Selected Works in Probability and Statistics

For futher volumes:
http://www.springer.com/series/8556

Sara van de Geer · Marten Wegkamp
Editors

Selected Works of Willem van Zwet

Springer

Editors
Sara van de Geer
Seminar für Statistik
HG G 24.1
Rämistrasse 101
8092 Zürich, Switzerland
geer@stat.math.ethz.ch

Marten Wegkamp
Department of Mathematics
and Department of Statistics
Cornell University
Ithaca, NY 14853-4201, USA
marten.wegkamp@cornell.edu

ISBN 978-1-4939-5111-6 ISBN 978-1-4614-1314-1 (eBook)
DOI 10.1007/978-1-4614-1314-1
Springer New York Dordrecht Heidelberg London

Printed on acid-free paper

Springer is a part of Springer Science+Business Media (www.springer.com)

To Willem

Preface to the Series

Springer's Selected Works in Probability and Statistics series offers scientists and scholars the opportunity of assembling and commenting upon major classical works in probability and statistics, and honors the work of distinguished scholars in probability and statistics. Each volume contains the original papers, original commentary by experts on the subject's papers, and relevant biographies and bibliographies.

Springer is committed to maintaining the volumes in the series with free access on SpringerLink, as well as to the distribution of print volumes. The full text of the volumes is available on SpringerLink with the exception of a small number of articles for which links to their original publisher is included instead. These publishers have graciously agreed to make the articles freely available on their websites. The goal is maximum dissemination of this material.

The subjects of the volumes have been selected by an editorial board consisting of Anirban DasGupta, Peter Hall, Jim Pitman, Michael Sörensen, and Jon Wellner.

Preface

With this collections volume, some of the important works of Willem van Zwet are moved to the front layers of modern statistics. We have made a selection based on discussions with Willem, and aiming at a representative sample. The result is a collection of papers that the new generations of statisticians should not be denied. They are here to stay, to enjoy and to form the basis for further research.

We have grouped the papers into six themes. The first three papers give an impression of the broad scope of statistics. One of its core business is as in all mathematics: classification, characterization, and unification. The third paper here discusses M- and Z-estimators, which have their modern face nowadays in non- and semi-parametric models.

The next theme concerns asymptotic theory. We cite Lucien Le Cam ([1]) *"If you need to use asymptotic arguments, don't forget to let your number of observations tend to infinity"*. Asymptotic statistics is indeed a subtle area involving much more than only pointwise limit theorems. The papers in this volume cover nonparametric tests as well as semi-parametric estimation, putting down the fundamentals for asymptotic efficiency in such models.

A very important, but sometimes notoriously technical topic, is second order approximations. With his co-authors, Willem deals with this topic in an impressingly elegant way. The beauty of concepts in this area is evolving further, for example by the formalization of the distance of distributions to the normal distribution. Within this theme, this volume contains the original contribution of Sergey Bobkov, Gennadiy Chistyakov and Friedrich Götze exposing the limits of near-normality.

Willem was very much intrigued by the bootstrap. It is often used without worrying about its validity, whereas Willem's intuition said its all round applicability is very questionable. This turned out to be a mind twisting and exciting issue: see the papers in this theme.

There is the modeling, the analysis of the model, and the statistical estimation. In the applications theme, we see all three aspects together. It shows that even though there are many sophisticated probabilistic models around, one still may have to start from scratch when looking at a particular real life problem. This is difficult hard work, but the final result is complete and beautiful.

Although statistics is not often associated with mathematical conjectures, it actually generates many. These are often questions in theoretical probability. The challenge to prove or disprove conjectures deserves its prominent place in statistics, and gives rise to fascinating storytelling.

This volume serves as basic reference for fundamental statistical theory, and at the same time reveals some of its history. We hope the unique mix will show the adventurous aspects of our profession, and that it will be an inspiration to all!

Zürich,
June 2011

Sara van de Geer
Marten Wegkamp

References

1. L. Le Cam (1990). Maximum likelihood: an introduction. *International Statistical Review/Revue Internationale de Statistique*, **58**, 153–171.

Acknowledgements

This series of selected works is possible only because of the efforts and cooperation of many people, societies, and publishers. The series editors originated the series and directed its development. The volume editors spent a great deal of time compiling the previously published material and the contributors provided comments on the significance of the papers. The societies and publishers who own the copyright to the original material made the volumes possible and affordable by their generous cooperation:

Allerton Press
Bernoulli Society
Institute of Mathematical Statistics
Indian Statistical Institute
John Wiley and Sons
Springer Science+Business Media
Statistica Sinica

Contents

Contributors

Peter J. Bickel
Department of Statistics, 367 Evans Hall. Berkeley, CA 94710-3860, USA
e-mail: bickel@stat.berkeley.edu

Sergey G. Bobkov
School of Mathematics, University of Minnesota, 127 Vincent Hall,
206 Church St. S.E., Minneapolis, MN 55455, USA
e-mail: bobkov@math.umn.edu

Gennadiy P. Chistyakov
Fakultät für Mathematik, Universität Bielefeld, Postfach 100131,
33501 Bielefeld, Germany
e-mail: chistyak@math.uni-bielefeld.de

Friedrich Götze
Fakultät für Mathematik, Universität Bielefeld, Postfach 100131,
33501 Bielefeld, Germany
e-mail: goetze@math.uni-bielefeld.de

Chris A.J. Klaassen
Korteweg-de Vries Institute for Mathematics, University of Amsterdam,
P.O. Box 94248, 1090 GE Amsterdam, The Netherlands
e-mail: C.A.J.Klaassen@uva.nl

Nelly Litvak
Department of Applied Mathematics, University of Twente,
P.O. Box 217, 7500 AE Enschede, The Netherlands
e-mail: n.litvak@ewi.utwente.nl

David M. Mason
Statistics Program, University of Delaware,
213 Townsend Hall, Newark, DE 19716, USA
e-mail: davidm@udel.edu

Jon A. Wellner
Department of Statistics, University of Washington, Seattle,
WA 98195-4322, USA
e-mail: jaw@stat.washington.edu

Biography of Willem van Zwet

Marten Wegkamp

Willem R. van Zwet was born in Leiden, the Netherlands, in 1934. He obtained a Masters degree in Mathematics at the University of Leiden in 1959. After his military service, Willem decided to continue his studies in statistics. The Mathematics Centre in Amsterdam was at that time the only place in the Netherlands with a proper statistics program. This centre had been founded after the war, in 1946. The first head of the Statistics Department was the mathematician David van Dantzig. His successor, Jan Hemelrijk, appointed Willem as sous-chef of the department in 1961. During daytime, Willem taught classes and did consulting work at the Centre. During the late hours, he worked on his thesis. With Jan Hemelrijk as advisor, Willem graduated in 1964 with a Ph.D. in Mathematics at the University of Amsterdam.

In 1965, he was appointed Associate Professor of Statistics at the University of Leiden. He spent the first semester as Associate Professor at the University of Oregon. Willem was promoted to Full Professor in 1968 and he remained in Leiden until his retirement in 1999. During 1990–1996, he visited the University of North Carolina at Chapel Hill on a regular basis as the William Newman Professor. He was a frequent visitor, and Miller Professor in 1997, of the University of California at Berkeley.

Willem is known for his pertinent contributions in various areas of mathematical statistics. This book is an homage to his scientific work. But Willem is also known as a talented and tireless organizer. The interview in Beran and Fisher (2009) paints an excellent picture of his academic life, and filled with many humorous anecdotes, makes for a recommended read. Statistics was still in its infancy in the early seventies in Europe and his service for the statistics community in his native Netherlands and worldwide are truly remarkable.

For instance, Willem served as member and chair of the European Regional Committee of the Institute of Mathematical Statistics (1969–1980) that organized the European Meetings of Statisticians. In 1972, he organized the first Lunteren Stochastics conference and he remained, until 1999, an organizer of this successful meeting, that continues to be held each Fall in Lunteren, the Netherlands. He was president of the Institute of Mathematical Statistics (1991–1992) and the Bernoulli Society for Mathematical Statistics and Probability (1987–1989), and vice-president

M. Wegkamp
Department of Mathematics and Department of Statistical Science
Cornell University, Ithaca, NY
e-mail: marten.wegkamp@cornell.edu

(1985–1989) and president (1997–1999) of the International Statistical Institute. Willem was Associate Editor (1972–1980) and Editor (1986–1988) of the Annals of Statistics and Editor-in-Chief of Bernoulli (2000–2003). He was the director of the Thomas Stieltjes Institute of Mathematics in the Netherlands (1992–1999), and founding director of the European research institute EURANDOM (1997–2000). Other activities included Dean of the School of Mathematics and Natural Sciences of the University of Leiden (1982–1984), chair of the scientific council and member of the board of the Mathematics Centre at Amsterdam (1983–1996) and the Leiden University Fund (1993–2005), member of the Board of Directors of the American Statistical Association (1993–1995) and member of the Corporation and the Board of NISS (1993–2002).

Fortunately, Willem's many scientific and organizational efforts are well recognized. He is a Fellow of the Institute of Mathematical Statistics (1972) and the American Statistical Association (1988), Honorary Fellow Royal Statistical Society (1978) and Honorary member of the International Statistical Institute (1999) and Netherlands Statistical Society (2000). He presented the Hotelling Lectures at the University of North Carolina (1988), the Wald Memorial Lectures (1992), and the Bahadur Lectures at the University of Chicago (2005). He is a member of the Royal Netherlands Academy of Sciences (1979) and the Academia Europaea (1990), and an honorary doctor of Charles University at Prague (1997). He received the Van Dantzig Medal of the Netherlands Society for Statistics and Operations Research (1970), the Bernoulli Medal (Tashkent, 1986), the Peace Medal of Charles University (1988), the Médaille de la Ville de Paris (1989), the Adolphe Quételet Medal of the International Statistical Institute (1993), the Certificate of Appreciation of the American Statistical Association (1995), the AKZO-Nobel Award (1996), and the Alexander von Humboldt Research Prize (2006). Perhaps the most prominent recognition happened in 1996, when Queen Beatrix of the Netherlands made him a Knight in the Order of the Netherlands Lion.

Reference

1. R.J. Beran and N.I. Fisher. An Evening Spent with Bill van Zwet. Statistical Science 24(1), 87–115, 2009.

Publications of Willem R. van Zwet

References

1. Convex Transformations of Random Variables, Math. Centre Tracts 7 (1964), Mathematisch Centrum, Amsterdam.
2. Convex transformations: A new approach to skewness and kurtosis, Statist. Neerlandica 18 (1964), 433-441.
3. De waarde van hydroxychloroquine (Plaquenil) voor de behandeling van chronische discoide lupus erythematodes (with J.H. Kraak, W.G. van Ketel and J.R. Prakken), Nederl. Tijdschr. Geneesk. 109 (1965), 461-469.
4. Bias in estimation from type I censored samples, Statist Neerlandica 20 (1966), 143-148.
5. On mixtures of distributions (with W. Molenaar), Ann. Math. Statist. 37 (1966), 281-283.
6. Het hachelijke oordeel, Statist. Neerlandica 12 (1967), 117-130.
7. On the combination of independent test statistics (with J. Oosterhoff), Ann. Math. Statist. 38 (1967), 659-680.
8. Host discrimination in pseudocoila bochei (with K. Bakker, S.N. Bagchee and E. Meelis), Ent Exp. & Appl. 10 (1967), 295-311.
9. An inequality for expected values of sample quantiles, Ann. Math. Statist. 38 (1967), 1817-1821.
10. On convexity preserving families of probability distributions, Statist. Neerlandica 22 (1968), 23-32.
11. Stemmingen zonder winnaar (with R. J. In 't Veld), Statist. Neerlandica 23 (1969), 269-276.
12. Asymptotic properties of isotonic estimators for the generalized failure rate function (with R.E. Barlow), Bull. Int. Statist. Inst. 43 (1969), 252-253.
13. Asymptotic properties of isotonic estimators for the generalized failure rate function. Part I: Strong consistency (with R.E. Barlow), Nonparametric Techniques in Statistical Inference, M.L. Puri editor, Cambridge University Press (1970), 159-173.
14. Some remarks on t he two-armed bandit (with J. Fabius), Ann. Math. Statist. 41 (1970), 1906-1916.
15. Grondbegrippen van de Waarschijnlijkheidsrekening (with J. Fabius), Math. Centre Syllabus 10 (1970), Mathematisch Centrum, Amsterdam.
16. Comparison of several nonparametric estimators of the failure rate function (with R.E. Barlow), Operations Research and Reliability, D. Grouchko editor, Gordon & Breach, New York (1971), 375-399.
17. The likelihood ratio test for the multinomial distribution (with J. Oosterhoff), Proceedings 6th Berkeley Symp. on Math. Statist and Probability 1 (1972), 31-49.
18. Asymptotic normality of nonparametric tests for independence (with F.H. Ruymgaart and G.R. Shorack), Ann. Math. Statist. 43 (1972), 1122-1135.
19. Asymptotic expansions for the power of distribution free tests in the one-sample problem (with W. Albers and P.J. Bickel), Ann. Statist. 3 (1976), 108-156.

20. Asymptotic expansions for the distribution functions of linear combinations of order statistics, Statistical Decision Theory and Related Topics II, S.S. Gupta and D.S. Moore editors, Academic Press, New York (1977), 421-438.
21. A proof of Kakutani's conjecture on random subdivision of longest intervals, Ann. Probability 6 (1978), 133-137.
22. Asymptotic expansions for the power of distribution free tests in the two-sample problem (with P.J. Bickel), Ann. Statist. 6 (1978), 937-1004.
23. Mean, median, mode, II, Statist. Neerlandica 33 (1979), 1-5.
24. A note on contiguity and Hellinger distance (with J. Oosterhoff), Contributions to Statistics (Jaroslau Hájek Memorial Volume), J. Jureckova editor, Academia, Prague (1979), 157-166.
25. The Edgeworth expansion for linear combinations of uniform order statistics, Proc. 2nd Prague Symp. on Asymptotic Statistics, P. Mandl and M. Huskova editors, North Holland, Amsterdam (1979), 93-101.
26. On a theorem of Hoeffding (with P.J. Bickel), Asymptotic Theory of Statistical Tests and Estimation, I.M. Chakravarti editor, Academic Press, New York (1980), 307-324.
27. A strong law for linear combinations of order statistics, Ann. Probability 8 (1980), 986-990.
28. On efficiency of first and second order (with P.J. Bickel and D.M. Chibisov), International Statistical Review 49 (1981), 169-175.
29. An asymptotic expansion for the distribution of the logarithm of the likelihood ratio (with D.M. Chibisov), Proc. Third Intern. Vilnius Conference on Probability Theory and Math. Statist. II, Akademia Nauk. USSR (1981), 55-56
30. The Berry-Esseen bound for U-statistics (with R. Helmers), Statistical Decision Theory and Related Topics III , Vol. 1, S.S. Gupta and J .O. Berger editors, Academic Press, New York (1982), 497-512.
31. On the Edgeworth expansion for the simple linear rank statistic, Nonparametric Statistical Inference, Coll. Math. Soc. Janos Bolyai, Budapest (1982), 889-909.
32. An inequality for random replacement sampling plans, Festschrift for Erich Lehmann; P.J. Bickel, K.A. Doksum and J.L. Hodges, Jr. editors, Wadsworth, Belmont (1982), 441-448.
33. Ranks and order statistics, Recent Advances in Statistics, Papers in Honor of Herman Chernoff; M.H. Rizvi, J. Rustagi and D. Siegmund editors, Academic Press, New York (1983), 407-422.
34. A Berry-Esseen bound for symmetric statistics, Z. Wahrsch. Verw. Gebiete 66 (1984), 425-440.
35. On the Edgeworth expansion for the logarithm of the likelihood ratio, I (with D.M. Chibisov), Teor. Veroyatnost i Primenen. 29 (1984), 417-439.
36. On the Edgeworth expansion for the logarithm of the likelihood ratio, II (with D.M. Chibisov), Asymptotic Statistics II, Proc. Third Prague Symp. Asympt. Statist, P. Mandl and M. Huskova editors, Elsevier Science Publishers, Amsterdam (1984), 451-461.
37. Van de Hulst on robust statistics: A historical note, Statist. Neerlandica 39 (1985), 81-95.
38. A simple analysis of third-order efficiency of estimates (with P.J. Bickel and F. Götze), Proc. Berkeley Conference in Honor of Jerzy Neyman and Jack Kiefer, II, L.M. LeCam and R.A. Olshen editors, Wadsworth, Monterey (1985), 749-768.
39. On estimating a parameter and its score function (with C.A.J. Klaassen), Proc. Berkeley Conference in Honor of Jerzy Neyman and Jack Kiefer, II, L.M. LeCam and R.A. Olshen editors, Wadsworth, Monterey (1985), 827-840.
40. The Edgeworth expansion for U-statistics of degree two (with P.J. Bickel and F. Götze), Ann. Statist. 14 (1986), 1463-1484.
41. A note on the strong approximation to the renewal process (with D.M. Mason), Pub. Inst. Univ. Paris XXXII, fasc. 1-2 (1987), 81-91.
42. A refinement of the KMT inequality for t he uniform empirical process (with D.M. Mason), Ann. Probability 15 (1987), 871-884.
43. On estimating a parameter and its score function, II (with C.A.J. Klaassen and A.W. van der Vaart), Statistical Decision Theory and Related Topics IV, 2, S.S. Gupta and J.O. Berger editors, Springer, New York (1988), 281-288.

44. Modelling the growth of a batch culture of plant cells: a corpuscular approach (with M.C.M. de Gunst, P.A.A. Harkes, J. Val and K.R. Libbenga), Enzyme Microb. Technol. 12 (1990), 61-71.
45. Andrei Nikolaevich Kolmogorov, Jaarboek 1989 Koninklijke Nederlandse Akademie van Wetenschappen, North Holland, Amsterdam (1990), 166-171.
46. A non-Markovian model for cell population growth: speed of convergence and central limit theorem (with M.C.M. de Gunst), Stoch. Processes and Appl. 41 (1992), 297-324.
47. Steekproeven uit steekproeven: De Baron van Munchhausen aan het werk? Verslag Afd. Nat, Koninklijke Nederlandse Akademie van Wetenschappen, 102 (5) (1993), 49-54.
48. Comment on double-blind refereeing, Statist. Sci. 8 (1993), 327-330.
49. A non-Markovian model for cell population growth: tail behavior and duration of the growth process (with M.C.M. de Gunst), Ann. Appl. Probability 3 (1993), 1112-1144.
50. The asymptotic distribution of point charges on a conducting sphere, Statistical Decision Theory and Related Topics, V, S.S. Gupta and J. O. Berger editors, Springer, New York (1993), 427-430.
51. Wassily Hoeffding's work in the sixties (with J. Oosterhoff), The Collected Works of Wassily Hoeffding, N.I. Fisher and P.K. Sen editors, Springer Verlag, New York (1994), 3-15.
52. Detecting differences in delay vector distributions (with C. Diks, F. Takens and J. de Goede), Phys. Rev. E. 53 (1996), 2169-2176.
53. Resampling: consistency of substitution estimators (with H. Putter), Ann. Statist. 24 (1996), 2297-2318.
54. Resampling fewer than n observations: Gains, losses, and remedies for losses (with P.J. Bickel and F. Götze), Statist. Sinica 7 (1997), 1-31.
55. On a set of the first category (with H. Putter), Festschrift for Lucien he Cam, D. Pollard, E. Torgersen and G.L. Yang, editors, Springer Verlag, New York (1997), 315-323.
56. An Edgeworth expansion for symmetric statistics (with V. Bentkus and F. Götze), Ann. Statist. 25 (1997), 851-896.
57. Comment on "Improving the relevance of economic statistics" by Yves Franchet, Economic Statistics: Accuracy, Timeliness and Relevance, Z. Kenessey, editor, US Department of Commerce, Bureau of Economic Analysis, Washington D.C. (1997), 23-26.
58. On the shape theorem for the supercritical contact process (with M. Fiocco), Prague Stochastics '98, M. Huskova, P. Lachout and J.A. Visek, editors, Union of Czech Mathematicians and Physicists, Prague (1998), 569-573.
59. Empirical Edgeworth expansions for symmetric statistics (with H. Putter), Ann. Statist 26 (1998), 1540-1569.
60. Discussion of "Towards systems of social statistics" by I.P. Fellegi and M. Wolfson, Bull. Intern. Statist. Inst LVII, 3 (1998), 96-97.
61. Introducing EURANDOM (with W.J.M. Senden), Statist. Neerlandica 53 (1999), 1-4.
62. No complaints so far, Leiden University Press, Leiden (1999), pp.15; also appeared in Nieuw Arch. Wisk. 17 (1999), 268-279.
63. A remark on consistent estimation (with E.W. van Zwet), Math. Methods of Statistics 8 (1999), 277-284.
64. Asymptotic efficiency of inverse estimators (with A.CM. van Rooij and F.H. Ruymgaart), Teor. Veroyatnost i Primenen 44 (1999), 826- 844.
65. Opening address of the 52nd Session of the International Statistical Institute at Helsinki, Bull. Intern. Statist. Inst. LVIII (3), 57-60.
66. David van Dantzig en de ontwikkeling van de stochastiek in Nederland, Uitbeelden in wiskunde, Proc. Symp. Van Dantzig 2000, G. Alberts & H. Blauwendraat editors, Centre for Mathematics and Computer Science, Amsterdam (2000), 99-121.
67. Statistics for the contact process (with M. Fiocco), Statistica Neerlandica 56 (2002), 243-251.
68. Een bemoeial met smaak: Interview Mike Keane (with G. Alberts), Nieuw Arch. Wisk. 5/3 (2002), 141-145.

69. Minimax invariant estimation of a continuous distribution function under entropy loss (with L. Mohammadi), Metrika 56 (2002), 31-42.
70. Decaying correlations for the supercritical contact process conditioned on survival (with M. Fiocco), Bernoulli 9 (2003), 763-781.
71. Parameter estimation for the supercritical contact process (with M. Fiocco), Bernoulli 9 (2003), 1071-1092.
72. Weak convergence results for the Kakutani interval splitting procedure (with R. Pyke), Ann. Probab. 32 (2004), 881-902.
73. On the minimal travel time needed to collect n items on a circle (with N. Litvak), Ann. Appl. Probab. 14 (2004), 881-902.
74. Niets nieuws onder de zon, STAtOR 5-2 (2004), 21-29.
75. Performance of control charts for specific alternative hypotheses (with A. Di Bucchianico, M. Huskova and P. Klasterecky), COMPSTAT 2004, Proceedings in Computational Statistics, J. Antoch editor, Physica Verlag, Heidelberg, New York (2004), 903-910.
76. Pythagoras voor de Tweede Kamer: Interview met Hans de Rijk, Cor Baayen en Jack van Lint (with Gerard Alberts), Nieuw Arch. Wisk. 5/5 (2004), 212-217.
77. Maximum likelihood estimation for the contact process (with M. Fiocco), Festschrift for Herman Rubin, Institute of Mathematical Statistics Lecture Notes-Monograph Series 45 (2004), 309-318.
78. Wassily Hoeffding 1914-1991 (with Nicholas I. Fisher), Biographical Memoirs of the National Academy of Sciences 86, The National Academies Press, Washington, D.C. (2005), 1-21.
79. Statistics and the law, ISI 2005, 55th Session of the International Statistical Institute, Sydney, Australia (2005), ISBN: I 877040 28 2.
80. Wetenschap of Maatschappij: In memoriam Guus Zoutendijk (with L.C.M. Kallenberg), Nieuw Arch. Wisk. 5/6 (2005), 193-196.
81. Maximum likelihood for the fully observed contact process (with M. Fiocco), Jour. Comp. and Appl. Math. 186 (2006), 117-129.
82. An expansion for a discrete non-lattice distribution (with F. Götze), Frontiers in Statistics, J. Fan & H.L. Koul editors, Imperial College London (2006), 257-274.
83. EURANDOM: a decade of European statistics (with W.Th.F. den Hollander and W.J.M. Senden), Statistica Neerlandica 62 (2008), 256-265.
84. Remembering Wassily Hoeffding (with N.I. Fisher), Statist. Sci. 23 (2008), 536-547.

Part I
Three Fundamental Statistics Papers

Convex transformations: A new approach to skewness and kurtosis *)

by W. R. van Zwet **)

UDC 519.2

Samenvatting

In dit artikel worden een tweetal orde-relaties voor waarschijnlijkheidsverdelingen voorgesteld, die – beter dan de klassieke maten gebaseerd op derde en vierde momenten – aangeven wanneer een verdeling een grotere scheefheid of kurtosis bezit dan een andere verdeling. Voorts wordt een aantal karakteriseringen en toepassingen van deze orde-relaties behandeld. Bewijzen worden in dit artikel niet gegeven; deze zijn te vinden in een meer uitgebreide, aan dit onderwerp gewijde studie [5].

1. Introduction

Every statistician will have at least an intuitive idea of what is meant by the concepts of „skewness" and „kurtosis" of a probability distribution and he will be aware of the fact that these should play an important role in applications. He will also probably feel vaguely dissatisfied with the existing measures for these concepts, i.e. the standardized third and fourth central moments, and indeed there are at least two perfectly good reasons for this uneasy feeling.

The first one is that, according to these measures, any pair of probability distributions that possess finite fourth moments may be compared as to skewness and kurtosis, whereas one feels that pairs of such distributions exist that are quite incomparable in these respects. The second reason is that, to the author's knowledge, very few interesting applications of any generality exist. It is fairly obvious that both disadvantages are closely related: the reason for the apparent lack of applications is precisely the fact that comparison of probability distributions on the basis of these measures is so often meaningless.

From the above it will be clear that at the root of the trouble lies the fact that these measures impose a simple ordering – i.e. an ordering where every pair of elements are comparable – on too large a class of probability distributions. Rather than restricting ourselves to considering smaller classes of distributions we shall try and find a more satisfactory approach by considering partial orderings – i.e. orderings where not every pair of distributions are necessarily comparable – to replace the classical measures.

*) Rapport S 328 van de afdeling Mathematische Statistiek van het Mathematisch Centrum te Amsterdam; Lezing gehouden op de Statistische Dag 1964.

**) Sous-chef voor mathematische statistiek van het Mathematisch Centrum te Amsterdam.

S. van de Geer and M. Wegkamp (eds.), *Selected Works of Willem van Zwet*, Selected Works in Probability and Statistics, DOI 10.1007/978-1-4614-1314-1_1, © Springer Science+Business Media, LLC 2012

It will be shown in this paper that two partial order relations exist that seem to cover our intuitive ideas about skewness and kurtosis. These order relations will not only be seen to imply the ordering according to the classical measures, but also to be so much stronger than the classical orderings as to permit meaningful applications. No proofs will be given in this paper; they may be found in [5], where a more extensive study of the subject is made. Part of the material presented here and in [5] was previously discussed in [4].

2. Notation

Let $\underline{x}$ be a non-degenerate real-valued random variable[1]) and let I be the smallest interval for which $P(\underline{x} \varepsilon I) = 1$. We define the distribution function F of $\underline{x}$ by

$$F(x) = \tfrac{1}{2} P(\underline{x} < x) + \tfrac{1}{2} P(\underline{x} \leqslant x)$$

and the expectation and central moments of $\underline{x}$ by

$$\mathscr{E}\underline{x} = \int_I x \, dF(x),$$

$$\sigma^2(\underline{x}) = \mu_2(\underline{x}) = \int_I (x - \mathscr{E}\underline{x})^2 \, dF(x), \text{ and}$$

$$\mu_k(\underline{x}) = \int_I (x - \mathscr{E}\underline{x})^k \, dF(x), \quad k = 3,4,\ldots,$$

where the right-hand sides denote STIELTJES integrals. We shall say that these expectations exist only if they are finite. The distribution given by F is said to be symmetrical about $x_0 \varepsilon I$ if

$$F(x_0 - x) + F(x_0 + x) = 1 \quad \text{for all real } x.$$

Let $\underline{x}_{1:n} \leqslant \underline{x}_{2:n} \leqslant \ldots \leqslant \underline{x}_{n:n}$ denote an ordered sample of size n from the distribution F; $\underline{x}_{i:n}$ is called the $i-th$ order statistic of a sample of size n from F. In the greater part of this paper we shall confine our attention to the class $\mathscr{F}$ of distribution functions F satisfying

(a) F is twice continuously differentiable on I;
(b) $F'(x) > 0$ on I;
(c) There exist integers i and n, $1 \leqslant i \leqslant n$, such that $\mathscr{E}\underline{x}_{i:n}$ exists.

For $F \varepsilon \mathscr{F}$ the inverse function G is uniquely defined on (0,1) by[2])

$$GF(x) = x \qquad \text{for } x \varepsilon I.$$

1) We denote random variables by underlining their symbols.

2) We shall usually not use brackets to denote composite functions and write GF and $GF(x)$ rather than $G(F(.))$ and $G(F(x))$.

We shall also be concerned with the subclass $\mathscr{S} \subset \mathscr{F}$ of symmetric distributions in $\mathscr{F}$.

When we consider simultaneously two random variables, $\underset{\sim}{x}$ and $\underset{\sim}{x}^*$, with distribution functions F and F^*, we shall adopt similar conventions and notations with regard to $\underset{\sim}{x}^*$ and F^*, and write: I^*, $\underset{\sim}{x}^*_{i;n}$ and G^*.

A real-valued function φ defined on I is said to be convex on I if for all $x_1, x_2 \varepsilon I$ and $0 \leqslant \lambda \leqslant 1$

$$\varphi(\lambda x_1 + (1-\lambda) x_2) \leqslant \lambda\varphi(x_1) + (1-\lambda)\varphi(x_2),$$

i.e. the graph of φ lies below any chord. We note that this definition implies continuity of φ on I, except perhaps at its endpoints, if these exist. A real-valued function φ on I is said to be antisymmetrical and concave-convex on I about $x_0 \varepsilon I$, if for all $x_0 - x \varepsilon I, x_0 + x \varepsilon I$,

$$\varphi(x_0 - x) + \varphi(x_0 + x) = 2\varphi(x_0),$$

and if φ is concave for $x \leqslant x_0$ and convex for $x \geqslant x_0$, $x \varepsilon I$; x_0 will be called a central point of φ.

3. Convex and concave-convex transformations

Suppose that φ is non-decreasing and convex on I and consider the random variables $\underset{\sim}{x}$ and $\varphi(\underset{\sim}{x})$. Apart from an overall linear change of scale such a transformation of the random variable $\underset{\sim}{x}$ to the random variable $\varphi(\underset{\sim}{x})$ effects a contraction of the lower part of the scale of measurement and an extension of the upper part. As, moreover, this deformation increases towards both ends of the scale, the transformation from $\underset{\sim}{x}$ to $\varphi(\underset{\sim}{x})$ produces what one intuitively feels to be an increased skewness to the right. The following theorem holds:

Theorem 3.1

If φ is a non-decreasing convex function on I, which is not constant on I, and if $\mu_{2k+1}(\underset{\sim}{x})$ and $\mu_{2k+1}(\varphi(\underset{\sim}{x}))$ exist, then

$$\frac{\mu_{2k+1}(\underset{\sim}{x})}{\sigma^{2k+1}(\underset{\sim}{x})} \leqslant \frac{\mu_{2k+1}(\varphi(\underset{\sim}{x}))}{\sigma^{2k+1}(\varphi(\underset{\sim}{x}))}, \quad \text{for } k = 1,2,\ldots.$$

It is intuitively equally appealing that a non-decreasing, antisymmetric and concave-convex transformation of a symmetrically distributed random variable should lead to an increased kurtosis of the distribution. We have:

Theorem 3.2

Let φ be a non-decreasing, antisymmetrical, concave-convex function on I, which is not constant on I, and let the distribution given by F be symmetrical about x_0, where x_0 denotes a central point of φ. Then, if $\mathscr{E}\varphi^{2k}(\underset{\sim}{x})$ exists,

$$\frac{\mu_{2k}(\underset{\sim}{x})}{\sigma^{2k}(\underset{\sim}{x})} \leqslant \frac{\mu_{2k}(\varphi(\underset{\sim}{x}))}{\sigma^{2k}(\varphi(\underset{\sim}{x}))}, \text{for } k = 2,3,\ldots.$$

4. Two weak-order relations

In the remaining part of this paper we shall confine our attention to distribution functions $F \varepsilon \mathscr{F}$; part of the results, however, remain valid without this restriction.

Returning to the theorems of section 3 we remark that they obviously continue to hold if one replaces $\varphi(\underset{\sim}{x})$ by any other random variable with the same distribution, i.e. they hold for any $\underset{\sim}{x}^*$ with distribution function F^* satisfying

$$F^*\varphi(x) = P(\underset{\sim}{x}^* \leqslant \varphi(x)) = P(\varphi(\underset{\sim}{x}) \leqslant \varphi(x)) = P(\underset{\sim}{x} \leqslant x) = F(x),$$

or $\qquad \varphi(x) = G^*F(x) \quad \text{on } I.$

We therefore define the following order relations on $\mathscr{F}$ and $\mathscr{S}$ respectively:

Definition 4.1

If $F, F^* \varepsilon \mathscr{F}$, then $F \underset{c}{<} F^*$ (or equivalently $F^* \underset{c}{>} F$) if and only if G^*F is convex on I.

Definition 4.2

If $F, F^* \varepsilon \mathscr{S}$, then $F \underset{s}{<} F^*$ (or equivalently $F^* \underset{s}{>} F$) if and only if G^*F is convex for $x > x_0$, $x \varepsilon I$, where x_0 denotes the point of symmetry of F.

We shall say in this case that F *c*-precedes or *s*-precedes F^*, or that F^* *c*-follows or *s*-follows F, and that the two are *c*-comparable or *s*-comparable. We shall also speak of *c*-ordering, *s*-ordering, *c*-comparison, *s*-comparison, etc., where the letters *c* and *s* stand for convex and symmetrical. According to the above the meaning of these definitions is clear: $F \underset{c}{<} F^*$ if and only if a random variable with distribution F may be transformed into one with distribution F^* by an increasing and convex transformation; for symmetrical distributions, $F \underset{s}{<} F^*$ if and only if this can be done by an increasing, antisymmetrical, concave-convex transformation. From the theorems of the preceding section the implications are also obvious: we have every right to say that $F \underset{c}{<} F^*$ implies that F^* has greater skewness to the right than F, whereas for symmetrical distributions $F \underset{s}{<} F^*$ implies that F^* has greater kurtosis than F.

Since it is easily seen that both order relations are reflexive ($F \prec F$) and transitive ($F \prec F^*$, $F^* \prec F^{**}$ implies $F \prec F^{**}$) they are weak orderings. If one defines an equivalence relation $\sim$ by

Definition 4.3

If $F, F^* \varepsilon \mathscr{F}$, then $F \sim F^*$ if and only if $F(x) = F^*(ax + b)$ for some constants $a > 0$ and b,

it is also easy to show that $F \sim F^*$ if and only if $F \underset{c}{<} F^*$ and $F^* \underset{c}{<} F$; for F, $F^* \varepsilon \mathscr{S}$ one finds that $F \sim F^*$ if and only if $F \underset{s}{<} F^*$ and $F^* \underset{s}{<} F$. Hence by passing to the collections $\overline{\mathscr{F}}$ and $\overline{\mathscr{S}}$ of equivalence classes one may define partial orderings ($F \prec F^*$, $F^* \prec F$ implies $F = F^*$) on $\overline{\mathscr{F}}$ and $\overline{\mathscr{S}}$ by ordering equivalence classes according to the c- and s-ordering of their representatives.

In statistical parlance the above asserts that c- and s-ordering are both independent of location and scale parameters. The classes $\overline{\mathscr{F}}$ and $\overline{\mathscr{S}}$ are the classes of types of laws belonging to $\mathscr{F}$ and $\mathscr{S}$. We may consequently restrict our attention to c- and s-comparison of standardized distribution functions.

Here we give only two examples of c- and s-ordering. The gamma distributions may be shown to be c-following one another with decreasing values of the parameter, whereas the symmetric beta distributions s-follow one another with increasing values of the parameter. Further examples may be found in [5].

5. Characterization theorems

In this section we give two theorems that provide a number of characterizations of the order relations $\underset{c}{<}$ and $\underset{s}{<}$ in terms of inequalities for expected values and odd moments of order statistics.

Theorem 5.1

Let R be a dense subset of (0,1). Then for $F, F^* \varepsilon \mathscr{F}$ the following statements are equivalent:

(1) $F \underset{c}{<} F^*$;

(2) $F(\mathscr{E}\underline{x}_{i:n}) \leqslant F^*(\mathscr{E}\underline{x}^*_{i:n})$ for all $n = 1,2,\ldots$ and $i = 1,2,\ldots,n$, for which $\mathscr{E}\underline{x}_{i:n}$ and $\mathscr{E}\underline{x}^*_{i:n}$ exist;

(3) $\dfrac{\mu_{2k+1}(\underline{x}_{i:n})}{\sigma^{2k+1}(\underline{x}_{i:n})} \leqslant \dfrac{\mu_{2k+1}(\underline{x}^*_{i:n})}{\sigma^{2k+1}(\underline{x}^*_{i:n})}$ for all $k = 1,2,\ldots, n = 1,2,\ldots$, and $i = 1,2,\ldots,n$, for which $\mu_{2k+1}(\underline{x}_{i:n})$ and $\mu_{2k+1}(\underline{x}^*_{i:n})$ exist;

(4) If i and n tend to infinity in such a way that $\lim \frac{i}{n} = r$, $r \varepsilon R$, then

$$\lim n\,(F^*(\mathscr{E}\underline{x}^*_{i:n}) - F(\mathscr{E}\underline{x}_{i:n})) \geqslant 0;$$

(5) If i and n tend to infinity in such a way that $\lim \frac{i}{n} = r$, $r \varepsilon R$, then for at least one value of $k = 1,2, \ldots$.

$$\lim \sqrt{\;}\left(\frac{\mu_{2k+1}(\underline{x}^*_{i:n})}{\sigma^{2k+1}(\underline{x}^*_{i:n})} - \frac{\mu_{2k+1}(\underline{x}_{i:n})}{\sigma^{2k+1}(\underline{x}_{i:n})}\right) \geqslant 0.$$

Two remarks should be made about this theorem. The first one is that for a given distribution F convexity would seem to be a rather heavy requirement to prove the inequalities of theorem 3.1. The equivalence of statements (1) and (5) of theorem 5.1 shows, however, that if these inequalities are to hold even for a single value of k and for the class of distributions of large sample order statistics from a given distribution, then convexity is necessary as well as sufficient. The second remark is that the equivalence of statements (2) and (4) and of (3) and (5) enable us to derive small sample inequalities from their large sample counterparts.

Theorem 5.2

Let R be a dense subset of $(\frac{1}{2}, 1)$. Then for $F, F^* \varepsilon \mathscr{S}$ the following statements are equivalent:

(1) $F \underset{s}{<} F^*$;

(2) $F(\mathscr{E}\underline{x}_{i:n}) \leqslant F^*(\mathscr{E}\underline{x}^*_{i:n})$ for all $n = 1,2, \ldots$ and $\frac{n+1}{2} \leqslant i \leqslant n$, for which $\mathscr{E}\underline{x}^*_{i:n}$ exists;

(3) If i and n tend to infinity in such a way that $\lim \frac{i}{n} = r$, $r \varepsilon R$, then

$$\lim n\,(F^*(\mathscr{E}\underline{x}^*_{i:n}) - F(\mathscr{E}\underline{x}_{i:n})) \geqslant 0;$$

(4) If i and n tend to infinity in such a way that $\lim \frac{i}{n} = r$, $\frac{1}{2} < r < 1$, then for all $k = 1,2, \ldots$

$$\lim \sqrt{n}\left(\frac{\mu_{2k+1}(\underline{x}^*_{i:n})}{\sigma^{2k+1}(\underline{x}^*_{i:n})} - \frac{\mu_{2k+1}(\underline{x}_{i:n})}{\sigma^{2k+1}(\underline{x}_{i:n})}\right) \geqslant 0;$$

(5) Statement (4) is valid for all $r \varepsilon R$ and at least one value of $k = 1,2, \ldots$.

We note that small sample inequalities concerning odd moments are lacking; the corresponding large sample result is given in statement (4). For $i \leqslant \frac{n+1}{2}$, $0 < r < \frac{1}{2}$ and R dense in $(0,\frac{1}{2})$, the inequalities of theorem 5.2 are of course reversed.

For large classes of distributions $F \varepsilon \mathcal{S}$ small sample inequalities between $F(\mathcal{E}\underline{x}_{i:n})$ and quantities of the type $\frac{i-\alpha}{n+1-2\alpha}$ may be obtained by s-comparison with a class of distribution functions for which the inverse functions G are incomplete beta functions. For these results we refer to [5], where one may also find still another characterization of the order relations $\underset{c}{<}$ and $\underset{s}{<}$ in terms of a measure of skewness based on the median.

6. Applications

Although it has been made clear that the relations $\underset{c}{<}$ and $\underset{s}{<}$ may be taken to indicate increasing skewness and kurtosis, we still have to demonstrate that these relations meet with more success in applications than the classical measures based on third and fourth moments. To this end three examples of comparison of distributions will be considered where skewness or kurtosis obviously play an important role.

The first example is taken from a paper by J. L. Hodges jr. and E. L. Lehmann [2]. They discuss the relative asymptotic efficiency $e_{W:N}(F)$ of Wilcoxon's two sample test W to the normal scores test N, for the case where the underlying distribution is of type F. Numerical evidence leads them to suppose that $e_{W:N}$ will increase as the tails of the underlying distribution grow heavier. Application of the relation $\underset{s}{<}$ to a formula for $e_{W:N}(F)$ given in [2] immediately yields the desired result:

Under certain regularity conditions, $F, F^* \varepsilon \mathcal{S}$ and $F \underset{s}{<} F^*$ implies $e_{W:N}(F) \leqslant \leqslant e_{W:N}(F^*)$.

The second example concerns a paper by H. Hotelling [3] where the behaviour of Student's test under non-standard conditions is studied. Let $\underline{x}_1, \underline{x}_2, \ldots, \underline{x}_n$ be a random sample from a distribution $F \varepsilon \mathcal{F}$, for which either $\mu = \mathcal{E}\underline{x}$ exists, or $F \varepsilon \mathcal{S}$; in the latter case we define μ by $F(\mu) = \frac{1}{2}$. Furthermore let

$$\underline{t}_n = \frac{\bar{\underline{x}} - \mu}{\underline{s}} \sqrt{n},$$

where $\bar{\underline{x}} = \frac{1}{n} \sum_{i=1}^{n} \underline{x}_i$ and $\underline{s}^2 = \frac{1}{n-1} \sum_{i=1}^{n} (\underline{x}_i - \bar{\underline{x}})^2$.

The probability that $\underline{t}_n$ will exceed a constant value t will be denoted by $P(\underline{t}_n > t \mid F)$ and we define

$$R_n(F) = \lim_{t \to \infty} \frac{P(\underline{t}_n > t \mid F)}{P(\underline{t}_n > t \mid \Phi)},$$

where Φ denotes the normal distribution function.

Suppose that, assuming the underlying distribution to be normal, one carries out STUDENT's right-sided test for the hypothesis $\mu \leqslant \mu_0$, whereas in fact F is not normal at all. Then obviously $R_n(F)$ denotes the limit of the ratio of the actual size and the assumed size of the test as both these sizes tend to zero. It may therefore serve to provide a rough idea of what to expect when the assumption of normality is violated.

For $n = 3$ numerical values found by HOTELLING for some symmetrical distributions seem to indicate – paradoxically enough at first sight – that $R_n(F)$ decreases as the tails of F become heavier. Making use of an expression for $R_n(F)$ given in [3] one easily shows this idea to be correct for s-ordered symmetric distributions, whereas a similar result may be proved for c-ordered distributions. In fact we have:

If $F, F^* \varepsilon \mathscr{F}$, and if either $\mathscr{E}\underline{x}, \mathscr{E}\underline{x}^*$ exist and $F \underset{c}{<} F^*$, or $F, F^* \varepsilon \mathscr{S}$ and $F \underset{s}{<} F^*$, then $R_n(F) \geqslant R_n(F^*)$ for $n = 2,3,\ldots$.

Finally we discuss the relative efficiency of sample median to sample mean in estimating the point of symmetry of a symmetric distribution. Let $\underline{x}_1, \underline{x}_2, \ldots, \underline{x}_n$ denote a random sample from a distribution $F \varepsilon \mathscr{S}$ with finite variance $\sigma^2(\underline{x})$, and suppose one wishes to estimate $\mathscr{E}\underline{x}$. Two unbiased estimates that are generally used are the sample median

$$\underline{x}_{\frac{n+1}{2}:n}$$

and the sample mean

$$\bar{\underline{x}}_n = \frac{1}{n} \sum_{i=1}^{n} \underline{x}_i,$$

where we have supposed n to be odd. The choice between them should depend on the ratio of their (small sample) efficiencies

$$r_n(F) = \frac{\operatorname{eff}\left(\underline{x}_{\frac{n+1}{2}:n}\right)}{\operatorname{eff}(\bar{\underline{x}}_n)} = \frac{\sigma^2(\underline{x})}{n\sigma^2\left(\underline{x}_{\frac{n+1}{2}:n}\right)}.$$

The following result is easily obtained:

For distributions $F, F^* \varepsilon \mathscr{S}$ having finite variances, $F \underset{s}{<} F^*$ implies $r_n(F) \leqslant$ $\leqslant r_n(F^*)$ for $n = 1, 3, 5, \ldots$.

This result supports the statement by G. W. BROWN and J. W. TUKEY [1] that „it is probable that the relative efficiencies of mean and median are greatly affected by the length of the tail".

References

[1] G. W. BROWN, J. W. TUKEY, Some distributions of sample means, Ann. Math. Statist. 17 (1946), 1-12.

[2] J. L. HODGES jr., E. L. LEHMANN, Comparison of normal scores and Wilcoxon tests, Proc. 4th Berkeley Symp. (1961) Vol. 1, 307-317.

[3] H. HOTELLING, The behavior of some standard statistical tests under nonstandard conditions, Proc. 4th Berkeley Symp. (1961) Vol. 1, 319-359.

[4] W. R. VAN ZWET, Two weak-order relations for distribution functions, Report S 303 (VP 17) Dept. of Math. Stat. (1962), Mathematisch Centrum, Amsterdam.

[5] W. R. VAN ZWET, Convex transformations of random variables, Math. Centre Tracts 7 (1964), Mathematisch Centrum, Amsterdam.

The Annals of Mathematical Statistics
1970, Vol. 41, No. 6, 1906–1916

SOME REMARKS ON THE TWO-ARMED BANDIT[1]

BY J. FABIUS AND W. R. VAN ZWET

University of Leiden and Mathematisch Centrum

1. Introduction and summary. In this paper we consider the following situation: An experimenter has to perform a total of N trials on two Bernoulli-type experiments E_1 and E_2 with success probabilities α and β respectively, where both α and β are unknown to him. The trials are to be carried out sequentially and independently, except that for each trial the experimenter may choose between E_1 and E_2, using the information obtained in all previous trials. The decisions on the part of the experimenter to use E_1 or E_2 in the successive trials may be randomized, i.e. for any trial he may use a chance mechanism in order to choose E_1 or E_2 with probabilities δ and $1-\delta$ respectively, where δ may depend on the decisions taken and the results obtained in the previous trials. A strategy Δ will be a set of such δ's, completely describing the experimenters behavior in every conceivable situation.

We assume the experimenter wants to maximize the number of successes. More precisely, we assume that he incurs a loss

$$L(\alpha,\beta,s) = N\max(\alpha,\beta)-s \tag{1.1}$$

if he scores a total of s successes. If he uses a strategy Δ, his expected loss is then given by the risk function

$$R(\alpha,\beta,\Delta) = N\max(\alpha,\beta)-E(S\mid\alpha,\beta,\Delta), \tag{1.2}$$

where S denotes the random number of successes obtained. Thus the risk of a strategy Δ equals the expected amount by which the number of successes the experimenter will obtain using Δ falls short of the number of successes he would score if he were clairvoyant and would use the more favorable experiment throughout the N trials. It is easy to see that $R(\alpha,\beta,\Delta)$ also equals $|\alpha-\beta|$ times the expected number of trials in which the less favorable experiment is performed under Δ.

We say that state $(m,k;n,l)$ is reached during the series of trials if in the first $m+n$ trials E_1 is performed m times, yielding k successes, and E_2 is performed n times, yielding l successes. Clearly, under a strategy Δ, the probability that this will happen is of the form

$$\pi_{\alpha,\beta,\Delta}(m,k;n,l) = p_\Delta(m,k;n,l)\alpha^k(1-\alpha)^{m-k}\beta^l(1-\beta)^{n-l}, \tag{1.3}$$

where $p_\Delta(m,k;n,l)$ depends on the state $(m,k;n,l)$ and the strategy Δ, but not on α and β. It is easy to show (e.g. by induction on N) that the class of all strategies is convex in the sense that there exists, for every pair of strategies Δ_1 and Δ_2 and for every $\lambda\in[0,1]$, a strategy Δ such that

$$p_\Delta(m,k;n,l) = \lambda p_{\Delta_1}(m,k;n,l)+(1-\lambda)p_{\Delta_2}(m,k;n,l) \tag{1.4}$$

Received November 1, 1968; revised September 3, 1969.

[1] Report S-399, Mathematisch Centrum, Amsterdam.

S. van de Geer and M. Wegkamp (eds.), *Selected Works of Willem van Zwet*, Selected Works in Probability and Statistics, DOI 10.1007/978-1-4614-1314-1_2, © Springer Science+Business Media, LLC 2012

for every state $(m, k; n, l)$. Moreover, this strategy Δ can always be taken to be such, that according to it the experimenter should base all his decisions exclusively on the numbers of successes and failures observed with E_1 and E_2, irrespective of the order in which these data became available. Denoting the class of all such strategies by $\mathscr{D}$ and remarking that $R(\alpha, \beta, \Delta)$ can be expressed in terms of the $\pi_{\alpha,\beta,\Delta}(m, k; n, l)$, we may conclude that $\mathscr{D}$ is an essentially complete class of strategies. We denote the probabilities δ constituting any strategy in $\mathscr{D}$ by $\delta(m, k; n, l)$: the probability with which the experimenter, having completed the first $m+n$ trials and thereby having reached state $(m, k; n, l)$, chooses E_1 for the next trial.

We note that if $p_\Delta(m, k; n, l) = 0$ for a state $(m, k; n, l)$, then $\delta(m, k; n, l)$ does not play any role in the description of Δ and may be assigned an arbitrary value without affecting the strategy. We shall say that any strategy Δ' such that $p_{\Delta'}(m, k; n, l) = p_\Delta(m, k; n, l)$ for all states $(m, k; n, l)$ constitutes a version of Δ.

Since we are considering a symmetric problem in the sense that it remains invariant when α and β are interchanged, it seems reasonable to consider strategies with a similar symmetry. Thus we are led to define the class $\mathscr{L}$ of all symmetric strategies: $\Delta \in \mathscr{L}$ iff $\Delta \in \mathscr{D}$ and $\delta(m, k; n, l) = 1-\delta(n, l; m, k)$ for all states $(m, k; n, l)$ with $p_\Delta(m, k; n, l) \neq 0$. Clearly, for $\Delta \in \mathscr{L}$,

$$(1.5) \qquad \delta(m, k; m, k) = \tfrac{1}{2} \quad \text{if} \quad p_\Delta(m, k; m, k) \geqq 0, \qquad \text{and}$$

$$(1.6) \qquad p_\Delta(m, k; n, l) = p_\Delta(n, l; m, k) \quad \text{for all states} \quad (m, k; n, l).$$

It follows that, for $\Delta \in \mathscr{L}$ and all (α, β),

$$(1.7) \qquad R(\alpha, \beta, \Delta) = R(\beta, \alpha, \Delta).$$

Among the contributions to the two-armed bandit problem the work of W. Vogel deserves special mention. Considering the same set-up we do, he discussed a certain subclass of the class $\mathscr{L}$ in [4], and obtained asymptotic bounds for the minimax risk for $N \to \infty$ in [5]. Since we shall not be concerned with asymptotics in this paper, we state the following result without a formal proof: The lower bound for the asymptotic minimax risk for $N \to \infty$ obtained by Vogel in [5] may be raised by a factor $2^{\frac{1}{2}}$. This is proved by applying the same method that was used in [5] to the optimal symmetric strategy for $\alpha+\beta = 1$ that was discussed in [4]. Combining this lower bound with the upper bound given in [5] we find that the asymptotic minimax risk must be between $0.265\,N^{\frac{1}{2}}$ and $0.376\,N^{\frac{1}{2}}$.

In Section 2 we study the Bayes strategies in $\mathscr{D}$. By means of a certain recurrence relation we arrive at a complete characterization of these strategies, thus generalizing D. Feldman's well-known result in [3] for the case where the experimenter knows the values of α and β except for their order. In addition we obtain expressions for the Bayes risk of any prior distribution. Using these results we proceed to derive in Section 3 certain monotonicity properties of $\delta(m, k; n, l)$ for any admissible strategy Δ in $\mathscr{D}$. Though these relations may seem intuitively evident, one does well to remember that the two-armed bandit problem has been shown to defy intuition in many aspects (cf. [2]). In Section 4 we prove the existence of an admissible symmetric minimax-risk strategy having the monotonicity properties just mentioned. This fact to some degree facilitates the search for minimax-risk strategies, but even so, the algebra involved becomes progressively more complicated with increasing N and seems to remain prohibitive already for N as small as 5.

2. Bayes strategies. For $\Delta \in \mathscr{D}$ we consider the expected number of successes $E(S \mid \alpha, \beta, \Delta)$ as a function of the $\delta(m, k; n, l)$. Clearly, the dependence on each $\delta(m, k; n, l)$ is linear. We denote the coefficient of $\delta(m, k; n, l)$ in $E(S \mid \alpha, \beta, \Delta)$ (and hence also in $-R(\alpha, \beta, \Delta)$) by $p_\Delta(m, k; n, l) c_{\alpha,\beta,\Delta}(m, k; n, l)$. If all $\delta(m, k; n, l)$ are strictly between 0 and 1, then all $p_\Delta(m, k; n, l)$ are positive and as a result all $c_{\alpha,\beta,\Delta}(m, k; n, l)$ are uniquely determined. Otherwise the $c_{\alpha,\beta,\Delta}(m, k; n, l)$ are defined by continuity.

THEOREM 1. *For any strategy* Δ *in* $\mathscr{D}$ *the functions* $c_{\alpha,\beta,\Delta}(m, k; n, l)$ *satisfy the following relations*

$$c_{\alpha,\beta,\Delta}(m,k;n,l) = (\alpha-\beta)\alpha^k(1-\alpha)^{m-k}\beta^l(1-\beta)^{n-l} \tag{2.1}$$

if $m+n = N-1$,

$$\begin{aligned} c_{\alpha,\beta,\Delta}(m,k;n,l) = {} & \delta(m+1,k+1;n,l)c_{\alpha,\beta,\Delta}(m+1,k+1;n,l) \\ & +\delta(m+1,k;n,l)c_{\alpha,\beta,\Delta}(m+1,k;n,l) \\ & +[1-\delta(m,k;n+1,l+1)]c_{\alpha,\beta,\Delta}(m,k;n+1,l+1) \\ & +[1-\delta(m,k;n+1,l)]c_{\alpha,\beta,\Delta}(m,k;n+1,l) \end{aligned} \tag{2.2}$$

if $m+n \leqq N-2$.

PROOF. By continuity it is obviously sufficient to consider the case where all $\delta(m, k; n, l)$ as well as α and β are strictly between 0 and 1. This ensures that expression (1.3) is positive for all states $(m, k; n, l)$. Hence the conditional expectation $e_{\alpha,\beta,\Delta}(m, k; n, l)$ of the total number of successes S under α, β and Δ given that the state $(m, k; n, l)$ is reached, exists. It is clearly a linear function of $\delta(m, k; n, l)$ and may thus be written in the form

$$e_{\alpha,\beta,\Delta}(m,k;n,l) = a_{\alpha,\beta,\Delta}(m,k;n,l)\,\delta(m,k;n,l) + b_{\alpha,\beta,\Delta}(m,k;n,l). \tag{2.3}$$

It follows that

$$c_{\alpha,\beta,\Delta}(m,k;n,l) = a_{\alpha,\beta,\Delta}(m,k;n,l)\,\alpha^k(1-\alpha)^{m-k}\beta^l(1-\beta)^{n-l}. \tag{2.4}$$

Dropping the subscripts α, β and Δ, we obtain, from the definition of $e(m, k; n, l)$,

$$\begin{aligned} e(m,k;n,l) = {} & \delta(m,k;n,l)[\alpha e(m+1,k+1;n,l) + (1-\alpha)\,e(m+1.k;n,l)] \\ & + [1-\delta(m,k;n,l)][\beta e(m,k;n+1,l+1) \\ & \qquad +(1-\beta)\,e(m,k;n+1,l)], \end{aligned} \tag{2.5}$$

and consequently

$$a(m,k;n,l) = \alpha e(m+1,k+1;n,l)+(1-\alpha)\,e(m+1,k;n,l) \tag{2.6}$$
$$\qquad -\beta e(m,k;n+1,l+1)-(1-\beta)\,e(m,k;n+1,l),$$

$$b(m,k;n,l) = \beta e(m,k;n+1,l+1)+(1-\beta)\,e(m,k;n+1,l). \tag{2.7}$$

If $m+n = N-1$, then (2.6) becomes $a(m, k; n, l) = \alpha-\beta$, and hence (2.1) follows from (2.4). On the other hand, rewriting (2.6) by means of (2.3) leads to

$$\begin{aligned} a(m,k;n,l) = {} & \alpha\delta(m+1,k+1;n,l)a(m+1,k+1;n,l) \\ & +(1-\alpha)\,\delta(m+1,k;n,l)a(m+1,k;n,l) \\ & +\beta[1-\delta(m,k;n+1,l+1)]a(m,k;n+1,l+1) \\ & +(1-\beta)[1-\delta(m,k;n+1,l)]a(m,k;n+1,l) \\ & +[\alpha b(m+1,k+1;n,l)+(1-\alpha)b(m+1,k;n,l) \\ & -\beta b(m,k;n+1,l+1) \\ & -(1-\beta)b(m,k;n+1,l)-\beta a(m,k;n+1,l+1) \\ & -(1-\beta)a(m,k;n+1,l)], \end{aligned}$$

where for $m+n = N-2$ the last expression between square brackets vanishes as one easily verifies using (2.6) and (2.7). This result, combined with (2.4), gives (2.2).

Let μ be a prior distribution on the closed unit square. For a strategy $\Delta\in\mathscr{D}$,

$$\rho(\mu,\Delta) = \int R(\alpha,\beta,\Delta)\,d\mu(\alpha,\beta) \tag{2.8}$$

denotes the average risk of Δ against μ. If we define

$$\gamma_{\mu,\Delta}(m,k;n,l) = \int c_{\alpha,\beta,\Delta}(m,k;n,l)\,d\mu(\alpha,\beta), \tag{2.9}$$

then $-p_\Delta(m, k; n, l)\gamma_{\mu,\Delta}(m, k; n, l)$ is the coefficient of $\delta(m, k; n, l)$ in $\rho(\mu, \Delta)$. It follows that any strategy Δ that has $\delta(m, k; n, l) = 1$ whenever $\gamma_{\mu,\Delta}(m, k; n, l) > 0$ and $\delta(m, k; n, l) = 0$ whenever $\gamma_{\mu,\Delta}(m, k; n, l) < 0$, minimizes $\rho(\mu, \Delta)$ for fixed μ and is therefore a Bayes strategy against μ. This may be seen by successively finding the optimal $\delta(m, k; n, l)$ for $m+n = N-1, N-2, \cdots, 0$, and noting that for $m+n = \nu$ these optimal values do not depend on the values of $\delta(m, k; n, l)$ for $m+n < \nu$. Conversely, every Bayes strategy against μ has a version with $\delta(m, k; n, l) = 1$ (or 0) whenever $\gamma_{\mu,\Delta}(m, k; n, l) > 0$ (or < 0).

THEOREM 2. *Let μ be a prior distribution on the closed unit square and let $\gamma_\mu(m, k; n, l)$ be defined by*

$$\gamma_\mu(m,k;n,l) = \int(\alpha-\beta)\alpha^k(1-\alpha)^{m-k}\beta^l(1-\beta)^{n-l}\,d\mu(\alpha,\beta) \tag{2.10}$$

if $m+n = N-1$,

$$\gamma_\mu(m,k;n,l) = \gamma_\mu^{+}(m+1,k+1;n,l)+\gamma_\mu^{+}(m+1,k;n,l) \tag{2.11}$$
$$\qquad -\gamma_\mu^{-}(m,k;n+1,l+1)-\gamma_\mu^{-}(m,k;n+1,l)$$

for $m+n \leqq N-2$, *where* x^+ *and* x^- *denote* $\max(0, x)$ *and* $\max(0, -x)$ *respectively. Then* $\Delta \in \mathscr{D}$ *is a Bayes strategy against* μ *if and only if it has a version with* $\delta(m, k; n, l) = 1$ *whenever* $\gamma_\mu(m, k; n, l) > 0$ *and* $\delta(m, k; n, l) = 0$ *whenever* $\gamma_\mu(m, k; n, l) < 0$.

PROOF. According to the remarks preceding the theorem, Δ is Bayes against μ iff it has a version for which $\delta(m, k; n, l) = 1$ (or 0) if $\gamma_{\mu,\Delta}(m, k; n, l) > 0$ (or < 0). Integrating (2.1) and (2.2) with respect to μ and substituting the values of the $\delta(m, k; n, l)$ we find that for this version of Δ, $\gamma_{\mu,\Delta}(m, k; n, l)$ equals $\gamma_\mu(m, k; n, l)$ as defined by (2.10) and (2.11) for all states.

We note that D. Feldman's characterization of the Bayes strategies in $\mathscr{D}$ against a prior distribution μ, which puts mass ξ and $1-\xi$ at points (α_0, β_0) and (β_0, α_0) respectively (cf. [3]), may be formulated as follows: Δ in $\mathscr{D}$ is Bayes against μ iff it has a version for which $\delta(m, k; n, l) = 1$ whenever $\eta_\mu(m, k; n, l) > 0$ and $\delta(m, k; n, l) = 0$ whenever $\eta_\mu(m, k; n, l) < 0$ where

$$\eta_\mu(m, k; n, l) = \xi\alpha_0^{\,k}(1-\alpha_0)^{m-k}\beta_0^{\,l}(1-\beta_0)^{n-l} - (1-\xi)\alpha_0^{\,l}(1-\alpha_0)^{n-l}\beta_0^{\,k}(1-\beta_0)^{m-k}$$

for all states $(m, k; n, l)$. It follows that $\operatorname{sgn}\eta_\mu(m, k; n, l) = \operatorname{sgn}\gamma_\mu(m, k; n, l)$ for all states $(m, k; n, l)$ and all μ of the type considered by Feldman. This fact may also be verified by a direct, though somewhat tedious argument.

To conclude this section we consider the Bayes risk $\rho(\mu)$ of an arbitrary prior distribution μ. This is defined as the average risk $\rho(\mu, \Delta)$ of any Bayes strategy Δ against μ, or equivalently, $\rho(\mu) = \inf_{\Delta \in \mathscr{D}} \rho(\mu, \Delta)$.

THEOREM 3. *For any prior distribution* μ,

$$\rho(\mu) = N\int \frac{|\alpha-\beta|}{2}\, d\mu(\alpha, \beta) - \sum_{m=0}^{N-1}\sum_{n=0}^{N-m-1}\sum_{k=0}^{m}\sum_{l=0}^{n} \frac{\binom{m+n}{n}\binom{m}{k}\binom{n}{l}}{2^{m+n+1}}|\gamma_\mu(m, k; n, l)|$$

$$= N\int (\alpha-\beta)^+\, d\mu(\alpha, \beta) - \textstyle\sum_{n=0}^{N-1}\sum_{l=0}^{n}\binom{n}{l}\gamma_\mu^{+}(0, 0; n, l)$$

$$= N\int (\alpha-\beta)^-\, d\mu(\alpha, \beta) - \textstyle\sum_{m=0}^{N-1}\sum_{k=0}^{m}\binom{m}{k}\gamma_\mu^{-}(m, k; 0, 0).$$

PROOF. Let $\Delta \in \mathscr{D}$ be Bayes against μ. Without loss of generality we may restrict attention to a version of Δ which has the property described in Theorem 2. For any such version and any state $(m, k; n, l)$ with $m+n \leqq N-1$ we have

$$\gamma_{\mu,\Delta}(m, k; n, l) = \gamma_\mu(m, k; n, l),$$

$$(\delta(m, k; n, l) - \tfrac{1}{2})\gamma_\mu(m, k; n, l) = \tfrac{1}{2}|\gamma_\mu(m, k; n, l)|,$$

$$\delta(m, k; n, l)\gamma_\mu(m, k; n, l) = \gamma_\mu^{+}(m, k; n, l),$$

$$-(1-\delta(m, k; n, l))\gamma_\mu(m, k; n, l) = \gamma_\mu^{-}(m, k; n, l).$$

Consequently for any state $(m, k; n, l)$ with $m+n \leqq N-1$ we obtain the following equalities, using (2.5) and the fact that $\gamma_{\mu,\Delta}(m,k;n,l)$ and hence $\gamma_\mu(m,k;n,l)$ equals the coefficient of $\delta(m,k;n,l)$ in the first member:

$$\int \alpha^k(1-\alpha)^{m-k}\beta^l(1-\beta)^{n-l}\, e_{\alpha,\beta,\Delta}(m,k;n,l)\,d\mu(\alpha,\beta)$$

$$= \tfrac{1}{2}|\gamma_\mu(m,k;n,l)|$$
$$+\tfrac{1}{2}\int \alpha^{k+1}(1-\alpha)^{m-k}\beta^l(1-\beta)^{n-l}\, e_{\alpha,\beta,\Delta}(m+1,k+1;n,l)\,d\mu(\alpha,\beta)$$
$$+\tfrac{1}{2}\int \alpha^{k}(1-\alpha)^{m-k+1}\beta^l(1-\beta)^{n-l}\, e_{\alpha,\beta,\Delta}(m+1,k;n,l)\,d\mu(\alpha,\beta)$$
$$+\tfrac{1}{2}\int \alpha^{k}(1-\alpha)^{m-k}\beta^{l+1}(1-\beta)^{n-l}\, e_{\alpha,\beta,\Delta}(m,k;n+1,l+1)\,d\mu(\alpha,\beta)$$
$$+\tfrac{1}{2}\int \alpha^{k}(1-\alpha)^{m-k}\beta^{l}(1-\beta)^{n-l+1}\, e_{\alpha,\beta,\Delta}(m,k;n+1,l)\,d\mu(\alpha,\beta)$$

$$(2.12)\qquad = \gamma_\mu^{\,+}(m,k;n,l)$$
$$+\int \alpha^{k}(1-\alpha)^{m-k}\beta^{l+1}(1-\beta)^{n-l}\, e_{\alpha,\beta,\Delta}(m,k;n+1,l+1)\,d\mu(\alpha,\beta)$$
$$+\int \alpha^{k}(1-\alpha)^{m-k}\beta^{l}(1-\beta)^{n-l+1}\, e_{\alpha,\beta,\Delta}(m,k;n+1,l)\,d\mu(\alpha,\beta)$$

$$= \gamma_\mu^{\,-}(m,k;n,l)$$
$$+\int \alpha^{k+1}(1-\alpha)^{m-k}\beta^l(1-\beta)^{n-l}\, e_{\alpha,\beta,\Delta}(m+1,k+1;n,l)\,d\mu(\alpha,\beta)$$
$$+\int \alpha^{k}(1-\alpha)^{m-k+1}\beta^l(1-\beta)^{n-l}\, e_{\alpha,\beta,\Delta}(m+1,k;n,l)\,d\mu(\alpha,\beta).$$

Observing that by definition $E(S\,|\,\alpha,\beta,\Delta) = e_{\alpha,\beta,\Delta}(0,0;0,0)$ and $e_{\alpha,\beta,\Delta}(m,k;n,l) = k+l$ for any state $(m,k;n,l)$ with $m+n = N$, we arrive at the three desired expressions by repeated application of the corresponding versions of (2.12).

3. Admissible strategies. For the type of problem considered in this paper every admissible strategy is also a Bayes strategy. In the sequel we shall, however, need a slightly stronger result. We shall say that a prior distribution is nonmarginal if, for some $\varepsilon > 0$, it assigns probability 1 to the set

$$(3.1)\qquad Q_\varepsilon = \{(\alpha,\beta)\,|\,|\alpha-\beta|\,\alpha(1-\alpha)\beta(1-\beta) \geqq \varepsilon, 0<\alpha<1, 0<\beta<1\}.$$

THEOREM 4. *Every admissible strategy $\Delta \in \mathscr{D}$ is Bayes against a nonmarginal prior distribution.*

PROOF. Let Δ be any strategy which is not Bayes against any nonmarginal prior. It is sufficient to show that Δ is not admissible.

For any sufficiently small $\varepsilon_i > 0$, consider the restricted problem where the parameter space is reduced to the set $A_i = Q_{\varepsilon_i}$ as defined by (3.1). Since A_i is compact, the assertion that every admissible strategy is Bayes remains true for the restricted problem. By our assumption Δ is not Bayes, and therefore not admissible in the new problem. It follows that there exists a strategy Δ_i that is Bayes against a prior distribution μ_i on A_i and for which $R(\alpha,\beta,\Delta_i) \leqq R(\alpha,\beta,\Delta)$ for all $(\alpha,\beta) \in A_i$. By a standard procedure we may select a sequence $\varepsilon_i \searrow 0$ and corresponding μ_i and

Δ_i such that the strategies Δ_i converge to a strategy Δ_0 in the sense that $\delta_i(m, k; n, l)$ converges to $\delta_0(m, k; n, l)$ for every state $(m, k; n, l)$. Obviously

$$R(\alpha, \beta, \Delta_0) \leqq R(\alpha, \beta, \Delta) \qquad \text{for all} \quad \alpha, \beta \in [0, 1]$$

since the inequality must hold on every A_i and both functions are continuous.

Since Δ_i converges to Δ_0 there exists a positive integer j for which Δ_j has the following properties:

(a) For all states with $\delta_0(m, k; n, l) = 0$, $\delta_j(m, k; n, l) \neq 1$;
(b) For all states with $\delta_0(m, k; n, l) = 1$, $\delta_j(m, k; n, l) \neq 0$;
(c) For all states with $0 < \delta_0(m, k; n, l) < 1$, $0 < \delta_j(m, k; n, l) < 1$.

This implies that $\delta_0(m, k; n, l) = \delta_j(m, k; n, l)$ for every state with $\delta_j(m, k; n, l) = 0$ or 1. Recalling that Δ_j is Bayes against μ_j and noting that this property can not be destroyed by changing only those $\delta_j(m, k; n, l)$ that are strictly between 0 and 1, we find that Δ_0 is Bayes against the prior distribution μ_j on A_j. As Δ is not Bayes against μ_j by our assumption, the inequality $R(\alpha, \beta, \Delta_0) \leqq R(\alpha, \beta, \Delta)$ on the closed unit square must be strict for at least one point (α, β) and the inadmissibility of Δ follows.

We are now in a position to prove a theorem that provides some insight in the structure of admissible strategies.

THEOREM 5. *If μ is a nonmarginal prior distribution and $m+n \leqq N-2$, then*

$$\gamma_\mu(m, k; n+1, l+1) < \gamma_\mu(m+1, k+1; n, l) \tag{3.2}$$

$$\gamma_\mu(m+1, k; n, l) < \gamma_\mu(m, k; n+1, l) \tag{3.3}$$

PROOF. For $m+n = N-2$, (2.10) yields

$$\gamma_\mu(m+1, k+1; n, l) - \gamma_\mu(m, k; n+1, l+1)$$
$$= \int (\alpha-\beta)^2 \alpha^k (1-\alpha)^{m-k} \beta^l (1-\beta)^{n-l} \, d\mu(\alpha, \beta),$$

which is strictly positive since μ is nonmarginal. In the same way one shows that (3.3) is satisfied for $m+n = N-2$.

Next we suppose that the theorem is valid for $m+n = \nu$, where $0 < \nu \leqq N-2$, and we assume $m+n = \nu-1$. By (2.11) we have then

$$\gamma_\mu(m+1, k+1; n, l) - \gamma_\mu(m, k; n+1, l+1)$$
$$\begin{aligned} &= [\gamma_\mu^+(m+2, k+2; n, l) - \gamma_\mu^+(m+1, k+1; n+1, l+1)] \\ &\quad + [\gamma_\mu^+(m+2, k+1; n, l) - \gamma_\mu^+(m+1, k; n+1, l+1)] \\ &\quad + [\gamma_\mu^-(m, k; n+2, l+2) - \gamma_\mu^-(m+1, k+1; n+1, l+1)] \\ &\quad + [\gamma_\mu^-(m, k; n+2, l+1) - \gamma_\mu^-(m+1, k+1, n+1, l)] \geqq 0 \end{aligned}$$

since by hypothesis each of these four expressions is nonnegative. Equality can occur only if all four expressions vanish. However, the first and the third one can vanish only if $\gamma_\mu(m+1, k+1; n+1, l+1) < 0$ and $\geqq 0$ respectively, and hence inequality (3.2) is strict.

Similarly (3.3) follows from

$$\begin{aligned}\gamma_\mu(m, k; n+1, l) - \gamma_\mu(m+1, k; n, l) &= [\gamma_\mu{}^+(m+1, k+1; n+1, l) - \gamma_\mu{}^+(m+2, k+1; n, l)] \\ &\quad + [\gamma_\mu{}^+(m+1, k; n+1, l) - \gamma_\mu{}^+(m+2, k; n, l)] \\ &\quad + [\gamma_\mu{}^-(m+1, k; n+1, l+1) - \gamma_\mu{}^-(m, k; n+2, l+1)] \\ &\quad + [\gamma_\mu{}^-(m+1, k; n+1, l) - \gamma_\mu{}^-(m, k; n+2, l)] \geqq 0\end{aligned}$$

and the fact that the first expression in square brackets can vanish only if $\gamma_\mu(m+2, k+1; n, l) < 0$ and the third one only if $\gamma_\mu(m+1, k; n+1, l+1) \geqq 0$, which would imply $\gamma_\mu(m+2, k+1; n, l) > 0$.

COROLLARY 1. *Every admissible strategy $\Delta \in \mathscr{D}$ has a version for which*

$$\delta(m, k; n+1, l+1) \leqq \delta(m+1, k+1; n, l) \tag{3.4}$$

$$\delta(m+1, k; n, l) \leqq \delta(m, k; n+1, l) \tag{3.5}$$

for all $m+n \leqq N-2$, where in each of these inequalities at least one member equals 0 or 1.

PROOF. By Theorem 4, Δ is Bayes against a nonmarginal prior μ, and as a result the theorem is proved by applying Theorem 5 and Theorem 2.

COROLLARY 2. *Every admissible strategy $\Delta \in \mathscr{D}$ has a version for which*

$$\delta(m, k; n, l)[1 - \delta(m+1, k+1; n, l)][1 - \delta(m+1, k; n, l)] = 0 \tag{3.6}$$

$$[1 - \delta(m, k; n, l)]\,\delta(m, k; n+1, l+1)\,\delta(m, k; n+1, l) = 0 \tag{3.7}$$

for all $m+n \leqq N-2$.

PROOF. As before, we let μ denote the nonmarginal prior of Theorem 4 and consider the version of Δ having $\delta(m, k; n, l) = 1$ (or 0) whenever $\gamma_\mu(m, k; n, l) > 0$ (or < 0). If (3.6) were false for this version, then $\gamma_\mu(m, k; n, l) \geqq 0$, $\gamma_\mu(m+1, k+1; n, l) \leqq 0$ and $\gamma_\mu(m+1, k; n, l) \leqq 0$. The second of these inequalities implies $\gamma_\mu(m, k; n+1, l+1) < 0$ by Theorem 5, and hence (2.11) shows that $\gamma_\mu(m, k; n, l) < 0$, which contradicts the first inequality.

Similarly, if (3.7) were false, then $\gamma_\mu(m, k; n, l) \leqq 0$, $\gamma_\mu(m, k; n+1, l+1) \geqq 0$ and $\gamma_\mu(m, k; n+1, l) \geqq 0$. The second inequality implies $\gamma_\mu(m+1, k+1; n, l) > 0$ by Theorem 5, and hence $\gamma_\mu(m, k; n, l) > 0$ by (2.11), which contradicts the first inequality.

Intuitively one might expect some further monotonicity relations, like e.g. (i): $\delta(m, k; n, l) \leqq \delta(m+1, k+1; n, l)$ and (ii): $\delta(m, k; n, l) \leqq \delta(m, k+1; n, l)$, for any reasonable strategy in $\mathscr{D}$. However, (i) is nothing but another version of Bradt, Johnson and Karlin's principle of staying on a winner (cf. [2]), which they showed not to be generally true for all Bayes strategies in $\mathscr{D}$. In fact, (i) and (ii) do not even hold for all admissible strategies in $\mathscr{D}$ as one can see from the example given in [2]: The Bayes strategies in $\mathscr{D}$ for the case $N = 2$ against the prior distribution μ, which puts mass .8 in (.1, 0) and mass .2 in (.9, 1), are precisely those strategies in $\mathscr{D}$ for which $\delta(0, 0; 0, 0) = 1$, $\delta(1, 1; 0, 0) = 0$, and $\delta(1, 0; 0, 0) = 1$. Thus there is an essentially unique and hence admissible Bayes strategy against μ, which violates (i) and (ii).

For admissible strategies, which are also symmetric, Corollary 1 takes the following more explicit form.

COROLLARY 3. *Every admissible strategy $\Delta \in \mathscr{L}$ has a version for which*

$$\delta(m, k; n, l) = 1, \qquad \delta(n, l; m, k) = 0 \tag{3.8}$$

whenever $m + n \leqq N - 1$, $k \geqq 1$, $m - k \leqq n - l$ and $(m, k; n, l) \neq (n, l; m, k)$.

PROOF. For the version of Δ that satisfies Corollary 1 we find by repeated application of (3.4) and (3.5) $\delta(m, k; n, l) \geqq \delta(m-k+l, l; n+k-l, k) \geqq \delta(n, l; m, k)$ where at least one of the extreme members must be 0 or 1. Since their sum equals 1 if $p_\Delta(m, k; n, l) \neq 0$, (3.8) will hold in this case. If $p_\Delta(m, k; n, l) = 0$, then by (1.6) we also have $p_\Delta(n, l; m, k) = 0$ and choosing $\delta(m, k; n, l) = 1$ and $\delta(n, l; m, k) = 0$ merely leads to another version of Δ.

We conclude this section by remarking that Corollaries 1, 2 and 3 obviously continue to hold if, instead of admissibility, we require that Δ be Bayes against a nonmarginal prior.

4. Symmetric minimax-risk strategies.

THEOREM 6. *There is a minimax-risk strategy which is admissible and belongs to $\mathscr{L}$.*

PROOF. The class $\mathscr{D}$, with the topology induced by the notion of convergence introduced in the proof of Theorem 4, is compact. The existence of a minimax-risk strategy in $\mathscr{D}$ is a well-known consequence of this. Moreover, the class $\mathscr{D}^*$ of all minimax-risk strategies in $\mathscr{D}$ is easily seen to be closed. Thus, if ν denotes Lebesgue measure on the unit square, there is a strategy $\Delta_1 \in \mathscr{D}^*$ such that $\rho(\nu, \Delta_1) = \min_{\Delta \in \mathscr{D}^*} \rho(\nu, \Delta)$. This follows from the continuity of $\rho(\nu, \cdot)$. Let $\Delta_2 \in \mathscr{D}$ be defined by $\delta_2(m, k; n, l) = 1 - \delta_1(n, l; m, k)$ for all states $(m, k; n, l)$. Then $p_{\Delta_2}(m, k; n, l) = p_{\Delta_1}(n, l; m, k)$ for all states, and hence $R(\alpha, \beta, \Delta_2) = R(\beta, \alpha, \Delta_1)$ for all (α, β), so that $\Delta_2 \in \mathscr{D}^*$. By convexity we now may construct a strategy $\Delta \in \mathscr{D}$ satisfying (1.4) with $\lambda = \frac{1}{2}$. It follows that $R(\alpha, \beta, \Delta) = \frac{1}{2}R(\alpha, \beta, \Delta_1) + \frac{1}{2}R(\alpha, \beta, \Delta_2)$ for all (α, β), and hence $\Delta \in \mathscr{D}^*$. Finally we define $\Delta^* \in \mathscr{L}$ by

$$\delta^*(m, k; n, l) = \tfrac{1}{2}\delta(m, k; n, l) + \tfrac{1}{2}[1 - \delta(n, l; m, k)]$$

for all states. The construction of Δ implies that $p_{\Delta^*}(m, k; n, l) = p_\Delta(m, k; n, l)$ for all states, and hence $\Delta^* \in \mathscr{D}^* \cap \mathscr{L}$.

In order to show that Δ^* is also admissible, we first remark that any strategy outside $\mathscr{D}^*$ has at some point (α, β) strictly larger risk than Δ^*, because Δ^* has minimax-risk. On the other hand, going through the steps leading to the construction of Δ^* once more, one easily verifies that $\rho(\nu, \Delta_1) = \rho(\nu, \Delta_2) = \rho(\nu, \Delta) = \rho(\nu, \Delta^*)$, so that $\rho(\nu, \Delta^*) \leqq \rho(\nu, \Delta')$ for any $\Delta' \in \mathscr{D}^*$. But because of the continuity of $R(\cdot, \cdot, \Delta)$, this implies that also within $\mathscr{D}^*$ there is no strategy improving on Δ^*, and thus the proof is complete.

The above proof really consists of two separate arguments mixed together. The first one is quite standard (cf. e.g. Theorem 8.6.4. in [1] and shows the existence of a symmetric minimax-risk strategy. The second argument, yielding admissibility, exploits an idea of Wald ([6] page 102). By the same argument, replacing $\mathscr{D}^*$ by the class of all Bayes strategies against any given prior distribution μ, one can prove the existence of an admissible Bayes strategy against μ.

Theorem 6 together with Corollaries 1, 2 and 3 yields

COROLLARY 4. *There is an admissible symmetric minimax-risk strategy which obeys* (3.4) *through* (3.8).

For $N = 1$ or 2, (1.5) and (3.8) uniquely determine a symmetric strategy. It follows from Corollary 4 and Corollary 3 that this strategy has minimax risk and is in fact the only admissible strategy in $\mathscr{L}$. For $N \geqq 3$ the situation rapidly becomes more complicated. In order to find a symmetric minimax-risk strategy Δ_0 satisfying (3.4) through (3.8) one first has to find a general expression for the risk function $R(\alpha, \beta, \Delta)$ of an arbitrary symmetric strategy Δ satisfying (3.8). Then, with the aid of (3.4) through (3.7), one has to solve the remaining $\delta(m, k; n, l)$ directly using the minimax property.

To accomplish the first step of computing $R(\alpha, \beta, \Delta)$ for an arbitrary symmetric strategy, one may proceed recursively. This is especially useful if one wants to find $R(\alpha, \beta, \Delta)$ for a number of values of N. If $X_\nu = 1 - Y_\nu = 1$ or 0 according to whether E_1 or E_2 is carried out on the νth trial ($\nu = 1, 2, \cdots, N$), then $R(\alpha, \beta, \Delta)$, being equal to $|\alpha - \beta|$ multiplied by the expected number of times the experimenter uses the less favorable experiment, is given by

$$R(\alpha, \beta, \Delta) = \tfrac{1}{2}N\,|\alpha - \beta| - \tfrac{1}{2}(\alpha - \beta)\textstyle\sum_{\nu=1}^{N} E(X_\nu - Y_\nu \mid \alpha, \beta, \Delta). \tag{4.1}$$

Remembering the definition of $\pi_{\alpha,\beta,\Delta}(m, k; n, l)$, we have

$$E(X_\nu - Y_\nu \mid \alpha, \beta, \Delta) = \textstyle\sum \pi_{\alpha,\beta,\Delta}(m, k; n, l)[2\delta(m, k; n, l) - 1], \tag{4.2}$$

where the summation is extended over all states $(m, k; n, l)$ with $m + n = \nu - 1$, and where the $\pi_{\alpha,\beta,\Delta}(m, k; n, l)$ can be computed recursively by means of

$$\begin{aligned} \pi_{\alpha,\beta,\Delta}(m, k; n, l) = {} & \alpha\delta(m-1, k-1; n, l)\pi_{\alpha,\beta,\Delta}(m-1, k-1; n, l) \\ & + (1-\alpha)\,\delta(m-1, k; n, l)\,\pi_{\alpha,\beta,\Delta}(m-1, k; n, l) \\ & + \beta[1 - \delta(m, k; n-1, l-1)]\,\pi_{\alpha,\beta,\Delta}(m, k; n-1, l-1) \\ & + (1-\beta)[1 - \delta(m, k; n-1, l)]\,\pi_{\alpha,\beta,\Delta}(m, k; n-1, l) \end{aligned} \tag{4.3}$$

starting from

$$\begin{aligned} \pi_{\alpha,\beta,\Delta}(0, k; 0, l) &= 1 \qquad \text{if } k = l = 0; \\ &= 0 \qquad \text{otherwise.} \end{aligned} \tag{4.4}$$

The work involved may be reduced somewhat by means of the relation

$$\pi_{\alpha,\beta,\Delta}(m,k;n,l)=\pi_{\alpha,\beta,\Delta}(n,l;m,k), \tag{4.5}$$

which is a consequence of (1.3) and (1.6).

For $N=3$, only $\delta(2,1;0,0)$ remains undetermined by the requirement that Δ be symmetric and must satisfy (3.8), and one finds

$$R(\alpha,\beta,\Delta)=\tfrac{3}{2}|\alpha-\beta|-\tfrac{1}{2}(\alpha-\beta)^2\{1+\delta(2,1;0,0)+[1-\delta(2,1;0,0)](\alpha+\beta)\}.$$

After a little algebra one sees that Δ_0 must have $\delta(2,1;0,0)=1$ and that $R(\alpha,\beta,\Delta_0)$ attains its maximum $M(\Delta_0)=\frac{9}{16}$ when $|\alpha-\beta|=\frac{3}{4}$.

For $N=4$ only $\delta(2,1;0,0)$, $\delta(3,1;0,0)$ and $\delta(3,2;0,0)$ are to be determined and

$$\begin{aligned}R(\alpha,\beta,\Delta)=2|\alpha-\beta|-\tfrac{1}{2}(\alpha-\beta)^2\{&(\alpha^2+\beta^2+3\alpha\beta-\alpha-\beta+3)-\delta(2,1;0,0)\alpha\beta\\&-\delta(3,2;0,0)[1+\delta(2,1;0,0)](\alpha^2+\beta^2+\alpha\beta-\alpha-\beta)\\&+\delta(3,1;0,0)\,\delta(2,1;0,0)(\alpha^2+\beta^2+\alpha\beta-2\alpha-2\beta+1)\}.\end{aligned}$$

Using (3.6), one finds after lengthy calculations that Δ_0 must have $\delta(2,1;0,0)=\frac{4}{5}$, $\delta(3,1;0,0)=\frac{1}{2}$ and $\delta(3,2;0,0)=1$, so that the risk function of Δ_0 is given by

$$R(\alpha,\beta,\Delta_0)=2|\alpha-\beta|-\tfrac{17}{10}(\alpha-\beta)^2+\tfrac{1}{5}(\alpha-\beta)^4$$

and attains its maximum $M(\Delta_0)=.617$ when $|\alpha-\beta|=.654$. For larger values of N the number of $\delta(m,k;n,1)$ that have to be determined increases rapidly, and consequently the algebra involved becomes distressingly complicated.

REFERENCES

[1] BLACKWELL, D. and GIRSHICK, M. A. (1954). *Theory of Games and Statistical Decisions.* Wiley, New York.

[2] BRADT, R. N., JOHNSON, S. M. and KARLIN, S. (1956). On sequential designs for maximizing the sum of n observations. *Ann. Math. Statist.* **27** 1060–1074.

[3] FELDMAN, D. (1962). Contributions to the "two-armed bandit" problem. *Ann. Math. Statist.* **33** 847–856.

[4] VOGEL, W. (1960a). Ein Irrfahrten-Problem und seine Anwendung auf die Theorie der sequentiellen Versuchs-Pläne, *Arch. Math.* **11** 310–320.

[5] VOGEL, W. (1960b). An asymptotic minimax theorem for the two-armed bandit problem. *Ann. Math. Statist.* **31** 444–451.

[6] WALD, A. (1950). *Statistical Decision Functions.* Wiley, New York.

VAN DE HULST ON ROBUST STATISTICS: A HISTORICAL NOTE

by W.R. van Zwet

Abstract This paper provides a discussion of an unpublished set of notes written in 1942 by the Dutch astronomer H.C. VAN DE HULST. In these notes VAN DE HULST derives the asymptotic variances of *M*-estimators as well as trimmed means and concludes that the asymptotic variance of what is now called HUBER's estimator is the same as that of a trimmed mean. This conclusion is usually ascribed to BICKEL (1965). A letter written by D. VAN DANTZIG in 1943 providing a critical evaluation of VAN DE HULST's results, adds interest to this suprisingly early contribution to the theory of robust statistics.

Key words: *Robust estimation, estimation of location, trimmed mean, M-estimator, history of statistics.*

1 Introduction

It is generally agreed that the history of modern mathematical statistics in the Netherlands begins with the work of VAN DANTZIG. Originally a pure mathematician, VAN DANTZIG turned to statistics and probability during the second world war. After the war, he did outstanding work in these areas and almost single-handedly educated an entire generation of mathematical statisticians and probabilists. He was a tireless promoter of applied mathematics and one of the founders of the Mathematisch Centrum at Amsterdam. Those who didn't know this rather formidable man, should read HEMELRIJK's (1959) obituary as well as some of VAN DANTZIG's papers listed there.

Though mathematical statistics was more or less unknown territory for Dutch mathematicians at the time VAN DANTZIG entered the field, this was certainly not the case for Dutch physicists and astronomers. It is the purpose of this historical note to show that, in fact, they knew quite a bit about the subject at a very early date. In particular, I shall discuss an unpublished set of notes on what is now called robust statistics, written in 1942 by VAN DE HULST, then an astronomy student at Utrecht and presently professor emeritus of theoretical astronomy at Leiden. Almost as fascinating as the notes themselves is the correspondence about the results between VAN DE HULST and VAN DANTZIG, and the remarks that VAN DANTZIG pencilled in the margins of the notebook. Asked for his opinion, VAN DANTZIG complained at great length about the lack of mathematical rigor, but finally relented somewhat and tried to encourage VAN DE HULST to continue working on statistical problems. But things turned out differently. In 1944 VAN DE HULST predicted the 21 cm radio spectral line of hydrogen, which eventually led to the birth of radio astronomy. From there, he went on to a brilliant career in theoretical astronomy and never bothered to publish his investigation of robust statistics. I'm endebted to professor VAN DE HULST for mentioning his work to me and for making the notebook and the ensuing correspondence available. Thanks

S. van de Geer and M. Wegkamp (eds.), *Selected Works of Willem van Zwet*, Selected Works in Probability and Statistics, DOI 10.1007/978-1-4614-1314-1_3, © Springer Science+Business Media, LLC 2012

also go to Professor S.M. STIGLER for drawing my attention to the work of DANIELL (1920).

2 A problem of Hertzsprung

In the issue of May 20, 1942, of the Bulletin of the Astronomical Institutes of the Netherlands, E. HERTZSPRUNG, director of the Observatory at Leiden, describes a sampling experiment to determine the variance of the trimmed mean. In connection with the determination of relative proper motions of stars in the Pleiades, HERTZSPRUNG discusses how one should assign weights to the observed values to account for differences in quality of the observations. He writes:
"The simplest way to deal with exorbitant observations is to reject them. In order to avoid special rules for onesided rejection the easy way of symmetrical rejection of the largest deviations to each side may be considered. The first question is then: How much is, in the case of Gaussian distribution of errors, the weight of the result diminished by a priori symmetrical rejection of outstanding observations? As the mathematical treatment of this question appears to be laborious beyond the needs mentioned above I gave preference to an empirical answer. On each of 12534 slips of paper was written with two decimals a deviation from zero in units of the mean error, in such a way that these deviations showed a Gaussian distribution. Thus 50 slips were marked with .00, 50 with +.01, 50 with -.01 etc.. Of these slips somewhat more than 1000 times 24 were picked out arbitrarily. Such 24 slips were in each case arranged according to the size of the deviation and mean squares of the sums of 24-x deviations calculated after symmetrical rejection of $x = 0,2,4, ..., 22$ extreme values."

This paragraph should warm a statistician's heart, except that he may feel slightly uneasy about "somewhat more than 1000" replications. And he has reason to feel uneasy: "Of all these samples of 24 exactly 1000 were picked out in such a way that the sum of all 24 deviations ($x = 0$) fairly well showed a Gaussian distribution with a mean square of 24."

From a theoretical point of view, this ruins a perfectly good sampling experiment, as VAN DANTZIG was quick to point out, especially since no further information is supplied. There is no way of assessing the accuracy of the estimated variances any more. On the other hand, if we assume that this data cleaning was done sensibly, there seems to be no reason, a priori, why the estimates should be much worse than they would have been otherwise.

We need some notation. $X_1, X_2, .., X_n$ will denote independent and identically distributed random variables with mean zero, finite variance and a common density f, which is symmetric about zero. The standard normal density will be denoted by ϕ. Let $X_{1:n} < X_{2:n} < ... < X_{n:n}$ be the ordered sample and define the trimmed means and their variances by

$$\overline{X}_{n,k} = \frac{1}{n-2k} \sum_{i=k+1}^{n-k} X_{i:n} \qquad (1)$$

$$\sigma^2_{n,k} = E\overline{X}^2_{n,k}. \tag{2}$$

For $f=\phi$, $n=24$ and $k=0,1,2, ..., 11$, Hertzsprung estimates the quantities $n\sigma^2_{n,k}=\sigma^2_{n,k}/\sigma^2_{n,0}$ by the corresponding ratios of the sampling variances of 1000 replications. His results are given in Table 1.

Table 1. HERTZSPRUNG's estimates of $n\sigma^2_{n,k}$ for $f=\phi$, $n=24$, $k=0,1, ... ,11$.

k	$s^2_{n,k}/s^2_{n,0}$	k	$s^2_{n,k}/s^2_{n,0}$	k	$s^2_{n,k}/s^2_{n,0}$
0	1.000	4	1.095	8	1.283
1	1.013	5	1.139	9	1.345
2	1.037	6	1.184	10	1.407
3	1.069	7	1.232	11	1.489

Commenting on these numbers, HERTZSPRUNG writes: "While the cancelling of two arbitrary observations out of 24 diminishes the weight from 24 to 22 the symmetrical rejection of the two largest deviations leaves a weight of nearly 23.7 for the mean of the rest. Hence there is not much reason for hesitation to do so, while the question is still left open as to how false the assigned weights must be in order to obtain an increase of weight by the procedure considered." The word "weight" is used for the reciprocal of the variance and what we have here is a plea for the 5% - trimmed mean!

Finally, HERTZSPRUNG notes that the formula

$$n\sigma^2_{n,k} = 1+.53(2k/n)^{3/2} \tag{3}$$

fits the data in Table 1 quite well. Since the median has asymptotic variance $\pi/(2n)$ in the normal case, he proposes to replace .53 in (3) by $\pi/2-1=$ .57.The entire paper doesn't take more than one page.

3 M-estimators

After attending a talk given by HERTZSPRUNG about his sampling experiment, VAN DE HULST decided to try and treat the problem of finding $\sigma^2_{n,k}$ mathematically. In letters to HERTZSPRUNG of April 15 and June 10, 1942, he computes values of $\sigma^2_{24,11}$ and $\sigma^2_{24,1}$. Since these computations also occur in his notes, I shall return to them later. In the weeks that followed, he apparently made quick progress and he wrote down his results in a notebook dated July 1942; there is a supplement dated October 1942. The notes are written in Dutch and are entitled "Over een probleem uit de waarschijnlijkheidsrekening" (On a problem in probability theory). In discussing these notes and other writings, I shall change the notation and terminology to one that is more common in statistics nowadays. Whenever direct quotes occur, the translation is mine.

Since the stated purpose of the notes is to compute the variance of the trimmed mean, it is surprising that the author should start with the asymptotic variance of an M-estimator:

"*Theorem.*

If n observations X_i (n very large) are distributed according to the symmetric probability (density) f, and one determines the number M by

$$\sum_{i=1}^{n} \psi(X_i - M) = 0, \tag{4}$$

where ψ is some odd function, then

$$EM^2 = \frac{\int \psi^2(x) f(x) dx}{n\{\int f(x) d\psi(x)\}^2}. \text{"} \tag{5}$$

VAN DE HULST calls this result well known and refers to page 470 *ff* of F. ZERNIKE's (1928) chapter on Probability Theory and Mathematical Statistics in Volume III of the Handbuch der Physik. The amazing implication is that the concept of an M-estimator as well as the expression for its asymptotic variance were known to physicists as early as 1928 and it seems worthwhile to take a look at this reference.

The fact that ZERNIKE, a Dutch physicist and a Nobel laureate a quarter of a century later, was asked to write on probability and statistics for the Handbuch, shows that he was considered an authority in this field by his colleagues, and indeed he did give a very interesting and up-to-date account of the area. His treatment of M-estimators starts with assuming a large number of observations and a symmetric error density f because (translating his German text) "for a skew error density a sharp distinction between systematic and random errors is not possible" and continues: "The commonly used best (summary) value of n observations is the arithmetic mean, determined by the equation $\Sigma(X_i - \bar{X}) = 0$. One considers the generalization of this equation $\Sigma\psi(X_i - M) = 0$, where ψ is an odd function of the argument $(x - M)$. This equation can be interpreted as follows: M is the mean of the X-values computed with weights $\psi/(X - M)$, i.e. with a symmetric weight function. If the number of X-values in every interval would be exactly equal to the (theoretical frequencies) calculated from the error law, then one would find $M = m$ (the expectation of X). From the statistical deviations of these numbers, one calculates for the deviation of M

$$\sigma^2(M) = \frac{\int \psi^2(x) f(x) dx}{n\{\int f(x) d\psi(x)\}^2}, \tag{6}$$

and for special choices of the function ψ this formula leads easily to the following results:

Arithmetic mean $\psi(x) = x$ $\sigma^2 = n^{-1}\sigma^2(x)$
Median $\psi(x) = sgn(x)$ $\sigma^2 = (4nf^2(0))^{-1}$
Best determination $\psi(x) = f'(x)/f(x)$ $\sigma^2 = (n\int(f'(x))^2/f(x)dx)^{-1}$."

ZERNIKE then goes on to provide the variances of these three estimators for four selected error densities. There is no further proof and regularity conditions are not mentioned. He does point out that for the median, the function ψ is not continuous and that the integral in the denominator of (6) should be interpreted as a Stieltjes integral. The interpretation of an M-estimator as a weighted mean with random weights is still part of the folklore in this field (cf. HUBER (1981), p. 44). The original of the result (6) is not clear as ZERNIKE doesn't provide any references but doesn't claim the result as his own either. What is clear, is that up-to-date knowledge of statistics did exist in the Netherlands in 1928.

But let us return to the notes of VAN DE HULST. Since there is no proof of the theorem in ZERNIKE's review paper, VAN DE HULST gives one. He prefaces his proof by the remark that the won't be bothered with details, and perhaps this is just as well. It doesn't make much sense to discuss regularity conditions if it isn't even clear what the conclusion of the theorem ought to be. As it stands, it is a statement about the limit of the variance of $n^{1/2}M$, as opposed to the variance of the limit distribution of $n^{1/2}M$. I doubt that, at the time, many people knew there was a difference between the two, and it certainly wasn't a distinction that was commonly made. We now realize, however, that the former type of result is usually harder to prove, but also less relevant than the latter, because asymptotic theory and small sample approximations concern distributions rather than moments.

Let us then ignore this distinction and look at VAN DE HULST's proof. He starts with the one-term Taylor expansion

$$0 = \sum\psi(X_i - M) = \sum\psi(X_i) - M\sum\psi'(X_i) \tag{7}$$

and rewrites it as

$$M = \frac{\sum\psi(X_i)}{\sum\psi'(X_i)}. \tag{8}$$

From here on one could argue that $n^{-1/2}\Sigma\psi(X_i)$ is asymptotically normal with mean zero and variance $\int\psi^2 f$ and that $n^{-1}\Sigma\psi'(X_i)$ converges in probability to $\int\psi' f$ if both integrals are finite. It follows by Slutsky's theorem that (8) implies that $n^{1/2}M$ is asymptotically normal with mean zero and variance given by (5). The convergence of the variance of $n^{1/2}M$ would take a bit more work.

Instead, VAN DE HULST follows a more devious route that was much travelled in those days. He discretizes the X's by partitioning the real line into a large number of small intervals. If N_j of the X's fall in the j-th interval and ψ_j and ψ_j' are values which ψ and ψ' assume somewhere in this interval, then approximately

$$M = \frac{\sum N_j\psi_j}{\sum N_j\psi_j'}. \tag{9}$$

Another Taylor expansion with respect to the N_j around EN_j, followed by the computation of variances and covariances for the multinomial distribution and a passage to the limit, produces. (5).

Of course it is not important how one proceeds from (8) on. The main problem is what happened to the remainder term in (7). To VAN DE HULST this might be a detail not to be worried about, but for VAN DANTZIG it was too much to swallow. In his letter of February 19, 1943, he writes: "It is really a pity that your discussion is so inexact, as this takes away much of the value of the several nice ideas that it contains. For instance, in the "proof" on page 3, I can understand the passage to $0=\Sigma\psi(X_i-M)=\Sigma\psi(X_i)-M\Sigma\psi'(X_i)$ only if (1) $\psi(x)$ is differentiable and (2) M is small, whereas $\psi''(x)$ remains bounded (even then I don't see how one can know in advance that the term involving M^2 can be neglected, since it is the calculation of M^2 we are concerned with)."

VAN DANTZIG is clearly right in pointing out that this kind of proof will work only for smooth functions ψ and I'm sure that VAN DE HULST readily agreed. It is interesting, however, that VAN DANTZIG had difficulty seeing how one shows that M is small, i.e. that M is a consistent estimator. If, in addition to its smoothness, one simply assumes ψ to be nondecreasing and strictly increasing on a set of positive probability under f, then $n^{-1}\Sigma\psi(X_i-m)$ is a continuous and nonincreasing function of m and $\int\psi(x-m)f(x)dx$ is strictly decreasing in m in a neighborhood of zero. The consistency of M now follows from the law of large numbers in the same way as in CRAMÉR's (1945) proof of the consistency of the maximum likelihood estimator. Of course CRAMÉR's book had not yet appeared, and apparently this argument was not yet generally known.

Having completed his proof, VAN DE HULST mentions the special cases of the mean and the median discussed by ZERNIKE, but again VAN DANTZIG is not impressed: "But under no circumstances can I understand how this proof can be applied to $\psi(x)=sgn(x)$. What is the meaning of $M=\Sigma\psi(X_i)/\Sigma\psi'(X_i)$ here? Therefore, these considerations don't prove anything for the case of the median."

Let us put these mathematical objections aside for a moment and take stock of what has been archieved so far. An expression has been derived for the asymptotic variance of an M-estimator of location. The proof is rather shaky, but we know today that the expression does indeed hold in great generality, including the case of the sample median. The reason is, that it is the smoothness of $\lambda(m)=E\psi(X-m)$ which is important rather than the smoothness of ψ. However, the original purpose was to find the asymptotic variance of the trimmed mean. Of course the two extreme cases of the trimmed mean, the sample median and the untrimmed mean, are also M-estimators but their asymptotic variances were already known. VAN DE HULST writes: "One wonders whether the intermediate cases (of the trimmed mean) ... can also be written in the form $\Sigma\psi(X_i-M)=0$ by choosing an appropriate function ψ. However, this is not the case and the result obtained above can not he used directly. Nevertheless it turns out that with a minor modification in the above proof, this case can also be treated."

4 Trimmed means

As before, let $X_1, ..., X_n$ be independent and identically distributed with mean zero, finite variance and a common density f, which is symmetric about zero. The distribution function corresponding to f is denoted by F. Consider the trimmed mean $\bar{X}_{n,k}$ and its variance $\sigma^2_{n,k}$ defined by (1) and (2). Denote the trimming fraction on each side by α and let a be the upper α-point of F, thus

$$\alpha = \frac{k}{n}, \quad F(a) = 1-\alpha. \tag{10}$$

Define the function ψ_0 by

$$\psi_0(x) = \begin{cases} -a & \text{if } x<-a, \\ x & \text{if } -a \leqslant x \leqslant a, \\ a & \text{if } x>a. \end{cases} \tag{11}$$

VAN DE HULST shows that if $n \to \infty$ and α remains bounded away from $\frac{1}{2}$, then

$$n\sigma^2_{n,k} \sim (1-2\alpha)^{-2} \int \psi_0^2(x) f(x)dx, \tag{12}$$

where $\sim$ denotes asymptotic equality. I shall try to give a simplified version of VAN DE HULST's argument which, I hope, still retains the original flavor. An entirely different proof of (12) was given earlier in DANIELL (1920), but this paper seems to have gone completely unnoticed (cf. STIGLER (1973)).

Define the interval $I = (X_{k:n}, X_{n-k:n}$ and let F_n be the empirical distribution function. We can write

$$\bar{X}_{n,k} = (1-2\alpha)^{-1} \int_I x dF_n(x). \tag{13}$$

Neglecting lower order terms, we have

$$\begin{aligned} \alpha - F_n(-a) &= F_n(X_{k:n}) - F_n(-a) \sim (X_{k:n}+a)f(a), \\ 1-\alpha-F_n(a) &= F_n(X_{n-k:n}) - F_n(a) \sim (X_{n-k:n}-a)f(a). \end{aligned} \tag{14}$$

This is intuitively clear, but not entirely trivial to prove rigorously. It is called the BAHADUR representation after BAHADUR (1966), who proved that the remainder term is $O(n^{-3/4}\log n)$ almost surely. Similarly,

$$\int_I x dF_n(x) - \int_{-a}^{a} x dF_n(x) \sim (X_{k:n}+X_{n-k:n})af(a). \tag{15}$$

Combining (13), (14), (15) and (11), we find

$$\bar{X}_{n,k} \sim (1-2\alpha)^{-1} \int \psi_0(x) dF_n(x) = (1-2\alpha)^{-1} n^{-1} \sum \psi_0(X_i), \tag{16}$$

and with a little bit of luck this implies (12).

Since $\int f d\psi_0 = (1-2\alpha)$, a comparison of (6) and (12) shows that the asymptotic variances of the α-trimmed mean and the M-estimator with $\psi = \psi_0$ (which is known as HUBER's estimator) coincide. Of course VAN DE HULST is pleased to note this and it confirms his intuition that M-estimators would have something to do with the problem. On the other hand, however, he finds the agreement of the two variances rather fortuitous.

Photograph 1. Part of a page in VAN DE HULST's notebook where he concludes that the trimmed mean has the same asymptotic variance as HUBER's estimator. Note the drawing of the influence curve of these two estimators, which is almost a symbol of robust statistics. The scribbled line at the bottom is VAN DANTZIG's.

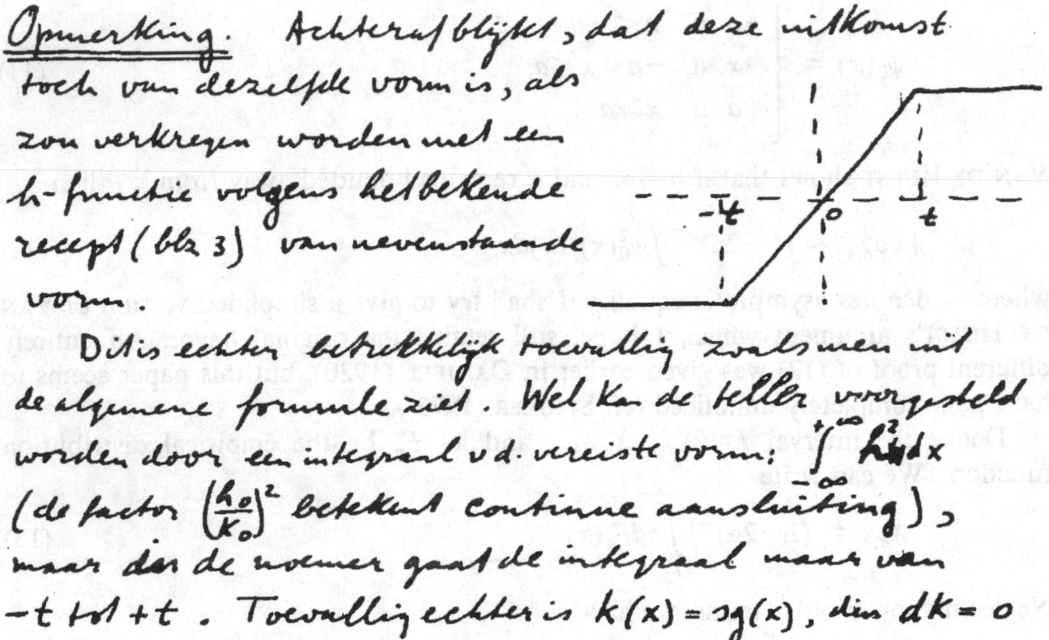
Opmerking. Achteraf blijkt, dat deze uitkomst toch van dezelfde vorm is, als zou verkregen worden met een h-functie volgens het bekende recept (blz 3) van nevenstaande vorm.

Dit is echter betrekkelijk toevallig zoals men uit de algemene formule ziet. Wel kan de teller voorgesteld worden door een integraal v.d. vereiste vorm: $\int_{-\infty}^{+\infty} h^2 dx$ (de factor $\left(\frac{h_0}{k_0}\right)^2$ betekent continue aansluiting), maar dan de noemer gaat de integraal maar van $-t$ tot $+t$. Toevallig echter is $k(x) = sg(x)$, dus $dk = 0$ hierbuiten. Men mag dus ook hier de grenzen door $-\infty$ en $+\infty$ vervangen. Kan steeds als functie w buiten interval door nul vervangen wordt.

A closer inspection of his argument, however, shows that the agreement goes much further than he may have realized. As I remarked below (8), we have $n^{-1}\Sigma\psi'(X_i) \sim \int \psi' f$ and hence

$$M \sim \frac{\Sigma\psi(X_i)}{n\int \psi'(x) f(x) dx}. \tag{17}$$

Choosing $\psi=\psi_0$, we find for HUBER's estimator H,

$$H \sim (1-2\alpha)^{-1} n^{-1} \sum \psi_0(X_i), \tag{18}$$

and together, (16) and (18) imply that $n^{1/2}(\bar{X}_{n,k}-H)$ tends to zero in probability. Thus HUBER's estimator and the α-trimmed mean are asymptotically equivalent estimators and the equality of the asymptotic variances is merely a consequence of this.

There is an interesting discussion connected with this result. When robust estimators began to be investigated, it was felt by some, that throwing away observations, as one does when computing a trimmed mean, is perhaps overdoing things a little. Wouldn't it be better to move these observations towards the center of the sample and replace them by $X_{k+1:n}$ and $X_{n-k:n}$ rather than deleting them outright? This led to the introduction of the so-called Winsorized mean

$$W_{n,k} = \frac{1}{n}\left[\sum_{i=k+1}^{n-k} X_{i:n} + k(X_{k+1:n}+X_{n-k:n})\right]. \tag{19}$$

Later, after HUBER's estimator had been introduced and found to perform well, representation (18) and the shape of ψ_0 as given by (11) seemed to suggest that the same thing is going on here: the X_i are replaced by $\psi_0(X_i)$, which moves the outlying observations towards the center. On the strength of this, it was assumed that the Winsorized mean would mimic HUBER's estimator and share its good performance to a greater extent than the trimmed mean. This argument is not really very convincing because, in view of the symmetry of f, the effect of moving observations towards the center depends very much on the exact number of observations moved and the exact positions they are moved to. Nevertheless, this idea was rather generally accepted until BICKEL (1965) proved that the asymptotic distribution of the trimmed mean is the same as that of HUBER's estimator. In HUBER's words (cf. HUBER (1981), p. 59): "This exemplifies how unreliable our intuition can be; we know now ... that the trimmed mean does not throw away all of the information sitting in the discarded observations, but that it does exactly what the Winsorized mean was supposed to do." Of course one can agree with this sentiment in general, but after seeing VAN DE HULST's notes, one is inclined to add that this depends very much on whose intuition one is talking about!

For $f=\phi$, (12) reduces to

$$n\sigma_{n,k}^2 \sim (1-2\alpha)^{-2}\{(1-2\alpha)-2a\phi(a)+2\alpha a^2\} \tag{20}$$

and for $n=24$, these values are given in Table 2.

Table 2. Asymptotic values of $n\sigma_{n,k}^2$ for $f=\phi$, $n=24$, $k=0, 1, \ldots, 11$.

k	$n\sigma_{n,k}^2$	k	$n\sigma_{n,k}^2$	k	$n\sigma_{n,k}^2$
0	1.000	4	1.114	8	1.294
1	1.021	5	1.153	9	1.352
2	1.048	6	1.195	10	1.417
3	1.079	7	1.242	11	1.483

These figures seem to agree reasonably well with HERTZSPRUNG's empirical data in Table 1. Commenting on VAN DE HULST's further efforts to improve the asymptotic approximations, VAN DANTZIG writes: "On the whole, I don't think it makes sense to try to explain the small deviations between your calculations and HERTZSPRUNG's results; in my opinion, the agreement is too good to be true." He argues that, since HERTZSPRUNG's data have been "doctored", one certainly can't attach "any value to the third, and perhaps even to the second decimal of the empirical data".

In view of the agreement which is "too good to be true", it is perhaps not surprising that VAN DANTZIG also feels that (20) is "almost certainly incorrect". The reason for his doubts is interesting. Apart from the argument for (12) and (20) that we have just sketched, VAN DE HULST also provides a second proof of (20), which is precisely the proof that one would give today: Given $X_{k:n}$ and $X_{n-k+1:n}$, the trimmed mean is distributed as an ordinary sample mean and the conditional second moment is easily calculated. Taking the expectation with respect to the bivariate normal limit distribution $(X_{k:n}+a)$ and $(X_{n-k+1:n}-a)$, one obtains (20). VAN DANTZIG argues, mistakenly, that second order terms of this bivariate distribution should also play a role and give rise to additional terms in the final result. The fact that the two proofs produce the same result, merely leads him to conclude that the first proof is probably incorrect too.

Van Dantzig's letter from which I have quoted repeatedly, was typical for the person I believe he was. His criticism was very much to the point and mathematically correct, except for his doubts about the validity of (20). It was offered in a matter-of-fact way and it probably didn't occur to him at first, that it might have a rather devastating effect on the receiver of the letter. He was genuinely trying to help and felt hat the best way to do this, was to explain his views as clearly as possible.

Photograph 2. End of VAN DANTZIG's letter to VAN DE HULST.

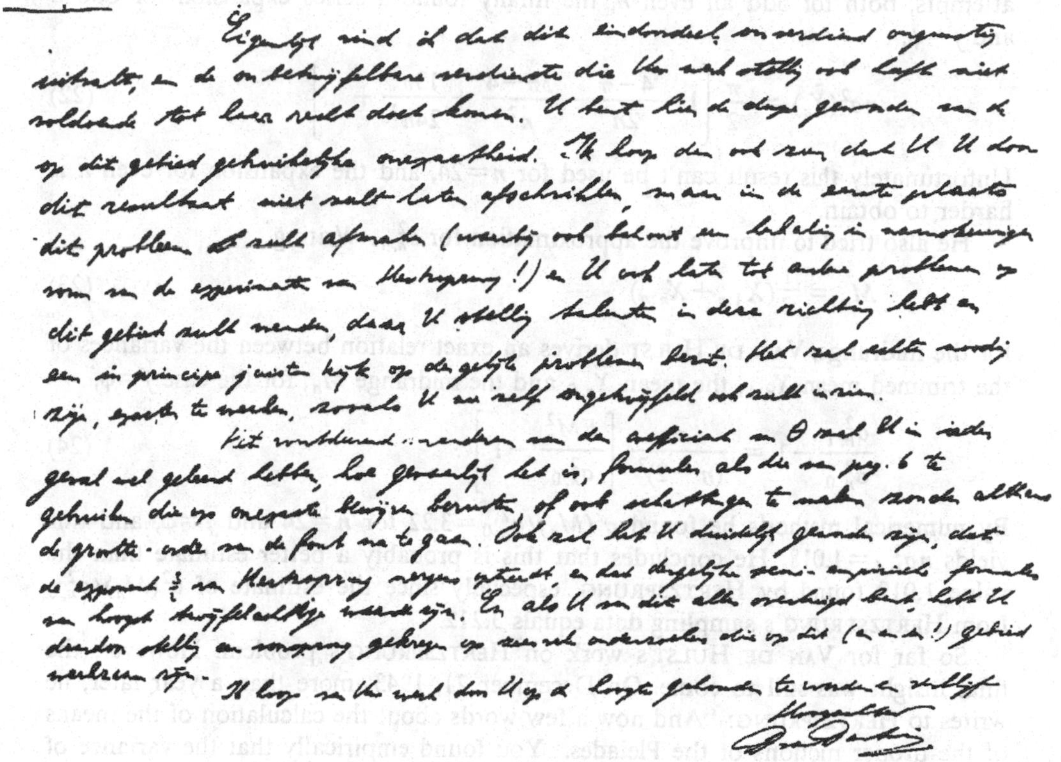

Eigenlijk vind ik dat dit eindoordeel onverdiend ongunstig uitvalt, en de onbetwijfelbare verdiensten die Uw werk stellig wel heeft niet voldoende tot hun recht doet komen. U bent hier de dupe geworden van de op dit gebied gebruikelijke onexactheid. Ik hoop dan ook zeer, dat U U door dit resultaat niet zult laten afschrikken, maar in de eerste plaats dit probleem ook zult afmaken (waartoe ook behoort een herhaling in vermeerdering van de experimenten van Hertzsprung!) en U ook later tot andere problemen op dit gebied zult wenden, daar U stellig talenten in deze richting hebt en een in principe juisten kijk op dergelijke problemen bezit. Het zal echter nodig zijn, exacter te werken, zooals U nu zelf ongetwijfeld ook zult inzien.

Uit het vervolg van de coefficient van θ zult U in ieder geval wel geleerd hebben, hoe gevaarlijk het is, formules als die van pag. 6 te gebruiken die op onexacte kunstgrepen berusten, of ook schattingen te maken zonder althans de grootte-orde van de fout na te gaan. Ook zal het U duidelijk geworden zijn, dat de exponent $\frac{3}{2}$ van Hertzsprung nergens uit volgt en dat dergelijke semi-empirische formules van hoogst twijfelachtige waarde zijn. En als U van dit alles overtuigd bent, hebt U daardoor stellig een verrijking gekregen op veel onderzoekers die op dit (en ander!) gebied werkzaam zijn.

Ik hoop van Uw werk door U op de hoogte gehouden te worden en verblijf

Hoogachtend

D. van Dantzig

I imagine that it was when rereading the letter, that he felt that perhaps he had been a bit too severe. He was basically a kind person and he could certainly recognize talent when he saw it. He added: "On second thought I find that my final opinion has turned out undeservedly unfavorable and that it doesn't do justice to the indubitable merit of your work. ... I do hope that you won't be discouraged by this result and that you'll first finish this problem ... and then turn to other problems in this area. You definitely have talents in this direction and in principle you have the right way of looking at such problems." Of course this praise was interspersed with admonitions to work more rigorously!

5 Other problems

As was mentioned in section 3, VAN DE HULST started out by studying the median $\tilde{X}_n$, which he defined in the usual way as

$$\tilde{X}_n = \begin{cases} X_{\frac{1}{2}(n+1):n} & = \bar{X}_{n,\frac{1}{2}(n-1)} \quad \text{if } n \text{ is odd,} \\ \frac{1}{2}(X_{\frac{1}{2}n:n} + X_{\frac{1}{2}n+1:n}) & = \bar{X}_{n,\frac{1}{2}n-1} \quad \text{if } n \text{ is even.} \end{cases} \tag{21}$$

It was known that for $f=\phi$, the asymptotic variance of $n^{1/2}\bar{X}_n$ equals $\frac{1}{2}\pi$. VAN DE HULST tried to calculate a correction term of order n^{-1}. After several unsuccessful attempts, both for odd an even n, the finally found a series expansion for odd n and $f=\phi$,

$$n\sigma^2(\bar{X}_n) = \frac{\pi}{2}\left[1-\frac{4-\pi}{2n}-\frac{3\pi-4}{n^2}+\frac{13\pi^2}{24n^2}+\ldots\right]. \tag{22}$$

Unfortunately this result can't be used for $n=24$, and the expansion for even n is harder to obtain.

He also tried to improve the approximation for $\sigma^2_{n,1}$. Writing

$$M_n = \frac{1}{2}(X_{1:n}+X_{n:n}) \tag{23}$$

for the midrange, VAN DE HULST derives an exact relation between the variances of the trimmed mean $\bar{X}_{n,1}$, the mean $\bar{X}_{n,0}$ and the midrange M_n, for the case $f=\phi$,

$$\frac{\sigma^2_{n,1}}{\sigma^2_{n,0}}-1 = \frac{4}{(n-2)^2}\left[\frac{EM_n^2}{\sigma^2_{n,0}}-1\right]. \tag{24}$$

By numerical methods he found $\sigma^2(M_n)/\sigma^2_{n,0}=3.22$ for $n=24$ and $f=\phi$, and this yields $n\sigma^2_{n,1}=1.018$. He concludes that this is probably a better estimate than the value 1.013 found by HERTZSPRUNG, especially since the estimate of $\sigma^2(M_n)/\sigma^2_{n,0}$ from HERTZSPRUNG's sampling data equals 3.212.

So far for VAN DE HULST's work on HERTZSPRUNG's problem. However, the final insight was still to come. On December 21, 1943, more than a year later, he writes to HERTZSPRUNG: "And now a few words about the calculation of the means of the proper motions of the Pleiades. You found empirically that the variance of the trimmed mean of 24 observations with a Gaussian distribution depends on $\alpha=k/n$ according to curve (a) (an increasing function of α is shown). I later found practically the same result by computation. In connection with the things you showed me on December 1 last, I discovered yet another possibility. It is possible to apply my formula (Formula (12) not to a Gaussian distribution, but to the true distribution of the measurement errors, including the so-called outliers. In a fictitious example I obtained curve (b) (a function of α which first decreases rapidly, and then increases more slowly, is shown). On this curve one can read off precisely how much the variance of the mean decreases, if one rejects one or more observations symmetrically. The location of the minimum in this example indicates, that it is best to reject about 25% of the observations, that is 3 on both sides out of 24!" He then proposes to estimate the distribution of the errors in HERTZSPRUNG's sampling experiment and find the optimal trimming percentage.

Presumably things never got to that point, but the computation to which VAN DE HULST refers is attached to the notebook. It concerns an error distribution with a range from -4.24 to +4.24 and his conclusion is that about 10% should be trimmed on both sides. Here he is definitely on the road that TUKEY would take in

1949!

6 Can we do better now?

It may be of interest to consider briefly how we would deal with this problem today and ask whether we could do essentially better. I believe the answer is: not very much. Of course we can now prove these results rigorously and in great generality. Also, we have sufficient computing power at our disposal to compute the required quantities exactly, if need be. It is a good thing that VAN DE HULST didn't have that possibility, because he wouldn't have discovered anything if he had.

The first thing we can do is to compute the asymptotic variance of HERTZSPRUNG's estimator of $n\sigma^2_{n,k}$,

$$\frac{s^2_{n,k}}{s^2_{n,0}} = \frac{\sum_{j=1}^{N}\{\overline{X}^{(j)}_{n,k}\}^2}{\sum_{j=1}^{N}\{\overline{X}^{(j)}_{n,0}\}^2}, \tag{25}$$

where $\overline{X}^{(j)}_{n,0}$ and $\overline{X}^{(j)}_{n,k}$, $j=1, \ldots, N$, denote the means and trimmed means of the $N=1000$ samples of size $n=24$. For $f=\phi$ we have, as n, $N\to\infty$,

$$\sigma^2\left[\frac{s^2_{n,k}}{s^2_{n,0}}\right] = \frac{4n\sigma^2_{n,k}}{N}(n\sigma^2_{n,k}-1)+\Theta(\frac{1}{Nn}+\frac{1}{N^2}), \tag{26}$$

where we may replace $n\sigma^2_{n,k}$ by the right-hand side of (20). Of course, one could also use $ns^2_{n,k}$ as an estimate for $n\sigma^2_{n,k}$, and then we find, for $f=\phi$,

$$\sigma^2(ns^2_{n,k}) = \frac{2n^2\sigma^4_{n,k}}{N}+\Theta(\frac{1}{Nn}+\frac{1}{N^2}). \tag{27}$$

This confirms what was already intuitively obvious, that HERTZSPRUNG's estimator is better than $ns^2_{n,k}$, especially for small values of $\alpha=k/n$, when $n\sigma^2_{n,k}$ is close to 1.

Another thing we can do is to calculate a second order approximation for $n\sigma^2_{n,k}$. Such approximations, including a term of order n^{-1}, can be found in HELMERS (1982) for general linear combinations of order statistics. However, use of these general formulas involves very lengthy computations and in this simple case, it is easier to start from scratch. In the notation of Section 4 we find, for $f=\phi$ and α bounded away from $\frac{1}{2}$,

$$n\sigma^2_{n,k} = \frac{1}{(1-2\alpha)^2}\left[(1-2\alpha)-2a\phi(a)+2\alpha a^2+n^{-1}\{-2\alpha(1-\alpha)\frac{a}{\phi(a)} + 2\alpha^2(1-\alpha)(\frac{a}{\phi(a)})^2+\alpha^2(1-2\alpha)\frac{1}{(\phi(a))^2}\}\right]+\Theta(n^{-2}). \tag{28}$$

For easy comparison Table 3 provides HERTZSPRUNG's estimates of $n\sigma^2_{n,k}$ taken from Table 1, the asymptotic standard deviation of these estimates computed from (26) and (20), the first order approximation of $n\sigma^2_{n,k}$ taken from Table 2 and finally the second order approximation of $n\sigma^2_{n,k}$ computed from (28).

Table 3. HERTZSPRUNG's estimate, its standard deviation and approximate values of $n\sigma^2_{n,k}$ for $f=\phi$, $n=24$, $k=0,1,\ldots,11$.

k	$s^2_{n,k}/s^2_{n,0}$	$\sigma(s^2_{n,k}/s^2_{n,0})$	$n\sigma^2_{n,k}$ first order	$n\sigma^2_{n,k}$ second order
0	1.000	0	1.000	1.000
1	1.013	.009	1.021	1.017
2	1.037	.014	1.048	1.043
3	1.069	.019	1.079	1.073
4	1.095	.023	1.114	1.106
5	1.139	.027	1.153	1.143
6	1.184	.031	1.195	1.185
7	1.232	.035	1.242	1.230
8	1.283	.039	1.294	1.280
9	1.345	.044	1.352	1.335
10	1.407	.049	1.417	1.397
11	1.489	.054	1.483	1.459

Inspection of this table shows that the agreement between HERTZSPRUNG's estimates and the asymptotic values is closer than one would expect; the difference never exceeds the standard deviation of the estimate and is considerably smaller in most cases. Perhaps HERTZSPRUNG's data cleaning worked rather well!

The second order approximation looks somewhat better than the first order approximation, except at the bottom of the table. This is as it should be. The first order result (20) is valid for all values of $\alpha=k/n$, but the second order result (28) holds only if α is bounded away from 1/2 as n tends to infinity. For values of α close to 1/2, it can therefore be expected to give a bad approximation. For $k=11$, it would be better to use the second order approximation for the median for even sample size, which VAN DE HULST was unable to find. For even n and $f=\phi$, it is

$$n\sigma^2(\tilde{X}_n) = \frac{\pi}{2}(1-\frac{6-\pi}{2n})+\mathcal{O}(n^{-2}), \tag{29}$$

and for $n=24$, this yields 1.477 .

References

BAHADUR, R.R. (1966) , A note on quantiles in large samples, *Ann. Math. Statist. 37*, 577-580.

BICKEL, P.J. (1965), On some robust estimates of location, *Ann. Math. Statist. 36*, 847-858.

CRAMÉR, H. (1945), *Mathematical Methods of Statistics*, Princeton University Press, Princeton.

DANIELL, P.J. (1920), Observations weighted according to order, *American Journal of Mathematics 42*, 222-236.

HELMERS, R. (1982), *Edgeworth Expansions for Linear Combinations of Order Statistics*, Mathematical Centre Tracts 105, Mathematisch Centrum, Amsterdam.

HEMELRIJK, J. (1959), In memoriam Prof.Dr. D. van Dantzig, *Statistica Neerlandica 13*, 416-432.

HERTZSPRUNG, E. (1942), On the symmetrical rejection of extreme observations, *Bulletin of the Astronomical Institutes of the Netherlands IX (349)*, 285-286.

HUBER, P.J. (1981), *Robust Statistics*, Wiley, New York.

STIGLER, S.M. (1973), Simon Newcomb, Percy Daniell, and the history of robust estimation 1885-1920, *J. Amer. Statist. Assoc. 68*, 872-879.

ZERNIKE, F. (1928), Wahrscheinlichkeitsrechnung und mathematische Statistik. In: *Handbuch der Physik*, Bd III, *Mathematische Hilfsmittel in der Physik*, Kap. 12, Springer, Berlin.

Dept. of Mathematics and Computer Science
University of Leiden
P.O. Box 9512
2300 RA Leiden
The Netherlands

Chapter 4
Discussion of three statistics papers by Willem van Zwet

Jon A. Wellner

Abstract I discuss three statistics papers of Willem van Zwet: [12], [5], and [15].

4.1 Introduction

I first met Willem at the 2nd conference on *Statistical decision theory and related topics*, held at Purdue University in May 1976. After a brief discussion over dinner on the topic of my dissertation (concerning certain limit theorems for linear combinations of order statistics), Willem tactfully pointed out that perhaps I had missed some interesting problems of a somewhat more fundamental nature concerning strong laws for such linear combinations. This brief conversation lead to [16]. Willem himself beautifully improved my results in [14] as discussed by David Mason elsewhere in this volume.

Beyond giving good advice, Willem is well-known to many for his story-telling abilities, both in his papers and over a beer in a corner at Oberwolfach. The three papers discussed here provide ample evidence of the former (with hints of the latter, especially in [15]). The reader interested in more of the latter should consult [1].

4.2 Paper 1.

The first of these three papers, *Convex transformations: a new approach to skewness and kurtosis,* is based on [13]. It gives a wonderfully clear exposition of partial orderings for distribution functions (or their associated random variables) which

Jon A. Wellner
Department of Statistics, University of Washington, Seattle, WA 98195-4322 e-mail: jaw@stat.washington.edu Supported in part by NSF Grant DMS-0804587, and by NI-AID grant 2R01 AI291968-04

S. van de Geer and M. Wegkamp (eds.), *Selected Works of Willem van Zwet*, Selected Works in Probability and Statistics, DOI 10.1007/978-1-4614-1314-1_4,

"cover our intuitive ideas about skewness and kurtosis", and have a variety of further statistical applications.

Briefly, for distribution functions F, F^* which are twice continuously differentiable on some interval I with $F'(x) > 0$ on I, $F \overset{c}{<} F^*$ if and only if G^*F is convex on I where G^* is the inverse function of F^* defined by $G^*F^*(x) = x$. Similarly, for the subclass of all distribution functions as above which are symmetric about some point x_0, $F \overset{s}{<} F^*$ if and only if G^*F is convex for $x \geq x_0, x \in I$, where x_0 is the (common) point of symmetry. It is easily seen from the forward and inverse probability integral transformations that if X has distribution function F, then $X^* \equiv G^*F(X)$ has distribution function F^*, and hence it is natural to write $X \overset{c}{<} X^*$ whenever $F \overset{c}{<} F^*$.

For $X \sim F$ and a positive integer k let

$$\begin{aligned}
\gamma_{2k+1}(F) &\equiv \gamma_{2k+1}(X) \equiv \frac{E(X-EX)^{2k+1}}{\sigma^{2k+1}(F)}, \\
\gamma_{2k}(F) &\equiv \gamma_{2k}(X) \equiv \frac{E(X-EX)^{2k}}{\sigma^{2k}(F)}, \quad \text{where} \\
\sigma^2(F) &\equiv \sigma^2(X) \equiv E(X-EX)^2,
\end{aligned}$$

assuming that the expectations exist. Thus $\gamma_1(F)$ is the classical skewness of F and $\gamma_2(F)$ is the kurtosis of F. The paper [12] starts with the basic results

$$\begin{aligned}
&\gamma_{2k+1}(X) \leq \gamma_{2k+1}(\varphi(X)) \text{ for any convex function } \varphi, \text{ and} \\
&\gamma_{2k}(X) \leq \gamma_{2k}(\varphi(X)) \text{ for any convex, odd about } x_0, \text{ function } \varphi \\
&\qquad\qquad \text{if } X \sim F \text{ symmetric about } x_0.
\end{aligned}$$

These results are discussed heuristically and used to motivate the definitions of $F \overset{c}{<} F^*$ and $F \overset{s}{<} F^*$. A natural choice of φ is exactly G^*F.

Willem himself writes about this paper:

> This is a short summary of my dissertation. Over the years, the Centrum voor Wiskunde en Informatica (Center for Mathematics and Computer Science) at Amsterdam has sold 800 copies. The reason is that the topic is revisited every ten years or so. Among other things, the thesis deals with a partial ordering of one-dimensional probability distributions that produces an increasing skewness to the right, and discusses a few simple consequences of this ordering. Nowadays I would formulate this as an ordering in terms of the fatness of the tail rather than in terms of skewness. The subject will doubtless enjoy yet another lifetime due to the current interest in heavy tails by queuing theory folks, financial mathematics people, et cetera. The thesis is now out of print, but it should be available in the libraries of some statistics departments.

Despite the recent comprehensive book [10], Willem's paper and his thesis [13] remain gems of the stochastic orderings literature.

Here is a conjecture related to van Zwet's $\overset{s}{<}$ ordering:

Conjecture: Let X have Chernoff's distribution as described in [6]; this distribution arises as the limit distribution in a variety of problems involving monotone nonparametric function estimation. Let Z be a random variable with a standard normal distribution (with mean zero, variance 1). Both X and Z have distributions symmetric about 0. I conjecture that $X \overset{s}{<} \sigma Z$ and that $f_X(t) = h(t)\varphi(t/\sigma)/\sigma$ with $h-$log-concave if $\sigma \geq .52$.

4.3 Paper 2.

The "two-armed bandit problem", apparently introduced in [8], is as follows: you are presented with a slot machine with two arms. One arm yields a payoff of \$1 with probability α and the other arm yields a payoff of \$1 with probability β. The rub is that you do not know which arm is connected with these probabilities, and you also don't know the values of α and β. The goal is to maximize your expected winnings in N successive pulls of one or the other of the two arms. Alternatively, if you are very patient and have lots of time to play the machine, you may have the goal of maximizing your limiting average expected winnings as N is allowed to become large. It has been known since [8] that there exist strategies achieving the latter goal: if X_k denotes the winnings from play k, then there is decision rule or strategy for choosing one or the other of the two arms so that

$$\frac{1}{N}\sum_{k=1}^{N} X_k \to \max\{\alpha,\beta\} \quad \text{as } N \to \infty$$

with probability one; see e.g. [7] and [4]. Finding optimal strategies for finite N is somewhat more difficult, but perhaps more important for a variety of real problems. If you have played both arms by step $m < N$, then playing the arm which has yielded the smaller winnings so far results in sub-optimal winnings in the next step, but a strategy involving always "playing the winner" can also be sub-optimal, as was shown by [3]. The results of these authors prompt Fabius and van Zwet to write:

> Though these relations may seem intuitively evident, one does well to remember that the two-armed bandit problem has been shown to defy intuition in many aspects (cf. [3]).

Fabius and van Zwet formulate the two-armed bandit problem in a general decision theoretic setting allowing randomized decision rules and an arbitrary prior distribution for (α, β) on $[0,1]^2$. They proceed by characterizing the class of all Bayes rules, and show (Theorem 4) that every admissible strategy is Bayes agains a "non-marginal prior distribution" π. They give an explicit example showing that "... there is an essentially unique and hence admissible Bayes strategy against π which violates (the monotonicity requirements) (i) and (ii) (of "play the winner" rules)... ", thereby reconfirming the results of [3]. Fabius and van Zwet go on to provide wonderfully explicit calculations of minimax symmetric rules and risk for $N = 3$ and $N = 4$.

For further development of these problems and themes, see [4], [9], and the survey by [7].

4.4 Paper 3.

In this delightful historical article Willem reviews the work of the Dutch astronomer Van de Hulst on the behavior of trimmed means, and the wonderful interactions between Van de Hulst and the imminent Dutch mathematician and statistician D. van Dantzig.

In connection with this paper Willem writes:

> At some time during the early 1980s I gave a talk for a general sciences audience. As many scientists routinely remove outliers from their data, I thought it might be useful to speak about trimmed means and what happens if you use them. In the talk I showed them the derivation of the asymptotic variance of the trimmed mean. There was a spirited discussion afterwards. To my utter surprise, Van de Hulst – a world-famous astronomer from Leiden – shows up in my office a few days later carrying a small notebook written in 1942 that contained precisely this asymptotic result. From a mathematical point of view, the proof left something to be desired, but the right ideas were all there. In 1942 he apparently knew all about M-estimators too, and this knowledge goes back to Nobel laureate Zernike in 1928. In his 1942 notes Van de Hulst also showed that what is now known as Huber's estimator has the same asymptotic variance as the trimmed mean. After Huber's estimator had been introduced, statisticians first believed that its asymptotic variance would coincide with that of the Winsorized mean until Bickel proved Van de Hulst's result in 1965 ([2]). Van de Hulst is justifiably pleased by the recognition provided in this paper and has shown it to all of his astronomy friends!.

Willem's article sketches the theory of "M-estimators" that was apparently well-known in the Dutch astronomy community in the 1930's and 1940's and that was used as a starting point by Van de Hulst in his investigations. A proof of the theorem concerning the asymptotic variance of an "M-estimator" was not given in the known reference, so Van de Hulst provided one. But van Dantzig felt that Van de Hulst's proof was not "rigorous". Willem provides a fascinating commentary on the interactions between the two scientists, with a very readable introduction to the theory (translated into modern notation and terminology), including connections between "M-estimators" (or "Z- estimators" as they are renamed slightly in [11]) and trimmed means via Bahadur's representation theorem for quantiles. In the last section of the paper Willem's intimate familiarity with second order expansions and correction terms comes into play in an elegant and subtle re-analysis of the results of Van de Hulst and empirical data concerning trimmed means from Hertzsprung.

I commend the article to the reader as a superb example of Willem at his storytelling best!

References

1. Beran, R. J. and Fisher, N. I. (2009). An evening spent with Bill van Zwet. *Statist. Sci.* **24** 87–115.
2. Bickel, P. J. (1965). On some robust estimates of location. *Ann. Math. Statist.* **36** 847–858.
3. Bradt, R. N., Johnson, S. M. and Karlin, S. (1956). On sequential designs for maximizing the sum of n observations. *Ann. Math. Statist.* **27** 1060–1074.
4. Duflo, M. (1996). *Algorithmes Stochastiques, Mathématiques & Applications* **23**, Springer-Verlag, Berlin.
5. Fabius, J. and van Zwet, W. R. (1970). Some remarks on the two-armed bandit. *Ann. Math. Statist.* **41** 1906–1916.
6. Groeneboom, P. and Wellner, J. A. (2001). Computing Chernoff's distribution. *J. Comput. Graph. Statist.* **10** 388–400.
7. Pemantle, R. (2007). A survey of random processes with reinforcement. *Probab. Surv.* **4** 1–79 (electronic).
8. Robbins, H. (1952). Some aspects of the sequential design of experiments. *Bull. Amer. Math. Soc.* **58** 527–535.
9. Rosenberger, W. F. (1996). New directions in adaptive designs. *Statistical Science* **11** 137–149.
10. Shaked, M. and Shanthikumar, J. G. (2007). *Stochastic Orders.* Springer Series in Statistics, Springer, New York.
11. van der Vaart, A. W. and Wellner, J. A. (1996). *Weak Convergence and Empirical Processes. With Applications to Statistics* Springer Series in Statistics, Springer-Verlag, New York.
12. van Zwet, W. R. (1964a). Convex transformations: A new approach to skew- ness and kurtosis. *Statistica Neerlandica* **18** 433–441.
13. van Zwet, W. R. (1964b). *Convex transformations of Random Variables.* Vol. 7 of Mathematical Centre Tracts. Mathematisch Centrum, Amsterdam.
14. van Zwet, W. R. (1980). A strong law for linear functions of order statistics. *Ann. Probab.* **8** 986–990.
15. van Zwet, W. R. (1985). van de Hulst on robust statistics: a historical note. *Statist. Neerlandica 39* 81–95.
16. Wellner, J. A. (1977). A Glivenko-Cantelli theorem and strong laws of large numbers for functions of order statistics. *Ann. Statist.* **5** 473–480.

Part II
Asymptotic Statistics

The Annals of Mathematical Statistics
1972, Vol. 43, No. 4, 1122–1135

ASYMPTOTIC NORMALITY OF NONPARAMETRIC TESTS FOR INDEPENDENCE[1]

BY F. H. RUYMGAART, G. R. SHORACK[2] AND W. R. VAN ZWET

Mathematisch Centrum, Amsterdam and University of Leiden

Asymptotic normality of linear rank statistics for testing the hypothesis of independence is established under fixed alternatives. A generalization of a result of Bhuchongkul [1] is obtained both with respect to the conditions concerning the orders of magnitude of the score functions and with respect to the smoothness conditions on these functions.

1. Introduction. For each n let $(X_1, Y_1), \cdots, (X_n, Y_n)$ be a random sample from a continuous bivariate distribution function (df) $H(x, y)$ having marginal dfs $F(x)$ and $G(y)$. The bivariate empirical df based on this sample is denoted by H_n. With respect to the n random variables (rvs) $X_i(Y_i)$ corresponding to the first (second) coordinates, the empirical df is denoted by $F_n(G_n)$, the ith order statistic by $X_{in}(Y_{in})$ and the rank of $X_i(Y_i)$ by $R_i(Q_i)$. All samples are defined on a single probability space $(\Omega, \mathscr{A}, P)$.

The rank statistics most commonly used to test the independence hypothesis $H = F.G$, are of the linear type

$$T_n = n^{-1} \sum_{i=1}^{n} a_n(R_i) b_n(Q_i) ,$$

where $a_n(i)$, $b_n(i)$ are real numbers for $i = 1, \cdots, n$ (see Hájek and Šidák [6]). A suitably standardized version of T_n will be (see also Bhuchongkul [1])

$$n^{\frac{1}{2}}(T_n - \mu) = n^{\frac{1}{2}}[\textstyle\iint J_n(F_n)K_n(G_n)\, dH_n - \mu] ; \tag{1.1}$$

here

$$J_n(s) = a_n(i) , \qquad K_n(s) = b_n(i) , \tag{1.2}$$

for $(i-1)/n < s \leqq i/n$ and $i = 1, \cdots, n$, and

$$\mu = \textstyle\iint J(F)K(G)\, dH , \tag{1.3}$$

for some functions J and K on $(0, 1)$ that can be thought of as limits of the score functions J_n and K_n.

In order to summarize the main results of this paper let us introduce the function

$$r = [I(1-I)]^{-1} \quad \text{on} \quad (0, 1) , \tag{1.4}$$

where I is the identity function on the unit interval. Under the hypothesis and under contiguous alternatives, asymptotic normality of (1.1) may be proved for score functions J and K of order $r^{\frac{1}{2}-\delta}$ for some $\delta > 0$ (see Hájek and Šidák [6]).

Received September 25, 1970; revised October 29, 1971.

[1] Report SW 5a/71 of the Department of Mathematical Statistics, Mathematisch Centrum, Amsterdam.

[2] This research was partially supported by NSF Contract No. GP-13739.

S. van de Geer and M. Wegkamp (eds.), *Selected Works of Willem van Zwet*, Selected Works in Probability and Statistics, DOI 10.1007/978-1-4614-1314-1_5, © Springer Science+Business Media, LLC 2012

Jogdeo [7] establishes asymptotic normality under the hypothesis of a statistic more general than T_n; the growth condition on his score functions in the case of T_n is $r^{\frac{1}{4}-\delta}$. By an approach analogous to that of Chernoff and Savage [3] for the two-sample problem, Bhuchongkul [1] proves asymptotic normality under fixed alternatives provided the score functions are of the order $\log r$ (see Section 2). The main purpose of this paper is to relax these conditions to $r^{\frac{1}{4}-\delta}$ in general and $r^{\frac{1}{2}-\delta}$ for a special class of dfs H.

In Theorem 2.1 the asymptotic normality of (1.1) is established for rather smooth score functions with orders of magnitude not exceeding r^a and r^b, where the numbers a and b satisfy the relations $a = (\frac{1}{2} - \delta)/p_0$ and $b = (\frac{1}{2} - \delta)/q_0$ for some $0 < \delta < \frac{1}{2}$ and some $p_0, q_0 > 1$ with $p_0^{-1} + q_0^{-1} = 1$. No condition other than continuity is imposed on the df H. The theorem is stronger than Theorem 1 of Bhuchongkul [1]. The proof is based on Hölder's inequality in the form

$$\int\int |\phi(F)\psi(G)| \, dH \leq [\int |\phi|^p \, dI]^{1/p}[\int |\psi|^q \, dI]^{1/q} , \tag{1.5}$$

where ϕ and ψ are functions on (0, 1), dI denotes Lebesgue measure restricted to the unit interval and $p, q > 1$ satisfy $p^{-1} + q^{-1} = 1$.

Theorem 2.2 gives asymptotic normality of (1.1) under much weaker conditions on the score functions. Here these functions are allowed to be of order r^a and r^b, where $a = b = \frac{1}{2} - \delta$ for some $0 < \delta < \frac{1}{2}$. The price for this is a condition on the df H, keeping it in some sense similar to the null hypothesis. This condition is

$$dH \leq C[r(F)r(G)]^{\delta/2} \, dF \, dG , \tag{1.6}$$

with fixed constants $C \geq 1$ and $0 < \delta < \frac{1}{2}$. Mathematically, (1.6) allows a direct factorization of the left-hand integral in (1.5) which is more efficient than Hölder's inequality. Intuitively, this condition prevents the large (small) X's from occurring in the same pair as large (small) Y's with too high a probability. Condition (1.6) trivially holds under the null hypothesis. More generally it is also satisfied if H can be written as a polynomial in its marginals F and G. This class of distributions was introduced by Lehmann [9] and the special case where $H = FG[1 + \alpha(1 - F)(1 - G)]$ for $-1 \leq \alpha \leq 1$ was considered by Gumbel [5]. Finally (1.6) holds for all bivariate normal distributions with a sufficiently small correlation coefficient (use Lemma 2 on page 166 of Feller [4] to see that (1.6) holds for a correlation coefficient between $-\delta/(2 - \delta)$ and $\delta/(2 - \delta)$).

2. Statement of the theorems. Each of the theorems below establishes the asymptotic normality

$$n^{\frac{1}{2}}(T_n - \mu) \to_d N(0, \sigma^2) \qquad \text{as } n \to \infty , \tag{2.1}$$

of (1.1); here μ and σ^2 are finite and are given by (1.3) and (3.10) respectively.

Let $\mathscr{H}$ denote the class of all continuous bivariate dfs H, and let $\mathscr{H}_{C\delta}$ denote the subclass that satisfies (1.6) for fixed $C \geq 1$ and $0 < \delta < \frac{1}{2}$.

To prove (2.1) for general H in $\mathscr{H}$ we require a strong boundedness condition on the score functions.

ASSUMPTION 2.1. The functions J and K are continuous on $(0, 1)$; each is differentiable except at an at most finite number of points, and in the open intervals between these points the derivatives are continuous. The functions J_n, K_n, J, K satisfy $|J_n| \leqq Dr^a$, $|K_n| \leqq Dr^b$ and

$$|J^{(i)}| \leqq Dr^{a+i}, \qquad |K^{(i)}| \leqq Dr^{b+i} \qquad \text{for } i = 0, 1,$$

where defined on $(0, 1)$. Here D is a positive constant and a and b satisfy

$$a = (\tfrac{1}{2} - \delta)/p_0, \quad b = (\tfrac{1}{2} - \delta)/q_0 \tag{2.2}$$

for some $0 < \delta < \frac{1}{2}$ and some $p_0, q_0 > 1$ with $p_0^{-1} + q_0^{-1} = 1$.
In proving (2.1) for the more restrictive class $\mathscr{H}_{C\delta}$ we only require a weak boundedness condition on the score functions.

ASSUMPTION 2.2. Assumption 2.1 holds with

$$a = b = \tfrac{1}{2} - \delta \tag{2.3}$$

for some $0 < \delta < \frac{1}{2}$.

We also need a condition on the convergence of J_n, K_n to J, K. Define

$$B_{0n} = n^{\frac{1}{2}} \textstyle\iint_{\Delta_n} [J_n(F_n)K_n(G_n) - J(F_n)K(G_n)]\, dH_n, \tag{2.4}$$

$$B_{0n}^* = n^{\frac{1}{2}} \textstyle\iint [J_n(F_n)K_n(G_n) - J(F_n^*)K(G_n^*)]\, dH_n, \tag{2.5}$$

where

$$\Delta_n = \Delta_{n1} \times \Delta_{n2} \quad \text{with} \quad \Delta_{n1} = [X_{1n}, X_{nn}) \quad \text{and} \quad \Delta_{n2} = [Y_{1n}, Y_{nn}), \tag{2.6}$$

$$F_n^* = [n/(n+1)]F_n, \qquad G_n^* = [n/(n+1)]G_n, \tag{2.7}$$

ASSUMPTION 2.3. Either (a) $B_{0n} \to_p 0$ as $n \to \infty$, or (b) $B_{0n}^* \to_p 0$ as $n \to \infty$.
This assumption is very general, but may occasionally be difficult to verify. However, most examples are special cases of Remarks 2.1 and 2.2 below.

REMARK 2.1. If the scores of (1.2) satisfy $a_n(i) = J(i/(n+1))$ and $b_n(i) = K(i/(n+1))$ for $1 \leqq i \leqq n$ for some functions J and K, then Assumption 2.3 (b) holds uniformly for H in $\mathscr{H}$. (In this case $B_{0n}^* = 0$ for all n.)

REMARK 2.2. Suppose that J and K are increasing and twice differentiable on $(0, 1)$, and that $|J^{(i)}| \leqq Dr^{a+i}$ and $|K^{(i)}| \leqq Dr^{b+i}$ for $i = 0, 1, 2$ where $D > 0$ and a and b satisfy (2.2). Let the scores $a_n(i)$ and $b_n(i)$ of (1.2) be the expectations of the ith order statistics of samples of size n from populations whose dfs are the inverse functions of J and K respectively. Then Assumption 2.1 holds and Assumption 2.3 (a) holds uniformly for all H in $\mathscr{H}$. (This statement generalizes Theorem 2 of [1] and the proof may be given in the same way. It relies mainly on the fact that $\sum_{i=1}^{n-1} |a_n(i) - J(i/n)| = O(n^a)$ and $\sum_{i=1}^{n-1} |b_n(i) - K(i/n)| = O(n^b)$, which follows from formulas (7.14) and (7.24) of [3] with $\alpha = a$ and $\alpha = b$ respectively.)

THEOREM 2.1. *If H is in $\mathscr{H}$ and if Assumptions* 2.1 *and* 2.3 *are satisfied, then the asymptotic normality* (2.1) *holds. Given any subclass $\mathscr{H}'$ of $\mathscr{H}$ such that As-*

sumption 2.3 holds uniformly for H in $\mathscr{H}'$ and such that $\sigma^2 = \sigma^2(H)$ is bounded away from 0 on $\mathscr{H}'$, the convergence in (2.1) is uniform for H in $\mathscr{H}'$.

Note that (2.2) is satisfied if $a = b = \frac{1}{4} - \varepsilon$ for some $0 < \varepsilon < \frac{1}{4}$ (take $p_0 = q_0 = 2$ and $\delta = 2\varepsilon$). Thus Theorem 2.1 allows a rate of growth $r^{\frac{1}{4}-\varepsilon}$ for the score functions J and K and $r^{\frac{5}{4}-\varepsilon}$ for their derivatives. In Theorem 1 of [1] these rates are $r^{\frac{1}{2}-\varepsilon}$ and r respectively; in fact the latter condition reduces the rate for J and K to $\log r$. Moreover in [1] the score functions are assumed to be twice differentiable throughout the unit interval.

THEOREM 2.2. *Fix $C \geqq 1$ and $0 < \delta < \frac{1}{2}$. If H is in $\mathscr{H}_{C\delta}$ and if Assumptions 2.2 and 2.3 are satisfied, then the asymptotic normality (2.1) holds. Given any subclass $\mathscr{H}'_{C\delta}$ of $\mathscr{H}_{C\delta}$ such that Assumption 2.3 holds uniformly for H in $\mathscr{H}'_{C\delta}$ and such that $\sigma^2 = \sigma^2(H)$ is bounded away from 0 on $\mathscr{H}'_{C\delta}$, the convergence in (2.1) is uniform for H in $\mathscr{H}'_{C\delta}$.*

3. Proof of the theorems: Asymptotic normality of the leading terms. Let $F^{-1}(s) = \inf\{x: F(x) \geqq s\}$ and $G^{-1}(t) = \inf\{y: G(y) \geqq t\}$; these definitions imply $F(F^{-1}) = G(G^{-1}) = I$. The random functions $F_n(F^{-1})$ and $G_n(G^{-1})$ are with probability 1 the empirical dfs of the sets of independent uniform (0, 1) rvs $F(X_1), \cdots, F(X_n)$ and $G(Y_1), \cdots, G(Y_n)$ respectively. Define the empirical processes $U_n = n^{\frac{1}{2}}[F_n(F^{-1}) - I]$ and $V_n = n^{\frac{1}{2}}[G_n(G^{-1}) - I]$ on [0, 1]. With probability 1 these processes satisfy $U_n(F) = n^{\frac{1}{2}}(F_n - F)$ and $V_n(G) = n^{\frac{1}{2}}(G_n - G)$ on $(-\infty, \infty)$. All of the above remarks follow from the fact that

$$(3.1) \qquad P(\Omega_0) = P(\{\omega : F_n(F^{-1}(F)) = F_n,\ G_n(G^{-1}(G)) = G_n \quad \text{for all} \quad x, y \quad \text{and} \quad n\}) = 1\,.$$

Without loss of generality we shall prove Theorems 2.1 and 2.2 in the case where both J and K fail to have a derivative at just one point, say at s_1 and t_1 respectively. For small positive γ define the sets

$$(3.2) \qquad S_{\gamma 1} = [F^{-1}(\gamma), F^{-1}(s_1 - \gamma)] \cup [F^{-1}(s_1 + \gamma), F^{-1}(1 - \gamma)]\,.$$
$$S_{\gamma 2} = [G^{-1}(\gamma), G^{-1}(t_1 - \gamma)] \cup [G^{-1}(t_1 + \gamma), G^{-1}(1 - \gamma)]\,,$$

$$(3.3) \qquad \Omega_{\gamma n} = \{\omega : \sup|F_n - F| < \gamma/2,\ \sup|G_n - G| < \gamma/2\}\,.$$

Let $S_\gamma = S_{\gamma 1} \times S_{\gamma 2}$ be the product set in the plane and let $\chi(\Omega_{\gamma n})$ denote the indicator function of $\Omega_{\gamma n}$. For ω in $\Omega_0 \cap \Omega_{\gamma n}$ the mean value theorem gives

$$n^{\frac{1}{2}}J(F_n) = n^{\frac{1}{2}}J(F) + U_n(F)J'(\Phi_n)$$

for all x in $\Delta_{n1} \cap S_{\gamma 1}$. In the above formula the function Φ_n is defined by $\Phi_n = F + \theta(F_n - F)$, where $\theta = \theta(\omega, x, n)$ is a number between 0 and 1. Thus with probability 1 (using Assumption 2.3 (a))

$$(3.4) \qquad n^{\frac{1}{2}}(T_n - \mu) = \textstyle\sum_{i=1}^{3} A_{in} + \sum_{i=0}^{2} B_{in} + \sum_{i=3}^{7} B_{\gamma in} + B_{8n} + C_n\,,$$

where B_{0n} is defined in (2.4) and where

$$A_{1n} = n^{\frac{1}{2}} \iint J(F)K(G)d(H_n - H),$$
$$A_{2n} = \iint U_n(F)J'(F)K(G)\,dH, \qquad A_{3n} = \iint V_n(G)J(F)K'(G)\,dH,$$
$$B_{1n} = n^{\frac{1}{2}} \iint_{\Delta_n^c} J_n(F_n)K_n(G_n)\,dH_n, \qquad B_{2n} = -n^{\frac{1}{2}} \iint_{\Delta_n^c} J(F)K(G)\,dH_n,$$
$$B_{\gamma 3n} = \chi(\Omega_{\gamma n}^c)\{n^{\frac{1}{2}} \iint_{\Delta_n} [J(F_n) - J(F)]K(G)dH_n - A_{2n}\},$$
$$B_{\gamma 4n} = \chi(\Omega_{\gamma n})n^{\frac{1}{2}} \iint_{\Delta_n \cap S_\gamma^c} [J(F_n) - J(F)]K(G)\,dH_n,$$
$$B_{\gamma 5n} = \chi(\Omega_{\gamma n}) \iint_{\Delta_n \cap S_\gamma} U_n(F)[J'(\Phi_n) - J'(F)]K(G)\,dH_n,$$
$$B_{\gamma 6n} = \chi(\Omega_{\gamma n}) \iint_{\Delta_n \cap S_\gamma} U_n(F)J'(F)K(G)d(H_n - H),$$
$$B_{\gamma 7n} = -\chi(\Omega_{\gamma n}) \iint_{\Delta_n^c \cup S_\gamma^c} U_n(F)J'(F)K(G)\,dH,$$
$$B_{8n} = n^{\frac{1}{2}} \iint_{\Delta_n} J(F)[K(G_n) - K(G)]dH_n - A_{3n},$$
$$C_n = n^{\frac{1}{2}} \iint_{\Delta_n} [J(F_n) - J(F)][K(G_n) - K(G)]\,dH_n.$$

Let us note that

$$\textstyle\sum_{i=3}^{7} B_{\gamma in} = n^{\frac{1}{2}} \iint_{\Delta_n} [J(F_n) - J(F)]K(G)dH_n - A_{2n},$$

which is symmetric to B_{8n}. For this reason B_{8n} will not be treated in the sequel.

We now proceed to prove the asymptotic normality of the A-terms. Let us start with the very useful remark that if a and b satisfy (2.2), then for $i = 1$ and 2 we can find numbers $p_i, q_i > 1$ satisfying $p_i^{-1} + q_i^{-1} = 1$ and

$$(a + \tfrac{1}{2} + \delta/2)p_1 < 1, \qquad bq_1 < 1, \qquad ap_2 < 1, \qquad (b + \tfrac{1}{2} + \delta/2)q_2 < 1. \tag{3.5}$$

As to the first pair of inequalities, we have $a + \frac{1}{2} + \delta/2 + b = 1 - \delta/2$ and consequently $a + \frac{1}{2} + \delta/2 < 1 - \delta/2$ (the numbers a and b are strictly positive). Now choose $p_1 = (a + \frac{1}{2} + 3\delta/4)^{-1}$ and let $q_1 = (1 - p_1^{-1})^{-1}$. Then $(a + \frac{1}{2} + \delta/2)p_1 < 1$ and $bq_1 = (\frac{1}{2} - a - \delta)/(\frac{1}{2} - a - 3\delta/4) < 1$. The second pair of inequalities can be obtained in the same way.

The rv A_{1n} can be written in the form

$$A_{1n} = n^{-\frac{1}{2}} \textstyle\sum_{i=1}^{n} A_{1in}, \tag{3.6}$$

where $A_{1in} = J(F(X_i))K(G(Y_i)) - \mu$ are independent and identically distributed (i.i.d.) with mean zero. Under Assumption 2.1 application of (1.5) with $p = p_0$ and $q = q_0$ shows that the rv A_{1in} has a finite absolute moment of order $2 + \delta_0$ for some $\delta_0 > 0$. The same conclusion holds under Assumption 2.2 for H in $\mathscr{H}_{C\delta}$ as may be seen by applying (1.6). Moreover this moment will be uniformly bounded above for H within $\mathscr{H}(\mathscr{H}_{C\delta})$.

Because

$$U_n(F) = n^{-\frac{1}{2}} \textstyle\sum_{i=1}^{n} (\phi_{X_i} - F),$$

where

$$\phi_{X_i}(x) = 0 \quad \text{if} \quad x < X_i \qquad \text{and} \qquad \phi_{X_i}(x) = 1 \quad \text{if} \quad x \geqq X_i, \tag{3.7}$$

we have

$$A_{2n} = n^{-\frac{1}{2}} \textstyle\sum_{i=1}^{n} A_{2in}, \tag{3.8}$$

where the $A_{2in} = \iint (\phi_{X_i} - F)J'(F)K(G)\,dH$ are i.i.d. with mean zero. Under Assumptions 2.1 or 2.2 we have

$$|A_{2in}| \leq D^2 r^{\frac{1}{2}-\delta/4}(F(X_i)) \iint r^{a+\frac{1}{2}+\delta/4}(F)r^b(G)\,dH\,.$$

For some $\delta_1 > 0$ the random part of this upper bound possesses an absolute moment of order $2 + \delta_1$ which is uniformly bounded above for H in $\mathscr{H}$. Under Assumption 2.1 the nonrandom integral is seen to be uniformly bounded above for H in $\mathscr{H}$ by application of (1.5) with $p = p_1$ and $q = q_1$ as in (3.5). Uniform boundedness of this integral for H in $\mathscr{H}_{c\delta}$ holds under Assumption 2.2, as may be shown by application of (1.6).

Analogously we can write

$$A_{3n} = n^{-\frac{1}{2}} \textstyle\sum_{i=1}^{n} A_{3in}\,, \tag{3.9}$$

where $A_{3in} = \iint (\phi_{Y_i} - G)J(F)K'(G)\,dH$ are i.i.d. with mean zero. Again for $\delta_1 > 0$ this rv has a finite absolute moment of order $2 + \delta_1$ which is uniformly bounded for H in $\mathscr{H}(\mathscr{H}_{c\delta})$. This time use (1.5) with $p = p_2$ and $q = q_2$ as in (3.5).

Combining (3.6), (3.8) and (3.9) we get $\sum_{i=1}^{3} A_{in} \to_d N(0, \sigma^2)$ as $n \to \infty$. The variance σ^2 is given by (see [1])

$$\sigma^2 = \mathrm{Var}\,[J(F(X))K(G(Y)) + \iint (\phi_X - F)J'(F)K(G)\,dH + \iint (\phi_Y - G)J(F)K'(G)\,dH]\,, \tag{3.10}$$

with ϕ defined in (3.7).

Since we have shown that an absolute moment of order larger than 2 exists and is uniformly bounded on $\mathscr{H}(\mathscr{H}_{c\delta})$, and because the variance is uniformly bounded away from zero on $\mathscr{H}'(\mathscr{H}'_{c\delta})$, the established convergence in distribution is uniform for H in $\mathscr{H}'(\mathscr{H}'_{c\delta})$ by Esséen's theorem (see e.g. [3], Section 4).

4. Some lemmas. We start with a number of lemmas to be used in the proofs of both Theorem 2.1 and Theorem 2.2.

LEMMA 4.1. *For any* $\zeta \geq 0$ *the function* r^ζ *is symmetric about* $\frac{1}{2}$, *decreasing on* $(0, \frac{1}{2}]$ *and has the property that for each* β *in* $(0, 1)$ *there exists a constant* $M = M_\beta$ *such that* $r^\zeta(\beta s) \leq Mr^\zeta(s)$ *for* $0 < s \leq \frac{1}{2}$ *and* $r^\zeta(1 - \beta(1 - s)) \leq Mr^\zeta(s)$ *for* $\frac{1}{2} < s < 1$.

PROOF. On $(0, \frac{1}{2}]$ we have $r^\zeta(\beta s) = (\beta s)^{-\zeta}(1 - \beta s)^{-\zeta} \leq \beta^{-\zeta} r^\zeta(s)$. A similar argument applies to the interval $(\frac{1}{2}, 1)$. □

LEMMA 4.2. *For each* ω *let* $\tilde{\Phi}_n = \tilde{\Phi}_{n\omega}$ *and* $\tilde{\Psi}_n = \tilde{\Psi}_{n\omega}$ *be functions on* $\Delta_{n1} = \Delta_{n1\omega}$ *and* $\Delta_{n2} = \Delta_{n2\omega}$ *respectively (see* (2.6)), *satisfying* $\min(F, F_n) \leq \tilde{\Phi}_n \leq \max(F, F_n)$ *and* $\min(G, G_n) \leq \tilde{\Psi}_n \leq \max(G, G_n)$ *where defined. Then uniformly for* $n = 1, 2, \cdots$ *and* $H \in \mathscr{H}$:

(i) $\sup_{\Delta_{n1}} r^\zeta(\tilde{\Phi}_n)r^{-\zeta}(F) = O_p(1)$ *for each* $\zeta \geq 0$;
(ii) $\sup_{\Delta_{n2}} r^\eta(\tilde{\Psi}_n)r^{-\eta}(G) = O_p(1)$ *for each* $\eta \geq 0$;
(iii) $\sup_{(-\infty,\infty)} |U_n(F)|r^{\frac{1}{2}-\tau}(F) = O_p(1)$ *for each* $\tau > 0$.

PROOF. (i) From formula (3.1) and e.g. from [11], Lemma A.3 it follows that for each $\varepsilon > 0$ there exists a constant $\beta = \beta_\varepsilon$ in (0, 1) such that

$$(4.1)\qquad P(\Omega_n) = P(\{\beta F \leqq F_n \leqq 1 - \beta(1-F) \text{ on } \Delta_{n1}\}) > 1 - \varepsilon\,,$$

for all n and uniformly in all continuous F. Because of the definition of $\tilde{\Phi}_n$ we have $\beta F \leqq \tilde{\Phi}_n \leqq 1 - \beta(1-F)$ on Δ_{n1}. By Lemma 4.1 this implies that for some constant $M_{\zeta\varepsilon}$ we have $r^\zeta(\tilde{\Phi}_n) \leqq M_{\zeta\varepsilon} r^\zeta(F)$ for x in Δ_{n1} on the set Ω_n.

(ii) This is analogous to (i).

(iii) This follows immediately from Lemma 2.2 of Pyke and Shorack [10]. □

For each positive integer k we define a function I_k on [0, 1] by

$$(4.2)\qquad I_k(0) = 0\,, \qquad I_k(s) = (i-1)/k \quad \text{for} \quad (i-1)/k < s \leqq i/k\,, \qquad i = 1, \cdots, k\,,$$

LEMMA 4.3. *As $k, n \to \infty$, $\sup_{(-\infty,\infty)} |U_n(I_k(F)) - U_n(F)| \to_p 0$ uniformly in all continuous F.*

PROOF. Note that $\sup_{-\infty<x<\infty} |U_n(I_k(F)) - U_n(F)| = \sup_{0\leqq s\leqq 1} |U_n(I_k) - U_n|$, which is no longer dependent on F. The U_n-processes converge weakly to a tied-down Wiener process U_0 (see e.g. Billingsley [2]). In Pyke and Shorack [10] these U_n- and U_0-processes are replaced by $\tilde{U}_n$- and $\tilde{U}_0$-processes defined on a single new probability space $(\tilde{\Omega}, \tilde{\mathscr{A}}, \tilde{P})$ and having the same finite dimensional distributions as the original processes (see also Skorokhod [12]). These new processes satisfy $\sup |\tilde{U}_n - \tilde{U}_0| \to_{\text{a.s.}} 0$ and hence also $\sup |\tilde{U}_n(I_k) - \tilde{U}_0(I_k)| \to_{\text{a.s.}} 0$ uniformly in k, as $n \to \infty$. Now $\sup |\tilde{U}_n(I_k) - \tilde{U}_n| \leqq \sup |\tilde{U}_n - \tilde{U}_0| + \sup |\tilde{U}_0 - \tilde{U}_0(I_k)| + \sup |\tilde{U}_0(I_k) - \tilde{U}_n(I_k)|$. For almost every $\tilde{\omega}$ the function $\tilde{U}_0$ is uniformly continuous on [0, 1] so that $\sup |\tilde{U}_0 - \tilde{U}_0(I_k)| \to_{\text{a.s.}} 0$ as $k \to \infty$. This proves that $\sup |\tilde{U}_n(I_k) - \tilde{U}_n| \to_{\text{a.s.}} 0$ for $k, n \to \infty$. This last result implies the convergence in probability of the lemma. □

Let ν and λ be the random indices $1 \leqq \nu(\omega), \lambda(\omega) \leqq n$ such that

$$(4.3)\qquad X_\nu = X_{nn} \quad \text{and} \quad Y_\lambda = Y_{nn}\,.$$

LEMMA 4.4. *As $n \to \infty$, $P(\{\alpha_n \leqq F(X_\nu) \leqq 1 - \alpha_n\} \cap \{\alpha_n \leqq G(Y_\nu) \leqq 1 - \alpha_n\}) \to 1$ uniformly for H in $\mathscr{H}$ provided only $\alpha_n = o(n^{-1})$.*

PROOF. The probability of the complementary event is bounded above by $2\alpha_n{}^n + 2[1 - (1-\alpha_n)^n] \to 0$ as $n \to \infty$, independently of H in $\mathscr{H}$. □

We conclude this section with some lemmas needed for Theorem 2.2.

LEMMA 4.5. *As $n \to \infty$, $P(\{Y_\nu = Y_{nn}\}) \to 0$ uniformly for H in $\mathscr{H}_{C\delta}$.*

PROOF. $P(\{Y_\nu = Y_{nn}\}) = P(\bigcup_{i=1}^n \{(X_i, Y_i) = (X_{nn}, Y_{nn})\}) = n \iint H^{n-1}\, dH$. Note that for all x, y we have $H(x,y) \leqq F(x)$ and $H(x,y) \leqq G(y)$. Letting $n_0 = (n-1)/2$ and applying (1.6) we obtain

$$\begin{aligned} n \iint H^{n-1}(x,y)\, dH(x,y) &\leqq n \iint F^{n_0}(x) G^{n_0}(y)\, dH(x,y) \\ &\leqq Cn[\textstyle\int_0^1 I^{n_0} r^{\delta/2}\, dI]^2 \\ &= Cn[\Gamma(n_0 - \delta/2 + 1)\Gamma(1 - \delta/2)/\Gamma(n_0 - \delta + 2)]^2 \\ &\leqq C_1 n n_0{}^{\delta-2} = O(n^{\delta-1}) \to 0 \end{aligned}$$

as $n \to \infty$, because $0 < \delta < \frac{1}{2}$. Here C_1 is a constant depending on C and δ only; hence the convergence is uniform for H in $\mathscr{H}_{C\delta}$. □

LEMMA 4.6. *As $n \to \infty$, $P(\{\gamma_n \leq G(Y_\nu) \leq 1 - \gamma_n\}) \to 1$ uniformly for H in $\mathscr{H}_{C\delta}$ provided $\gamma_n \leq an^{-\delta}$ for some positive constant a.*

PROOF. This probability equals $1 - P(\{G(Y_\nu) < \gamma_n\}) - P(\{G(Y_\nu) > 1 - \gamma_n\})$ for n larger than $(2a)^{1/\delta}$. Because of the independence of the sample elements, application of (1.6) gives

$$\begin{aligned} P(\{G(Y_\nu) < \gamma_n\}) &= nP([\textstyle\bigcap_{i=1}^{n-1} \{F(X_i) \leq F(X_n)\}] \cap \{G(Y_n) < \gamma_n\}) \\ &= n \textstyle\int_{-\infty}^{\infty} \int_{-\infty}^{G^{-1}(\gamma_n)} F^{n-1}(x_n)\, dH(x_n, y_n) \\ &\leq Cn[\textstyle\int_0^1 I^{n-1} r^{\delta/2}\, dI][\int_0^{\gamma_n} r^{\delta/2}\, dI] \\ &= C_1 n[\Gamma(n - \delta/2)/\Gamma(n + 1 - \delta)] n^{-\delta+\delta^2/2} \\ &\leq C_2 n n^{-1+\delta/2} n^{-\delta+\delta^2/2} = C_2 n^{-\delta/2+\delta^2/2} \to 0 \end{aligned}$$

as $n \to \infty$, because $-\delta/2 + \delta^2/2 < 0$ for $0 < \delta < \frac{1}{2}$. Here C_1 and C_2 are constants depending on C, δ and a only; hence the convergence is uniform for H in $\mathscr{H}_{C\delta}$. □

5. Proof of the theorems: Asymptotic negligibility of the remainder terms under Assumption 2.3(a). Let us start with a further decomposition of C_n, which can be seen to be the sum of

$$\begin{aligned} C_{\gamma 1n} &= \chi(\Omega_{\gamma n}^c) n^{\frac{1}{2}} \textstyle\iint_{\Delta_n} [J(F_n) - J(F)][K(G_n) - K(G)]\, dH_n\,, \\ C_{\gamma 2n} &= \chi(\Omega_{\gamma n}) n^{\frac{1}{2}} \textstyle\iint_{\Delta_n \cap S_\gamma^c} [J(F_n) - J(F)] K(G_n)\, dH_n\,, \\ C_{\gamma 3n} &= -\chi(\Omega_{\gamma n}) n^{\frac{1}{2}} \textstyle\iint_{\Delta_n \cap S_\gamma^c} [J(F_n) - J(F)] K(G)\, dH_n\,, \\ C_{\gamma 4n} &= \chi(\Omega_{\gamma n}) \textstyle\iint_{\Delta_n \cap S_\gamma} U_n(F) J'(\Phi_n)[K(G_n) - K(G)]\, dH_n\,. \end{aligned}$$

From this we see that $B_{\gamma 4n}$ and $C_{\gamma 3n}$ cancel out. The asymptotic negligibility of the other B- and C-terms will be given as corollaries to the lemmas of the previous section.

COROLLARY 5.1. *As $n \to \infty$, $B_{1n} \to_p 0$ uniformly for H in $\mathscr{H}(\mathscr{H}_{C\delta})$.*

PROOF. The rv B_{1n} is bounded by $\sum_{i=1}^{3} B_{1in}$ where

$$\begin{aligned} B_{11n} &= n^{\frac{1}{2}} |J_n(1)| \textstyle\iint_{\{X_{nn}\} \times \Delta_{n2}} |K_n(G_n(y))|\, dH_n(x, y)\,, \\ B_{12n} &= n^{\frac{1}{2}} |J_n(1) K_n(1)| \textstyle\iint_{\{(X_{nn}, Y_{nn})\}} dH_n(x, y)\,, \\ B_{13n} &= n^{\frac{1}{2}} |K_n(1)| \textstyle\iint_{\Delta_{n1} \times \{Y_{nn}\}} |J_n(F_n(x))|\, dH_n(x, y)\,. \end{aligned}$$

Under the assumptions of Theorem 2.1 we have at once that the sum of these terms is of order $O(n^{-\frac{1}{2}+a+b}) = O(n^{-\delta}) \to 0$ as $n \to \infty$, uniformly for H in $\mathscr{H}$.

Under the assumptions of Theorem 2.2 first consider B_{11n}. By Assumption 2.2, $|K_n(G_n(y))| \leq Dr^b(G_n(y))$. Application of Lemma 4.2 (ii) with $\Psi_n = G_n$ and $\eta = b$ gives the existence of a constant M such that $\Omega_{1n} = \{r^b(G_n) \leq Mr^b(G) \text{ on } \Delta_{n2}\}$ has probability larger than $1 - \varepsilon$ uniformly for $n = 1, 2, \cdots$ and all continuous H. Also

$$\chi(\Omega_{1n})B_{11n} \leqq DMn^{-\frac{1}{2}}|J_n(1)|r^b(G(Y_\nu)),$$

where ν is defined by (4.3). Set $\gamma_n = n^{-\frac{1}{2}}|J_n(1)|$ and note that by (1.2) and Assumption 2.2 we have $\gamma_n \leqq D_1 n^{-\delta}$ for some constant $D_1 \geqq D$. Let $\Omega_{2n} = \{\gamma_n \leqq G(Y_\nu) \leqq 1 - \gamma_n\}$. Then

$$\chi(\bigcap_{i=1}^2 \Omega_{in})B_{11n} \leqq DM\gamma_n^{1-b}(1-\gamma_n)^{-b} = O(n^{-\delta(1-b)}) \to 0$$

as $n \to \infty$. Applying Lemma 4.6 we see that $P(\bigcap_{i=1}^2 \Omega_{in}) > 1 - 2\varepsilon$ for n large enough, uniformly for H in $\mathscr{H}_{C\delta}$. A symmetric argument can be given for B_{13n}.

For the rv B_{12n} use Lemma 4.5 to see that the set on which this rv may assume a nonzero value has probability converging to zero as $n \to \infty$, uniformly for H in $\mathscr{H}_{C\delta}$. □

COROLLARY 5.2. *As $n \to \infty$, $B_{2n} \to_p 0$ uniformly for H in $\mathscr{H}(\mathscr{H}_{C\delta})$.*

PROOF. The rv B_{2n} is bounded by $\sum_{i=1}^2 B_{2in}$ where

$$B_{21n} = D^2 n^{-\frac{1}{2}} r^a(F(X_\nu))r^b(G(Y_\nu)),$$
$$B_{22n} = D^2 n^{-\frac{1}{2}} r^a(F(X_\lambda))r^b(G(Y_\lambda)),$$

with ν and λ defined by (4.3).

Under the assumptions of Theorem 2.1 consider $\Omega_{1n} = \{\alpha_n \leqq F(X_\nu) \leqq 1 - \alpha_n\} \cap \{\alpha_n \leqq G(Y_\nu) \leqq 1 - \alpha_n\}$, with $\alpha_n = n^{a+b-\frac{3}{2}}$. Note that $n\alpha_n \to 0$. Then

$$\chi(\Omega_{1n})B_{21n} \leqq D^2 n^{-\frac{1}{2}}\alpha_n^{-a-b}(1-\alpha_n)^{-a-b}$$
$$= D^2(n\alpha_n)^{1-a-b}(1-\alpha_n)^{-a-b} \to 0$$

as $n \to \infty$. Lemma 4.4 gives that $P(\Omega_{1n}) \to 1$ as $n \to \infty$, uniformly for H in $\mathscr{H}$. The same argument applies for the rv B_{22n}.

Under the assumptions of Theorem 2.2 consider

$$\Omega_{2n} = \{\beta_n \leqq F(X_\nu) \leqq 1 - \beta_n\} \cap \{\gamma_n \leqq G(Y_\nu) \leqq 1 - \gamma_n\},$$

with $\beta_n = (n \log n)^{-1}$ and $\gamma_n = n^{-\delta}$. Then by (2.3)

$$\chi(\Omega_{2n})B_{21n} \leqq D^2 n^{-\frac{1}{2}}\beta_n^{-a}\gamma_n^{-b}(1-\beta_n)^{-a}(1-\gamma_n)^{-b} \to 0$$

as $n \to \infty$. By Lemmas 4.4 and 4.6 we see that $P(\Omega_{2n}) \to 1$ as $n \to \infty$, uniformly for H in $\mathscr{H}_{C\delta}$. The rv B_{22n} can be treated in the same way. □

COROLLARY 5.3. *For fixed γ, $B_{\gamma 3n} \to_p 0$ and $C_{\gamma 1n} \to_p 0$ as $n \to \infty$, uniformly for H in $\mathscr{H}$.*

PROOF. $P(\Omega_{\gamma n}^c) \to 0$ uniformly for H in $\mathscr{H}$ by the Glivenko-Cantelli theorem and because the distribution of $\sup |F_n - F|$ does not depend on H in $\mathscr{H}$. □

COROLLARY 5.4. *For fixed γ, $B_{\gamma 5n} \to_p 0$ and $C_{\gamma 4n} \to_p 0$ as $n \to \infty$, uniformly for H in $\mathscr{H}$.*

PROOF. According to Lemma 4.2 (iii) with $\tau = \frac{1}{2}$, for given $\varepsilon > 0$ there exists a constant M such that $\Omega_n = \{\sup |U_n(F)| \leqq M\}$ has probability larger than $1-\varepsilon$ for all n and H in $\mathscr{H}$. Also

$$\chi(\Omega_n)|B_{\gamma 5n}| \leqq M \sup_{\Delta_{n1}\cap S_{\gamma 1}} |J'(\Phi_n) - J'(F)| \sup_{S_{\gamma 2}} |K(G)| \, .$$

The function $K(G)$ is bounded on $S_{\gamma 2}$ and the bound does not depend on H in $\mathscr{H}$. The function J' is uniformly continuous on $[\gamma/2, s_1 - \gamma/2] \cup [s_1 + \gamma/2, 1-\gamma/2]$. Since $|\Phi_n - F| \leqq |F_n - F|$ where Φ_n is defined, the Glivenko-Cantelli theorem yields $\sup_{\Delta_{n1}\cap S_{\gamma 1}} |J'(\Phi_n) - J'(F)| \to_p 0$ uniformly for H in $\mathscr{H}$. A similar argument may be used for $C_{\gamma 4n}$. □

Corollary 5.5. *For fixed* γ, $B_{\gamma 6n} \to_p 0$ *as* $n \to \infty$, *uniformly for* H *in* $\mathscr{H}$.

Proof. For arbitrary k we have (see (4.2)) $|B_{\gamma 6n}| \leqq \sum_{i=1}^{3} B_{\gamma 6ikn}$, where

$$B_{\gamma 61kn} = \iint_{\Delta_n\cap S_\gamma} |U_n(F)J'(F)K(G) - U_n(I_k(F))J'(I_k(F))K(I_k(G))| \, dH_n \, ,$$
$$B_{\gamma 62kn} = |\iint_{\Delta_n\cap S_\gamma} U_n(I_k(F))J'(I_k(F))K(I_k(G)) \, d(H_n - H)| \, ,$$
$$B_{\gamma 63kn} = \iint_{\Delta_n\cap S_\gamma} |U_n(F)J'(F)K(G) - U_n(I_k(F))J'(I_k(F))K(I_k(G))| \, dH \, .$$

Let us first consider $B_{\gamma 61kn}$ and $B_{\gamma 63kn}$, which are both bounded by the supremum of the integrand over the set S_γ. Let an arbitrary $\varepsilon > 0$ be given. Application of Lemma 4.3 gives the existence of constants $\eta_{kn} \to 0$ as $k, n \to \infty$, such that $\Omega_{kn} = \{\sup |U_n(F) - U_n(I_k(F))| \leqq \eta_{kn}\}$ has probability larger than $1-\varepsilon$ for all k, n and all H in $\mathscr{H}$. Note that on $([\gamma, s_1 - \gamma] \cup [s_1 + \gamma, 1 - \gamma]) \times ([\gamma, t_1 - \gamma] \cup [t_1 + \gamma, 1 - \gamma])$ the function $J'(s)K(t)$ is bounded, say by a constant M_γ, and uniformly continuous. By Lemma 4.2 (iii) with $\tau = \frac{1}{2}$, there exists a constant M such that $\Omega_n = \{\sup |U_n(F)| \leqq M\}$ has probability larger than $1-\varepsilon$. Let us finally write $\zeta_{k\gamma} = \max_{S_\gamma} |J'(F)K(G) - J'(I_k(F))K(I_k(G))|$, which tends to zero as $k \to \infty$, uniformly for H in $\mathscr{H}$. Hence for $i = 1, 3$

$$\chi(\Omega_{kn} \cap \Omega_n) B_{\gamma 6ikn} \leqq \eta_{kn} M_\gamma + M\zeta_{k\gamma} \to 0$$

as $k, n \to \infty$ for fixed γ. Because $P(\Omega_{kn} \cap \Omega_n) > 1 - 2\varepsilon$ uniformly for H in $\mathscr{H}$ we may conclude that $B_{\gamma 61kn} \to_p 0$ and $B_{\gamma 63kn} \to_p 0$ uniformly for H in $\mathscr{H}$, as $k, n \to \infty$.

Let us next consider $B_{\gamma 62kn}$ for a fixed value k. For each ω in Ω_n the integrand in the expression for this rv is a simple step function assuming a value $a_{ijkn}(\omega)$ on the rectangle

$$R_{ijkn} = (F^{-1}((i-1)/k), F^{-1}(i/k)] \times (G^{-1}((j-1)/k), G^{-1}(j/k)] \cap S_\gamma \cap \Delta_n \, ,$$

for $i = 1, \cdots, k$ and $j = 1, \cdots, k$. Because $|a_{ijkn}| \leqq M(M_\gamma + \zeta_{k\gamma})$ on Ω_n, we have

$$\chi(\Omega_n)B_{\gamma 62kn} = |\textstyle\sum_{i=1}^{k} \sum_{j=1}^{k} a_{ijkn} \iint_{R_{ijkn}} d(H_n - H)|$$
$$\leqq 4k^2 M(M_\gamma + \zeta_{k\gamma}) \sup |H_n - H| \to_p 0$$

as $n \to \infty$, uniformly for H in $\mathscr{H}$. Here Theorem 1-m of Kiefer [8] is used. The conclusion of the corollary follows by straightforward combination of these results. □

COROLLARY 5.6. *As $\gamma \downarrow 0$ and $n \to \infty$, $B_{\gamma 7n} \to_p 0$ and $C_{\gamma 2n} \to_p 0$, uniformly for H in $\mathscr{H}(\mathscr{H}_{c\delta})$.*

PROOF. Let $\varepsilon > 0$ be given and let us first consider $B_{\gamma 7n}$. By Lemma 4.2 (iii), taking $\tau = \delta/4$, there exists a constant M_1 such that $\Omega_{1n} = \{|U_n(F)| \leq M_1 r^{-\frac{1}{2}+\delta/4}(F)\}$ has probability larger than $1 - \varepsilon$ for all n and H in $\mathscr{H}$. From Assumption 2.1 (Assumption 2.2) it may be seen that

$$\chi(\Omega_{1n})|B_{\gamma 7n}| \leq D^2 M_1 \iint_{\Delta_n{}^c \cup S_\gamma{}^c} r^{a+\frac{1}{2}+\delta/4}(F) r^b(G)\, dH \,. \tag{5.1}$$

Next consider $C_{\gamma 2n}$. By Assumption 2.1 (Assumption 2.2) we have $|K(G_n)| \leq Dr^b(G_n)$ on Δ_{n2} and application of Lemma 4.2 (ii) with $\tilde{\Psi}_n = G_n$ and $\eta = b$ gives the existence of a constant M_2 such that $\Omega_{2n} = \{r^b(G_n) \leq M_2 r^b(G)$ on $\Delta_{n2}\}$ has probability larger than $1 - \varepsilon$ for all n and H in $\mathscr{H}$. Take an arbitrary ω in Ω and let us first consider those values of x in Δ_{n1} for which the open random interval between the points $F(x)$ and $F_n(x)$ does not contain s_1. Then by continuity of J on the closed and differentiability on the open interval, the mean value theorem can be applied; it follows from Assumption 2.1 (Assumption 2.2) that

$$n^{\frac{1}{2}}|J(F_n) - J(F)| = |U_n(F)J'(\Phi_{0n})| \leq D|U_n(F)|r^{a+1}(\Phi_{0n}) \,.$$

For those values of x in Δ_{n1} for which the open random interval between the points $F(x)$ and $F_n(x)$ does contain s_1, the mean value theorem can be applied stepwise, since J is continuous on the closed interval and differentiable on the two open intervals between $F(x)$, $F_n(x)$ and s_1. We thus get the estimate

$$n^{\frac{1}{2}}|J(F_n) - J(F)| \leq |U_n(F)| \textstyle\sum_{i=1}^{2} |J'(\Phi_{in})| \leq D|U_n(F)| \sum_{i=1}^{2} r^{a+1}(\Phi_{in})$$

by Assumption 2.1 (Assumption 2.2). Where defined on Δ_{n1}, both Φ_{0n} and Φ_{1n}, Φ_{2n} lie between F and F_n. By Lemma 4.2 (i), taking $\zeta = a + 1$, there exists a constant M_3 such that $\Omega_{3n} = \{\max_{i=0,1,2} r^{a+1}(\Phi_{in}) \leq M_3 r^{a+1}(F)$ where defined on $\Delta_{n1}\}$ has probability larger than $1 - \varepsilon$ for all n and H in $\mathscr{H}$. Combining these results we have

$$E(\chi(\textstyle\bigcap_{i=1}^{3} \Omega_{in})|C_{\gamma 2n}|) \leq 2D^2 M_1 M_2 M_3 \iint_{S_\gamma{}^c} r^{a+\frac{1}{2}+\delta/4}(F) r^b(G)\, dH \,. \tag{5.2}$$

From (5.1) and (5.2) it is clear that the corollary is proved if we show that the integral on the right in (5.1) converges to zero as $\gamma \downarrow 0$ and $n \to \infty$, uniformly for H in $\mathscr{H}(\mathscr{H}_{c\delta})$. For this purpose we start with the integral

$$\iint_{S_\gamma{}^c} r^{a+\frac{1}{2}+\delta/4}(F) r^b(G)\, dH \,, \tag{5.3}$$

and note that $S_\gamma{}^c \subset (S_{\gamma 1}^c \times (-\infty, \infty)) \cup ((-\infty, \infty) \times S_{\gamma 2}^c)$. Under Assumption 2.1, by application of (1.5) with $p = p_1$ and $q = q_1$ as in (3.5), we find that (5.3) is bounded uniformly for H in $\mathscr{H}$ by

$$\begin{aligned}\text{(5.4)}\quad &[\textstyle\int_{(0,\gamma)\cup(s_1-\gamma,s_1+\gamma)\cup(1-\gamma,1)} r^{(a+\frac{1}{2}+\delta/4)p_1}\, dI]^{1/p_1}[\int r^{bq_1}\, dI]^{1/q_1} \\ &\quad + [\textstyle\int r^{(a+\frac{1}{2}+\delta/4)p_1} dI]^{1/p_1}[\int_{(0,\gamma)\cup(t_1-\gamma,t_1+\gamma)\cup(1-\gamma,1)} r^{bq_1}\, dI]^{1/q_1} \,.\end{aligned}$$

Since by (3.5) both exponents of the function r are smaller than 1, the dominated convergence theorem implies convergence of (5.4) to zero as $\gamma \downarrow 0$. Under Assumption 2.2 and for H in $\mathscr{H}_{C\delta}$, by an application of (1.6) we see that (5.3) is bounded uniformly for H in $\mathscr{H}_{C\delta}$ by

$$(5.5)\qquad C[\int_{(0,\gamma)\cup(s_1-\gamma,s_1+\gamma)\cup(1-\gamma,1)} r^{1-\delta/4}\,dI][\int r^{\frac{1}{2}-\delta/2}\,dI] + C[\int r^{1-\delta/4}\,dI][\int_{(0,\gamma)\cup(t_1-\gamma,t_1+\gamma)\cup(1-\gamma,1)} r^{\frac{1}{2}-\delta/2}\,dI],$$

which by the dominated convergence theorem converges to zero as $\gamma \downarrow 0$. Hence under the assumptions of Theorem 2.1 (Theorem 2.2) a value $\tilde{\gamma}$ of γ can be chosen such that (5.3) is smaller than ε for all H in $\mathscr{H}(\mathscr{H}_{C\delta})$ provided $\gamma \leqq \tilde{\gamma}$. For this $\tilde{\gamma}$ there exists an index $\tilde{n} = \tilde{n}_{\tilde{\gamma}}$ such that $P(\{\Delta_n \supset S_{\tilde{\gamma}}\}) > 1 - \varepsilon$ uniformly for H in $\mathscr{H}$, provided $n \geqq \tilde{n}$. It follows that under the assumptions of Theorem 2.1 (Theorem 2.2) the integral on the right in (5.1) is smaller than ε with probability larger than $1 - \varepsilon$ uniformly for H in $\mathscr{H}(\mathscr{H}_{C\delta})$ for all $\gamma \leqq \tilde{\gamma}$ and all $n \geqq \tilde{n}$. □

In order to show how the results of these corollaries can be combined to complete the proof of Theorems 2.1 and 2.2, let an arbitrary $\varepsilon > 0$ be given. First use Corollary 5.6 to choose a fixed γ and an index n_1 to ensure $P(\{|B_{\gamma 7n}|, |C_{\gamma 2n}| \leqq \varepsilon\}) > 1 - \varepsilon$ for all $n > n_1$. Next use Assumption 2.3 (a) and Corollaries 5.1–5.5 to choose for the above fixed γ an index $n_2 = n_{2\gamma} > n_1$ such that $P(\{|B_{in}|, |B_{\gamma jn}|, |C_{\gamma kn}| \leqq \varepsilon$ for $i = 0, 1, 2$; $j = 3, 5, 6$; $k = 1, 4\}) > 1 - \varepsilon$ for $n > n_2$. This implies that the probability that the sum of all these second order terms does not exceed 10ε is larger than $1 - 2\varepsilon$ uniformly for H in $\mathscr{H}(\mathscr{H}_{C\delta})$, as $n > n_2$.

6. Replacing Assumption 2.3(a) **by Assumption** 2.3(b). We shall now suppose that Assumption 2.3 (b) holds. Again the theorems will be considered only in the case where J and K fail to have a derivative at one point, s_1 and t_1 respectively. The proof is based on an analogue of (3.4). We shall need both the empirical processes and the processes $U_n^*(F) = n^{\frac{1}{2}}(F_n^* - F)$, $V_n^*(G) = n^{\frac{1}{2}}(G_n^* - G)$. Instead of the set $\Omega_{\gamma n}$ we shall use $\Omega^*_{\gamma n} = \{\omega : \sup|F_n^* - F| < \gamma/2,\ \sup|G_n^* - G| < \gamma/2\}$. The role of Δ_n will be taken over by its closure $\bar{\Delta}_n = \bar{\Delta}_{n1} \times \bar{\Delta}_{n2} = [X_{1n}, X_{nn}] \times [Y_{1n}, Y_{nn}]$. Because integration over $\bar{\Delta}_n$ with respect to dH_n is the same as integration over the entire plane, we now have the simpler decomposition

$$(6.1)\qquad n^{\frac{1}{2}}(T_n - \mu) = \textstyle\sum_{i=1}^3 A_{in} + B^*_{0n} + \sum_{i=1}^5 B^*_{\gamma in} + B^*_{6n} + C_n^*,$$

with probability 1. Here B^*_{0n} is defined in (2.5), the A-terms are as given in Section 3 and

$$B^*_{\gamma 1n} = \chi(\Omega^{*c}_{\gamma n})\{n^{\frac{1}{2}} \textstyle\iint [J(F_n^*) - J(F)]K(G)\,dH_n - A_{2n}\},$$
$$B^*_{\gamma 2n} = \chi(\Omega^*_{\gamma n}) n^{\frac{1}{2}} \textstyle\iint_{S_\gamma{}^c} [J(F_n^*) - J(F)]K(G)\,dH_n,$$
$$B^*_{\gamma 3n} = \chi(\Omega^*_{\gamma n}) \textstyle\iint_{S_\gamma} U_n^*(F)[J'(\Phi_n^*) - J'(F)]K(G)\,dH_n,$$
$$B^*_{\gamma 4n} = \chi(\Omega^*_{\gamma n}) \textstyle\iint_{\bar{\Delta}_n \cap S_\gamma} U_n^*(F)J'(F)K(G)\,d(H_n - H),$$
$$B^*_{\gamma 5n} = \chi(\Omega^*_{\gamma n})\{\textstyle\iint_{\bar{\Delta}_n \cap S_\gamma} U_n^*(F)J'(F)K(G)\,dH - A_{2n}\},$$
$$B^*_{6n} = n^{\frac{1}{2}} \textstyle\iint J(F)[K(G_n^*) - K(G)]\,dH_n - A_{3n},$$
$$C_n^* = n^{\frac{1}{2}} \textstyle\iint [J(F_n^*) - J(F)][K(G_n^*) - K(G)]\,dH_n.$$

The function Φ_n^* arises from application of the mean value theorem and lies strictly between F and F_n^* where defined. The analogues of B_{1n} and B_{2n} are missing in this decomposition; this essentially simplifies the proof of the theorems. However, if one tries to prove the validity of Assumption 2.3 (b) when Assumption 2.3 (a) is given to hold, problems similar to those connected with B_{1n} and B_{2n} recur.

Only the second order terms differ from those in (3.4). For their asymptotic negligibility we need the following modifications of Lemma 4.2.

LEMMA 6.1. *For each* ω *let* $\tilde{\Phi}_n^* = \tilde{\Phi}_{n\omega}^*$ *and* $\tilde{\Psi}_n^* = \tilde{\Psi}_{n\omega}^*$ *be functions on* $\bar{\Delta}_{n1} = \bar{\Delta}_{n1\omega}$ *and* $\bar{\Delta}_{n2} = \bar{\Delta}_{n2\omega}$ *respectively, satisfying* $\min(F, F_n^*) \leqq \tilde{\Phi}_n^* \leqq \max(F, F_n^*)$ *and* $\min(G, G_n^*) \leqq \tilde{\Psi}_n^* \leqq \max(G, G_n^*)$ *where defined. Then, uniformly for* $n = 1, 2, \cdots$ *and* $H \in \mathscr{H}$:

(i) $\sup_{\bar{\Delta}_{n1}} r^{\zeta}(\tilde{\Phi}_n^*) r^{-\zeta}(F) = O_p(1)$ *for each* $\zeta \geqq 0$;
(ii) $\sup_{\bar{\Delta}_{n2}} r^{\eta}(\tilde{\Psi}_n^*) r^{-\eta}(G) = O_p(1)$ *for each* $\eta \geqq 0$.

PROOF. It suffices to prove (i). Let us first show that for each $\varepsilon > 0$ there exists a $\beta = \beta_\varepsilon$ in $(0, 1)$ such that $P(\{\beta F \leqq F_n^* \leqq 1 - \beta(1 - F) \text{ on } \bar{\Delta}_{n1}\}) > 1 - \varepsilon$, for all n and uniformly in all continuous F. By (4.1) and because $\frac{1}{2} \leqq n/(n+1) \leqq 1$, we only have to prove that $P(\{n/(n+1) \leqq 1 - \beta[1 - F(X_{nn})]\}) > 1 - \varepsilon$ for β small enough. Because the $F(X_i)$ are independent uniform (0, 1) rvs, this probability equals $1 - \{1 - 1/[\beta(n+1)]\}^n > 1 - \varepsilon$ for all n and uniformly in all continuous F, provided $\beta = \beta_\varepsilon$ is chosen sufficiently small. The proof can be concluded in the same way as that of Lemma 4.2. □

LEMMA 6.2. *Uniformly in all continuous F we have:*

(i) $\sup_{\bar{\Delta}_{n1}} |U_n^*(F) - U_n(F)| r^{\frac{1}{2}-\rho}(F) \to_p 0$ *as* $n \to \infty$, *for each* $\rho > 0$;
(ii) $\sup_{\bar{\Delta}_{n1}} |U_n^*(F)| r^{\frac{1}{2}-\tau}(F) = O_p(1)$ *uniformly for* $n = 1, 2, \cdots$, *for each* $\tau > 0$.

PROOF. (i) Note that $|U_n^*(F) - U_n(F)| r^{\frac{1}{2}-\rho}(F) < n^{-\frac{1}{2}} r^{\frac{1}{2}-\rho}(F)$ and that for any fixed $\beta \in (0, 1)$ we have $r^{\frac{1}{2}-\rho}(\beta/n) = r^{\frac{1}{2}-\rho}(1 - \beta/n) = O(n^{\frac{1}{2}-\rho})$. Because the $F(X_i)$ are independent uniform rvs, given an arbitrary $\varepsilon > 0$ we can choose a $\beta = \beta_\varepsilon$ in (0, 1) such that $P(\{\beta/n \leqq F(X_{1n}) \leqq F(X_{nn}) \leqq 1 - \beta/n\}) > 1 - \varepsilon$ for all n and uniformly for all continuous F. Part (i) follows from a combination of these results. (ii) follows from (i) and Lemma 4.2 (iii). □

The proof that the sum of the B^*- and C^*-terms converges in probability to zero can be given by a method quite similar to that of Section 5, by using Lemmas 6.1, 6.2 instead of Lemma 4.2.

REFERENCES

[1] BHUCHONGKUL, S. (1964). A class of nonparametric tests for independence in bivariate populations. *Ann. Math. Statist.* **35** 138–149.
[2] BILLINGSLEY, P. (1968). *Convergence of Probability Measures.* Wiley, New York.
[3] CHERNOFF, H. and I. R. SAVAGE (1958). Asymptotic normality and efficiency of certain nonparametric test statistics. *Ann. Math. Statist.* **29** 972–994.

[4] FELLER, W. (1957). *An Introduction to Probability Theory and its Applications* **1**. Wiley, New York.

[5] GUMBEL, E. J. (1958). Multivariate distributions with given margins and analytical examples. *Bull. Inst. Internat. Statist.* **37-3** 363–373.

[6] HÁJEK, J. and Z. ŠIDÁK (1967). *Theory of Rank Tests*. Academic Press, New York.

[7] JOGDEO, K. (1968). Asymptotic normality in nonparametric methods. *Ann. Math. Statist.* **39** 905–922.

[8] KIEFER, J. (1961). On large deviations of the empiric df of vector chance variables and a law of the iterated logarithm. *Pacific J. Math.* **11** 649–660.

[9] LEHMANN, E. L. (1953). The power of rank tests. *Ann. Math. Statist.* **24** 23–43.

[10] PYKE, R. and G. R. SHORACK (1968). Weak convergence of a two-sample empirical process and a new approach to Chernoff-Savage theorems. *Ann. Math. Statist.* **39** 755–771.

[11] SHORACK, G. R. (1972). Functions of order statistics. *Ann. Math. Statist.* **43** 412–427.

[12] SKOROKHOD, A. V. (1956). Limit theorems for stochastic processes. *Theor. Probability Appl.* **1** 261–290.

A NOTE ON CONTIGUITY AND HELLINGER DISTANCE*

by

J. OOSTERHOFF (1), AND W. R. VAN ZWET (2)

1. Introduction

For $n = 1, 2, \ldots$ let $(\mathscr{X}_{n1}, \mathscr{A}_{n1}), \ldots, (\mathscr{X}_{nn}, \mathscr{A}_{nn})$ be arbitrary measurable spaces. Let P_{ni} and Q_{ni} be probability measures defined on $(\mathscr{X}_{ni}, \mathscr{A}_{ni})$, $i = 1, \ldots, n$; $n = 1, 2, \ldots$, and let $P_n^{(n)} = \prod_{i=1}^{n} P_{ni}$ and $Q_n^{(n)} = \prod_{i=1}^{n} Q_{ni}$ $(n = 1, 2, \ldots)$ denote the product probability measures. For each i and n let X_{ni} be the identity map from $\mathscr{X}_{ni}$ onto $\mathscr{X}_{ni}$. Then P_{ni} and Q_{ni} represent the two possible distributions of the random element X_{ni} as well as the probability measures of the underlying probability space. Obviously $X_{n1}, \ldots, X_{nn}$ are independent under both $P_n^{(n)}$ and $Q_n^{(n)}$ $(n = 1, 2, \ldots)$.

The sequence $\{Q_n^{(n)}\}$ is said to be contiguous with respect to the sequence $\{P_n^{(n)}\}$ if $\lim_{n\to\infty} P_n^{(n)}(A_n) = 0$ implies $\lim_{n\to\infty} Q_n^{(n)}(A_n) = 0$ for any sequence of measurable sets A_n. This one-sided contiguity notion is denoted by $\{Q_n^{(n)}\} \lhd \{P_n^{(n)}\}$ (the notation is due to H. Witting & G. Nölle [7]). The sequences $\{P_n^{(n)}\}$ and $\{Q_n^{(n)}\}$ are said to be contiguous with respect to each other if both $\{Q_n^{(n)}\} \lhd \{P_n^{(n)}\}$ and $\{P_n^{(n)}\} \lhd \{Q_n^{(n)}\}$. This two-sided contiguity concept we denote by $\{P_n^{(n)}\} \lhd \rhd \{Q_n^{(n)}\}$.

The main purpose of this note is to characterize contiguity of product probability measures in terms of their marginals. To this end we introduce the Hellinger distance $H(P, Q)$ between two probability measures P and Q on the same σ-field, defined by

$$(1.1)\qquad H(P, Q) = \left\{\int (p^{1/2} - q^{1/2})^2 \, d\mu\right\}^{1/2} = \left\{2 - 2\int p^{1/2} q^{1/2} \, d\mu\right\}^{1/2},$$

where $p = dP/d\mu$, $q = dQ/d\mu$ and μ is any σ-finite measure dominating $P + Q$. This metric is independent of the choice of μ and satisfies $0 \leq H(P, Q) \leq 2^{1/2}$.

---

* Report SW 36/75 Mathematisch Centrum, Amsterdam

AMS (MOS) subject classification scheme (1970): 62E20

KEY WORDS & PHRASES: *asymptotic normality, contiguity, Hellinger distance, log likelihood ratio.*

Defining the total variation distance of P and Q by

$$\|P - Q\| = \sup |P(A) - Q(A)|\,, \tag{1.2}$$

where the supremum is taken over all measurable sets A, we have the following inequalities (Le Cam [4])

$$\tfrac{1}{2}H^2(P, Q) \leqq \|P - Q\| \leqq H(P, Q)\,. \tag{1.3}$$

The Hellinger distances of the product measures and of their marginals are connected by the relationship

$$H^2(P_n^{(n)}, Q_n^{(n)}) = 2 - 2\prod_{i=1}^{n}\{1 - \tfrac{1}{2}H^2(P_{ni}, Q_{ni})\}\,. \tag{1.4}$$

For further reference we first mention two easy results, viz.

$$\sum_{i=1}^{n} H^2(P_{ni}, Q_{ni}) = o(1) \quad \text{for} \quad n \to \infty \Rightarrow \{P_n^{(n)}\} \triangleleft\triangleright \{Q_n^{(n)}\}\,, \tag{1.5}$$

and

$$\{Q_n^{(n)}\} \triangleleft \{P_n^{(n)}\} \Rightarrow \sum_{i=1}^{n} H^2(P_{ni}, Q_{ni}) = O(1) \quad \text{for} \quad n \to \infty\,. \tag{1.6}$$

The proof of (1.5) is an immediate consequence of the string of implications

$$\sum_{i=1}^{n} H^2(P_{ni}, Q_{ni}) = o(1) \Rightarrow \sum_{i=1}^{n} \log\{1 - \tfrac{1}{2}H^2(P_{ni}, Q_{ni})\} = o(1)$$

$$\Rightarrow H^2(P_n^{(n)}, Q_n^{(n)}) = o(1) \Rightarrow \|P_n^{(n)} - Q_n^{(n)}\| = o(1) \Rightarrow \{P_n^{(n)}\} \triangleleft\triangleright \{Q_n^{(n)}\}\,.$$

To prove (1.6) suppose that $\limsup_{n\to\infty} H(P_n^{(n)}, Q_n^{(n)}) = 2^{1/2}$. Then by (1.3) $\limsup_{n\to\infty} \|P_n^{(n)} - Q_n^{(n)}\| = 1$ in contradiction to $\{Q_n^{(n)}\} \triangleleft \{P_n^{(n)}\}$. Thus $\limsup_{n\to\infty} H^2(P_n^{(n)}, Q_n^{(n)}) < 2$, therefore $\liminf_{n\to\infty} \prod_{i=1}^{n}\{1 - \tfrac{1}{2}H^2(P_{ni}, Q_{ni})\} > 0$ and hence $\limsup_{n\to\infty} \sum_{i=1}^{n} H^2(P_{ni}, Q_{ni}) < \infty$ and the proof is complete.

It can be shown by counterexamples that in (1.5) the condition cannot be weakened to $\sum_{i=1}^{n} H^2(P_{ni}, Q_{ni}) = O(1)$, and that in (1.6) the conclusion cannot be strengthened to $\sum_{i=1}^{n} H^2(P_{ni}, Q_{ni}) = o(1)$, for $n \to \infty$. Hence there remains a gap between the sufficient condition and the necessary condition for contiguity in (1.5) and (1.6) respectively. In section 2 we obtain conditions which are both sufficient and necessary for contiguity of the product measures by adding another condition to $\sum_{i=1}^{n} H^2(P_{ni}, Q_{ni}) = O(1)$.

In many applications asymptotic normality of the log likelihood ratio statistic Λ_n (see (3.1)) plays an important part. Since

$$\mathscr{L}(\Lambda_n \mid P_n^{(n)}) \to_w \mathscr{N}(-\tfrac{1}{2}\sigma^2; \sigma^2) \quad \text{implies} \quad \{P_n^{(n)}\} \lhd\rhd \{Q_n^{(n)}\}$$

(cf. Hájek & Šidák [1], Le Cam [2], [3], [4], Roussas [6]), we have to impose stronger conditions on the marginals P_{ni} and Q_{ni} to ensure the asymptotic normality of Λ_n. Some sufficient (and almost necessary) conditions for the asymptotic normality of Λ_n, which are clearly stronger than those in section 2, are given in section 3. These conditions are closely related to some earlier results of Le Cam [3], [4].

2. Contiguity of product measures

We begin by noting the following useful implication:

$$\text{(2.1)} \qquad \{Q_n^{(n)}\} \lhd \{P_n^{(n)}\} \Rightarrow [\lim_{n\to\infty} \sum_{i=1}^{n} P_{ni}(A_{ni}) = 0 \Rightarrow \lim_{n\to\infty} \sum_{i=1}^{n} Q_{ni}(A_{ni}) = 0]$$

for any collection of measurable sets A_{ni}. For suppose $\lim_{n\to\infty} \sum_{i=1}^{n} P_{ni}(A_{ni}) = 0$. Then $\lim_{n\to\infty} P_n^{(n)}(\bigcup_{i=1}^{n} A_{ni}) = 0$, hence by contiguity $\lim_{n\to\infty} Q_n^{(n)}(\bigcup_{i=1}^{n} A_{ni}) = 1 - \lim_{n\to\infty} \prod_{i=1}^{n} (1 - Q_{ni}(A_{ni}))$ $= 0$ and therefore $\lim_{n\to\infty} \sum_{i=1}^{n} Q_{ni}(A_{ni}) = 0$.

Now let μ_{ni} be a σ-finite measure on $(\mathscr{X}_{ni}, \mathscr{A}_{ni})$ dominating $P_{ni} + Q_{ni}$ and write $p_{ni} = \mathrm{d}P_{ni}/\mathrm{d}\mu_{ni}$ and $q_{ni} = \mathrm{d}Q_{ni}/\mathrm{d}\mu_{ni}$ ($i = 1, \ldots, n$; $n = 1, 2, \ldots$). The main result of this section is

Theorem 1. $\{Q_n^{(n)}\} \lhd \{P_n^{(n)}\}$ *iff*

$$\text{(2.2)} \qquad \limsup_{n\to\infty} \sum_{i=1}^{n} H^2(P_{ni}, Q_{ni}) < \infty$$

and

$$\text{(2.3)} \qquad \lim_{n\to\infty} \sum_{i=1}^{n} Q_{ni}(q_{ni}(X_{ni})/p_{ni}(X_{ni}) \geqq c_n) = 0 \quad \textit{whenever} \quad c_n \to \infty\,.$$

Proof. First assume that (2.2) and (2.3) are satisfied. Write

$$L_{ni} = q_{ni}(X_{ni})/p_{ni}(X_{ni})\,, \quad i = 1, \ldots, n\,; \quad n = 1, 2, \ldots\,,$$

and consider $\prod_{i=1}^{n} L_{ni}$. It is easily shown (cf. Le Cam [4], Roussas [6]) that $\{Q_n^{(n)}\} \lhd$ $\lhd \{P_n^{(n)}\}$ is equivalent to tightness of the sequence of distributions $\{\mathscr{L}(\prod_{i=1}^{n} L_{ni} \mid Q_n^{(n)});$

$n = 1, 2, \ldots\}$. The tightness of this set of distributions can also be expressed in the more convenient form

(2.4) $$\lim_{n\to\infty} Q_n^{(n)}\Big(\prod_{i=1}^n L_{ni} \geqq k_n\Big) = 0 \quad \text{whenever} \quad k_n \to \infty\,.$$

Hence we have to prove (2.4). Let $0 < k_n \to \infty$. Let $0 < c_n \to \infty$ be real numbers to be chosen in the sequel. If 1_A denotes the indicator function of the set A, we have by (2.3) and Markov's inequality for $n \to \infty$

$$Q_n^{(n)}\Big(\prod_{i=1}^n L_{ni} \geqq k_n\Big)$$

$$\leqq Q_n^{(n)}\Big(\prod_{i=1}^n L_{ni} \geqq k_n \wedge L_{ni} < c_n \quad \text{for} \quad i = 1, \ldots, n\Big) + Q_n^{(n)}\Big(\bigcup_{i=1}^n \{L_{ni} \geqq c_n\}\Big)$$

$$\leqq Q_n^{(n)}\Big(\prod_{i=1}^n L_{ni}^{1/2}\, 1_{(0,c_n)}(L_{ni}) \geqq k_n^{1/2}\Big) + \sum_{i=1}^n Q_{ni}(L_{ni} \geqq c_n)$$

$$\leqq k_n^{-1/2} \prod_{i=1}^n \int_{q_{ni} < c_n p_{ni}} q_{ni}^{3/2} p_{ni}^{-1/2}\, d\mu_{ni} + o(1)\,.$$

Since for all $c_n \geqq 1$

$$\int_{q_{ni} < c_n p_{ni}} q_{ni}^{3/2} p_{ni}^{-1/2}\, d\mu_{ni}$$

$$\leqq \int_{q_{ni} < c_n p_{ni}} q_{ni}\, d\mu_{ni} + \int_{q_{ni} < c_n p_{ni}} q_{ni} p_{ni}^{-1/2} (q_{ni}^{1/2} - p_{ni}^{1/2})\, d\mu_{ni}$$

$$\leqq 1 + \int_{q_{ni} < c_n p_{ni}} q_{ni}^{1/2} p_{ni}^{-1/2} (q_{ni}^{1/2} - p_{ni}^{1/2})^2\, d\mu_{ni} + \int_{q_{ni} < c_n p_{ni}} q_{ni}^{1/2} (q_{ni}^{1/2} - p_{ni}^{1/2})\, d\mu_{ni}$$

$$\leqq 1 + c_n^{1/2} \int (q_{ni}^{1/2} - p_{ni}^{1/2})^2\, d\mu_{ni} + 1 - \int q_{ni}^{1/2} p_{ni}^{1/2}\, d\mu_{ni}$$

$$- \int_{q_{ni} \geqq c_n p_{ni}} q_{ni}^{1/2} (q_{ni}^{1/2} - p_{ni}^{1/2})\, d\mu_{ni} \leqq 1 + (c_n^{1/2} + \tfrac{1}{2})\, H^2(P_{ni}, Q_{ni})\,,$$

it follows that

$$\limsup_{n\to\infty} Q_n^{(n)}\Big(\prod_{i=1}^n L_{ni} \geqq k_n\Big)$$

$$\leqq \limsup_{n\to\infty} k_n^{-1/2} \prod_{i=1}^n \{1 + (c_n^{1/2} + \tfrac{1}{2})\, H^2(P_{ni}, Q_{ni})\}$$

$$\leqq \limsup_{n\to\infty} k_n^{-1/2} \exp\Big\{(c_n^{1/2} + \tfrac{1}{2}) \sum_{i=1}^n H^2(P_{ni}, Q_{ni})\Big\}\,.$$

Choosing c_n in such a way that $c_n = o((\log k_n)^2)$ for $n \to \infty$, (2.2) implies $Q_n^{(n)}(\prod_{i=1}^{n} L_{ni} \geqq k_n) = o(1)$ for $n \to \infty$ and (2.4) is established.

Conversely, suppose that $\{Q_n^{(n)}\} \lhd \{P_n^{(n)}\}$. Since (1.6) implies that (2.2) is satisfied, it remains to prove (2.3). Let $0 < c_n \to \infty$ and consider the inequality, valid for $c_n \geqq 4$,

$$\sum_{i=1}^{n} \int_{q_{ni} \geqq c_n p_{ni}} p_{ni}\, d\mu_{ni} \leqq c_n^{-1/2} \sum_{i=1}^{n} \int_{q_{ni} \geqq c_n p_{ni}} p_{ni}^{1/2} q_{ni}^{1/2}\, d\mu_{ni}$$

$$= c_n^{-1/2} \left\{ \sum_{i=1}^{n} \int_{q_{ni} \geqq c_n p_{ni}} p_{ni}^{1/2}(q_{ni}^{1/2} - p_{ni}^{1/2})\, d\mu_{ni} + \sum_{i=1}^{n} \int_{q_{ni} \geqq c_n p_{ni}} p_{ni}\, d\mu_{ni} \right\}$$

$$\leqq c_n^{-1/2} \left\{ \sum_{i=1}^{n} \int_{q_{ni} \geqq c_n p_{ni}} (q_{ni}^{1/2} - p_{ni}^{1/2})^2\, d\mu_{ni} + \sum_{i=1}^{n} \int_{q_{ni} \geqq c_n p_{ni}} p_{ni}\, d\mu_{ni} \right\}$$

$$\leqq c_n^{-1/2} \left\{ \sum_{i=1}^{n} H^2(P_{ni}, Q_{ni}) + \sum_{i=1}^{n} \int_{q_{ni} \geqq c_n p_{ni}} p_{ni}\, d\mu_{ni} \right\}.$$

Since by (2.2) $c_n^{-1/2} \sum_{i=1}^{n} H^2(P_{ni}, Q_{ni}) \to 0$ for $n \to \infty$, it follows that $\lim_{n\to\infty} \sum_{i=1}^{n} P_n(L_{ni} \geqq c_n) = 0$. Hence (2.1) implies that $\lim_{n\to\infty} \sum_{i=1}^{n} Q_{ni}(L_{ni} \geqq c_n) = 0$ and the proof of the theorem is complete. □

Corollary 1. $\{P_n^{(n)}\} \lhd\rhd \{Q_n^{(n)}\}$ *iff* (2.2) *and* (2.3) *are satisfied and*

$$\text{(2.5)} \qquad \lim_{n\to\infty} \sum_{i=1}^{n} P_{ni}(p_{ni}(X_{ni})/q_{ni}(X_{ni}) \geqq c_n) = 0 \quad \textit{whenever} \quad c_n \to \infty\,.$$

In connection with contiguity Hellinger distance seems to be a more appropriate metric than total variation distance. Note that from (1.3) and (1.6) we immediately obtain the implication

$$\text{(2.6)} \qquad \{Q_n^{(n)}\} \lhd \{P_n^{(n)}\} \Rightarrow \sum_{i=1}^{n} \|P_{ni} - Q_{ni}\|^2 = O(1) \quad \text{for} \quad n \to \infty\,,$$

where again the order term cannot be strenghtened to $o(1)$. However, $\sum_{i=1}^{n} \|P_{ni} - Q_{ni}\|^2 = O(1)$ is too weak a condition to replace (2.2) in Theorem 1. On the other hand we cannot strengthen this condition to $\sum_{i=1}^{n} \|P_{ni} - Q_{ni}\|^r = O(1)$ for some $r < 2$, since $\{Q_n^{(n)}\} \lhd \{P_n^{(n)}\}$ does not necessarily imply $\sum_{i=1}^{n} \|P_{ni} - Q_{ni}\|^r = O(1)$ for any positive $r < 2$. The following example serves to illustrate these points.

Example. Let μ_{ni} denote Lebesgue measure on (0,1), let $p_{ni} = 1_{(0,1)}$ and let $q_{ni} = (1 + n^{-1/2})\, 1_{(0,1-n^{-1/2})} + n^{-1/2}\, 1_{[1-n^{-1/2},1)}$, $i = 1, \ldots, n$; $n = 1, 2, \ldots$. Then $\sum_{i=1}^{n} \|P_{ni} - Q_{ni}\|^2 = (1 - n^{-1/2})^2 \leqq 1$ and (2.3) is trivially satisfied since q_{ni}/p_{ni} is uniformly bounded. But $\{Q_n^{(n)}\} \lhd \{P_n^{(n)}\}$ does not hold because $\sum_{i=1}^{n} H^2(P_{ni}, Q_{ni}) = 2n\{1 - \int q_{ni}^{1/2}\, d\mu_{ni}\} = 2n\{1 - (1 + n^{-1/2})^{1/2}(1 - n^{-1/2}) - n^{-3/4}\} = n^{1/2}(1 + o(1))$ for $n \to \infty$.

Taking $q_{ni} = (1 + n^{-1/2})\, 1_{(0,1/2)} + (1 - n^{-1/2})\, 1_{[1/2,1)}$ for all i and n, we have $\{Q_n^{(n)}\} \lhd \{P_n^{(n)}\}$ since (2.3) is satisfied and $\sum_{i=1}^{n} H^2(P_{ni}, Q_{ni}) = 2n\{1 - \frac{1}{2}(1 + n^{-1/2})^{1/2} - \frac{1}{2}(1 - n^{-1/2})^{1/2}\} = \frac{1}{4} + o(1)$ for $n \to \infty$. However, in this case $\sum_{i=1}^{n} \|P_{ni} - Q_{ni}\|^r = n(\frac{1}{2}n^{-1/2})^r \to \infty$ for $n \to \infty$ if $r < 2$.

3. Asymptotic normality of Λ_n

Define

$$\text{(3.1)} \qquad \Lambda_n = \sum_{i=1}^{n} \log \{q_{ni}(X_{ni})/p_{ni}(X_{ni})\}\,, \quad n = 1, 2, \ldots.$$

Note that, with probability one, Λ_n is well-defined under $P_n^{(n)}$, although Λ_n may assume the value $-\infty$ with positive probability under $P_n^{(n)}$.

In our search for necessary and sufficient conditions for the weak convergence $\mathscr{L}(\Lambda_n \mid P_n^{(n)}) \to_w \mathscr{N}(-\frac{1}{2}\sigma^2; \sigma^2)$ in terms of the marginal distributions of the X_{ni} we shall confine ourselves to the case where the summands in (3.1) satisfy the traditional u.a.n. condition (cf. Loève [5]).

Theorem 2. *For any* $\sigma \geqq 0$.

$$\text{(3.2)} \qquad \mathscr{L}(\Lambda_n \mid P_n^{(n)}) \to_w \mathscr{N}(-\tfrac{1}{2}\sigma^2; \sigma^2)$$

and

$$\text{(3.3)} \qquad \lim_{n\to\infty} \max_{1 \leq i \leq n} P_{ni}(|\log \{q_{ni}(X_{ni})/p_{ni}(X_{ni})\}| \geqq \varepsilon) = 0$$

for every $\varepsilon > 0$ *iff for every* $\varepsilon > 0$

$$\text{(3.4)} \qquad \lim_{n\to\infty} \sum_{i=1}^{n} H^2(P_{ni}, Q_{ni}) = \tfrac{1}{4}\sigma^2\,,$$

$$\lim_{n\to\infty} \sum_{i=1}^{n} Q_{ni}(q_{ni}(X_{ni})/p_{ni}(X_{ni}) \geqq 1 + \varepsilon) = 0\,, \tag{3.5}$$

$$\lim_{n\to\infty} \sum_{i=1}^{n} P_{ni}(p_{ni}(X_{ni})/q_{ni}(X_{ni}) \geqq 1 + \varepsilon) = 0\,, \tag{3.6}$$

or equivalently, iff (3.4) *holds and for every* $\varepsilon > 0$

$$\lim_{n\to\infty} \sum_{i=1}^{n} \int_{|q_{ni}-p_{ni}|\geqq \varepsilon p_{ni}} (q_{ni}^{1/2} - p_{ni}^{1/2})^2 \, d\mu_{ni} = 0\,. \tag{3.7}$$

Proof. To simplify the notation we write $r_{ni} = q_{ni}/p_{ni}$. We first show that (3.5) and (3.6) are equivalent to (3.7). From

$$\sum_{i=1}^{n} \int_{|q_{ni}-p_{ni}|\geqq \varepsilon p_{ni}} (q_{ni}^{1/2} - p_{ni}^{1/2})^2 \, d\mu_{ni}$$

$$= \sum_{i=1}^{n} \left\{ \int_{r_{ni}\geqq 1+\varepsilon} q_{ni}(1 - r_{ni}^{-1/2})^2 \, d\mu_{ni} + \int_{r_{ni}\leqq 1-\varepsilon} p_{ni}(1 - r_{ni}^{1/2})^2 \, d\mu_{ni} \right\}$$

we obtain the double inequality

$$\{1 - (1+\varepsilon)^{-1/2}\}^2 \sum_{i=1}^{n} Q_{ni}(r_{ni}(X_{ni}) \geqq 1 + \varepsilon)$$

$$+ \{1 - (1-\varepsilon)^{1/2}\}^2 \sum_{i=1}^{n} P_{ni}(r_{ni}^{-1}(X_{ni}) \geqq (1-\varepsilon)^{-1})$$

$$\leqq \sum_{i=1}^{n} \int_{|q_{ni}-p_{ni}|\geqq \varepsilon p_{ni}} (q_{ni}^{1/2} - p_{ni}^{1/2})^2 \, d\mu_{ni}$$

$$\leqq \sum_{i=1}^{n} Q_{ni}(r_{ni}(X_{ni}) \geqq 1 + \varepsilon) + \sum_{i=1}^{n} P_{ni}(r_{ni}^{-1}(X_{ni}) \geqq (1-\varepsilon)^{-1})$$

and the equivalence of (3.5) and (3.6) to (3.7) is immediate.

Next we note that both (3.2), (3.3) and (3.4), (3.5), (3.6) imply $\{P_n^{(n)}\} \lhd\rhd \{Q_n^{(n)}\}$ (cf. Corollary 1).

The remainder of the proof relies on the normal convergence theorem (cf. Loève [5]). According to an equivalent form of this theorem (3.2) and (3.3) are equivalent to

$$\lim_{n\to\infty} \sum_{i=1}^{n} P_{ni}(|\log r_{ni}(X_{ni})| \geqq \delta) = 0 \quad \text{for every} \quad \delta > 0\,, \tag{3.8}$$

$$\lim_{\delta\downarrow 0} \lim_{n\to\infty} \sum_{i=1}^{n} \int_{|\log r_{ni}|\leqq \delta} (\log r_{ni}) \, dP_{ni} = -\tfrac{1}{2}\sigma^2\,, \tag{3.9}$$

$$\lim_{\delta\downarrow 0} \lim_{n\to\infty} \sum_{i=1}^{n} \left\{ \int_{|\log r_{ni}|\leqq \delta} (\log r_{ni})^2 \, dP_{ni} - \left(\int_{|\log r_{ni}|\leqq \delta} (\log r_{ni}) \, dP_{ni} \right)^2 \right\} = \sigma^2\,. \tag{3.10}$$

By the contiguity of $\{P_n^{(n)}\}$ and $\{Q_n^{(n)}\}$ and (2.1) the condition (3.8) is equivalent to (3.5) and (3.6) and hence to (3.7). Henceforth we assume (3.7), (3.8) and $\{P_n^{(n)}\} \lhd\rhd \{Q_n^{(n)}\}$. We still have to show that (3.4) is equivalent to (3.9) and (3.10).

Let $0 < \delta < 1$. For $|\log r_{ni}| \leq \delta$ we have the expansion

$$\text{(3.11)} \qquad \begin{aligned} \log r_{ni} &= 2 \log\{1 + (q_{ni}^{1/2} - p_{ni}^{1/2})\, p_{ni}^{-1/2}\} \\ &= 2(q_{ni}^{1/2} - p_{ni}^{1/2})\, p_{ni}^{-1/2} - (q_{ni}^{1/2} - p_{ni}^{1/2})^2\, p_{ni}^{-1}(1 + \varrho_{ni\delta}) \end{aligned}$$

with $|\varrho_{ni\delta}| < 2\delta$. Thus

$$\begin{aligned} &\int_{|\log r_{ni}| \leq \delta} (\log r_{ni})\, p_{ni}\, d\mu_{ni} \\ &= -2\int_{|\log r_{ni}| \leq \delta} (q_{ni}^{1/2} - p_{ni}^{1/2})^2\, d\mu_{ni} + \int_{|\log r_{ni}| \leq \delta} (q_{ni} - p_{ni})\, d\mu_{ni} \\ &\quad - \int_{|\log r_{ni}| \leq \delta} \varrho_{ni\delta}(q_{ni}^{1/2} - p_{ni}^{1/2})^2\, d\mu_{ni}\,. \end{aligned}$$

Since by (3.7)

$$\lim_{n\to\infty} \left\{ \sum_{i=1}^{n} \int_{|\log r_{ni}| \leq \delta} (q_{ni}^{1/2} - p_{ni}^{1/2})^2\, d\mu_{ni} - \sum_{i=1}^{n} H^2(P_{ni}, Q_{ni}) \right\} = 0$$

and by (3.8), $\{P_n^{(n)}\} \lhd\rhd \{Q_n^{(n)}\}$ and (2.1)

$$\sum_{i=1}^{n} \int_{|\log r_{ni}| \leq \delta} (q_{ni} - p_{ni})\, d\mu_{ni} = -\sum_{i=1}^{n} \int_{|\log r_{ni}| > \delta} (q_{ni} - p_{ni})\, d\mu_{ni} \to 0$$

for $n \to \infty$, we have

$$\text{(3.12)} \qquad \begin{aligned} &\lim_{\delta\downarrow 0} \limsup_{n\to\infty} \left| \sum_{i=1}^{n} \int_{|\log r_{ni}| \leq \delta} (\log r_{ni})\, dP_{ni} + 2\sum_{i=1}^{n} H^2(P_{ni}, Q_{ni}) \right| \\ &\leq \lim_{\delta\downarrow 0} \limsup_{n\to\infty} 2\delta \sum_{i=1}^{n} H^2(P_{ni}, Q_{ni}) = 0\,, \end{aligned}$$

where we have used (1.6). Similarly,

$$\text{(3.13)} \qquad \begin{aligned} &\lim_{\delta\downarrow 0} \limsup_{n\to\infty} \sum_{i=1}^{n} \left\{ \int_{|\log r_{ni}| \leq \delta} (\log r_{ni})\, dP_{ni} \right\}^2 \\ &\leq \lim_{\delta\downarrow 0} \limsup_{n\to\infty} \delta \sum_{i=1}^{n} \left| \int_{|\log r_{ni}| \leq \delta} (\log r_{ni})\, dP_{ni} \right| \\ &\leq \lim_{\delta\downarrow 0} \limsup_{n\to\infty} \delta(2 + 2\delta) \sum_{i=1}^{n} H^2(P_{ni}, Q_{ni}) = 0\,. \end{aligned}$$

Finally (3.11) implies that for $|\log r_{ni}| \leqq \delta < 1$

$$(\log r_{ni})^2 = 4(q_{ni}^{1/2} - p_{ni}^{1/2})^2 p_{ni}^{-1} + \bar{\varrho}_{ni\delta}(q_{ni}^{1/2} - p_{ni}^{1/2})^2 p_{ni}^{-1}$$

with $|\bar{\varrho}_{ni\delta}| < 10\delta$. Hence, in view of (3.7) and (1.6),

$$\text{(3.14)} \qquad \lim_{\delta \downarrow 0} \limsup_{n\to\infty} \left| \sum_{i=1}^{n} \int_{|\log r_{ni}| \leqq \delta} (\log r_{ni})^2 \, dP_{ni} - 4 \sum_{i=1}^{n} H^2(P_{ni}, Q_{ni}) \right| = 0 \, .$$

The equivalence of (3.4) to (3.9) and (3.10) is now an immediate consequence of (3.12), (3.13) and (3.14). The theorem is proved. □

In the one sample case where, for each n, $X_{n1}, \ldots, X_{nn}$ are identically distributed, condition (3.3) is implied by (3.2) and Theorem 2 slightly simplifies. This remains true in the k sample case ($k \geqq 2$) provided all sample sizes tend to infinity.

The first part of the proof of Theorem 2 also shows that the conditions (2.3) and (2.5) in Corollary 1 may be replaced by the single condition

$$\lim_{n\to\infty} \sum_{i=1}^{n} \int_{|q_{ni} - p_{ni}| \geqq c_n p_{ni}} (q_{ni}^{1/2} - p_{ni}^{1/2})^2 \, d\mu_{ni} = 0 \quad \text{whenever} \quad c_n \to \infty \, .$$

The proof of Theorem 2 could also be given in a more roundabout way. Introducing the r.v.'s

$$W_{ni} = 2\{q_{ni}(X_{ni})/p_{ni}(X_{ni})\}^{1/2} - 2 \, , \quad i = 1, \ldots, n \, ; \quad n = 1, 2, \ldots ,$$

one shows that $\mathscr{L}(\sum_{i=1}^{n} W_{ni} \mid P_n^{(n)}) \to_w \mathscr{N}(-\frac{1}{4}\sigma^2; \sigma^2)$ iff $\mathscr{L}(\Lambda_n \mid P_n^{(n)}) \to_w \mathscr{N}(-\frac{1}{2}\sigma^2; \sigma^2)$, provided the respective u.a.n. conditions are satisfied. It is then not difficult to prove that the weak convergence of $\sum_{i=1}^{n} W_{ni}$ and the u.a.n. condition on the summands are equivalent to (3.4) and (3.7). In this proof (3.7) appears as the Lindeberg condition in the central limit theorem applied to $\sum_{i=1}^{n} W_{ni}$.

The equivalence of both weak convergence results has first been proved by Le Cam ([3], [4]). The initial assumptions $\lim_{n\to\infty} \sup_{1\leqq i\leqq n} H^2(P_{ni}, Q_{ni}) = 0$ and $\limsup_{n\to\infty} \|P_n^{(n)} - Q_n^{(n)}\| < 1$ made by Le Cam are not restrictive since they are implied by our condition (3.7) and the contiguity of $\{P_n^{(n)}\}$ and $\{Q_n^{(n)}\}$, respectively. One part of this proof is also contained in Hájek & Šidák [1].

References

[1] HÁJEK, J. - ŠIDÁK, Z. (1967). "Theory of rank tests". Academic Press, New York.

[2] LE CAM, L. (1960). Locally asymptotically normal families of distributions. *Univ. California Publ. Statist.*, **3**, 37—98, University of California Press.

[3] LE CAM, L. (1966). Likelihood functions for large numbers of independent observations. *Research papers in statistics (Festschrift for J. Neyman)*, 167—187, F. N. David (ed.), Wiley, New York.

[4] LE CAM, L. (1969). Théorie asymptotique de la décision statistique. *Les Presses de l'Université de Montréal.*

[5] LOÈVE, M. (1963). "Probability theory (3rd ed.)". Van Nostrand, New York.

[6] ROUSSAS, G. G. (1972). Contiguity of probability measures: some applications in statistics, *Cambridge University Press.*

[7] WITTING, H. - NÖLLE G. (1970). "Angewandte mathematische Statistik". Teubner, Stuttgart.

(1) CATH. UNIV. NYMEGEN, NYMEGEN, THE NETHERLANDS

(2) CENTRAAL REKENINSTITUUT DER RIJKSUNIVERSITEIT, LEIDEN, THE NETHERLANDS

Received October 1975

ON ESTIMATING A PARAMETER AND ITS SCORE FUNCTION

C. A. J. KLAASSEN AND W. R. VAN ZWET University of Leiden

ABSTRACT: We consider the problem of estimating a real-valued parameter θ in the presence of an abstract nuisance parameter η, such as an unknown distributional shape. Attention is restricted to the case in which the *score functions* for θ and η are orthogonal, so that fully asymptotically efficient estimation is not a priori impossible. For fixed sample size, we provide a bound of Cramér-Rao type. The bound differs from the classical one for known η by a term involving the integrated mean square error of an estimator of a multiple of the score function for θ for the case where θ is known. This implies that an estimator of θ can only perform well over a class of shapes η if it is possible to estimate the score function for θ accurately over this class.

Key Words and Phrases: Adaptation, score function, Cramér-Rao inequality, semiparametric models

1980 Mathematics Subject Classification: Primary, 62F11; secondary, 62F35, 62G05.

Research supported in part by Office of Naval Research Contract N 00014-80-C-0163.

From *Proceedings of the Berkeley Conference in Honor of Jerzy Neyman and Jack Kiefer*, Volume II, Lucien M. Le Cam and Richard A. Olshen, eds.,

S. van de Geer and M. Wegkamp (eds.), *Selected Works of Willem van Zwet*, Selected Works in Probability and Statistics, DOI 10.1007/978-1-4614-1314-1_7,

1. AN INEQUALITY OF CRAMÉR-RAO TYPE

Let $X_1, \ldots, X_N$ be independent and identically distributed (i.i.d.) random variables with a common density $f(\cdot; \eta, \theta)$ with respect to a σ-finite measure μ on $\mathbb{R}$. The parameter of interest θ belongs to an open subset Θ of $\mathbb{R}$, and the nuisance parameter η ranges over an arbitrary set H. For unknown η and θ, it is required to estimate θ, which is done by means of an estimator $T_N = T_N(X_1, \ldots, X_N)$ for some measurable function $T_N : \mathbb{R}^N \to \mathbb{R}$. We are interested in the variance of T_N under $f(\cdot; \eta, \theta)$. We shall write $P_{\eta\theta}$, $E_{\eta\theta}$, and $\sigma^2_{\eta\theta}$ for probabilities, expectations, and variances under this model. The indicator function of a set B will be denoted by 1_B.

Throughout, we shall make the following regularity assumptions on the model and on the estimators to be considered. The first set of assumptions concerns differentiability in quadratic mean of the square root of the density with respect to θ. We assume that for every (η, θ) there exists a function $\tau(\cdot; \eta, \theta)$ such that

$$E_{\eta\theta}\tau^2(X_1; \eta, \theta) > 0, \tag{1.1}$$

$$\lim_{\theta'\to\theta} E_{\eta\theta}\left[\frac{f^{\frac12}(X_1; \eta, \theta') - f^{\frac12}(X_1; \eta, \theta)}{(\theta' - \theta)f^{\frac12}(X_1; \eta, \theta)} - \frac{1}{2}\tau(X_1; \eta, \theta)\right]^2 = 0, \tag{1.2}$$

$$\lim_{\theta'\to\theta} \frac{P_{\eta\theta'}(f(X_1; \eta, \theta) = 0)}{(\theta' - \theta)^2} = 0. \tag{1.3}$$

Clearly this defines $\tau(\cdot; \eta, \theta)$ a.e. $[P_{\eta\theta}]$ and ensures that $E_{\eta\theta}\tau^2(X_1; \eta, \theta) < \infty$. We complete the definition of τ by requiring arbitrarily that

$$\tau(x; \eta, \theta) = 0 \text{ if } f(x; \eta, \theta) = 0. \tag{1.4}$$

Note that an equivalent formulation of (1.1)-(1.3) is

$$\int \rho^2(x; \eta, \theta)\,d\mu(x) > 0, \tag{1.5}$$

$$\lim_{\theta'\to\theta} \int \left[\frac{f^{\frac12}(x; \eta, \theta') - f^{\frac12}(x; \eta, \theta)}{(\theta' - \theta)} - \frac{1}{2}\rho(x; \eta, \theta)\right]^2 d\mu(x) = 0, \tag{1.6}$$

where ρ is of the form

$$\rho(x;\, \eta,\, \theta) = \tau(x;\, \eta,\, \theta) f^{\frac{1}{2}}(x;\, \eta,\, \theta), \tag{1.7}$$

i.e., $\rho = 0$ if $f = 0$.

For fixed η, the function $\tau(\cdot;\, \eta,\, \theta)$ is called the score function for θ, and if f is differentiable in the ordinary sense, it coincides with $\partial \log f(\cdot;\, \eta,\, \theta)/\partial\theta$ a.e. $[P_{\eta\theta}]$. For known η, the Fisher information concerning θ that is contained in a single observation X_1 is defined by

$$I_\eta(\theta) = E_{\eta\theta}\tau^2(X_1;\, \eta,\, \theta) = \int \rho^2(x;\, \eta,\, \theta)\,d\mu(x) \in (0,\, \infty). \tag{1.8}$$

Our second set of assumptions concerns the estimator T_N. We assume that for every (η, θ)

$$E_{\eta\theta} T_N = \chi(\eta,\, \theta) \in (-\infty,\, \infty) \tag{1.9}$$

and that if $E_{\eta\theta} T_N^2 < \infty$ for a certain (η, θ), then T_N^2 is uniformly integrable with respect to $P_{\eta\theta'}$, for all θ' in a neighborhood of θ. Thus, for some $\varepsilon > 0$,

$$\lim_{C\to\infty} \sup_{|\theta' - \theta| < \varepsilon} E_{\eta\theta'} T_N^2 1_{\{|T_N| \geq C\}} = 0. \tag{1.10}$$

Under the assumptions made so far, the Cramér-Rao inequality for known η is valid for T_N, so

$$\sigma^2_{\eta\theta}(T_N) \geq \frac{\{\dot{\chi}(\eta,\, \theta)\}^2}{N I_\eta(\theta)} \tag{1.11}$$

where $\dot{\chi}(\eta, \theta) = \partial\chi(\eta, \theta)/\partial\theta$. Define the function $J(\cdot;\, \eta,\, \theta)$ by

$$J(x;\, \eta,\, \theta) = \frac{\dot{\chi}(\eta,\, \theta)}{I_\eta(\theta)} \tau(x;\, \eta,\, \theta) \tag{1.12}$$

and let

$$S_N(\eta,\, \theta) = \frac{1}{N}\sum_{i=1}^{N} J(X_i;\, \eta,\, \theta). \tag{1.13}$$

We note that (1.11) is a consequence of the orthogonality of $S_N(\eta, \theta)$ and $T_N - S_N(\eta, \theta)$, which yields

$$\begin{aligned}\sigma^2_{\eta\theta}(T_N) &= \sigma^2_{\eta\theta}(S_N(\eta,\, \theta)) + \sigma^2_{\eta\theta}(T_N - S_N(\eta,\, \theta)) \\ &= \frac{\{\dot{\chi}(\eta,\, \theta)\}^2}{N I_\eta(\theta)} + \sigma^2_{\eta\theta}(T_N - S_N(\eta,\, \theta)).\end{aligned} \tag{1.14}$$

But this implies, in addition, that $\sigma^2_{\eta\theta}(T_N)$ can come close to the Cramér-Rao bound (1.11) only if $T_N - \chi(\eta, \theta)$ is close to $S_N(\eta, \theta)$ under $P_{\eta\theta}$. However, if H is a large set in some function space, say, then $T_N - \chi(\eta, \theta)$ can obviously not mimic $S_N(\eta, \theta)$ arbitrarily well for all $\eta \in H$, and consequently, $\sigma^2_{\eta\theta}(T_N)$ cannot come arbitrarily close to the Cramér-Rao bound for all $\eta \in H$ simultaneously. Since we are considering the case in which $\eta \in H$ is unknown, it should be possible to improve on (1.11).

Let us turn this argument around for a moment. If T_N performs well as an estimator of θ—or rather of $\chi(\eta, \theta)$, which may include a bias term—for all $\eta \in H$, then $T_N - \chi(\eta, \theta)$ must resemble $S_N(\eta, \theta)$ under $P_{\eta\theta}$ for every $\eta \in H$ and $\theta \in \Theta$. It would seem therefore that $T_N - \chi(\eta, \theta)$ must contain information about the unknown function $J(\cdot; \eta, \theta)$. Let us try to extract this information. For every fixed $\theta \in \Theta$, let $\psi(X_1; \theta)$ be a sufficient statistic for X_1 with respect to the remaining parameter $\eta \in H$. According to the factorization theorem this means that

$$f(x; \eta, \theta) = g(\psi(x; \theta); \eta, \theta) \cdot h(x; \theta) \quad \text{a.e.} \quad [\mu] \tag{1.15}$$

for appropriately chosen g and h. Suppose, moreover, that for all (η, θ),

$$E_{\eta\theta}(\tau(X_1; \eta, \theta)|\psi(X_1; \theta)) = 0 \quad \text{a.e.} \quad [P_{\eta\theta}]. \tag{1.16}$$

Then, for $i = 1, \ldots, N$, we have

$$\begin{aligned} &E_{\eta\theta}(S_N(\eta, \theta)|\psi(X_j; \theta) \text{ for } j \neq i;\ X_i = x) \\ &- E_{\eta\theta}(S_N(\eta, \theta)|\psi(X_j; \theta) \text{ for } j \neq i;\ \psi(X_i; \theta) = \psi(x; \theta)) \\ &= \frac{1}{N} J(x; \eta, \theta) \quad \text{a.e. } [P_{\eta\theta}]. \end{aligned} \tag{1.17}$$

Since $T_N - \chi(\eta, \theta)$ resembles $S_N(\eta, \theta)$ under $P_{\eta\theta}$, we can hope that $NE_{\eta\theta}(T_N|\psi(X_j; \theta) \text{ for } j \neq i;\ X_i = x) - NE_{\eta\theta}(T_N|\psi(X_j; \theta) \text{ for } j \neq i;$ $\psi(X_i; \theta) = \psi(x; \theta))$ or rather its symmetrized version

$$\begin{aligned} J_N(x; \theta) = \sum_{i=1}^{N} \{&E_{\eta\theta}(T_N|\psi(X_j; \theta) \text{ for } j \neq i;\ X_i = x) \\ &- E_{\eta\theta}(T_N|\psi(X_j; \theta) \text{ for } j \neq i;\ \psi(X_i; \theta) \\ &= \psi(x; \theta))\} \end{aligned} \tag{1.18}$$

can serve as an estimator of $J(x; \eta, \theta)$. Note that since for each j, $\psi(X_j; \theta)$ is sufficient for X_j for fixed θ, J_N is indeed independent of η. For known θ it is therefore a legitimate estimator.

We shall prove the following result.

Theorem 1.1: *Suppose that for every* (η, θ) *assumptions* (1.1)-(1.3), (1.9), *and* (1.10) *are satisfied. For every fixed* θ, *let* $\psi(X_1; \theta)$ *be sufficient for* X_1 *with respect to* η *and let* (1.16) *hold for all* (η, θ). *Then, for every* (η, θ),

$$\sigma^2_{\eta\theta}(T_N) \geqslant \frac{\{\dot{\chi}(\eta, \theta)\}^2}{NI_\eta(\theta)} + \frac{1}{N} E_{\eta\theta} \int \{J_N(x; \theta) - J(x; \eta, \theta)\}^2 f(x; \eta, \theta) d\mu(x). \quad (1.19)$$

The theorem asserts that the Cramér-Rao bound may be improved by adding N^{-1} times the integrated mean square error (MSE) of the estimator J_N of the function J, which is an unknown multiple of the score function τ. It is unsatisfactory that the right-hand side of (1.19) depends on the choice of T_N. However, one can rephrase the theorem to assert only the existence of an estimator J_N such that (1.19) holds. The message of the theorem is then clear: The accuracy with which one can estimate θ for unknown η is delimited by how well one can do for known η on the one hand and how well one can estimate $J(\cdot; \eta, \theta)$ for known θ on the other. Clearly, the latter depends strongly on the class H. If $\tau(\cdot; \eta, \theta)$ runs through a large class of score functions as η ranges over H, then the integrated MSE of any estimator of J can be quite large, especially for some particularly irregular choices of τ. If τ is restricted to a smaller class of nicely behaved score functions as $\eta \in$ H, then the integrated MSE can be much smaller. Finally, if η is known, so that H consists of a single element, then $J(\cdot; \eta, \theta)$ can serve as an estimator of itself, and (1.19) reduces to the Cramér-Rao inequality.

In a sense, the result of Theorem 1.1 is not at all surprising. Adaptive estimators of a parameter for an unknown distributional shape are always based on some kind of preliminary estimate of the unknown score function followed by a good estimate of θ for the distributional shape corresponding to the estimated score function. For such estimators it is to be expected that a bound on their accuracy should involve both the accuracy of estimating θ for known η and that of estimating η for known θ. The novel aspect of Theorem 1.1, however, is that it is not assumed that the estimator T_N is based on a preliminary estimate of the score function, but that an estimate of J for known θ is derived from T_N. In effect we are saying that any successful adaptive estimation procedure must involve, either explicitly or implicitly, the estimation of the score function (or rather of J) and that because of this, the accuracy of estimating J enters into the lower bound for the variance of the adaptive estimator.

Though Theorem 1.1 is purely a finite sample result, it obviously has asymptotic implications. As an example, it clearly provides a finite sample analogue of a conjecture of Bickel (1982) which states, loosely speaking, that asymptotically fully efficient adaptive estimation is possible only if a consistent estimator of the score function exists.

In this connection the role of assumption (1.16) is of interest. It is well known [cf. Stein (1956), Bickel (1982), and Begun, Hall, Huang, and Wellner (1983)] that a necessary condition for asymptotically fully efficient adaptive estimation to be possible is that the two estimation problems—that of θ for known η and that of η for known θ—are, in a sense, asymptotically orthogonal. Since $\psi(X_1; \theta)$ is sufficient with respect to η for known θ, and $\tau(X_1; \eta, \theta)$ contains the information about θ locally for known η, assumption (1.16) is indeed an asymptotic orthogonality condition of this kind. In making this assumption, we are therefore restricting attention to the case in which fully asymptotically efficient estimation is not a priori impossible. In a way, this restriction is a reasonable one because without it, the Cramér-Rao inequality (1.11) is no longer a logical point of departure. In a companion paper, we intend to discuss the more general situation in which orthogonality is not necessarily present.

Even though it serves the same purpose, assumption (1.16) looks a bit different from the orthogonality conditions employed by other authors. Stein (1956) and Begun et al. (1983) define a class of score functions for η as the class of all limits, in the ordinary sense or in

L^2, of the form

$$\lim_{\nu\to\infty} \frac{\log f(\cdot;\, \eta_\nu,\, \theta) - \log f(\cdot;\, \eta,\, \theta)}{d(\eta_\nu,\, \eta)},$$

or

$$2 \lim_{\nu\to\infty} \frac{f^{\frac{1}{2}}(\cdot;\, \eta_\nu,\, \theta) - f^{\frac{1}{2}}(\cdot;\, \eta,\, \theta)}{d(\eta_\nu,\, \eta) f^{\frac{1}{2}}(\cdot;\, \eta,\, \theta)},$$

where d denotes an appropriately chosen distance, and $\lim d(\eta_\nu, \eta) = 0$. For all such "score functions" $\sigma(\cdot;\, \eta,\, \theta)$ of the form

$$E_{\eta\theta}\, \tau(X_1;\, \eta,\, \theta)\sigma(X_1;\, \eta,\, \theta) = 0. \tag{1.20}$$

Bickel (1982) considers all "score functions" $\sigma(\cdot;\, \eta,\, \theta)$ of the form

$$\frac{f(\cdot;\, \eta',\, \theta) - f(\cdot;\, \eta,\, \theta)}{f(\cdot;\, \eta,\, \theta)}$$

for $\eta' \in H$ and again requires (1.20), which now reduces to

$$E_{\eta\theta}\, \tau(X_1;\, \eta,\, \theta) = 0 \quad \text{for all } \eta' \in H. \tag{1.21}$$

Under an additional completeness assumption on the sufficient statistic $\psi(X_1;\, \theta)$, condition (1.16) in Theorem 1.1 can be replaced by a condition of the form (1.20) for an appropriate class of "score functions" σ. Since we are not concerned with asymptotics, in which only local properties count, there seems to be no need to introduce differentiation with respect to η to define our score functions. Bickel's definition, however, has the drawback that the expectation in (1.21) need not exist. To remedy this we consider all score functions of the form

$$\frac{f^{\frac{1}{2}}(\cdot;\, \eta',\, \theta) - f^{\frac{1}{2}}(\cdot;\, \eta,\, \theta)}{f^{\frac{1}{2}}(\cdot;\, \eta,\, \theta)}$$

for $\eta' \in H$ and require (1.20), which reduces to

$$\int \tau(x;\, \eta,\, \theta) f^{\frac{1}{2}}(x;\, \eta,\, \theta) f^{\frac{1}{2}}(x;\, \eta',\, \theta)\, d\mu(x) = 0 \quad \text{for all } \eta' \in H. \tag{1.22}$$

Of course we have to tailor the completeness assumption on $\psi(X_1;\, \theta)$ to this particular choice of score functions. Define densities

$$f(\cdot;\, \eta,\, \eta',\, \theta) = A(\eta,\, \eta',\, \theta) f^{\frac{1}{2}}(\cdot;\, \eta,\, \theta) f^{\frac{1}{2}}(\cdot;\, \eta',\, \theta)$$

for all η' in the set $H_{\eta\theta}$ where

$$A^{-1}(\eta,\, \eta',\, \theta) = \int f^{\frac{1}{2}}(x;\, \eta,\, \theta) f^{\frac{1}{2}}(x;\, \eta',\, \theta)\, d\mu(x) > 0.$$

We shall write $P_{\eta\eta'\theta}$ and $E_{\eta\eta'\theta}$ for probabilities and expectations under this model. For every fixed η and θ we assume that $\psi(X_1;\, \theta)$ is complete with respect to $\eta' \in H_{\eta\theta}$ under this model; i.e., if for some $(\eta,\, \theta)$ and for some measurable function m, $E_{\eta\eta'\theta} m(\psi(X_1;\, \theta)) = 0$ for all $\eta' \in H_{\eta\theta}$, then $P_{\eta\eta'\theta}(m(\psi(X_1;\, \theta)) = 0) = 1$ for all $\eta' \in H_{\eta\theta}$.

Theorem 1.2: *Suppose that for every* $(\eta,\, \theta)$ *assumptions* (1.1)-(1.3), (1.9), *and* (1.10) *are satisfied. For every fixed* θ, *let* $\psi(X_1;\, \theta)$ *be sufficient for* X_1 *with respect to* η; *for every fixed* $(\eta,\, \theta)$, *let* $\psi(X_1;\, \theta)$ *be complete with respect to* η' *under the model* $P_{\eta\eta'\theta}$. *Suppose, finally, that* (1.22) *holds for all* $(\eta,\, \eta',\, \theta)$. *Then, for every* $(\eta,\, \theta)$, *inequality* (1.19) *holds.*

In Section 2 we shall provide the proofs of Theorems 1.1 and 1.2. The most obvious example, i.e., the estimation of location for an unknown symmetric density, is briefly discussed in Section 3.

2. PROOF OF THE THEOREMS

Let

$$f_N(x;\, \eta,\, \theta) = \prod_{i=1}^{N} f(x_i;\, \eta,\, \theta)$$

denote the density of $X = (X_1,\, \ldots,\, X_N)$ with respect to the N-fold product measure μ_N taken at the point $x = (x_1,\, \ldots,\, x_N)$. Since N is fixed, a standard argument shows that (1.6) and (1.7)—or equivalently (1.2) and (1.3)—imply

$$\lim_{\theta' \to \theta} \int \left[\frac{f_N^{\frac{1}{2}}(x;\, \eta,\, \theta') - f_N^{\frac{1}{2}}(x;\, \eta,\, \theta)}{(\theta' - \theta)} - \frac{1}{2}\rho_N(x;\, \eta,\, \theta) \right]^2 d\mu_N(x) = 0, \quad (2.1)$$

where

$$\rho_N(x;\, \eta,\, \theta) = f_N^{\frac{1}{2}}(x;\, \eta,\, \theta) \sum_{i=1}^{N} \tau(x_i;\, \eta,\, \theta). \quad (2.2)$$

Suppose that $E_{\eta\theta} T_N^2 < \infty$ for a certain (η, θ). Take $\varepsilon > 0$ as in (1.10) and $|\theta' - \theta| < \varepsilon$. In view of (1.9) and (2.1),

$$\frac{\chi(\eta, \theta') - \chi(\eta, \theta)}{(\theta' - \theta)} = \int T_N(x) \cdot \frac{f_N^{\frac{1}{2}}(x; \eta, \theta') - f_N^{\frac{1}{2}}(x; \eta, \theta)}{(\theta' - \theta)}$$

$$\cdot \{f_N^{\frac{1}{2}}(x; \eta, \theta') + f_N^{\frac{1}{2}}(x; \eta, \theta)\} d\mu_N(x)$$

$$= \int T_N(x)\left\{\frac{1}{2}\rho_N(x; \eta, \theta) + \Delta_N(x; \eta, \theta, \theta')\right\}$$

$$\cdot \{f_N^{\frac{1}{2}}(x; \eta, \theta') + f_N^{\frac{1}{2}}(x; \eta, \theta)\} d\mu_N(x) \qquad (2.3)$$

with

$$\lim_{\theta' \to \theta} \int \Delta_N^2(x; \eta, \theta, \theta') d\mu_N(x) = 0. \qquad (2.4)$$

Because of (1.10), $E_{\eta\theta'} T_N^2$ is bounded for $|\theta' - \theta| < \varepsilon$ and by the Cauchy-Schwarz inequality

$$\lim_{\theta' \to \theta} \int T_N(x) \Delta_N(x; \eta, \theta, \theta') \cdot \{f_N^{\frac{1}{2}}(x; \eta, \theta') + f_N^{\frac{1}{2}}(x; \eta, \theta)\} d\mu_N(x) = 0. \qquad (2.5)$$

By another application of the Cauchy-Schwarz inequality combined with (1.6), (2.1), and (1.10),

$$\lim_{\theta' \to \theta} \int T_N(x) \rho_N(x; \eta, \theta) \cdot \{f_N^{\frac{1}{2}}(x; \eta, \theta') - f_N^{\frac{1}{2}}(x; \eta, \theta)\} d\mu_N(x) = 0. \qquad (2.6)$$

Together, (2.3), (2.5), (2.6), and (2.2) imply the existence of $\dot{\chi}(\eta, \theta)$ as well as

$$\dot{\chi}(\eta, \theta) = E_{\eta\theta} T_N \sum_{i=1}^{N} \tau(X_i; \eta, \theta). \qquad (2.7)$$

Repeating this argument with both T_N and χ replaced by 1, we find

$$E_{\eta\theta} \tau(X_1; \eta, \theta) = 0. \qquad (2.8)$$

Combining (2.7) and (2.8) we arrive at the decomposition (1.14).

To prove Theorem 1.1 it remains to study $\sigma_{\eta\theta}^2(T_N - S_N(\eta, \theta))$ for $S_N(\eta, \theta)$ as defined by (1.12) and (1.13). We begin by noting that

$$\sigma_{\eta\theta}^2(T_N - S_N(\eta, \theta)) \geqslant E_{\eta\theta}\, \sigma_{\eta\theta}^2(T_N - S_N(\eta, \theta) | \psi(X_1; \theta), \ldots, \psi(X_N; \theta)). \qquad (2.9)$$

Consider the conditional distribution of $X = (X_1, \ldots, X_N)$ given $\psi(X_1; \theta), \ldots, \psi(X_N; \theta)$. Under this conditional probability model, $X_1, \ldots, X_N$ are still i.i.d., and an application of Hájek's projection lemma [cf. Hájek (1968)] to this conditional setup yields

$$\sigma^2_{\eta\theta}(T_N - S_N(\eta, \theta) | \psi(X_1; \theta), \ldots, \psi(X_N; \theta))$$
$$\geq \sum_{i=1}^{N} \sigma^2_{\eta\theta}\{E_{\eta\theta}(T_N - S_N(\eta, \theta) | \psi(X_j; \theta) \text{ for } j \neq i; X_i) \\ | \psi(X_1; \theta), \ldots, \psi(X_N; \theta)\}. \quad (2.10)$$

It follows from (2.9), (2.10), and the inequality

$$\sum_{i=1}^{N} a_i^2 \geq N^{-1}\left(\sum_{i=1}^{N} a_i\right)^2$$

that

$$\sigma^2_{\eta\theta}(T_N - S_N(\eta, \theta)) \geq \sum_{i=1}^{N} E_{\eta\theta}\{E_{\eta\theta}(T_N - S_N(\eta, \theta) | \psi(X_j; \theta) \text{ for } j \neq i; X_i) \\ - E_{\eta\theta}(T_N - S_N(\eta, \theta) | \psi(X_1; \theta), \ldots, \psi(X_N; \theta))\}^2$$
$$= \sum_{i=1}^{N} E_{\eta\theta} \int \{E_{\eta\theta}(T_N - S_N(\eta, \theta) | \psi(X_j; \theta) \text{ for } j \neq i; X_i = x) \\ - E_{\eta\theta}(T_N - S_N(\eta, \theta) | \psi(X_j; \theta) \text{ for } j \neq i; \psi(X_i; \theta) = \psi(x; \theta))\}^2 \\ \cdot f(x; \eta, \theta) d\mu(x)$$
$$\geq N^{-1} E_{\eta\theta} \int \left\{ \sum_{i=1}^{N} [E_{\eta\theta}(T_N - S_N(\eta, \theta) | \psi(X_j; \theta) \text{ for } j \neq i; X_i = x) \right. \\ \left. - E_{\eta\theta}(T_N - S_N(\eta, \theta) | \psi(X_j; \theta) \text{ for } j \neq i; \psi(X_i; \theta) = \psi(x; \theta))] \right\}^2 \\ \cdot f(x; \eta, \theta) d\mu(x). \quad (2.11)$$

But since (1.16) implies (1.17), and in view of definition (1.18), we can write (2.11) as

$$\sigma^2_{\eta\theta}(T_N - S_N(\eta, \theta)) \\ \geq N^{-1} E_{\eta\theta} \int \{J_N(x; \theta) - J(x; \eta, \theta)\}^2 f(x; \eta, \theta) d\mu(x). \quad (2.12)$$

Theorem 1.1 now follows from (1.14) and (2.12).

To prove Theorem 1.2, we note that the factorization theorem, cf. (1.15), ensures that for fixed (η, θ), $\psi(X_1; \theta)$ is sufficient for X_1

with respect to $\eta' \in H_{\eta\theta}$ under the model $P_{\eta\eta'\theta}$. It follows that

$$E_{\eta\eta'\theta}(\tau(X_1;\ \eta,\ \theta)\,|\,\psi(X_1;\ \theta)) \tag{2.13}$$

is independent of $\eta' \in H_{\eta\theta}$. However, according to (1.22),

$$E_{\eta\eta'\theta}\{E_{\eta\eta'\theta}(\tau(X_1;\ \eta,\ \theta)\,|\,\psi(X_1;\ \theta))\} = 0 \tag{2.14}$$

for all $\eta' \in H_{\eta\theta}$, and the completeness assumption implies that the conditional expectation in (2.13) vanishes a.s. under $P_{\eta\eta'\theta}$ for every $\eta' \in H_{\eta\theta}$. Since (2.13) is independent of η', we can take $\eta' = \eta$, and (1.16) follows. Theorem 1.2 is now a consequence of Theorem 1.1.

3. ESTIMATING LOCATION UNDER SYMMETRY

Let H be the class of probability densities η with respect to Lebesgue measure on $\mathbb{R}$, which are symmetric about 0 and absolutely continuous with derivative η' and which possess a finite Fisher information

$$I_\eta = \int \left\{\frac{\eta'(x)}{\eta(x)}\right\}^2 \eta(x)dx < \infty. \tag{3.1}$$

Let $X_1, \ldots, X_N$ be i.i.d. with a common density $f(\cdot;\ \eta,\ \theta) = \eta(\cdot - \theta)$, where $\eta \in H$ and $\theta \in \mathbb{R}$ are both unknown. Under this model it is reasonable to estimate θ by a location equivariant estimator $T_N = T_N(X_1, \ldots, X_N)$, i.e., an estimator satisfying

$$T_N(x_1 + a, \ldots, x_N + a) = T_N(x_1, \ldots, x_N) + a \tag{3.2}$$

for all $x = (x_1, \ldots, x_N) \in \mathbb{R}^N$ and $a \in \mathbb{R}$. If we assume that $E_{\eta\theta} T_N^2 < \infty$, then

$$\chi(\eta,\ \theta) = E_{\eta\theta} T_N = \phi(\eta) + \theta, \tag{3.3}$$

so that $\dot{\chi}(\eta,\ \theta) \equiv 1$.

It is easy to see that for this model the regularity conditions (1.5)-(1.7), or equivalently (1.1)-(1.3), are satisfied with $\tau = -\eta'(\cdot - \theta)/\eta(\cdot - \theta)$. Clearly, assumptions (1.9) and (1.10) on T_N also hold. Choosing

$$\psi(x;\ \theta) = |x - \theta|, \tag{3.4}$$

we see that for fixed θ, $\psi(X_1;\ \theta)$ is sufficient for X_1 with respect to

$\eta \in H$. Since η'/η is an odd function and η is symmetric, we have

$$E_{\eta\theta}\left(\frac{\eta'(X_1 - \theta)}{\eta(X_1 - \theta)}\,\middle|\, |X_1 - \theta|\right) = E_{\eta 0}\left(\frac{\eta'(X_1)}{\eta(X_1)}\,\middle|\, |X_1|\right) = 0 \quad \text{a.s.} \tag{3.5}$$

in view of (3.1). Hence the assumptions of Theorem 1.1 are satisfied, and

$$\sigma^2_{\eta\theta}(T_N) \geqslant \frac{1}{NI_\eta} + \frac{1}{N} E_{\eta\theta} \int \{J_N(x;\,\theta) - J(x;\,\eta,\,\theta)\}^2 \eta(x - \theta)\,dx, \tag{3.6}$$

where

$$J(x;\,\eta,\,\theta) = -I_\eta^{-1} \frac{\eta'(x - \theta)}{\eta(x - \theta)}, \tag{3.7}$$

$$\begin{aligned} J_N(x;\,\theta) &= \sum_{i=1}^{N} \{E_{\eta\theta}(T_N \,|\, |X_j - \theta| \text{ for } j \neq i;\ X_i = x) \\ &\qquad - E_{\eta\theta}(T_N \,|\, |X_j - \theta| \text{ for } j \neq i;\ |X_i - \theta| = |x - \theta|)\} \\ &= \frac{1}{2} \sum_{i=1}^{N} \{E_{\eta\theta}(T_N \,|\, |X_j - \theta| \text{ for } j \neq i;\ X_i - \theta = x - \theta) \\ &\qquad - E_{\eta\theta}(T_N \,|\, |X_j - \theta| \text{ for } j \neq i;\ X_i - \theta = -(x - \theta))\}. \end{aligned} \tag{3.8}$$

Obviously, neither side of (3.6) depends on θ. We can therefore simplify (3.6) to

$$\sigma^2_{\eta 0}(T_N) \geqslant \frac{1}{NI_\eta} + \frac{1}{N} E_{\eta 0} \int \{J_N(x) - J(x;\,\eta)\}^2 \eta(x)\,dx, \tag{3.9}$$

where

$$J(x;\,\eta) = -I_\eta^{-1} \frac{\eta'(x)}{\eta(x)}, \tag{3.10}$$

$$\begin{aligned} J_N(x) = \frac{1}{2} \sum_{i=1}^{N} \{&E_{\eta 0}(T_N \,|\, |X_j| \text{ for } j \neq i;\ X_i = x) \\ &- E_{\eta 0}(T_N \,|\, |X_j| \text{ for } j \neq i;\ X_i = -x)\}. \end{aligned} \tag{3.11}$$

This result is given in Klaassen (1981), which also contains a discussion of the implications of inequality (3.9).

REFERENCES

Begun, J. M., Hall, W. J., Huang, W. M., and Wellner, J. A. (1983). Information and asymptotic efficiency in parametric-nonparametric models. *Ann. Statist. 11*, 432-452.

Bickel, P. J. (1982). On adaptive estimation. *Ann. Statist. 10*, 647-671.

Hájek, J. (1968). Asymptotic normality of simple linear rank statistics under alternatives. *Ann. Math. Statist. 39*, 325-346.

Klaassen, C. A. J. (1981). *Statistical Performance of Location Estimators.* Mathematical Centre Tract 133, Mathematisch Centrum, Amsterdam.

Stein, C. (1956). Efficient nonparametric testing and estimation. *Proc. Third Berkeley Symp. Math. Statist. and Probability 1*, 187-195. University of California Press, Berkeley.

ON ESTIMATING A PARAMETER AND ITS SCORE FUNCTION, II

C. A. J. KLAASSEN, A. W. VAN DER VAART AND W. R. VAN ZWET

Department of Mathematics
University of Leiden
Leiden, Netherlands

1. INTRODUCTION

A bound of Cramér-Rao type is provided for an estimator of a real-valued parameter θ in the presence of an abstract nuisance parameter η, such as an unknown distributional shape, on the basis of N i.i.d. observations. The bound consists of the reciprocal of the effective Fisher information in the sample, plus a term involving the integrated mean squared error of an estimator of a multiple of the so-called conditional score function for θ, for the case where θ is known. This implies that an estimator of θ can only perform well over a class of shapes η if it is possible to estimate the conditional score function for θ accurately over this class. For the special case where fully adaptive estimation may be possible, this result was given in a companion paper (Klaassen and van Zwet (1985)).

2. AN INEQUALITY OF CRAMÉR-RAO TYPE

Let $X_1, \ldots, X_N$ be independent and identically distributed (i.i.d.) random variables taking values in some measurable space $(\mathcal{X}, \mathcal{A})$, with a common density $f(\cdot;\eta,\theta)$ with respect to a σ-finite measure μ on $(\mathcal{X}, \mathcal{A})$. The parameter of interest θ belongs to an open subset Θ of R and the nuisance parameter η ranges over an arbitrary set H. For unknown η and θ, it is required to estimate θ and this is done by means of an estimator $T_N = T_N(X_1, \ldots, X_N)$ for some measurable function $T_N \colon \mathcal{X}^N \to \mathrm{R}$. We are interested in finding a lower bound for the variance of T_N under $f(\cdot;\eta,\theta)$. We shall write $P_{\eta\theta}$, $E_{\eta\theta}$ and $\sigma^2_{\eta\theta}$ for probabilities, expectations and variances under this model.

For every fixed $\theta \in \Theta$ and $j = 1, \ldots, N$, let $\psi(X_j;\theta)$ be a sufficient statistic for X_j with respect to $\eta \in H$. According to the factorization theorem this is equivalent to assuming that

$$f(x;\eta,\theta) = g(\psi(x;\theta);\eta,\theta)h(x;\theta) \quad \text{a.e. } [\mu], \tag{2.1}$$

where $g(\cdot;\eta,\theta)$ may be chosen to be the density of $\psi(X_1;\theta)$ with respect to a σ-finite measure ν_θ.

We shall assume that $f^{\frac{1}{2}}$ is differentiable in quadratic mean with respect

S. van de Geer and M. Wegkamp (eds.), *Selected Works of Willem van Zwet*, Selected Works in Probability and Statistics, DOI 10.1007/978-1-4614-1314-1_8, © Springer Science+Business Media, LLC 2012

to θ with a derivative which is not essentially zero, thus for every (η,θ)

$$\lim_{\theta'\to\theta}\int[(\theta'-\theta)^{-1}\{f^{\frac{1}{2}}(x;\eta,\theta')-f^{\frac{1}{2}}(x;\eta,\theta)\}-\frac{1}{2}\tau(x;\eta,\theta)f^{\frac{1}{2}}(x;\eta,\theta)]^2d\mu(x)=0, \tag{2.2}$$

$$I(\eta,\theta)=E_{\eta\theta}\tau^2(X_1;\eta,\theta)>0. \tag{2.3}$$

Obviously (2.2) implies that $I(\eta,\theta)<\infty$. Note that $\tau(\cdot\,;\eta,\theta)$ is simply an L_2-version of the classical score function for θ, $\partial\log f(x;\eta,\theta)/\partial\theta$; $I(\eta,\theta)$ is the Fisher information concerning θ which is contained in a single observation X_1 and measures how well θ can be estimated when η is known. However, since η is unknown, one expects the information concerning θ to be smaller. As discussed in Begun, Hall, Huang and Wellner (1983), the information loss results from a reduction of the score function.

First we define score functions in the η-direction. We shall say that $\beta(\cdot\,;\eta,\theta)$ is an η-score function if there exists a sequence $\eta_k\in H$ such that

$$\lim_{k\to\infty}\int[k\{f^{\frac{1}{2}}(x;\eta_k,\theta)-f^{\frac{1}{2}}(x;\eta,\theta)\}-\frac{1}{2}\beta(x;\eta,\theta)f^{\frac{1}{2}}(x;\eta,\theta)]^2d\mu(x)=0. \tag{2.4}$$

It is easy to see that, in view of (2.1), (2.4) implies that

$$\beta(x;\eta,\theta)=b(\psi(x;\theta);\eta,\theta)\quad\text{a.e. }[P_{\eta\theta}], \tag{2.5}$$

where b satisfies

$$\lim_{k\to\infty}\int[k\{g^{\frac{1}{2}}(v;\eta_k,\theta)-g^{\frac{1}{2}}(v;\eta,\theta)\}-\frac{1}{2}b(v;\eta,\theta)g^{\frac{1}{2}}(v;\eta,\theta)]^2d\nu_\theta(v)=0, \tag{2.6}$$

so that b is an η-score function for the model $\{g(\cdot\,;\eta,\theta)\colon\ \eta\in H,\theta\in\Theta\}$. Let $B(\eta,\theta)$ denote the set of all η-score functions for the original model — i.e. functions β for which (2.4)–(2.5) hold for an appropriate sequence $\eta_k\in H$ — and let $\tilde{B}(\eta,\theta)$ be the closure in L_2 of the linear span of $B(\eta,\theta)$.

The effective score function τ_E for θ in the presence of the nuisance parameter η, is defined as

$$\tau_E(x;\eta,\theta)=\tau(x;\eta,\theta)-b_E(\psi(x;\theta);\eta,\theta), \tag{2.7}$$

where $b_E(\psi(x;\theta);\eta,\theta)$ is the L_2-projection of τ on $\tilde{B}(\eta,\theta)$, thus

$$\begin{aligned}I_E(\eta,\theta)&=E_{\eta\theta}\{\tau(X_1;\eta,\theta)-b_E(\psi(X_1;\theta);\eta,\theta)\}^2\\&=\min_{\beta\in\tilde{B}(\eta,\theta)}\ E_{\eta\theta}\{\tau(X_1;\eta,\theta)-\beta(X_1)\}^2.\end{aligned} \tag{2.8}$$

$I_E(\eta,\theta)$ is the effective Fisher information, which measures how well θ can be estimated when η is unknown (cf. Begun et al. (1983), but note that we do not assume that $B(\eta,\theta)$ itself is a linear space).

Let $C(\eta,\theta)$ denote the set of all square-integrable functions $b(\psi(x;\theta))$ with $E_{\eta\theta}\ b(\psi(X_1;\theta))=0$. In the special case where $\tilde{B}(\eta,\theta)=C(\eta,\theta)$, $b_E(v;\eta,\theta)$ equals the conditional expectation $E_{\eta\theta}(\tau(X_1;\eta,\theta)|\psi(X_1;\theta)=v)$ and τ_E and I_E

reduce to

$$\tau_C(x;\eta,\theta) = \tau(x;\eta,\theta) - E_{\eta\theta}(\tau(X_1;\eta,\theta)|\psi(X_1;\theta) = \psi(x;\theta)), \tag{2.9}$$

$$I_C(\eta,\theta) = E_{\eta\theta}\tau_C^2(X_1;\eta,\theta), \tag{2.10}$$

which are called the conditional score function and the conditional Fisher information for θ. In general, however, $\tilde{B}(\eta,\theta)$ may be a proper subset of $C(\eta,\theta)$, and hence

$$I_C(\eta,\theta) \le I_E(\eta,\theta) \le I(\eta,\theta) \tag{2.11}$$

as is clear from figure 1. Of course, we still have $\tau_C = \tau_E$ and $I_C = I_E$ if $E_{\eta\theta}(\tau(X_1;\eta,\theta)|\psi(X_1;\theta) = \psi(\cdot\,;\theta))$ happens to be in $\tilde{B}(\eta,\theta)$.

So far we have discussed various aspects of the model. Concerning the estimator T_N, we assume that, whenever $E_{\eta\theta}\, T_N^2 < \infty$ for a certain (η,θ), then

$$\sup_{(\eta',\theta')\in A_\varepsilon} E_{\eta'\theta'}T_N^2 < \infty \tag{2.12}$$

for some $\varepsilon > 0$, where

$$A_\varepsilon = \{(\eta',\theta') \colon \int |f(x;\eta',\theta') - f(x;\eta,\theta)|d\mu(x) < \varepsilon\} \tag{2.13}$$

consists of parameter values "close" to (η,θ). For simplicity we shall also assume that T_N is an unbiased estimator of θ, i.e. for all (η,θ),

$$E_{\eta\theta}T_N = \theta. \tag{2.14}$$

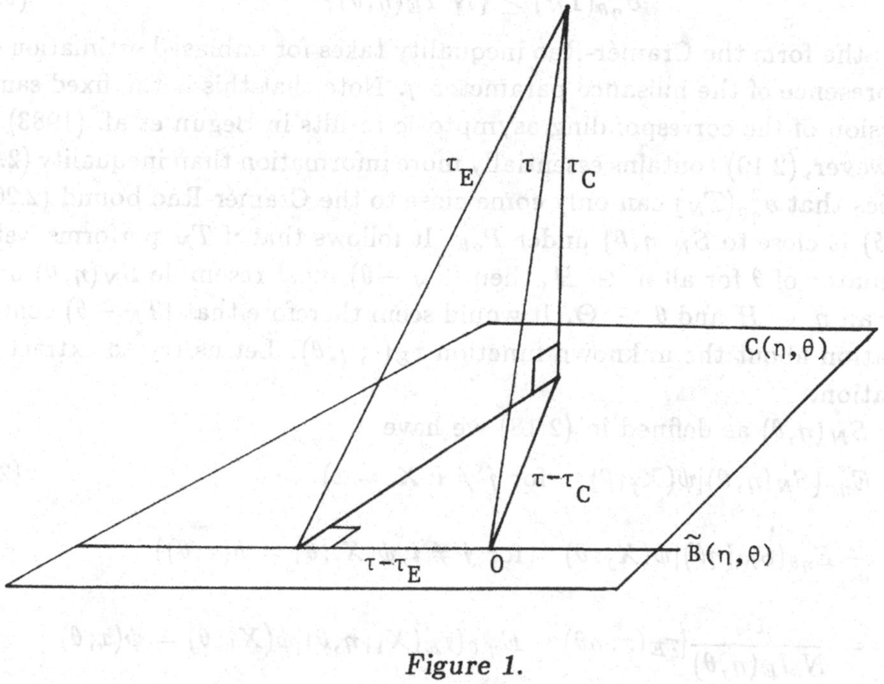

Figure 1.

The role of assumption (2.12)–(2.13) is to ensure that (2.14) implies that

$$E_{\eta\theta}T_N \sum_{j=1}^{N} \tau(X_j;\eta,\theta) = 1, \tag{2.15}$$

$$E_{\eta\theta}T_N \sum_{j=1}^{N} \beta(X_j) = 0 \quad \text{for all } \beta \in \tilde{B}(\eta,\theta), \tag{2.16}$$

and hence in particular

$$E_{\eta\theta}T_N \sum_{j=1}^{N} \tau_E(X_j;\eta,\theta) = 1 \tag{2.17}$$

in view of (2.7)–(2.8). If we define

$$S_N(\eta,\theta) = \frac{1}{NI_E(\eta,\theta)} \sum_{j=1}^{N} \tau_E(X_j;\eta,\theta) \tag{2.18}$$

then (2.17) asserts that $S_N(\eta,\theta)$ and $T_N - S_N(\eta,\theta)$ are uncorrelated under $P_{\eta\theta}$. As a consequence

$$\begin{aligned} \sigma^2_{\eta\theta}(T_N) &= \sigma^2_{\eta\theta}(S_N(\eta,\theta)) + \sigma^2_{\eta\theta}(T_N - S_N(\eta,\theta)) \\ &= \{N\ I_E(\eta,\theta)\}^{-1} + \sigma^2_{\eta\theta}(T_N - S_N(\eta,\theta)), \end{aligned} \tag{2.19}$$

and hence

$$\sigma^2_{\eta\theta}(T_N) \geq \{N\ I_E(\eta,\theta)\}^{-1} \tag{2.20}$$

which is the form the Cramér-Rao inequality takes for unbiased estimation of θ in the presence of the nuisance parameter η. Note that this is the fixed sample size version of the corresponding asymptotic results in Begun et al. (1983).

However, (2.19) contains essentially more information than inequality (2.20). It implies that $\sigma^2_{\eta\theta}(T_N)$ can only come close to the Cramér-Rao bound (2.20) if $(T_N - \theta)$ is close to $S_N(\eta,\theta)$ under $P_{\eta\theta}$. It follows that if T_N performs well as an estimator of θ for all $\eta \in H$, then $(T_N - \theta)$ must resemble $S_N(\eta,\theta)$ under $P_{\eta\theta}$ for all $\eta \in H$ and $\theta \in \Theta$. It would seem therefore that $(T_N - \theta)$ contains information about the unknown function $\tau_E(\cdot;\eta,\theta)$. Let us try to extract this information.

For $S_N(\eta,\theta)$ as defined in (2.18) we have

$$\begin{aligned} &E_{\eta\theta}(S_N(\eta,\theta)|\psi(X_j;\theta) \quad \text{for } j \neq i; X_i = x) \\ &- E_{\eta\theta}(S_N(\eta\theta)|\psi(X_j;\theta) \quad \text{for } j \neq i; \psi(X_i;\theta) = \psi(x;\theta)) \\ &= \frac{1}{N\ I_E(\eta,\theta)}[\tau_E(x;\eta\theta) - E_{\eta,\theta}(\tau_E(X_1;\eta,\theta)|\psi(X_1;\theta) = \psi(x;\theta))] \end{aligned} \tag{2.21}$$

$$= \frac{1}{N\, I_E(\eta,\theta)} \tau_C(x;\eta,\theta)$$

in view of (2.7) and (2.9). If $(T_N - \theta)$ resembles $S_N(\eta,\theta)$ under $P_{\eta\theta}$, we may hope that

$$J_N(x;\theta) = \sum_{i=1}^{N} \{E_{\eta\theta}(T_N|\psi(X_j;\theta) \quad \text{for } j \neq i; X_i = x) \qquad (2.22)$$
$$- E_{\eta\theta}(T_N|\psi(X_j;\theta) \quad \text{for } j \neq i; \psi(X_i;\theta) = \psi(x;\theta))\}$$

can serve as an estimator of

$$J(x;\eta,\theta) = \frac{1}{I_E(\eta,\theta)} \tau_C(x;\eta,\theta). \qquad (2.23)$$

Note that since for each j, $\psi(X_j;\theta)$ is sufficient for X_j with respect to $\eta \in H$ for fixed $\theta \in \Theta$, J_N is indeed independent of η. For known θ, it is therefore a legitimate estimator of J.

We shall prove the following result.

Theorem 2.1. *Suppose that assumptions* (2.1)–(2.3) *and* (2.12)–(2.14) *are satisfied for every* (η,θ). *Then, for every* (η,θ),

$$\sigma^2_{\eta\theta}(T_N) \geq \frac{1}{N\, I_E(\eta\theta)} + \frac{1}{N} E_{\eta\theta} \int \{J_N(x;\theta) - J(x;\eta,\theta)\}^2 f(x;\eta,\theta)\, d\mu(x). \quad (2.24)$$

The theorem asserts that the Cramér-Rao bound (2.20) may be improved by adding N^{-1} times the integrated mean squared error (MSE) of the estimator J_N of the function J, which is an unknown multiple of the conditional score function τ_C. For practical purposes it is unsatisfactory that the right-hand side of (2.24) depends on the choice of T_N. However, one may obviously rephrase the theorem to assert only the existence of an estimator J_N such that (2.24) holds. The message of the theorem is then clear: the accuracy with which one can estimate θ for unknown η is delimited by the effective Fisher information on the one hand and by how well one can estimate $J(\cdot\,;\eta,\theta)$ for known θ on the other. Clearly the latter depends heavily on the class H. If $J(\cdot\,;\eta,\theta)$ runs through a large class of functions as η ranges over H, then the integrated MSE of any estimator of J may be quite large, especially for particularly irregular choices of J. If J is restricted to a smaller class of nicely behaved functions as $\eta \in H$, then the integrated MSE may be much smaller. Finally, if η is known so that H consists of a single element, then $J(\cdot\,;\eta,\theta)$ can serve as an estimator of itself and (2.24) reduces to the Cramér-Rao inequality (2.20).

In a sense, the result of the theorem is not surprising. Estimation of a parameter θ for an unknown distributional shape is based typically on a preliminary estimate of an unknown score function followed by a good estimate of θ for the distributional shape corresponding to the estimated score function.

For such estimators a result like (2.24) is to be expected. The interesting aspect of the theorem, however, is that it is not assumed that the estimator T_N is based on a preliminary estimate of a score function, but that an estimate of J for known θ is derived from T_N. In effect we are saying that a successful estimation procedure for θ must involve — either explicitly or implicitly — the estimation of J and that because of this, the accuracy of estimating J enters into the lower bound for the variance of the estimator of θ.

Although the theorem is purely a finite sample result, it obviously has asymptotic implications. An asymptotic analogue would imply that effective estimation of θ, i.e.

$$\{N\ I_E(\eta,\theta)\}^{\frac{1}{2}}(T_N-\theta) \xrightarrow{\mathcal{D}} N(0,1), \tag{2.25}$$

is possible only if the function J can be estimated consistently with respect to integrated MSE for known θ. In this context it is unsatisfactory that J involves the conditional score function τ_C rather than the effective score function τ_E and, indeed, Klaassen (1987) has shown that a somewhat stronger version of (2.25) does entail consistent estimation of $\tau_E/I_E(\eta,\theta)$.

Of course this discrepancy disappears if $\tau_C=\tau_E$, i.e. if the function $E_{\eta\theta}(\tau(X_1;\eta,\theta)|\psi(X_1;\theta)=\psi(x;\theta))$ is an element of $\tilde{B}(\eta,\theta)$. This situation is rather common and examples, including non-i.i.d. models, are given by van der Vaart (1986), who also explicitly constructs an effective estimator of θ based on a preliminary consistent estimator of τ_C for such models.

An even more special case occurs if $\tau_C=\tau$, so that $I_E=I$ and $J=\tau/I$. Now (2.24) provides a finite sample analogue of the statement that fully adaptive estimation of θ is possible only if τ/I can be estimated consistently. This situation was discussed in the companion paper Klaassen and van Zwet (1985).

3. PROOF OF THE THEOREM

The proof resembles that of theorem 1.1 in Klaassen and van Zwet (1985). Let

$$f_N(x)=\prod_{j=1}^{N} f(x_j;\eta,\theta)$$

denote the density of $X=(X_1,\ldots,X_N)$ with respect to the N-fold product measure μ^N taken at the point $x=(x_1,\ldots,x_N)$ and write

$$\frac{f_N^{\frac{1}{2}}(x;\eta,\theta')-f_N^{\frac{1}{2}}(x;\eta,\theta)}{(\theta'-\theta)}=\frac{1}{2}\rho_N(x;\eta,\theta)+\Delta_N(x;\eta,\theta,\theta'), \tag{3.1}$$

with

$$\rho_N(x;\eta,\theta) = f_N^{\frac{1}{2}}(x;\eta,\theta)\sum_{i=1}^{N} \tau(x_i;\eta,\theta). \tag{3.2}$$

Since N is fixed, a standard argument shows that (2.2) implies

$$\lim_{\theta'\to\theta} \int \Delta_N^2(x;\eta,\theta,\theta')d\mu_N(x) = 0. \tag{3.3}$$

In view of (2.14) we have

$$1 = \int T_N(x)\frac{f_N^{\frac{1}{2}}(x;\eta,\theta') - f_N^{\frac{1}{2}}(x;\eta,\theta)}{(\theta'-\theta)}\{f_N^{\frac{1}{2}}(x;\eta,\theta') + f_N^{\frac{1}{2}}(x;\eta,\theta)\}d\mu_N(x) \tag{3.4}$$

$$= \int T_N(x)\{\frac{1}{2}\rho_N(x;\eta,\theta) + \Delta_N(x;\eta,\theta,\theta')\}\{f_N^{\frac{1}{2}}(x;\eta,\theta') + f_N^{\frac{1}{2}}(x;\eta,\theta)\}d\mu_N(x).$$

If $E_{\eta\theta}T_N^2 = \infty$, there is nothing to prove. Suppose therefore that $E_{\eta\theta}T_N^2 < \infty$. Since (2.2) ensures that

$$\lim_{\theta'\to\theta} \int |f(x;\eta,\theta') - f(x;\eta,\theta)|d\mu(x) = 0,$$

(2.12) and (2.13) yield

$$\limsup_{\theta'\to\theta} E_{\eta\theta'}T_N^2 < \infty. \tag{3.5}$$

Together, (3.3), (3.5) and the Cauchy-Schwarz inequality show that

$$\lim_{\theta'\to\theta} \int T_N(x)\Delta_N(x;\eta,\theta,\theta')\{f_N^{\frac{1}{2}}(x;\eta,\theta') + f_N^{\frac{1}{2}}(x;\eta,\theta)\}d\mu_N(x) = 0, \tag{3.6}$$

$$|\int T_N(x)\rho_N(x;\eta,\theta)\{f_N^{\frac{1}{2}}(x;\eta,\theta') - f_N^{\frac{1}{2}}(x;\eta,\theta)\}d\mu_N(x)|$$

$$\leq \{C^2\int \{f_N^{\frac{1}{2}}(x;\eta,\theta') - f_N^{\frac{1}{2}}(x;\eta,\theta)\}2d\mu_N(x)\int \rho_N^2(x;\eta,\theta)d\mu_N(x)\}^{\frac{1}{2}} \tag{3.7}$$

$$+\{\int T_N^2(x)\{f_N^{\frac{1}{2}}(x;\eta,\theta') - f_N^{\frac{1}{2}}(x;\eta,\theta)\}^2d\mu_N(x)\int_{\{|T_N|>C\}} \rho_N^2(x;\eta,\theta)d\mu_N(x)\}^{\frac{1}{2}}$$

for every $C > 0$. As θ' tends to θ, the first term on the right tends to zero for every C in view of (2.2). Since $E_{\eta\theta}T_N^2 < \infty$ and $E_{\eta\theta}\tau^2(X_1;\eta,\theta) < \infty$, the second term converges to zero as $C \to \infty$. It follows that the left-hand side of (3.7) converges to zero, and together with (3.4) and (3.6) this proves (2.15). A similar argument produces (2.16) and (2.19) follows.

It remains to show that

$$\sigma_{\eta\theta}^2(T_N - S_N(\eta,\theta)) \geq \frac{1}{N}E_{\eta\theta}\int\{J_N(x;\theta) - J(x;\eta,\theta)\}^2 f(x;\eta,\theta)d\mu(x). \tag{3.8}$$

To see this, we copy the argument leading from (2.9) to (2.11) in Klaassen and van Zwet (1985), even though $S_N(\eta,\theta)$ is defined differently in that paper. We find

$$\sigma^2_{\eta\theta}(T_N - S_N(\eta,\theta)) \tag{3.9}$$

$$\geq N^{-1}E_{\eta\theta}\int\{\sum_{i=1}^{N}[E_{\eta\theta}(T_N - S_N(\eta,\theta)|\psi(X_j;\theta) \text{ for } j\neq i; X_i = x)$$

$$- E_{\eta\theta}(T_N - S_N(\eta,\theta)|\psi(X_j;\theta) \text{ for } j\neq i; \psi(X_i;\theta) = \psi(x;\theta))]\}^2 f(x;\eta,\theta)d\mu(x).$$

In view of (2.21)–(2.23), (3.9) is identical to (3.8) and the proof complete.

ACKNOWLEDGMENTS

Research was supported in part by the Office of Naval Research Contract N00014-80-C-0163.

BIBLIOGRAPHY

Begun, J. M., Hall, W. J., Huang, W. M. and Wellner, J. A. (1983). Information and asymptotic efficiency in parametric-nonparametric models. *Ann. Statist.* **11**, 432–452.

Klaassen, C. A. J. (1987). Consistent estimation of the influence function of locally asymptotically linear estimators. *Ann. Statist.*, to appear.

Klaassen, C. A. J. and van Zwet, W. R. (1985). On estimating a parameter and its score function. *Proc. Berkeley Conference in Honor of Jerzy Neyman and Jack Kiefer*, L. Le Cam and R. A. Olshen, editors, Vol. II, 827–839.

van der Vaart, A. W. (1986). Estimating a real parameter in a class of semi-parametric models. Submitted to *Ann. Statist.*

Volume 8, No. 2 (1999), pp. 277–284 Allerton Press, Inc.

MATHEMATICAL METHODS OF STATISTICS

A REMARK ON CONSISTENT ESTIMATION

E. W. VAN ZWET[1] AND W. R. VAN ZWET[2]

[1]Department of Mathematics, University of Utrecht
P.O. Box 80010, 3508 TA Utrecht, The Netherlands
vanzwet@math.ruu.nl

[2]Department of Mathematics, University of Leiden
P.O. Box 9512, 2300 RA Leiden, The Netherlands
vanzwet@wi.leidenuniv.nl

In this paper we re-examine two auxiliary results in Putter and van Zwet [7]. Viewed in a new light these results provide some insight in two related phenomena, to wit consistency of estimators and local asymptotic equivariance. Though technically quite different, our conclusions will be similar to those in Beran [1] and LeCam and Yang [5].

Key words: non-parametric models, consistency, local asymptotic equivariance, convolution theorem, sets of the first category.

AMS 1991 Subject Classification: Primary 62G05, 62G20.

1. Introduction

Let $(\mathcal{X}, \mathcal{A})$ be a measurable space and $\mathcal{P}$ a collection of probability measures on $(\mathcal{X}, \mathcal{A})$. Let Π be a topology on $\mathcal{P}$, so that $(\mathcal{P}, \Pi)$ is a topological space. Finally, let $X_1, X_2, \ldots$ denote a sequence of i.i.d. random variables with values in $\mathcal{X}$ and (unknown) common distribution $P \in \mathcal{P}$.

For $N = 1, 2, \ldots$, we consider a map $\tau_N : (\mathcal{P}, \Pi) \to (\mathcal{R}, \rho)$, where $(\mathcal{R}, \rho)$ is a metric space. Both spaces $(\mathcal{P}, \Pi)$ and $(\mathcal{R}, \rho)$ are equipped with the σ-algebras of Borel sets $\mathcal{B}(\mathcal{P}, \Pi)$ and $\mathcal{B}(\mathcal{R}, \rho)$, which are generated by the open sets in $(\mathcal{P}, \Pi)$ and $(\mathcal{R}, \rho)$ respectively. Probability distributions on these spaces are probability measures on the Borel sets, and are induced by measurable maps from $(\mathcal{X}^\infty, \mathcal{A}^\infty, P^\infty)$ to $(\mathcal{P}, \mathcal{B}(\mathcal{P}, \Pi))$ or $(\mathcal{R}, \mathcal{B}(\mathcal{R}, \rho))$. We assume throughout that each τ_N is measurable.

S. van de Geer and M. Wegkamp (eds.), *Selected Works of Willem van Zwet*, Selected Works in Probability and Statistics, DOI 10.1007/978-1-4614-1314-1_9, © Springer Science+Business Media, LLC 2012

Having observed the i.i.d. sample $X_1, \ldots, X_N$ with common distribution $P \in \mathcal{P}$, our aim is to estimate the somewhat abstract $\mathcal{R}$-valued "parameter" $\tau_N(P)$. For a measurable map $t_N : \mathcal{X}^N \to \mathcal{R}$, let $T_N = t_N(X_1, \ldots, X_N)$ be an estimator of $\tau_N(P)$ based on $X_1, \ldots, X_N$. We shall say that T_N is a *consistent estimator* of $\tau_N(P)$ for $P \in \mathcal{P}$ if

$$(1.1) \qquad \rho(T_N, \tau_N(P)) \to^P 0, \qquad \text{for every } P \in \mathcal{P},$$

where $\to^P$ indicates convergence in probability under P as $N \to \infty$. The more formally inclined reader should view this expression as shorthand for the correct but laborious statement that the sequence $\{T_N\}_{N=1}^{\infty}$ is a consistent sequence of estimators of the sequence $\{\tau_N(P)\}_{N=1}^{\infty}$. If we wish to stress the role of the metric ρ in (1.1) we call T_N ρ-consistent.

In many applications the topology Π on $\mathcal{P}$ will be metrized by the Hellinger metric H, so that $(\mathcal{P}, H)$ is a metric space. Recall that for $P, Q \in \mathcal{P}$ with densities f and g with respect to a common σ-finite measure μ on $(\mathcal{X}, \mathcal{A})$, the Hellinger distance H of P and Q is defined as

$$(1.2) \qquad H(P,Q) = \left\{ \int_{\mathcal{X}} (f^{1/2} - g^{1/2})^2 \, d\mu \right\}^{1/2} .$$

Many results in asymptotic statistics do not hold for all underlying distributions $P \in \mathcal{P}$, but only for $P \in \mathcal{P} \setminus D$, where the exceptional set D is in some sense small compared to $\mathcal{P}$. For a finite-dimensional parametric family $\mathcal{P} = \{P_\theta : \theta \in \Theta\}$ with $\Theta \subset \mathbb{R}^k$, we may identify $\mathcal{P}$ with Θ, and the exceptional subset of Θ will typically be small in the sense that it has Lebesgue measure zero. On the more general topological spaces of distributions $(\mathcal{P}, \Pi)$ that we consider in this paper, there is no obvious analogue for Lebesgue measure for which "small" sets can naturally be described as sets of measure zero. It is therefore hardly surprising that in the present set-up, the exceptional set D will be small in a topological sense: D will be a set of the first category in $(\mathcal{P}, \Pi)$. We recall that a *set of the first category* is a countable union of nowhere dense sets, and that a set is *nowhere dense* in $(\mathcal{P}, \Pi)$ if its closure does not contain an open set in $(\mathcal{P}, \Pi)$.

When thinking of exceptional sets $D \subset \mathcal{P}$ of the first category as "small", a word of caution may not be amiss. Even for parametric families $\mathcal{P} = \{P_\theta : \theta \in \Theta\}$ with $\Theta \subset \mathbb{R}^k$, such sets D may correspond to subsets of the parameter space of positive Lebesgue measure and one has to impose regularity conditions to make sure that this phenomenon does not occur. By allowing exceptional sets of the first category, one is – in a sense – sweeping some technical difficulties under the rug in exchange for cleaner statements of results. Of course one also has the added generality of dealing with arbitrary $\mathcal{P}$ rather than parametric families, but again this comes at a price. With increasing complexity of the space $\mathcal{P}$, the concept of a set of the first category gradually looses its significance. For instance, as long as $(\mathcal{P}, \Pi)$ is topologically complete, we know that the complement of a set of the first category is everywhere dense, but in incomplete spaces a set of the first category may equal the entire space. This gradual loss of significance corresponds precisely to what one would expect: as the complexity of the model $\mathcal{P}$ increases, many results will hold in decreasing generality. All in all, we feel that exceptional sets of the first category may well be the proper concept in much of asymptotic statistical theory.

The framework introduced so far was used to discuss consistency of bootstrap estimators in Putter and van Zwet [7]. Two of the auxiliary results obtained in that paper are re-examined in the present note from a somewhat different point of view, to wit consistency of estimators in general and the related subject of local asymptotic equivariance. Though technically quite different, our conclusions will be similar in spirit to those in Beran [1] and LeCam and Yang [5].

2. Consistency

The parameter sequence $\{\tau_N(P)\}$ will be called *locally asymptotically constant* (LAC) at $P \in \mathcal{P}$ if for every $C > 0$,

$$\lim_{N\to\infty} \sup_{\{Q\in\mathcal{P}:\ H(P,Q)\leq CN^{-1/2}\}} \rho(\tau_N(Q), \tau_N(P)) = 0. \tag{2.1}$$

If τ_N converges to a limit τ pointwise in $(\mathcal{R}, \rho)$, then the LAC property is obviously equivalent to the statement that the limit of $\tau_N(P_N)$ is the same for any sequence P_N contained in shrinking Hellinger balls of radii of order $N^{-1/2}$ and centered at P. The following proposition is simply a re-statement of Lemma 5.3 in Putter and van Zwet [7].

Proposition 2.1. *Let the topology* Π *on* $\mathcal{P}$ *be metrized by Hellinger metric* H *and suppose that*

(i) *For every* $N = 1, 2, \ldots$, *the map* $\tau_N\colon (\mathcal{P}, \Pi) \to (\mathcal{R}, \rho)$ *is continuous;*

(ii) *There exists a* ρ*-consistent estimator* $T_N = t_N(X_1, \ldots, X_N)$ *of* $\tau_N(P)$.

Then there exists a set D *of the first category in* $(\mathcal{P}, \Pi)$ *such that* $\{\tau_N(P)\}$ *is LAC at every* $P \in \mathcal{P} \setminus D$.

Let P^N and Q^N denote the joint distribution of $X_1, \ldots, X_N$ under the models P and Q respectively. We have

$$1 - \tfrac{1}{2}H^2(Q^N, P^N) = (1 - \tfrac{1}{2}H^2(Q, P))^N \tag{2.2}$$

and hence, if P and Q are at Hellinger distance of order $N^{-1/2}$, the N-dimensional distributions P^N and Q^N are at Hellinger distance of order 1. This is the case where the two models are contiguous and, for large N, the joint distributions of $X_1, \ldots, X_N$ are essentially different under P and Q. Yet, in order to be continuous and estimable, the parameter $\tau_N(P)$ has to be essentially the same under both models for large N and "most" P. Hence, if, e.g., $\tau_N(P) = \theta(P^N)$ is a parameter of the distribution P^N which varies non-trivially with its argument P^N, then one must expect that $\tau_N(P)$ will not be estimable.

A case of particular interest is as follows. Consider a sequence of random variables $Y_N = y_N(X_1, \ldots, X_N; P)$, where y_N is a measurable map from $\mathcal{X}^N \times \mathcal{P}$ to a *separable* metric space $(\mathcal{Y}, \eta)$. Let $\mathcal{R}$ be the space of all probability distributions on $(\mathcal{Y}, \eta)$ equipped with Prohorov's metric ρ which is defined for distributions $R_1, R_2 \in \mathcal{R}$ by

$$\rho(R_1, R_2) = \inf\{\varepsilon > 0\colon R_1(A) \leq R_2(A^\varepsilon) + \varepsilon, \quad \text{for all } A \in \mathcal{B}(\mathcal{Y}, \eta)\}, \tag{2.3}$$

where A^ϵ is an ϵ-neighborhood of A. Since $(\mathcal{Y}, \eta)$ is separable, ρ metrizes weak convergence of probability measures in $\mathcal{R}$ (c.f. Dudley [3], Section 11.3). Note that the separability of $(\mathcal{Y}, \eta)$ also implies that $(\mathcal{R}, \rho)$ is separable (Billingsley [2], p. 239). The parameter to be estimated is

$$\tau_N(P) = \mathcal{L}_P(Y_N), \tag{2.4}$$

the probability distribution of Y_N under P. Obviously τ_N maps $(\mathcal{P}, H)$ into $(\mathcal{R}, \rho)$. This is the estimation problem that the bootstrap is designed to solve.

We shall say that Y_N is *locally asymptotically distributionfree* (LAD) if its distribution τ_N satisfies (2.1) with ρ taken to be the Prohorov metric. Obviously, the conclusion of Proposition 2.1 may now be interpreted to assert that Y_N is LAD.

At first sight the dependence of $Y_N = y_N(X_1, \ldots, X_N; P)$ on P may seem somewhat unusual. It is introduced to allow us to standardize the statistic of interest and this standardization may depend on P. If $(\mathcal{Y}, \eta)$ is a Euclidean space $\mathbb{R}^k$, one may · instance wish to study an estimator $Z_N = z_N(X_1, \ldots, X_N)$ of a k-dimensional rameter $\zeta(P)$ and in this case the distribution $\tau_N(P)$ of $Y_N = a_N(Z_N - \zeta(P))$ is distribution of interest for some normalizing sequence of constants a_N. Perhaps even more compelling reason for allowing Y_N to depend on P is that otherwise s distribution will typically not be estimable in cases of interest.

To see this, suppose that $Y_N = y_N(X_1, \ldots, X_N)$ does not depend on P. In the first place this ensures continuity of each τ_N in view of (2.2) and because $\rho(\tau_N(Q), \tau_N(P))$ will tend to zero if $H(Q^N, P^N)$ does. Hence assumption (i) of Proposition 2.1 is automatically satisfied and we obtain

Corollary 2.1. *Let $\tau_N(P)$ be the law of $Y_N = y_N(X_1, \ldots, X_N)$, where Y_N does not depend on P. If there exists a Prohorov-consistent estimator of $\tau_N(P)$, then there exists a set D of the first category in $(\mathcal{P}, H)$ such that Y_N is LAD at every $P \in \mathcal{P} \setminus D$.*

In other words, the distribution $\tau_N(P)$ of $Y_N = y_N(X_1, \ldots, X_N)$ can only be estimated Prohorov-consistently on the basis of $X_1, \ldots, X_N$ if Y_N is LAD at "most" P. But we already noted that if P and Q are at Hellinger distance of order $N^{-1/2}$, then P^N and Q^N will be essentially different N-dimensional distributions. Hence the LAD property of Y_N at a given P_0 indicates that the distribution of $Y_N = y_N(X_1, \ldots, X_N)$ is insensitive to significant changes of the distribution P^N of the random vector $(X_1, \ldots, X_N)$ in all possible directions around P_0^N. Such a statistic is not of much interest since no statistical procedure based on it will enable us to distinguish between significantly different contiguous models.

Let us therefore return to the case where $Y_N = y_N(X_1, \ldots, X_N; P)$ does indeed depend on P. If $\tau_N(P)$ is to be continuous as well as estimable, then clearly the role of this dependence of Y_N on P must be to offset the change in the distribution of the statistic that would otherwise occur for changes in the underlying distribution P of Hellinger distance of order $N^{-1/2}$. To arrive at the proper dependence of Y_N on P that will ensure the LAD property which is necessary for the estimability of $\tau_N(P)$, one starts with the statistic of interest $Z_N = z_N(X_1, \ldots, X_N)$ and asks what the influence of a contiguous change in the distribution of the X_i will be on the distribution of Z_N. This is often relatively easy as the statistic is supposedly constructed for some purpose related to distinguishing between the different models $P \in \mathcal{P}$ and the effect of contiguous changes of P on its distribution in usually known to first order. The second step is to construct a dependence on P that will destroy this first order change. This step is basically intelligent guesswork.

Another approach to this problem is to propose a particular dependence of Y_N on P and then try to find out in which cases this produces the desired result of making the distribution $\tau_N(P)$ of Y_N continuous and LAC by checking (2.1). The prime example of this was alluded to above: if $(\mathcal{Y}, \eta)$ is a Euclidean space $\mathbb{R}^k$ and the k-vector Z_N is a consistent estimator of a k-dimensional parameter $\zeta(P)$, then we may choose

$$Y_N = N^{1/2}(Z_N - \zeta(P)), \quad Z_N = z_N(X_1, \ldots, X_N). \tag{2.5}$$

This transformation will certainly remove at least part of the first order effect of the change in P on the distribution of the statistic Z_N. Moreover, the distribution $\tau_N(P)$ of Y_N is clearly of much interest for statistical purposes. To make sure that the normalization in (2.5) makes sense we assume that

(2.6) the sequence $\{\tau_N(P)\}$ is tight for every $P \in \mathcal{P}$.

The problem of estimating this distribution $\tau_N(P)$ was studied by Efron [4] who proposed the so-called naïve bootstrap as an estimator.

Let us assume that ζ is continuous. Then the continuity of each individual τ_N is also guaranteed. We saw earlier that for every fixed N the distribution of Z_N is continuous and this continues to hold for the distribution of $N^{1/2}Z_N$. Similarly the continuity of ζ implies the continuity of $N^{1/2}\zeta$ for fixed N. Hence assumption (i) of Proposition 2.1 may be replaced by the requirement that $\zeta\colon (\mathcal{P}, H) \to \mathbb{R}^k$ is continuous.

In the present case where Y_N is given by (2.5), the LAC property of $\tau_N(P)$ – or equivalently the LAD property of Y_N – is usually expressed by saying that Z_N is a *regular estimator* of $\zeta(P)$ or that Z_N is *locally asymptotically equivariant* (LAE). This terminology is usually reserved for the case where the distribution $\tau_N(P)$ of Y_N converges weakly to a limit distribution, but we shall adopt this terminology also in the slightly weaker case where (2.6) holds. Proposition 2.1 now reduces to

Corollary 2.2. *Let $(\mathcal{Y}, \eta) = \mathbb{R}^k$, let $\zeta\colon (\mathcal{P}, H) \to \mathbb{R}^k$, let Z_N and Y_N as defined in* (2.5) *be random variables with values in $\mathbb{R}^k$ and suppose that* (2.3), (2.4) *and* (2.6) *hold. If ζ is continuous and there exists a Prohorov-consistent estimator of $\tau_N(P)$, then there exists a set D of the first category in $(\mathcal{P}, H)$ such that Z_N is LAE at every $P \in \mathcal{P} \setminus D$.*

Interest in LAE estimators is motivated by the convolution theorem. In the present setting we shall use slight modification of the version of this theorem given in van der Vaart [8]. First of all a path in $\mathcal{P}$ through a fixed point $P \in \mathcal{P}$ is a map $t \to P_t$ from some interval $(0, \varepsilon)$ into $\mathcal{P}$ such that

$$\int \left[t^{-1}(dP_t^{1/2} - dP^{1/2}) - \tfrac{1}{2} g\, dP^{1/2}\right]^2 \to 0 \quad \text{as} \quad t \downarrow 0 \tag{2.7}$$

for some $g \in L_2(P)$. The collection of all such functions g constitutes the *tangent cone* $T(P)$ at the point P. Second, we shall assume that the functional $\zeta : (\mathcal{P}, H) \to \mathbb{R}^k$ is *differentiable* in the sense that there exists a vector-valued function $\dot{\zeta}_P \in L_2(P)^k$ such that for every path P_t through P

$$t^{-1}(\zeta(P_t) - \zeta(P)) \to \int \dot{\zeta}_P g dP. \tag{2.8}$$

With these definitions we have a version of the

Convolution Theorem. *In the set-up of Corollary 2.2 with* (2.3), (2.4) *and* (2.6) *being satisfied, suppose that* ζ *is differentiable and* Z_N *is LAE. Then for every* P *for which the tangent cone* $T(P)$ *contains a point in the interior of its linear span,*

$$\lim_{N\to\infty} \sup_{\{Q\in\mathcal{P}:\, H(P,Q)\le CN^{-1/2}\}} \rho(\tau_N(Q), M_N(P) * N(0, \Sigma(\zeta, P))) = 0. \tag{2.9}$$

Here $\{M_N(P)\}$ denotes a tight and LAE sequence of probability distributions on $\mathbb{R}^k$ which only depends on P, $*$ denotes convolution and $N(0, \Sigma(\zeta, P))$ denotes the k-variate normal distribution with mean 0 and the Cramér–Rao bound $\Sigma(\zeta, P)$ for estimating ζ as the covariance matrix.

Combining Corollary 2.2 and the convolution theorem we find

Corollary 2.3. *Consider the set-up of Corollary 2.2 with* (2.3), (2.4) *and* (2.6) *being satisfied. Suppose that* ζ *is differentiable and that there exists a Prohorov-consistent estimator of* $\tau_N(P)$. *Then there exists a set* D *of the first category in* $(\mathcal{P}, H)$ *such that* (2.9) *holds for all* $P \in \mathcal{P} \setminus D$ *for which the tangent cone* $T(P)$ *contains an interior point of its linear span.*

Now let us step back for a moment and examine what we have shown. In a large non-parametric model $\mathcal{P}$, the tangent cone $T(P)$ will contain an interior point of its linear span for many points $P \in \mathcal{P}$. In fact, $T(P)$ may well be a linear space for all interior points $P \in \mathcal{P}$. In such cases we find roughly speaking that if we estimate a differentiable functional $\zeta(P)$ at rate $N^{-1/2}$ by a statistic $Z_N = z_N(X_1, \ldots, X_N)$, then at "most" points $P \in \mathcal{P}$ it is not possible to estimate the distribution $\tau_N(P)$ of $Y_N = N^{1/2}(Z_N - \zeta(P))$ consistently, unless this distribution behaves asymptotically like $M_N(P) * N(0, \Sigma(\zeta, P))$ under all sequences of underlying distributions inside Hellinger balls of radius $N^{-1/2}$ around P.

Let us compare Corollary 2.3 with the results obtained in Beran [1]. Theorem 2.1 in Beran [1] deals with the parametric case where $\mathcal{P} = \{P_\theta : \theta \in \Theta\}$ with $\Theta \subset \mathbb{R}^k$ under the assumption that $\tau_N(P)$ converges to a limit distribution, the support of which satisfies a technical condition. Otherwise, the conditions are more or less similar to ours: a LAN assumption and Fréchet-differentiability of ζ. Among other things this theorem asserts that if the parametric bootstrap estimates $\tau_N(P)$ Prohorov-consistently, then (2.9) will hold for all $\theta \in \Theta$. There is no exceptional set where the conclusion does not hold. The important differences with Corollary 2.3 in the present paper are that we do not assume convergence but only tightness of $\tau_N(P)$, but that we do have an exceptional set of the first category. Incidentally this set of the first category corresponds to a set of Lebesgue measure zero in the parameter space (cf. Putter [6], Theorem 2.6.1). The fact that we discuss estimability of $\tau_N(P)$ as opposed to consistency of a particular estimator – the parametric bootstrap – is a minor matter.

Theorem 2.2 in Beran [1] deals with general non-parametric families $\mathcal{P}$. This is achieved by replacing the class $\mathcal{P}$ of all possible distributions on $(\mathcal{X}, \mathcal{A})$ by the class $\mathcal{P}_0$ of all discrete distributions with finite support, which is dense in $\mathcal{P}$. This reduces the problem to a finite-dimensional one which can be treated as before. Instead of an exceptional set of the first category, the exceptional set is now $\mathcal{P} \setminus \mathcal{P}_0$. Again the main difference with Corollary 2.3 above is the requirement of weak convergence of $\tau_N(P)$ instead of tightness. We shall return to this matter in Section 3.

There is also an interesting connection with van Zwet [9]. One of the conclusions of that paper is that, roughly speaking, the naïve bootstrap can only be consistent if Y_N is either asymptotically normal or asymptotically distributionfree (or of course a sum of variables of both types). This seems to come very close to Corollary 2.3.

3. Local Asymptotic Constancy

In this section we do not assume throughout that the topological space $(\mathcal{P}, \Pi)$ is metrizable. We shall say that the sequence $\tau_N : (\mathcal{P}, \Pi) \to (\mathcal{R}, \rho)$ is *equicontinuous* at $P \in \mathcal{P}$ if for every $\varepsilon > 0$ there exists a neighborhood U_ε of P such that $Q \in U_\varepsilon$ implies that

$$\sup_N \rho(\tau_N(Q), \tau_N(P)) \leq \varepsilon. \tag{3.1}$$

The following proposition is a re-statement of Lemma 4.2 in Putter and van Zwet [7].

Proposition 3.1. *Suppose that*

(i) *For every* $N = 1, 2, \ldots$, *the map* $\tau_N : (\mathcal{P}, \Pi) \to (\mathcal{R}, \rho)$ *is continuous;*

(ii) τ_N *converges pointwise to a limit* τ.

Then there exists a set D *of the first category in* $(\mathcal{P}, \Pi)$ *such that the sequence* τ_N *is equicontinuous at every* $P \in \mathcal{P} \setminus D$.

A comparison with the results in Section 2 shows that the convergence $\tau_N(P) \to \tau(P)$ for all $P \in \mathcal{P}$ makes a great deal of difference for the estimability of $\tau_N(P)$. Obviously, equicontinuity of τ_N on $\mathcal{P} \setminus D$ implies that if we can find a consistent estimator $\hat{P}_N$ of P, then $\tau_N(P)$ can be estimated consistently for all $P \in \mathcal{P} \setminus D$. But a comparison with Proposition 2.1 also shows that if $\tau_N \to \tau$, we no longer need the estimability of $\tau_N(P)$ to conclude that $\{\tau_N(P)\}$ is LAC at every $P \in \mathcal{P} \setminus D$. In fact, we now have more because if Π is metrized by the Hellinger metric, (3.1) ensures that for any $\delta_N \downarrow 0$,

$$\lim_{N \to \infty} \sup_{\{Q \in \mathcal{P} :\, H(P,Q) \leq \delta_N\}} \rho(\tau_N(Q), \tau_N(P)) = 0, \tag{3.2}$$

which is much stronger than (2.1).

References

[1] R. J. Beran, *Diagnosing bootstrap success*, Ann. Inst. Statist. Math., 49 (1997), 1–24.

[2] P. Billingsley, *Convergence of Probability Measures*, Wiley, New York, 1968.

[3] R. M. Dudley, *Real Analysis and Probability*, Wadsworth, Belmont, 1989.

[4] B. Efron, B. (1979), *The bootstrap: Another look at the jackknife*, Ann. Statist., 7 (1979), 1–26.

[5] L. M. Le Cam and G. L. Yang, *Asymptotics in Statistics: Some Basic Concepts*, Springer, New York, 1990.

[6] H. Putter, *Consistency of Resampling Methods*, PhD thesis, University of Leiden, 1994.

[7] H. Putter and W. R. van Zwet, *Resampling: Consistency of substitution estimators*, Ann. Statist., 24 (1996), 2297–2318.

[8] A. W. van der Vaart, *On the asymptotic information bound*, Ann. Statist., 17 (1989), 1487–1500.

[9] W. R. van Zwet, *Resampling: The jackknife and the naïve bootstrap*, unpublished manuscript, 1995.

[*Received January* 1999]

Chapter 10
Finite samples and asymptotics

Chris A.J. Klaassen

Abstract Willem van Zwet is a scientist and a scholar with a broad spectrum of research interests. This is reflected by the five papers in this section, which study very different fundamental problems and which have four of his PhD students and his youngest son as coauthor.

10.1 Introduction

Willem van Zwet is a scientist and a scholar with a broad spectrum of research interests within statistics and probability theory with the stress on statistics. This is clear from his list of publications. This breadth is obvious also from the selection of papers we discuss in this section: asymptotic normality of rank test statistics, Hellinger distance and contiguity, estimation of parameters and score functions, and consistency and asymptotic equivariance. Willem has tried to instill this attitude of broad interest towards research also into his PhD students.

As a thesis advisor Willem van Zwet has been unparalleled. He didn't suggest standard problems to his students, but he has boosted their careers by tackling interesting, fundamental, relevant problems. Most of his PhD students have become full professor in statistics (12 out of 16, so far). Actually the five papers in this section all have been written with scientific children, i.e. PhD students, as coauthors, except for the most recent one, which has been written in collaboration with his youngest biological son Erik.

One of the distinctions between these five papers is that three of them belong to the realm of asymptotic statistics, whereas the other two are of the finite sample type. Research in mathematical statistics is or should be motivated mainly by

Chris A.J. Klaassen
Korteweg-de Vries Institute for Mathematics, University of Amsterdam, P.O. Box 94248, 1090 GE Amsterdam, The Netherlands
e-mail: C.A.J.Klaassen@uva.nl

S. van de Geer and M. Wegkamp (eds.), *Selected Works of Willem van Zwet*, Selected Works in Probability and Statistics, DOI 10.1007/978-1-4614-1314-1_10, © Springer Science+Business Media, LLC 2012

real life problems where data have to be interpreted. Since data always come in finite amounts, the core business of mathematical statistics should be finite sample statistics. Typically however, finite sample problems are difficult to handle mathematically. That is why one resorts to approximations, which are obtained typically by letting the sample size tend to infinity. This results in asymptotic statistics. This approach is applied quite often; actually, to such an extent that asymptotic statistics has become the norm, or at least the benchmark at which different techniques are compared.

10.2 Asymptotic Normality of Nonparametric Tests for Independence

David van Dantzig (1900–1959) is one of the initiators of the Mathematical Centre in Amsterdam. It was founded in February 1947 as a non-profit institute aiming at the promotion of pure mathematics and its applications. The topologist Van Dantzig intended to contribute to the reconstruction of Dutch society after the second world war by stimulating the study and application of mathematical statistics. In his philosophy the statistician should choose for his data a statistical model with the weakest of possible assumptions. Therefore, a main theme at the Mathematical Centre, the cradle of much of Dutch mathematical statistics, has been the study of nonparametric and rank procedures during the first decades of its existence. These methods had developed strongly, also internationally, with the monograph of Hájek and Šidák (1967) as a milestone. The present paper fits perfectly well in this tradition. It presents conditions for asymptotic normality for a class of rank statistics used for testing independence, and it seems to be one of the last ones in a series of papers that present weaker and weaker conditions for asymptotic normality to hold. We mention Wald and Wolfowitz (1944), Noether (1949), Hoeffding (1951), Hájek (1961), and Bhuchongkul (1964).

Let $(X_1,Y_1),\dots,(X_n,Y_n)$ be independent and identically distributed random vectors with continuous distribution function $H(\cdot,\cdot)$ on $\mathbb{R}^2$ and marginal distribution functions $F(\cdot)$ and $G(\cdot)$. Locally most powerful rank tests of the null hypothesis of independence $H(\cdot,\cdot)=F(\cdot)G(\cdot)$ are of the type

$$\begin{aligned} T_n &= \frac{1}{n}\sum_{i=1}^{n} J_n\left(\frac{R_i}{n}\right) K_n\left(\frac{Q_i}{n}\right) = \frac{1}{n}\sum_{i=1}^{n} J_n\left(F_n(X_i)\right) K_n\left(G_n(Y_i)\right) \\ &= \int\!\!\int J_n\left(F_n(x)\right) K_n\left(G_n(y)\right) dH_n(x,y), \qquad (10.1) \end{aligned}$$

where R_i is the rank of X_i among $X_1,\dots,X_n$, where Q_i is the rank of Y_i among $Y_1,\dots,Y_n$, and where $F_n(\cdot)$, $G_n(\cdot)$, and $H_n(\cdot,\cdot)$ are the empirical versions of $F(\cdot)$, $G(\cdot)$, and $H(\cdot,\cdot)$, respectively. Assuming existence of functions $J(\cdot)$ and $K(\cdot)$ on the unit interval such that

$$\int\int [J_n(F_n(x))K_n(G_n(y)) - J(F_n(x))K(G_n(y))]\,dH_n(x,y) = o_P\left(\frac{1}{\sqrt{n}}\right) \quad (10.2)$$

holds and with the notation

$$\mu = \int\int J(F(x))K(G(y))\,dH(x,y) \quad (10.3)$$

we note heuristically that

$$\begin{aligned}
T_n - \mu &= \int\int J(F_n(x))K(G_n(y))\,dH_n(x,y) \\
&\quad - \int\int J(F(x))K(G(y))\,dH(x,y) + o_P\left(\frac{1}{\sqrt{n}}\right) \\
&= \int\int \{J(F_n(x)) - J(F(x))\}K(G_n(y))\,dH_n(x,y) \\
&\quad + \int\int J(F(x))\{K(G_n(y)) - K(G(y))\}\,dH_n(x,y) \quad (10.4) \\
&\quad + \int\int J(F(x))K(G(y))\,d\{H_n(x,y) - H(x,y)\} + o_P\left(\frac{1}{\sqrt{n}}\right) \\
&= \int\int \{F_n(x) - F(x)\}J'(F(x))K(G(y))\,dH(x,y) \\
&\quad + \int\int \{G_n(y) - G(y)\}J(F(x))K'(G(y))\,dH(x,y) \\
&\quad + \int\int J(F(x))K(G(y))\,d\{H_n(x,y) - H(x,y)\} + o_P\left(\frac{1}{\sqrt{n}}\right)
\end{aligned}$$

holds. Consequently, it is intuitively clear that asymptotic normality of $\sqrt{n}(T_n - \mu)$ holds with limit variance as in (3.10) of the paper, and that this asymptotic normality will hold uniformly over appropriate classes of distribution functions $H(\cdot,\cdot)$. However, the technical difficulties are many, especially since the authors have been aiming at minimal conditions. Actually they have replaced (10.4) above by an expression with 13 terms, 10 of which had to be shown to be asymptotically negligible.

Since $H(\cdot,\cdot)$ is not necessarily equal to $F(\cdot)G(\cdot)$, the asymptotic normality is shown under so-called fixed alternatives to the hypothesis of independence. A well-known example is the Van der Waerden normal scores rank correlation coefficient with $J(u) = K(u) = \Phi^{-1}(u) \approx \sqrt{-2\log[u(1-u)]}$ and $J'(u) = K'(u) = 1/\phi(\Phi^{-1}(u)) \approx \sqrt{2\pi}\,[u(1-u)]^{-1}, 0 < u < 1$. Clearly conditions are needed on the behavior of $J(u)$ and $K(u)$ as u comes close to 0 or 1, in order for (10.4) to be valid and for the asymptotic normality to hold. Ruymgaart, Shorack, and Van Zwet improve on the conditions of Bhuchongkul (1964), and for the case $J(\cdot) = K(\cdot)$ they need that $|J(u)|\,[u(1-u)]^{1/4-\delta}$ and $|J'(u)|\,[u(1-u)]^{5/4-\delta}$ are bounded for some positive δ, thus incorporating the Van der Waerden normal scores rank correlation coefficient.

Frits Ruymgaart, Willem's third PhD student, has generalized these results in Ruymgaart (1974) and in his PhD thesis, Ruymgaart (1973), to the case where the score functions $J(\cdot)$ and $K(\cdot)$ may have discontinuities. It is very likely that this is the reason why Ruymgaart (1974) has gotten many more citations in the literature than the article under discussion. In any case, these papers are still being cited because of their relevance to the study of semiparametric copula models. For example the Van der Waerden normal scores rank correlation coefficient is semiparametrically efficient in the normal copula model; see Klaassen and Wellner (1997). In the normal copula model one assumes that if all components of a random vector are transformed into normal random variables, then the resulting random vector has a multivariate normal distribution. It was noted by Li (2000) that this normal copula model was in use for the pricing of credit default swaps. This practice has been blamed for the global financial crisis of 2007–2009. Motivated by risk management problems, the study of copula models has led to research on a generalization of T_n from (10.1), namely

$$T_n = \frac{1}{n}\sum_{i=1}^{n} J\left(\frac{R_i}{n}, \frac{Q_i}{n}\right) = \frac{1}{n}\sum_{i=1}^{n} J\left(F_n(X_i), G_n(Y_i)\right); \tag{10.5}$$

see e.g. Fermanian, Radulovic, and Wegkamp (2004) and Schmidt and Stadtmüller (2006). We may conclude that Ruymgaart, Shorack, and Van Zwet have added a technically complicated, but thorough and useful result to the statistical literature, which is a milestone in a long development.

10.3 A Note on Contiguity and Hellinger Distance

Consider two sequences of probability measures (P_n) and (Q_n) defined on a common sequence of measurable spaces $(\mathscr{X}_n, \mathscr{A}_n)$. The sequence of probability measures (Q_n) is called *contiguous* with respect to the sequence (P_n), if for every sequence $(A_n), A_n \in \mathscr{A}_n$, the convergence $P_n(A_n) \to 0$ implies $Q_n(A_n) \to 0$, notation $(Q_n) \lhd (P_n)$. The two-sided version of this fundamental concept has been introduced by Le Cam (1960), and advertised and applied by Roussas (1972). Jaroslav Hájek visited Lucien Le Cam in Berkeley and popularized the concept in Hájek and Šidák (1967) calling the most important results Le Cam's first, second, and third lemma.

Willem van Zwet and his first PhD student, his contemporary Kobus Oosterhoff, were the first to geometrize contiguity for product measures, and they chose the Hellinger distance as a natural metric for this. Their results have been published as Oosterhoff and Van Zwet (1979) in the Hájek Memorial Volume, which is just the proper place for this paper, given Hájek's interest in the topic.

Let $P_n = P_n^{(n)}$ be the product measure $P_n^{(n)} = \prod_{i=1}^n P_{ni}$ and $Q_n = Q_n^{(n)} = \prod_{i=1}^n Q_{ni}$ with $(\mathscr{X}_n, \mathscr{A}_n)$ the product space of $(\mathscr{X}_{ni}, \mathscr{A}_{ni})$, $i = 1, \ldots, n$. Let μ_{ni} be a σ-finite measure on $(\mathscr{X}_{ni}, \mathscr{A}_{ni})$ dominating both P_{ni} and Q_{ni}. Denote the densities of P_{ni} and Q_{ni} by p_{ni} and q_{ni}, respectively, and write $H(P_{ni}, Q_{ni})$ for the Hellinger distance of P_{ni} and Q_{ni},

$$H(P_{ni},Q_{ni}) = \sqrt{\int \left(p_{ni}^{1/2} - q_{ni}^{1/2}\right)^2 d\mu_{ni}}. \tag{10.6}$$

It is easy to see, as Oosterhoff and Van Zwet (1979) show, that

$$\lim_{n\to\infty} \sum_{i=1}^{n} H^2(P_{ni},Q_{ni}) = 0 \Longrightarrow (Q_n^{(n)}) \lhd (P_n^{(n)})$$
$$\Longrightarrow \limsup_{n\to\infty} \sum_{i=1}^{n} H^2(P_{ni},Q_{ni}) < \infty \tag{10.7}$$

holds. As the one-sided contiguity is an 'asymmetric' property and the Hellinger distance is symmetric, the gap between the contiguity in the middle term and the left hand or the right hand side of (10.7) can be closed only by an asymmetric condition. Indeed, with the additional notation

$$H_c(P_{ni},Q_{ni}) = \sqrt{\int_{|q_{ni}-p_{ni}|\geq cp_{ni}} \left(p_{ni}^{1/2} - q_{ni}^{1/2}\right)^2 d\mu_{ni}} \tag{10.8}$$

the main result of Oosterhoff and Van Zwet (1979) may be formulated as follows,

$$(Q_n^{(n)}) \lhd (P_n^{(n)}) \Longleftrightarrow \begin{cases} \limsup_{n\to\infty} \sum_{i=1}^{n} H^2(P_{ni},Q_{ni}) < \infty, \\ \lim_{c\to\infty} \limsup_{n\to\infty} \sum_{i=1}^{n} H_c^2(P_{ni},Q_{ni}) = 0. \end{cases} \tag{10.9}$$

The second result from this paper is related to the First Lemma of Le Cam. Let Λ_n be the loglikelihood ratio

$$\Lambda_n = \sum_{i=1}^{n} \log\left(q_{ni}(X_{ni})/p_{ni}(X_{ni})\right). \tag{10.10}$$

The First Lemma of Le Cam implies

$$\Lambda_n \xrightarrow{P_n^{(n)}}_w \mathscr{N}(-\frac{1}{2}\sigma^2,\sigma^2) \Longrightarrow (Q_n^{(n)}) \lhd (P_n^{(n)}). \tag{10.11}$$

Oosterhoff and Van Zwet succeeded in formulating necessary and sufficient conditions for this convergence to normality of the loglikelihood ratio in terms of Hellinger distances as follows. For any $\sigma \geq 0$ we have

$$\left.\begin{array}{l} \Lambda_n \xrightarrow{P_n^{(n)}}_w \mathscr{N}(-\frac{1}{2}\sigma^2,\sigma^2), \\ \lim_{\varepsilon\downarrow 0} \limsup_{n\to\infty} \max_{1\leq i\leq n} H_\varepsilon^2(P_{ni},Q_{ni}) = 0 \end{array}\right\} \Longleftrightarrow$$
$$\begin{cases} \lim_{n\to\infty} \sum_{i=1}^{n} H^2(P_{ni},Q_{ni}) = \frac{1}{4}\sigma^2, \\ \lim_{\varepsilon\downarrow 0} \limsup_{n\to\infty} \sum_{i=1}^{n} H_\varepsilon^2(P_{ni},Q_{ni}) = 0. \end{cases} \tag{10.12}$$

In order to stress the relation to the Hellinger distance we have chosen to formulate (2.3) of Theorem 1 and the uniform asymptotic negligibility condition (3.3) of Theorem 2 of Oosterhoff and Van Zwet (1979) in terms of $H_c(P_{ni}, Q_{ni})$ from (10.8); cf. (3.7) of ibid. To see the equivalence of (2.3) of Theorem 1 with the second condition at the right hand side of (10.9), we note that for $c > 2$

$$\begin{aligned}(1-c^{-1/2})^{-2}H_c^2(P_{ni}, Q_{ni}) &\le Q_{ni}\,(q_{ni}(X_{ni}) \ge c\,p_{ni}(X_{ni})) \\ &\le (1-c^{-1})^{-2}H_c^2(P_{ni}, Q_{ni}) \qquad (10.13)\end{aligned}$$

holds. To derive the equivalence of the uniform asymptotic negligibility condition (3.3) of Theorem 2 with the second condition at the left hand side of (10.12) we note that for $\varepsilon > 0$

$$\begin{aligned}&(\sqrt{1+\varepsilon}-1)^2\, P_{ni}\,(|\,q_{ni}(X_{ni}) - p_{ni}(X_{ni})\,| \ge \varepsilon\, p_{ni}(X_{ni})) \\ &\le H_\varepsilon^2(P_{ni}, Q_{ni}) \qquad (10.14) \\ &\le P_{ni}\,(|\,q_{ni}(X_{ni}) - p_{ni}(X_{ni})\,| \ge \varepsilon\, p_{ni}(X_{ni})) \\ &\quad + Q_{ni}\,(|\,q_{ni}(X_{ni}) - p_{ni}(X_{ni})\,| \ge \varepsilon\, p_{ni}(X_{ni}))\end{aligned}$$

holds, and we use (10.11) to obtain that the uniform asymptotic negligibility condition

$$\lim_{\varepsilon\downarrow 0}\limsup_{n\to\infty}\max_{1\le i\le n} P_{ni}\,(|\,q_{ni}(X_{ni}) - p_{ni}(X_{ni})\,| \ge \varepsilon\, p_{ni}(X_{ni})) = 0 \qquad (10.15)$$

implies

$$\lim_{\varepsilon\downarrow 0}\limsup_{n\to\infty}\max_{1\le i\le n} Q_{ni}\,(|\,q_{ni}(X_{ni}) - p_{ni}(X_{ni})\,| \ge \varepsilon\, p_{ni}(X_{ni})) = 0. \qquad (10.16)$$

It is well known that contiguity holds in regular parametric models for i.i.d. random variables. Let $\{P(\theta) : \theta \in \Theta\}, \Theta \subset \mathbb{R}^k$, be a collection of distributions that have densities $p(\theta)$ with respect to some σ-finite measure μ. The most important condition for regularity is the existence of a score function $\dot{\ell}(\theta) \in L_2^k(P(\theta))$ such that the map $\theta \mapsto p^{1/2}(\theta)$ is continuously Fréchet differentiable in $L_2(\mu)$ as follows,

$$|h|^{-2}\int\left(p^{1/2}(\theta+h) - p^{1/2}(\theta) - \frac{1}{2}h^T\dot{\ell}(\theta)p^{1/2}(\theta)\right)^2 d\mu \to 0,\, h\to 0, \qquad (10.17)$$

with the map $\theta \mapsto \dot{\ell}(\theta)p^{1/2}(\theta)$ from Θ to $L_2^k(P(\theta))$ continuous. In these regular models Local Asymptotic Normality holds and via (10.11) this yields the contiguity

$$\{P^n(\theta_n + n^{-1/2}t_n)\} \lhd \{P^n(\theta_n)\}, \qquad (10.18)$$

as $n\to\infty$, $\theta_n \to \theta$, and $t_n \to t$ for fixed θ and t, and hence the corresponding mutual contiguity; see e.g. Section 2.1 of Bickel et al. (1993). To circumvent the cumbersome proof of Local Asymptotic Normality, one might use the characterization (10.9) of contiguity in order to prove (10.18) as follows. By (a) from the proof of

Lemma A.9.5 of Bickel et al. (1993) we see that (10.17) holds uniformly for θ in compacts. Consequently,

$$n\int\left(p^{1/2}(\theta_n+n^{-1/2}t_n)-p^{1/2}(\theta_n)-\frac{1}{2}n^{-1/2}t_n^T\dot{\ell}(\theta_n)p^{1/2}(\theta_n)\right)^2 d\mu\to 0 \quad (10.19)$$

holds as $n\to\infty$, $\theta_n\to\theta$, and $t_n\to t$ for fixed θ and t. It follows that the first condition at the right hand side of (10.9) is satisfied. To prove that the second condition at the right hand side of (10.9) is satisfied as well, we first note that for $c\geq 1$

$$\begin{aligned}\int_{p(\theta+h)\geq cp(\theta)}&\left(p^{1/2}(\theta+h)-p^{1/2}(\theta)\right)^2 d\mu\\ &\leq 3\int_{p(\theta+h)\geq cp(\theta)}\left(p^{1/2}(\theta+h)-p^{1/2}(\theta)-\frac{1}{2}h^T\dot{\ell}(\theta)p^{1/2}(\theta)\right)^2 d\mu\\ &+3\int_{p(\theta+h)\geq cp(\theta)}\left(p^{1/2}(\theta-h)-p^{1/2}(\theta)+\frac{1}{2}h^T\dot{\ell}(\theta)p^{1/2}(\theta)\right)^2 d\mu \quad (10.21)\\ &+3\int_{p(\theta+h)\geq cp(\theta)}p(\theta)d\mu\end{aligned}$$

and

$$\int_{p(\theta+h)\geq cp(\theta)}\left\{\left(p^{1/2}(\theta+h)-p^{1/2}(\theta)\right)^2-\left(c^{1/2}-1\right)^2 p(\theta)\right\}d\mu\geq 0 \quad (10.21)$$

hold, and hence for sufficiently large values of c

$$\begin{aligned}\int_{p(\theta+h)\geq cp(\theta)}&\left(p^{1/2}(\theta+h)-p^{1/2}(\theta)\right)^2 d\mu\\ &\leq 3\left[1-3(\sqrt{c}-1)^{-2}\right]^{-1}\times \quad (10.22)\\ &\int_{p(\theta+h)\geq cp(\theta)}\left\{\left(p^{1/2}(\theta+h)-p^{1/2}(\theta)-\frac{1}{2}h^T\dot{\ell}(\theta)p^{1/2}(\theta)\right)^2\right.\\ &\qquad\left.+\left(p^{1/2}(\theta-h)-p^{1/2}(\theta)+\frac{1}{2}h^T\dot{\ell}(\theta)p^{1/2}(\theta)\right)^2\right\}d\mu. \quad (10.23)\end{aligned}$$

To complete the proof of the contiguity in (10.18), we note that the second condition at the right hand side of (10.9) follows from (10.19) and (10.22) by the substitutions $\theta=\theta_n$ and $h=n^{-1/2}t_n$.

One of the simplest examples of nonregular parametric models for i.i.d. random variables where contiguity may be determined easily via (10.9), is the location family for the exponential distribution. With μ_{ni} Lebesgue measure on $(\mathbb{R},\mathscr{B})$ and

$$p_{ni}(x)=e^{-x}\mathbf{1}_{(0,\infty)}(x),\quad q_{ni}(x)=p_{ni}(x-\frac{t}{n}),\quad x\in\mathbb{R}, \quad (10.24)$$

some computation shows

$$\limsup_{n\to\infty} \sum_{i=1}^{n} H^2(P_{ni}, Q_{ni}) = |t| \tag{10.25}$$

and

$$\lim_{c\to\infty} \limsup_{n\to\infty} \sum_{i=1}^{n} H_c^2(P_{ni}, Q_{ni}) = |t| \, \mathbf{1}_{[t<0]}, \tag{10.26}$$

which by (10.9), i.e. Theorem 1 of Oosterhoff and Van Zwet (1979), implies the one-sided contiguity from the left hand side of (10.9), but only for t positive.

More tedious computations are necessary in case of triangular densities

$$p_{ni}(x) = (1 - |x|) \vee 0, \quad q_{ni}(x) = (1 - |x - \frac{\sigma}{\sqrt{n \log n}}|) \vee 0, \quad x \in \mathbb{R}. \tag{10.27}$$

In my master thesis, Klaassen (1974), written under supervision of Kobus Oosterhoff, the conditions at the right hand side of (10.12) have been checked, and hence Theorem 2 of Oosterhoff and Van Zwet (1979) is applicable here.

The main result, Theorem 1 of ibid., is called by Jacod and Shiryaev (1987, p. 576) 'the first general contiguity result'. It has been generalized to nonindependent and continuous time cases, and it has been one of the roots of Jacod and Shiryaev (1987).

Furthermore, the paper is cited in several publications as a reference for contiguity. We mention Strasser (1985), Bickel, Klaassen, Ritov, and Wellner (1993), Cabaña and Cabaña (1997) and Bose, Gangopadhyay, and Goswami (2007). The results are explicitly used in e.g. Khmaladze (1988), Eubank (2000), Pfanzagl (2000), Putter and Young (2001), and Ferger (2001).

We conclude that Oosterhoff and Van Zwet (1979) is a fundamental paper, which presents useful tools for verifying contiguity.

10.4 On Estimating a Parameter and its Score Function On Estimating a Parameter and its Score Function, II

In the early seventies several semiparametrically efficient estimators for the symmetric location model have been constructed. These estimators were called adaptive because they adapt to the unknown underlying symmetric density $f(\cdot)$ of the errors in the observations in such a way that they attain the asymptotic variance bound of $1/I(f)$ with $I(f)$ the Fisher information for location. After a chat with Peter Huber about these estimators, Willem van Zwet suggested me, his PhD student, to study these estimators and to show that there is some loss somewhere. This has been a very stimulating research program and it resulted in my thesis Klaassen (1981). The articles under discussion are generalizations to much more general models of the inequality in Theorem 3.2.1 of ibid. for the symmetric location case. These inequalities improve the Cramér-Rao lower bound for unbiased estimators of the parameter of interest by adding a multiple of the integrated mean squared error of an estimator, given the parameter of interest, of the efficient influence function. In this way they state that the parameter of interest can be estimated accurately only if, given the parameter of interest, the efficient influence function can be estimated accurately as well. So, if the parameter of interest is the only unknown parameter, this additional term vanishes and the inequality reduces to the ordinary Cramér-Rao inequality.

The charm of these inequalities is that they are finite sample results. They have been proved for i.i.d. situations where, given the parameter of interest, there exists a sufficient statistic with respect to the nuisance parameter. All these inequalities are based on a conditional version of Projection Lemma 4.1 of Hájek (1968). An asymptotic version of these results does not need these sufficient statistics and it states that asymptotically efficient estimation of the parameter of interest is possible if and only if the efficient influence function can be estimated consistently; for a generalization of this, see Klaassen (1987).

The first paper is applicable in situations where adaptive estimation should be possible, i.e. where the semiparametrically efficient influence function is the same as the efficient influence function for the case that the nuisance parameter is known. The second paper studies the general semiparametric situation. Still another PhD student of Willem is a coauthor here, namely Aad van der Vaart. In chapter 5 of his PhD thesis Van der Vaart (1988), he continues research on models of the above type. There he constructs asymptotically efficient estimators for semiparametric models with a sufficient statistic with respect to the nuisance parameter.

10.5 A Remark on Consistent Estimation

A fundamental rule of thumb in statistics states that 'substituting unknown parameters in statistical procedures by estimators of them, yields appropriate procedures.' Consequently, if one is simulating the distribution of a statistic and the distribution of the underlying random variables is unknown, one may replace the latter distribution by an estimator of it, like the empirical. The resulting bootstrap was introduced by Efron (1979). Clearly, an important question is :'When does the bootstrap work and when it doesn't?' For some important classes of situations the validity of the bootstrap was proved by Bickel and Freedman (1981), who also presented some counter-examples.

Hein Putter, writing his doctoral thesis Putter (1994) under supervision of Willem van Zwet, has studied the question in the setting of a general substitution estimator. In Putter and Van Zwet (1996) they write 'This is commonly called a "plug-in estimator," but this expression is of the same sad grammatical level as "see-through clothes."' However, more importantly in the context of the paper under discussion they prove that substitution estimators work under all underlying distributions, except for a 'small' subset within the set of underlying distributions metrized by Hellinger distance, namely for a subset of the first category.

Under local asymptotic normality Beran (1997) has proved that the bootstrap fails in estimating the distribution of an estimator consistently, precisely at those parameter values at which the bootstrapped estimator is not locally asymptotically equivariant. He also has shown that the set of these parameter values has Lebesgue measure 0. At the points where an estimator is locally asymptotically equivariant or regular, the Hájek-LeCam convolution theorem holds; Hájek (1970).

Willem van Zwet and his second son Erik have used results from Putter and Van Zwet (1996) in order to prove within a very general framework that, if the distribution of an estimator can be estimated consistently in the Prohorov metric, then there exists a subset of the first category within the set of underlying distributions metrized by Hellinger distance, such that the estimator is locally asymptotically equivariant outside this subset of the first category. In Van Zwet and Van Zwet (1999) they prove under somewhat stricter conditions, but still within the same very general framework, that for the same subset of the first category as above the distribution of the estimator has a locally asymptotically uniform convolution structure. We conclude that Willem and Erik van Zwet have generalized the connection between validness of the bootstrap and local asymptotic equivariance and convolution structure as noted by Beran (1997), to the general i.i.d. case in an elegant way.

References

1. Beran, R.J. (1997), Diagnosing bootstrap success, *Ann. Inst. Statist. Math.* **49**, 1–14.
2. Bhuchongkul, S. (1964), A class of nonparametric tests for independence in bivariate populations, *Ann. Math. Statist.* **35**, 138–149.
3. Bickel, P.J. and Freedman, D.A. (1981), Some Asymptotic Theory for the Bootstrap, *Ann. Statist.* **9**, 1196-1217.
4. Bickel, P.J., Klaassen, C.A.J., Ritov, Y., and Wellner, J.A. (1993), *Efficient and Adaptive Estimation in Semiparametric Models*, Johns Hopkins Univ. Press, Baltimore; reprint (1998), Springer, New York.
5. Bose, A., Gangopadhyay, S., and Goswami, A. (2007), A note on random coin tossing, *Indag. Math. (N.S.)* **18**, 405–416.
6. Cabaña, A. and Cabaña, E.M. (1997), Transformed empirical processes and modified Kolmogorov-Smirnov tests for multivariate distributions, *Ann. Statist.* **25**, 2388–2409.
7. Efron, B. (1979), Bootstrap Methods: Another Look at the Jackknife, *Ann. Statist.* **7**, 1–26.
8. Eubank, R.L. (2000), Testing for no effect by cosine series methods, *Scand. J. Statist.* **27**, 747–763.
9. Ferger, D. (2001), Analysis of change-point estimators under the null hypothesis, *Bernoulli* **7**, 487–506.
10. Fermanian, J.-D., Radulovic, D., and Wegkamp, M. (2004), Weak convergence of empirical copula processes, *Bernoulli* **10**, 847–860.
11. Hájek, J. (1961), Some extensions of the Wald-Wolfowitz-Noether theorem, *Ann. Math. Statist.* **32**, 506–523.
12. Hájek, J. (1968), Asymptotic normality of simple linear rank statistics under alternatives, *Ann. Math. Statist.* **39**, 325–346.
13. Hájek, J. (1970), A characterization of limiting distributions of regular estimates, *Z. Wahrscheinlichkeitstheorie und Verw. Gebiete* **14**, 323–330.

14. Hájek, J., and Šidák, Z. (1967), *Theory of rank tests*, Academic Press, New York, Academia Publishing House of the Czechoslovak Academy of Sciences, Prague.
15. Hoeffding, W. (1951), A combinatorial central limit theorem, *Ann. Math. Statist.* **22**, 558-566.
16. Jacod, J. and Shiryaev, A.N. (1987), *Limit Theorems for Stochastic Processes*, Springer, Berlin.
17. Khmaladze, E.V. (1988), An innovation approach to goodness-of-fit tests, *Ann. Statist.* **16**, 1503–1516.
18. Klaassen, C.A.J. (1974), Applications of Hellinger distance in hypothesis testing (in Dutch), Master thesis, University of Nijmegen.
19. Klaassen, C.A.J. (1981). *Statistical Performance of Location Estimators*, Mathematical Centre Tracts 133, Mathematical Centre, Amsterdam.
20. Klaassen, C.A.J. (1987). Consistent estimation of the influence function of locally asymptotically linear estimators. *Ann. Statist.* **15**, 1548–1562.
21. Klaassen, C.A.J. and Van Zwet, W.R. (1985), On estimating a parameter and its score function, *Proceedings of the Berkeley Conference in Honor of Jerzy Neyman and Jack Kiefer (Berkeley, Calif., 1983)*, L.M. Le Cam and R.A. Olshen (eds.), Wadsworth Statist./Probab. Ser., Wadsworth, Belmont, CA, **II**, 827–839.
22. Klaassen, C.A.J., Van der Vaart, A.W., and Van Zwet, W.R. (1988), On estimating a parameter and its score function, II, *Statistical Decision Theory and Related Topics IV, (West Lafayette, Ind., 1986)*, S.S. Gupta and J.O. Berger (eds.), Springer, New York, **2**, 281–288.
23. Klaassen, C.A.J. and Wellner, J.A. (1997), Efficient estimation in the bivariate normal copula model: normal margins are least favourable, *Bernoulli* **3**, 55–77.
24. Le Cam, L.M. (1960), Locally asymptotically normal families of distributions. Certain approximations to families of distributions and their use in the theory of estimation and testing hypotheses, *Univ. California Publ. Statist.* **3**, 37–98.
25. Li, D.X. (2000), On Default Correlation: A Copula Function Approach, *Journal of Fixed Income* **9**, 43-54.
26. Noether, G.E. (1949), On a theorem of Wald and Wolfowitz, *Ann. Math. Statist.* **20**, 455–458.
27. Oosterhoff, J. and Van Zwet, W.R. (1979), A note on contiguity and Hellinger distance. *Contributions to Statistics, Jaroslav Hájek Memorial Volume*, J. Jurečková (ed.), Reidel, Dordrecht, 157–166.
28. Pfanzagl, J. (2000), On local uniformity for estimators and confidence limits, *J. Statist. Plann. Inf.* **84**, 27–53.
29. Putter, H. (1994), *Consistency of Resampling Methods*, PhD thesis, University of Leiden.
30.
31. Putter, H. and Van Zwet, W.R. (1996), Resampling: consistency of substitution estimators, *Ann. Statist.* **24**, 2297–2318.
32. Putter, H. and Young, G.A. (2001), On the effect of covariance function estimation on the accuracy of Kriging predictors, *Bernoulli* **7**, 421–438.
33. Roussas, G.G. (1972), *Contiguity of Probability Measures: Some Applications in Statistics*, Cambridge Tracts in Mathematics and Mathematical Physics, **63**, Cambridge University Press, London-New York.
34. Ruymgaart, F.H. (1973), *Asymptotic theory of rank tests for independence*, Mathematical Centre Tracts **43**, Mathematisch Centrum, Amsterdam.
35. Ruymgaart, F.H. (1974), Asymptotic normality of nonparametric tests for independence, *Ann. Statist.* **2**, 892–910.
36. Ruymgaart, F.H., Shorack, G.R., and Van Zwet, W.R. (1972), Asymptotic normality of nonparametric tests for independence, *Ann. Math. Statist.* **43**, 1122–1135.
37. Schmidt, R. and Stadtmüller, U. (2006), Non-parametric estimation of tail dependence, *Scand. J. Statist.* **33**, 307–335.
38. Strasser, H. (1985), *Mathematical Theory of Statistics: Statistical Experiments and Asymptotic Decision Theory*, de Gruyter Studies in Mathematics **7**, de Gruyter, Berlin.
39. Van der Vaart, A.W. (1988), *Statistical Estimation in Large Parameter Spaces*, CWI Tracts 44, CWI, Amsterdam.

40. Van Zwet, E.W. and Van Zwet, W.R. (1999), A remark on consistent estimation, *Johann Pfanzagl—on the occasion of his 70th birthday, Math. Methods Statist.*, **8**, 277–284.
41. Wald, A. and Wolfowitz, J. (1944), Statistical tests based on permutations of the observations, *Ann. Math. Statist.* **15**, 358–372.

Part III
Second Order Asymptotics

The Annals of Statistics
1978, Vol. 6, No. 5, 937-1004

ASYMPTOTIC EXPANSIONS FOR THE POWER OF DISTRIBUTIONFREE TESTS IN THE TWO-SAMPLE PROBLEM[1]

BY P. J. BICKEL[2] AND W. R. VAN ZWET[3]

University of California, Berkeley and University of Leiden

Asymptotic expansions are established for the power of distributionfree tests in the two-sample problem. These expansions are then used to obtain deficiencies in the sense of Hodges and Lehmann for distributionfree tests with respect to their parametric competitors and for the estimators of shift associated with these tests.

1. Introduction. Let $X_1, X_2, \cdots, X_N$, $N = m + n$, be independent random variables such that $X_1, \cdots, X_m$ are identically distributed with common distribution function F and density f and $X_{m+1}, \cdots, X_N$ are identically distributed with distribution function G and density g. For $N = 2, 3, \cdots$ and $0 < \varepsilon \leqq m/N \leqq 1 - \varepsilon < 1$, consider the problem of testing the hypothesis $F = G$ against a sequence of alternatives that is contiguous to the hypothesis. The level α of the sequence of tests is fixed in (0, 1). Standard tests for this two-sample problem are linear rank tests and permutation tests and expressions for the limiting powers of such tests are well known. In this paper we shall establish asymptotic expansions to order N^{-1} for the powers π_N of such tests, i.e., expressions of the form $\pi_N = c_0 + c_1 N^{-\frac{1}{2}} + c_{2,N} N^{-1} + o(N^{-1})$. Of course this involves finding similar expansions for the distribution function of the test statistic under the hypothesis as well as under contiguous alternatives. For simplicity we shall eventually limit our discussion to contiguous location alternatives. Extension of the results to general contiguous alternatives is straightforward but messy.

A number of authors have computed formal expansions for the distributions of various two-sample rank statistics without proof of their validity. Their purpose was to obtain better numerical approximations for the critical value of the test statistic and the power of the test than can be provided by the usual normal approximation. For an account of this work we refer to a review paper of Bickel

Received April 1976; revised March 1977.

[1] Report SW 38/76 (revised) Mathematisch Centrum, Amsterdam.

[2] Research supported by the National Science Foundation, Grant MPS-73-08698, by the Office of Naval Research, Contract N00014-75-C-0444, and by the Netherlands Organization for Pure Scientific Research.

[3] Research supported by the National Science Foundation, Grant GP-29123, and by the Office of Naval Research, Contracts N00014-69-A-0200-1036 and N00014-75-C-0444.

AMS 1970 *subject classifications.* Primary 62G10, 62G20; Secondary 60F05.

Key words and phrases. Distributionfree tests, linear rank tests, permutation test, Hodges-Lehmann estimators, power, contiguous alternatives, asymptotic expansions, Edgeworth expansions, deficiency, rejective sampling, sampling without replacement.

S. van de Geer and M. Wegkamp (eds.), *Selected Works of Willem van Zwet*, Selected Works in Probability and Statistics, DOI 10.1007/978-1-4614-1314-1_11, © Springer Science+Business Media, LLC 2012

(1974), which incidentally also contains a short preview of the present paper including a brief description of the expansion of the distribution function of the two-sample linear rank statistic under the hypothesis (cf. Corollary 2.1 in the present paper). This result was also proved independently by Robinson (1977). An earlier proof by Rogers (1971) for the special case of the two-sample Wilcoxon statistic under the hypothesis unfortunately appears to contain a nontrivial error.

We shall not discuss the numerical aspects of the expansions we obtain but we shall concentrate on a rather delicate type of asymptotic comparison of the power functions of various parametric and nonparametric tests. Consider two sequences of tests $\{T_N\}$ and $\{T_N'\}$ for the same hypothesis at the same fixed level α. Let $\pi_N(\theta_N)$ and $\pi_N'(\theta_N)$ denote the powers of these tests against the same sequence of contiguous alternatives parametrized by a parameter θ. If T_N is more powerful than T_N' we search for a number $k_N = N + d_N$ such that $\pi_N(\theta_N) = \pi'_{k_N}(\theta_N)$. Here k_N and d_N are treated as continuous variables, the power π_N' being defined for real N by linear interpolation between consecutive integers. The quantity d_N was named the deficiency of $\{T_N'\}$ with respect to T_N by Hodges and Lehmann (1970), who introduced this concept and initiated its study. Of course, in many cases of interest d_N is analytically intractable and one can only study its asymptotic behavior as N tends to infinity.

Suppose that for $N \to \infty$, the ratio N/k_N tends to a limit e, the asymptotic relative efficiency of $\{T_N'\}$ with respect to $\{T_N\}$. If $0 < e < 1$, we have $d_N \sim (e^{-1} - 1)N$ and further asymptotic information about d_N is not particularly revealing. On the other hand, if $e = 1$, the asymptotic behavior of d_N (which may now be anything from $o(1)$ to $o(N)$) does provide important additional information. Of special interest is the case where d_N tends to a finite limit.

Asymptotic expansions for the power of the type we discussed above are precisely what is needed for an asymptotic evaluation of d_N. With the aid of such expansions we arrive at the following results. Let F be a distribution function with density f, let b be a positive real number and define $\theta_N = bN^{-\frac{1}{2}}$. Consider the problem of testing the hypothesis (F, F) against the sequence of simple alternatives $(F(\cdot + \Delta_N \theta_N), F(\cdot - (1 - \Delta_N)\theta_N))$ at level α. Let d_N denote the deficiency of the locally most powerful rank test with respect to the most powerful test for this problem. For the rank test the power is independent of Δ_N but for the most powerful test it is not and we choose Δ_N in such a way that the power of the most powerful test is minimal. Under certain regularity conditions on F we establish an expansion for d_N with remainder $o(1)$. To indicate the qualitative behavior of d_N it suffices to note that the expansion is of the form

$$d_N = \frac{1}{\int \Psi_1^2(t)\,dt} \sum_{j=1}^{N} \sigma^2(\Psi_1(U_{j:N})) + \bar{d}_{N,0} + o(1) \tag{1.1}$$

where $\Psi_1 = f'(F^{-1})/f(F^{-1})$, σ^2 indicates a variance, $U_{j:N}$ denotes the jth order statistic of a sample of size N from a uniform distribution on (0, 1) and

$\tilde{d}_{N,0} = O(1)$. Alternatively we may write

$$(1.2) \qquad d_N = \frac{1}{\int \Psi_1^2(t)\,dt} \int_{N^{-1}}^{1-N^{-1}} (\Psi_1'(t))^2 t(1-t)\,dt + \tilde{d}_{N,0} + o(1) \\ + O(N^{-\frac{1}{2}} \int_{N^{-1}}^{1-N^{-1}} (\Psi_1'(t))^2 \{t(1-t)\}^{\frac{1}{2}}\,dt)\,,$$

where Ψ_1' is the derivative of Ψ_1. If we replace the exact scores $-E\Psi_1(U_{j:N})$ in the locally most powerful rank test by the corresponding approximate scores $-\Psi_1(j/(N+1))$, then (1.1) changes to

$$(1.3) \qquad d_N = \frac{1}{\int \Psi_1^2(t)\,dt} \sum_{j=1}^{N} E\{\Psi_1(U_{j:N}) - \Psi_1(j/(N+1))\}^2 + \tilde{d}_{N,0} + o(1)$$

and (1.2) continues to hold. Thus the asymptotic behavior of d_N is governed by that of the first term in these expansions and under the conditions imposed, all we can say is that it is $o(N^{\frac{1}{2}})$ but not $o(1)$. Typically, however, it will be $O(1)$ or only slightly larger than that. By taking F to be a normal distribution we find that the deficiency of both the normal scores test and van der Waerden's test with respect to the test based on the difference of the sample means for contiguous normal location alternatives is asymptotic to $\log \log N$. For logistic shift alternatives the deficiency of Wilcoxon's test with respect to the most powerful parametric test tends to a finite limit. Turning to distributionfree tests other than rank tests, we find that for contiguous normal location alternatives the deficiency of the permutation test based on the sample means with respect to Student's test tends to zero for $N \to \infty$.

If the locally most powerful rank test for shift has nondecreasing scores, then there exists a corresponding Hodges–Lehmann estimator of shift in the two-sample problem (cf. Hodges and Lehmann (1963)). There is a similar correspondence between the locally most powerful parametric test for shift and the maximum likelihood estimator of shift in the two-sample problem. We shall exploit this correspondence to obtain asymptotic expansions for the distribution functions of these estimators. We shall show that, when suitably defined, the deficiency of the Hodges–Lehmann estimator associated with the locally most powerful rank test with respect to the maximum likelihood estimator is asymptotically equivalent to the deficiency of the parent tests for $\alpha = \frac{1}{2}$.

This paper is thus the natural counterpart of Albers, Bickel and van Zwet (ABZ) (1976) where exactly the same programme is carried out for the one-sample problem. Without exception the results are also qualitatively the same but contrary to what one might think at first sight, this in itself is rather surprising. Of course there is a strong similarity between the one- and two-sample cases but there is also one major difference. In the nonparametric one-sample location problem the underlying distribution is always symmetric both under the hypothesis and under the alternative. Because of this symmetry, the power expansions for contiguous location alternatives do not contain a term of order

$N^{-\frac{1}{2}}$ for any of the parametric or nonparametric tests considered. Since attention is restricted to sequences of tests $\{T_N\}$ and $\{T_N'\}$ with asymptotic relative efficiency 1, the leading terms of the power expansions coincide and these expansions must therefore be of the form $\pi_N = c_0 + c_{2,N}N^{-1} + o(N^{-1})$ and $\pi_N' = c_0 + c'_{2,N}N^{-1} + o(N^{-1})$. In the comparison of rank tests T_N' with parametric tests T_N it is found that the deficiency d_N is of the order of $N(\pi_N - \pi_N') = (c_{2,N} - c'_{2,N}) + o(1) = o(N^{\frac{1}{2}})$. In the two-sample problem, however, the underlying distributions are not required to be symmetric and as a result the power expansions do in general contain a term of order $N^{-\frac{1}{2}}$. It is not clear a priori that this term should be the same in each expansion and because d_N is again of the order of $N(\pi_N - \pi_N')$, one should expect d_N to be of the order $N^{\frac{1}{2}}$. It turns out, however, that for the most powerful test, the locally most powerful test, the locally most powerful rank test and its approximate scores analogue, the term of order $N^{-\frac{1}{2}}$ in the power expansion for contiguous location alternatives is in fact the same for each of these four tests. Borrowing a phrase from Pfanzagl (1977) who noted the same phenomenon for the (asymmetric) parametric one-sample problem, first order efficiency apparently implies second order efficiency in these cases. It follows that again d_N is of the order of $(c_{2,N} - c'_{2,N})$ and since $c_{2,N}$ and $c'_{2,N}$ exhibit precisely the same asymptotic behavior as in the one-sample case, our deficiency results are qualitatively the same as in ABZ (1976). The reader should note that Pfazagl's concept of second order efficiency which in general implies $d_N = o(N^{\frac{1}{2}})$, is different from Rao's concept of second order efficiency as discussed in Efron (1975), which is more in the nature of $d_N = o(1)$. This difference in terminology is not as illogical as it may seem because Rao's concept is related to the asymptotic performance of an estimator M_N as measured by the asymptotic variance of $N^{\frac{1}{2}}M_N$ and expansions for this quantity are typically in powers of N^{-1} rather than $N^{-\frac{1}{2}}$.

Throughout this paper we shall draw heavily on the techniques developed for the one-sample case in ABZ (1976) but several new difficulties appear that make the two-sample case essentially more complicated. The main source of trouble is the occurrence of terms of order $N^{-\frac{1}{2}}$ in our expansions. Not only do they make the actual computation of the expansions much more laborious, but their presence also poses a number of technical problems that are hard to handle under the conditions imposed, which are comparable to those in ABZ (1976). Another complicating factor is that the distribution theory for the two-sample rank statistic is more involved than for its one-sample counterpart. In the one-sample case a conditioning argument reduces the rank statistic to a weighted sum of independent Bernoulli random variables. A similar argument in the two-sample case leads to the much less manageable random variable indicated below.

In Section 2 we point out that for arbitrary F and G, the conditional distribution of the two-sample linear rank statistic given the order statistics of the combined sample is the same as the distribution of the sample sum in a rejective sampling scheme. We establish an expansion for the distribution function of such a sample sum which may be of interest in its own right. As a corollary we obtain an expansion for the distribution function of the rank statistic under the hypothesis. In Section 3 we return to general F and G and obtain an unconditional expansion for the distribution function of the rank statistic. We specialize to contiguous location alternatives in Section 4 and derive an expansion for the power of the rank test. In Section 5 we deal with the important case where the scores are exact or approximate scores generated by a smooth function. The permutation test based on the sample means is discussed in Section 6. The results on deficiencies of distributionfree tests are contained in Section 7. Section 8 is devoted to estimators. Some technical results are dealt with in the appendix.

2. An expansion for the conditional distribution of two-sample rank statistics and its application to rejective sampling.

Let $X_1, X_2, \cdots, X_N$, $N = m + n$, be independent random variables (rv's) such that $X_1, \cdots, X_m$ are identically distributed (i.d.) with common distribution function (df) F and density f and $X_{m+1}, \cdots, X_N$ are i.d. with common df G and density g. Let $Z_1 < Z_2 < \cdots < Z_N$ denote the order statistics of $X_1, \cdots, X_N$, define the antiranks $D_1, D_2, \cdots, D_N$ by $X_{D_j} = Z_j$ and let

$$V_j = 1 \quad \text{if} \quad m + 1 \leqq D_j \leqq N \qquad (2.1)$$
$$= 0 \quad \text{otherwise.}$$

For a specified vector of scores $a = (a_1, a_2, \cdots, a_N)$ define a two-sample rank statistic by

$$T = \sum_{j=1}^N a_j V_j\,. \qquad (2.2)$$

Our aim is to obtain an asymptotic expansion as $N \to \infty$ for the distribution of T for suitable sequences of pairs of df's (F_N, G_N), arrays of scores $\{a_{j,N}\}$, $1 \leqq j \leqq N$, and sample sizes (m_N, n_N). As in Albers, Bickel and van Zwet (ABZ) (1976) we shall suppress dependence on N whenever possible and formally present our results in terms of error bounds for fixed, but arbitrary, values of N.

Under the null-hypothesis that $F = G$,

$$P(V_1 = v_1, \cdots, V_N = v_N) = \frac{1}{\binom{N}{n}}$$

for any vector $(v_1, \cdots, v_N)$ with m coordinates equal to 0 and n coordinates equal to 1. In general, conditional on $Z = (Z_1, \cdots, Z_N)$,

$$P(V_1 = v_1, \cdots, V_N = v_N \mid Z) = c^{-1}(P) \prod_{j=1}^N P_j^{v_j}(1 - P_j)^{1-v_j}\,, \qquad (2.3)$$

where

$$P_j = \frac{\lambda g(Z_j)}{(1 - \lambda)f(Z_j) + \lambda g(Z_j)}\,, \qquad (2.4)$$

(2.5) $$\lambda = \frac{n}{N},$$

(2.6) $$c(P) = \sum \prod_{j=1}^{N} P_j^{w_j}(1 - P_j)^{1-w_j},$$

and the summation is over all vectors $(w_1, \cdots, w_N)$ consisting of m zeros and n ones.

Let $W_1, W_2, \cdots, W_N$ be independent rv's with $P(W_j = 1) = 1 - P(W_j = 0) = p_j$, $1 \leq j \leq N$. Suppose that

(2.7) $$\begin{aligned} p_j &= 0 \quad \text{for at most } m \text{ indices } j \\ p_j &= 1 \quad \text{for at most } n \text{ indices } j \end{aligned}$$

and consider the conditional distribution of $\sum a_j W_j$ given that $\sum W_j = n$. Note that if we replace $p = (p_1, \cdots, p_N)$ by $P = (P_1, \cdots, P_N)$, then this is the distribution of T given Z. For general p this distribution is of interest in its own right since $\sum a_j W_j$ given $\sum W_j = n$ is the sample sum we obtain when we use a rejective sampling scheme with parameters $p_1, \cdots, p_N$ in selecting a sample of size n from the sampling frame $\{a_1, a_2, \cdots, a_N\}$ (see Hájek (1964) for details).

Define

(2.8) $$\rho(t, p) = E(\exp\{itN^{-\frac{1}{2}} \sum_{j=1}^{N} a_j(W_j - p_j)\} \mid \sum_{j=1}^{N} W_j = n),$$

(2.9) $$R(x, p) = P(N^{-\frac{1}{2}} \sum_{j=1}^{N} a_j(W_j - p_j) \leq x \mid \sum_{j=1}^{N} W_j = n).$$

Our program for obtaining an Edgeworth expansion for the df of T parallels in part that of ABZ (1976). We obtain a formula for ρ. From this formula we obtain an expansion for ρ which we can rigorously translate into an Edgeworth expansion for R. Because of the connection with rejective sampling we isolate this result as the only theorem in this section. In the next section we proceed with our main program and obtain an expansion for the df of T by replacing p by P and taking the expectation of the resulting expression. We begin with

LEMMA 2.1. *Define*

(2.10) $$\psi(s, t, p) = \exp\{isN^{-\frac{1}{2}} \sum_{j=1}^{N} (p_j - \lambda)\} \prod_{j=1}^{N} [p_j \exp\{iN^{-\frac{1}{2}}(1-p_j)(s+a_j t)\} + (1 - p_j)\exp\{-iN^{-\frac{1}{2}} p_j(s + a_j t)\}],$$

(2.11) $$\nu(t, p) = \int_{-\pi N^{\frac{1}{2}}}^{\pi N^{\frac{1}{2}}} \psi(s, t, p)\, ds,$$

(2.12) $$c(p) = \sum \prod_{j=1}^{N} p_j^{w_j}(1 - p_j)^{1-w_j},$$

where the last summation is over all vectors $(w_1, \cdots, w_N)$ *consisting of m zeros and n ones. Then, if* (2.7) *is satisfied,*

(2.13) $$\rho(t, p) = \frac{1}{2\pi c(p) N^{\frac{1}{2}}} \int_{-\pi N^{\frac{1}{2}}}^{\pi N^{\frac{1}{2}}} \psi(s, t, p)\, ds = \frac{\nu(t, p)}{\nu(0, p)}.$$

PROOF. Begin with the identity

$$\begin{aligned} E(\exp\{iN^{-\frac{1}{2}}[s \textstyle\sum (W_j - p_j) + t \sum a_j(W_j - p_j)]\}) \\ = \textstyle\sum_{k=0}^{N} E(\exp\{itN^{-\frac{1}{2}} \sum a_j(W_j - p_j)\} \mid \sum W_j = k) \\ \times P(\textstyle\sum W_j = k) \exp\{isN^{-\frac{1}{2}}(k - \sum p_j)\}. \end{aligned}$$

Because the system $\{(2\pi N^{\frac{1}{2}})^{-1}\exp(iksN^{-\frac{1}{2}}): k = 0, \pm 1, \cdots\}$ is orthonormal on $[-\pi N^{\frac{1}{2}}, \pi N^{\frac{1}{2}}]$ this implies

$$\rho(t, p) = (2\pi N^{\frac{1}{2}} P(\sum W_j = n))^{-1} \int_{-\pi N^{\frac{1}{2}}}^{\pi N^{\frac{1}{2}}} \exp\{isN^{-\frac{1}{2}} \sum (p_j - \lambda)\} \times E(\exp\{iN^{-\frac{1}{2}} \sum (s + a_j t)(W_j - p_j)\})\, ds\,.$$

Elementary considerations now yield (2.13). □

Note that if $p_j = \lambda$ for all j (which corresponds to the null-hypothesis in the two-sample problem) our formula agrees with that of Erdös and Rényi for random sampling without replacement (cf. Rényi (1970), page 462). In fact their result motivated our approach.

In our asymptotic study of ψ, ν and ρ we shall repeatedly come across the following functions of p.

$$\omega(p) = N^{-\frac{1}{2}} \textstyle\sum_{j=1}^N (p_j - \lambda)\,, \tag{2.14}$$

$$\sigma^2(p) = N^{-1} \textstyle\sum_{j=1}^N p_j(1 - p_j)\,, \tag{2.15}$$

$$\bar{a}(p) = \textstyle\sum_{j=1}^N p_j(1 - p_j)a_j / \sum_{j=1}^N p_j(1 - p_j)\,, \tag{2.16}$$

$$\begin{aligned}\tau^2(p) &= N^{-1} \textstyle\sum_{j=1}^N p_j(1 - p_j)(a_j - \bar{a}(p))^2 \\ &= N^{-1} \textstyle\sum_{j=1}^N p_j(1 - p_j)a_j^2 - \sigma^2(p)\bar{a}^2(p)\,,\end{aligned} \tag{2.17}$$

$$\kappa_{3,i}(p) = N^{-1} \textstyle\sum_{j=1}^N p_j(1 - p_j)(1 - 2p_j)(a_j - \bar{a}(p))^i\,, \qquad i = 0, 1, 2, 3\,, \tag{2.18}$$

$$\kappa_{4,i}(p) = N^{-1} \textstyle\sum_{j=1}^N p_j(1 - p_j)(1 - 6p_j + 6p_j^2)(a_j - \bar{a}(p))^i\,, \qquad i = 0, 1, \cdots, 4\,. \tag{2.19}$$

In this notation we shall suppress the dependence on p when this is convenient. Let l denote Lebesgue measure on R^1 and define

$$\gamma(\varepsilon, \zeta, p) = l\{x : \exists j\ |x - a_j| < \zeta, \varepsilon \leq p_j \leq 1 - \varepsilon\}\,. \tag{2.20}$$

LEMMA 2.2. *Suppose that positive numbers* c, C, δ *and* ε *exist such that*

$$\tau^2(p) \geq c\,, \qquad \frac{1}{N} \textstyle\sum_{j=1}^N a_j^4 \leq C\,, \tag{2.21}$$

$$\gamma(\varepsilon, \zeta, p) \geq \delta N \zeta \qquad \text{for some} \quad \zeta \geq N^{-\frac{3}{2}} \log N\,. \tag{2.22}$$

Then there exist positive numbers b, B *and* β *depending only on* c, C, δ *and* ε *such that*

$$|\psi(s, t, p)| \leq BN^{-\beta \log N} \tag{2.23}$$

for all pairs (s, t) *such that* $|s| \leq \pi N^{\frac{1}{2}}$, $|t| \leq bN^{\frac{3}{2}}$ *and either* $|s| \geq \log(N + 1)$ *or* $|t| \geq \log(N + 1)$.

PROOF.

$$\begin{aligned}|\psi(s, t, p)| &= \textstyle\prod_{j=1}^N [1 - 2p_j(1 - p_j)\{1 - \cos(N^{-\frac{1}{2}}(s + a_j t))\}]^{\frac{1}{2}} \\ &\leq \exp\{-\textstyle\sum_{j=1}^N p_j(1 - p_j)[\frac{1}{2}N^{-1}(s + a_j t)^2 - \frac{1}{24}N^{-2}(s + a_j t)^4]\} \\ &\leq \exp\{-\tfrac{1}{2}[\tau^2 t^2 + \sigma^2(s + \bar{a}t)^2] \\ &\qquad + \tfrac{1}{12}N^{-1}[N^{-1} \textstyle\sum_{j=1}^N (a_j - \bar{a})^4 t^4 + (s + \bar{a}t)^4]\}\,.\end{aligned} \tag{2.24}$$

Now (2.21) ensures that

$$\sigma^2(p) \geqq N\tau^4(p)/\textstyle\sum_{j=1}^N a_j^4 \geqq C^{-1}c^2 , \tag{2.25}$$

$$|\bar{a}(p)| \leqq [N^{-1} \textstyle\sum_{j=1}^N a_j^4]^{\frac{1}{4}}/\sigma^2(p) \leqq c^{-2}C^{\frac{5}{4}} , \tag{2.26}$$

and by (2.21), (2.24), (2.25) and (2.26) we conclude that there exist positive b_1, B and β depending only on c and C such that for $|s| \leqq b_1 N^{\frac{1}{2}}$ and $|t| \leqq b_1 N^{\frac{1}{2}}$

$$|\psi(s, t, p)| \leqq B \exp\{-\beta(s^2 + t^2)\} . \tag{2.27}$$

Next note that (2.25) and (2.21) imply that the number of indices j for which $p_j(1 - p_j) \geqq \frac{1}{2}c^2/C$ is at least $2Nc^2/C$ and the number of j for which $|a_j| \leqq (C/c)^{\frac{1}{2}}$ is at least $N - Nc^2/C$. Hence the number of indices j for which $|a_j| \leqq (C/c)^{\frac{1}{2}}$ and $p_j(1 - p_j) \geqq \frac{1}{2}c^2/C$ is at least Nc^2/C. Put $b_2 = \frac{1}{2}b_1(c/C)^{\frac{1}{2}}$ and we see that if $b_1 N^{\frac{1}{2}} \leqq |s| \leqq \pi N^{\frac{1}{2}}$ and $|t| \leqq b_2 N^{\frac{1}{2}}$, then for at least Nc^2/C indices j

$$[1 - 2p_j(1 - p_j)\{1 - \cos(N^{-\frac{1}{2}}(s + a_j t))\}] \leqq 1 - c^2C^{-1}\left\{1 - \cos\left(\frac{b_1}{2}\right)\right\} .$$

Combining this with (2.27) we see that it only remains to be shown that positive numbers b, B and β exist depending only on c, C, δ and ε and such that (2.23) holds for $|s| \leqq \pi N^{\frac{1}{2}}$ and $(b_1 \wedge b_2)N^{\frac{1}{2}} \leqq |t| \leqq bN^{\frac{3}{2}}$. For this we can appeal to the corresponding part of the proof of Lemma 2.2 in ABZ (1976) with only minor modifications. □

Define functions $\mu_k(p)$, $1 \leqq k \leqq 6$, and $A_k(p)$, $0 \leqq k \leqq 6$, by

$$\mu_1 = \frac{\omega}{\sigma^2} , \qquad \mu_2 = \frac{1}{\sigma^2} - \frac{\omega^2}{\sigma^4} , \qquad \mu_3 = \frac{3\omega}{\sigma^4} - \frac{\omega^3}{\sigma^6} ,$$

$$\mu_4 = \frac{3}{\sigma^4} - \frac{6\omega^2}{\sigma^6} + \frac{\omega^4}{\sigma^8} , \qquad \mu_5 = \frac{15\omega}{\sigma^6} - \frac{10\omega^3}{\sigma^8} + \frac{\omega^5}{\sigma^{10}} , \tag{2.28}$$

$$\mu_6 = \frac{15}{\sigma^6} - \frac{45\omega^2}{\sigma^8} + \frac{15\omega^4}{\sigma^{10}} - \frac{\omega^6}{\sigma^{12}} ,$$

$$A_0 = 1 + \frac{N^{-\frac{1}{2}}}{6}\kappa_{3,0}\mu_3 + \frac{N^{-1}}{72}(3\kappa_{4,0}\mu_4 - \kappa_{3,0}^2\mu_6) ,$$

$$A_1 = -\frac{N^{-\frac{1}{2}}}{2}\kappa_{3,1}\mu_2 + \frac{N^{-1}}{12}(2\kappa_{4,1}\mu_3 - \kappa_{3,0}\kappa_{3,1}\mu_5) ,$$

$$A_2 = -\frac{N^{-\frac{1}{2}}}{2}\kappa_{3,2}\mu_1 + \frac{N^{-1}}{24}\{-6\kappa_{4,2}\mu_2 + (2\kappa_{3,0}\kappa_{3,2} + 3\kappa_{3,1}^2)\mu_4\} , \tag{2.29}$$

$$A_3 = \frac{N^{-\frac{1}{2}}}{6}\kappa_{3,3} + \frac{N^{-1}}{36}\{-6\kappa_{4,3}\mu_1 + (\kappa_{3,0}\kappa_{3,3} + 9\kappa_{3,1}\kappa_{3,2})\mu_3\} ,$$

$$A_4 = \frac{N^{-1}}{24}\{\kappa_{4,4} - (2\kappa_{3,1}\kappa_{3,3} + 3\kappa_{3,2}^2)\mu_2\} ,$$

$$A_5 = -\frac{N^{-1}}{12}\kappa_{3,2}\kappa_{3,3}\mu_1 ,$$

$$A_6 = \frac{N^{-1}}{72}\kappa_{3,3}^2 ,$$

where we have suppressed the dependence on p. We shall show that

$$\tilde{\nu}(t, p) = \frac{(2\pi)^{\frac{1}{2}}}{\sigma(p)} \exp\left\{-\frac{\omega^2(p)}{2\sigma^2(p)} - \frac{\tau^2(p)t^2}{2} - i\omega(p)\bar{a}(p)t\right\} \sum_{k=0}^{6} A_k(p)(it)^k \tag{2.30}$$

is an asymptotic expansion for $\nu(t, p)$.

LEMMA 2.3. *Suppose that positive numbers c, C, δ and ε exist such that* (2.21) *and* (2.22) *are satisfied. Then there exist positive numbers b, B and β depending only on c, C, δ and ε such that for $|t| \leqq bN^{\frac{3}{2}}$,*

$$|\nu(t, p) - \tilde{\nu}(t, p)| \leqq B\left[(N^{-\frac{3}{2}} + N^{-\frac{5}{4}}|t|^5) \exp\left\{-\frac{ct^2}{8}\right\} + N^{-\beta \log N}\right]. \tag{2.31}$$

PROOF. In this proof b, b_i, B_i, β_i and N_0 denote appropriately chosen positive numbers depending only on c, C, δ and ε.

Arguing as in the proof of Theorem 2.1 in ABZ (1976) we find by Taylor expansion of $\log \psi$ that if $|s + a_j t| \leqq \frac{1}{2}\pi N^{\frac{1}{2}}$ for all j, then

$$\begin{aligned} \psi(s, t, p) &= \exp\left\{i\omega s - \frac{\tau^2 t^2}{2} - \frac{\sigma^2(s + \bar{a}t)^2}{2}\right. \\ &\quad - \frac{iN^{-\frac{3}{2}}}{6} \sum p_j(1 - p_j)(1 - 2p_j)(s + a_j t)^3 \\ &\quad \left. + \frac{N^{-2}}{24} \sum p_j(1 - p_j)(1 - 6p_j + 6p_j^2)(s + a_j t)^4 + M_1(s, t, p)\right\}, \end{aligned} \tag{2.32}$$

where

$$\begin{aligned} |M_1(s, t, p)| &\leqq C_1 N^{-\frac{5}{2}} \sum |s + a_j t|^5 \\ &\leqq 16C_1(N^{-\frac{5}{2}}|t|^5 \sum |a_j - \bar{a}|^5 + N^{-\frac{3}{2}}|s + \bar{a}t|^5) \end{aligned}$$

for some absolute constant C_1. Now (2.21) and (2.26) imply that $N^{-1} \sum |a_j - \bar{a}|^3$, $N^{-1} \sum |a_j - \bar{a}|^4$, $N^{-\frac{1}{4}} \max |a_j|$ and $N^{-\frac{5}{4}} \sum |a_j - \bar{a}|^5$ are bounded. Using (2.21) and (2.25) we find that for all $|s| \leqq b_1 N^{\frac{1}{2}}$ and $|t| \leqq b_1 N^{\frac{1}{4}}$

$$\frac{N^{-\frac{3}{2}}}{6} \sum |s + a_j t|^3 + \frac{N^{-2}}{24} \sum (s + a_j t)^4 + |M_1(s, t, p)| \leqq \frac{\tau^2 t^2 + \sigma^2(s + \bar{a}t)^2}{4}.$$

Hence further expansion of part of the exponential in (2.32) shows that

$$\psi(s, t, p) = \tilde{\psi}(s, t, p) + M_2(s, t, p) \tag{2.33}$$

for $|s| \leqq b_1 N^{\frac{1}{2}}$ and $|t| \leqq b_1 N^{\frac{1}{4}}$, where

$$\begin{aligned} \tilde{\psi}(s, t, p) &= \exp\left\{i\omega s - \frac{\tau^2 t^2}{2} - \frac{\sigma^2(s + \bar{a}t)^2}{2}\right\} \\ &\quad \times \left[1 - \frac{iN^{-\frac{3}{2}}}{6} \sum p_j(1 - p_j)(1 - 2p_j)(s + a_j t)^3\right. \\ &\quad + \frac{N^{-2}}{24} \sum p_j(1 - p_j)(1 - 6p_j + 6p_j^2)(s + a_j t)^4 \\ &\quad \left. - \frac{N^{-3}}{72} \left(\sum p_j(1 - p_j)(1 - 2p_j)(s + a_j t)^3\right)^2\right], \end{aligned} \tag{2.34}$$

$$(2.35) \qquad |M_2(s, t, p)| \leq (N^{-\frac{3}{2}} + N^{-\frac{5}{4}}|t|^5) M_3(t, s + \bar{a}t) \exp\left\{-\frac{\tau^2 t^2 + \sigma^2(s + \bar{a}t)^2}{4}\right\}$$

and M_3 is a polynomial in t and $(s + \bar{a}t)$ of fixed degree with coefficients depending only on c and C. Therefore, for $|t| \leq b_1 N^{\frac{1}{4}}$,

$$(2.36) \qquad \int_{-b_1 N^{\frac{1}{2}}}^{b_1 N^{\frac{1}{2}}} |\psi(s, t, p) - \tilde{\psi}(s, t, p)|\, ds \leq B_1 (N^{-\frac{3}{2}} + N^{-\frac{5}{4}}|t|^5) \exp\left\{-\frac{ct^2}{8}\right\}.$$

Next we show that for $|t| \leq b_1 N^{\frac{1}{4}}$,

$$(2.37) \qquad \int_{b_1 N^{\frac{1}{2}} \leq |s| \leq \pi N^{\frac{1}{2}}} |\psi(s, t, p)|\, ds \leq B_2 N^{-\beta_2 \log N},$$

$$(2.38) \qquad \int_{|s| \geq b_1 N^{\frac{1}{2}}} |\tilde{\psi}(s, t, p)|\, ds \leq B_3 N^{-\beta_3 \log N}.$$

For $N \geq N_0$, (2.37) is a consequence of Lemma 2.2 and since $|\psi| \leq 1$ we can choose B_2 so that (2.37) holds for all N. Because for all s and t

$$(2.39) \qquad |\tilde{\psi}(s, t, p)| \leq \exp\left\{-\frac{\tau^2 t^2 + \sigma^2(s + \bar{a}t)^2}{2}\right\} M_4(t, s + \bar{a}t)$$

where M_4 is a polynomial depending only on c and C, (2.38) follows. Combining (2.11), (2.36), (2.37) and (2.38) we see that for $|t| \leq b_1 N^{\frac{1}{4}}$

$$(2.40) \qquad \begin{aligned} &|\nu(t, p) - \textstyle\int_{-\infty}^{\infty} \tilde{\psi}(s, t, p)\, ds| \\ &\quad \leq B_4 \left[(N^{-\frac{3}{2}} + N^{-\frac{5}{4}}|t|^5) \exp\left\{-\frac{ct^2}{8}\right\} + N^{-\beta_4 \log N} \right]. \end{aligned}$$

A direct application of Lemma 2.2, the fact that $|\psi| \leq 1$ and (2.39) show that we can choose B_4 and β_4 so that (2.40) continues to hold for $b_1 N^{\frac{1}{4}} \leq |t| \leq b N^{\frac{3}{2}}$ with b as in Lemma 2.2.

It remains to be shown that for all s and t

$$(2.41) \qquad \tilde{\nu}(t, p) = \int_{-\infty}^{\infty} \tilde{\psi}(s, t, p)\, ds.$$

This follows by straightforward but tedious computation using the fact that

$$\begin{aligned} (2\pi)^{-\frac{1}{2}} \int_{-\infty}^{\infty} \left(\frac{z}{\sigma(p)} + \frac{i\omega(p)}{\sigma^2(p)}\right)^k e^{-\frac{1}{2}z^2}\, dz &= \mu_k(p) \qquad \text{for even } k \\ &= i\mu_k(p) \qquad \text{for odd } k. \end{aligned}$$

□

We now turn to our asymptotic expansion for rejective sampling. For $1 \leq k \leq 6$, define functions $Q_k(p)$ by

$$(2.42) \qquad \begin{aligned} Q_1 &= -\frac{N^{-\frac{1}{2}}}{2}\kappa_{3,1}\mu_2 + \frac{N^{-1}}{6}\left[\kappa_{4,1}\mu_3 - 3\kappa_{3,0}\kappa_{3,1}\left(2\frac{\omega}{\sigma^6} - \frac{\omega^3}{\sigma^8}\right)\right], \\ Q_2 &= -\frac{N^{-\frac{1}{2}}}{2}\kappa_{3,2}\mu_1 + \frac{N^{-1}}{8}\left[-2\kappa_{4,2}\mu_2 + 2\kappa_{3,0}\kappa_{3,2}\left(\frac{1}{\sigma^4} - \frac{\omega^2}{\sigma^6}\right) + \kappa_{3,1}^2\mu_4\right], \\ Q_3 &= \frac{N^{-\frac{1}{2}}}{6}\kappa_{3,3} + \frac{N^{-1}}{12}\left[-2\kappa_{4,3}\mu_1 + 3\kappa_{3,1}\kappa_{3,2}\mu_3\right], \\ Q_k &= A_k, \quad k = 4, 5, 6. \end{aligned}$$

Let Φ and ϕ denote the standard normal df and its density and let H_k denote the Hermite polynomial of degree k, thus

$$(2.43)\qquad H_0(x) = 1\,, \quad H_1(x) = x\,, \quad H_2(x) = x^2 - 1\,, \quad H_3(x) = x^3 - 3x\,,$$
$$H_4(x) = x^4 - 6x^2 + 3\,, \quad H_5(x) = x^5 - 10x^3 + 15x\,.$$

We shall show that expansions for (2.8) and (2.9) are given by

$$(2.44)\qquad \tilde{\rho}(t, p) = \exp\left\{-\frac{\tau^2(p)t^2}{2} - i\omega(p)\bar{a}(p)t\right\}[1 + \textstyle\sum_{k=1}^{6} Q_k(p)(it)^k]\,,$$

$$(2.45)\qquad \tilde{R}(x, p) = \Phi\left(\frac{x + \omega(p)\bar{a}(p)}{\tau(p)}\right) - \phi\left(\frac{x + \omega(p)\bar{a}(p)}{\tau(p)}\right)\sum_{k=1}^{6}\frac{Q_k(p)}{(\tau(p))^k} \times H_{k-1}\left(\frac{x + \omega(p)\bar{a}(p)}{\tau(p)}\right).$$

Note that $\tilde{\rho}$ is the Fourier–Stieltjes transform of $\tilde{R}$, i.e., $\tilde{\rho}(t, p) = \int e^{itx}\, d\tilde{R}(x, p)$.

THEOREM 2.1. *Suppose that positive numbers c, C, D, δ and ε exist such that (2.21) and (2.22) are satisfied and*

$$(2.46)\qquad |\omega(p)| \leq D\,.$$

Then there exist positive numbers N_0 and B depending only on c, C, D, δ and ε such that for $N \geq N_0$, $R(x, p)$ is well defined and

$$(2.47)\qquad \sup_x |R(x, p) - \tilde{R}(x, p)| \leq BN^{-\frac{5}{4}}\,.$$

PROOF. In this proof b, B_i, β, η and N_0 denote appropriately chosen positive numbers depending only on c, C, D, δ and ε.

By (2.21), (2.25), (2.26), (2.46) and Lemma 2.3 we have for $N \geq N_0$,

$$(2.48)\qquad |\tilde{\nu}(0, p)| \geq \eta\,, \quad |\nu(0, p) - \tilde{\nu}(0, p)| \leq \frac{\eta}{2}\,,$$

so that $|\nu(0, p)| \geq \eta/2 > 0$. In the first place it follows that for $N \geq N_0$, $c(p) > 0$ and hence (2.7) is satisfied and $R(x, p)$ is properly defined. We assume that $N \geq N_0$ and we shall show that, with b as in Lemma 2.3,

$$(2.49)\qquad \int_{-bN^{\frac{3}{4}}}^{bN^{\frac{3}{4}}} \left|\frac{\rho(t, p) - \tilde{\rho}(t, p)}{t}\right| dt \leq B_1 N^{-\frac{5}{4}}\,.$$

By Esseen's smoothing lemma (Esseen (1945)) this suffices to prove the theorem because $\tilde{R}(-\infty, p) = 0$, $\tilde{R}(\infty, p) = 1$ and the derivative of $\tilde{R}$ with respect to x is bounded.

By (2.21), (2.25), (2.26) and (2.46), $\tilde{\rho}$ has a bounded derivative with respect to t. Also

$$\left|\frac{d\rho(t, p)}{dt}\right| \leq N^{-\frac{1}{2}} E(|\textstyle\sum a_j(W_j - p_j)|\,|\, \sum W_j = n) \leq N^{-\frac{1}{2}} \sum |a_j| \leq C^{\frac{1}{4}} N^{\frac{1}{2}}\,.$$

Since $\rho(0, p) = \tilde{\rho}(0, t) = 1$, it follows that

$$(2.50)\qquad \int_{-N^{-2}}^{N^{-2}} \left|\frac{\rho(t, p) - \tilde{\rho}(t, p)}{t}\right| dt \leq B_2 N^{-\frac{3}{2}}\,.$$

Next we note that (2.21), (2.25) and (2.26) ensure that for all t

$$(2.51) \qquad |\tilde{\nu}(t, p)| \leqq B_3 \exp\left\{-\frac{ct^2}{4}\right\}.$$

Together with (2.13), (2.48) and Lemma 2.3 this implies that for $|t| \leqq bN^{\frac{3}{2}}$

$$(2.52) \qquad \begin{aligned}\left|\rho(t, p) - \frac{\tilde{\nu}(t, p)}{\tilde{\nu}(0, p)}\right| &\leqq \frac{2}{\eta}|\nu(t, p) - \tilde{\nu}(t, p)| + \frac{2}{\eta^2}|\tilde{\nu}(t, p)|\,|\nu(0, p) - \tilde{\nu}(0, p)| \\ &\leqq B_4\left[(N^{-\frac{3}{2}} + N^{-\frac{5}{4}}|t|^5)\exp\left\{-\frac{ct^2}{8}\right\} + N^{-\beta \log N}\right].\end{aligned}$$

Again with the aid of (2.21), (2.25) (2.26) and (2.46) one can easily check that, for $1 \leqq k \leqq 6$, Q_k is obtained from A_k/A_0 by expanding the denominator and discarding all terms of order $N^{-\frac{3}{2}}$, i.e., that $|Q_k - A_k/A_0| \leqq B_5 N^{-\frac{3}{2}}$. It follows that

$$(2.53) \qquad \left|\tilde{\rho}(t, p) - \frac{\tilde{\nu}(t, p)}{\tilde{\nu}(0, p)}\right| \leqq B_6 N^{-\frac{3}{2}} \exp\left\{-\frac{ct^2}{4}\right\}$$

and combined with (2.52) this yields

$$(2.54) \qquad \int_{N^{-2} \leqq |t| \leqq bN^{\frac{3}{2}}} \left|\frac{\rho(t, p) - \tilde{\rho}(t, p)}{t}\right| dt \leqq B_7(N^{-\frac{3}{2}} \log N + N^{-\frac{5}{4}}) \leqq B_8 N^{-\frac{5}{4}}.$$

Together with (2.50) this proves (2.49) and the theorem. □

Two remarks should be made with regard to Theorem 2.1. The first one concerns condition (2.46) that does not occur in the preceding lemmas. The meaning of this condition is perhaps obscured by the fact that we make it do some odd jobs in the proof for which it is not really needed. We use it to show that (2.7) is satisfied for $N \geqq N_0$, but (2.25) ensures that the number of indices j with $p_j = 0$ (or $p_j = 1$) cannot exceed $m - C^{-1}c^2N + |\omega(p)|N^{\frac{1}{2}}$ (or $n - C^{-1}c^2N + |\omega(p)|N^{\frac{1}{2}}$) so that $|\omega(p)| \leqq C^{-1}c^2N^{\frac{1}{2}}$ already implies (2.7) for all N. Condition (2.46) is also used to obtain (2.50), but in (2.50) we may replace N^{-2} by an arbitrarily high power of N^{-1} without doing any damage to the proof, and then the trivial bound $|\omega(p)| \leqq N^{\frac{1}{2}}$ suffices. Finally we note that since

$$(2.55) \qquad \min(\lambda, 1 - \lambda) \geqq \sigma^2(p) - N^{-\frac{1}{2}}|\omega(p)|,$$

(2.46) forces λ to be bounded away from 0 and 1 for large N, which is obviously important although it does not show up explicitly in the proof. However, here $|\omega(p)| \leqq \frac{1}{2}C^{-1}c^2N^{\frac{1}{2}}$ would be sufficient.

The basic function of assumption (2.46), however, is to avoid a large (or intermediate) deviation situation that the condition $\sum W_j = n$ would get us into if $\omega(p) = N^{-\frac{1}{2}}(E \sum W_j - n)$ would not be bounded. Technically speaking this is reflected in the proof at the point where (2.46) is used to show that $\nu(0, p)$ is bounded away from zero. Also (2.46) ensures that (2.45) provides an expansion in powers of $N^{-\frac{1}{2}}$ to the required order.

To see what happens when condition (2.46) is relaxed, we prefer not to try to adapt the proof of Theorem 2.1 but to answer this question more directly by remarking that the conditional distribution of $\sum a_j W_j$ given $\sum W_j = n$ remains unchanged if we replace p by $\bar{p}$ where $\bar{p}_j/(1-\bar{p}_j) = \xi p_j/(1-p_j)$ $1 \leqq j \leqq N$, for some $0 \leqq \xi \leqq \infty$. If (2.7) is satisfied there exists a unique ξ for which $\sum \bar{p}_j = N\lambda$. Since $\omega(\bar{p}) = 0$ it follows that if (2.21) and (2.22) are satisfied with p replaced by $\bar{p}$, then (2.47) holds with $\tilde{R}(x, \bar{p})$ instead of $\tilde{R}(x, p)$. Of course the snag is that in general $\bar{p}$ can only be expressed analytically in terms of p as an infinite series. However, if $\omega(p) = O(N^{\alpha})$ for some $\alpha < \frac{1}{2}$, then a finite number of terms of this series will yield the required degree of accuracy and an explicit expansion for $R(x, p)$ can be obtained. If $\alpha = 0$ this is expansion (2.45) but for $0 < \alpha < \frac{1}{2}$ more terms have to be included.

The second remark concerns the remainder $O(N^{-\frac{5}{4}})$ of our expansion. It is clear that by requiring that $\sum |a_j|^5 \leqq CN$ in Theorem 2.1 one obtains $|R - \tilde{R}| \leqq BN^{-\frac{3}{2}} \log(N+1)$. Of course the "natural" order of the remainder is $O(N^{-\frac{3}{2}})$ and the factor $\log(N+1)$ is due only to technical difficulties in finding the conditional expectation of $\sum a_j W_j$ given $\sum W_j = n$.

The special case $p_j = \lambda$, $1 \leqq j \leqq N$, which is random sampling without replacement, is worth singling out because it corresponds to the null-hypothesis in the two-sample problem. Let $\bar{\lambda}$ denote the vector $(\lambda, \cdots, \lambda)$. For $p = \bar{\lambda}$, (2.45) simplifies to

$$\begin{aligned}\tilde{R}(x, \bar{\lambda}) = \Phi\left(\frac{x}{\tau(\bar{\lambda})}\right) - \frac{\phi(x/\tau(\bar{\lambda}))}{\lambda(1-\lambda)}\Bigg[\frac{\lambda(1-\lambda)}{2N} H_1\left(\frac{x}{\tau(\bar{\lambda})}\right) \\ + \frac{\{\lambda(1-\lambda)\}^{\frac{1}{2}}(1-2\lambda)}{6} \frac{\sum (a_j - a_{\bullet})^3}{\{\sum (a_j - a_{\bullet})^2\}^{\frac{3}{2}}} H_2\left(\frac{x}{\tau(\bar{\lambda})}\right) \\ + \left\{\frac{1 - 6\lambda + 6\lambda^2}{24} \frac{\sum (a_j - a_{\bullet})^4}{\{\sum (a_j - a_{\bullet})^2\}^2} - \frac{(1-2\lambda)^2}{8N}\right\} H_3\left(\frac{x}{\tau(\bar{\lambda})}\right) \\ + \frac{(1-2\lambda)^2}{72} \frac{\{\sum (a_j - a_{\bullet})^3\}^2}{\{\sum (a_j - a_{\bullet})^2\}^3} H_5\left(\frac{x}{\tau(\bar{\lambda})}\right)\Bigg],\end{aligned} \tag{2.56}$$

where

$$\tau^2(\bar{\lambda}) = \frac{\lambda(1-\lambda)}{N} \sum_{j=1}^{N} (a_j - a_{\bullet})^2, \tag{2.57}$$

$$a_{\bullet} = \bar{a}(\bar{\lambda}) = \frac{1}{N} \sum_{j=1}^{N} a_j. \tag{2.58}$$

Define, with l denoting Lebesgue measure on R^1,

$$\gamma(\zeta) = l\{x : \exists j \ |x - a_j| < \zeta\}. \tag{2.59}$$

For $p = \bar{\lambda}$, Theorem 2.1 yields

COROLLARY 2.1. *Suppose that positive numbers c, C, δ and ε exist such that*

$$\varepsilon \leqq \lambda \leqq 1 - \varepsilon, \tag{2.60}$$

$$\frac{1}{N} \sum_{j=1}^{N} (a_j - a_{\bullet})^2 \geqq c, \qquad \frac{1}{N} \sum_{j=1}^{N} a_j^4 \leqq C, \tag{2.61}$$

(2.62) $$\gamma(\zeta) \geqq \delta N\zeta \quad \textit{for some} \quad \zeta \geqq N^{-\frac{3}{4}} \log N .$$

Then there exists $B > 0$ depending only on c, C, δ and ε such that

$$\sup_x |R(x, \bar{\lambda}) - \tilde{R}(x, \bar{\lambda})| \leqq BN^{-\frac{5}{4}} .$$

Note that there is considerable further simplification in (2.56) if we either have almost equal sample sizes, i.e., $\lambda = \frac{1}{2} + O(N^{-\frac{1}{2}})$, or antisymmetric scores, i.e., $a_j + a_{N-j+1}$ is constant for all j. The latter happens for the locally most powerful rank test against shift alternatives when the underlying distribution is symmetric. In either case the H_2 and H_5 terms disappear so that the correction to the leading normal term is of order N^{-1} only and is due solely to a correction to the variance, the H_1 term, and a kurtosis correction corresponding to H_3.

3. An unconditional expansion. We encounter several difficulties on the way to a usable unconditional expansion:

(i) the distribution of Z is awkward to handle analytically;

(ii) as in ABZ (1976), the random variables obtained by substituting P for p in $\bar{\rho}$ or $\tilde{R}$ are generally not summable;

(iii) again as in ABZ (1976), final simplification is not possible with our present techniques unless we assume that the sequence of alternatives is contiguous to the hypothesis as $N \to \infty$.

In this section we shall deal with the first two difficulties. Although we do not assume contiguity we shall be governed in the form of our expansion, which will involve polynomials in $(P_j - \lambda)$, in the number of terms that we calculate and in what we relegate to the remainder by the consideration that we expect $P_j = \lambda + O_P(N^{-\frac{1}{2}})$ and $\sum (P_j - \lambda) = O_P(1)$.

Recall that we assumed that $X_1, \cdots, X_N$ are independent, $X_1, \cdots, X_m$ having common density f and $X_{m+1}, \cdots, X_N$ having density g. We shall write P for probabilities and E for expectations calculated under this model. In addition we need to consider an auxiliary model where $X_1, \cdots, X_N$ are i.i.d. with common density $h = (1 - \lambda)f + \lambda g$ and df $H = (1 - \lambda)F + \lambda G$. We shall write P_H for probabilities, E_H for expectations and σ_H^2 for variances calculated under this second model.

To simplify our notation we assume from this point on that

(3.1) $$\textstyle\sum_{j=1}^N a_j = 0 .$$

Since $T = \sum (a_j - a_\cdot)V_j + na_\cdot$ it is obvious how all expansions need to be modified if (3.1) does not hold.

We meet difficulty (i) through

LEMMA 3.1.

(3.2) $$E \exp\{itN^{-\frac{1}{2}}T\} = \frac{E_H \nu(t, P) \exp\{itN^{-\frac{1}{2}} \sum a_j P_j\}}{2\pi N^{\frac{1}{2}} B_{N,n}(\lambda)} ,$$

where

$$B_{N,n}(\lambda) = \binom{N}{n}\lambda^n(1-\lambda)^{N-n}\,.$$

PROOF. Under our original model the density of Z at the point $z = (z_1, \cdots, z_N)$ with $z_1 < z_2 < \cdots < z_N$ is given by

$$\sum \prod_{j=1}^m f(z_{i_j}) \prod_{j=m+1}^N g(z_{i_j})\,,$$

where the sum ranges over all permutations $i_1, \cdots, i_N$ of $1, \cdots, N$. Under our second model this density is

$$N! \prod_{j=1}^N [(1-\lambda)f(z_j) + \lambda g(z_j)]\,.$$

By the Radon–Nikodym theorem and Lemma 2.1,

$$\begin{aligned} E\exp\{itN^{-\frac12}T\} &= E\,\frac{\nu(t,P)}{\nu(0,P)}\exp\{itN^{-\frac12}\textstyle\sum a_jP_j\} \\ &= E_H\,\frac{\nu(t,P)}{\nu(0,P)}\exp\{itN^{-\frac12}\textstyle\sum a_jP_j\}\sum\prod_{j=1}^m \frac{f(Z_{i_j})}{h(Z_{i_j})}\prod_{j=m+1}^N\frac{g(Z_{i_j})}{h(Z_{i_j})}\frac{1}{N!} \\ &= [B_{N,n}(\lambda)]^{-1}E_H\,\frac{\nu(t,P)}{\nu(0,P)}\exp\{itN^{-\frac12}\textstyle\sum a_jP_j\}c(P)\,, \end{aligned}$$

where c is defined by (2.6) or (2.12). The lemma follows from (2.11) and (2.13). □

Lemma 3.1 shows that we are concerned with $\tilde{\nu}$ rather than $\tilde{\rho}$, but since $\tilde{\nu}$ as a function of P is no more summable than $\tilde{\rho}$, we still have to face difficulty (ii). We do this by showing that $\tilde{\nu}$ may be replaced by a summable function ν^* outside a set that will later be seen to have sufficiently small probability. Define

$$\nu^*(t,p) = \left[\frac{2\pi}{\lambda(1-\lambda)}\right]^{\frac12}\exp\left\{-\frac{\lambda(1-\lambda)}{2N}\sum a_j{}^2t^2\right\}\sum_{k=0}^6 A_k{}^*(p)(it)^k\,, \tag{3.3}$$

where

$$\begin{aligned} A_0{}^*(p) &= 1 + \frac{1}{2\lambda(1-\lambda)N}\Big[\sum(p_j-\lambda)^2 - \{\sum(p_j-\lambda)\}^2 \\ &\qquad - \frac{1-\lambda+\lambda^2}{6}\Big]\,, \\ A_1{}^*(p) &= N^{-\frac32}\sum a_jp_j\Big[1-\frac{1-2\lambda}{\lambda(1-\lambda)}\sum(p_j-\lambda)\Big]\,, \\ A_2{}^*(p) &= \frac{(1-2\lambda)}{2N}\sum a_j{}^2(p_j-\lambda) - \frac{\sum a_j{}^2}{2N^2} \\ &\qquad \times[(1-2\lambda)\sum(p_j-\lambda) - \lambda(1-\lambda)] \\ &\qquad - \frac{1}{2N}\sum a_j{}^2(p_j-\lambda)^2 - \frac{(1-2\lambda)^2}{2\lambda(1-\lambda)N^2}\{\sum a_jp_j\}^2\,, \qquad (3.4) \\ A_3{}^*(p) &= \frac{N^{-\frac32}}{6}\Big[\lambda(1-\lambda)(1-2\lambda)\sum a_j{}^3 + (1-6\lambda+6\lambda^2)\sum a_j{}^3(p_j-\lambda) \\ &\qquad - \frac{3}{N}(1-2\lambda)^2\sum a_j{}^2\sum a_jp_j\Big]\,, \end{aligned}$$

$$A_4^*(p) = \frac{\lambda(1-\lambda)(1-6\lambda+6\lambda^2)}{24N^2} \sum a_j^4 - \frac{\lambda(1-\lambda)(1-2\lambda)^2}{8N^3} \{\sum a_j^2\}^2$$
$$+ \frac{(1-2\lambda)^2}{8N^2} \{\sum a_j^2(p_j - \lambda)\}^2 ,$$

$$A_5^*(p) = \frac{N^{-\frac{5}{2}}}{12} \lambda(1-\lambda)(1-2\lambda)^2 \sum a_j^3 \sum a_j^2(p_j - \lambda) ,$$

$$A_6^*(p) = \frac{\lambda^2(1-\lambda)^2(1-2\lambda)^2}{72N^3} \{\sum a_j^3\}^2 .$$

LEMMA 3.2. *Suppose that* (3.1) *holds and positive numbers c, C and ε exist such that* (2.21) *is satisfied and*

$$\varepsilon \leqq \lambda \leqq 1 - \varepsilon . \tag{3.5}$$

Then there exist positive numbers B and β depending only on c, C and ε such that

$$|\tilde{\nu}(t, p) - \nu^*(t, p)| \leqq B \exp\{-\beta t^2\}[\{N^{-\frac{3}{2}} + N^{-\frac{5}{4}}|t|\}\{1 + N \sum (p_j - \lambda)^4\} + N^{-\frac{3}{4}}\{\sum (p_j - \lambda)^4\}] . \tag{3.6}$$

PROOF. For simplicity we make use of order symbols in this proof and $O(x)$ will denote a quantity that is bounded by $B_1|x|$ where B_1 depends only on c, C and ε.

Suppose first that $|\omega(p)| > 1$. Then (2.21) and (3.5) are easily seen to imply that $|\nu^*(t, p)| = O(\omega^2(p) \exp\{-\varepsilon(1-\varepsilon)ct^2/4\})$, whereas for $\tilde{\nu}(t, p)$ we have the bound (2.51). The right-hand side of (3.6), however, contains a term $BN^{\frac{1}{4}}\omega^4(p) \exp\{-\beta t^2\}$ so that the lemma is trivial for $|\omega(p)| > 1$.

We therefore assume that $|\omega(p)| \leqq 1$. Noting that $\sigma^2(p)$ is bounded away from zero (cf. (2.25)), we expand σ^{-2}, $\bar{a}$, τ^2 and $\kappa_{r,i}$ about the point $p_j = \lambda$, $1 \leqq j \leqq N$, using elementary inequalities to bound the remainders in terms of N and

$$M_1 = N^{-1} \sum (p_j - \lambda)^4 , \qquad M_2 = N^{-1}|\sum (p_j - \lambda)| .$$

We find

$$\frac{1}{\sigma^2(p)} = \frac{1}{\lambda(1-\lambda)}\left[1 - \frac{(1-2\lambda)}{\lambda(1-\lambda)N} \sum (p_j - \lambda) + \frac{1}{\lambda(1-\lambda)N} \sum (p_j - \lambda)^2\right]$$
$$+ O(M_1 + M_2^2)$$
$$= \frac{1}{\lambda(1-\lambda)} + O(M_1^{\frac{1}{2}} + M_2) ,$$

$$\bar{a}(p) = \frac{(1-2\lambda)}{\lambda(1-\lambda)N} \sum a_j p_j + O(M_1^{\frac{1}{2}}) = O(M_1^{\frac{1}{4}}) ,$$

$$\tau^2(p) = \frac{\lambda(1-\lambda)}{N} \sum a_j^2 + \frac{(1-2\lambda)}{N} \sum a_j^2(p_j - \lambda) - \frac{1}{N} \sum a_j^2(p_j - \lambda)^2$$
$$- \frac{(1-2\lambda)^2}{\lambda(1-\lambda)N^2} \{\sum a_j p_j\}^2 + O(M_1^{\frac{3}{4}}) ,$$

$$\kappa_{3,0}(p) = \lambda(1-\lambda)(1-2\lambda) + O(M_1^{1/2} + M_2)\,,$$
$$\kappa_{3,1}(p) = -\frac{2\lambda(1-\lambda)}{N} \sum a_j p_j + O(M_1^{1/2})\,,$$
$$\kappa_{3,2}(p) = \frac{\lambda(1-\lambda)(1-2\lambda)}{N} \sum a_j^2 + O(M_1^{1/4})\,,$$
$$\kappa_{3,3}(p) = \frac{\lambda(1-\lambda)(1-2\lambda)}{N} \sum a_j^3 + \frac{(1-6\lambda+6\lambda^2)}{N} \sum a_j^3(p_j - \lambda)$$
$$- \frac{3(1-2\lambda)^2}{N^2} \sum a_j^2 \sum a_j p_j + O(N^{1/4}M_1^{1/2}) \quad (= O(1))\,,$$
$$\kappa_{4,0}(p) = \lambda(1-\lambda)(1-6\lambda+6\lambda^2) + O(M_1^{1/4})\,,$$
$$\kappa_{4,1}(p) = O(M_1^{1/4})\,,$$
$$\kappa_{4,2}(p) = \frac{\lambda(1-\lambda)(1-6\lambda+6\lambda)^2}{N} \sum a_j^2 + O(M_1^{1/4})\,,$$
$$\kappa_{4,3}(p) = O(1)\,,$$
$$\kappa_{4,4}(p) = \frac{\lambda(1-\lambda)(1-6\lambda+6\lambda^2)}{N} \sum a_j^4 + O(N^{1/4}M_1^{1/4}) \quad (= O(1))\,.$$

To illustrate computations involved we present the argument for $\kappa_{3,3}$. By (2.21), the result for $\bar{a}(p)$ and the fact that $0 \leqq M_1 \leqq 1$, we have

$$\begin{aligned}\kappa_{3,3}(p) &= N^{-1} \sum p_j(1-p_j)(1-2p_j)a_j^3 \\ &\quad - 3N^{-1}\bar{a}(p) \sum p_j(1-p_j)(1-2p_j)a_j^2 + O(M_1^{1/2}) \\ &= N^{-1}\lambda(1-\lambda)(1-2\lambda) \sum a_j^3 + N^{-1}(1-6\lambda+6\lambda^2) \sum a_j^3(p_j-\lambda) \\ &\quad - 3N^{-2}(1-2\lambda)^2 \sum a_j^2 \sum a_j p_j \\ &\quad + O(M_1^{1/2} + N^{-1} \sum |a_j|^3(p_j-\lambda)^2 + N^{-1}M_1^{1/4} \sum a_j^2|p_j-\lambda|)\,.\end{aligned}$$

Hölder's inequality and (2.21) imply that

$$N^{-1}M_1^{1/4} \sum a_j^2|p_j-\lambda| \leqq N^{-1}M_1^{1/4}(\sum |a_j|^{8/3})^{3/4}(NM_1)^{1/4} = O(M_1^{1/2})\,,$$
$$\begin{aligned}N^{-1} \sum |a_j|^3(p_j-\lambda)^2 &\leqq N^{-1}(CN)^{1/4} \sum a_j^2(p_j-\lambda)^2 \leqq C^{1/4}N^{-3/4}(NM_1 \sum a_j^4)^{1/2} \\ &= O(N^{1/4}M^{1/2})\,.\end{aligned}$$

As $\bar{a}(p)$ is bounded, $\kappa_{3,3}(p)$ is obviously also $O(1)$. Note that the atypical order of the remainder $O(N^{1/4}M_1^{1/2})$ originates from the term $O(N^{-1} \sum |a_j|^3(p_j-\lambda)^2)$ where we have to sacrifice a factor $O(N^{-1/4})$ in order to apply Hölder's inequality and (2.21). The same thing occurs for $\kappa_{4,4}(p)$.

For $\mu_k(p)$ defined by (2.28) we find

$$\mu_1(p) = \frac{1}{\lambda(1-\lambda)N^{1/2}} \sum (p_j - \lambda) + O(M_1^{1/2} + N^{1/2}M_2^2)\,,$$
$$\mu_2(p) = \frac{1}{\lambda(1-\lambda)} + O(M_1^{1/2} + M_2 + NM_2^2)\,,$$

$$\mu_3(p) = \frac{3}{\lambda^2(1-\lambda)^2 N^{\frac{1}{2}}} \sum (p_j - \lambda) + O(M_1^{\frac{1}{2}} + NM_2^2)\,,$$

$$\mu_4(p) = \frac{3}{\lambda^2(1-\lambda)^2} + O(M_1^{\frac{1}{2}} + M_2 + NM_2^2)\,,$$

$$\mu_5(p) = O(N^{\frac{1}{2}}M_2)\,,$$

$$\mu_6(p) = \frac{15}{\lambda^3(1-\lambda)^3} + O(M_1^{\frac{1}{2}} + M_2 + NM_2^2)\,.$$

Straightforward but tedious calculation now yields

$$\begin{aligned}
&\textstyle\sum_{k=0}^{6} A_k(p)(it)^k \\
&= \left[1 + \frac{(1-2\lambda)}{2\lambda(1-\lambda)N} \sum (p_j - \lambda) - \frac{(1-\lambda+\lambda^2)}{12\lambda(1-\lambda)N}\right] \\
&\quad + \frac{\sum a_j p_j}{N^{\frac{1}{2}}}\, it - \frac{\sum a_j^2}{2N^2}[(1-2\lambda)\sum (p_j-\lambda) - \lambda(1-\lambda)](it)^2 \\
&\quad + \frac{1}{6N^{\frac{3}{2}}}\Big[\lambda(1-\lambda)(1-2\lambda)\sum a_j^3 + (1-6\lambda+6\lambda^2)\sum a_j^3(p_j-\lambda) \\
&\quad - \frac{3(1-2\lambda)^2}{N}\sum a_j^2 \sum a_j p_j\Big](it)^3 \\
&\quad + \frac{\lambda(1-\lambda)}{24N^2}\left[(1-6\lambda+6\lambda^2)\sum a_j^4 - \frac{3(1-2\lambda)^2}{N}\{\sum a_j^2\}^2\right](it)^4 \\
&\quad + \frac{\lambda^2(1-\lambda)^2(1-2\lambda)^2}{72N^3}\{\sum a_j^3\}^2(it)^6 + O((|t|^3+t^4)[N^{-\frac{5}{4}} + N^{-\frac{1}{4}}M_1^{\frac{1}{2}}] \\
&\quad + (1+t^6)[N^{-\frac{3}{2}} + N^{-\frac{1}{2}}M_1^{\frac{1}{2}} + N^{\frac{1}{2}}M_2^2])\,.
\end{aligned} \tag{3.7}$$

Next we expand the remaining factor in (2.30). Because both $\tau^2(p)$ and its leading term $\lambda(1-\lambda)N^{-1}\sum a_j^2$ are bounded away from zero, there exists $\beta > 0$ depending only on c, C and D, such that

$$\begin{aligned}
&\frac{(2\pi)^{\frac{1}{2}}}{\sigma(p)} \exp\left\{-\frac{\omega^2(p)}{2\sigma^2(p)} - \frac{\tau^2(p)t^2}{2} - i\omega(p)\bar{a}(p)t\right\} \\
&= \left[\frac{2\pi}{\lambda(1-\lambda)}\right]^{\frac{1}{2}} \exp\left\{-\frac{\lambda(1-\lambda)}{2N}\sum a_j^2 t^2\right\} \\
&\quad \times \Big[1 - \frac{1}{2\lambda(1-\lambda)N}\{(1-2\lambda)\sum (p_j-\lambda) - \sum (p_j-\lambda)^2 + \{\sum (p_j-\lambda)\}^2\} \\
&\quad - \frac{(1-2\lambda)}{\lambda(1-\lambda)N^{\frac{3}{2}}}\sum (p_j-\lambda)\sum a_j p_j (it) + \frac{1}{2N}\Big\{(1-2\lambda)\sum a_j^2(p_j-\lambda) \\
&\quad - \sum a_j^2(p_j-\lambda)^2 - \frac{(1-2\lambda)^2}{\lambda(1-\lambda)N}\{\sum a_j p_j\}^2\Big\}(it)^2 \\
&\quad + \frac{(1-2\lambda)^2}{8N^2}\{\sum a_j^2(p_j-\lambda)\}^2(it)^4\Big] \\
&\quad + O(\exp\{-\beta t^2\}[N^{-\frac{3}{2}} + N^{\frac{1}{2}}M_1 + N^{\frac{5}{2}}M_2^4])\,.
\end{aligned}$$

Multiplication by (3.7) yields (3.6). □

Here is our first unconditional expansion. Define

$$\rho(t) = E\exp\{itN^{-\frac{1}{2}}T\}, \tag{3.8}$$

$$\begin{aligned}\rho^*(t) = \exp\left\{-\frac{\lambda(1-\lambda)}{2N}\sum a_j^2 t^2\right\} E_H\Bigg[\exp\{itN^{-\frac{1}{2}}\sum a_j P_j\} \\ \times\left\{1 + \frac{1}{2\lambda(1-\lambda)N}(\sum (P_j-\lambda)^2 - \{\sum (P_j-\lambda)\}^2) \right. \\ \left. + \sum_{k=1}^{6} A_k^*(P)(it)^k\right\}\Bigg].\end{aligned} \tag{3.9}$$

LEMMA 3.3. *Suppose that* (3.1) *holds and that positive numbers* c, C, δ, δ' *and* ε *exist with* $\delta' < \min(\frac{1}{2}, \delta/2, c^2C^{-1}/4)$ *and such that* (2.62) *is satisfied and*

$$\frac{1}{N}\sum a_j^2 \geq c, \qquad \frac{1}{N}\sum a_j^4 \leq C, \tag{3.10}$$

$$P_H\left(\varepsilon \leq \frac{\lambda g(X_1)}{h(X_1)} \leq 1-\varepsilon\right) \geq 1-\delta'. \tag{3.11}$$

Then there exist positive numbers b, B, β_1 *and* β_2 *depending only on* c, C, δ, δ' *and* ε *such that for* $|t| \leq bN^{\frac{3}{2}}$,

$$|\rho(t) - \rho^*(t)| \leq B\Bigg[\exp\{-\beta_1 t^2\}(N^{-\frac{3}{2}} + N^{-\frac{3}{2}}|t|) \times\left\{1 + N^2 E_H\left(\frac{g(X_1)}{h(X_1)} - 1\right)^4\right\} + N^{-\beta_2 \log N}\Bigg]. \tag{3.12}$$

PROOF. In this proof we again use O symbols that are uniform for fixed c, C, δ, δ' and ε. Note that $E_H\{g(X_1)/h(X_1)\} = 1$, so that (3.11) and Markov's inequality ensure that $\min(\lambda, 1-\lambda) \geq \varepsilon(1-\delta')$.

Take a number $\delta'' \in (\delta', \min(\frac{1}{2}, \delta/2, c^2C^{-1}/4)$ and define the event E by

$$\begin{aligned}E &= \{\varepsilon \leq P_j \leq 1-\varepsilon \text{ for at least } (1-\delta'')N \text{ indices } j\} \\ &= \left\{\varepsilon \leq \frac{\lambda g(X_j)}{h(X_j)} \leq 1-\varepsilon \text{ for at least } (1-\delta'')N \text{ indices } j\right\}.\end{aligned}$$

Applying an exponential bound for binomial probabilities (cf. Okamoto (1958)) we find that (3.11) implies

$$P_H(E) \geq 1 - \exp\{-2N(\delta''-\delta')^2\}. \tag{3.13}$$

Because λ and $(1-\lambda)$ are bounded away from 0, the same is true for $N^{\frac{1}{2}}B_{N,n}(\lambda)$. Also, (2.10) and (2.11) imply that $|\nu(t,p)| \leq 2\pi N^{\frac{1}{2}}$ for all t and p. Hence application of Lemma 3.1 shows that

$$\rho(t) = \frac{E_H \nu(t,P)\exp\{itN^{-\frac{1}{2}}\sum a_j P_j\}\chi_E}{2\pi N^{\frac{1}{2}}B_{N,n}(\lambda)} + O(\exp\{-N(\delta''-\delta')^2\}), \tag{3.14}$$

where χ_E denotes the indicator of E.

Since $\delta'' < \delta/2$, (2.62) ensures the validity of (2.22) on the set E with δ replaced by $\delta - 2\delta''$. If $\sum'$ denotes summation over those indices j for which $P_j \notin [\varepsilon, 1-\varepsilon]$ and k denotes the number of these indices, then $k \leqq \delta'' N$ on E and as a result

$$\tau^2(P) \geqq \frac{\varepsilon(1-\varepsilon)}{N}\left[\sum_{j=1}^N (a_j - \bar a(P))^2 - \sum{}' (a_j - \bar a(P))^2\right]$$
$$\geqq \frac{\varepsilon(1-\varepsilon)}{N}\left[\sum_{j=1}^N a_j^2 + N\{\bar a(P)\}^2 - 2\sum{}' a_j^2 - 2k\{\bar a(P)\}^2\right]$$
$$\geqq \frac{\varepsilon(1-\varepsilon)}{N}\left[cN - 2\{k\sum{}' a_j^4\}^{\frac12}\right] \geqq \varepsilon(1-\varepsilon)[c - 2\{\delta'' C\}^{\frac12}] > 0$$

on E, because $\delta'' < \min(\frac12, c^2C^{-1}/4)$.

We have shown that on the set E, a and P satisfy the conditions on a and p in Lemmas 2.3 and 3.2. Combining (3.14), (2.31) and (3.6) we obtain

$$\begin{aligned}\rho(t) = {}& \frac{E_H \nu^*(t, P)\exp\{itN^{-\frac12}\sum a_j P_j\}\chi_E}{2\pi N^{\frac12} B_{N,n}(\lambda)} \\ & + O(N^{-\beta_2 \log N} + \exp\{-\beta_1 t^2\}[\{N^{-\frac32} + N^{-\frac32}|t|\} \\ & \times \{1 + NE_H \sum (P_j - \lambda)^4\} + N^{-\frac32}E_H\{\sum (P_j - \lambda)\}^4]) \end{aligned} \tag{3.15}$$

for $|t| \leqq bN^{\frac32}$, where b, β_1 and β_2 depend on c, C, δ, δ' and ε only.

Because of (3.13) and the fact that $\nu^*(t, p) = O(N)$, (3.15) remains valid if we delete χ_E. Using

$$2\pi N^{\frac12} B_{N,n}(\lambda) = \left[\frac{2\pi}{\lambda(1-\lambda)}\right]^{\frac12}\left(1 - \frac{1-\lambda+\lambda^2}{12\lambda(1-\lambda)N}\right) + O(N^{-2})$$

one easily verifies that in (3.15) the first term on the right may be replaced by $\rho^*(t)$ without changing the order of the remainder. Since

$$E_H \sum (P_j - \lambda)^4 = E_H \sum \left(\frac{\lambda g(X_i)}{h(X_i)} - \lambda\right)^4 = \lambda^4 N E_H \left(\frac{g(X_1)}{h(X_1)} - 1\right)^4,$$

$$E_H\{\sum (P_j - \lambda)\}^4 = \lambda^4 E_H\left\{\sum\left(\frac{g(X_i)}{h(X_i)} - 1\right)\right\}^4 \leqq 3\lambda^4 N^2 E_H\left(\frac{g(X_1)}{h(X_1)} - 1\right)^4,$$

the proof of the lemma is complete. □

Define

$$\pi_j = E_H P_j, \quad \pi = (\pi_1, \cdots, \pi_N). \tag{3.16}$$

In the remaining part of this section we obtain a further expansion for $\rho(t)$ and convert this expansion into one for the df of T. Although we still do not assume contiguity, we shall be guided in what terms we include in the remainder by the fact that under contiguous alternatives we expect $(P_j - \pi_j)$ to behave roughly like $O_{P_H}(N^{-1})$. Let

$$K(x) = \Phi(x) - \phi(x)\sum_{k=0}^5 \alpha_k H_k(x), \tag{3.17}$$

where Φ and ϕ denote the standard normal df and its density, the Hermite polynomials H_k are given by (2.43) and

$$
\begin{aligned}
\alpha_0 &= \frac{\sum a_j \pi_j}{\{\lambda(1-\lambda)\sum a_j^2\}^{\frac12} N}\,, \\
\alpha_1 &= \frac{\sigma_H^2(\sum a_j P_j) - \sum a_j^2 E_H(P_j-\lambda)^2 + (1-2\lambda)\sum a_j^2(\pi_j-\lambda)}{2\lambda(1-\lambda)\sum a_j^2} \\
&\quad - \frac{(1-2\lambda)^2\{\sum a_j\pi_j\}^2}{2\lambda^2(1-\lambda)^2 N \sum a_j^2} + \frac{1}{2N}\,, \\
\alpha_2 &= ([\lambda(1-\lambda)(1-2\lambda)\sum a_j^3 + (1-6\lambda+6\lambda^2)\sum a_j^3(\pi_j-\lambda) \\
&\quad - 3(1-2\lambda)^2 N^{-1}\sum a_j^2 \sum a_j\pi_j])/(6\{\lambda(1-\lambda)\sum a_j^2\}^{\frac32})\,, \\
\alpha_3 &= \big(\lambda(1-\lambda)(1-6\lambda+6\lambda^2)\sum a_j^4 - 3\lambda(1-\lambda)(1-2\lambda)^2 N^{-1}\{\sum a_j^2\}^2 \\
&\quad + 3(1-2\lambda)^2\{\sum a_j^2(\pi_j-\lambda)\}^2\big)/(24\{\lambda(1-\lambda)\sum a_j^2\}^2)\,, \\
\alpha_4 &= \frac{(1-2\lambda)^2\sum a_j^3\sum a_j^2(\pi_j-\lambda)}{12\{\lambda(1-\lambda)\}^{\frac32}\{\sum a_j^2\}^{\frac52}}\,, \\
\alpha_5 &= \frac{(1-2\lambda)^2\{\sum a_j^3\}^2}{72\lambda(1-\lambda)\{\sum a_j^2\}^3}\,.
\end{aligned} \tag{3.18}
$$

THEOREM 3.1. *Suppose that* (3.1) *holds and that positive numbers* c, C, δ *and* ε *exist such that* (3.10) *and* (2.62) *are satisfied and*

$$\varepsilon \leqq \lambda \leqq 1-\varepsilon\,. \tag{3.19}$$

Then there exists $B>0$ *depending only on* c, C, δ *and* ε *such that*

$$
\begin{aligned}
\sup_x \left| P\left(\frac{T}{\{\lambda(1-\lambda)\sum a_j^2\}^{\frac12}} \leqq x\right) - K\left(x - \frac{\sum a_j\pi_j}{\{\lambda(1-\lambda)\sum a_j^2\}^{\frac12}}\right)\right| \\
\leqq B\left\{N^{-\frac54} + N^{\frac34} E_H\left(\frac{g(X_1)}{h(X_1)} - 1\right)^4 + N^{-\frac12}[\sum\{E_H|P_j-\pi_j|^3\}^{\frac49}]^{\frac94}\right\}.
\end{aligned} \tag{3.20}
$$

PROOF. In this proof B_i and β_i denote appropriately chosen positive numbers depending only on c, C, δ and ε. We shall have to consider the rv

$$U = N^{-\frac12}\sum a_j(P_j - \pi_j) \tag{3.21}$$

and we note that

$$
\begin{aligned}
E_H|U|^3 &\leqq N^{-\frac32}[\sum |a_j|\{E_H|P_j-\pi_j|^3\}^{\frac13}]^3 \\
&\leqq C^{\frac34} N^{-\frac34}[\sum\{E_H|P_j-\pi_j|^3\}^{\frac49}]^{\frac94}\,.
\end{aligned} \tag{3.22}
$$

Since $\sup_x(1+|K(x)|) \leqq B_1(1+E_H U^2) \leqq B_1(2+E_H|U|^3)$ we may assume without loss of generality that $E_H|U|^3 \leqq 1$, because otherwise (3.20) is satisfied trivially for $B = 3B_1C^{\frac34}$. Hence $\sup_x(1+|K(x)|) \leqq 3B_1$ and similar bounds $|\alpha_k| \leqq B_2(1+E_H U^2) \leqq 3B_2$ and $\sup_x|K'(x)| \leqq 3B_3$ hold for $\alpha_0, \cdots, \alpha_5$ and for the derivative K' of K.

Take $\delta' = \min(\frac14, \delta/4, c^2C^{-1}/8)$. In view of $1+|K| \leqq 3B_1$ it is again no loss of generality to assume that $E_H(g(X_1)/h(X_1)-1)^4 \leqq \delta'\varepsilon^4/16$, because otherwise

(3.22) with $B = 48B_1/(\delta'\varepsilon^4)$ is trivially true. Hence by (3.19) and Markov's inequality

$$P_H\left(\tfrac{1}{2}\varepsilon \leqq \frac{\lambda g(X_1)}{h(X_1)} \leqq 1 - \tfrac{1}{2}\varepsilon\right) \geqq P_H\left(\left|\frac{g(X_1)}{h(X_1)} - 1\right| \leqq \tfrac{1}{2}\varepsilon\right) \geqq 1 - \delta',$$

so that the conditions of Lemma 3.3 are satisfied and (3.12) holds.

The proof hinges on the expansion

$$\exp\{itN^{-\frac{1}{2}} \textstyle\sum a_j P_j\} = \exp\{itN^{-\frac{1}{2}} \sum a_j \pi_j\}[1 + itU + \frac{1}{2}(itU)^2] + O(|tU|^3)$$

and its truncation to fewer terms. We apply this expansion to (3.9) and in the resulting expression we replace P by π wherever this is possible without giving rise to remainder terms that would be awkward to handle at this point. Using elementary inequalities to separate out and bound those parts of the remainder that depend on the $(P_j - \lambda)$ rather than on the $(P_j - \pi_j)$, we arrive at

$$(3.23)\qquad |\rho^*(t) - \bar\rho(t)| \leqq B_4|t| \exp\{-\beta_3 t^2\}\Big[N^{-\frac{3}{2}} + N^{\frac{1}{2}}E_H\left(\frac{g(X_1)}{h(X_1)} - 1\right)^4 + E_H|U|^3$$
$$+ N^{-1}E_H|U \textstyle\sum a_j^2(P_j - \pi_j)| + N^{-2}E_H\{\sum a_j^2(P_j - \pi_j)\}^2\Big],$$

$$(3.24)\qquad \bar\rho(t) = \exp\left\{itN^{-\frac{1}{2}} \textstyle\sum a_j\pi_j - t^2\dfrac{\lambda(1-\lambda)}{2N}\sum a_j^2\right\}$$
$$\times\left[1 + \textstyle\sum_{k=1}^{6} \alpha_{k-1}\left(\dfrac{\lambda(1-\lambda)\sum a_j^2}{N}\right)^{\frac{1}{2}k}(it)^k\right].$$

Because $\max |a_j| \leqq (CN)^{\frac{1}{2}}$ we find by the same reasoning as in (3.22),

$$N^{-1}E_H|U \textstyle\sum a_j^2(P_j - \pi_j)| + N^{-2}E_H\{\sum a_j^2(P_j - \pi_j)\}^2$$
$$\leqq B_5 N^{-\frac{5}{4}}E_H\{\textstyle\sum |a_j(P_j - \pi_j)|\}^2$$
$$\leqq B_5 N^{-\frac{5}{4}}[1 + E_H\{\textstyle\sum |a_j(P_j - \pi_j)|\}^3]$$
$$\leqq B_5 N^{-\frac{5}{4}} + B_6 N^{-\frac{1}{2}}[\textstyle\sum \{E_H|P_j - \pi_j|^3\}^{\frac{4}{9}}]^{\frac{9}{4}}.$$

Together with (3.22) this shows that (3.23) may be reduced to

$$(3.25)\qquad |\rho^*(t) - \bar\rho(t)| \leqq B_7|t| \exp\{-\beta_3 t^2\}\Big\{N^{-\frac{5}{4}} + N^{\frac{1}{2}}E_H\left(\frac{g(X_1)}{h(X_1)} - 1\right)^4$$
$$+ N^{-\frac{1}{2}}[\textstyle\sum \{E_H|P_j - \pi_j|^3\}^{\frac{4}{9}}]^{\frac{9}{4}}\Big\}.$$

As $\alpha_0, \cdots, \alpha_5$ are bounded and $N^{-\frac{1}{2}}|\sum' a_j\pi_j| \leqq C^{\frac{1}{2}}N^{\frac{1}{2}}$, we have $|\bar\rho'(t)| \leqq B_8 N^{\frac{1}{2}}$ for all t. Since $|\rho'(t)| \leqq N^{-\frac{1}{2}}E|T| \leqq C^{\frac{1}{2}}N^{\frac{1}{2}}$ for all t and $\rho(0) = \bar\rho(0) = 1$,

$$(3.26)\qquad |\rho(t) - \bar\rho(t)| \leqq B_9 N^{\frac{1}{2}}|t| \qquad \text{for all } t.$$

Combining Lemma 3.3, (3.25) and (3.26) we find

$$\int_{-bN^{\frac{3}{2}}}^{bN^{\frac{3}{2}}} \left|\frac{\rho(t) - \bar\rho(t)}{t}\right| dt$$
$$(3.27)\qquad \leqq B_9 N^{-\frac{3}{2}} + \int_{N^{-2} \leqq |t| \leqq bN^{\frac{3}{2}}} \left|\frac{\rho(t) - \bar\rho(t)}{t}\right| dt$$
$$\leqq B_{10}\left\{N^{-\frac{5}{4}} + N^{\frac{3}{4}}E_H\left(\frac{g(X_1)}{h(X_1)} - 1\right)^4 + N^{-\frac{1}{2}}[\textstyle\sum \{E_H|P_j - \pi_j|^3\}^{\frac{4}{9}}]^{\frac{9}{4}}\right\}.$$

Now $\bar{\rho}(t)$ is the Fourier–Stieltjes transform of $K(\{N^{\frac{1}{2}}x - \sum a_j \pi_j\}\{\lambda(1-\lambda) \sum a_j^2\}^{-\frac{1}{2}})$ as a function of x. This is a function of bounded variation assuming the values 0 and 1 at $-\infty$ and $+\infty$ and having a derivative that is bounded by $3B_3 c^{-\frac{1}{2}}\{\varepsilon(1-\varepsilon)\}^{-1}$ in absolute value. It follows from the smoothing lemma (Esseen (1945)) that

$$\sup_x \left| P(N^{-\frac{1}{2}}T \leq x) - K\left(\frac{N^{\frac{1}{2}}x - \sum a_j \pi_j}{\{\lambda(1-\lambda) \sum a_j^2\}^{\frac{1}{2}}}\right) \right|$$

is bounded above by the right-hand side of (3.20). A change of scale completes the proof. □

Theorem 3.1 provides the basic expansion for the distribution of T under contiguous alternatives. Only first and second moments of functions of order statistics remain to be determined. In Section 4 we shall be concerned with a further simplification of the expansion and a precise evaluation of the order of the remainder. With regard to this remainder we are in a seemingly less favorable position than we were at the same stage in the one-sample problem (cf. ABZ (1976), Theorem 2.3), because the third remainder term in (3.20) is larger than the corresponding term in the one-sample case by a factor $N^{\frac{1}{2}}$. This is due to the appearance of the remainder term $N^{-1}E_H|U \sum a_j^2(P_j - \pi_j)|$ that does not occur for the one-sample statistic. It will turn out, however, that we shall need only a slightly stronger condition than before to show that the remainder is still $O(N^{-\frac{5}{4}})$.

The conditions of Theorem 3.1 concern only the sample ratio λ and the scores a. There are no assumptions about the underlying densities f and g but this is merely a trick; obviously something like contiguity is needed to make the expansion meaningful in the sense that the remainder is at all small. With regard to the conditions on the scores, (3.10) acts as a safeguard against too rapid growth and (2.62) ensures that the a_j do not cluster too much around too few points, thus preventing a too pronounced lattice character of the distribution of T, as was pointed out in ABZ (1976). It was also noted there that in the important case of exact scores $a_j = EJ(U_{j:N})$, with $U_{1:N} < U_{2:N} < \cdots < U_{N:N}$ order statistics from the uniform distribution on (0, 1), both (3.10) and (2.62) will be satisfied for all N with fixed c, C and δ if J is a continuously differentiable, nonconstant function on (0, 1) with $\int J^4 < \infty$. The same is true for approximate scores $a_j = J(j/(N+1))$ provided that J is monotone near 0 and 1.

4. Contiguous location alternatives. The analysis in this section will be carried out for contiguous location alternatives rather than for contiguous alternatives in general. The general case can be treated in much the same way as the location case, but the conditions as well as the results become more involved.

We recall some assumptions and notation from Section 3 of ABZ (1976). Let F be a df with a density f that is positive on R^1 and four times differentiable with derivatives $f^{(i)}$, $i = 1, \cdots, 4$. Define

$$\psi_i = \frac{f^{(i)}}{f}, \qquad i = 1, \cdots, 4, \tag{4.1}$$

and suppose that positive numbers ε' and C' exist such that for

$$(4.2)\qquad m_1 = 6\,, \quad m_2 = 3\,, \quad m_3 = \tfrac{4}{3}\,, \quad m_4 = 1\,,$$
$$\sup\{\textstyle\int_{-\infty}^{\infty} |\psi_i(x+y)|^{m_i} f(x)\,dx : |y| \leq \varepsilon'\} \leq C'\,, \qquad i = 1, \cdots, 4\,.$$

So far, we have studied the distribution of T under the assumption that $X_1, \cdots, X_N$ are independent, $X_1, \cdots, X_m$ having common df F and $X_{m+1}, \cdots, X_N$ having df G. We now add the assumptions that

$$(4.3)\qquad G(x) = F(x - \theta)$$

for all x and that

$$(4.4)\qquad 0 \leq \theta \leq DN^{-\frac{1}{2}}$$

for some $D > 0$. Probabilities under this particular model will still be denoted by P. Note that (4.2), (4.3) and (4.4) together imply contiguity.

In Section 3 we also introduced an auxiliary model where $X_1, \cdots, X_N$ are supposed to be i.i.d. with common df $H = (1 - \lambda)F + \lambda G$. In view of (4.3) this common df now becomes $H(x) = (1 - \lambda)F(x) + \lambda F(x - \theta)$. Probabilities, expectations and variances under this model will be denoted by P_H, E_H and σ_H^2 as before. Similarly, P_F, E_F and σ_F^2 will indicate probabilities, expectations and variances under a third model where $X_1, \cdots, X_N$ are i.i.d. with common df F. Note that for $\theta = 0$ these three models coincide.

Define

$$(4.5)\qquad \tilde{K}(x) = \Phi(x) - \phi(x) \textstyle\sum_{k=0}^{5} \tilde{\alpha}_k H_k(x)\,,$$

where

$$\tilde{\alpha}_0 = \frac{1}{6}\left(\frac{\lambda(1-\lambda)}{\sum a_j^2}\right)^{\frac{1}{2}} [3(1-2\lambda)\theta^2 \textstyle\sum a_j E_F \psi_2(Z_j) - 6N^{-1}\theta \sum a_j E_F \psi_1(Z_j)$$
$$- \theta^3 \textstyle\sum a_j E_F\{(1 - 3\lambda + 3\lambda^2)\psi_3(Z_j) - 6\lambda(1-\lambda)\psi_1(Z_j)\psi_2(Z_j)$$
$$+ 3\lambda(1-\lambda)\psi_1^3(Z_j)\}]\,,$$

$$\tilde{\alpha}_1 = \frac{1}{8\sum a_j^2}[-4(1-2\lambda)\theta \textstyle\sum a_j^2 E_F\psi_1(Z_j) + 2(1-2\lambda)^2\theta^2 \sum a_j^2 E_F \psi_2(Z_j)$$
$$- 4\lambda(1-\lambda)\theta^2 \textstyle\sum a_j^2 E_F \psi_1^2(Z_j) + 4\lambda(1-\lambda)\theta^2\sigma_F^2(\sum a_j \psi_1(Z_j))$$
$$- 4(1-2\lambda)^2 N^{-1}\theta^2\{\textstyle\sum a_j E_F \psi_1(Z_j)\}^2$$
$$+ \lambda(1-\lambda)(1-2\lambda)^2\theta^4\{\textstyle\sum a_j E_F \psi_2(Z_j)\}^2] + \frac{1}{2N}\,,$$

$$(4.6)\qquad \tilde{\alpha}_2 = \frac{1}{12\{\lambda(1-\lambda)\}^{\frac{1}{2}}(\sum a_j^2)^{\frac{3}{2}}}[2(1-2\lambda)\textstyle\sum a_j^3 - 2(1 - 6\lambda + 6\lambda^2)$$
$$\times \theta \textstyle\sum a_j^3 E_F \psi_1(Z_j) + 6(1-2\lambda)^2 N^{-1}\theta \sum a_j^2 \sum a_j E_F \psi_1(Z_j)$$
$$- 3\lambda(1-\lambda)(1-2\lambda)^2\theta^3 \textstyle\sum a_j^2 E_F \psi_1(Z_j) \sum a_j E_F \psi_2(Z_j)]\,,$$

$$\tilde{\alpha}_3 = \frac{1}{24\lambda(1-\lambda)(\sum a_j^2)^2}\{(1 - 6\lambda + 6\lambda^2)\textstyle\sum a_j^4 + 3\lambda(1-\lambda)(1-2\lambda)^2\theta^2$$
$$\times \{\textstyle\sum a_j^2 E_F \psi_1(Z_j)\}^2 + 2\lambda(1-\lambda)(1-2\lambda)^2\theta^2 \sum a_j^3 \sum a_j E_F \psi_2(Z_j)]$$
$$- \frac{(1-2\lambda)^2}{8\lambda(1-\lambda)N}\,,$$

$$\tilde{\alpha}_4 = -\frac{(1-2\lambda)^2\theta \sum a_j^3 \sum a_j^2 E_F\psi_1(Z_j)}{12\{\lambda(1-\lambda)\}^{\frac{1}{2}}(\sum a_j^2)^{\frac{5}{2}}},$$

$$\tilde{\alpha}_5 = \frac{(1-2\lambda)^2(\sum a_j^3)^2}{72\lambda(1-\lambda)(\sum a_j^2)^3},$$

and let

$$(4.7)\qquad \eta = -\left(\frac{\lambda(1-\lambda)}{\sum a_j^2}\right)^{\frac{1}{2}} \theta \sum a_j E_F\psi_1(Z_j).$$

We shall show that $\tilde{K}(x-\eta)$ is an expansion for the df of $\{\lambda(1-\lambda)\sum a_j^2\}^{-\frac{1}{2}}T$. The expansion will be established in Theorem 4.1 and an evaluation of the order of the remainder will be given in Theorem 4.2.

Let $\pi(F, \theta)$ denote the power of the one-sided level α test based on T for the hypothesis $F = G$ against the alternative $G(x) = F(x-\theta)$. Suppose that

$$(4.8)\qquad \varepsilon'' \leq \alpha \leq 1 - \varepsilon'',$$

for some $\varepsilon'' > 0$. We shall prove that an expansion for $\pi(F, \theta)$ is given by

$$(4.9)\qquad \tilde{\pi}(F, \theta) = 1 - \Phi(u_\alpha - \eta) + \phi(u_\alpha - \eta) \textstyle\sum_{k=0}^{5} \tilde{\beta}_k H_k(u_\alpha - \eta),$$

where $u_\alpha = \Phi^{-1}(1-\alpha)$ is the upper α-point of the standard normal distribution and

$$\tilde{\beta}_0 = \tilde{\alpha}_0 - \frac{(1-2\lambda)\sum a_j^3}{6\{\lambda(1-\lambda)\}^{\frac{1}{2}}(\sum a_j^2)^{\frac{3}{2}}}(u_\alpha^2 - 1) + 2\tilde{\alpha}_5(2u_\alpha^3 - 5u_\alpha) - \frac{u_\alpha}{2N}$$
$$- \left\{\frac{(1-6\lambda+6\lambda^2)\sum a_j^4}{24\lambda(1-\lambda)(\sum a_j^2)^2} - \frac{(1-2\lambda)^2}{8\lambda(1-\lambda)N}\right\}(u_\alpha^3 - 3u_\alpha),$$

$$(4.10)\qquad \tilde{\beta}_1 = \tilde{\alpha}_1 + \tilde{\alpha}_5(u_\alpha^2 - 1)^2 - \frac{(1-2\lambda)^2}{12(\sum a_j^2)^2}\theta^2 \sum a_j^3 \sum a_j E_F\psi_2(Z_j)(u_\alpha^2 - 1),$$

$$\tilde{\beta}_2 = \tilde{\alpha}_2 - \tilde{\alpha}_4(u_\alpha^2 - 1),$$
$$\tilde{\beta}_3 = \tilde{\alpha}_3 - 2\tilde{\alpha}_5(u_\alpha^2 - 1),$$
$$\tilde{\beta}_k = \tilde{\alpha}_k \quad \text{for } k = 4, 5.$$

THEOREM 4.1. *Suppose that* (3.1) *and* (4.3) *hold and that positive numbers c, C, C′, D, δ, ε and ε′ exist such that* (3.10), (2.62), (3.19), (4.2) *and* (4.4) *are satisfied. Define*

$$(4.11)\qquad M = N^{-\frac{5}{4}} + N^{-\frac{1}{2}}\theta^3[\sum \{E_F|\psi_1(Z_j) - E_F\psi_1(Z_j)|^3\}^{\frac{4}{3}}]^{\frac{3}{4}} + N^{-\frac{3}{4}}\theta^3[\sum \{E_F(\psi_2(Z_j) - E_F\psi_2(Z_j))^2\}^{\frac{2}{3}}]^{\frac{3}{4}}.$$

Then there exists $B > 0$ depending only on c, C, C′, D, δ, ε and ε′ such that

$$(4.12)\qquad \sup_x \left| P\left(\frac{T}{\{\lambda(1-\lambda)\sum a_j^2\}^{\frac{1}{2}}} \leq x\right) - \tilde{K}(x-\eta)\right| \leq BM.$$

If, in addition, (4.8) *is satisfied there exists $B' > 0$ depending only on c, C, C′, D, δ, ε, ε′ and ε″ such that*

$$(4.13)\qquad |\pi(F, \theta) - \tilde{\pi}(F, \theta)| \leq B'M.$$

PROOF. The proof of (4.12) hinges on Taylor expansion with respect to θ of the moments under P_H of functions of $P = (P_1, \cdots, P_N)$ occurring in expansion (3.20). Since both H and P depend on θ the argument is highly technical and laborious and it is therefore given in the appendix. Theorem 3.1, Corollary A.1, (A.12) and (A.13) immediately yield (4.12).

The one-sided level α test based on T rejects the hypothesis if $T\{\lambda(1-\lambda)\sum a_j^2\}^{-\frac{1}{2}} \geqq \xi_\alpha$ with possible randomization if equality occurs. Using (4.12) for $\theta = 0$ (or Corollary 2.1), (3.10), (3.19) and (4.8) we easily show that

$$\xi_\alpha = u_\alpha + \frac{(1-2\lambda)\sum a_j^3}{6\{\lambda(1-\lambda)\}^{\frac{1}{2}}(\sum a_j^2)^{\frac{3}{2}}}(u_\alpha^2 - 1) - 2\tilde{\alpha}_5(2u_\alpha^3 - 5u_\alpha) + \frac{u_\alpha}{2N} + \left\{\frac{(1-6\lambda+6\lambda^2)\sum a_j^4}{24\lambda(1-\lambda)(\sum a_j^2)^2} - \frac{(1-2\lambda)^2}{8\lambda(1-\lambda)N}\right\}(u_\alpha^3 - 3u_\alpha) + O(N^{-\frac{3}{2}}), \tag{4.14}$$

where, in this proof, $O(x)$ denotes a quantity bounded by $B_1|x|$ with B_1 depending only on c, C, C', D, δ, ε, ε' and ε''. Because of (4.12),

$$\pi(F, \theta) = 1 - \tilde{K}(\xi_\alpha - \eta) + O(M).$$

Using (4.14), (4.8) and the bounds provided by Corollary A.1, we now expand $\tilde{K}(\xi_\alpha - \eta)$ about the point $(u_\alpha - \eta)$ and arrive at (4.13). □

Define

$$\Psi_i(t) = \psi_i(F^{-1}(t)) = \frac{f^{(i)}(F^{-1}(t))}{f(F^{-1}(t))}, \qquad i = 1, \cdots, 4. \tag{4.15}$$

THEOREM 4.2. *Let M be defined by* (4.11) *and suppose that positive numbers D, C and δ exist such that* (4.4) *is satisfied and that $|\Psi_1'(t)| \leqq C\{t(1-t)\}^{-\frac{5}{4}+\delta}$ and $|\Psi_2'(t)| \leqq C\{t(1-t)\}^{-\frac{3}{2}+\delta}$. Then there exist $B > 0$ depending only D, C and δ such that*

$$M \leqq BN^{-\frac{5}{4}}.$$

PROOF. The proof is similar to that of Corollary A2.1 in ABZ (1976). To deal with the second term of M we take $h = \Psi_1$ and replace $\frac{4}{3}$ by $\frac{5}{4}$ in the proof of that corollary. For the third term of M we take $h = \Psi_2$, replace $\frac{4}{3}$ by $\frac{3}{2}$, appeal to condition R_2 instead of R_3 and otherwise proceed as in the proof of Corollary A2.1 of ABZ (1976). □

5. Exact and approximate scores. A further simplification of the expansions in Section 4 may obtained if we make certain smoothness assumptions about the scores a_j. Consider a continuous function J on (0, 1) and let $U_{1:N} < U_{2:N} < \cdots < U_{N:N}$ denote order statistics of a sample of size N from the uniform distribution on (0, 1). For $N = 1, 2, \cdots$ we define the exact scores generated by J by

$$a_j = a_{j,N} = EJ(U_{j:N}), \qquad j = 1, \cdots, N, \tag{5.1}$$

and the approximate scores generated by J by

$$a_j = a_{j,N} = J\left(\frac{j}{N+1}\right), \qquad j = 1, \cdots, N. \tag{5.2}$$

For exact scores and general J Theorem 5.1 will provide expansions for the df of T under contiguous location alternatives of type F and for the power of the rank test against these alternatives. In Theorem 5.2 we consider the special case $J = -\Psi_1$, with Ψ_1 as in (4.15), for exact as well as approximate scores. Note that the exact scores generated by $-\Psi_1$ define the locally most powerful rank test.

As in Section 4 of ABZ (1976) it is now no longer feasible to keep the order of the remainder in our expansions down to $O(N^{-\frac{3}{2}})$ and we shall be content with $o(N^{-1})$. Also as in ABZ (1976) we shall formulate the results in this section for a fixed scores generating function J and a fixed df F, leaving the construction of uniformity classes to the reader.

DEFINITION 5.1. $\mathscr{J}$ is the class of functions J on $(0, 1)$ that are twice continuously differentiable and nonconstant on $(0, 1)$ and satisfy

$$\int_0^1 J(t)\,dt = 0\,, \tag{5.3}$$

$$\lim_{t\to 0,1} \{t(1-t)\}^{\frac{7}{6}} J'(t) = 0\,, \tag{5.4}$$

$$\limsup_{t\to 0,1} t(1-t)\left|\frac{J''(t)}{J'(t)}\right| < \tfrac{3}{2}\,. \tag{5.5}$$

$\mathscr{F}$ is the class of df's F on R^1 with positive and four times differentiable densities f and such that, for $\psi_i = f^{(i)}/f$, $\Psi_i = \psi_i(F^{-1})$, $m_1 = 6$, $m_2 = 3$, $m_3 = \frac{4}{3}$, $m_4 = 1$,

$$\limsup_{y\to 0} \int_{-\infty}^{\infty} |\psi_i(x+y)|^{m_i} f(x)\,dx < \infty\,, \qquad i = 1, \cdots, 4\,, \tag{5.6}$$

$$\limsup_{t\to 0,1} t(1-t)\left|\frac{\Psi_1''(t)}{\Psi'(t)}\right| < \tfrac{3}{2}\,. \tag{5.7}$$

Note that one can argue as in the proof of Corollary A2.1 of ABZ (1976) to show that, in conjunction with (5.5), condition (5.4) is weaker than the assumption $\int J^6(t)\,dt < \infty$. Define

$$\begin{aligned}
\bar{\alpha}_0 = \tfrac{1}{6}\left(\frac{\lambda(1-\lambda)}{N\int J^2(t)\,dt}\right)^{\frac{1}{2}} &[3(1-2\lambda)N\theta^2 \int J(t)\Psi_2(t)\,dt - 6\theta \int J(t)\Psi_1(t)\,dt \\
&- N\theta^3 \int J(t)\{(1-3\lambda+3\lambda^2)\Psi_3(t) - 6\lambda(1-\lambda)\Psi_1(t)\Psi_2(t) \\
&+ 3\lambda(1-\lambda)\Psi_1^3(t)\}\,dt]\,,
\end{aligned}$$

$$\begin{aligned}
\bar{\alpha}_1 = \frac{1}{8\int J^2(t)\,dt} &[-4(1-2\lambda)\theta \int J^2(t)\Psi_1(t)\,dt \\
&+ 2(1-2\lambda)^2\theta^2 \int J^2(t)\Psi_2(t)\,dt - 4\lambda(1-\lambda)\theta^2 \int J^2(t)\Psi_1^2(t)\,dt \\
&+ 4\lambda(1-\lambda)\theta^2 \iint J(s)J(t)\Psi_1'(s)\Psi_1'(t)[s\wedge t - st]\,ds\,dt \\
&- 4(1-2\lambda)^2\theta^2\{\int J(t)\Psi_1(t)\,dt\}^2 \\
&+ \lambda(1-\lambda)(1-2\lambda)^2 N\theta^4\{\int J(t)\Psi_2(t)\,dt\}^2] + \frac{1}{2N}\,,
\end{aligned} \tag{5.8}$$

$$\bar{\alpha}_2 = \frac{1}{12\{\lambda(1-\lambda)N\}^{\frac{1}{2}}\{\int J^2(t)\,dt\}^{\frac{3}{2}}}[2(1-2\lambda)\int J^3(t)\,dt$$
$$-2(1-6\lambda+6\lambda^2)\theta\int J^3(t)\Psi_1(t)\,dt$$
$$+6(1-2\lambda)^2\theta\int J^2(t)\,dt\int J(t)\Psi_1(t)\,dt$$
$$-3\lambda(1-\lambda)(1-2\lambda)^2N\theta^3\int J^2(t)\Psi_1(t)\,dt\int J(t)\Psi_2(t)\,dt],$$

$$\bar{\alpha}_3 = \frac{1}{24\lambda(1-\lambda)N\{\int J^2(t)\,dt\}^2}[(1-6\lambda+6\lambda^2)\int J^4(t)\,dt$$
$$+3\lambda(1-\lambda)(1-2\lambda)^2N\theta^2\{\int J^2(t)\Psi_1(t)\,dt\}^2$$
$$+2\lambda(1-\lambda)(1-2\lambda)^2N\theta^2\int J^3(t)\,dt\int J(t)\Psi_2(t)\,dt] - \frac{(1-2\lambda)^2}{8\lambda(1-\lambda)N},$$

$$\bar{\alpha}_4 = -\frac{(1-2\lambda)^2\theta}{12\{\lambda(1-\lambda)N\}^{\frac{1}{2}}}\frac{\int J^3(t)\,dt\int J^2(t)\Psi_1(t)\,dt}{\{\int J^2(t)\,dt\}^{\frac{5}{2}}},$$

$$\bar{\alpha}_5 = \frac{(1-2\lambda)^2}{72\lambda(1-\lambda)N}\frac{\{\int J^3(t)\,dt\}^2}{\{\int J^2(t)\,dt\}^3},$$

$$\bar{K}_1(x) = \Phi(x) - \phi(x)\Big[\textstyle\sum_{k=0}^5 \bar{\alpha}_k H_k(x)$$
$$+\tfrac{1}{2}\left(\frac{\lambda(1-\lambda)}{N\int J^2(t)\,dt}\right)^{\frac{1}{2}}\theta\Big\{2\textstyle\sum_{j=1}^N \mathrm{Cov}\,(J(U_{j:N}),\Psi_1(U_{j:N})) \tag{5.9}$$
$$-\frac{\int J(t)\Psi_1(t)\,dt}{\int J^2(t)\,dt}\textstyle\sum_{j=1}^N \sigma^2(J(U_{j:N}))\Big\}\Big],$$

$$\bar{K}_2(x) = \Phi(x) - \phi(x)\Big[\textstyle\sum_{k=0}^5 \bar{\alpha}_k H_k(x)$$
$$+\tfrac{1}{2}\left(\frac{\lambda(1-\lambda)}{N\int J^2(t)\,dt}\right)^{\frac{1}{2}}\theta\Big\{2\int_{N^{-1}}^{1-N^{-1}} J'(t)\Psi_1'(t)t(1-t)\,dt \tag{5.10}$$
$$-\frac{\int J(t)\Psi_1(t)\,dt}{\int J^2(t)\,dt}\int_{N^{-1}}^{1-N^{-1}}(J'(t))^2t(1-t)\,dt\Big\}\Big],$$

$$\bar{\eta} = -\left(\frac{\lambda(1-\lambda)N}{\int J^2(t)\,dt}\right)^{\frac{1}{2}}\theta\int J(t)\Psi_1(t)\,dt, \tag{5.11}$$

where all integrals are over (0, 1) unless otherwise indicated. We shall show that $\bar{K}_1(x-\bar{\eta})$ and $\bar{K}_2(x-\bar{\eta})$ are expansions for the df of $\{\lambda(1-\lambda)\sum a_j^2\}^{-\frac{1}{2}}T$ for exact scores. Furthermore let

$$\bar{\beta}_0 = \bar{\alpha}_0 - \frac{(1-2\lambda)}{6\{\lambda(1-\lambda)N\}^{\frac{1}{2}}}\frac{\int J^3(t)\,dt}{\{\int J^2(t)\,dt\}^{\frac{3}{2}}}(u_\alpha^2-1) + 2\bar{\alpha}_5(2u_\alpha^3-5u_\alpha) - \frac{u_\alpha}{2N}$$
$$-\left\{\frac{(1-6\lambda+6\lambda^2)}{24\lambda(1-\lambda)N}\frac{\int J^4(t)\,dt}{\{\int J^2(t)\,dt\}^2} - \frac{(1-2\lambda)^2}{8\lambda(1-\lambda)N}\right\}(u_\alpha^3-3u_\alpha),$$

$$\bar{\beta}_1 = \bar{\alpha}_1 + \bar{\alpha}_5(u_\alpha^2-1)^2 - \frac{(1-2\lambda)^2\theta^2}{12}\frac{\int J^3(t)\,dt\int J(t)\Psi_2(t)\,dt}{\{\int J^2(t)\,dt\}^2}(u_\alpha^2-1), \tag{5.12}$$

$$\bar{\beta}_2 = \bar{\alpha}_2 - \bar{\alpha}_4(u_\alpha^2 - 1)\,,$$
$$\bar{\beta}_3 = \bar{\alpha}_3 - 2\bar{\alpha}_5(u_\alpha^2 - 1)\,,$$
$$\bar{\beta}_k = \bar{\alpha}_k \quad \text{for} \quad k = 4, 5\,,$$

$$\bar{\pi}_i(F, \theta) = 1 - \bar{K}_i(u_\alpha - \bar{\eta}) + \phi(u_\alpha - \bar{\eta}) \textstyle\sum_{k=0}^{5} (\bar{\beta}_k - \bar{\alpha}_k) H_k(u_\alpha - \bar{\eta})\,, \qquad i = 1, 2\,, \tag{5.13}$$

i.e., $\bar{\pi}_i(F, \theta)$ equals $1 - \bar{K}_i(u_\alpha - \bar{\eta})$ with $\bar{\alpha}_k$ replaced by $\bar{\beta}_k$, $k = 0, \cdots, 3$.

THEOREM 5.1. *Let* $F \in \mathscr{F}$, $J \in \mathscr{J}$, $a_j = EJ(U_{j:N})$ *for* $j = 1, \cdots, N$, $G(x) = F(x - \theta)$, $0 \leqq \theta \leqq DN^{-\frac{1}{2}}$, $\varepsilon \leqq \lambda \leqq 1 - \varepsilon$ *and* $\varepsilon' \leqq \alpha \leqq 1 - \varepsilon'$ *for positive* D, ε *and* ε'. *Then, for every fixed* F, J, D, ε *and* ε', *there exist positive numbers* B, $\delta_1, \delta_2, \cdots$ *such that* $\lim_{N\to\infty} \delta_N = 0$ *and for every* N

$$\sup_x \left| P\left(\frac{T}{\{\lambda(1-\lambda)\sum a_j^2\}^{\frac{1}{2}}} \leqq x\right) - \bar{K}_1(x - \bar{\eta})\right| \leqq \delta_N N^{-1}\,, \tag{5.14}$$

$$\sup_x \left| P\left(\frac{T}{\{\lambda(1-\lambda)\sum a_j^2\}^{\frac{1}{2}}} \leqq x\right) - \bar{K}_2(x - \bar{\eta})\right| \leqq \delta_N N^{-1} + BN^{-\frac{3}{2}} \int_{N^{-1}}^{1-N^{-1}} |J'(t)|(|J'(t)| + |\Psi_1'(t)|)\{t(1-t)\}^{\frac{1}{2}}\, dt\,, \tag{5.15}$$

$$|\pi(F, \theta) - \bar{\pi}_1(F, \theta)| \leqq \delta_N N^{-1}\,, \tag{5.16}$$

$$|\pi(F, \theta) - \bar{\pi}_2(F, \theta)| \leqq \delta_N N^{-1} + BN^{-\frac{3}{2}} \int_{N^{-1}}^{1-N^{-1}} |J'(t)|(|J'(t)| + |\Psi_1'(t)|)\{t(1-t)\}^{\frac{1}{2}}\, dt\,. \tag{5.17}$$

PROOF. In the first part of the proof we shall not need requirement (5.4) but only the weaker assumption $\int J^4(t)\, dt < \infty$. We proceed as in the proof of Theorem 4.1 in ABZ (1976), drawing heavily on the results in Appendix 2 of ABZ (1976). Note that these results remain valid in the present context even though the definition of the functions Ψ_i is slightly different here. Throughout the proof we shall make use of O and o symbols that are uniform for fixed F, J, D, ε and ε'.

Because $\sum a_j = N \int J(t)\, dt = 0$ and in view of the remark made at the end of Section 3, the assumptions of Theorem 4.1 are satisfied. The proof of Corollary A2.1 of ABZ (1976) shows that (5.6) and (5.7) imply that

$$\Psi_1'(t) = o(\{t(1-t)\}^{-\frac{7}{6}}) \qquad \text{for} \quad t \to 0, 1\,. \tag{5.18}$$

Hence, because of (5.7), $\Psi_1''(t) = o(\{t(1-t)\}^{-\frac{13}{6}})$ and $\Psi_1(t) = o(\{t(1-t)\}^{-\frac{1}{6}})$ for $t \to 0, 1$. Since $f(F^{-1})$ has a summable derivative Ψ_1 on $(0, 1)$, $f(F^{-1})$ must have limits at 0 and 1; as f is positive on R^1, these limits must be equal to 0. It follows that $f(F^{-1}(t)) = o(\{t(1-t)\}^{\frac{5}{6}})$ for $t \to 0, 1$. Combining these facts with the identity $\Psi_2'(t) = \Psi_1''(t) f(F^{-1}(t)) + 3\Psi_1(t)\Psi_1'(t)$, we find that

$$\Psi_2'(t) = o(\{t(1-t)\}^{-\frac{4}{3}}) \qquad \text{for} \quad t \to 0, 1\,. \tag{5.19}$$

Thus the assumptions of Theorem 4.2 are also satisfied and we can take the expansions of Section 4 as a starting point for proving Theorem 5.1.

In $\tilde{\alpha}_0, \cdots, \tilde{\alpha}_5, \tilde{\beta}_0, \cdots, \tilde{\beta}_5$ defined by (4.6) and (4.10) we may replace E_F, σ_F^2 and $\psi_i(Z_j)$ by E, σ^2 and $\psi_i(F^{-1}(U_{j:N})) = \Psi_i(U_{j:N})$ without changing anything. Next, arguing as in Corollary A2.2 of ABZ (1976), we see that for all sums of the form $\sum a_j^k$ and $\sum a_j^k Eh(U_{j:N})$ occurring in $\tilde{\alpha}_0, \cdots, \tilde{\alpha}_5, \tilde{\beta}_0, \cdots, \tilde{\beta}_5$ we may write

$$\frac{1}{N} \sum a_j^k = \int J^k(t)\, dt + o(1)\,, \tag{5.20}$$

$$\frac{1}{N} \sum a_j^k Eh(U_{j:N}) = \int J^k(t) h(t)\, dt + o(1)\,, \tag{5.21}$$

and also

$$\frac{1}{N} \sigma^2\left(\sum a_j \Psi_1(U_{j:N})\right) = \iint J(s)J(t)\Psi_1'(s)\Psi_1'(t)[s \wedge t - st]\, ds\, dt + o(1)\,. \tag{5.22}$$

We note that $\bar{\alpha}_0, \cdots, \bar{\alpha}_5, \bar{\beta}_0, \cdots, \bar{\beta}_5$ are obtained from $\tilde{\alpha}_0, \cdots, \tilde{\alpha}_5, \tilde{\beta}_0, \cdots, \tilde{\beta}_5$ by replacing every expression of the form (5.20)—(5.22) by the corresponding integral on the right in (5.20)—(5.22). Since $\int J^2(t)\, dt > 0$, we know that for those terms in $\tilde{\alpha}_0, \cdots, \tilde{\alpha}_5, \tilde{\beta}_0, \cdots, \tilde{\beta}_5$ that are $O(N^{-1})$, this substitution can only introduce errors that are $o(N^{-1})$.

The first terms in $\tilde{\alpha}_0$, $\tilde{\alpha}_1$ and $\tilde{\alpha}_2$ as well as the second term in $\tilde{\beta}_0$ are generally not $O(N^{-1})$ but only $O(N^{-\frac{1}{2}})$, and here the substitution of integrals for sums gives rise to more complicated remainder terms. This creates problems we did not encounter in the one-sample case where certain symmetries prohibit the occurrence of $O(N^{-\frac{1}{2}})$ terms. We have

$$\frac{1}{N} \sum a_j^2 = \int J^2(t)\, dt - \frac{1}{N} \sum \sigma^2(J(U_{j:N}))\,,$$

$$\frac{1}{N} \sum a_j^3 = \int J^3(t)\, dt - \frac{1}{N} \sum \operatorname{Cov}(J(U_{j:N}), J^2(U_{j:N})) - \frac{1}{N} \sum EJ(U_{j:N})\sigma^2(J(U_{j:N}))\,,$$

$$\frac{1}{N} \sum a_j E\Psi_2(U_{j:N}) = \int J(t)\Psi_2(t)\, dt - \frac{1}{N} \sum \operatorname{Cov}(J(U_{j:N}), \Psi_2(U_{j:N}))$$

$$\frac{1}{N} \sum a_j^2 E\Psi_1(U_{j:N}) = \int J^2(t)\Psi_1(t)\, dt - \frac{1}{N} \sum \operatorname{Cov}(J^2(U_{j:N}), \Psi_1(U_{j:N})) - \frac{1}{N} \sum E\Psi_1(U_{j:N})\sigma^2(J(U_{j:N}))\,.$$

By (A2.22) in ABZ (1976), $N^{-\frac{3}{2}} \sum \sigma^2(J(U_{j:N})) = o(N^{-1})$. It follows that for $k = 0, \cdots, 5$,

$$\tilde{\alpha}_k - \bar{\alpha}_k = o(N^{-1}) + O(M_1)\,, \qquad \tilde{\beta}_k - \bar{\beta}_k = o(N^{-1}) + O(M_1)\,, \tag{5.23}$$

$$\begin{aligned} M_1 = (1 - 2\lambda)N^{-\frac{3}{2}}[&|\textstyle\sum \operatorname{Cov}(J(U_{j:N}), J^2(U_{j:N}))| \\ &+ |\textstyle\sum EJ(U_{j:N})\sigma^2(J(U_{j:N}))| + |\textstyle\sum \operatorname{Cov}(J(U_{j:N}), \Psi_2(U_{j:N}))| \\ &+ |\textstyle\sum \operatorname{Cov}(J^2(U_{j:N}), \Psi_1(U_{j:N}))| + |\textstyle\sum E\Psi_1(U_{j:N})\sigma^2(J(U_{j:N}))|]\,. \end{aligned} \tag{5.24}$$

By (A2.17), (A2.22) and (A2.23) in ABZ (1976) we have

$$\eta = \bar{\eta} + \tfrac{1}{2}\left(\frac{\lambda(1-\lambda)}{N \int J^2(t)\,dt}\right)^{\frac{1}{2}} \theta \Big[2 \sum \operatorname{Cov}\,(J(U_{j:N}), \Psi_1(U_{j:N}))$$

$$- \frac{\int J(t)\Psi_1(t)\,dt}{\int J^2(t)\,dt} \sum \sigma^2(J(U_{j:N}))\Big] + o(N^{-1}) \qquad (5.25)$$

$$= \bar{\eta} + o(N^{-\frac{1}{2}}) .$$

Hence, uniformly in x,

$$\tilde{K}(x-\eta) = \Phi(x-\bar{\eta}) - \phi(x-\bar{\eta})[(\eta-\bar{\eta}) + \textstyle\sum_{k=0}^{5} \bar{\alpha}_k H_k(x-\bar{\eta})]$$
$$+ o(N^{-1}) + O(M_1)$$
$$= \tilde{K}_1(x-\bar{\eta}) + o(N^{-1}) + O(M_1) ,$$

and similarly

$$\tilde{\pi}(F,\theta) = \tilde{\pi}_1(F,\theta) + o(N^{-1}) + O(M_1) .$$

It follows that, in order to prove (5.14) and (5.16), it suffices to show that $M_1 = o(N^{-1})$. Since (5.15) and (5.17) are immediate consequences of (5.14) and (5.16) on the one hand and (A2.22) and (A2.23) in ABZ (1976) on the other, the proof of the theorem will then be complete.

At this point we finally need condition (5.4) rather than the weaker assumption $\int J^4(t)\,dt < \infty$. Using (5.4), (5.18) and (5.19) and proceeding as in the proof of Corollary A2.1 in ABZ (1976), we find that each term of M_1 is

$$o(N^{-\frac{3}{2}} \textstyle\int_{N^{-1}}^{1-N^{-1}} \{t(1-t)\}^{-\frac{3}{2}}\,dt) = o(N^{-1}) . \qquad (5.26)$$

□

REMARK. In the above we have stressed the fact that the only reason for requiring (5.4) rather than assuming $\int J^4(t)\,dt < \infty$ is that we have to show that $M_1 = o(N^{-1})$. However, there are special cases of interest where $\int J^4(t)\,dt < \infty$ suffices. If either $\lambda = \frac{1}{2}$, or f is a symmetric density and $J(t)$ is antisymmetric about $t = \frac{1}{2}$, then $M_1 = 0$. Less trivially, since $\int J^4(t)\,dt < \infty$ and (5.5) imply that $J'(t) = o(\{t(1-t)\}^{-\frac{5}{4}})$, we can follow the reasoning leading to (5.26) while retaining the factor $(1-2\lambda)$, to arrive at

$$M_1 = o(|1-2\lambda| N^{-\frac{3}{2}} \textstyle\int_{N^{-1}}^{1-N^{-1}} \{t(1-t)\}^{-\frac{7}{4}}\,dt) = o(|1-2\lambda| N^{-\frac{3}{4}}) . \qquad (5.27)$$

Hence in the special cases where either $\lambda = \frac{1}{2} + O(N^{-\frac{1}{4}})$, or f is a symmetric density and J is antisymmetric about the point $\frac{1}{2}$, the conclusions of Theorem 5.1 will hold if condition (5.4) is replaced by the assumption $\int J^4(t)\,dt < \infty$. Comparison with ABZ (1976) shows that in these special cases the conditions under which Theorem 5.1 holds are essentially the same as the conditions of the comparable Theorem 4.1 in ABZ (1976) for the one-sample problem. This is not surprising as one may think of the one-sample case under contiguous alternatives as a two-sample situation with $\lambda = \frac{1}{2} + O_P(N^{-\frac{1}{2}})$.

We now turn to the special case $J = -\Psi_1$. For $F \in \mathscr{F}$ we obtain by partial integration

$$\begin{aligned}(5.28)\qquad &\int \Psi_1(t)\Psi_2(t)\,dt = \tfrac{1}{2}\int \Psi_1^3(t)\,dt\,,\\ &\int \Psi_1^2(t)\Psi_2(t)\,dt = \tfrac{2}{3}\int \Psi_1^4(t)\,dt\,,\\ &\int \Psi_1(t)\Psi_3(t)\,dt = \tfrac{2}{3}\int \Psi_1^4(t)\,dt - \int \Psi_2^2(t)\,dt\,,\\ &\int\int \Psi_1(s)\Psi_1(t)\Psi_1'(s)\Psi_1'(t)[s\wedge t - st]\,ds\,dt = \tfrac{1}{4}\int \Psi_1^4(t)\,dt - \tfrac{1}{4}\left(\int \Psi_1^2(t)\,dt\right)^2.\end{aligned}$$

Substitution of $J = -\Psi_1$ and application of (5.28) considerably simplifies the expressions (5.8) and (5.12) for $\bar{\alpha}_k$ and $\bar{\beta}_k$. Note that $\bar{\eta}$ defined by (5.11) reduces to

$$(5.29)\qquad \eta^* = \theta\{\lambda(1-\lambda)N\int \Psi_1^2(t)\,dt\}^{\frac{1}{2}}\,.$$

The expressions for $\bar{\alpha}_k$ and $\bar{\beta}_k$ simplify somewhat further if we express θ in terms of η^* throughout. Finally we rearrange the terms in $\sum \bar{\alpha}_k H_k(x-\eta^*)$ and $\sum \bar{\beta}_k H_k(u_\alpha - \eta^*)$ according to the integrals involved and substitute the explicit expressions (2.43) for the Hermite polynomials H_k. In this way we find after laborious but straightforward calculations that for $J = -\Psi_1$,

$$\begin{aligned}(5.30)\qquad &\Phi(x-\bar{\eta}) - \phi(x-\bar{\eta})\textstyle\sum_{k=0}^5 \bar{\alpha}_k H_k(x-\bar{\eta}) = L_0(x)\,,\\ &1-\Phi(u_\alpha-\bar{\eta}) + \phi(u_\alpha-\bar{\eta})\textstyle\sum_{k=0}^5 \bar{\beta}_k H_k(u_\alpha-\bar{\eta}) = \pi_0^*(F,\theta)\,,\end{aligned}$$

where

$$\begin{aligned}L_0(x) = {}&\Phi(x-\eta^*)\\ &-\frac{\phi(x-\eta^*)}{288}\Bigg[\frac{24(1-2\lambda)}{\{\lambda(1-\lambda)N\}^{\frac{1}{2}}}\frac{\int \Psi_1^3(t)\,dt}{\{\int \Psi_1^2(t)\,dt\}^{\frac{3}{2}}}\\ &\times\{-2(x^2-1) - 2\eta^*x + \eta^{*2}\} + \frac{4}{\lambda(1-\lambda)N}\frac{\int \Psi_1^4(t)\,dt}{\{\int \Psi_1^2(t)\,dt\}^2}\\ &\times\{3(1-6\lambda+6\lambda^2)(x^3-3x+\eta^*(x^2-1)) - 3(1-5\lambda+5\lambda^2)\eta^{*2}x\\ &+5(1-3\lambda+3\lambda^2)\eta^{*3}\} - \frac{48}{\lambda(1-\lambda)N}\frac{\int \Psi_2^2(t)\,dt}{\{\int \Psi_1^2(t)\,dt\}^2}\\ &\times(1-3\lambda+3\lambda^2)\eta^{*3} + \frac{(1-2\lambda)^2}{\lambda(1-\lambda)N}\frac{\{\int \Psi_1^3(t)\,dt\}^2}{\{\int \Psi_1^2(t)\,dt\}^3}\\ &\times\{4(x^5-10x^3+15x) + 4\eta^*(x^4-6x^2+3) - 8\eta^{*2}(x^3-3x)\\ &-4\eta^{*3}(x^2-1) + 5\eta^{*4}x - \eta^{*5}\} + \frac{144x}{N} + \frac{36}{\lambda(1-\lambda)N}\\ &\times\{-(1-2\lambda)^2(x^3-3x+\eta^*x^2) + \eta^* + (1-5\lambda+5\lambda^2)\eta^{*2}x\\ (5.31)\qquad &+(1-3\lambda+3\lambda^2)\eta^{*3}\}\Bigg]\,,\end{aligned}$$

$$\begin{aligned}\pi_0^*(F,\theta) = {}&1-\Phi(u_\alpha-\eta^*)\\ &+\frac{\eta^*\phi(u_\alpha-\eta^*)}{288}\Bigg[\frac{24(1-2\lambda)}{\{\lambda(1-\lambda)N\}^{\frac{1}{2}}}\frac{\int \Psi_1^3(t)\,dt}{\{\int \Psi_1^2(t)\,dt\}^{\frac{3}{2}}}(-2u_\alpha+\eta^*)\\ &+\frac{4}{\lambda(1-\lambda)N}\frac{\int \Psi_1^4(t)\,dt}{\{\int \Psi_1^2(t)\,dt\}^2}\{3(1-6\lambda+6\lambda^2)(u_\alpha^2-1)\end{aligned}$$

$$- 3(1-5\lambda+5\lambda^2)\eta^* u_\alpha + 5(1-3\lambda+3\lambda^2)\eta^{*2}\}$$
$$- \frac{48}{\lambda(1-\lambda)N} \frac{\int \Psi_2^2(t)\,dt}{\{\int \Psi_1^2(t)\,dt\}^2}(1-3\lambda+3\lambda^2)\eta^{*2}$$
$$+ \frac{(1-2\lambda)^2}{\lambda(1-\lambda)N} \frac{\{\int \Psi_1^3(t)\,dt\}^2}{\{\int \Psi_1^2(t)\,dt\}^3}\{-8(2u_\alpha^2-1)$$
$$+ 4\eta^*(u_\alpha^3+3u_\alpha) - 8\eta^{*2}(u_\alpha^2-1) + 5\eta^{*3}u_\alpha - \eta^{*4}\}$$
$$+ \frac{36}{\lambda(1-\lambda)N}\{-(1-2\lambda)^2 u_\alpha^2 + 1 + (1-5\lambda+5\lambda^2)\eta^* u_\alpha$$
$$+ (1-3\lambda+3\lambda^2)\eta^{*2}\}\Big] .$$

Define

$$L_1(x) = L_0(x) + \frac{\eta^*\phi(x-\eta^*)}{2N\int \Psi_1^2(t)\,dt} \textstyle\sum_{j=1}^N \sigma^2(\Psi_1(U_{j:N})) ,$$
$$L_2(x) = L_0(x) + \frac{\eta^*\phi(x-\eta^*)}{2N\int \Psi_1^2(t)\,dt} \int_{N^{-1}}^{1-N^{-1}} (\Psi_1'(t))^2 t(1-t)\,dt ,$$
$$L_3(x) = L_0(x) + \frac{\eta^*\phi(x-\eta^*)}{2N\int \Psi_1^2(t)\,dt} \textstyle\sum_{j=1}^N E\Big(\Psi_1(U_{j:N}) - \Psi_1\Big(\frac{j}{N+1}\Big)\Big)^2 ,$$
$$\pi_1^*(F,\theta) = \pi_0^*(F,\theta) - \frac{\eta^*\phi(u_\alpha-\eta^*)}{2N\int \Psi_1^2(t)\,dt} \textstyle\sum_{j=1}^N \sigma^2(\Psi_1(U_{j:N})) , \tag{5.32}$$
$$\pi_2^*(F,\theta) = \pi_0^*(F,\theta) - \frac{\eta^*\phi(u_\alpha-\eta^*)}{2N\int \Psi_1^2(t)\,dt} \int_{N^{-1}}^{1-N^{-1}} (\Psi_1'(t))^2 t(1-t)\,dt ,$$
$$\pi_3^*(F,\theta) = \pi_0^*(F,\theta) - \frac{\eta^*\phi(u_\alpha-\eta^*)}{2N\int \Psi_1^2(t)\,dt} \textstyle\sum_{j=1}^N E\Big(\Psi_1(U_{j:N}) - \Psi_1\Big(\frac{j}{N+1}\Big)\Big)^2 .$$

Note that (5.9), (5.10), (5.11), (5.13), (5.30) and (5.31) imply that for $J = -\Psi_1$, $\bar{K}_i(x-\bar{\eta}) = L_i(x)$ and $\bar{\pi}_i(F,\theta) = \pi_i^*(F,\theta)$ for $i = 1, 2$. The expansions L_3 and π_3^* are connected only with approximate scores that were not considered so far.

THEOREM 5.2. *Let $F \in \mathscr{F}$, $J = -\Psi_1$, $G(x) = F(x-\theta)$, $0 \leqq \theta \leqq DN^{-\frac{1}{2}}$, $\varepsilon \leqq \lambda \leqq 1-\varepsilon$ and $\varepsilon' \leqq \alpha \leqq 1-\varepsilon'$ for positive D, ε and ε'. Then, for every fixed F, D, ε and ε', there exist positive numbers B, $\delta_1, \delta_2, \cdots$ with $\lim_{N\to\infty}\delta_N = 0$ such that the following statements hold for every N.*

(i) *For exact scores $a_j = -E\Psi_1(U_{j:N})$,*

$$\sup_x \Big| P\Big(\frac{T}{\{\lambda(1-\lambda)\sum a_j^2\}^{\frac{1}{2}}} \leqq x\Big) - L_1(x)\Big| \leqq \delta_N N^{-1} , \tag{5.33}$$
$$\sup_x \Big| P\Big(\frac{T}{\{\lambda(1-\lambda)\sum a_j^2\}^{\frac{1}{2}}} \leqq x\Big) - L_2(x)\Big| \leqq \delta_N N^{-1} + BN^{-\frac{3}{2}} \int_{N^{-1}}^{1-N^{-1}} (\Psi_1'(t))^2 \{t(1-t)\}^{\frac{1}{2}}\,dt , \tag{5.34}$$
$$|\pi(F,\theta) - \pi_1^*(F,\theta)| \leqq \delta_N N^{-1} , \tag{5.35}$$
$$|\pi(F,\theta) - \pi_2^*(F,\theta)| \leqq \delta_N N^{-1} + BN^{-\frac{3}{2}} \int_{N^{-1}}^{1-N^{-1}} (\Psi_1'(t))^2 \{t(1-t)\}^{\frac{1}{2}}\,dt ; \tag{5.36}$$

(ii) *For approximate scores* $a_j = -\Psi_1(j/(N+1))$,

$$(5.37)\qquad \sup_x \left| P\left(\frac{T - \lambda \sum a_j}{\{\lambda(1-\lambda)\sum a_j^2\}^{\frac{1}{2}}} \leq x\right) - L_3(x)\right| \leq \delta_N N^{-1},$$

$$(5.38)\qquad \sup_x \left| P\left(\frac{T - \lambda \sum a_j}{\{\lambda(1-\lambda)\sum a_j^2\}^{\frac{1}{2}}} \leq x\right) - L_2(x)\right| \leq \delta_N N^{-1} + BN^{-\frac{3}{2}} \int_{N^{-1}}^{1-N^{-1}} (\Psi_1'(t))^2\{t(1-t)\}^{\frac{1}{2}}\, dt,$$

$$(5.39)\qquad |\pi(F,\theta) - \pi_3^*(F,\theta)| \leq \delta_N N^{-1}$$

and (5.36) *continues to hold.*

PROOF. For $F \in \mathscr{F}$, Ψ_1 is not constant on (0, 1), $\int \Psi_1(t)\,dt = 0$ and Ψ_1^6 is summable. In view of the remark following Definition 5.1, this implies that $J \in \mathscr{J}$. We have already noted that $K_i(x-\eta) = L_i(x)$ and $\bar{\pi}_i(F,\theta) = \pi_i^*(F,\theta)$ for $i = 1, 2$, if $J = -\Psi_1$. Part (i) of the theorem is therefore an immediate consequence of Theorem 5.1.

To prove part (ii) we retrace the proof of Theorem 5.1 for $J = -\Psi_1$ and approximate scores $a_j = -\Psi_1(j/(N+1))$. The first difficulty we encounter is that in general $\sum a_j \neq 0$. However, Lemma A2.3 of ABZ (1976), (5.7) and (5.18) yield

$$(5.40)\qquad a_\cdot = \frac{1}{N}\sum_{j=1}^N a_j = -\int_0^1 \Psi_1(t)\,dt + O\left(N^{-1}\int_{N^{-1}}^{1-N^{-1}} |\Psi_1'(t)|\,dt\right) = o(N^{-\frac{5}{6}}),$$

and one easily verifies that the conditions of Theorem 4.1 hold for the reduced scores $a_j - a_\cdot$. Since the assumptions of Theorem 4.2 are also satisfied, we have

$$(5.41)\qquad \sup_x \left| P\left(\frac{T - \lambda \sum a_j}{\{\lambda(1-\lambda)\sum (a_j - a_\cdot)^2\}^{\frac{1}{2}}} \leq x\right) - \hat{K}(x-\hat{\eta})\right| = O(N^{-\frac{5}{4}}),$$

where $\hat{K}$ and $\hat{\eta}$ are obtained from $\tilde{K}$ and η by replacing a_j by $a_j - a_\cdot$ throughout. Because, by (3.10) and (5.40),

$$(5.42)\qquad \sum (a_j - a_\cdot)^2 = \sum a_j^2(1 + o(N^{-\frac{5}{3}})),$$

we can change the norming constant $\sum (a_j - a_\cdot)^2$ of T in (5.41) back to $\sum a_j^2$ with impunity. As $\int \Psi_1(t)\,dt = 0$, (5.42) also ensures that $|\hat{\eta} - \eta| = o(N^{-\frac{5}{3}})$. Finally (A2.16) of ABZ (1976) and (5.18) imply that $\sigma_F^2(\sum a_j \psi_1(Z_j)) = O(N)$ for $J = -\Psi_1$ and, together with (5.42), (3.10), (5.6) and (5.40), this yields $\sup_x |\hat{K}(x) - \tilde{K}(x)| = o(N^{-\frac{4}{3}})$. Combining these results we find

$$(5.43)\qquad \sup_x \left| P\left(\frac{T - \lambda \sum a_j}{\{\lambda(1-\lambda)\sum a_j^2\}^{\frac{1}{2}}} \leq x\right) - \tilde{K}(x-\eta)\right| = O(N^{-\frac{5}{4}})$$

and similarly

$$(5.44)\qquad |\pi(F,\theta) - \bar{\pi}(F,\theta)| = O(N^{-\frac{5}{4}}).$$

The remainder of the proof parallels that of Theorem 5.1 for the special case $J = -\Psi_1$. We replace all sums as well as $\sigma^2(\sum a_j \Psi_1(U_{j:N}))$ by the appropriate integrals. The reasoning of Corollary A2.2 of ABZ (1976) shows that for those terms in the expansions that are $O(N^{-1})$, this substitution will only lead to errors that are $o(N^{-1})$. For the $O(N^{-\frac{1}{2}})$ terms the error committed is $O(M_1) + O(M_2)$, where M_1 is given by (5.24) with $J = -\Psi_1$ and M_2 originates from the difference between exact and approximate scores. It was shown in the proof of Theorem 5.1 that $M_1 = o(N^{-1})$. With regard to M_2, (5.7), Lemma A2.3 of ABZ (1976), (5.18) and (5.19) imply that, uniformly in j,

$$\left|\{E\Psi_1(U_{j:N})\}^k - \Psi_1{}^k\left(\frac{j}{N+1}\right)\right|$$

$$= O(N^{-1}) + o\left(N^{-1}\left\{\frac{j(N-j+1)}{(N+1)^2}\right\}^{-1-k/6}\right) \tag{5.45}$$

$$|E\Psi_1(U_{j:N})| = O(1) + o\left(\left\{\frac{j(N-j+1)}{(N+1)^2}\right\}^{-\frac{1}{6}}\right),$$

$$|E\Psi_2(U_{j:N})| = O(1) + o\left(\left\{\frac{j(N-j+1)}{(N+1)^2}\right\}^{-\frac{1}{3}}\right),$$

where $k = 1, 2, 3$. It follows that M_2 is of the form (5.26) and is therefore $o(N^{-1})$.

It remains to replace η by η^*. Because of (5.7), (5.18) and Lemma A2.3 of ABZ (1976), $N^{-1}\sum \sigma^2(\Psi_1(U_{j:N})) = o(N^{-\frac{1}{2}})$, and in view of (5.45),

$$\frac{1}{N}\sum \Psi_1\left(\frac{j}{N+1}\right)E\Psi_1(U_{j:N}) - \int \Psi_1^2(t)\,dt$$

$$= -\frac{1}{N}\sum E\Psi_1(U_{j:N})\left[\Psi_1(U_{j:N}) - \Psi_1\left(\frac{j}{N+1}\right)\right] = o(N^{-\frac{1}{2}}),$$

$$\frac{1}{N}\sum \Psi_1^2\left(\frac{j}{N+1}\right) - \int \Psi_1^2(t)\,dt = -\frac{1}{N}\sum\left[E\Psi_1^2(U_{j:N}) - \Psi_1^2\left(\frac{j}{N+1}\right)\right]$$
$$= o(N^{-\frac{1}{2}}).$$

Hence, for $J = -\Psi_1$,

$$\eta = \eta^* - \frac{\eta^*}{2N\int \Psi_1^2(t)\,dt}\sum E\left\{\Psi_1(U_{j:N}) - \Psi_1\left(\frac{j}{N+1}\right)\right\}^2 + o(N^{-\frac{1}{2}}) \tag{5.46}$$
$$= \eta^* + o(N^{-\frac{1}{2}}),$$

and a comparison with (5.25) for $J = -\Psi_1$ show that (5.37) and (5.39) will hold if L_3 and $\pi_3{}^*$ can be obtained from L_1 and $\pi_1{}^*$ by replacing $\sum \sigma^2(\Psi_1(U_{j:N}))$ by $\sum E\{\Psi_1(U_{j:N}) - \Psi_1(j/(N+1))\}^2$. Since this is true, (5.37) and (5.39) are proved. The validity of (5.38) and (5.36) for approximate scores is a consequence of (5.37), (5.39) and Corollary A2.2 of ABZ (1976). The proof of the theorem is complete. □

At this point it is appropriate to repeat some remarks made in ABZ (1976). The correspondence between expansions (5.34) and (5.38) and the fact that

(5.36) holds for both exact and approximate scores seem to be typical for the case $J = -\Psi_1$. In the general case where $J \neq -\Psi_1$, expansions (5.15) and (5.17) will not hold for approximate scores even if T is replaced by $T - \lambda \sum a_j$ in (5.15). A second remark is that the growth conditions on J' and Ψ_1' implicit in our assumptions (viz. (5.4) and (5.18)) do not guarantee that the right-hand side in (5.15), (5.17), (5.34), (5.36) and (5.38) is indeed $o(N^{-1})$ as is our aim. For this we would need $J'(t) = o(\{t(1-t)\}^{-1})$ and $\Psi_1'(t) = o(\{t(1-t)\}^{-1})$. This may explain the presence of the remaining expansions in Theorems 5.1 and 5.2, which are less explicit but do have remainder $o(N^{-1})$ under the conditions stated. Note that their presence in Theorem 5.2 also indicates that even for $J = -\Psi_1$, expansions for exact and approximate scores are not necessarily identical to $o(N^{-1})$. Finally we should point out that similar expansions with remainder $o(N^{-1})$ might have been given in Theorem 4.2 of ABZ (1976) where they were unfortunately omitted.

We conclude this section with a few examples of the power expansions in Theorems 5.1 and 5.2. First we consider the powers $\pi_W(\Phi, \theta)$ and $\pi_W(\Lambda, \theta)$ of Wilcoxon's two-sample test (W) against normal and logistic location alternatives $(\Phi(x), \Phi(x-\theta))$ and $(\Lambda(x), \Lambda(x-\theta))$ respectively, where $\Lambda(x) = (1+\exp\{-x\})^{-1}$ and $\theta = O(N^{-\frac{1}{2}})$. We find

$$
\begin{aligned}
\pi_W(\Phi, \theta) = 1 - \Phi(u_\alpha - \tilde{\eta}) + \frac{\tilde{\eta}\phi(u_\alpha - \tilde{\eta})}{N} \\
\times \Bigg[-\tfrac{1}{2} - \frac{37 - 217\lambda + 217\lambda^2}{20\lambda(1-\lambda)}(u_\alpha^2 - 1) \\
+ \frac{1}{\lambda(1-\lambda)}\left\{\frac{3^{\frac{1}{2}}}{6} + \frac{67 - 437\lambda + 437\lambda^2}{20}\right\} u_\alpha \tilde{\eta} \\
- \frac{1}{\lambda(1-\lambda)}\left\{\frac{3^{\frac{1}{2}}}{6} + \frac{\pi}{36} + \frac{29 - 219\lambda + 219\lambda^2}{20}\right\} \tilde{\eta}^2 \\
+ \frac{(1 - 6\lambda + 6\lambda^2)}{\lambda(1-\lambda)} \frac{6 \arctan 2^{\frac{1}{2}}}{\pi}\{u_\alpha^2 - 1 - 2u_\alpha\tilde{\eta} + \tilde{\eta}^2\}\Bigg] \\
+ o(N^{-1}),
\end{aligned}
\tag{5.47}
$$

where $\tilde{\eta} = (3\lambda(1-\lambda)N/\pi)^{\frac{1}{2}}\theta$, and

$$
\begin{aligned}
\pi_W(\Lambda, \theta) = 1 - \Phi(u_\alpha - \eta^*) + \frac{\eta^*\phi(u_\alpha - \eta^*)}{N} \\
\times \Bigg[-\tfrac{1}{2} - \frac{1 - \lambda + \lambda^2}{20\lambda(1-\lambda)}(u_\alpha^2 - 1) + \frac{1 - 5\lambda + 5\lambda^2}{20\lambda(1-\lambda)} u_\alpha \eta^* \\
- \frac{1 - 3\lambda + 3\lambda^2}{20\lambda(1-\lambda)}\eta^{*2}\Bigg] + o(N^{-1}),
\end{aligned}
\tag{5.48}
$$

where $\eta^* = (\lambda(1-\lambda)N/3)^{\frac{1}{2}}\theta$.

As a second example we compute expansions for the powers $\pi_{NS}(\Phi, \theta)$ and $\pi_{NS}(\Lambda, \theta)$ of the two-sample normal scores test against the normal and logistic location alternatives described above. One of the integrals occurring in this computation is

$$\int_{N^{-1}}^{1-N^{-1}} \frac{t(1-t)}{\{\phi(\Phi^{-1}(t))\}^2}\,dt = 2\int_0^{\Phi^{-1}(1-N^{-1})} \frac{\Phi(x)(1-\Phi(x))}{\phi(x)}\,dx \tag{5.49}$$

and since its asymptotic evaluation is not entirely trivial, we provide some details. Let γ denote Euler's constant

$$\gamma = \lim_{k\to\infty}\left\{\sum_{j=1}^{k}\frac{1}{j} - \log k\right\} = 0.577216\ldots \tag{5.50}$$

and note that (cf. Ryshik and Gradstein (1957), page 197)

$$\int_0^\infty e^{-u}\log u\,du = -\gamma\,, \tag{5.51}$$

$$\int_0^\infty \phi(u)\log u\,du = -\tfrac{1}{4}\log 2 - \tfrac{1}{4}\gamma\,. \tag{5.52}$$

To evaluate (5.49) we begin by writing for $z > 0$

$$\begin{aligned}\int_0^z \frac{1-\Phi(x)}{\phi(x)}\,dx &= \int_0^z dx\int_x^\infty e^{-\frac{1}{2}(y-x)(y+x)}\,dy = \tfrac{1}{2}\int_0^\infty du\int_u^{2z+u} e^{-\frac{1}{2}uv}\,dv \\ &= \int_0^\infty \frac{1}{u}\,e^{-\frac{1}{2}u^2}(1-e^{-zu})\,du\,.\end{aligned} \tag{5.53}$$

It follows from (5.53) and (5.51) that for $z \to \infty$,

$$\begin{aligned}\int_0^z \frac{1-\Phi(x)}{\phi(x)}\,dx &= \int_0^{z^{-\frac{1}{2}}} \frac{1}{u}(1-e^{-zu})\,du + \int_{z^{-\frac{1}{2}}}^\infty \frac{1}{u}\,e^{-\frac{1}{2}u^2}\,du + o(1) \\ &= \int_0^{z^{\frac{1}{2}}} \frac{1}{u}(1-e^{-u})\,du + \tfrac{1}{2}\int_{(2z)^{-1}}^\infty \frac{1}{u}\,e^{-u}\,du + o(1) \\ &= \tfrac{1}{2}\log z - \int_0^{z^{\frac{1}{2}}} e^{-u}\log u\,du + \tfrac{1}{2}\log(2z) \\ &\quad + \tfrac{1}{2}\int_{(2z)^{-1}}^\infty e^{-u}\log u\,du + o(1) \\ &= \log z + \tfrac{1}{2}\log 2 + \tfrac{1}{2}\gamma + o(1)\,.\end{aligned} \tag{5.54}$$

Similarly (5.53), (5.51) and (5.52) imply that

$$\begin{aligned}\int_0^\infty \frac{(1-\Phi(x))^2}{\phi(x)}\,dx &= \int_0^\infty \phi(x)\,dx\int_0^\infty \frac{1}{u}\,e^{-\frac{1}{2}u^2}(1-e^{-xu})\,du \\ &= \int_0^\infty \frac{1}{u}\{\tfrac{1}{2}e^{-\frac{1}{2}u^2} - (1-\Phi(u))\}\,du \\ &= \int_0^\infty \log u\{\tfrac{1}{2}ue^{-\frac{1}{2}u^2} - \phi(u)\}\,du \\ &= \tfrac{1}{4}\log 2 + \tfrac{1}{4}\int_0^\infty e^{-x}\log x\,dx - \int_0^\infty \phi(u)\log u\,du \\ &= \tfrac{1}{2}\log 2\,.\end{aligned} \tag{5.55}$$

Since $\log\Phi^{-1}(1-N^{-1}) = \frac{1}{2}\log\log N + \frac{1}{2}\log 2 + o(1)$ for $N \to \infty$, (5.49), (5.54) and (5.55) imply that

$$\int_{N^{-1}}^{1-N^{-1}} \frac{t(1-t)}{\{\phi(\Phi^{-1}(t))\}^2}\,dt = \log\log N + \log 2 + \gamma + o(1)\,. \tag{5.56}$$

With the aid of (5.56) we find

$$\text{(5.57)}\qquad \pi_{NS}(\Phi, \theta) = 1 - \Phi(u_\alpha - \eta^*) + \frac{\eta^*\phi(u_\alpha - \eta^*)}{N}[-\tfrac{1}{2}\log\log N - \tfrac{1}{2}\log 2 - \tfrac{1}{2}\gamma + \tfrac{1}{2} - \tfrac{1}{4}(u_\alpha^2 - 1)] + o(N^{-1}),$$

where now $\eta^* = \{\lambda(1-\lambda)N\}^{\frac{1}{2}}\theta$, and

$$\text{(5.58)}\qquad \begin{aligned}\pi_{NS}(\Lambda, \theta) = 1 - \Phi(u_\alpha - \bar\eta) + \frac{\bar\eta\phi(u_\alpha - \bar\eta)}{N}\Big[&\tfrac{1}{2}\log\log N + \tfrac{1}{2}\log 2 + \tfrac{1}{2}\gamma \\ &- \tfrac{3}{2} - \frac{1 - 3\lambda + 3\lambda^2}{12\lambda(1-\lambda)}(u_\alpha^2 - 1) \\ &+ \left\{\frac{3^{\frac{1}{2}}(1-2\lambda)^2}{4\lambda(1-\lambda)} - \frac{\pi}{6} - \frac{4 - 21\lambda + 21\lambda^2}{12\lambda(1-\lambda)}\right\}u_\alpha\bar\eta \\ &+ \left\{\frac{6(1 - 5\lambda + 5\lambda^2)}{\lambda(1-\lambda)}\arctan 2^{\frac{1}{2}} - \frac{3^{\frac{1}{2}}(1-2\lambda)^2}{4\lambda(1-\lambda)}\right. \\ &\left. - \frac{11\pi(1 - 5\lambda + 5\lambda^2)}{6\lambda(1-\lambda)} + \frac{5 - 21\lambda + 21\lambda^2}{12\lambda(1-\lambda)}\right\}\bar\eta^2\Big] \\ &+ o(N^{-1}),\end{aligned}$$

where now $\bar\eta = \{\lambda(1-\lambda)N/\pi\}^{\frac{1}{2}}\theta$. Note that Theorem 5.2 ensures that expansion (5.57) is also valid for van der Waerden's two-sample test which is based on the approximate scores $a_j = \Phi^{-1}(j/(N+1))$.

It may be useful to remark here that an integral similar to (5.56) also occurs in ABZ (1976), formula (4.25) on page 130, where its asymptotic behavior is determined numerically. However, the numerically computed value is incorrect and in formulas (4.25) and (6.8) in ABZ (1976) the number $\frac{1}{2}\log 2 + 0.05832\cdots$ should be replaced by $\frac{1}{2}\gamma = 0.288608\cdots$ (cf. the correction note in this issue).

6. The permutation test based on the sample means. In ABZ (1976) two results were given for permutation tests in the one-sample problem. The first of these is an asymptotic expansion for the power of the locally most powerful permutation test against contiguous shift alternatives. Secondly it was shown that the difference between the powers of the permutation test based on the sum of transformed observations $\sum j(X_i)$ and Student's test applied to $j(X_1), \cdots, j(X_N)$ is $o(N^{-1})$ for a large class of alternatives.

In the present paper we shall forego the two-sample analogue of the first mentioned result; the expansion can be obtained in a straightforward manner in much the same way as in the one-sample case but the computations will be extremely tedious. We shall concentrate on the comparison with Student's test. For simplicity we take j to be the identity, thus comparing the two-sample permutation test based on the sample means with Student's two-sample test. Also, we restrict attention to contiguous location alternatives.

As before, we assume that $X_1, \cdots, X_N$ are independent, $X_1, \cdots, X_m$ having common df F and $X_{m+1}, \cdots, X_N$ having common df $G(x) = F(x - \theta)$; $Z = (Z_1, \cdots, Z_N)$ denotes the vector of order statistics of $X_1, \cdots, X_N$. We wish to test the hypothesis $\theta = 0$ against the alternative $\theta > 0$ at a fixed level $\alpha \in (0, 1)$. We denote probabilities and expectations under the alternative by P and E, and under the hypothesis by P_F and E_F. Note that we do not assume that F has a density, as we did in the previous sections.

The permutation test rejects the hypothesis if

$$\sum_{i=m+1}^N X_i \geqq \xi_\alpha(Z) , \tag{6.1}$$

possibly with randomization if equality occurs. Here $\xi_\alpha(Z)$ is chosen in such a way that

$$P_F(\sum_{i=m+1}^N X_i \geqq \xi_\alpha \mid Z) = \alpha \quad \text{a.s.} \tag{6.2}$$

with an obvious modification if there is randomization. If F is known, Student's test rejects the hypothesis if

$$\mathcal{T} = \frac{\{\lambda(1-\lambda)N(N-2)\}^{\frac{1}{2}}(X_\cdot^{(2)} - X_\cdot^{(1)})}{[\sum_{i=1}^m (X_i - X_\cdot^{(1)})^2 + \sum_{i=m+1}^N (X_i - X_\cdot^{(2)})^2]^{\frac{1}{2}}} \geqq t_\alpha , \tag{6.3}$$

where

$$X_\cdot^{(1)} = \frac{1}{m} \sum_{i=1}^m X_i , \qquad X_\cdot^{(2)} = \frac{1}{n} \sum_{i=m+1}^N X_i .$$

Here t_α depends on F, N, λ and α and is chosen in such a way that the test has level α. Again there may be randomization. Let $\pi_{Pe}(F, \theta)$ and $\pi_{St}(F, \theta)$ denote the power against the alternative $(F, F(\cdot - \theta))$ of the tests (6.1) and (6.3) respectively.

THEOREM 6.1. *Suppose that positive numbers c, C, D, ε, ε', δ and $r > 8$ exist such that F^{-1} is differentiable on an interval of length at least δ where*

$$\frac{d}{dt} F^{-1}(t) \geqq c , \tag{6.4}$$

and such that $\int |x|^r \, dF(x) \leqq C$, $0 \leqq \theta \leqq DN^{-\frac{1}{2}}$, $\varepsilon \leqq \lambda \leqq 1 - \varepsilon$ and $\varepsilon' \leqq \alpha \leqq 1 - \varepsilon'$. Then there exist $B > 0$ depending only on c, C, D, ε, ε' and δ, and $\beta > 0$ depending only on r such that

$$|\pi_{Pe}(F, \theta) - \pi_{St}(F, \theta)| \leqq BN^{-1-\beta} . \tag{6.5}$$

PROOF. We shall draw heavily on the proof of Theorem 5.2 in ABZ (1976). The only essentially new problem is caused again by the occurrence of a term of order $N^{-\frac{1}{2}}$ in the expansions. The O symbols in this proof are uniform for fixed c, C, D, ε, ε' and δ. Since both tests are location invariant we may assume without loss of generality that $\int x \, dF(x) = 0$.

We begin by collecting some results on moments that will be needed throughout the proof. Define

$$\beta = \min\left(\frac{r-8}{2r+8}, \tfrac{1}{4}\right), \tag{6.6}$$

$$X_{\bullet} = Z_{\bullet} = \frac{1}{N}\sum_{i=1}^{N} X_i = \frac{1}{N}\sum_{j=1}^{N} Z_j, \tag{6.7}$$

and note that $EX_N^k = E(X_1 + \theta)^k$ and that $EX_1 = 0$. Proceeding as in the proof of Theorem 5.2 in ABZ (1976) we see that

$$\frac{c^2\delta^3}{12} \leqq EX_1^2 \leqq [EX_1^4]^{\frac{1}{2}} \leqq C^{2/r}, \tag{6.8}$$

and that, uniformly on a set of probability $1 - O(N^{-1-\beta})$ under P as well as under P_F,

$$\frac{1}{N}\sum_{i=1}^{m} X_i^k = (1-\lambda)EX_1^k + O(N^{-\beta}), \qquad k = 1, \cdots, 4, \tag{6.9}$$

$$\frac{1}{N}\sum_{i=m+1}^{N} X_i^k = \lambda EX_1^k + O(N^{-\beta}), \qquad k = 1, \cdots, 4, \tag{6.10}$$

$$\frac{1}{N}\sum_{i=1}^{m} (X_i - X_{\bullet})^k = (1-\lambda)EX_1^k + O(N^{-\beta}), \qquad k = 2, \cdots, 4, \tag{6.11}$$

$$\frac{1}{N}\sum_{i=m+1}^{N} (X_i - X_{\bullet})^k = \lambda EX_1^k + O(N^{-\beta}), \qquad k = 2, \cdots, 4. \tag{6.12}$$

For $k = 1$, (6.9) and (6.10) are insufficient for our purposes. Arguing as in (5.13) in ABZ (1976) for $\tau = N^{-\frac{3}{8}}$, we find

$$\frac{1}{N}\sum_{i=1}^{m} X_i = O(N^{-\frac{3}{8}}), \qquad \frac{1}{N}\sum_{i=m+1}^{N} X_i = O(N^{-\frac{3}{8}}), \tag{6.13}$$

uniformly with probability $1 - O(N^{-1-\beta})$ under both P and P_F.

We shall also have to consider the quantity $l\{x : \exists\, i\ |x - X_i| < \zeta\}$ for some $\zeta \geqq N^{-\frac{3}{2}} \log N$, where l denotes Lebesgue measure. Borrowing from the proof of Theorem 5.2 in ABZ (1976) again, we find that for $\zeta = N^{-\frac{3}{2}} \log N$,

$$l\{x : \exists\, i\ \ |x - X_i| < \zeta\} \geqq \frac{\delta N \zeta}{6} \tag{6.14}$$

with probability $1 - O(N^{-1-\beta})$ both under P and under P_F. Let E_1 be a set on which (6.9)—(6.14) hold uniformly, with $P(E_1) = 1 - O(N^{-1-\beta})$ and $P_F(E_1) = 1 - O(N^{-1-\beta})$.

Under the hypothesis P_F and conditional on Z the df of

$$N^{-\frac{1}{2}}\left(\sum_{i=m+1}^{N} X_i - \lambda \sum_{j=1}^{N} Z_j\right)$$

equals $R(x, p)$ defined in (2.9) with $p_j = \lambda$ and $a_j = Z_j$ for $j = 1, \cdots, N$. Hence Corollary 2.1 provides an expansion for this conditional df that holds uniformly on any set where the $a_j = Z_j$ satisfy (2.61) and (2.62) for some fixed positive c, C and δ, and in view of (6.8)—(6.14) such a set is contained in E_1. Since $\varepsilon' \leqq \alpha \leqq 1 - \varepsilon'$, this yields an expansion for $\xi_\alpha(Z)$. We find (cf. (4.14))

$$\begin{aligned}
&\frac{\xi_\alpha(Z) - \lambda \sum Z_j}{[\lambda(1-\lambda)\sum (Z_j - Z_\cdot)^2]^{\frac{1}{2}}} \\
&\quad = u_\alpha + \frac{(1-2\lambda)\sum (Z_j - Z_\cdot)^3}{6\{\lambda(1-\lambda)\}^{\frac{1}{2}}[\sum (Z_j - Z_\cdot)^2]^{\frac{3}{2}}}(u_\alpha^2 - 1) \\
&\quad - \frac{(1-2\lambda)^2[\sum (Z_j - Z_\cdot)^3]^2}{36\lambda(1-\lambda)[\sum (Z_j - Z_\cdot)^2]^3}(2u_\alpha^3 - 5u_\alpha) + \frac{u_\alpha}{2N} \qquad (6.15)\\
&\quad + \left\{\frac{(1-6\lambda+6\lambda^2)\sum (Z_j - Z_\cdot)^4}{24\lambda(1-\lambda)[\sum (Z_j - Z_\cdot)^2]^2} - \frac{(1-2\lambda)^2}{8\lambda(1-\lambda)N}\right\}(u_\alpha^3 - 3u_\alpha) \\
&\quad + O(N^{-\frac{3}{2}}),
\end{aligned}$$

uniformly on E_1.

Next we start to compute under the alternative P. We have $P(E_1) = 1 - O(N^{-1-\beta})$and on E_1 we can use (6.8), (6.11) and (6.12) to replace the random terms of order $O(N^{-1})$ on the right in (6.15) by constants. In this way we arrive at

$$\pi_{P_\epsilon}(F, \theta) = P(T^* \geqq \xi_\alpha^* + U_1 + O(N^{-1-\beta})) + O(N^{-1-\beta}), \tag{6.16}$$

where the first remainder term depends on Z but may now be taken to be uniformly $O(N^{-1-\beta})$, and where

$$T^* = \frac{\sum_{i=m+1}^N X_i - \lambda \sum Z_j}{[\lambda(1-\lambda)\sum (Z_j - Z_\cdot)^2]^{\frac{1}{2}}}, \tag{6.17}$$

$$\begin{aligned}
\xi_\alpha^* &= u_\alpha + \frac{(1-2\lambda)EX_1^3}{6\{\lambda(1-\lambda)N\}^{\frac{1}{2}}(EX_1^2)^{\frac{3}{2}}}(u_\alpha^2 - 1) \\
&\quad - \frac{(1-2\lambda)^2(EX_1^3)^2}{36\lambda(1-\lambda)N(EX_1^2)^3}(2u_\alpha^3 - 5u_\alpha) + \frac{u_\alpha}{2N} \qquad (6.18)\\
&\quad + \left\{\frac{(1-6\lambda+6\lambda^2)EX_1^4}{24\lambda(1-\lambda)N(EX_1^2)^2} - \frac{(1-2\lambda)^2}{8\lambda(1-\lambda)N}\right\}(u_\alpha^3 - 3u_\alpha),
\end{aligned}$$

$$U_1 = \frac{(1-2\lambda)(u_\alpha^2 - 1)}{6\{\lambda(1-\lambda)\}^{\frac{1}{2}}}\left\{\frac{\sum (Z_j - Z_\cdot)^3}{[\sum (Z_j - Z_\cdot)^2]^{\frac{3}{2}}} - \frac{N^{-\frac{1}{2}}EX_1^3}{(EX_1^2)^{\frac{3}{2}}}\right\}. \tag{6.19}$$

The basic problem is now to show that the rv U_1 originating from the $O(N^{-\frac{1}{2}})$ term in (6.15), may be omitted in (6.16). Since U_1 is a rv of order N^{-1}, this problem is nontrivial. We shall show that because U_1 depends only on Z and is approximately centered, a cancellation occurs which makes its contribution to (6.16) of negligible order. Several methods of proof are possible. We choose one that does not require any additional assumptions.

In (6.16), P may be replaced by P_F if X_i is replaced by $X_i + \theta$ for $i = m+1, \cdots, N$, which transforms T^* and U_1 into $T^*(\theta)$ and $U_1(\theta)$, say. On the set E_1, (6.8)—(6.13) ensure that we can expand $T^*(\theta)$ and $U_1(\theta)$ about T^* and

U_1. Replacing rv's by their expected values if the difference is of negligible order, a simple calculation shows that under P_F we have, uniformly on the set E_1,

$$T^*(\theta) = T^*\left[1 - \frac{3\lambda(1-\lambda)N\theta^2}{2\sum(Z_j - Z_\cdot)^2}\right] - T^{*2}\frac{\{\lambda(1-\lambda)\}^{\frac{1}{2}}\theta}{[\sum(Z_j - Z_\cdot)^2]^{\frac{1}{2}}} + \frac{\{\lambda(1-\lambda)\}^{\frac{1}{2}}N\theta}{[\sum(Z_j - Z_\cdot)^2]^{\frac{1}{2}}} - \frac{\{\lambda(1-\lambda)\}^{\frac{3}{2}}N^{\frac{1}{2}}\theta^3}{2(EX_1^2)^{\frac{3}{2}}} + O(N^{-1-\beta}), \tag{6.20}$$

$$U_1(\theta) = U_1 + O(N^{-1-\beta}). \tag{6.21}$$

Another easy calculation where we use (6.8)—(6.13) to bound the terms in (6.20) and (6.21) and to replace rv's by their expected values whenever possible, and where we note that $\xi_\alpha^* = u_\alpha + O(N^{-\frac{1}{2}})$, shows that uniformly on E_1, the inequality $T^*(\theta) \geqq \xi_\alpha^* + U_1(\theta) + O(N^{-1-\beta})$ is equivalent to $T^* \geqq \xi_\alpha^*(\theta) - U_0 + U_1 + O(N^{-1-\beta})$, where

$$\xi_\alpha^*(\theta) = \xi_\alpha^* - \frac{\{\lambda(1-\lambda)N\}^{\frac{1}{2}}\theta}{(EX_1^2)^{\frac{1}{2}}} + \frac{\{\lambda(1-\lambda)\}^{\frac{1}{2}}\theta}{(NEX_1^2)^{\frac{1}{2}}}u_\alpha^2 - \frac{\lambda(1-\lambda)\theta^2}{2EX_1^2}u_\alpha, \tag{6.22}$$

$$U_0 = \{\lambda(1-\lambda)\}^{\frac{1}{2}}N\theta\left\{\frac{1}{[\sum(Z_j - Z_\cdot)^2]^{\frac{1}{2}}} - \frac{1}{(NEX_1^2)^{\frac{1}{2}}}\right\}. \tag{6.23}$$

Since $P_F(E_1) = 1 - O(N^{-1-\beta})$, this implies

$$\begin{aligned}\pi_{P_e}(F,\theta) &= P_F(T^*(\theta) \geqq \xi_\alpha^* + U_1(\theta) + O(N^{-1-\beta})) + O(N^{-1-\beta}) \\ &= P_F(T^* \geqq \xi_\alpha^*(\theta) - U_0 + U_1 + O(N^{-1-\beta})) + O(N^{-1-\beta}),\end{aligned} \tag{6.24}$$

where the first remainder term in the last member depends on $X_1, \cdots, X_N$ but is uniformly $O(N^{-1-\beta})$.

Since U_0 and U_1 depends on $X_1, \cdots, X_N$ only through Z, we can compute $P_F(T^* \leqq \xi_\alpha^*(\theta) - U_0 + U_1 + O(N^{-1-\beta}))$ by taking the expectation under P_F of the conditional df of T^* given Z under P_F evaluated at the point $\xi_\alpha^*(\theta) - U_0 + U_1 + O(N^{-1-\beta})$. Corollary 2.1 provides an expansion for the conditional df of T^* given Z under P_F that is valid uniformly on E_1, and $P_F(E_1) = 1 - O(N^{-1-\beta})$. Combining these facts and simplifying as much as possible with the aid of (6.8)—(6.13) (note that (6.8), (6.11) and (6.12) imply that $U_0 = O(N^{-\beta})$ and $U_1 = O(N^{-\frac{1}{2}-\beta})$) we find

$$\begin{aligned}\pi_{P_e}(F,\theta) = 1 &- E_F\Phi(\xi_\alpha^*(\theta) - U_0 + U_1) + \frac{(1-2\lambda)}{6\{\lambda(1-\lambda)\}^{\frac{1}{2}}} \\ &\times E_F\left[\frac{\sum(Z_j - Z_\cdot)^3}{[\sum(Z_j - Z_\cdot)^2]^{\frac{3}{2}}}\phi(\xi_\alpha^*(\theta) - U_0)H_2(\xi_\alpha^*(\theta) - U_0)\right] \\ &+ \phi(\xi_\alpha^*(\theta))\left[\frac{1}{2N}\xi_\alpha^*(\theta) + \left\{\frac{(1-6\lambda+6\lambda^2)}{24\lambda(1-\lambda)N}\frac{EX_1^4}{(EX_1^2)^2}\right.\right. \\ &\left. - \frac{(1-2\lambda)^2}{8\lambda(1-\lambda)N}\right\}H_3(\xi_\alpha^*(\theta)) + \frac{(1-2\lambda)^2}{72\lambda(1-\lambda)N}\frac{(EX_1^3)^2}{(EX_1^2)^3} \\ &\left.\times H_5(\xi_\alpha^*(\theta))\right] + O(N^{-1-\beta}).\end{aligned} \tag{6.25}$$

Thus we see that the contribution of U_1 to the expansion for $\pi_{Pe}(F, \theta)$ is restricted to its contribution to $-E_F\Phi(\xi_\alpha^*(\theta) - U_0 + U_1)$. On the set E_1 we have $U_1 = \tilde{U}_1 + M$, where

$$\tilde{U}_1 = \frac{(1-2\lambda)(u_\alpha^2 - 1)}{6\{\lambda(1-\lambda)N\}^{\frac{1}{2}}} \left\{\left(\frac{\sum (Z_j - Z_.)^3}{NEX_1^3} - 1\right) - \tfrac{3}{2}\left(\frac{\sum (Z_j - Z_.)^2}{NEX_1^2} - 1\right)\right\},$$

and

$$N^{\frac{1}{2}}M = O\left(\left\{\frac{\sum (Z_j - Z_.)^2}{NEX_1^2} - 1\right\}^2 + \left|\frac{\sum (Z_j - Z_.)^2}{NEX_1^2} - 1\right| \cdot \left|\frac{\sum (Z_j - Z_.)^3}{NEX_1^3} - 1\right|\right)$$

uniformly on E_1; also $U_0 = O(N^{-\beta})$ uniformly on E_1. Let χ_{E_1} denote the indicator of E_1. Then, because $P_F(E_1) = 1 - O(N^{-1-\beta})$,

$$\begin{aligned}
E_F\Phi(\xi_\alpha^*(\theta) - U_0 + U_1) &= E_F\Phi(\xi_\alpha^*(\theta) - U_0\chi_{E_1} + \tilde{U}_1 + M\chi_{E_1}) + O(N^{-1-\beta}) \\
&= E_F\Phi(\xi_\alpha^*(\theta) - U_0\chi_{E_1}) + E_F\phi(\xi_\alpha^*(\theta) - U_0\chi_{E_1})\tilde{U}_1 \\
&\quad + O(N^{-1-\beta} + E_F\{\tilde{U}_1^2 + |M|\chi_E\}) \\
&= E_F\Phi(\xi_\alpha^*(\theta) - U_0) + \phi(\xi_\alpha^*(\theta))E_F\tilde{U}_1 \\
&\quad + O(N^{-1-\beta} + E_F\{N^{-\beta}|\tilde{U}_1| + \tilde{U}_1^2 + |M|\chi_E\}).
\end{aligned}$$

Noting that $\sum (Z_j - Z_.)^k = \sum (X_i - X_.)^k$, $E_F X_i = 0$ and $E_F|X_i|^r \leq C$ for some $r > 8$, one easily verifies that $E_F\tilde{U}_1 = O(N^{-\frac{3}{2}})$, $E_F\tilde{U}_1^2 = O(N^{-2})$ and $E_F|M|\chi_E = O(N^{-\frac{3}{2}})$. It follows that

$$E_F\Phi(\xi_\alpha^*(\theta) - U_0 + U_1) = E_F\Phi(\xi_\alpha^*(\theta) - U_0) + O(N^{-1-\beta}), \tag{6.26}$$

and hence U_1 may be omitted in (6.25) because its contribution is of negligible order. Retracing our steps back to (6.16) we conclude that the same must be true there, so that

$$\pi_{Pe}(F, \theta) = P(T^* \geq \xi_\alpha^* + O(N^{-1-\beta})) + O(N^{-1-\beta}). \tag{6.27}$$

The remainder of the proof parallels that of Theorem 5.2 in ABZ (1976). Let $\tilde{T}$ be Student's statistic as defined in (6.3). The inequality $T^* \geq a$ is algebraically equivalent to $\tilde{T} \geq a\{(N-2)/(N-a^2)\}^{\frac{1}{2}}$ on the set where $\sum (X_i - X_.)^2 \neq 0$ and provided that $a^2 < N$. Since $\sum (X_i - X_.)^2 \neq 0$ on E_1 for sufficiently large N and $\varepsilon' \leq \alpha \leq 1 - \varepsilon'$, this implies that

$$\pi_{Pe}(F, \theta) = P\left(\tilde{T} \geq \xi_\alpha^* + \frac{u_\alpha^3 - 2u_\alpha}{2N} + O(N^{-1-\beta})\right) + O(N^{-1-\beta}). \tag{6.28}$$

In the same way as in the proof of Theorem 5.2 in ABZ (1976) we show that

$$\sup_t P(t \leq \tilde{T} \leq t + O(N^{-1-\beta})) = O(N^{-1-\beta}) \tag{6.29}$$

and hence

$$\pi_{Pe}(F, \theta) = P\left(\tilde{T} \geq \xi_\alpha^* + \frac{u_\alpha^3 - 2u_\alpha}{2N}\right) + O(N^{-1-\beta}). \tag{6.30}$$

Now ξ_α^* depends only on N, λ, α and F but not on θ, and arguing as in the proof of Theorem 5.2 in ABZ (1976) we find that this together with $\pi_{Pe}(F, 0) = \alpha$ ensures that

$$t_\alpha = \xi_\alpha^* + \frac{u_\alpha^3 - 2u_\alpha}{2N} + O(N^{-1-\beta}), \tag{6.31}$$

with t_α defined as in (6.3). Combination of (6.29), (6.30) and (6.31) completes the proof. □

Although we have conducted the proof in such a way as to avoid actually establishing expansions for $\pi_{Pe}(F, \theta)$ and $\pi_{St}(F, \theta)$, the excursion from (6.16) to (6.26) and back has, in fact, brought us rather close to obtaining such expansions. Suppose that the conditions of Theorem 6.1 are satisfied but drop the assumption $\int x\,dF(x) = 0$ that was made in the proof merely for convenience. Define

$$\tilde{\eta} = \frac{\{\lambda(1-\lambda)N\}^{\frac{1}{2}}\theta}{\sigma(X_1)}, \tag{6.32}$$

$$\kappa_3(F) = \frac{E(X_1 - EX_1)^3}{\sigma^3(X_1)}, \qquad \kappa_4(F) = \frac{E(X_1 - EX_1)^4}{\sigma^4(X_1)} - 3, \tag{6.33}$$

where all moments are computed under F since only X_1 is involved. A relatively straightforward computation starting with (6.25) and (6.26) yields

$$\begin{aligned}\pi_{Pe}(F, \theta) = 1 - \Phi(u_\alpha - \tilde{\eta}) + \frac{\tilde{\eta}\phi(u_\alpha - \tilde{\eta})}{72}\Bigg[&\frac{12(1-2\lambda)\kappa_3(F)}{\{\lambda(1-\lambda)N\}^{\frac{1}{2}}}(\tilde{\eta} - 2u_\alpha)\\ &+ \frac{(1-2\lambda)^2\kappa_3^2(F)}{\lambda(1-\lambda)N}(-\tilde{\eta}^4 + 5u_\alpha\tilde{\eta}^3 - 8u_\alpha^2\tilde{\eta}^2 + 4u_\alpha^3\tilde{\eta} + 8\tilde{\eta}^2\\ &- 24u_\alpha\tilde{\eta} + 20u_\alpha^2 - 10)\\ &+ \frac{3\kappa_4(F)}{\lambda(1-\lambda)N}\{-(1 - 3\lambda + 3\lambda^2)(\tilde{\eta}^2 - 3)\\ &+ 3(1 - 5\lambda + 5\lambda^2)u_\alpha\tilde{\eta} - 3(1 - 6\lambda + 6\lambda^2)u_\alpha^2\}\\ &- \frac{18u_\alpha^2}{N}\Bigg] + O(N^{-1-\beta}),\end{aligned} \tag{6.34}$$

where β is given by (6.6) and the O symbol is uniform for fixed c, C, D, ε, ε' and δ. Theorem 6.1 ensures that the same expansion is valid for $\pi_{St}(F, \theta)$.

The case where F is normal is perhaps of most interest because both tests are then asymptotically efficient. Since Φ satisfies the stronger regularity conditions needed to replace β by $\frac{1}{2}$ we find in this case

$$\begin{aligned}\pi_{Pe}(\Phi, \theta) &= \pi_{St}(\Phi, \theta) + O(N^{-\frac{3}{2}})\\ &= 1 - \Phi(u_\alpha - \eta^*) - \frac{u_\alpha^2\eta^*\phi(u_\alpha - \eta^*)}{4N} + O(N^{-\frac{3}{2}}),\end{aligned} \tag{6.35}$$

where $\eta^* = \{\lambda(1-\lambda)N\}^{\frac{1}{2}}\theta$.

7. Deficiencies of distributionfree tests. In analogy to the one-sample case we want to compare the distributionfree tests discussed so far to the best parametric tests for the two-sample problem when the hypothesis and the alternative are both simple. The situation is more complicated than in the one-sample case because of the shift invariance of the distributionfree tests involved. Let $X_1, \cdots, X_N$ be independent and let (F, G) denote the hypothesis that $X_1, \cdots, X_m$ have common df F and $X_{m+1}, \cdots, X_N$ have common df G. For fixed F and θ and varying $\Delta \in R^1$, consider the simple hypothesis H_F and the simple alternative $K_{F,\theta,\Delta}$, where

$$H_F = (F, F), \qquad K_{F,\theta,\Delta} = (F(\cdot + \Delta\theta), F(\cdot - (1-\Delta)\theta)).$$

The shift invariance of the distributionfree tests ensures that their power against $K_{F,\theta,\Delta}$ is independent of Δ, so that it was sufficient to consider only alternatives with $\Delta = 0$ in the preceding sections. Note that the form of the locally most powerful rank test against $K_{F,\theta,\Delta}$ is also independent of Δ. However, the envelope power $\pi^+(F, \theta, \Delta)$, i.e., the power of the most powerful level α test of H_F against $K_{F,\theta,\Delta}$, does depend on Δ and the "right" Δ against which comparisons should be made is thus the value Δ_0 that minimizes the envelope power. It is given to first order by $\Delta_0 \sim \lambda$. For values of Δ whose asymptote is different there is not even an asymptotically efficient shift invariant test, so that the deficiency of a shift invariant test with respect to the best test is not of much interest in this case. Of course we shall have to provide a more precise asymptotic evaluation of Δ_0 because we are concerned with second order terms.

Suppose that F is a fixed df with density f that is positive and five times differentiable on R^1. The most powerful level α test for H_F against $K_{F,\theta,\Delta}$ rejects H_F for large values of the statistic

$$S_{\theta,\Delta} = \sum_{i=1}^{m} \log \frac{f(X_i + \Delta\theta)}{f(X_i)} + \sum_{i=m+1}^{N} \log \frac{f(X_i - (1-\Delta)\theta)}{f(X_i)}.$$

This statistic is a sum of independent rv's and we can therefore obtain an Edgeworth expansion for its df under H_F and under $K_{F,\theta,\Delta}$ and hence for the power $\pi^+(F, \theta, \Delta)$ by proceeding in the classical manner and expanding the cumulants of the statistic. In this expansion for $\pi^+(F, \theta, \Delta)$ we minimize with respect to Δ. We shall give each of these expansions but we omit the tedious computations.

Define Ψ_i by (4.15) for $i = 1, \cdots, 5$, and take

$$\begin{aligned}
&\tilde{\pi}^+(F, \theta, \Delta)\\
&= 1 - \Phi(u_\alpha - \tilde{\eta}) - \frac{\tilde{\eta}\phi(u_\alpha - \tilde{\eta})}{288}\Bigg[\frac{24}{N^{\frac12}} \frac{\tau_3 \int \Psi_1^3(t)\,dt}{\{\tau_2 \int \Psi_1^2(t)\,dt\}^{\frac32}}(-2u_\alpha + \tilde{\eta})\\
&\quad + \frac{4}{N}\frac{\tau_4 \int \Psi_1^4(t)\,dt}{\{\tau_2 \int \Psi_1^2(t)\,dt\}^2}\{-3(u_\alpha^2 - 1) + 3\tilde{\eta}u_\alpha - 2\tilde{\eta}^2\}\\
&\quad + \frac{12}{N}\frac{\tau_4 \int \Psi_2^2(t)\,dt}{\{\tau_2 \int \Psi_1^2(t)\,dt\}^2}\tilde{\eta}^2 + \frac{1}{N}\frac{\{\tau_3 \int \Psi_1^3(t)\,dt\}^2}{\{\tau_2 \int \Psi_1^2(t)\,dt\}^3}\\
&\quad \times \{8(2u_\alpha^2 - 1) - 4\tilde{\eta}(u_\alpha^3 + 3u_\alpha) + \tilde{\eta}^2(8u_\alpha^2 + 1) - 5\tilde{\eta}^3 u_\alpha + \tilde{\eta}^4\}\\
&\quad - \frac{36}{N}\frac{\tau_4}{\tau_2^2}(-u_\alpha^2 + 1 + \tilde{\eta}_\alpha)\Bigg],
\end{aligned} \tag{7.1}$$

where

$$\tilde{\eta} = [N\{(1-\lambda)\Delta^2 + \lambda(1-\Delta)^2\} \textstyle\int \Psi_1{}^2(t)\,dt]^{\frac{1}{2}}\theta\,, \tag{7.2}$$

$$\tau_k = (1-\lambda)\Delta^k + \lambda(\Delta - 1)^k\,, \qquad k = 2, 3, 4\,. \tag{7.3}$$

LEMMA 7.1. *Let F satisfy (5.6) for $m_i = 5/i$, $i = 1, \cdots, 5$, and suppose that positive numbers D, D', ε and ε' exist such that $0 \leqq \theta \leqq DN^{-\frac{1}{2}}$, $|\Delta\theta| \leqq D'N^{-\frac{1}{2}}$, $\varepsilon \leqq \lambda \leqq 1 - \varepsilon$ and $\varepsilon' \leqq \alpha \leqq 1 - \varepsilon'$. Then there exists $B > 0$ depending only on F, D, D', ε and ε' such that*

$$|\pi^+(F, \theta, \Delta) - \tilde{\pi}^+(F, \theta, \Delta)| \leqq BN^{-\frac{3}{2}}\,. \tag{7.4}$$

PROOF. Under the conditions of the lemma we find that under $H_F = (F, F)$

$$\begin{aligned}
P_F\Big(&\frac{S_{\theta,\Delta} + \frac{1}{2}\tilde{\eta}^2}{\tilde{\eta}} \leq x\Big) \\
&= \Phi(x) - \frac{\phi(x)}{288}\Big[\frac{24}{N^{\frac{1}{2}}}\frac{\tau_3 \int \Psi_1{}^3(t)\,dt}{\{\tau_2 \int \Psi_1{}^2(t)\,dt\}^{\frac{3}{2}}}\{2(x^2-1) - 3\tilde{\eta}x + \tilde{\eta}^2\} \\
&\quad + \frac{4}{N}\frac{\tau_4 \int \Psi_1{}^4(t)\,dt}{\{\tau_2 \int \Psi_1{}^2(t)\,dt\}^2}\{3(x^3 - 3x) - 6\tilde{\eta}(x^2-1) + 5\tilde{\eta}^2x - 2\tilde{\eta}^3\} \\
&\quad + \frac{12}{N}\frac{\tau_4 \int \Psi_2{}^2(t)\,dt}{\{\tau_2 \int \Psi_1{}^2(t)\,dt\}^2}\{-\tilde{\eta}^2x + \tilde{\eta}^3\} \\
&\quad + \frac{1}{N}\frac{\{\tau_3 \int \Psi_1{}^3(t)\,dt\}^2}{\{\tau_2 \int \Psi_1{}^2(t)\,dt\}^3}\{4(x^5 - 10x^3 + 15x) - 12\tilde{\eta}(x^4 - 6x^2 + 3) \\
&\quad + 13\tilde{\eta}^2(x^3 - 3x) - 6\tilde{\eta}^3(x^2-1) + \tilde{\eta}^4x\} \\
&\quad + \frac{36\tau_4}{N\tau_2{}^2}\{-(x^3-3x) + 2\tilde{\eta}(x^2-1) - \tilde{\eta}^2x\}\Big] + O(N^{-\frac{3}{2}})\,,
\end{aligned}$$

whereas under $K_{F,\theta,\Delta}$,

$$\begin{aligned}
P\Big(&\frac{S_{\theta,\Delta} - \frac{1}{2}\tilde{\eta}^2}{\tilde{\eta}} \leq x\Big) \\
&= \Phi(x) - \frac{\phi(x)}{288}\Big[\frac{24}{N^{\frac{1}{2}}}\frac{\tau_3 \int \Psi_1{}^3(t)\,dt}{\{\tau_2 \int \Psi_1{}^2(t)\,dt\}^{\frac{3}{2}}}\{2(x^2-1) + 3\tilde{\eta}x + \tilde{\eta}^2\} \\
&\quad + \frac{4}{N}\frac{\tau_4 \int \Psi_1{}^4(t)\,dt}{\{\tau_2 \int \Psi_1{}^2(t)\,dt\}^2}\{3(x^3 - 3x) + 6\tilde{\eta}(x^2-1) + 5\tilde{\eta}^2x + 2\tilde{\eta}^3\} \\
&\quad + \frac{12}{N}\frac{\tau_4 \int \Psi_2{}^2(t)\,dt}{\{\tau_2 \int \Psi_1{}^2(t)\,dt\}^2}\{-\tilde{\eta}^2x - \tilde{\eta}^3\} \\
&\quad + \frac{1}{N}\frac{\{\tau_3 \int \Psi_1{}^3(t)\,dt\}^2}{\{\tau_2 \int \Psi_1{}^2(t)\,dt\}^3}\{4(x^5 - 10x^3 + 15x) + 12\tilde{\eta}(x^4 - 6x^2 + 3) \\
&\quad + 13\tilde{\eta}^2(x^3 - 3x) + 6\tilde{\eta}^3(x^2-1) + \tilde{\eta}^4x\} \\
&\quad + \frac{36\tau_4}{N\tau_2{}^2}\{-(x^3-3x) - 2\tilde{\eta}(x^2-1) - \tilde{\eta}^2x\}\Big] + O(N^{-\frac{3}{2}})\,.
\end{aligned}$$

The remainder terms $O(N^{-\frac{3}{2}})$ are uniform for fixed F, D, D' ε and ε'. Together these expansions yield (7.4). □

For large values of $|\Delta\theta N^{\frac{1}{2}}|$, both $\pi^+(F, \theta, \Delta)$ and $\tilde{\pi}^+(F, \theta, \Delta)$ will come close to 1 as $N \to \infty$. It follows that an asymptotic expansion for the value Δ_0 that minimizes $\pi^+(F, \theta, \Delta)$ may be obtained by minimizing $\tilde{\pi}^+(F, \theta, \Delta)$ instead. This yields

$$\Delta_0 = \lambda + \frac{\{\lambda(1-\lambda)\}^{\frac{1}{2}}}{4N^{\frac{1}{2}}} \frac{\int \Psi_1^3(t)\,dt}{\{\int \Psi_1^2(t)\,dt\}^{\frac{3}{2}}}(-2u_\alpha + \eta^*) + O(N^{-1}) \tag{7.5}$$

with η^* as in (5.29). Since the derivative of $\tilde{\eta}$ with respect to Δ vanishes at $\Delta = \lambda$, (7.5) is sufficient to determine $\tilde{\eta}$ and $\tilde{\pi}^+$ for $\Delta = \Delta_0$ up to a remainder $O(N^{-\frac{3}{2}})$. Noting that indeed $|\Delta_0\theta| = O(N^{-\frac{1}{2}})$, we substitute (7.5) for Δ in (7.1)—(7.3) and neglecting terms that are $O(N^{-\frac{3}{2}})$, we find that $\tilde{\pi}^+(F, \theta, \Delta_0)$ reduces to

$$\begin{aligned}
\tilde{\pi}^+(F, \theta) = 1 - \Phi(u_\alpha - \eta^*) &+ \frac{\eta^*\phi(u_\alpha - \eta^*)}{288} \\
&\times \Bigg[\frac{24(1-2\lambda)}{\{\lambda(1-\lambda)N\}^{\frac{1}{2}}} \frac{\int \Psi_1^3(t)\,dt}{\{\int \Psi_1^2(t)\,dt\}^{\frac{3}{2}}}(-2u_\alpha + \eta^*) \\
&+ \frac{4(1-3\lambda+3\lambda^2)}{\lambda(1-\lambda)N} \frac{\int \Psi_1^4(t)\,dt}{\{\int \Psi_1^2(t)\,dt\}^2} \\
&\times \{3(u_\alpha^2 - 1) - 3\eta^* u_\alpha + 2\eta^{*2}\} \\
&- \frac{12(1-3\lambda+3\lambda^2)}{\lambda(1-\lambda)N} \frac{\int \Psi_2^2(t)\,dt}{\{\int \Psi_1^2(t)\,dt\}^2}\eta^{*2} - \frac{9}{N}\frac{\{\int \Psi_1^3(t)\,dt\}^2}{\{\int \Psi_1^2(t)\,dt\}^3} \\
&\times (2u_\alpha - \eta^*)^2 + \frac{(1-2\lambda)^2}{\lambda(1-\lambda)N}\frac{\{\int \Psi_1^3(t)\,dt\}^2}{\{\int \Psi_1^2(t)\,dt\}^3}\{-8(2u_\alpha^2 - 1) \\
&+ 4\eta^*(u_\alpha^3 + 3u_\alpha) - \eta^{*2}(8u_\alpha^2 + 1) + 5\eta^{*3}u_\alpha - \eta^{*4}\} \\
&+ \frac{36(1-3\lambda+3\lambda^2)}{\lambda(1-\lambda)N}\{-(u_\alpha^2 - 1) + \eta^* u_\alpha\}\Bigg]
\end{aligned} \tag{7.6}$$

with η^* as in (5.29). Summarizing, we have

LEMMA 7.2. *Let F satisfy* (5.6) *for $m_i = 5/i$, $i = 1, \cdots, 5$, and suppose that positive numbers D, ε and ε' exist such that $0 \leq \theta \leq DN^{-\frac{1}{2}}$, $\varepsilon \leq \lambda \leq 1 - \varepsilon$ and $\varepsilon' \leq \alpha \leq 1 - \varepsilon'$. Then there exists $B > 0$ depending only on F, D, ε and ε' such that*

$$|\pi^+(F, \theta, \Delta_0) - \tilde{\pi}^+(F, \theta)| \leq BN^{-\frac{3}{2}}. \tag{7.7}$$

For the same testing problem Theorem 5.2 provides an expansion for the power $\pi(F, \theta)$ of the locally most powerful rank test. Together, Theorem 5.2 and Lemma 7.2 enable us to find an asymptotic expression for the deficiency d_N of the locally most powerful rank test with respect to the most powerful parametric test for H_F against K_{F,θ,Δ_0}. To ensure that F satisfies the assumptions of both Theorem 5.2 and Lemma 7.2, we require that $F \in \mathscr{F}_1$, where

DEFINITION 7.1. $\mathscr{F}_1$ is the class of df's F on R^1 with positive and five times differentiable densities f and such that (5.6) is satisfied for $i = 1, \cdots, 5$ with $m_1 = 6$, $m_2 = 3$, $m_3 = \frac{5}{3}$, $m_4 = \frac{5}{4}$, $m_5 = 1$, and such that (5.7) holds.

Furthermore, we define

$$
\begin{aligned}
\tilde{d}_{N,0} = \frac{1}{48}\Bigg[&\frac{4\int \Psi_1^4(t)\,dt}{\{\int \Psi_1^2(t)\,dt\}^2}\{3(u_\alpha^2 - 1) - 2\eta^* u_\alpha\} \\
&- \frac{4(1-3\lambda+3\lambda^2)}{\lambda(1-\lambda)}\left\{\frac{\int \Psi_1^4(t)\,dt - 3\int \Psi_2^2(t)\,dt}{\{\int \Psi_1^2(t)\,dt\}^2} + 3\right\}\eta^{*2} \\
&- \frac{3\{\int \Psi_1^3(t)\,dt\}^2}{\{\int \Psi_1^2(t)\,dt\}^3}(2u_\alpha - \eta^*)^2 - \frac{3(1-2\lambda)^2}{\lambda(1-\lambda)}\frac{\{\int \Psi_1^3(t)\,dt\}^2}{\{\int \Psi_1^2(t)\,dt\}^3}\eta^{*2} \\
&- 12\{u_\alpha^2 + 3 - 2\eta^* u_\alpha\}\Bigg], \qquad (7.8)
\end{aligned}
$$

$$\tilde{d}_{N,1} = \tilde{d}_{N,0} + \frac{1}{\int \Psi_1^2(t)\,dt}\sum_{j=1}^N \sigma^2(\Psi_1(U_{j:N}))\,,$$

$$\tilde{d}_{N,2} = \tilde{d}_{N,0} + \frac{1}{\int \Psi_1^2(t)\,dt}\int_{N^{-1}}^{1-N^{-1}} (\Psi_1'(t))^2 t(1-t)\,dt\,,$$

$$\tilde{d}_{N,3} = \tilde{d}_{N,0} + \frac{1}{\int \Psi_1^2(t)\,dt}\sum_{j=1}^N E\left(\Psi_1(U_{j:N}) - \Psi_1\left(\frac{j}{N+1}\right)\right)^2,$$

where Ψ_i and η^* are given by (4.15) and (5.29) and $U_{j:N}$ is the jth order statistic of a sample of size N from the uniform distribution on (0, 1).

THEOREM 7.1. *Let d_N be the deficiency of the locally most powerful rank test with respect to the most powerful test for testing H_F against K_{F,θ,Δ_0} on the basis of $X_1, \cdots, X_N$ and at level α. Suppose that $F \in \mathscr{F}_1$, $cN^{-\frac{1}{2}} \leq \theta \leq CN^{-\frac{1}{2}}$, $\varepsilon \leq \lambda \leq 1-\varepsilon$ and $\varepsilon' \leq \alpha \leq 1-\varepsilon'$ for positive c, C, ε and ε'. Then, for every fixed F, c, C, ε and ε', there exist positive numbers $B, \delta_1, \delta_2, \cdots$ with $\lim_{N\to\infty}\delta_N = 0$ such that*

$$|d_N - \tilde{d}_{N,1}| \leq \delta_N\,, \qquad (7.9)$$

$$|d_N - \tilde{d}_{N,2}| \leq \delta_N + BN^{-\frac{1}{2}}\int_{N^{-1}}^{1-N^{-1}} (\Psi_1'(t))^2\{t(1-t)\}^{\frac{1}{2}}\,dt\,. \qquad (7.10)$$

If in the above the locally most powerful rank test is replaced by the rank test with the corresponding approximate scores $a_j = -\Psi_1(j/(N+1))$ then

$$|d_N - \tilde{d}_{N,3}| \leq \delta_N \qquad (7.11)$$

and (7.10) *continues to hold.*

PROOF. Let us first consider the locally most powerful rank test and show that the expansions (5.35) and (7.7) yield (7.9). The conditions of the theorem ensure that η^*, $\{\lambda(1-\lambda)\}^{-1}$ and u_α are bounded. As $\mathscr{F}_1 \subset \mathscr{F}$, (5.18) holds and the reasoning leading up to (5.46) gives

$$
\begin{aligned}
N^{-1}\sum_{j=1}^N \sigma^2(\Psi_1(U_{j:N})) &\leq N^{-1}\sum_{j=1}^N E\left\{\Psi_1(U_{j:N}) - \Psi_1\left(\frac{j}{N+1}\right)\right\}^2 \\
&= o(N^{-\frac{3}{2}})\,. \qquad (7.12)
\end{aligned}
$$

In view of these remarks we find from (5.35), (5.32) and (5.31) that the power

$\pi(F, \theta)$ of the locally most powerful rank test satisfies

$$\pi(F,\theta) = 1 - \Phi(u_\alpha - \eta^*) + \frac{\eta^*\phi(u_\alpha - \eta^*)}{12} \frac{(1-2\lambda)}{\{\lambda(1-\lambda)N\}^{\frac{1}{2}}} \frac{\int \Psi_1^3(t)\,dt}{\{\int \Psi_1^2(t)\,dt\}^{\frac{3}{2}}}(-2u_\alpha + \eta^*) + o(N^{-\frac{1}{2}}) . \tag{7.13}$$

From Lemma 7.2 and (7.6) it is clear that $\pi^+(F, \theta, \Delta_0)$ also equals the right-hand side of (7.13). Since d_N is obtained by replacing N and η^* by $(N + d_N)$ and $\eta^*(1 + d_N N^{-1})^{\frac{1}{2}}$ in $\pi(F, \theta)$ and equating the result to $\pi^+(F, \theta, \Delta_0)$, and since η^* is bounded away from zero, we find that $d_N = o(N^{\frac{1}{2}})$.

Having obtained this crude bound for d_N we study the effect of the substitution of $(N + d_N)$ and $\eta^*(1 + d_N N^{-1})^{\frac{1}{2}}$ for N and η^* a bit more carefully. The effect on $\pi_0^*(F, \theta)$ as given in (5.31) is obviously the addition of a term

$$\frac{\eta^*\phi(u_\alpha - \eta^*)}{2N} d_N + o(N^{-\frac{5}{4}}) ; \tag{7.14}$$

to prove that this remains true for $\pi_1^*(F, \theta)$ in (5.32) it is clearly sufficient to show that

$$\frac{1}{N} \sum_{j=1}^{N} \sigma^2(\Psi_1(U_{j:N})) = \frac{1}{N+1} \sum_{j=1}^{N+1} \sigma^2(\Psi_1(U_{j:N+1})) + o(N^{-\frac{5}{4}}) . \tag{7.15}$$

Once this has been established, (5.35) and (7.7) imply that an expansion for d_N may be obtained by equating (7.14) to $\tilde{\pi}^+(F, \theta) - \pi_1^*(F, \theta) + o(N^{-1})$ and an easy computation yields (7.9).

To prove (7.15), we let $b_{j,N}$ denote the density of $U_{j:N}$ and we note the well-known recurrence relation $(N + 1)b_{j,N} = jb_{j+1,N+1} + (N - j + 1)b_{j,N+1}$. We have

$$\begin{aligned}\sigma^2(\Psi_1(U_{j:N})) &= \frac{j}{N+1} E\{\Psi_1(U_{j+1:N+1}) - E\Psi_1(U_{j:N})\}^2 \\ &\quad + \frac{N-j+1}{N+1} E\{\Psi_1(U_{j:N+1}) - E\Psi_1(U_{j:N})\}^2 \\ &= \frac{j}{N+1}\sigma^2(\Psi_1(U_{j+1:N+1})) + \frac{N-j+1}{N+1}\sigma^2(\Psi_1(U_{j:N+1})) \\ &\quad + \frac{j(N-j+1)}{(N+1)^2}\{E[\Psi_1(U_{j+1:N+1}) - \Psi_1(U_{j:N+1})]\}^2 .\end{aligned}$$

Summation on j gives

$$\begin{aligned}\frac{1}{N} \sum_{j=1}^{N} \sigma^2(\Psi_1(U_{j:N})) &= \frac{1}{N+1} \sum_{j=1}^{N+1} \sigma^2(\Psi_1(U_{j:N+1})) \\ &\quad + \sum_{j=1}^{N} \frac{j(N-j+1)}{N(N+1)^2}\{E[\Psi_1(U_{j+1:N+1}) - \Psi_1(U_{j:N+1})]\}^2 .\end{aligned} \tag{7.16}$$

By Fubini's theorem and (5.18),

$$\begin{aligned}|E[\Psi_1(U_{j+1:N+1}) - \Psi_1(U_{j:N+1})]| &\leqq E \int_{U_{j:N+1}}^{U_{j+1:N+1}} |\Psi_1'(t)|\, dt \\ &= \int_0^1 |\Psi_1'(t)| P(U_{j:N+1} \leqq t < U_{j+1:N+1})\, dt \\ &= \binom{N+1}{j} \int_0^1 |\Psi_1'(t)| t^j (1-t)^{N+1-j}\, dt \\ &\leqq M\left(\frac{j}{N+1}\right) \frac{(N+1)^{\frac{4}{3}}}{\{j(N+1-j)\}^{\frac{7}{6}}},\end{aligned}$$

where M is a bounded function on (0, 1) with $\lim_{t\to 0,1} M(t) = 0$. Hence

$$\sum_{j=1}^{N} \frac{j(N-j+1)}{N(N+1)^2} \{E[\Psi_1(U_{j+1:N+1}) - \Psi_1(U_{j:N+1})]\}^2$$
$$= o\left(N^{-2} \int_{N^{-1}}^{1-N^{-1}} \{t(1-t)\}^{-\frac{4}{3}}\, dt\right) = o(N^{-\frac{5}{3}}).$$

Together with (7.16), this proves (7.15) and establishes expansion (7.9).

For the rank test based on the approximate scores the proof that (5.39) and (7.7) yield expansion (7.11) proceeds in the same way as above, the only difference being that instead of (7.15) we now show that

$$\frac{1}{N} \sum_{j=1}^{N} E\left(\Psi_1(U_{j:N}) - \Psi_1\left(\frac{j}{N+1}\right)\right)^2 \tag{7.17}$$
$$= \frac{1}{N+1} \sum_{j=1}^{N+1} E\left(\Psi_1(U_{j:N+1}) - \Psi_1\left(\frac{j}{N+2}\right)\right)^2 + o(N^{-\frac{5}{3}}).$$

Using the recurrence relation for $b_{j,N}$ again, we find after some arithmetic

$$\frac{1}{N} \sum_{j=1}^{N} E\left(\Psi_1(U_{j:N}) - \Psi_1\left(\frac{j}{N+1}\right)\right)^2$$
$$\begin{aligned}&= \frac{1}{N+1} \sum_{j=1}^{N+1} E\left(\Psi_1(U_{j:N+1}) - \Psi_1\left(\frac{j}{N+2}\right)\right)^2 \\ &\quad + \frac{2}{N+1} \sum_{j=1}^{N+1} \left\{\Psi_1\left(\frac{j}{N+2}\right) - \frac{j-1}{N} \Psi_1\left(\frac{j-1}{N+1}\right) \right. \\ &\quad \left. - \frac{N-j+1}{N} \Psi_1\left(\frac{j}{N+1}\right)\right\} E\left(\Psi_1(U_{j:N+1}) - \Psi_1\left(\frac{j}{N+2}\right)\right) \\ &\quad + \frac{1}{N+1} \sum_{j=1}^{N+1} \left\{\frac{j-1}{N}\left(\Psi_1\left(\frac{j}{N+2}\right) - \Psi_1\left(\frac{j-1}{N+1}\right)\right)^2 \right. \\ &\quad \left. + \frac{N-j+1}{N}\left(\Psi_1\left(\frac{j}{N+2}\right) - \Psi_1\left(\frac{j}{N+1}\right)\right)^2\right\}.\end{aligned} \tag{7.18}$$

Now (5.18) ensures that

$$\left|\Psi_1\left(\frac{j}{N+2}\right) - \Psi_1\left(\frac{j-1}{N+1}\right)\right| \leqq \tilde{M}\left(\frac{j}{N+2}\right) \frac{(N-j+2)}{(N+2)^2} \left\{\frac{j(N-j+2)}{(N+2)^2}\right\}^{-\frac{7}{6}}$$
$$\text{for} \quad j = 2, \cdots, N+1,$$

$$\left|\Psi_1\left(\frac{j}{N+2}\right) - \Psi_1\left(\frac{j}{N+1}\right)\right| \leqq \tilde{M}\left(\frac{j}{N+2}\right) \frac{j}{(N+2)^2} \left\{\frac{j(N-j+2)}{(N+2)^2}\right\}^{-\frac{7}{6}}$$
$$\text{for} \quad j = 1, \cdots, N,$$

where $\tilde{M}$ is a bounded function on $(0, 1)$ with $\lim_{t\to 0,1} \tilde{M}(t) = 0$. Similarly, (5.7), Lemma A2.3 in ABZ (1976) and (5.18) imply that

$$\left|E\Psi_1(U_{j:N+1}) - \Psi_1\left(\frac{j}{N+2}\right)\right| \leq \tilde{M}_2\left(\frac{j}{N+2}\right) N^{-1} \left\{\frac{j(N-j+2)}{(N+2)^2}\right\}^{-\frac{7}{8}} \quad \text{for} \quad j = 1, \cdots, N+1 .$$

It follows that both the second and third terms on the right in (7.18) are

$$o\left(N^{-2} \textstyle\int_{N^{-1}}^{1-N^{-1}} \{t(1-t)\}^{-\frac{7}{4}}\, dt\right) = o(N^{-\frac{5}{3}}) ,$$

which proves (7.17) and therefore (7.11).

Finally, the validity of expansion (7.10) for exact as well as approximate scores is a simple consequence of (7.9) and (7.11) and the fact that Theorem 5.2 clearly implies that both $\sum \sigma^2(\Psi_1(U_{j:N}))$ and $\sum E(\Psi_1(U_{j:N}) - \Psi_1(j/(N+1)))^2$ equal

$$\textstyle\int_{N^{-1}}^{1-N^{-1}} (\Psi_1'(t))^2 t(1-t)\, dt + o(1) + O\left(N^{-\frac{1}{2}} \int_{N^{-1}}^{1-N^{-1}} (\Psi_1'(t))^2 \{t(1-t)\}^{\frac{1}{2}}\, dt\right) .$$

This completes the proof of the theorem. □

Like Theorems 5.1 and 5.2, Theorem 7.1 presents us with a choice between an expansion with remainder $o(1)$ and one which is more explicit but may have a remainder of larger order under the conditions of the theorem. If $\Psi_1'(t) = o(\{t(1-t)\}^{-1})$ for $t \to 0, 1$, then $d_N = \tilde{d}_{N,2} + o(1)$ for exact as well as approximate scores and expansion (7.10) is obviously preferable. This appears to be the most common case. However, if $\Psi_1'(t)$ is of exact order $\{t(1-t)\}^{-1}$, then (7.10) yields only

$$d_N = \frac{\int_{N^{-1}}^{1-N^{-1}} (\Psi_1'(t))^2 t(1-t)\, dt}{\int_0^1 \Psi_1^2(t)\, dt} + O(1) = O(\log N) .$$

Finally, if $\Psi_1'(t) \sim \{t(1-t)\}^{-1-\delta}$ for $t \to 0, 1$ and some $0 < \delta < \frac{1}{6}$, then (7.10) reduces to $d_N = O(N^{2\delta})$.

In general, all we can say under the conditions of Theorem 7.1 is that

$$\text{(7.19)} \qquad d_N = \frac{\sum \sigma^2(\Psi_1(U_{j:N}))}{\int \Psi_1^2(t)\, dt} + O(1) = O\left(\textstyle\int_{N^{-1}}^{1-N^{-1}} (\Psi_1'(t))^2 t(1-t)\, dt\right) = o(N^{\frac{1}{3}})$$

for exact scores, and that

$$\text{(7.20)} \qquad d_N = \frac{\sum E(\Psi_1(U_{j:N}) - \Psi_1(j/(N+1)))^2}{\int \Psi_1^2(t)\, dt} + O(1)$$
$$= O\left(\textstyle\int_{N^{-1}}^{1-N^{-1}} (\Psi_1'(t))^2 t(1-t)\, dt\right) = o(N^{\frac{1}{3}})$$

for approximate scores. Even this result, however, is rather surprising because one might have expected these deficiencies to be of the order $N^{\frac{1}{2}}$. The reason that they are of smaller order than $N^{\frac{1}{2}}$ is of course that the power expansions for the rank tests in Theorem 5.2 and for the most powerful test in Lemma 7.2 agree not only in their leading terms of order 1 but also in their second order terms of order $N^{-\frac{1}{2}}$. It is only in third order terms that differences begin to show up. Borrowing a phrase from Pfanzagl (1977), who noted the same phenomenon in the parametric one-sample problem, first order efficiency apparently implies second order efficiency in the cases considered. Note that results very similar to (7.19) and (7.20) were obtained for one-sample rank tests in ABZ (1976). In that case, however, there is no cause for surprise because certain symmetries that are present in the nonparametric one-sample problem ensure that there is no term of order $N^{-\frac{1}{2}}$ in any of the power expansions. Finally we should perhaps point out that in the present two-sample case, the fact that we have evaluated the envelope power for Δ_0 as given in (7.5) instead of for the conventional choice $\Delta = \lambda$, is of no consequence for these considerations. For $\Delta = \lambda$ the term involving $(2u_\alpha - \eta^*)^2$ should simply be omitted from (7.6) and (7.8) and this does not influence the qualitative behavior of $\bar{\pi}^+$ or $\bar{d}_{N,i}$.

To provide some examples of Theorem 7.1 we compute the expansion (7.10) for the special case where F is the logistic df $\Lambda(x) = (1 + e^{-x})^{-1}$ or the normal df Φ. The computations resemble those at the end of Section 5. Suppose that $c \leqq \theta N^{\frac{1}{2}} \leqq C$, $\varepsilon \leqq \lambda \leqq 1 - \varepsilon$ and $\varepsilon' \leqq \alpha \leqq 1 - \varepsilon'$ for positive c, C, ε and ε'. As both examples concern symmetric distributions for which $\int \Psi_1^3(t)\,dt = 0$, the second order term in (7.5) vanishes so that we may take $\Delta_0 = \lambda$ in both cases. For $F = \Lambda$ we are therefore concerned with the problem of testing the hypothesis (Λ, Λ) against the alternative $(\Lambda(\cdot + \lambda\theta), \Lambda(\cdot - (1 - \lambda)\theta))$ and d_N denotes the deficiency of Wilcoxon's two-sample test with respect to the most powerful test for this problem. We find

$$(7.21) \qquad d_N = \tfrac{1}{20}\left[4u_\alpha^2 + 16 + 4\eta^* u_\alpha + \frac{1 - 3\lambda + 3\lambda^2}{\lambda(1 - \lambda)}\eta^{*2}\right] + o(1)$$

with $\eta^* = \{\lambda(1 - \lambda)N/3\}^{\frac{1}{2}}\theta$. In this example d_N remains bounded as $N \to \infty$.

In the second example we consider the testing problem (Φ, Φ) versus $(\Phi(\cdot + \lambda\theta), \Phi(\cdot - (1 - \lambda)\theta))$. Now d_N is the deficiency of the two-sample normal scores test (or van der Waerden's two-sample test) with respect to the most powerful test based on the difference of the sample means. We obtain

$$(7.22) \qquad d_N = \log\log N + \tfrac{1}{2}(u_\alpha^2 - 3) + \log 2 + \gamma + o(1)\,,$$

where γ denotes Euler's constant (cf. (5.50)). Now $d_N \sim \log\log N \to \infty$ as $N \to \infty$. Note that there is no dependence on θ or λ in this expansion.

So far in this section we have compared distributionfree tests to the most powerful test for a simple hypothesis against a simple alternative. However, all distributionfree tests occurring in this paper—rank tests as well as the permutaion test discussed in Section 6—are invariant under changes of location and scale. It would therefore be more realistic to compare these tests to the uniformly most powerful location and scale invariant test, if such a test exists. For the two-sample normal location problem Student's test answers this description and its power would therefore be a more suitable basis for comparison than the envelope power. For the problem of testing (Φ, Φ) against $(\Phi(\cdot + \lambda\theta), \Phi(\cdot - (1-\lambda)\theta))$ the power of the most powerful test equals $1 - \Phi(u_\alpha - \eta^*)$ with $\eta^* = \{\lambda(1-\lambda)N\}^{\frac{1}{2}}\theta$. Assuming again that $c \leqq \theta N^{\frac{1}{2}} \leqq C$, $\varepsilon \leqq \lambda \leqq 1 - \varepsilon$ and $\varepsilon' \leqq \alpha \leqq 1 - \varepsilon'$ for positive c, C, ε and ε', the power of Student's two-sample test is given by (6.35) and its deficiency with respect to the most powerful test is therefore equal to $\frac{1}{2}u_\alpha^2 + o(1)$. It follows from (7.22) that the deficiency of the two-sample normal scores test (or van der Waerden's two-sample test) with respect to Student's two-sample test for the normal location problem is given by

$$d_N = \log\log N - \tfrac{3}{2} + \log 2 + \gamma + o(1)\,, \tag{7.23}$$

where now the expansion does not even depend on α. Since both tests are location invariant, (7.23) also denotes the deficiency for testing (Φ, Φ) against $(\Phi, \Phi(\cdot - \theta))$.

We conclude this section by comparing the permutation test discussed in Section 6 to Student's test. Theorem 7.2 is an immediate consequence of Theorem 6.1, expansion (6.34) and (6.8).

THEOREM 7.2. *Suppose that positive numbers c, c', C, D, ε, ε', δ and $r > 8$ exist such that the conditions of Theorem 6.1 are satisfied and that $\theta \geqq c'N^{-\frac{1}{2}}$. Let d_N denote the deficiency of the permutation test based on the sample means with respect to Student's two-sample test for testing (F, F) against $(F, F(\cdot - \theta))$ on the basis of $X_1, \cdots, X_N$ and at level α. Then there exist $B > 0$ depending only on c, c', C, D, ε, ε' and δ, and $\beta > 0$ depending only on r such that*

$$d_N \leqq BN^{-\beta}\,. \tag{7.24}$$

The case $F = \Phi$ is of course of most interest because then the theorem asserts that for the normal location problem there exists a distributionfree test whose deficiency with respect to the best location and scale invariant test tends to zero. We note that the remark at the end of Section 6 implies that in this case (7.24) may be replaced by $d_N \leqq BN^{-\frac{1}{2}}$. For $F \neq \Phi$ the theorem merely shows how closely the permutation test resembles Student's test with the correct significance level for F.

8. Expansions and deficiencies for related estimators. Let $X_1, \cdots, X_N$ be independent and let (F, G) denote the hypothesis that $X_1, \cdots, X_m$ have common df F and $X_{m+1}, \cdots, X_N$ have common df G. Let $T = T(X_1, \cdots, X_N)$ be the rank statistic given by (2.2) and suppose that the scores a_j are nondecreasing in $j = 1, \cdots, N$. Define the statistic M by

$$\begin{aligned} M(X_1, &\cdots, X_N) \\ &= \tfrac{1}{2}\sup\{t : T(X_1, \cdots, X_m, X_{m+1} - t, \cdots, X_N - t) > \lambda \textstyle\sum a_j\} \\ &\quad + \tfrac{1}{2}\inf\{t : T(X_1, \cdots, X_m, X_{m+1} - t, \cdots, X_N - t) < \lambda \textstyle\sum a_j\}\,. \end{aligned} \tag{8.1}$$

Under the model $(F, F(\cdot - \mu))$, M was proposed as an estimator of μ by Hodges and Lehmann (1963). They showed that the normal approximation to the power of the level $\frac{1}{2}$ test based on T for contiguous location alternatives can be used to establish asymptotic normality of M. In the same way we shall show that a power expansion yields an expansion for the df of $N^{\frac{1}{2}}(M - \mu)$. Note that we do not make the assumption of Hodges and Lehmann (1963) that the distribution of T under (F, F) is symmetric about $\lambda \sum a_j$, which occurs, e.g., when either $\lambda = \frac{1}{2}$ or the scores are antisymmetric. As a result the power expansion involved will be for the test based on T at level $\bar{\alpha} = \frac{1}{2} + O(N^{-\frac{1}{2}})$ rather than at level $\frac{1}{2}$, but for our deficiency computations this will not make any difference. We shall restrict attention to the case where T is the statistic of the locally most powerful rank test or its approximate scores analogue, so that the a_j will be exact or approximate scores generated by the score function $-\Psi_1$, with Ψ_1 as in (4.15). To ensure that the scores are nondecreasing we require that the density f of F is strongly unimodal, i.e., that $\log f$ is concave.

Let $\mathscr{F}$ be given by Definition 5.1, let $\pi(\alpha, F, \theta)$ denote the power of the level α right-sided test based on T against the alternative $(F, F(\cdot - \theta))$ and define

$$\bar{\alpha} = \frac{1}{2} + \frac{(1-2\lambda)}{6\{2\pi\lambda(1-\lambda)N\}^{\frac{1}{2}}} \frac{\int \Psi_1^3(t)\,dt}{\{\int \Psi_1^2(t)\,dt\}^{\frac{3}{2}}} . \tag{8.2}$$

Furthermore define, with Ψ_i as in (4.15),

$$\begin{aligned} \bar{L}_0(x) = \Phi(x) - \frac{\phi(x)}{288}\Bigg[& \frac{24(1-2\lambda)}{\{\lambda(1-\lambda)N\}^{\frac{1}{2}}} \frac{\int \Psi_1^3(t)\,dt}{\{\int \Psi_1^2(t)\,dt\}^{\frac{3}{2}}}(x^2+2) \\ & - \frac{4}{\lambda(1-\lambda)N} \frac{\int \Psi_1^4(t)\,dt}{\{\int \Psi_1^2(t)\,dt\}^2} \\ & \times \{5(1-3\lambda+3\lambda^2)x^3 - 3(1-6\lambda+6\lambda^2)x\} \\ & + \frac{48(1-3\lambda+3\lambda^2)}{\lambda(1-\lambda)N} \frac{\int \Psi_2^2(t)\,dt}{\{\int \Psi_1^2(t)\,dt\}^2} x^3 \\ & + \frac{(1-2\lambda)^2}{\lambda(1-\lambda)N} \frac{\{\int \Psi_1^3(t)\,dt\}^2}{\{\int \Psi_1^2(t)\,dt\}^3}(x^5 - 4x^3 - 12x) \\ & - \frac{36}{\lambda(1-\lambda)N}\{(1-3\lambda+3\lambda^2)x^3 + x\}\Bigg], \end{aligned} \tag{8.3}$$

$$\bar{L}_1(x) = \bar{L}_0(x) - \frac{x\phi(x)}{2N\int \Psi_1^2(t)\,dt} \sum_{j=1}^N \sigma^2(\Psi_1(U_{j:N})) ,$$

$$\bar{L}_2(x) = \bar{L}_0(x) - \frac{x\phi(x)}{2N\int \Psi_1^2(t)\,dt} \int_{N^{-1}}^{1-N^{-1}} (\Psi_1'(t))^2 t(1-t)\,dt ,$$

$$\bar{L}_3(x) = \bar{L}_0(x) - \frac{x\phi(x)}{2N\int \Psi_1^2(t)\,dt} \sum_{j=1}^N E\left(\Psi_1(U_{j:N}) - \Psi_1\left(\frac{j}{N+1}\right)\right)^2 .$$

Probabilities under the model $(F, F(\cdot - \mu))$ are denoted by $P_{F,\mu}$.

THEOREM 8.1. *Suppose that $F \in \mathscr{F}$, that f is strongly unimodal and that either $a_j = -E\Psi_1(U_{j:N})$ for $j = 1, \cdots, N$, or $a_j = -\Psi(j/(N+1))$ for $j = 1, \cdots, N$. Let ε and C be positive numbers and suppose that $\varepsilon \leqq \lambda \leqq 1-\varepsilon$. Then there exist positive* *numbers* $B, \delta_1, \delta_2, \cdots$, *with* $\lim_{N\to\infty} \delta_N = 0$, *which depend only on* F, ε *and* C, *such that*

$$\sup_{|\xi|\le C} |P_{F,\mu}(N^{\frac{1}{2}}(M-\mu) \le \xi) - \{1-\pi(\bar{\alpha}, F, -\xi N^{-\frac{1}{2}})\}| \le \delta_N N^{-1} \tag{8.4}$$

and such that the following statements hold:

(i) *for exact scores* $a_j = -E\Psi_1(U_{j:N})$,

$$\sup_{|x|\le C} |P_{F,\mu}(\{\lambda(1-\lambda)N \textstyle\int \Psi_1^2(t)\,dt\}^{\frac{1}{2}}(M-\mu) \le x) - \bar{L}_1(x)| \le \delta_N N^{-1}, \tag{8.5}$$

$$\begin{aligned} \sup_{|x|\le C} |P_{F,\mu}(\{\lambda(1-\lambda)N \textstyle\int \Psi_1^2(t)\,dt\}^{\frac{1}{2}}(M-\mu) \le x) - \bar{L}_2(x)| \\ \le \delta_N N^{-1} + BN^{-\frac{3}{2}} \textstyle\int_{N^{-1}}^{1-N^{-1}} (\Psi_1'(t))^2\{t(1-t)\}^{\frac{1}{2}}\,dt\,; \end{aligned} \tag{8.6}$$

(ii) *for approximate scores* $a_j = -\Psi_1(j/(N+1))$,

$$\sup_{|x|\le C} |P_{F,\mu}(\{\lambda(1-\lambda)N \textstyle\int \Psi_1^2(t)\,dt\}^{\frac{1}{2}}(M-\mu) \le x) - \bar{L}_3(x)| \le \delta_N N^{-1} \tag{8.7}$$

and (8.6) *continues to hold.*

Proof. In view of (8.1) we have for $\theta = -\xi N^{-\frac{1}{2}}$,

$$P_{F,\mu}(N^{\frac{1}{2}}(M-\mu) \le \xi) = P_{F,\theta}(M \le 0),$$
$$P_{F,\theta}(T < \lambda \textstyle\sum a_j) \le P_{F,\theta}(M \le 0) \le P_{F,\theta}(T \le \lambda \sum a_j).$$

For $\theta = -\xi N^{-\frac{1}{2}}$ and $\varepsilon' \le \alpha \le 1-\varepsilon'$, the conditions of Theorem 5.2 are satisfied except, of course, that $\theta < 0$ if $\xi > 0$. However, the theorem remains valid for $|\theta| \le DN^{-\frac{1}{2}}$; it was formulated for positive θ merely because we were discussing one-sided tests against one-sided alternatives at that point. It follows that $P_{F,\theta}(T = \lambda \sum a_j) = o(N^{-1})$ uniformly for $|\xi| \le C$, so that

$$\begin{aligned} P_{F,\mu}(N^{\frac{1}{2}}(M-\mu) \le \xi) &= P_{F,\theta}(T \le \lambda \textstyle\sum a_j) + o(N^{-1}) \\ &= 1-\pi(\alpha, F, -\xi N^{-\frac{1}{2}}) + o(N^{-1}), \end{aligned} \tag{8.8}$$

where α is the level of the test that rejects if $T > \lambda \sum a_j$. Noting that $\sum a_j = 0$ for exact scores, we find from (5.33) and (5.37) for $x = \eta^* = 0$, that $\alpha = \bar{\alpha} + o(N^{-1})$. In view of (5.35) and (5.39) this yields $\pi(\alpha, F, -\xi N^{-\frac{1}{2}}) = \pi(\bar{\alpha}, F, -\xi N^{-\frac{1}{2}}) + o(N^{-1})$ uniformly for $|\xi| \le C$ and together with (8.8) this proves (8.4). The remainder of the theorem follows from (8.8) and expansions (5.33), (5.34), (5.37) and (5.38) with x and η^* replaced by 0 and $-x$. □

The natural parametric competitor of M as an estimator of μ is of course the maximum likelihood estimator M'. Under the model $K_{F,\mu,\Delta} = (F(\cdot + \Delta\mu), F(\cdot - (1-\Delta)\mu))$, $M' = M_\Delta'$ is the solution of

$$\Delta \textstyle\sum_{i=1}^m \psi_1(X_i + \Delta M') - (1-\Delta) \sum_{i=m+1}^N \psi_1(X_i - (1-\Delta)M') = 0 \tag{8.9}$$

with $\psi_1 = f'/f$ as in (4.1). Note that, in contrast to M, the estimator M_Δ' as well as its distribution under $K_{F,\mu,\Delta}$ depend on Δ.

The df of M_Δ' under $K_{F,\mu,\Delta}$ is connected with the power of the locally most powerful test for $H_F = (F, F)$ against $K_{F,\theta,\Delta}$. For $\theta > 0$, this test rejects H_F for large values of the statistic

$$S_\Delta = \Delta \sum_{i=1}^m \psi_1(X_i) - (1-\Delta) \sum_{i=m+1}^N \psi_1(X_i) . \tag{8.10}$$

Let $\pi(\alpha, F, \theta, \Delta)$ denote the power against $K_{F,\theta,\Delta}$ of this right-sided test at level α. Suppose that F is a fixed df with density f that is positive and five times differentiable on R^1 and define

$$\begin{aligned}
&\tilde{\pi}(\alpha, F, \theta, \Delta) \\
&\quad = 1 - \Phi(u_\alpha - \tilde{\eta}) - \frac{\tilde{\eta}\phi(u_\alpha - \tilde{\eta})}{288}\Bigg[\frac{24}{N^{\frac{1}{2}}} \frac{\tau_3 \int \Psi_1^3(t)\,dt}{\{\tau_2 \int \Psi_1^2(t)\,dt\}^{\frac{3}{2}}}(-2u_\alpha + \tilde{\eta}) \\
&\quad + \frac{4}{N} \frac{\tau_4 \int \Psi_1^4(t)\,dt}{\{\tau_2 \int \Psi_1^2(t)\,dt\}^2}\{-3(u_\alpha^2 - 1) + 3\tilde{\eta}u_\alpha - 5\tilde{\eta}^2\} \\
&\quad + \frac{48}{N} \frac{\tau_4 \int \Psi_2^2(t)\,dt}{\{\tau_2 \int \Psi_1^2(t)\,dt\}^2}\tilde{\eta}^2 + \frac{1}{N} \frac{\{\tau_3 \int \Psi_1^3(t)\,dt\}^2}{\{\tau_2 \int \Psi_1^2(t)\,dt\}^3} \\
&\quad \times \{8(2u_\alpha^2 - 1) - 4\tilde{\eta}(u_\alpha^3 + 3u_\alpha) + 8\tilde{\eta}^2(u_\alpha^2 - 1) - 5\tilde{\eta}^3 u_\alpha + \tilde{\eta}^4\} \\
&\quad + \frac{36\tau_4}{N\tau_2^2}\{(u_\alpha^2 - 1) - \tilde{\eta}u_\alpha - \tilde{\eta}^2\}\Bigg],
\end{aligned} \tag{8.11}$$

where $\tilde{\eta}$ and τ_k are given by (7.2) and (7.3).

LEMMA 8.1. *Let F satisfy* (5.6) *for $m_i = 5/i$, $i = 1, \cdots, 5$, and suppose that positive numbers D, D', ε and ε' exist such that $|\theta| \leqq DN^{-\frac{1}{2}}$, $|\Delta\theta| \leqq D'N^{-\frac{1}{2}}$, $\varepsilon \leqq \lambda \leqq 1 - \varepsilon$ and $\varepsilon' \leqq \alpha \leqq 1 - \varepsilon'$. Then there exists $B > 0$ depending only on F, D, D', ε and ε' such that*

$$|\pi(\alpha, F, \theta, \Delta) - \tilde{\pi}(\alpha, F, \theta, \Delta)| \leqq BN^{-\frac{3}{2}} . \tag{8.12}$$

PROOF. The proof proceeds in the same manner as that of Lemma 7.1 and again we omit the details. Under the conditions of the lemma we find that under $K_{F,\theta,\Delta}$,

$$\begin{aligned}
&P\left(\frac{S_\Delta}{\{N\tau_2 \int \Psi_1^2(t)\,dt\}^{\frac{1}{2}}} \leqq x\right) \\
&\quad = \Phi(x - \tilde{\eta}) - \frac{\phi(x - \tilde{\eta})}{288}\Bigg[\frac{24}{N^{\frac{1}{2}}} \frac{\tau_3 \int \Psi_1^3(t)\,dt}{\{\tau_2 \int \Psi_1^2(t)\,dt\}^{\frac{3}{2}}}\{2(x^2 - 1) + 2\tilde{\eta}x - \tilde{\eta}^2\} \\
&\quad + \frac{4}{N} \frac{\tau_4 \int \Psi_1^4(t)\,dt}{\{\tau_2 \int \Psi_1^2(t)\,dt\}^2}\{3(x^3 - 3x) + 3\tilde{\eta}(x^2 - 1) - 3\tilde{\eta}^2 x + 5\tilde{\eta}^3\} \\
&\quad - \frac{48}{N} \frac{\tau_4 \int \Psi_2^2(t)\,dt}{\{\tau_2 \int \Psi_1^2(t)\,dt\}^2}\tilde{\eta}^3 + \frac{1}{N} \frac{\{\tau_3 \int \Psi_1^3(t)\,dt\}^2}{\{\tau_2 \int \Psi_1^2(t)\,dt\}^3}\{4(x^5 - 10x^3 + 15x) \\
&\quad + 4\tilde{\eta}(x^4 - 6x^2 + 3) - 8\tilde{\eta}^2(x^3 - 3x) - 4\tilde{\eta}^3(x^2 - 1) + 5\tilde{\eta}^4 x - \tilde{\eta}^5\} \\
&\quad + \frac{36\tau_4}{N\tau_2^2}\{-(x^3 - 3x) - \tilde{\eta}(x^2 - 1) + \tilde{\eta}^2 x + \tilde{\eta}^3\}\Bigg] + O(N^{-\frac{3}{2}}) .
\end{aligned} \tag{8.13}$$

The remainder term is uniformly $O(N^{-\frac{3}{2}})$ for fixed F, D, D', ε and ε'. This expansion yields (8.12). □

Note that the expansions (8.12) and (8.13) are valid also for negative values of θ, but that the right-sided test considered here is not locally most powerful against these alternatives.

If the conditions of Lemma 8.1 are fulfilled and if, moreover, f is strongly unimodal so that ψ_1 is nonincreasing, then we can establish the connection between $\pi(\alpha, F, \theta, \Delta)$ and the df of M_Δ' by arguing as in the proof of Theorem 8.1. Writing $P_{F,\mu,\Delta}$ for probabilities under $K_{F,\mu,\Delta}$ and taking $\theta = -\xi N^{-\frac{1}{2}}$ we find that

$$P_{F,\theta,\Delta}(S_\Delta < 0) \leqq P_{F,\mu,\Delta}(N^{\frac{1}{2}}(M_\Delta' - \mu) \leqq \xi) \leqq P_{F,\theta,\Delta}(S_\Delta \leqq 0).$$

In view of (8.12) and (8.13) this implies that uniformly for $|\xi|, |x| \leqq C$,

$$P_{F,\mu,\Delta}(N^{\frac{1}{2}}(M_\Delta' - \mu) \leqq \xi) = 1 - \pi(\tilde{\alpha}, F, \theta, \Delta) + O(N^{-\frac{3}{2}}), \tag{8.14}$$

$$\tilde{\alpha} = \tfrac{1}{2} - \frac{1}{6\{2\pi N\}^{\frac{1}{2}}} \frac{\tau_3 \int \Psi_1^3(t)\,dt}{\{\tau_2 \int \Psi_1^2(t)\,dt\}^{\frac{3}{2}}}, \tag{8.15}$$

$$\begin{aligned} P_{F,\mu,\Delta}(\{\tau_2 N \textstyle\int \Psi_1^2(t)\,dt\}^{\frac{1}{2}}(M_\Delta' - \mu) \leqq x) &= \Phi(x) - \frac{\phi(x)}{288}\Bigg[-\frac{24}{N^{\frac{1}{2}}} \frac{\tau_3 \int \Psi_1^3(t)\,dt}{\{\tau_2 \int \Psi_1^2(t)\,dt\}^{\frac{3}{2}}}(x^2 + 2) \\ &\quad - \frac{4}{N} \frac{\tau_4 \int \Psi_1^4(t)\,dt}{\{\tau_2 \int \Psi_1^2(t)\,dt\}^2}(5x^3 - 3x) + \frac{48}{N} \frac{\tau_4 \int \Psi_2^2(t)\,dt}{\{\tau_2 \int \Psi_1^2(t)\,dt\}^2} x^3 \\ &\quad + \frac{1}{N} \frac{\{\tau_3 \int \Psi_1^3(t)\,dt\}^2}{\{\tau_2 \int \Psi_1^2(t)\,dt\}^3}(x^5 - 4x^3 - 12x) - \frac{36\tau_4}{N\tau_2^2}(x^3 + x)\Bigg] \\ &\quad + O(N^{-\frac{3}{2}}). \end{aligned} \tag{8.16}$$

We have already remarked that the df of $(M_\Delta' - \mu)$ under $K_{F,\mu,\Delta}$ depends on Δ and thus the same problem arises that we encountered in Section 7, viz. to determine the "right" Δ for which M and M' should be compared. It is easy to see from (8.16) that the value $\Delta = \Delta^0$ that is least favorable for M' in the sense that it minimizes (maximizes) $P_{F,\mu,\Delta}(\{N\lambda(1-\lambda) \int \Psi_1^2(t)\,dt\}^{\frac{1}{2}}(M_\Delta' - \mu) \leqq x)$ for positive (negative) x is given by

$$\Delta^0 = \lambda - \frac{\{\lambda(1-\lambda)\}^{\frac{1}{2}}}{4N^{\frac{1}{2}}} \frac{\int \Psi_1^3(t)\,dt}{\{\int \Psi_1^2(t)\,dt\}^{\frac{3}{2}}} \frac{x^2 + 2}{x} + O(N^{-1}).$$

However, we shall not take $\Delta = \Delta^0$ as a basis for comparing M and M' but we shall simply choose $\Delta = \lambda$ instead. We advance three reasons for doing so. The reader who does not find these reasons sufficiently compelling should realize that we are merely granting the maximum likelihood estimator a slight additional advantage.

(i) The second order term of Δ^0 depends on x just as the second order term of Δ_0 in (7.5) depends on θ. This did not deter us from choosing $\Delta = \Delta_0$ as a basis for comparison in Section 7, but we feel the situation is slightly different there. In Section 7 we were comparing with envelope power and in general this means comparing with a different most powerful test for each alternative (θ, Δ). This being so, there seems to be little reason not to choose the least favorable testing problem for each value of θ, i.e., to take $\Delta = \Delta_0$. All we are doing is locating a curve $(\theta, \Delta_0(\theta))$ of least favorable alternatives in the set of all alternatives (θ, Δ) and comparing with envelope power on that curve only.

Our attitude would have been different, however, if in Section 7 we had been comparing with the power of the locally most powerful test rather than with the envelope power. The locally most powerful test is of course independent of θ (cf. (8.10)) and for every fixed Δ we would therefore be comparing with a single fixed test for all θ. In this case it would still be reasonable to choose $\Delta = \lambda$ which is least favorable to first order, but if $\int \Psi_1^3(t)\,dt \neq 0$, it would seem to be rather extreme to compute the power of the locally most powerful test at each θ for $\Delta = \Delta_0 = \Delta_0(\theta)$ which is least favorable to second order in this case too. After all, for every fixed Δ there would be a single locally most powerful test that does better than that for all values of θ except the one for which $\Delta_0(\theta) = \Delta$. It is precisely for such sets of alternatives (Δ fixed, θ unknown) that the locally most powerful test is designed and it seems unrealistic to assess its performance only for a different one-parameter set of alternatives $(\theta, \Delta_0(\theta))$.

The present problem for the maximum likelihood estimator is of course very similar to the one for the locally most powerful test. Again the choice $\Delta = \Delta^0$ depending on x appears to be rather extreme because for every Δ the df of the maximum likelihood estimator is more concentrated around μ than this choice would indicate at all but at most two points.

(ii) Even though, in general, the distribution of M_Δ' under $K_{F,\mu,\Delta}$ is not symmetric about μ, most reasonable measures of dispersion are built around the distribution of $|M_\Delta' - \mu|$ rather than $(M_\Delta' - \mu)$. It is clear from (8.16) that $P_{F,\mu,\Delta}(\{N\lambda(1-\lambda)\int \Psi_1^2(t)\,dt\}^{\frac{1}{2}}|M_\Delta' - \mu| \leq x)$ is minimized by $\Delta = \lambda + O(N^{-1})$; it is also obvious from (8.16) that it makes no difference for our asymptotic results if we take $\Delta = \lambda$ instead (cf. the remark following (7.5)). Hence $\Delta = \lambda$ is the "right" choice of Δ for our asymptotic comparison of M and M', provided that the comparison is made on the basis of the distributions of $|M - \mu|$ and $|M' - \mu|$.

(iii) Our final argument is the rather more pedestrain one that any choice of Δ other than $\Delta = \lambda + o(N^{-\frac{1}{2}})$ would to a certain extent destroy the simplicity of the main results in this section. We shall elaborate points (ii) and (iii) after proving Theorem 8.3.

We now substitute $\Delta = \lambda$ in (8.14)—(8.16) and find that $\tilde{\alpha}$ reduces to $\bar{\alpha}$ as defined in (8.2) and that the expansion on the right in (8.16) becomes

$$
\begin{aligned}
L^*(x) = \Phi(x) - \frac{\phi(x)}{288}\Bigg[& \frac{24(1-2\lambda)}{\{\lambda(1-\lambda)N\}^{\frac{1}{2}}} \frac{\int \Psi_1^3(t)\,dt}{\{\int \Psi_1^2(t)\,dt\}^{\frac{3}{2}}}(x^2+2) \\
& - \frac{4(1-3\lambda+3\lambda^2)}{\lambda(1-\lambda)N} \frac{\int \Psi_1^4(t)\,dt}{\{\int \Psi_1^2(t)\,dt\}^2}(5x^3-3x) + \frac{48(1-3\lambda+3\lambda^2)}{\lambda(1-\lambda)N} \\
& \times \frac{\int \Psi_2^2(t)\,dt}{\{\int \Psi_1^2(t)\,dt\}^2}x^3 + \frac{(1-2\lambda)^2}{\lambda(1-\lambda)N} \frac{\{\int \Psi_1^3(t)\,dt\}^2}{\{\int \Psi_1^2(t)\,dt\}^3}(x^5-4x^3-12x) \\
& - \frac{36(1-3\lambda+3\lambda^2)}{\lambda(1-\lambda)N}(x^3+x)\Bigg]. \qquad (8.17)
\end{aligned}
$$

We have proved

THEOREM 8.2. *Suppose that F satisfies (5.6) for $m_i = 5/i$, $i = 1, \cdots, 5$ and that f is strongly unimodal. Let ε and C be positive numbers and suppose that $\varepsilon \leqq \lambda \leqq 1 - \varepsilon$. Then there exists $B > 0$ depending only on F, ε and C, such that*

$$\sup_{|\xi| \leq C} |P_{F,\mu,\lambda}(N^{\frac{1}{2}}(M_\lambda' - \mu) \leqq \xi) - \{1 - \pi(\tilde{\alpha}, F, -\xi N^{-\frac{1}{2}}, \lambda)\}| \leqq BN^{-\frac{3}{2}}, \tag{8.18}$$

$$\sup_{|x| \leq C} |P_{F,\mu,\lambda}(\{\lambda(1-\lambda)N \textstyle\int \Psi_1^2(t)\, dt\}^{\frac{1}{2}}(M_\lambda' - \mu) \leqq x) - L^*(x)| \leqq BN^{-\frac{3}{2}}. \tag{8.19}$$

There is no unique natural measure to assess the performance of the estimators M and M_λ' on the basis of the expansions (8.5)—(8.7) and (8.19) and consequently there is no unique natural definition of the deficiency of M with respect to M_λ' either. Let us, for a moment, indicate the dependence on the sample size N in our notation and write M_N and $M'_{\lambda,N}$ for M and M_λ'. For any real ξ we define the deficiency $D_N(\xi)$ of the sequence of estimators $\{M_N\}$ with respect to the estimator $M'_{\lambda,N}$ by equating the df's of $(M_{N+D_N} - \mu)$ under $P_{F,\mu,\lambda}$ (or $P_{F,\mu}$) and of $(M'_{\lambda,N} - \mu)$ under $P_{F,\mu,\lambda}$ at the point $\xi N^{-\frac{1}{2}}$, thus

$$P_{F,\mu}(M_{N+D_N} - \mu \leqq \xi N^{-\frac{1}{2}}) = P_{F,\mu,\lambda}(M'_{\lambda,N} - \mu \leqq \xi N^{-\frac{1}{2}}), \tag{8.20}$$

with the usual convention that the probability on the left is defined by linear interpolation for nonintegral values of $N + D_N$. Of course, one will normally not be inclined to judge the performance of $\{M_N\}$ with respect to $M'_{\lambda,N}$ on the basis of $D_N(\xi)$ for one value of ξ only, but rather on the behavior of $D_N(\xi)$ as a function of ξ. In our asymptotic study this will not make any difference because the expansions for $D_N(\xi)$ will be found to be independent of ξ.

Turning to the corresponding tests, we let $d_N(\alpha, \theta)$ denote the deficiency in the usual sense of the locally most powerful rank test (or its approximate scores version) with respect to the locally most powerful test for the problem of testing $H_F = (F, F)$ against $K_{F,\theta,\lambda} = (F(\cdot + \lambda\theta), F(\cdot - (1-\lambda)\theta))$ at level α. Since we shall be concerned with negative as well as positive values of θ, we note that for positive (negative) θ the tests involved reject H_F for large (small) values of the statistics given in (2.2) and (8.10), where the scores in (2.2) are exact or approximate scores generated by $-\Psi_1$.

Let $\mathscr{F}_1$ be given by Definition 7.1 and define

$$\tilde{D}_{N,1} = -\frac{1}{4}\frac{\int \Psi_1^4(t)\, dt}{\{\int \Psi_1^2(t)\, dt\}^2} - \frac{3}{4} + \frac{1}{\int \Psi_1^2(t)\, dt} \textstyle\sum_{j=1}^N \sigma^2(\Psi_1(U_{j:N})),$$

$$\tilde{D}_{N,2} = -\frac{1}{4}\frac{\int \Psi_1^4(t)\, dt}{\{\int \Psi_1^2(t)\, dt\}^2} - \frac{3}{4} + \frac{1}{\int \Psi_1^2(t)\, dt} \int_{N^{-1}}^{1-N^{-1}} (\Psi_1'(t))^2 t(1-t)\, dt, \tag{8.21}$$

$$\tilde{D}_{N,3} = -\frac{1}{4}\frac{\int \Psi_1^4(t)\, dt}{\{\int \Psi_1^2(t)\, dt\}^2} - \frac{3}{4} + \frac{1}{\int \Psi_1^2(t)\, dt} \textstyle\sum_{j=1}^N E\left(\Psi_1(U_{j:N}) - \Psi_1\left(\frac{j}{N+1}\right)\right)^2.$$

THEOREM 8.3. *Let $d_N(\alpha, \theta)$ be the deficiency of the locally most powerful rank test with respect to the locally most powerful test for testing H_F against $K_{F,\theta,\lambda}$ at* *level α. Let $D_N(\xi)$ be the deficiency of the Hodges–Lehmann estimator associated with the locally most powerful rank test with respect to the maximum likelihood estimator for estimating μ under $K_{F,\mu,\lambda}$. Suppose that $F \in \mathscr{F}_1$ and that f is strongly unimodal. Let c, C and ε be positive numbers and suppose that $c \leqq |\xi| \leqq C$ and $\varepsilon \leqq \lambda \leqq 1 - \varepsilon$. Then there exist positive numbers $B, \delta_1, \delta_2, \cdots$, with $\lim_{N\to\infty} \delta_N = 0$, which depend only on F, c, C and ε, such that*

$$|D_N(\xi) - d_N(\tfrac{1}{2}, -\xi N^{-\frac{1}{2}})| \leqq \delta_N , \tag{8.22}$$

$$|D_N(\xi) - \tilde{D}_{N,1}| \leqq \delta_N , \tag{8.23}$$

$$|D_N(\xi) - \tilde{D}_{N,2}| \leqq \delta_N + BN^{-\frac{1}{2}} \textstyle\int_{N^{-1}}^{1-N^{-1}} (\Psi_1'(t))^2 \{t(1-t)\}^{\frac{1}{2}}\, dt . \tag{8.24}$$

If in the locally most powerful rank test and in the associated estimator, the exact scores are replaced by the approximate scores $a_j = -\Psi_1(j/(N+1))$, then (8.22) *and* (8.24) *remain valid and* (8.23) *is replaced by*

$$|D_N(\xi) - \tilde{D}_{N,3}| \leqq \delta_N . \tag{8.25}$$

PROOF. Since $\mathscr{F}_1 \subset \mathscr{F}$, the conditions of Theorems 8.1 and 8.2 are satisfied and (8.5)—(8.7) and (8.19) provide expansions for the df's of the estimators considered. Substituting the appropriate expansions in (8.20) and proceeding exactly as in the proof of Theorem 7.1, we arrive at (8.23) and (8.24) for the estimator associated with the locally most powerful rank test and at (8.24) and (8.25) for its approximate scores version.

Turning to the corresponding tests, (8.4) and (8.18) clearly imply that for negative values of ξ the computation for obtaining an expansion for $d_N(\bar{\alpha}, -\xi N^{-\frac{1}{2}})$ is precisely the same as for $D_N(\xi)$. In view of (8.23) and (8.25) this computation determines the deficiency up to $o(1)$ and hence

$$|D_N(\xi) - d_N(\bar{\alpha}, -\xi N^{-\frac{1}{2}})| = o(1) , \qquad \text{for} \quad -C \leqq \xi \leqq -c . \tag{8.26}$$

For positive ξ, $d_N(\alpha, -\xi N^{-\frac{1}{2}})$ refers to testing for negative shift and therefore to the left-sided tests rather than the right-sided tests whose powers appear in (8.4) and (8.18). Since the powers of the left- and right-sided versions of a test sum to 1 if their significance levels do, we find

$$|D_N(\xi) - d_N(1-\bar{\alpha}, -\xi N^{-\frac{1}{2}})| = o(1) , \qquad \text{for} \quad c \leqq \xi \leqq C . \tag{8.27}$$

Note that (8.26) and (8.27) hold for exact as well as approximate scores and that the remainder terms are uniformly $o(1)$ for fixed F, c, C and ε.

It remains to show that $\bar{\alpha}$ may be replaced by $\frac{1}{2}$ in (8.26) and (8.27). If we take $\Delta = \lambda$ in the power expansion for the locally most powerful test in Lemma 8.1 and compare the result with the power expansion for the most powerful test in Lemma 7.2, we see that the terms of orders 1 and $N^{-\frac{1}{2}}$ agree and that in the terms of order N^{-1} only certain coefficients differ. Moreover, for $\Delta = \lambda$ the conditions of Lemma 8.1 are identical with those of Lemma 7.2. This means that if we replace the most powerful test by the locally most powerful test in Theorem 7.1, then the theorem will remain valid if some of the coefficients in

$\tilde{d}_{N,0}$ are changed. Thus, under the conditions of Theorem 7.1 there exists, for exact as well as approximate scores, an expansion for $d_N(\alpha, \theta)$ with a bounded derivative with respect to α and a remainder term $o(1)$. This statement remains correct for $-CN^{-\frac{1}{2}} \leqq \theta \leqq -cN^{-\frac{1}{2}}$ because the power expansions in Lemma 8.1 and Theorem 5.2 are valid for negative θ too (cf. the remark in the proof of Theorem 8.1) so that the only change in the expansion for $d_N(\alpha, \theta)$ is a change of sign of u_α to account for the switch from the right-sided to the left-sided tests. Noting that $c \leqq |\xi| \leqq C$ and that $\bar{\alpha} = \frac{1}{2} + O(N^{-\frac{1}{2}})$ we find that we may indeed replace $\bar{\alpha}$ by $\frac{1}{2}$ in (8.26) and (8.27) without affecting the right-hand side and its uniformity for fixed F, c, C and ε. This proves (8.22) and the theorem. □

A number of comments should be made at this point. First of all we recall remarks (ii) and (iii) in our discussion earlier in this section concerning the choice of Δ for which M and M' should be compared. Suppose we define deficiencies $D_N'(\xi)$ by

$$P_{F,\mu}(|M_{N+D_N'} - \mu| \leqq \xi N^{-\frac{1}{2}}) = P_{F,\mu,\Delta}(|M'_{\Delta,N} - \mu| \leqq \xi N^{-\frac{1}{2}})$$

for that value of Δ that minimizes the right-hand side. In view of remark (ii), Theorem 8.3 implies that $D_N'(\xi)$ is also asymptotically equivalent to the $\tilde{D}_{N,\xi}$. Thus our results can be thought of as corresponding exactly to those of ABZ (1976) where deficiencies are defined in terms of a positive quantile of the symmetrically distributed centered estimators in the one-sample problem. Since the deficiency is asymptotically independent of the value of ξ, we obtain the same answers for deficiencies based on reasonable functionals of the distributions of $N^{\frac{1}{2}}|M - \mu|$ and $N^{\frac{1}{2}}|M' - \mu|$, such as the asymptotic second moment. This agrees with what was found in the one-sample case in Albers (1974).

The choice $\Delta = \lambda$ is less obvious in equation (8.20) which defines $D_N(\xi)$. In remark (iii) we pointed out that if we would not choose $\Delta = \lambda + o(N^{-\frac{1}{2}})$, then our results would become essentially more complicated. The first source of trouble is the difference of the significance levels $\bar{\alpha}$ and $\tilde{\alpha}$ given by (8.2) and (8.15). Except in the trivial case where $\int \Psi_1^3(t)\, dt = 0$, we find that $(\tilde{\alpha} - \bar{\alpha})$ is of the order of $N^{-\frac{1}{2}}(\Delta - \lambda)$ and a change of the order of $N^{-\frac{1}{2}}(\Delta - \lambda)$ in the level of significance of one of the two tests produces a change of the same order in its power. Unless $\Delta - \lambda = o(N^{-\frac{1}{2}})$ such an effect is not negligible for our purposes and this means that it would no longer be true that the deficiency for the estimators is asymptotically equivalent in the sense of (8.22) to the deficiency of the parent tests at the same level. In fact a correction term of the order of $N^{\frac{1}{2}}(\Delta - \lambda)$ would have to be introduced in (8.22) to ensure its validity. Note that there is no contradiction here with the fact that in the proof of Theorem 8.3 we could change $\bar{\alpha}$ to $\frac{1}{2}$ with impunity, because there we were concerned with the same change of level for both tests simultaneously. A second unpleasant consequence of choosing $\Delta = \Delta^0$ (or even $\Delta = \lambda + bN^{-\frac{1}{2}}$ with b independent of x) would be that the expansions for $D_N(\xi)$ would no longer be independent of

ξ. By taking $\Delta = \Delta^0$, we would therefore destroy at one stroke the two most striking features of Theorem 8.3.

Next we note that upon formal substitution of $\alpha = \frac{1}{2}$ and $\theta = 0$ the expansions for d_N in Theorem 7.1 reduce to the expansions for $D_N(\xi)$ in Theorem 8.3. This shows that for every $\xi \neq 0$, $D_N(\xi)$ is nonnegative for sufficiently large N.

In the proof of Theorem 8.3 we indicated how one can obtain expansions for the deficiency of the locally most powerful rank test or its approximate scores analogue with respect to the locally most powerful test. At that point there was no need to produce these expansions, but we shall do so now because they may be of independent interest. The simplest way to describe these results is the following. In the formulation of Theorem 7.1 change the words "most powerful test" to "locally most powerful test" and K_{F,θ,Δ_0} to $K_{F,\theta,\lambda}$; change $\tilde{d}_{N,0}$ in (7.8) to

$$(8.28) \qquad \tilde{d}_{N,0} = \tfrac{1}{48}\left[\frac{4\int \Psi_1^4(t)\,dt}{\{\int \Psi_1^2(t)\,dt\}^2}\{3(u_\alpha^2 - 1) - 2\eta^* u_\alpha\} - 12(u_\alpha^2 + 3 - 2\eta^* u_\alpha)\right].$$

With these changes Theorem 7.1 holds. When comparing the expansions for d_N in (7.8) with those based on (8.28) we see that the expansions in (7.8) consist of three parts. The term involving $(2u_\alpha - \eta^*)^2$ is due to the fact that comparisons with the most powerful test were made for $\Delta = \Delta_0$ rather than $\Delta = \lambda$ (cf. the discussion following Theorem 7.1). The other terms involving η^{*2} represent the deficiency of the locally most powerful test with respect to the most powerful test for $\Delta = \lambda$. The remaining terms are due to the transition from the locally most powerful test to the two rank tests. All four tests are efficient to second order, i.e., for each pair the deficiency is $o(N^{\frac{1}{2}})$, and the reason for this is that the terms of orders 1 and $N^{-\frac{1}{2}}$ are the same in all four power expansions (cf. the discussion following Theorem 7.1).

We conclude with one example of Theorem 8.3. For estimating μ in the normal location model $(\Phi(\cdot + \lambda\mu), \Phi(\cdot - (1 - \lambda)\mu))$, the deficiency of either one of the Hodges–Lehmann estimators associated with the normal scores test and with van der Waerden's test with respect to the difference of the sample means is given by

$$(8.29) \qquad D_N(\xi) = \log\log N - \tfrac{3}{2} + \log 2 + \gamma + o(1),$$

where γ is Euler's constant as in (5.50). Note that this expansion is the same as expansion (7.23) for the deficiency of the normal scores test (or van der Waerden's test) with respect to Student's test for any α.

APPENDIX

Expansions for the contiguous location case. In this appendix we provide the tools for deriving Theorem 4.1 from Theorem 3.1. The quantities appearing in the expansion of Theorem 3.1 are expected values under P_H of functions of $P_1, \cdots, P_N$ and in the setup of Section 4 both H and $P_1, \cdots, P_N$ depend on θ.

Our task is to provide Taylor expansions in θ with error bounds for these quantities, thus reducing expectations E_H to expectations E_F while at the same time expanding the rv's involved. Since we are only concerned with the models P_H and P_F under the assumptions of Section 4, we suppose throughout that $X_1, \cdots, X_N$ are i.i.d. with common density h under P_H and f under P_F, where $h(x) = (1-\lambda)f(x) + \lambda f(x-\theta)$ and f is positive and four times differentiable on R^1. Define $\xi(x, t)$, $p(x, t)$ and $\tilde{p}(x, t)$ by

$$(1-\lambda)F(\xi(x,t)) + \lambda F(\xi(x,t)-t) = F(x)\,, \tag{A.1}$$

$$p(x,t) = \frac{\lambda f(x-t)}{(1-\lambda)f(x) + \lambda f(x-t)}\,. \tag{A.2}$$

$$\tilde{p}(x,t) = p(\xi(x,t),t)\,. \tag{A.3}$$

As in Appendix 1 of ABZ (1976), these functions are introduced because $\tilde{p}(Z_1,\theta), \cdots, \tilde{p}(Z_N,\theta)$ under P_F have the same joint distribution as $P_1, \cdots, P_N$ under P_H. Our main problem is therefore to expand $\tilde{p}(x, t)$ as a function of t around $t = 0$.

With $\psi_i = f^{(i)}/f$ as in (4.1), we define for $i = 1, \cdots, 4$,

$$\chi_i(x,t) = |\psi_i(\xi(x,t))| + |\psi_i(\xi(x,t)-t)| \tag{A.4}$$

and for any function q of two variables we write

$$q_{i,j}(x,t) = \frac{\partial^{i+j} q(x,t)}{\partial x^i\,\partial t^j}\,.$$

Then elementary but tedious computations yield

$$\begin{aligned}
&\tilde{p}(x,0) = \lambda\,,\\
&\tilde{p}_{0,1}(x,0) = -\lambda(1-\lambda)\psi_1(x)\,,\\
&\tilde{p}_{0,2}(x,0) = \lambda(1-\lambda)(1-2\lambda)\psi_2(x)\,,\\
&\tilde{p}_{0,3}(x,0) = -\lambda(1-\lambda)\{(1-3\lambda+3\lambda^2)\psi_3(x) - 6\lambda(1-\lambda)\psi_1(x)\psi_2(x)\\
&\qquad\qquad + 3\lambda(1-\lambda)\psi_1^3(x)\}\,,
\end{aligned} \tag{A.5}$$

$$\begin{aligned}
&|\tilde{p}_{0,1}| \leq b_1\chi_1\,,\\
&|\tilde{p}_{0,2}| \leq b_2(\chi_2 + \chi_1^2)\,,\\
&|\tilde{p}_{0,3}| \leq b_3(\chi_3 + \chi_2^{3/2} + \chi_1^3)\,,\\
&|\tilde{p}_{0,4}| \leq b_4(\chi_4 + \chi_3^{4/3} + \chi_2^2 + \chi_1^4)\,,
\end{aligned} \tag{A.6}$$

where $b_1, \cdots, b_4$ are positive constants.

Define $\pi_j = E_H P_j$ as in (3.16).

THEOREM A.1. *Suppose that positive numbers C, C' and ε' exist such that $\sum a_j^4 \leq CN$, $0 \leq \theta \leq \varepsilon'$ and (4.2) is satisfied. Then there exists $B > 0$ depending* *only on C, C′ and ε′ such that*

$$\begin{aligned}\sum a_j(\pi_j - \lambda) = \lambda(1-\lambda)\{&-\theta \sum a_j E_F \psi_1(Z_j) \\ &+ (1-2\lambda)\frac{\theta^2}{2} \sum a_j E_F \psi_2(Z_j) \\ &- \frac{\theta^3}{6} \sum a_j E_F[(1 - 3\lambda + 3\lambda^2)\psi_3(Z_j) \\ &- 6\lambda(1-\lambda)\psi_1(Z_j)\psi_2(Z_j) + 3\lambda(1-\lambda)\psi_1^3(Z_j)]\} + M_1 , \\ |M_1| \leqq BN^{\frac{5}{4}}\theta^4 ;\end{aligned} \tag{A.7}$$

$$\begin{aligned}\sum a_j^2(\pi_j - \lambda) = \lambda(1-\lambda)\Big\{&-\theta \sum a_j^2 E_F \psi_1(Z_j) \\ &+ (1-2\lambda)\frac{\theta^2}{2} \sum a_j^2 E_F \psi_2(Z_j)\Big\} + M_2 , \\ |M_2| \leqq BN^{\frac{5}{4}}\theta^3 ;\end{aligned} \tag{A.8}$$

$$\begin{aligned}&\sum a_j^3(\pi_j - \lambda) = -\lambda(1-\lambda)\theta \sum a_j^3 E_F \psi_1(Z_j) + M_3 , \\ &|M_3| \leqq BN^{\frac{13}{12}}\theta^2 ;\end{aligned} \tag{A.9}$$

$$\begin{aligned}&\sum a_j^2 E_H(P_j - \lambda)^2 = \lambda^2(1-\lambda)^2\theta^2 \sum a_j^2 E_F \psi_1^2(Z_j) + M_4 , \\ &|M_4| \leqq BN^{\frac{5}{4}}\theta^3 ;\end{aligned} \tag{A.10}$$

$$\begin{aligned}&\sigma_H^2(\sum a_j P_j) = \lambda^2(1-\lambda)^2\theta^2\sigma_F^2(\sum a_j \psi_1(Z_j)) + M_5 , \\ &|M_5| \leqq B\{N^2\theta^{\frac{24}{5}} + N\theta^{\frac{19}{5}}[E_F|\sum a_j(\psi_1(Z_j) - E_F\psi_1(Z_j))|^3]^{\frac{1}{3}} \\ &\qquad + \theta^3\sigma_F(\sum a_j \psi_1(Z_j))\sigma_F(\sum a_j \psi_2(Z_j)) + \theta^4\sigma_F^2(\sum a_j \psi_2(Z_j))\} ;\end{aligned} \tag{A.11}$$

$$E_H\left(\frac{\lambda g(X_1)}{h(X_1)} - \lambda\right)^4 \leqq B\theta^4 ; \tag{A.12}$$

$$[\sum \{E_H|P_j - \pi_j|^3\}^{\frac{4}{9}}]^{\frac{9}{4}} \leqq \theta^3[\sum \{E_F|\psi_1(Z_j) - E_F\psi_1(Z_j)|^3\}^{\frac{4}{9}}]^{\frac{9}{4}} + BN^{\frac{9}{4}}\theta^6 . \tag{A.13}$$

PROOF. Although the proof is very similar to that of Theorem A1.1 and the relevant part of Corollary A1.1 in ABZ (1976), there are additional complications due to the fact that now $\tilde{p}_{0,2}(x, 0) \not\equiv 0$. We begin by noting that the distribution of $\xi(X_1, t)$ under F is that of X_1 under $\lambda F(x) + (1-\lambda)F(x-t)$, so that (4.2) and (A.6) imply the existence of $B_1 > 0$ depending only on C' and such that

$$\sup \{E_F|\tilde{p}_{0,i}(X_1, \nu\theta)|^{m_i} : 0 \leqq \nu \leqq 1\} \leqq B_1 , \qquad i = 1, \cdots, 4 , \tag{A.14}$$

where $m_1 = 6$, $m_2 = 3$, $m_3 = \frac{4}{3}$, $m_4 = 1$.

Using Lemma A1.1 of ABZ (1976) together with $\sum a_j^4 \leqq CN$ and (A.14), we find that

$$\begin{aligned}|M_1| &\leqq \frac{\theta^4}{24} \sup \{\sum |a_j| E_F|\tilde{p}_{0,4}(Z_j, \nu\theta)| : 0 \leqq \nu \leqq 1\} \\ &\leqq \frac{(CN)^{\frac{1}{4}}\theta^4}{24} \sup \{N E_F|\tilde{p}_{0,4}(X_1, \nu\theta)| : 0 \leqq \nu \leqq 1\} \leqq \frac{B_1 C^{\frac{1}{4}}}{24} N^{\frac{5}{4}}\theta^4 ,\end{aligned}$$

$$|M_2| \leqq \frac{\theta^3}{6} \sup \{\sum a_j{}^2 E_F|\tilde{p}_{0,3}(Z_j, \nu\theta)| : 0 \leqq \nu \leqq 1\}$$

$$\leqq \frac{\theta^3}{6} (\sum a_j{}^8)^{\frac{1}{4}} \sup \{[N E_F|\tilde{p}_{0,3}(X_1, \nu\theta)|^{\frac{4}{3}}]^{\frac{3}{4}} : 0 \leqq \nu \leqq 1\} \leqq \frac{B_1{}^{\frac{3}{4}} C^{\frac{1}{2}}}{6} N^{\frac{5}{4}} \theta^3 ,$$

$$|M_3| \leqq \frac{\theta^2}{2} \sup \{\sum |a_j|^3 E_F|\tilde{p}_{0,2}(Z_j, \nu\theta)| : 0 \leqq \nu \leqq 1\}$$

$$\leqq \frac{\theta^2}{2} (\sum |a_j|^{\frac{9}{2}})^{\frac{2}{3}} (NB_1)^{\frac{1}{3}} \leqq \frac{B_1{}^{\frac{1}{3}} C^{\frac{3}{4}}}{2} N^{\frac{13}{12}} \theta^2 ,$$

$$|M_4| \leqq \frac{\theta^3}{6} \sup \{\sum a_j{}^2 E_F[2|\tilde{p}_{0,3}(Z_j, \nu\theta)| + 6|\tilde{p}_{0,1}(Z_j, \nu\theta)\tilde{p}_{0,2}(Z_j, \nu\theta)|] : 0 \leqq \nu \leqq 1\}$$

$$\leqq \frac{\theta^3}{6} [2(\sum a_j{}^8)^{\frac{1}{4}} (NB_1)^{\frac{3}{4}} + 6(\sum a_j{}^4)^{\frac{1}{2}} (NB_1)^{\frac{1}{2}}] \leqq (B_1{}^{\frac{1}{2}} + B_1{}^{\frac{3}{4}}) C^{\frac{1}{2}} N^{\frac{5}{4}} \theta^3 ,$$

$$E_H \left(\frac{\lambda g(X_1)}{h(X_1)} - \lambda \right)^4 \leqq \theta^4 \sup \{E_F \tilde{p}_{0,1}^4(X_1, \nu\theta) : 0 \leqq \nu \leqq 1\} \leqq B_1{}^{\frac{3}{4}} \theta^4 ,$$

which proves (A.7)—(A.10) and (A.12). To establish (A.13) we note that

$$|\tilde{p}(Z_j, \theta) - E_F \tilde{p}(Z_j, \theta)| \leqq \theta |\tilde{p}_{0,1}(Z_j, 0) - E_F \tilde{p}_{0,1}(Z_j, 0)|$$
$$+ \frac{\theta^2}{2} \int_0^1 2(1 - \nu)\{|\tilde{p}_{0,2}(Z_j, \nu\theta)| + E_F|\tilde{p}_{0,2}(Z_j, \nu\theta)|\} \, d\nu .$$

Hence

$$E_H|P_j - \pi_j|^3 \leqq \frac{\theta^3}{16} E_F|\psi_1(Z_j) - E_F \psi_1(Z_j)|^3 + 4\theta^6 \int_0^1 2(1 - \nu) E_F|\tilde{p}_{0,2}(Z_j, \nu\theta)|^3 \, d\nu ,$$

$$\sum \{E_H|P_j - \pi_j|^3\}^{\frac{4}{9}} \leqq \theta^{\frac{4}{3}} \sum \{E_F|\psi_1(Z_j) - E_F \psi_1(Z_j)|^3\}^{\frac{4}{9}} + 2(B_1 + 1) N \theta^{\frac{8}{3}} ,$$

and (A.13) follows.

It remains to prove (A.11). We have

$$\tilde{p}(x, t) - \lambda + \lambda(1 - \lambda) t \psi_1(x) - \tfrac{1}{2}\lambda(1 - \lambda)(1 - 2\lambda) t^2 \psi_2(x)$$
$$= \frac{t^2}{2} \int_0^1 2(1 - \nu)(\tilde{p}_{0,2}(x, \nu t) - \tilde{p}_{0,2}(x, 0)) \, d\nu = \frac{t^3}{6} \int_0^1 3(1 - \nu)^2 \tilde{p}_{0,3}(x, \nu t) \, d\nu ,$$

and as a result

$$(\tilde{p}(x, t) - \lambda + \lambda(1 - \lambda) t \psi_1(x) - \tfrac{1}{2}\lambda(1 - \lambda)(1 - 2\lambda) t^2 \psi_2(x))^2$$
$$\leqq \left| \frac{t^2}{2} \int_0^1 2(1 - \nu)(\tilde{p}_{0,2}(x, \nu t) - \tilde{p}_{0,2}(x, 0)) \, d\nu \right|^{\frac{6}{5}} \left| \frac{t^3}{6} \int_0^1 3(1 - \nu)^2 \tilde{p}_{0,3}(x, \nu t) \, d\nu \right|^{\frac{4}{5}}$$
$$\leqq |t|^{\frac{24}{5}} \{|\tfrac{1}{2} \int_0^1 2(1 - \nu)(\tilde{p}_{0,2}(x, \nu t) - \tilde{p}_{0,2}(x, 0)) \, d\nu|^3$$
$$+ |\tfrac{1}{6} \int_0^1 3(1 - \nu)^2 \tilde{p}_{0,3}(x, \nu t) \, d\nu|^{\frac{4}{3}}\}$$
$$\leqq |t|^{\frac{24}{5}} \int_0^1 \{|\tilde{p}_{0,2}(x, \nu t)|^3 + |\tilde{p}_{0,2}(x, 0)|^3 + |\tilde{p}_{0,3}(x, \nu t)|^{\frac{4}{3}}\} \, d\nu .$$

Similarly,

$$|\tilde{p}(x, t) - \lambda + \lambda(1 - \lambda) t \psi_1(x) - \tfrac{1}{2}\lambda(1 - \lambda)(1 - 2\lambda) t^2 \psi_2(x)|^{\frac{3}{2}}$$
$$\leqq |t|^{\frac{21}{5}} \int_0^1 \{|\tilde{p}_{0,2}(x, \nu t)|^3 + |\tilde{p}_{0,2}(x, 0)|^3 + |\tilde{p}_{0,3}(x, \nu t)|^{\frac{4}{3}}\} \, d\nu .$$

It follows that

$$\begin{aligned}
&\sigma_F^2(\sum a_j\{\tilde{p}(Z_j, \theta) + \lambda(1-\lambda)\theta\psi_1(Z_j) - \tfrac{1}{2}\lambda(1-\lambda)(1-2\lambda)\theta^2\psi_2(Z_j)\}) \\
&\quad \leqq N \sum a_j^2 E_F(\tilde{p}(X_1, \theta) - \lambda + \lambda(1-\lambda)\theta\psi_1(X_1) - \tfrac{1}{2}\lambda(1-\lambda)(1-2\lambda)\theta^2\psi_2(X_1))^2 \\
&\quad \leqq 3B_1 C^{\frac{1}{2}} N^2 \theta^{\frac{24}{5}} , \\
&|\mathrm{Cov}_F(\sum a_j\{\tilde{p}(Z_j, \theta) + \lambda(1-\lambda)\theta\psi_1(Z_j) \\
&\quad - \tfrac{1}{2}\lambda(1-\lambda)(1-2\lambda)\theta^2\psi_2(Z_j)\}, \sum a_j\psi_1(Z_j))| \\
&\qquad \leqq [E_F|\sum a_j\{\tilde{p}(Z_j, \theta) - \lambda + \lambda(1-\lambda)\theta\psi_1(Z_j) \\
&\qquad\qquad - \tfrac{1}{2}\lambda(1-\lambda)(1-2\lambda)\theta^2\psi_2(Z_j)\}|^{\frac{3}{2}}]^{\frac{2}{3}}[E_F|\sum a_j(\psi_1(Z_j) - E_F\psi_1(Z_j))|^3]^{\frac{1}{3}} \\
&\qquad \leqq [(\sum |a_j^3|)^{\frac{1}{2}} N E_F|\tilde{p}(X_1, \theta) - \lambda + \lambda(1-\lambda)\theta\psi_1(X_1) \\
&\qquad\qquad - \tfrac{1}{2}\lambda(1-\lambda)(1-2\lambda)\theta^2\psi_2(X_1)|^{\frac{3}{2}}]^{\frac{2}{3}}[E_F|\sum a_j(\psi_1(Z_j) - E_F\psi_1(Z_j))|^3]^{\frac{1}{3}} \\
&\qquad \leqq (3B_1)^{\frac{2}{3}} C^{\frac{1}{4}} N \theta^{\frac{14}{5}} [E_F|\sum a_j(\psi_1(Z_j) - E_F\psi_1(Z_j))|^3]^{\frac{1}{3}} , \\
&\qquad\quad |\mathrm{Cov}_F(\sum a_j\{\tilde{p}(Z_j, \theta) + \lambda(1-\lambda)\theta\psi_1(Z_j) \\
&\qquad\qquad\quad - \tfrac{1}{2}\lambda(1-\lambda)(1-2\lambda)\theta^2\psi_2(Z_j)\}, \sum a_j\psi_2(Z_j))| \\
&\qquad\qquad\qquad \leqq (3B_1)^{\frac{1}{2}} C^{\frac{1}{4}} N \theta^{\frac{12}{5}} \sigma_F(\sum a_j\psi_2(Z_j)) .
\end{aligned}$$

These inequalities ensure that there exists $B_2 > 0$ depending only on B_1 and C such that

$$\begin{aligned}
&|\sigma_H^2(\sum a_j P_j) - \sigma_F^2(\sum a_j\{\lambda(1-\lambda)\theta\psi_1(Z_j) - \tfrac{1}{2}\lambda(1-\lambda)(1-2\lambda)\theta^2\psi_2(Z_j)\})| \\
&\quad \leqq B_2\{N^2\theta^{\frac{24}{5}} + N\theta^{\frac{19}{5}}[E_F|\sum a_j(\psi_1(Z_j) - E_F\psi_1(Z_j))|^3]^{\frac{1}{3}} + N\theta^{\frac{22}{5}}\sigma_F(\sum a_j\psi_2(Z_j))\} .
\end{aligned}$$

Since $N\theta^{\frac{22}{5}}\sigma_F(\sum a_j\psi_2(Z_j)) \leqq N^2\theta^{\frac{24}{5}} + \theta^4\sigma_F^2(\sum a_j\psi_2(Z_j))$, (A.11) follows immediately and the proof of the theorem is complete. □

COROLLARY A.1. *Suppose that* (3.1) *and* (4.3) *hold and that positive numbers* c, C, C', D, ε *and* ε' *exist such that* (3.10), (3.19), (4.2) *and* (4.4) *are satisfied. Let* K, α_i, $\tilde{K}$, $\tilde{\alpha}_i$ *and* η *be defined by* (3.17), (3.18), (4.5), (4.6) *and* (4.7). *Then there exists* $B > 0$ *depending only on* c, C, C', D, ε *and* ε' *such that*

$$\begin{aligned}
&\sup_x \left| K\left(x - \frac{\sum a_j\pi_j}{\{\lambda(1-\lambda)\sum a_j^2\}^{\frac{1}{2}}}\right) - \tilde{K}(x-\eta)\right| \\
&\quad \leqq B\{N^{-\frac{5}{4}} + N^{-\frac{1}{2}}\theta^3[\sum \{E_F|\psi_1(Z_j) - E_F\psi_1(Z_j)|^3\}^{\frac{4}{9}}]^{\frac{9}{4}} \\
&\qquad + N^{-\frac{3}{4}}\theta^3[\sum \{E_F(\psi_2(Z_j) - E_F\psi_2(Z_j))^2\}^{\frac{3}{4}}]^{\frac{4}{3}}\} , \qquad \text{(A.15)}
\end{aligned}$$

$$\theta^2 \frac{|\sum a_j E_F\psi_2(Z_j)|}{(\sum a_j^2)^{\frac{1}{2}}} \leqq BN^{-\frac{1}{2}} , \qquad \theta\frac{|\sum a_j^2 E_F\psi_1(Z_j)|}{\sum a_j^2} \leqq BN^{-\frac{1}{2}} , \qquad \text{(A.16)}$$

$$\frac{|\sum a_j^3|}{(\sum a_j^2)^{\frac{3}{2}}} \leqq BN^{-\frac{1}{2}} ,$$

$$\theta^2\frac{\sigma_F^2(\sum a_j\psi_1(Z_j))}{\sum a_j^2} \leqq B\{N^{-1} + N^{-\frac{1}{4}}\theta^3[\sum \{E_F|\psi_1(Z_j) - E_F\psi_1(Z_j)|^3\}^{\frac{4}{9}}]^{\frac{9}{4}}\} , \qquad \text{(A.17)}$$

and all other terms occurring in $\tilde{\alpha}_0, \cdots, \tilde{\alpha}_5$ *are bounded in absolute value by* BN^{-1}.

PROOF. In this proof $O(x)$ will denote a quantity that is bounded by $B_1|x|$ with B_1 depending only on c, C, C', D, ε and ε'.

We begin by noting that (A.16) and the last statement in Corollary A.1 are immediate consequences of Hölder's inequality, (3.10), (4.2) and (4.4). Also

$$\begin{aligned} \theta^2\sigma_F^2\left(\sum a_j\psi_1(Z_j)\right) &\leqq 1 + \theta^3\sigma_F^3\left(\sum a_j\psi_1(Z_j)\right) \\ &\leqq 1 + \theta^3 E_F\left|\sum a_j(\psi_1(Z_j) - E_F\psi_1(Z_j))\right|^3 \\ &\leqq 1 + \theta^3\left[\sum |a_j|\{E_F|\psi_1(Z_j) - E_F\psi_1(Z_j)|^3\}^{\frac{1}{3}}\right]^3 \\ &\leqq 1 + \theta^3\left(\sum a_j^4\right)^{\frac{3}{4}}\left[\sum \{E_F|\psi_1(Z_j) - E_F\psi_1(Z_j)|^3\}^{\frac{4}{9}}\right]^{\frac{9}{4}}, \end{aligned} \tag{A.18}$$

and in view of (3.10) and (4.4), this implies (A.17). For later use we note that similarly

$$\sigma_F^2\left(\sum a_j\psi_2(Z_j)\right) \leqq C^{\frac{1}{2}}N^{\frac{1}{2}}\left[\sum \{E_F(\psi_2(Z_j) - E_F\psi_2(Z_j))^2\}^{\frac{2}{3}}\right]^{\frac{3}{2}}. \tag{A.19}$$

It remains to prove (A.15). Since (A.15) is trivially satisfied for $N < (D/\varepsilon')^2$, we may assume that $0 \leqq \theta \leqq \varepsilon'$ so that Theorem A.1 applies. Because of (3.1), $\sum a_j\pi_j = \sum a_j(\pi_j - \lambda)$. In view of the bounds obtained above, we can truncate expansions (A.7) and (A.8) to

$$\begin{aligned} \sum a_j\pi_j &= \lambda(1-\lambda)\left\{-\theta\sum a_j E_F\psi_1(Z_j) \right. \\ &\qquad \left. + (1-2\lambda)\frac{\theta^2}{2}\sum a_j E_F\psi_2(Z_j)\right\} + O(N\theta^3) \\ &= -\lambda(1-\lambda)\theta\sum a_j E_F\psi_1(Z_j) + O(N\theta^2) = O(N\theta), \end{aligned} \tag{A.20}$$

$$\sum a_j^2(\pi_j - \lambda) = -\lambda(1-\lambda)\theta\sum a_j^2 E_F\psi_1(Z_j) + O(1) = O(N^{\frac{1}{2}}). \tag{A.21}$$

Using (A.8)—(A.11), (A.20), (A.21), (3.10), (3.19) and (4.4) we expand $\alpha_0, \cdots, \alpha_5$ and find

$$\begin{aligned} \sup_x |K(x) - \hat{K}(x)| = O\Big(N^{-\frac{5}{4}} &+ \theta^{\frac{19}{5}}\left[E_F\left|\sum a_j(\psi_1(Z_j) - E_F\psi_1(Z_j))\right|^3\right]^{\frac{1}{3}} \\ &+ N^{-1}\theta^3\sigma_F\left(\sum a_j\psi_1(Z_j)\right)\sigma_F\left(\sum a_j\psi_2(Z_j)\right) \\ &+ N^{-1}\theta^4\sigma_F^2\left(\sum a_j\psi_2(Z_j)\right)\Big), \end{aligned} \tag{A.22}$$

where

$$\hat{K}(x) = \Phi(x) - \phi(x)\sum_{k=0}^{5}\hat{\alpha}_k H_k(x), \tag{A.23}$$

$$\begin{aligned} \hat{\alpha}_0 &= -\left(\frac{\lambda(1-\lambda)}{\sum a_j^2}\right)^{\frac{1}{2}} N^{-1}\theta\sum a_j E_F\psi_1(Z_j), \\ \hat{\alpha}_1 &= \tilde{\alpha}_1 - \frac{1}{8\sum a_j^2}\lambda(1-\lambda)(1-2\lambda)^2\theta^4\left\{\sum a_j E_F\psi_2(Z_j)\right\}^2, \\ \hat{\alpha}_2 &= \tilde{\alpha}_2 - \frac{\{\lambda(1-\lambda)\}^{\frac{1}{2}}}{4(\sum a_j^2)^{\frac{3}{2}}}(1-2\lambda)^2\theta^3\sum a_j^2 E_F\psi_1(Z_j)\sum a_j E_F\psi_2(Z_j), \\ \hat{\alpha}_3 &= \tilde{\alpha}_3 - \frac{1}{12(\sum a_j^2)^2}(1-2\lambda)^2\theta^2\sum a_j^3\sum a_j E_F\psi_2(Z_j), \\ \hat{\alpha}_k &= \tilde{\alpha}_k \qquad \text{for} \quad k = 4, 5, \end{aligned} \tag{A.24}$$

with $\tilde{\alpha}_k$ as given by (4.6). By applying elementary inequalities (A.22) may be simplified to

$$\text{(A.25)} \qquad \sup_x |K(x) - \hat{K}(x)| = O(N^{-\frac{5}{4}} + N^{-\frac{5}{4}}\theta^3 E_F|\sum a_j(\psi_1(Z_j) - E_F\psi_1(Z_j))|^3 + N^{-\frac{5}{4}}\theta^3\sigma_F^2(\sum a_j\psi_2(Z_j)))\,.$$

With the aid of (A.7), (A.20) and the bounds obtained in the first part of the proof we now expand $\hat{K}(x - \sum a_j\pi_j\{\lambda(1-\lambda)\sum a_j^2\}^{-\frac{1}{2}})$ about the point $(x - \eta)$ and obtain

$$\text{(A.26)} \qquad \sup_x \left|\hat{K}\left(x - \frac{\sum a_j \pi_j}{\{\lambda(1-\lambda)\sum a_j^2\}^{\frac{1}{2}}}\right) - \tilde{K}(x-\eta)\right| = O(N^{-\frac{5}{4}} + N^{-1}\theta^3\sigma_F^2(\sum a_j\psi_1(Z_j)))$$

with $\tilde{K}$ as given by (4.5). Combining (A.25), (A.26), (A.18) and (A.19) we see that (A.15) and Corollary A.1 are proved. □

REFERENCES

ALBERS, W. (1974). *Asymptotic Expansions and the Deficiency Concept in Statistics.* Mathematical Centre Tracts **58**, Amsterdam.

ALBERS, W., BICKEL, P. J. and VAN ZWET, W. R. (1976). Asymptotic expansions for the power of distribution free tests in the one-sample problem. *Ann. Statist.* **4** 108–156.

BICKEL, P. J. (1974). Edgeworth expansions in nonparametric statistics. *Ann. Statist.* **2** 1-20.

EFRON, B. (1975). Defining the curvature of a statistical problem (with applications to second order efficiency). *Ann. Statist.* **3** 1189–1217.

ESSEEN, C. F. (1945). Fourier analysis of distribution functions. A mathematical study of the Laplace-Gaussian law. *Acta Math.* **77** 1–125.

HÁJEK, J. (1964). Asymptotic theory of rejective sampling with varying probabilities from finite populations. *Ann. Math. Statist.* **35** 1491–1523.

HODGES, J. L. and LEHMANN, E. L. (1963). Estimates of location based on rank tests. *Ann. Math. Statist.* **34** 598-611.

HODGES, J. L. and LEHMANN, E. L. (1970). Deficiency. *Ann. Math. Statist.* **41** 783-801.

OKAMOTO, M. (1958). Some inequalities relating to the partial sum of binomial probabilities. *Ann. Inst. Statist. Math.* **10** 29–35.

PFANZAGL, J. (1977). First order efficiency implies second order efficiency. To appear in J. Hájek Memorial Volume, Academia, Prague; North-Holland, Amsterdam.

RÉNYI, A. (1970). *Probability Theory.* North-Holland, Amsterdam.

ROBINSON, J. (1978). An asymptotic expansion for samples from a finite population. *Ann. Statist.* **6** 1005–1011.

ROGERS, W. F. (1971). Exact null distributions and asymptotic expansions for rank test statistics. Technical Report 145, Departments of O.R. and Statistics, Stanford Univ.

RYSHIK, I. M. and GRADSTEIN, I. S. (1957). *Tables of series, products and integrals.* V.E.B. Deutscher Verlag der Wiss., Berlin.

DEPARTMENT OF STATISTICS
UNIVERSITY OF CALIFORNIA
BERKELEY, CALIFORNIA 94720

UNIVERSITY OF LEIDEN
DEPARTMENT OF MATHEMATICS
WASSENAARSEWEG 80
P.O. BOX 9512
2300 RA LEIDEN
THE NETHERLANDS

International Statistical Review, **49** (1981), pp. 169–175 Longman Group Limited/Printed in Great Britain

On Efficiency of First and Second Order

P. J. Bickel[1], D. M. Chibisov[2] and W. R. van Zwet[3]

Department of Statistics, University of California, Berkeley, U.S.A., Steklov Mathematical Institute, Moscow, U.S.S.R. and Department of Mathematics, University of Leiden, The Netherlands

Summary

It has been noted by a number of authors that if two tests are asymptotically efficient for the same testing problem, then typically their powers will not only agree to first but also to second order. A general result of this type was given by Pfanzagl (1979) in a paper entitled 'First order efficiency implies second order efficiency'. Because of their technical nature, however, these contributions give little insight into the nature of this phenomenon. The purpose of the present paper is to provide an intuitive understanding of the phenomenon by proving a simple theorem of this kind under mild assumptions.

Key words: Asymptotic efficiency of tests; Second order efficiency; Deficiency.

1 Introduction

For $N = 1, 2, \ldots$, we consider an experiment with outcome X_N taking values in an arbitrary sample space. Let $P_{N,0}$ and $P_{N,1}$ be two possible distributions of X_N, with densities $p_{N,0}$ and $p_{N,1}$ with respect to some dominating measure μ_N. We shall write $E_{N,0}$ and $E_{N,1}$ for expectations under $P_{N,0}$ and $P_{N,1}$ respectively. Define the logarithm of the likelihood ratio by

$$\Lambda_N = \log \frac{p_{N,1}(X_N)}{p_{N,0}(X_N)}$$

with the usual conventions for vanishing $p_{N,0}$ and/or $p_{N,1}$.

Now consider a sequence $\alpha_N \in (0, 1)$ and let $\phi_N(\Lambda_N, \alpha_N)$ denote the test function of the most powerful level-α_N test for $P_{N,0}$ against $P_{N,1}$; thus

$$\phi_N(\Lambda_N, \alpha_N) = \begin{cases} 0 & \text{for } \Lambda_N < c_N(\alpha_N), \\ 1 & \text{for } \Lambda_N > c_N(\alpha_N), \end{cases}$$

with

$$E_{N,0}\, \phi_N(\Lambda_N, \alpha_N) = \alpha_N, \quad E_{N,1}\, \phi_N(\Lambda_N, \alpha_N) = \pi_N^*(\alpha_N),$$

where $\pi_N^*(\alpha_N)$ is the maximum attainable power against $P_{N,1}$ at level α_N.

For $N = 1, 2, \ldots$, let Z_N be a random variable depending only on the outcome X_N of the Nth experiment and let $\psi_N(Z_N, \alpha_N)$ denote the test function of the level-α_N right-sided test based on the statistic Z_N, i.e.

$$\psi_N(Z_N, \alpha_N) = \begin{cases} 0 & \text{for } Z_N < d_N(\alpha_N), \\ 1 & \text{for } Z_N > d_N(\alpha_N). \end{cases}$$

[1] Research supported by the U.S. Office of Naval Research, Contract N00014-80-C-0163, the National Science Foundation, Grant MC S76 10238 A01, and by the Netherlands' Organization for Pure Scientific Research.
[2] Research supported by the Netherlands' Organization for Pure Scientific Research.
[3] Research supported by the U.S. Office of Naval Research, Contract N00014-80-C-0163.

We have

$$E_{N,0}\,\psi_N(Z_N,\alpha_N)=\alpha_N,\quad E_{N,1}\,\psi_N(Z_N,\alpha_N)=\pi_N(\alpha_N),$$

where $\pi_N(\alpha_N)$ is the power of this test against $P_{N,1}$.

For a sequence $\tau_N\in(0,1]$, we shall say that the sequence of level-α_N tests $\psi_N(Z_N,\alpha_N)$ is *τ_N-efficient* if, for $N\to\infty$,

$$\pi_N^*(\alpha_N)-\pi_N(\alpha_N)=o(\tau_N).$$

In a more usual terminology first and second order efficiency correspond to τ_N-efficiency with $\tau_N=1$ and $\tau_N=N^{-1/2}$ respectively.

Finally, let us define for $N=1,2,\ldots,$

$$\Delta_N=\begin{cases}0 & \text{if } Z_N=\Lambda_N=\pm\infty,\\ Z_N-\Lambda_N & \text{otherwise,}\end{cases}$$

and let us denote the indicator function of a set B by I_B.

Having established our notation, we now give an informal description of the phenomenon we wish to study. Let us think of N as denoting sample size, i.e. N is the number of independent random variables involved in the Nth testing problem. We are interested primarily in sequences of testing problems where $\alpha_N\geq\varepsilon$ and $\pi_N^*(\alpha_N)\leq 1-\varepsilon$ for some $\varepsilon>0$ and all N. Such sequences exist if it is impossible to discriminate perfectly between $P_{N,0}$ and $P_{N,1}$ even as $N\to\infty$ and this is true if $P_{N,0}$ and $P_{N,1}$ are contiguous. A sufficient condition for contiguity, and one which is often fulfilled in this case, is asymptotic normality of Λ_N both under $P_{N,0}$ and $P_{N,1}$. But if Λ_N is asymptotically normal, it will usually also be possible to obtain an Edgeworth expansion for its distribution function under $P_{N,0}$ and $P_{N,1}$ and this will yield a similar expansion for the power of the test based on Λ_N, viz.

$$\pi_N^*(\alpha_N)=c_0+c_1N^{-1/2}+o(N^{-1/2}).\tag{1.1}$$

Typically the remainder term on the right in (1.1) will be $O(N^{-1})$.

Suppose that the sequence of tests $\psi_N(Z_N,\alpha_N)$ is asymptotically efficient to first order, or 1-efficient in our terminology. For most statistical problems such 1-efficient tests abound. They are usually based on statistics Z_N that closely resemble Λ_N. Typically $\Delta_N=Z_N-\Lambda_N$ tends to zero in probability both under $P_{N,0}$ and $P_{N,1}$ and in the situation we have described so far, this suffices to ensure 1-efficiency. Of course, these 1-efficient tests can also be based on statistics Z_N which do not resemble Λ_N at all, because the test statistic associated with a test is by no means unique. However, we shall not be concerned with such alternative representations and suppose that $\Delta_N\to 0$ in $P_{N,0}$- and in $P_{N,1}$-probability.

Recalling that N is the sample size, we note that one often finds that a sequence of random variables Δ_N tending to zero in probability does so at the rate of $N^{-1/2}$. Thus, for 1-efficient tests, $N^{1/2}\Delta_N$ will typically be bounded in probability both under $P_{N,0}$ and $P_{N,1}$. Hence, one may expect to be able to establish Edgeworth expansions for the distribution functions of Z_N under $P_{N,0}$ and $P_{N,1}$, which differ from those for Λ_N only in the term of order $N^{-1/2}$ and in those of smaller order. This yields a similar expansion for the power of the test based on Z_N,

$$\pi_N(\alpha_N)=c_0+c_1'N^{-1/2}+o(N^{-1/2}),\tag{1.2}$$

where the remainder term on the right will typically be $O(N^{-1})$ or of slightly larger order. The fact that the leading terms in expansions (1.1) amd (1.2) have the same value c_0 reflects the 1-efficiency of the sequence $\psi_N(Z_N,\alpha_N)$. There would seem to be no reason *a priori* to expect that also $c_1=c_1'$, which would entail $N^{-1/2}$-efficiency or efficiency of second order.

However, in those cases where expansions (1.1) and (1.2) were explicitly computed, one does indeed find that $c_1=c_1'$ and hence that the sequence $\psi_N(Z_N,\alpha_N)$ is $N^{-1/2}$-efficient. This phenomenon was noticed by Pfanzagl (1973), (1975) and Chibisov (1974) for a number of tests for the parametric one-sample problem and by Bickel & van Zwet (1978) for rank tests for the nonparametric two-sample problem. Some tests for the one-sample problem for the case where nuisance parameters are present were considered by Chibisov (1973) and Pfanzagl (1974) and also found to be $N^{-1/2}$-efficient. Finally, it was shown by Pfanzagl (1979) that first-order efficiency forces second-order efficiency for a large class of one-sample tests in the presence of nuisance parameters. With an appropriate definition of efficiency a similar result was obtained for estimators.

In each of these contributions, $N^{-1/2}$-efficiency is established by imposing the conditions needed to obtain expansions (1.1) and (1.2) and then checking that these expansions are in fact identical. This method of proof coupled with its extreme technicality makes an intuitive understanding of the phenomenon rather difficult. The purpose of the present paper is to provide such an intuitive understanding by proving a simple theorem of this kind under rather mild assumptions. Since our aim is to provide insight rather than generality, we shall only be concerned with the simple hypothesis testing problem described above and avoid the technicalities inherent in the treatment of nuisance parameters and estimation problems, although extension to these situations is certainly possible. Having mentioned estimation, however, we should note that Rao's (1961, 1962) concept of second order efficiency of estimators as discussed by Efron (1975) and Ghosh, Sinha & Wieand (1980), refers to optimality up to $o(N^{-1})$ and would therefore correspond to N^{-1}-efficiency or third order efficiency in our terminology. This difference in terminology is not as illogical as it may seem because most results of these authors concern the performance of an estimator as measured by its risk relative to a symmetric loss function and expansions for this quantity typically do not contain a term of order $N^{-1/2}$, so that the term of order N^{-1} is indeed the second order term in this case.

In section 2 we present our result, discuss its meaning and explain why it is true. A formal proof of the theorem is given in section 3. Though this proof is straightforward, the non-mathematically inclined may wish to skip it.

2 Discussion of the result

We adopt the notation and conventions introduced in the previous section. In particular we recall that τ_N is an arbitrary sequence in (0, 1] and that first and second order efficiency correspond to τ_N-efficiency with $\tau_N = 1$ and $\tau_N = N^{-1/2}$ respectively.

Theorem. Suppose that

$$\liminf_N \alpha_N > 0, \tag{2.1}$$

and that there exists $A > 0$ such that for every $x_0 \in \mathbb{R}$, every $\gamma > 0$ and $N \to \infty$,

$$\sup_{x \le x_0} P_{N,0}(x - \tau_N^{1/2} \le \Delta_N \le x) = O(\tau_N^{1/2}), \tag{2.2}$$

$$E_{N,0}|\Delta_N| 1_{(\gamma\tau_N^{1/2}, A)}(|\Delta_N|) = o(\tau_N), \tag{2.3}$$

$$P_{N,0}(\Delta_N \ge A) = o(\tau_N), \tag{2.4}$$

$$P_{N,1}(\Delta_N \le -A) = o(\tau_N). \tag{2.5}$$

Then the sequence of tests $\psi_N(Z_N, \alpha_N)$ is τ_N-efficient.

Let us briefly discuss the conditions of this theorem. First, assumption (2.2) is clearly satisfied for any sequence τ_N if the distributions of Δ_N under $P_{N,0}$ possess uniformly bounded densities.

More generally, (2.2) will hold if the distribution functions F_N of Λ_N under $P_{N,0}$ can be approximated with a uniform error of order $\tau_N^{1/2}$ by distribution functions G_N with uniformly bounded densities, i.e. if $\sup|F_N(x) - G_N(x)| = O(\tau_N^{1/2})$. This is certainly the case when F_N has a normal approximation or an Edgeworth expansion with the required error.

If $\tau_N \to 0$, assumption (2.2) clearly implies that the distributions of Λ_N under $P_{N,0}$ do not tend to a degenerate limit. In view of this, conditions (2.3)–(2.4) serve to ensure that under $P_{N,0}$, $|\Delta_N|$ is small compared to the variation of Λ_N. Note that these conditions refer only to values of $|\Delta_N|$ which excede $\gamma\tau_N^{1/2}$ and that they are satisfied if the distribution of Δ_N under $P_{N,0}$ either assigns probability 1 to a set where $|\Delta_N| = o(\tau_N^{1/2})$, or has at most very small tails outside that range. Thus, roughly speaking, Δ_N is required to be $o(\tau_N^{1/2})$ under $P_{N,0}$; under $P_{N,1}$ condition (2.5) is even weaker, but this is to a certain extent artificial and is due to our efforts to replace conditions under $P_{N,1}$ as much as possible by conditions under $P_{N,0}$ which will usually be easier to verify. It follows that one cannot hope to say more about the differences between the distribution functions of Λ_N and Z_N under $P_{N,0}$ and $P_{N,1}$ than that they are $o(\tau_N^{1/2})$. One would therefore expect to be able to prove that $\pi_N^*(\alpha_N) - \pi_N(\alpha_N) = o(\tau_N^{1/2})$ but, somewhat surprisingly, the conclusion of the theorem is the stronger statement that $\pi_N^*(\alpha_N) - \pi_N(\alpha_N) = o(\tau_N)$. The condition that roughly Δ_N is $o(\tau_N^{1/2})$ under $P_{N,0}$ cannot essentially be improved. By taking $\Delta_N = \Delta\tau_N^{1/2}$ where Δ is independent of Λ_N for every N, one easily constructs examples where $\pi^*(\alpha_N) - \pi(\alpha_N)$ is of exact order τ_N.

A formal proof of the theorem will be given in section 3. At this point we shall be content to provide an intuitive explanation of the result by sketching the proof for the special case where there exist numbers δ_N such that for $N = 1, 2, \ldots,$

$$|\Delta_N| < \delta_N = o(\tau_N^{1/2}). \tag{2.6}$$

We should perhaps stress that a boundedness assumption like (2.6) is not likely to be fulfilled in concrete examples. It is made here merely to avoid technicalities at this stage and bring out the essential simplicity of the proof.

Let us write c_N and d_N for $c_N(\alpha_N)$ and $d_N(\alpha_N)$ respectively. Since the tests $\phi_N(\Lambda_N, \alpha_N)$ and $\psi_N(Z_N, \alpha_N)$ have the same level α_N, (2.6) clearly implies that we may assume that $|c_N - d_N| < \delta_N$. Invoking (2.6) once more, we see that if $\Lambda_N \geq c_N$ and $Z_N \leq d_N$, then $d_N - \delta_N \leq \Lambda_N \leq d_N + \delta_N$; the same conclusion holds if $\Lambda_N \leq c_N$ and $Z_N \geq d_N$. It follows that on the set where $\phi_N(\Lambda_N, \alpha_N) \neq \psi_N(Z_N, \alpha_N)$ we have $|\Lambda_N - d_N| \leq \delta_N$, and again because both tests have level α_N we find with the aid of (2.2),

$$\begin{aligned}
\pi_N^*(\alpha_N) - \pi_N(\alpha_N) &= E_{N,1}\{\phi_N(\Lambda_N, \alpha_N) - \psi_N(Z_N, \alpha_N)\} \\
&= E_{N,0}(e^{\Lambda_N} - e^{d_N})\{\phi_N(\Lambda_N, \alpha_N) - \psi_N(Z_N, \alpha_N)\} \\
&\leq e^{d_N}(e^{\delta_N} - 1)\, P_{N,0}(|\Lambda_N - d_N| \leq \delta_N) \\
&= O(\delta_N \tau_N^{1/2}) = o(\tau_N).
\end{aligned}$$

Note that we need to have d_N bounded above, but this is an easy consequence of (2.1) and (2.6). The above sketch should make it clear that the essential thing which makes the theorem work is that not only do $\phi_N(\Lambda_N, \alpha_N)$ and $\psi_N(Z_N, \alpha_N)$ resemble each other closely, but that also Λ_N is almost constant on the set where they differ.

Let us finally discuss the relevance of the theorem to the problem of first and second order efficiency. As was pointed out in section 1, first order efficiency of $\psi_N(Z_N, \alpha_N)$ will typically imply that $N^{1/2}\Delta_N$ is bounded in probability both under $P_{N,0}$ and $P_{N,1}$. Also, $E_{N,0}|N^{1/2}\Delta_N|^r$ and $E_{N,1}|N^{1/2}\Delta_N|^r$ will usually be bounded for some $r > 1$ and this (or even uniform integrability of $|N^{1/2}\Delta_N|$ under $P_{N,0}$ and $P_{N,1}$) is amply sufficient to ensure that assumptions (2.3)–(2.5) are satisfied for $\tau_N = N^{-1/2}$. But then the theorem ensures that $\psi_N(Z_N, \alpha_N)$ is efficient to second order under the very mild conditions (2.1) and (2.2) for $\tau_N = N^{-1/2}$.

An examination of our proof shows that if we replace (2.6) by

$$|\Delta_N| \leq M\tau_N^{1/2} \tag{2.7}$$

for $M < \infty$ and $N = 1, 2, \ldots$, then we obtain the conclusion

$$\pi_N^*(\alpha_N) - \pi_N(\alpha_N) = O(\tau_N). \tag{2.8}$$

Taking $\tau_N = N^{-1}$ we conclude that if $|N^{1/2}\Delta_N|$ is bounded and if, e.g., the distribution of Λ_N under $P_{N,0}$ tends to normality at the rate of $N^{-1/2}$, then $\pi_N^*(\alpha_N) - \pi_N(\alpha_N) = O(N^{-1})$. This means that the tests based on Z_N have a finite deficiency in the sense of Hodges & Lehmann (1970). That is, if we let $\alpha_N = \alpha \in (0, 1)$ for $N = 1, 2, \ldots$ and define N' to be the smallest integer for which $\pi_{N'}(\alpha) \geq \pi_N^*(\alpha)$ then $\limsup (N' - N) < \infty$.

Assumption (2.7) is of course not necessary to obtain (2.8) and $N^{-1/2}$-efficiency is frequently coupled with finite deficiency. However, the coupling is not inevitable. An example, in a nuisance parameter context, is provided by the normal scores test studied by Bickel & van Zwet (1978). Note that this bears out what we have said about the order of the remainder terms in (1.1) and (1.2).

3 Proof of the theorem

Take a sequence $\alpha_N \in (0, 1)$ satisfying (2.1) and write $c_N = c_N(\alpha_N)$ and $d_N = d_N(\alpha_N)$. If $d_N = -\infty$ for some N then

$$\begin{aligned} 1 - \pi_N(\alpha_N) &\leq P_{N,1}(Z_N = -\infty) \leq P_{N,1}(\Lambda_N = -\infty) + P_{N,1}(\Delta_N = -\infty) \\ &= P_{N,1}(\Delta_N = -\infty) = o(\tau_N) \end{aligned}$$

because of (2.5), and $\psi_N(Z_N, \alpha_N)$ is clearly τ_N-efficient. Take A as in (2.3)–(2.5). Then

$$\begin{aligned} \alpha_N &\leq P_{N,0}(Z_N \geq d_N) \leq P_{N,0}(\Lambda_N \geq d_N - A) + P_{N,0}(\Delta_N \geq A) \\ &\leq e^{-d_N + A}\, P_{N,1}(\Lambda_N \geq d_N - A) + o(\tau_N) \leq e^{-d_N + A} + o(\tau_N) \end{aligned}$$

because of (2.4). In view of (2.1) it is therefore no loss of generality to assume that for some $D < \infty$ and all N,

$$-\infty < d_N \leq D. \tag{3.1}$$

Define

$$\bar{\Delta}_N = \begin{cases} d_N - \Lambda_N & \text{if } \Lambda_N < d_N \leq Z_N \text{ or } Z_N \leq d_N < \Lambda_N, \\ 0 & \text{otherwise.} \end{cases}$$

Obviously,

$$\bar{\Delta}_N^+ \leq \Delta_N^+, \; \bar{\Delta}_N^- \leq \Delta_N^-, \tag{3.2}$$

where $x^+ = x \vee 0$ and $x^- = (-x) \vee 0$ denote the positive and negative parts of a number x.

Let α_N' be such that $d_N = c_N(\alpha_N')$. Then

$$\begin{aligned} \{\pi_N^*(\alpha_N) - \pi_N(\alpha_N)\} &+ \{\pi_N^*(\alpha_N') - \pi_N^*(\alpha_N) - e^{d_N}(\alpha_N' - \alpha_N)\} \\ &= \pi_N^*(\alpha_N') - \pi_N(\alpha_N) - e^{d_N}(\alpha_N' - \alpha_N) \\ &= E_{N,0}\{\phi_N(\Lambda_N, \alpha_N') - \psi_N(Z_N, \alpha_N)\}(e^{\Lambda_N} - e^{d_N}) \\ &= E_{N,0}|\phi_N(\Lambda_N, \alpha_N') - \psi_N(Z_N, \alpha_N)|\,|e^{\Lambda_N} - E^{d_N}| \\ &\leq e^{d_N} E_{N,0}|e^{-\bar{\Delta}_N} - 1|, \end{aligned}$$

since $\{\phi_N(\Lambda_N, \alpha'_N) - \psi_N(Z_N, \alpha_N)\}$ is nonnegative, or nonpositive, if $(\Lambda_N - d_N)$ is positive, or negative, and equals zero if $\bar{\Delta}_N = 0$ and $(\Lambda_N - d_N) \neq 0$. A similar argument yields

$$\pi_N^*(\alpha'_N) - \pi_N^*(\alpha_N) - e^{d_N}(\alpha'_N - \alpha_N) = E_{N,0}\{\phi_N(\Lambda_N, \alpha'_N) - \phi_N(\Lambda_N, \alpha_N)\}(e^{\Lambda_N} - e^{d_N}) \geq 0$$

and hence

$$0 \leq \pi_N^*(\alpha_N) - \pi_N(\alpha_N) \leq e^{d_N} E_{N,0} | e^{-\bar{\Delta}_N} - 1 |. \tag{3.3}$$

By (2.3) there exists a sequence $\gamma_N \downarrow 0$ such that

$$E_{N,0} | e^{-\Delta_N} - 1 | I_{(\gamma_N \tau_N^{1/2}, A)}(|\Delta_N|) = o(\tau_N)$$

for $N \to \infty$. In view of (3.2) this implies

$$E_{N,0} | e^{-\bar{\Delta}_N} - 1 | I_{(\gamma_N \tau_N^{1/2}, \infty]}(|\bar{\Delta}_N|) I_{[0,A)}(|\Delta_N|) = o(\tau_N). \tag{3.4}$$

Also (3.1) and (2.2) with $x_0 = D + 1$ yield

$$\begin{aligned} E_{N,0} | e^{-\bar{\Delta}_N} - 1 | I_{[0, \gamma_N \tau_N^{1/2}]}(|\bar{\Delta}_N|) &= O(\gamma_N \tau_N^{1/2} P_{N,0}(0 < |\bar{\Delta}_N| \leq \gamma_N \tau_N^{1/2})) \\ &= o(\tau_N^{1/2} P_{N,0}(0 < |\Lambda_N - d_N| \leq \tau_N^{1/2})) = o(\tau_N). \end{aligned} \tag{3.5}$$

If $\Delta_N \geq 0$ then $\bar{\Delta}_N \geq 0$ and because of (2.4)

$$E_{N,0} | e^{-\bar{\Delta}_N} - 1 | I_{[A,\infty]}(\Delta_N) \leq P_{N,0}(\Delta_N \geq A) = o(\tau_N). \tag{3.6}$$

If $\Delta_N \leq 0$ then $\bar{\Delta}_N \leq 0$ and (2.5) ensures that

$$\begin{aligned} e^{d_N} E_{N,0} | e^{-\bar{\Delta}_N} - 1 | I_{[-\infty,-A]}(\Delta_N) &\leq e^{d_N} E_{N,0} e^{-\bar{\Delta}_N} I_{[-\infty,-A]}(\Delta_N) I_{[-\infty,0)}(\bar{\Delta}_N) \\ &\leq E_{N,0} e^{\Lambda_N} I_{[-\infty,A]}(\Delta_N) = P_{N,1}(\Delta_N \leq -A) = o(\tau_N). \end{aligned} \tag{3.7}$$

Together (3.1) and (3.3)–(3.7) imply that $\pi_N^*(\alpha_N) - \pi_N(\alpha_N) = o(\tau_N)$ and the proof is complete.

References

Bickel, P.J. & van Zwet, W.R. (1978). Asymptotic expansions for the power of distribution free tests in the two-sample problem. *Ann. Statist.* **6**, 937–1004.

Chibisov, D.M. (1973). Asymptotic expansions for Neyman's $C(\alpha)$ tests. *Proc. Second Japan–USSR Symp. on Probability Theory*, ed. G. Maruyama and Yu. V. Prokhorov. *Lecture Notes in Mathematics* **330**, pp. 16–45. Springer, Berlin.

Chibisov, D.M. (1974). Asymptotic expansions for some asymptotically optimal tests. *Proc. Prague Symp. on Asymptotic Statistics*, ed. J. Hájek, **2**, 37–68.

Efron, B. (1975). Defining the curvature of a statistical problem (with applications to second order efficiency). *Ann. Statist.* **3**, 1189–1242.

Ghosh, J.K., Sinha, B.K. & Wieand, H.S. (1980). Second order efficiency of the MLE with respect to a bounded bowl-shaped loss function. *Ann. Statist.* **8**, 506–521.

Hodges, J.L. & Lehmann, E.L. (1970). Deficiency. *Ann. Math. Statist.* **41**, 783–801.

Pfanzagl, J. (1973). Asymptotic expansions related to minimum contrast estimators. *Ann. Statist.* **1**, 993–1026.

Pfanzagl, J. (1974). Asymptotically optimum estimation and test procedures. *Proc. Prague Symp. on Asymptotic Statistics*, ed. J. Hájek, **1**, 201–272.

Pfanzagl, J. (1975). On asymptotically complete classes. *Statistical Inference and Related Topics*, **2**, ed. M. L. Puri, pp. 1–43. Academic Press, New York.

Pfanzagl, J. (1979). First order efficiency implies second order efficiency. *Contributions to Statistics [J. Hájek Memorial Volume]*, ed. J. Jurečková, pp. 167–196. Academia, Prague,

Rao, C.R. (1961). Asymptotic efficiency and limiting information. *Proc. 4th Berkeley Symp.* **1**, 531–545.

Rao, C.R. (1962). Efficient estimates and optimum inference procedures in large samples (with discussion). *J. R. Statist. Soc.* B **24**, 46–72.

Résumé

Plusieurs auteurs ont remarqué, que, si deux tests sont asymptotiquement efficients pour le même problème de test statistique, leurs puissances s'accorderont normalement non seulement du premier mais aussi du deuxième ordre. Pfanzagl (1979) donna un résultat général de ce genre dans son article "First order efficiency implies second order efficiency". Cependant, à cause de leur structure technique, ces contributions ne donnent qu'une idée peu claire de la nature de ce phénomène. Le but de cet article-ci est d'établir une notion intuïtive du phénomène en démontrant un théorème simple de ce genre sous des conditions souples.

[*Paper received October* 1980, *revised January* 1981]

Z. Wahrscheinlichkeitstheorie verw. Gebiete
66, 425–440 (1984)

Zeitschrift für
Wahrscheinlichkeitstheorie
und verwandte Gebiete

A Berry-Esseen Bound for Symmetric Statistics

W.R. van Zwet*

University of Leiden, Dept. of Mathematics and Computer Science, Wassenaarseweg 80,
P.O. Box 9512, 2300 RA Leiden, The Netherlands

Summary. The rate of convergence of the distribution function of a symmetric function of N independent and identically distributed random variables to its normal limit is investigated. Under appropriate moment conditions the rate is shown to be $\mathcal{O}(N^{-\frac{1}{2}})$. This theorem generalizes many known results for special cases and two examples are given. Possible further extensions are indicated.

1. Introduction

During the past decade a good deal of effort has been devoted to extending the theory of Berry-Esseen bounds and Edgeworth expansions to more complicated sequences of random variables than normalized sums of independent and identically distributed (i.i.d.) random variables or vectors. From a statistical point of view, this study of higher order asymptotics for large classes of test statistics and estimators has proved extremely fruitful: it has yielded much that is significant for statistical theory as well as useful in practical applications. To the probabilist, however, most test statistics and estimators occurring in statistical theory appear to be strange artefacts, which are neither particularly interesting objects for study in themselves nor very promising starting points for developing a general probabilistic theory.

There is, perhaps, one exception which is the class of U-statistics introduced by Hoeffding (1948). Though it is usually studied for its statistical applications, it surely constitutes a large class of random variables which would seem to be a natural extension of sums of i.i.d. random variables. Let $X_1, X_2, \ldots$ be i.i.d. random variables and let h: $\mathbb{R}^k \to \mathbb{R}$ be a symmetric function of its k arguments. For $N \geqq k$, a U-statistic of degree k is defined as

$$U = \sum_{1 \leqq i_1 < i_2 < \ldots < i_k \leqq N} \sum h(X_{i_1}, X_{i_2}, \ldots, X_{i_k}) \tag{1.1}$$

* Research supported by the U.S. Office of Naval Research, Contract N 00014-80-C-0163

and the idea is to study its asymptotic behavior for a fixed h as $N\to\infty$. For $k=1$, we are back in the case of sums of i.i.d. random variables. As soon as $k\geqq 2$, the degree doesn't play an important role any more except, of course, for the fact that it stays fixed as $N\to\infty$. Many authors therefore discuss only the case of degree two, on the understanding that the case $k>2$ is similar. Let us follow this tradition for a moment and take

$$U=\sum_{1\leqq i<j\leqq N}\sum h(X_i, X_j), \tag{1.2}$$

where $h(x,y)=h(y,x)$. Assume that

$$Eh(X_1,X_2)=0, \quad Eh^2(X_1,X_2)<\infty, \tag{1.3}$$

and define

$$g(x)=E(h(X_1,X_2)|X_1=x), \quad \psi(x,y)=h(x,y)-g(x)-g(y), \tag{1.4}$$

$$\hat{U}=(N-1)\sum_{i=1}^{N} g(X_i), \quad \Delta=\sum_{1\leqq i<j\leqq N}\sum \psi(X_i,X_j). \tag{1.5}$$

Clearly, $E(\psi(X_1,X_2)|X_1)=0$ a.s. so that the random variables $g(X_i)$ and $\psi(X_i,X_j)$ are pairwise uncorrelated and since $U=\hat{U}+\Delta$,

$$\sigma^2(U)=\sigma^2(\hat{U})+\sigma^2(\Delta)=N(N-1)^2 Eg^2(X_1)+\tfrac{1}{2}N(N-1)E\psi^2(X_1,X_2). \tag{1.6}$$

If it is assumed that

$$Eg^2(X_1)>0, \tag{1.7}$$

then $\sigma^2(\hat{U})$ dominates the right-hand side of (1.6) and $U\sigma^{-1}(U)$ is asymptotically normal (cf. Hoeffding (1948)).

The speed of convergence to normality was investigated by a number of authors who proved in increasing generality that

$$\sup_x\left|P\left(\frac{U}{\sigma(U)}\leqq x\right)-\Phi(x)\right|=\mathcal{O}(N^{-\frac{1}{2}}), \tag{1.8}$$

where Φ denotes the standard normal distribution function (d.f.). Suppose that (1.3) and (1.7) are satisfied so that asymptotic normality is ensured. Bickel (1974) established the Berry-Esseen bound (1.8) under the additional assumption that h is bounded. Chan and Wierman (1977) and Callaert and Janssen (1978) successively reduced this assumption first to $Eh^4(X_1,X_2)<\infty$ and then to $E|h(X_1,X_2)|^3<\infty$. Helmers and Van Zwet (1982) showed that $E|g(X_1)|^3<\infty$ suffices. They also proved that the assumption $Eh^2(X_1,X_2)<\infty$ in (1.3) may be relaxed, provided $\sigma(U)$ is replaced by $\sigma(\hat{U})$ in (1.8). This need not concern us here, however, since we shall concentrate on the case of finite variance in the present paper.

Let us consider the more general case of a symmetric statistic. As before, let $X_1,\ldots,X_N$ be i.i.d. and let $\tau\colon \mathbb{R}^N\to\mathbb{R}$ be a symmetric function of its N arguments.

Define

$$T=\tau(X_1,\ldots,X_N) \tag{1.9}$$

and assume that

$$ET=0, \quad ET^2=1. \tag{1.10}$$

We wish to study the asymptotic behavior of T as $N \to \infty$. The difference with the previous problem is that then we were dealing with a kernel function h that remains fixed as $N \to \infty$, or perhaps with uniformity classes of such functions of a fixed degree k. Now the degree of the kernel τ equals the sample size N and both tend to infinity together.

Define

$$T_j = E(T|X_j), \quad \hat{T}_1 = \sum_{j=1}^{N} T_j, \tag{1.11}$$

then $\hat{T}_1$ and $(T - \hat{T}_1)$ are again uncorrelated. It follows that if $\sigma^2(T) \sim \sigma^2(\hat{T}_1)$ as $N \to \infty$ and the summands T_j satisfy the Lindeberg condition, then $T\sigma^{-1}(T)$ is asymptotically normal.

The aim of this paper is to prove the following theorem of Berry-Esseen type.

Theorem 1.1. *Suppose that* (1.10) *is satisfied and that positive numbers A and B exist such that*

$$E|E(T|X_1)|^3 \leqq AN^{-\frac{3}{2}}, \tag{1.12}$$

$$1 + E\{E(T|X_1, \ldots, X_{N-2})\}^2 - 2E\{E(T|X_1, \ldots, X_{N-1})\}^2 \leqq BN^{-3}. \tag{1.13}$$

Then

$$\sup_x |P(T \leqq x) - \Phi(x)| \leqq C(A+B)N^{-\frac{1}{2}}, \tag{1.14}$$

where C denotes a universal constant.

Note that although we have formulated the theorem as a uniform error bound for a fixed but arbitrary N and T, it is a purely asymptotic result because the constant C is not specified. It applies to sequences of symmetric statistics $T_N = \tau_N(X_{N,1}, \ldots, X_{N,N})$ where, for every fixed N, $X_{N,1}, \ldots, X_{N,N}$ are i.i.d. with a common d.f. F_N, provided (1.10), (1.12) and (1.13) are satisfied for every N and fixed values of A and B.

The theorem will be proved in Sects. 2 and 3. In Sect. 2 we collect some facts concerning L_2-projections and in Sect. 3 we provide a proof of the theorem based on these facts. Some examples and possible extensions are discussed in Sects. 4 and 5.

2. L_2-Projections

L_2-projections were introduced in statistics by Hoeffding (1948, 1961) and have been used effectively by many authors since then. Most recently Efron and Stein (1981) and Karlin and Rinott (1982) have used these orthogonal projections to establish certain variance inequalities. To indicate decomposition by repeated orthogonal projection, these authors have introduced the descriptive term *ANOVA-type decomposition*, but we prefer to speak of *Hoeffding's decomposition* instead. What follows are some simple and well-known facts concerning L_2-projections written down in an easy notation.

Let $X_1,\ldots,X_N$ be independent random variables and let $T=\tau(X_1,\ldots,X_N)$ have $ET^2<\infty$. Note that at this point we do *not* assume that $X_1,\ldots,X_N$ are identically distributed, that τ is symmetric in its N arguments, or that $ET=0$ and $ET^2=1$. Define $\Omega=\{1,2,\ldots,N\}$. For any $D\subset\Omega$, let

$$E(T\,|\,D)=E(T\,|\,X_i,\ i\in D) \tag{2.1}$$

denote the conditional expectation given all X_i with indices in D. Define

$$T_D=\sum_{A\subset D}(-1)^{|D|-|A|}E(T\,|\,A), \tag{2.2}$$

where the summation is over all subsets A of D, including the empty set, and $|\cdot|$ denotes the cardinality of a set. Of course $T_\phi=E(T\,|\,\phi)=ET$ a.s. and for convenience we shall write

$$T_j=T_{\{j\}}=E(T\,|\,X_j)-ET,\quad j=1,\ldots,N. \tag{2.3}$$

The basic property of T_D is that

$$E(T_D|D')=0\quad\text{a.s.}\quad\text{unless } D\subset D'. \tag{2.4}$$

To see this, write $C=D\cap D'$ and note that, if $|D|-|C|=k>0$,

$$E(T_D|D')=\sum_{A\subset D}(-1)^{|D|-|A|}E(T\,|\,A\cap C)=\sum_{B\subset C}E(T\,|\,B)\sum_{j=0}^{k}(-1)^{|D|-|B|-j}\binom{k}{j}=0\ \text{ a.s..}$$

It follows in particular that $ET_D=0$ if $D\neq\phi$ and that the random variables T_D, $D\subset\{1,\ldots,N\}$ are pairwise uncorrelated, i.e.

$$ET_D T_{D'}=0\quad\text{if } D\neq D'. \tag{2.5}$$

Since the order of the two operations in $E(T_D|D')$ may be interchanged with impunity, we have $E(T_D|D')=[E(T\,|\,D')]_D$. Hence (2.4) also yields that if T depends only on X_i for $i\in D'$, then

$$T_D=0\quad\text{a.s.}\quad\text{unless } D\subset D'. \tag{2.6}$$

For $m=0,1,\ldots,N$, let $\mathscr{L}_m$ denote the linear space of random variables with finite variance that is spanned by functions of at most m of the variables $X_1,\ldots,X_N$, thus

$$\mathscr{L}_m=\{Z:Z=\sum\ \sum_{1\leqq i_1<i_2<\ldots<i_m\leqq N}\psi_{i_1,\ldots,i_m}(X_{i_1},\ldots,X_{i_m}),EZ^2<\infty\}.$$

We define $\hat{T}_m$ to be the L_2-projection of T on $\mathscr{L}_m$ if $\hat{T}_m\in\mathscr{L}_m$ and $E(T-\hat{T}_m)^2$ is minimal, or equivalently, if $\hat{T}_m\in\mathscr{L}_m$ and $E(T-\hat{T}_m)Z=0$ for all $Z\in\mathscr{L}_m$. We have

$$\hat{T}_0=ET,\quad \hat{T}_1-\hat{T}_0=\sum_{j=1}^{N}T_j,\quad \hat{T}_m-\hat{T}_{m-1}=\sum_{|D|=m}T_D,\quad \hat{T}_N=T. \tag{2.7}$$

To check this, note that $\hat{T}_m \in \mathscr{L}_m$ and that $ET_D Z=0$ if $|D| \geqq m+1$ and $Z \in \mathscr{L}_m$ by (2.4). Hence we have Hoeffding's decomposition

$$T=\hat{T}_0+(\hat{T}_1-\hat{T}_0)+\ldots+(\hat{T}_N-\hat{T}_{N-1})=\sum_{D \subset \Omega} T_D \tag{2.8}$$

and since all terms are pairwise uncorrelated,

$$ET^2=\sum_{D \subset \Omega} ET_D^2. \tag{2.9}$$

If we apply (2.8) to $E(T|A)$ instead of T, (2.6) yields

$$E(T|A)=\sum_{D \subset A} T_D \tag{2.10}$$

which is the inverse of relation (2.2).

For $m=0,1,\ldots,N$, let us write

$$W_m=E(T|X_{m+1},\ldots,X_N), \tag{2.11}$$

$$T=\sum_{j=1}^{m} T_j+W_m+\Delta_m. \tag{2.12}$$

Clearly $\sum_{j=1}^{m} T_j+W_m$ is the best approximation of T in L_2 by a random variable which depends on $X_1,\ldots,X_m$ only through a sum of functions of each one of these variables separately. We shall need some information concerning the error Δ_m of this approximation. For $r=0,1,\ldots,N$, define

$$\Omega_r=\{1,2,\ldots,r\}, \quad \Omega_r^c=\Omega-\Omega_r=\{r+1,\ldots,N\}. \tag{2.13}$$

By (2.10) and (2.8),

$$W_m=\sum_{D \subset \Omega_m^c} T_D, \tag{2.14}$$

$$\Delta_0=0, \quad \Delta_m=\sum_{\substack{D \cap \Omega_m \neq \emptyset \\ |D| \geqq 2}} T_D=\sum_{\substack{k=1 \\ k+l \geqq 2}}^{m} \sum_{l=0}^{N-m} \sum_{\substack{A \subset \Omega_m \\ |A|=k}} \sum_{\substack{B \subset \Omega_m^c \\ |B|=l}} T_{A \cup B}. \tag{2.15}$$

Now let us assume that $X_1,\ldots,X_N$ are identically distributed, that $T=\tau(X_1,\ldots,X_N)$ is a symmetric function of these variables and that $ET=0$, $ET^2=1$, so that we are back in the situation of Sect. 1. Then (2.15) and (2.5) imply that

$$E\Delta_m^2=\sum_{r=2}^{N}\left\{\binom{N}{r}-\binom{N-m}{r}\right\} ET_{\Omega_r}^2, \quad m=0,1,\ldots,N. \tag{2.16}$$

If $D(E\Delta_m^2)=E\Delta_{m+1}^2-E\Delta_m^2$ and $D^{s+1}(E\Delta_m^2)=DD^s(E\Delta_m^2)$, then (2.16) yields

$$(-1)^{s+1} D^s(E\Delta_m^2)=\sum_{r=2}^{N-m}\binom{N-m-s}{r-s} ET_{\Omega_r}^2 \geqq 0, \quad s \geqq 1,$$

(cf. Karlin and Rinott (1982) who show that $EW_{N-m}^2 = 1-(N-m)ET_1^2 - E\Delta_{N-m}^2$ is absolutely monotone). In particular, $E\Delta_m^2$ is nondecreasing and concave for $m = 0, 1, \ldots, N$. Also

$$0 \leqq -D^2(E\Delta_0^2) = 2E\Delta_1^2 - E\Delta_2^2 = 2(1-ET_1^2 - EW_1^2) - (1-2ET_1^2 - EW_2^2)$$
$$= 1 + E\{E(T|X_1, \ldots, X_{N-2})\}^2 - 2E\{E(T|X_1, \ldots, X_{N-1})\}^2 \quad (2.17)$$

and under the conditions of Theorem 1.1 we therefore have

$$0 \leqq 2E\Delta_1^2 - E\Delta_2^2 = \sum_{r=2}^{N} \binom{N-2}{r-2} ET_{\Omega_r}^2 \leqq BN^{-3}. \quad (2.18)$$

It follows that

$$0 \leqq E\Delta_1^2 = \sum_{r=2}^{N} \binom{N-1}{r-1} ET_{\Omega_r}^2 \leqq BN^{-2}, \quad (2.19)$$

$$0 \leqq E\Delta_N^2 = \sum_{r=2}^{N} \binom{N}{r} ET_{\Omega_r}^2 \leqq \tfrac{1}{2} BN^{-1}, \quad (2.20)$$

$$0 \leqq E\Delta_m^2 \leqq mE\Delta_1^2 \leqq BmN^{-2}, \quad m = 0, \ldots, N, \quad (2.21)$$

because of the concavity of $E\Delta_m^2$.

So far we have implicitly assumed that the random variable T is real valued, but of course everything in this section goes through for complex valued T with appropriate modifications. In (2.5), $ET_D T_{D'}$, should be replaced by $ET_D \bar{T}_{D'}$, where $\bar{T}_{D'}$ denotes the complex conjugate of $T_{D'}$; furthermore, in all expectations of squares such as ET^2, ET_D^2, EW_m^2, $E\Delta_m^2$ etc., the squares should be replaced by their moduli $E|T^2|$, $E|T_D^2|$, $E|W_m^2|$, $E|\Delta_m^2|$ etc. Thus in particular (2.9) becomes

$$E|T^2| = \sum_{D \subset \Omega} E|T_D^2|. \quad (2.22)$$

3. Proof of Theorem 1.1

Let us agree to take $C \geqq 3$. For $1 \leqq N \leqq 3B$, we have $C(A+B)N^{-\frac{1}{2}} \geqq CBN^{-\frac{1}{2}} \geqq CN^{\frac{1}{2}}/3 \geqq 1$, so that (1.14) is trivially satisfied. We therefore assume that $N > 3B$.

In view of (2.12) and (2.20),

$$|ET_1^2 - N^{-1}| \leqq \tfrac{1}{2} BN^{-2} \leqq \frac{1}{6N} \quad (3.1)$$

and hence, under the conditions of the theorem,

$$A \geqq N^{\frac{3}{2}} E|T_1|^3 \geqq (NET_1^2)^{\frac{3}{2}} \geqq (1 - \tfrac{1}{2} BN^{-1})^{\frac{3}{2}} \geqq (\tfrac{5}{6})^{\frac{3}{2}}. \quad (3.2)$$

Let

$$\gamma(t) = Ee^{itT_1} \quad (3.3)$$

be the characteristic function of T_1. By (3.1) and (1.12),

$$\left|\gamma(t)-1+\frac{t^2}{2N}\right|\leqq\frac{1}{4}BN^{-2}t^2+\frac{1}{6}AN^{-\frac{3}{2}}|t|^3\leqq\frac{t^2}{6N}$$

for all $|t|\leqq H=\frac{1}{2}A^{-1}N^{\frac{1}{2}}$. For $|t|\leqq H$, we have $t^2\leqq(6/5)^3N/4\leqq\frac{1}{2}N$ and

$$0<1-\frac{2t^2}{3N}\leqq|\gamma(t)|\leqq1-\frac{t^2}{3N}\leqq\exp\left\{-\frac{t^2}{3N}\right\}, \tag{3.4}$$

$$\frac{t^2}{2N}\leqq1-|\gamma^2(t)|\leqq\frac{4t^2}{3N}. \tag{3.5}$$

Let

$$\psi(t)=Ee^{itT} \tag{3.6}$$

denote the characteristic function of T. According to Esseen's smoothing lemma (cf. Feller (1971), p. 538)

$$\sup_x|P(T\leqq x)-\Phi(x)|\leqq\frac{1}{\pi}\int_{-H}^{H}\left|\frac{\psi(t)-e^{-\frac{1}{2}t^2}}{t}\right|dt+\frac{4}{H}.$$

Define $h=\min(2N^{\frac{1}{2}},H)$ and let $C_1,C_2,\ldots$ denote universal constants throughout the proof. From (1.12), (3.1) and the proof of the classical Berry-Esseen theorem we conclude that

$$\int_{-h}^{h}\left|\frac{\gamma^N(t)-e^{-\frac{1}{2}t^2}}{t}\right|dt\leqq C_1AN^{-\frac{1}{2}}.$$

Because of (3.2)

$$\int_{|t|\geqq h}\left|\frac{e^{-\frac{1}{2}t^2}}{t}\right|dt\leqq\frac{1}{2e^2}N^{-\frac{1}{2}}\leqq AN^{-\frac{1}{2}}$$

and combining these results we find

$$\sup_x|P(T\leqq x)-\Phi(x)|\leqq\frac{1}{\pi}\int_{-h}^{h}\left|\frac{\psi(t)-\gamma^N(t)}{t}\right|dt$$
$$+\frac{1}{\pi}\int_{h\leqq|t|\leqq H}\left|\frac{\psi(t)}{t}\right|dt+C_2AN^{-\frac{1}{2}}. \tag{3.7}$$

To analyze $\psi(t)$ for $|t|\leqq h$, we employ decomposition (2.12) for $m=N$, i.e. $T=\hat{T}_1+\Delta_N$, to obtain

$$\psi(t)=Ee^{it\hat{T}_1}(1+it\Delta_N)+R_N=\gamma^N(t)+itEe^{it\hat{T}_1}\Delta_N+R_N, \tag{3.8}$$

$$|R_N|\leqq\tfrac{1}{2}t^2E\Delta_N^2\leqq\frac{Bt^2}{4N} \tag{3.9}$$

in view of (2.20). Similarly,

$$|tEe^{it\hat{T}_1}\Delta_N|\leqq|t|\{E\Delta_N^2\}^{\frac{1}{2}}\leqq(\tfrac{1}{2}B)^{\frac{1}{2}}|t|N^{-\frac{1}{2}}. \tag{3.10}$$

A more delicate analysis starts with noting that

$$
\begin{aligned}
Ee^{it\hat{T}_1}\Delta_N &= \sum_{k=2}^{N} \sum_{|D|=k} Ee^{it\hat{T}_1} T_D \\
&= \sum_{r=2}^{N} \binom{N}{r} \gamma^{N-r}(t) ET_{\Omega_r} \prod_{j=1}^{r} e^{itT_j} \\
&= \sum_{r=2}^{N} \binom{N}{r} \gamma^{N-r}(t) ET_{\Omega_r} \prod_{j=1}^{r} (e^{itT_j} - \gamma(t))
\end{aligned}
$$

where the final step follows from (2.4). For $2 \leqq r \leqq N$,

$$\binom{N}{r}^2 \leqq 6 \binom{N-2}{r-2} \binom{N+2}{r+2}$$

and since

$$E|e^{itT_j} - \gamma(t)|^2 = 1 - |\gamma^2(t)|, \tag{3.11}$$

repeated application of Schwarz's inequality yields

$$
\begin{aligned}
|Ee^{it\hat{T}_1}\Delta_N| &\leqq 6^{\frac{1}{2}} \sum_{r=2}^{N} \binom{N-2}{r-2}^{\frac{1}{2}} (ET_{\Omega_r}^2)^{\frac{1}{2}} \cdot \binom{N+2}{r+2}^{\frac{1}{2}} |\gamma^2(t)|^{\frac{1}{2}(N-r)} (1 - |\gamma^2(t)|)^{\frac{1}{2}r} \\
&\leqq \frac{6^{\frac{1}{2}}}{1 - |\gamma^2(t)|} \cdot \left[\sum_{r=2}^{N} \binom{N-2}{r-2} ET_{\Omega_r}^2 \right]^{\frac{1}{2}} \cdot \left[\sum_{r=2}^{N} \binom{N+2}{r+2} |\gamma^2(t)|^{N-r} (1 - |\gamma^2(t)|)^{r+2} \right]^{\frac{1}{2}} \\
&\leqq \frac{6^{\frac{1}{2}}}{1 - |\gamma^2(t)|} \left[\sum_{r=2}^{N} \binom{N-2}{r-2} ET_{\Omega_r}^2 \right]^{\frac{1}{2}}.
\end{aligned}
$$

Invoking (2.18) and (3.5), we see that for $|t| \leqq H$

$$|tEe^{it\hat{T}_1}\Delta_N| \leqq (24B)^{\frac{1}{2}} |t|^{-1} N^{-\frac{1}{2}}. \tag{3.12}$$

Combining (3.8), (3.9), (3.10) and (3.12) and then using (3.2), we arrive at

$$\int_{-h}^{h} \left| \frac{\psi(t) - \gamma^N(t)}{t} \right| dt \leqq (B + 8B^{\frac{1}{2}}) N^{-\frac{1}{2}} \leqq 6(A+B) N^{-\frac{1}{2}}. \tag{3.13}$$

It remains to consider $\psi(t)$ for $h \leqq |t| \leqq H$ in order to bound the second integral in (3.7). For any fixed $|t|$ in this interval we take

$$m = \left[\frac{3N \log N}{t^2} \right], \tag{3.14}$$

where $[x]$ denotes the integer part of x. For $|t| \geqq h$, we have $0 \leqq m \leqq N$, and using decomposition (2.12) for this value of m, we obtain

$$\psi(t) = E \exp\left\{ it \left(\sum_{j=1}^{m} T_j + W_m \right) \right\} \cdot (1 + it\Delta_m) + R_m, \tag{3.15}$$

$$|R_m| \leqq \tfrac{1}{2} t^2 E\Delta_m^2 \leqq \frac{Bmt^2}{2N^2} \leqq \frac{3B \log N}{2N} \tag{3.16}$$

because of (2.21). Since $|t| \leqq H$, (3.4) and (3.2) imply

$$\left|E\exp\left\{it\left(\sum_{j=1}^{m}T_j+W_m\right)\right\}\right|\leqq|\gamma(t)|^m\leqq\exp\left\{-\frac{mt^2}{3N}\right\}$$
$$\leqq\exp\left\{-\log N+\frac{t^2}{3N}\right\}\leqq N^{-1}\exp\left\{\frac{1}{12A^2}\right\}\leqq\frac{2A}{N}. \tag{3.17}$$

Let us define the complex valued random variable $Z=\exp\{itW_m\}$ which depends on $X_{m+1},\dots,X_N$ only. By (2.15) and two applications of (2.4),

$$E\exp\left\{it\left(\sum_{j=1}^{m}T_j+W_m\right)\right\}\Delta_m$$
$$=\sum_{\substack{k=1\\k+l\geqq 2}}^{m}\sum_{l=0}^{N-m}\sum_{\substack{A\subset\Omega_m\\|A|=k}}\sum_{\substack{B\subset\Omega_m^c\\|B|=l}}\gamma^{m-k}(t)\cdot E[T_{A\cup B}\prod_{j\in A}e^{itT_j}E(Z|B)]$$
$$=\sum_{\substack{k=1\\k+l\geqq 2}}^{m}\sum_{l=0}^{N-m}\sum_{\substack{A\subset\Omega_m\\|A|=k}}\sum_{\substack{B\subset\Omega_m^c\\|B|=l}}\gamma^{m-k}(t)\cdot E[T_{A\cup B}\prod_{j\in A}(e^{itT_j}-\gamma(t))Z_B]. \tag{3.18}$$

It follows from (2.22) and (2.6) that

$$\sum_{B\subset\Omega_m^c}E|Z_B^2|=E|Z^2|=1. \tag{3.19}$$

By Schwarz's inequality and (3.11),

$$E|T_{A\cup B}\prod_{j\in A}(e^{itT_j}-\gamma(t))Z_B|\leqq(ET_{A\cup B}^2)^{\frac{1}{2}}(1-|\gamma^2(t)|)^{\frac{1}{2}|A|}(E|Z_B^2|)^{\frac{1}{2}}$$

for every $A\subset\Omega_m$ and $B\subset\Omega_m^c$. Another application of Schwarz's inequality to the terms in (3.18) with $k=1$ and $k\geqq 2$ separately, followed by the use of (2.18) and (2.19) yields

$$\left|E\exp\left\{it\left(\sum_{j=1}^{m}T_j+W_m\right)\right\}\Delta_m\right|\leqq m|\gamma(t)|^{m-1}(1-|\gamma^2(t)|)^{\frac{1}{2}}$$
$$\cdot\left[\sum_{l=1}^{N-m}\sum_{\substack{B\subset\Omega_m^c\\|B|=l}}ET_{\Omega_{l+1}}^2\right]^{\frac{1}{2}}\left[\sum_{l=1}^{N-m}\sum_{\substack{B\subset\Omega_m^c\\|B|=l}}E|Z_B^2|\right]^{\frac{1}{2}}$$
$$+\left[\sum_{k=2}^{m}\sum_{l=0}^{N-m}\sum_{\substack{A\subset\Omega_m\\|A|=k}}\sum_{\substack{B\subset\Omega_m^c\\|B|=l}}\frac{k(k-1)}{m(m-1)}ET_{A\cup B}^2\right]^{\frac{1}{2}}$$
$$\cdot\left[\sum_{k=2}^{m}\sum_{\substack{A\subset\Omega_m\\|A|=k}}\frac{m(m-1)}{k(k-1)}|\gamma^2(t)|^{m-k}(1-|\gamma^2(t)|)^k\sum_{B\subset\Omega_m^c}E|Z_B^2|\right]^{\frac{1}{2}}$$
$$\leqq m|\gamma(t)|^{m-1}(1-|\gamma^2(t)|)^{\frac{1}{2}}\left[\sum_{r=2}^{N-m+1}\binom{N-m}{r-1}ET_{\Omega_r}^2\right]^{\frac{1}{2}}$$
$$+6^{\frac{1}{2}}\left[\sum_{r=2}^{N}\binom{N-2}{r-2}ET_{\Omega_r}^2\right]^{\frac{1}{2}}\left[\sum_{k=2}^{m}\binom{m+2}{k+2}|\gamma^2(t)|^{m-k}(1-|\gamma^2(t)|)^k\right]^{\frac{1}{2}}$$
$$\leqq B^{\frac{1}{2}}\left[\frac{m}{N}|\gamma(t)|^{m-1}(1-|\gamma^2(t)|)^{\frac{1}{2}}+6^{\frac{1}{2}}N^{-3/2}(1-|\gamma^2(t)|)^{-1}\right]. \tag{3.20}$$

Hence, by (3.4), (3.5), (3.14) and (3.2),

$$\begin{aligned}
&\left|tE\exp\left\{it\left(\sum_{j=1}^{m} T_j+W_m\right)\right\}\Delta_m\right| \\
&\leqq(3B)^{\frac{1}{2}}\left[2N^{-\frac{3}{2}}\log N\exp\left\{\frac{2t^2}{3N}\right\}+2^{\frac{3}{2}}N^{-\frac{1}{2}}|t|^{-1}\right] \\
&\leqq 5B^{\frac{1}{2}}[N^{-\frac{3}{2}}\log N+N^{-\frac{1}{2}}|t|^{-1}]
\end{aligned} \tag{3.21}$$

for $h\leqq|t|\leqq H$. Combining (3.15)-(3.17) and (3.21) and again using (3.2), we arrive at

$$\int_{h\leqq|t|\leqq H}\left|\frac{\psi(t)}{t}\right|dt\leqq\frac{3B(\log N)^2}{4N}+\frac{A\log N}{N}+\frac{5B^{\frac{1}{2}}(\log N)^2}{2N^{\frac{3}{2}}}+\frac{5B^{\frac{1}{2}}}{N^{\frac{3}{4}}}\leqq 7(A+B)N^{-\frac{1}{2}}. \tag{3.22}$$

Together (3.7), (3.13) and (3.22) establish Theorem 1.1. □

4. Examples

In this section we apply Theorem 1.1 to two special cases - U-statistics and linear functions of order statistics - to see whether we can obtain results comparable to the best available ones for these well-studied special cases.

Let $X_1,\ldots,X_N$ be i.i.d. random variables and let h be a function of $k(\leqq N)$ variables satisfying

$$Eh(X_1,\ldots,X_k)=0,\quad Eh^2(X_1,\ldots,X_k)<\infty. \tag{4.1}$$

Define the U-statistic U by (1.1), the function g by

$$g(x)=E(h(X_1,\ldots,X_k)|X_1=x) \tag{4.2}$$

and suppose that

$$Eg^2(X_1)>0,\quad E|g_1(X_1)|^3<\infty. \tag{4.3}$$

We shall show that Theorem 1.1 implies

Corollary 4.1. *There exists a universal constant C such that*

$$\sup_x\left|P\left(\frac{U}{\sigma(U)}\leqq x\right)-\Phi(x)\right|\leqq C\left[\frac{E|g(X_1)|^3}{\{Eg^2(X_1)\}^{\frac{3}{2}}}+\frac{(k-1)^2Eh^2(X_1,\ldots,X_k)}{Eg^2(X_1)}\right]N^{-\frac{1}{2}}$$

whenever $1\leqq k\leqq N$ *and provided* (4.1) *and* (4.3) *are satisfied.*

For $k=2$ this is the best result known for the case where $Eh^2(X_1,\ldots,X_k)<\infty$, as was pointed out in section 1. Since the assumption of finite variance is a natural limitation of the results in this paper, we conclude that Theorem 1.1 performs as well as might be expected for this special case. This is not really surprising, as Theorem 1.1 and its proof are modeled after the earlier work on U-statistics.

To prove the corollary, we begin by noting that (2.6) implies that

$$U_D=0\quad\text{if } |D|\geqq k+1. \tag{4.4}$$

For $r=0,1,\dots,k$, define

$$g_r(X_1,\dots,X_r)=(h(X_1,\dots,X_k))_{\Omega_r}=\sum_{A\subset\Omega_r}(-1)^{r-|A|}E(h(X_1,\dots,X_k)|A). \tag{4.5}$$

In particular, $g_0=0$ and $g_1=g$ as defined in (4.2). It follows from (2.9) that

$$Eh^2(X_1,\dots,X_k)=\sum_{r=0}^{k}\binom{k}{r}Eg_r^2(X_1,\dots,X_r). \tag{4.6}$$

Obviously, for $r=0,1,\dots,k$,

$$U_{\Omega_r}=\binom{N-r}{k-r}g_r(X_1,\dots,X_r) \tag{4.7}$$

and because of (2.7), (4.4) and (4.6) we have

$$E\hat{U}_1^2=NEU_1^2=N\binom{N-1}{k-1}^2 Eg^2(X_1), \tag{4.8}$$

$$E|U_1|^3=\binom{N-1}{k-1}^3 E|g(X_1)|^3, \tag{4.9}$$

$$\begin{aligned}\sum_{r=2}^{N}\binom{N-2}{r-2}EU_{\Omega_r}^2&=\sum_{r=2}^{k}\binom{N-2}{r-2}\binom{N-r}{k-r}^2 Eg_r^2(X_1,\dots,X_r)\\&=\binom{N}{k}\sum_{r=2}^{k}\frac{r(r-1)}{N(N-1)}\binom{N-r}{k-r}\cdot\binom{k}{r}Eg_r^2(X_1,\dots,X_r)\\&\leqq\binom{N-2}{k-2}^2 Eh^2(X_1,\dots,X_k).\end{aligned} \tag{4.10}$$

Define $T=U/\sigma(U)$, so that $ET^2=1$. Take

$$A=\frac{E|g(X_1)|^3}{\{Eg^2(X_1)\}^{\frac{3}{2}}},\qquad B=4(k-1)^2\frac{Eh^2(X_1,\dots,X_k)}{Eg^2(X_1)}. \tag{4.11}$$

By (4.8)–(4.10),

$$E|T_1|^3=\frac{E|U_1|^3}{\{EU^2\}^{\frac{3}{2}}}\leqq\frac{E|U_1|^3}{\{E\hat{U}_1^2\}^{\frac{3}{2}}}=AN^{-\frac{1}{2}},$$

$$\sum_{r=2}^{N}\binom{N-2}{r-2}ET_{\Omega_r}^2\leqq\frac{\binom{N-2}{k-2}^2 Eh^2(X_1,\dots,X_k)}{E\hat{U}_1^2}\leqq BN^{-3}. \tag{4.12}$$

Note that the results of these computations are correct also for $k=1$. In view of (2.17) and (2.18), it follows that assumptions (1.12) and (1.13) of Theorem 1.1 are satisfied with A and B as in (4.11). The corollary follows.

We now turn to our second example. Let $X_1, X_2, \ldots, X_N$ be i.i.d. random variables with a common distribution function F, which is not assumed to be continuous. Let $X_{(1)} \leqq X_{(2)} \leqq \ldots \leqq X_{(N)}$ denote the corresponding order statistics. For real numbers $c_1, c_2, \ldots, c_N$, we consider a normed linear function of order statistics

$$L = N^{-\frac{1}{2}} \sum_{j=1}^{N} c_j (X_{(j)} - EX_{(j)}). \tag{4.13}$$

Suppose that

$$E|X_1|^3 < \infty, \quad \sigma^2(L) > 0, \tag{4.14}$$

and let

$$\max_{1 \leqq j \leqq N} |c_j| = a, \quad N \max_{2 \leqq j \leqq N} |c_j - c_{j-1}| = b. \tag{4.15}$$

Theorem 1.1 implies

Corollary 4.2. *There exists a universal constant C such that*

$$\sup_x \left| P\left(\frac{L}{\sigma(L)} \leqq x\right) - \Phi(x) \right| \leqq C \left[\frac{a^3 E|X_1|^3}{\sigma^3(L)} + \frac{b^2 \{E|X_1|\}^2}{\sigma^2(L)}\right] N^{-\frac{1}{2}}$$

whenever (4.14) *and* (4.15) *are satisfied.*

If $\sigma^2(L)$ is bounded below and $E|X_1|^3$, a and b are bounded above as $N \to \infty$, then Corollary 4.2 provides a Berry-Esseen bound of order $N^{-\frac{1}{2}}$. In view of (4.15) we are then dealing with the case of smooth weights $c_1, \ldots, c_N$, but not necessarily smooth underlying distribution function F. For this case, the best result to date has been obtained by Helmers (1981; 1982) and this result is essentially equivalent to Corollary 4.2. Thus once again, Theorem 1.1 appears to perform in a satisfactory manner.

To prove corollary 4.2 we adopt some additional notation. For $n \leqq N$, $X_{1:n} \leqq X_{2:n} \leqq \ldots \leqq X_{n:n}$ will denote the order statistics corresponding to $X_1, X_2, \ldots, X_n$; we take $X_{0:n} = -\infty$, $X_{n+1:n} = +\infty$. We shall find it convenient to introduce i.i.d. random variables $U_1, U_2, \ldots, U_N$ with a common uniform distribution on (0,1) and pretend that $X_i = F^{-1}(U_i)$ for $i = 1, \ldots, N$. Clearly this does not affect the distribution of L. The rank of U_i among $U_1, \ldots. U_N$ will be denoted by R_i,

$$R_i = \sum_{k=1}^{N} 1_{(0, U_i]}(U_k),$$

and we define

$$K_1 = R_{N-1} \wedge R_N, \quad K_2 = R_{N-1} \vee R_N, \tag{4.16}$$

where $x \wedge y = \min(x, y)$ and $x \vee y = \max(x, y)$. Furthermore we let $b_{j,N}$ be the beta density

$$b_{j,N}(y) = \frac{N!}{(j-1)!\,(N-j)!} y^{j-1} (1-y)^{N-j}, \quad 0 < y < 1,$$

and we define the functions G, H and M by

$$G(x)=\int_{-\infty}^{x} F(y)\,dy, \quad H(x)=\int_{x}^{\infty}(1-F(y))\,dy, \quad M(x)=\int_{-\infty}^{x} F(y)(1-F(y))\,dy. \tag{4.17}$$

Obviously G, H and M are monotone and by (4.14), M is bounded. Finally we introduce the random variable

$$Z=L-E(L|U_1,\dots,U_{N-1})-E(L|U_1,\dots,U_{N-2},U_N)+E(L|U_1,\dots,U_{N-2}) \tag{4.18}$$

and note that

$$EZ^2=EL^2+E\{E(L|U_1,\dots,U_{N-2})\}^2-2E\{E(L|U_1,\dots,U_{N-1})\}^2. \tag{4.19}$$

Straightforward but somewhat tedious computations show that with probability 1

$$\begin{aligned} N^{\frac{1}{2}}L_1&=N^{\frac{1}{2}}E(L|U_1)\\ &=\frac{1}{N}\sum_{j=1}^{N} c_j\int_0^1\{1_{(0,U_1)}(y)-(1-y)\}\,b_{j,N}(y)\,dF^{-1}(y), \end{aligned} \tag{4.20}$$

$$\begin{aligned} N^{\frac{1}{2}}Z=&\sum_{j=1}^{N-1}(c_{j+1}-c_j)(M(X_{j:N-2})-M(X_{j-1:N-2}))\\ &-\sum_{j=1}^{K_1}(c_{j+1}-c_j)(G(X_{j:N})-G(X_{j-1:N}))\\ &+\sum_{j=K_2}^{N}(c_j-c_{j-1})(H(X_{j+1:N})-H(X_{j:N})). \end{aligned} \tag{4.21}$$

By (4.15), $\sum|c_j|\,b_{j,N}(y)\leqq aN$ and hence

$$\begin{aligned} N^{\frac{1}{2}}|L_1|&\leqq a\left\{\int_0^{U_1} y\,dF^{-1}(y)+\int_{U_1}^{1}(1-y)\,dF^{-1}(y)\right\}\\ &\leqq a\left\{|F^{-1}(U_1)|+\int_0^1|F^{-1}(y)|\,dy\right\}=a\{|X_1|+E|X_1|\}. \end{aligned} \tag{4.22}$$

Because of (4.15) and the monotonicity of M, G and H,

$$|Z|\leqq bN^{-\frac{3}{2}}[M(\infty)+G(X_{N-1}\wedge X_N)+H(X_{N-1}\vee X_N)]. \tag{4.23}$$

Define $T=L/\sigma(L)$. Combining (4.14), (4.22) and (4.23) we find after elementary calculations

$$E|T_1|^3\leqq\frac{4a^3E|X_1|^3}{\sigma^3(L)}N^{-\frac{3}{2}}, \tag{4.24}$$

$$\frac{EZ^2}{\sigma^2(L)}\leqq\frac{25b^2\{E|X_1|\}^2}{\sigma^2(L)}N^{-3}. \tag{4.25}$$

Corollary 4.2 follows from (4.19), (4.24), (4.25) and Theorem 1.1.

We should perhaps point out that (4.20) and (4.21) are valid under the sole assumption that $E|X_1|<\infty$ and can therefore be used to treat other cases than the one of smooth weights. Any set of assumptions ensuring that $E|T_1|^3 = \mathcal{O}(N^{-\frac{3}{2}})$ and $EZ^2/\sigma^2(L)=\mathcal{O}(N^{-3})$ as $N\to\infty$, will produce a Berry-Esseen bound of order $N^{-\frac{1}{2}}$. Smoothness of the underlying distribution function F can clearly replace smoothness of the weights c_j and intermediate versions are also possible.

5. Possible Extensions

Theorem 1.1 provides a Berry-Esseen bound for a symmetric function τ of i.i.d. random variables $X_1,\ldots,X_N$ under the relatively simple moment assumptions (1.12) and (1.13). For a particular case it may be laborious to check these assumptions, but the work involved is basically straightforward. The technical intricacies of the proof of a Berry-Esseen-type result have been dispensed with and what remains can be done by brute force. Of course this only makes sense up to a point: if too much brute force is needed, one may prefer to tackle the intricacies directly instead.

It would seem that this might be the deciding factor in judging how far the present result can usefully be generalized. There doesn't seem to be a reason, a priori, why one should need the symmetry of τ or the fact that $X_1,\ldots,X_N$ are identically distributed. Hoeffding's decomposition (2.9) works without these assumptions and it should be possible to adapt the remainder of the proof. In short, one should be able to generalize theorem 1.1 to arbitrary functions of independent random variables. Of course the assumptions needed to replace (1.12) and (1.13) will not look nearly as pleasant; worse still, they will probably be almost impossible to check in most nontrivial cases.

One would guess, however, that there is one slight but significant generalization that would still be feasible. This is the k-sample situation, where the independent random variables $X_1,\ldots,X_N$ are split into a fixed number (k) of groups. Within each group the variables are i.i.d. and τ is a symmetric function of the variables in such a group.

Another possible type of extension is to relax the moment assumptions $ET^2<\infty$ and $E|N^{\frac{1}{2}}T_1|^3<\infty$ by the following standard argument. Let $T=\tilde{T}+R$. If we have a Berry-Esseen bound for $\tilde{T}$,

$$\sup_x |P(\tilde{T}\leq x)-\Phi(x)|\leq cN^{-\frac{1}{2}} \tag{5.1}$$

and R satisfies

$$P(|R|\geq aN^{-\frac{1}{2}})\leq bN^{-\frac{1}{2}}, \tag{5.2}$$

then we have a Berry-Esseen bound for T,

$$\sup_x |P(T\leq x)-\Phi(x)|\leq (a+b+c)N^{-\frac{1}{2}}. \tag{5.3}$$

In principle, no moments of R - and therefore of T - are needed, but we note that (5.2) is often established with the aid of a moment of low order and the

Markov inequality. We have not incorporated this idea in Theorem 1.1 because it is well-known and may be applied ad hoc whenever needed.

The above argument may be used for other purposes than merely to relax the moment assumptions. As we have noted before (cf. (2.17) and (2.18)), assumption (1.13) of Theorem 1.1 is equivalent to

$$2E\Delta_1^2 - E\Delta_2^2 = \sum_{r=2}^{N} \binom{N-2}{r-2} ET_{\Omega_r}^2 \leqq BN^{-3}. \tag{5.4}$$

However, if we require that for some positive integer $N' \leqq N$,

$$E(T-\hat{T}_{N'})^2 = \sum_{r=N'+1}^{N} \binom{N}{r} ET_{\Omega_r}^2 \leqq BN^{-\frac{3}{2}}, \tag{5.5}$$

then

$$P(|T-\hat{T}_{N'}| \geqq N^{-\frac{1}{2}}) \leqq BN^{-\frac{1}{2}}$$

and by (5.3) and (3.2) the conclusion of Theorem 1.1 will hold for T if it holds for $\hat{T}_{N'}$. But for $\hat{T}_{N'}$ instead of T, assumption (5.4) reduces to

$$\sum_{r=2}^{N'} \binom{N-2}{r-2} ET_{\Omega_r}^2 \leqq BN^{-3} \tag{5.6}$$

because of (2.7), (2.6) and (2.4). It follows that (5.5) and (5.6) together may replace assumption (1.13) in Theorem 1.1.

We may even go one step further and replace assumption (5.6) in its turn by the requirement that for some N'' with $1 \leqq N'' \leqq N'$,

$$\sum_{r=N''+1}^{N'} \binom{N-1}{r-1} ET_{\Omega_r}^2 \leqq B(N \log N)^{-2}, \tag{5.7}$$

$$\sum_{r=2}^{N''} \binom{N-2}{r-2} ET_{\Omega_r}^2 \leqq BN^{-3}. \tag{5.8}$$

To see this, we go over the proof of Theorem 1.1 and find that the full force of assumption (5.4) (or (2.18)), as opposed to (2.19), is used only in (3.12) and (3.20). In both places, a strengthened version of (2.19), viz.

$$\sum_{r=2}^{N} \binom{N-1}{r-1} ET_{\Omega_r}^2 \leqq B(N \log N)^{-2} \tag{5.9}$$

would also have been sufficient. Alternatively, we could have required a mixture of (5.4) and (5.9), such as (5.8) combined with

$$\sum_{r=N''+1}^{N} \binom{N-1}{r-1} ET_{\Omega_r}^2 \leqq B(N \log N)^{-2}, \tag{5.10}$$

and the proof would still have gone through with minor modifications. Applying (5.10) to $\hat{T}_{N'}$ instead of T, we obtain (5.7).

Thus we have shown that (5.5), (5.7) and (5.8) together may replace assumption (1.13) in Theorem 1.1. These conditions may be substantially weaker than (1.13), especially if N' and N'' are taken to be of the order of $N^{\frac{1}{2}}(\log N)^{-2}$ and $(\log N)^2$ respectively. In general, however, these assumptions will be hard to check.

References

1. Bickel, P.J.: Edgeworth expansions in nonparametric statistics. Ann. Statist. **2**, 1–20 (1974)
2. Callaert, H., Janssen, P.: The Berry-Esseen theorem for *U*-statistics. Ann. Statist. **6**, 417–421 (1978)
3. Chan, Y.-K., Wierman, J.: On the Berry-Esseen theorem for *U*-statistics, Ann. Probability **5**, 136–139 (1977)
4. Efron, B., Stein, C.: The jackknife estimate of variance, Ann. Statist. **9**, 586–596 (1981)
5. Feller, W.: An Introduction to Probability Theory and Its Applications. Vol. II, 2nd Ed. New York: Wiley 1971
6. Helmers, R.: A Berry-Esseen theorem for linear combinations of order statistics. Ann. Probability **9**, 342–347 (1981)
7. Helmers, R.: Edgeworth Expansions for Linear Combinations of Order Statistics. Mathematical Centre Tracts 105. Mathematisch Centrum, Amsterdam (1982)
8. Helmers, R., Van Zwet, W.R.: The Berry-Esseen bound for *U*-statistics. Statistical Decision Theory and Related Topics, III Vol. 1, S.S. Gupta and J.O. Berger (eds.), 497–512. New York: Academic Press 1982
9. Hoeffding, W.: A class of statistics with asymptotically normal distributions. Ann. Math. Statist. **19**, 293–325 (1948)
10. Hoeffding, W.: The strong law of large numbers for *U*-statistics. Inst. of Statist., Univ. of North Carolina, Mimeograph Series No. 302 (1961)
11. Karlin, S., Rinott, Y.: Applications of ANOVA type decompositions for comparisons of conditional variance statistics including jackknife estimates. Ann. Statist. **10**, 485–501 (1982)

Received February 2, 1983

The Annals of Statistics
1986, Vol. 14, No. 4, 1463–1484

THE EDGEWORTH EXPANSION FOR *U*-STATISTICS OF DEGREE TWO

BY P. J. BICKEL,[1] F. GÖTZE[2] AND W. R. VAN ZWET[2]

University of California, Berkeley, University of Bielefeld and University of Leiden

An Edgeworth expansion with remainder $o(N^{-1})$ is established for a U-statistic with a kernel h of degree 2. The assumptions involved appear to be very mild; in particular, the common distribution of the summands $h(X_i, X_j)$ is not assumed to be smooth.

1. Introduction. Let $X_1, X_2, \dots, X_N$ be independent and identically distributed (i.i.d.) random variables assuming values in a measurable space $(\mathscr{X}, \mathscr{B})$ with a common distribution P_X. Let $h\colon \mathscr{X}\times\mathscr{X} \to \mathbb{R}$ be measurable and symmetric in its two arguments, i.e., $h(x, y) = h(y, x)$. For $N \geq 2$, a U-statistic of degree 2 is defined as

$$U_N = \sum_{i=1}^{N-1} \sum_{j=i+1}^{N} h(X_i, X_j). \tag{1.1}$$

Note that we do not follow the usual convention of dividing the sum in (1.1) by the number $\binom{N}{2}$ of its terms. Since our results concern the standardized version of U, this does not make any difference.

We assume throughout that

$$Eh(X_1, X_2) = 0, \qquad Eh^2(X_1, X_2) < \infty, \tag{1.2}$$

and define

$$g(x) = E\big(h(X_1, X_2)|X_1 = x\big), \qquad \psi(x, y) = h(x, y) - g(x) - g(y), \tag{1.3}$$

$$\hat{U}_N = (N-1)\sum_{i=1}^{N} g(X_i), \qquad \Delta_N = \sum_{i=1}^{N-1}\sum_{j=i+1}^{N} \psi(X_i, X_j), \tag{1.4}$$

so that

$$U_N = \hat{U}_N + \Delta_N. \tag{1.5}$$

Since $E(\psi(X_1, X_2)|X_1) = 0$ almost surely (a.s.), the random variables $g(X_i)$ and

Received January 1985; revised November 1985.
[1] Research supported by the U.S. Office of Naval Research, Contract N0014-80-C-1063 and by the Netherlands Organization for Pure Scientific Research.
[2] Research supported by the U.S. Office of Naval Research, Contract N00014-80-C-1063.
AMS 1980 *subject classifications*. Primary 62E20; secondary 60F05.
Key words and phrases. Edgeworth expansion, second order asymptotics, *U*-statistics.

S. van de Geer and M. Wegkamp (eds.), *Selected Works of Willem van Zwet*, Selected Works in Probability and Statistics, DOI 10.1007/978-1-4614-1314-1_14, © Springer Science+Business Media, LLC 2012

$\psi(X_i, X_j)$, $1 \le i < j \le N$, are pairwise uncorrelated and hence

$$\begin{aligned}\sigma_N^2 &= \sigma^2(U_N) = \sigma^2(\hat{U}_N) + \sigma^2(\Delta_N) \\ &= N(N-1)^2 Eg^2(X_1) + \tfrac{1}{2}N(N-1)E\psi^2(X_1, X_2).\end{aligned} \tag{1.6}$$

If it is assumed that

$$\sigma_g^2 = Eg^2(X_1) > 0, \tag{1.7}$$

then $\sigma^2(\hat{U}_N)$ dominates the right-hand side of (1.6) and $\sigma_N^{-1}U_N$ is asymptotically normal as $N \to \infty$ [cf. Hoeffding (1948), where U-statistics were introduced].

The speed of convergence to normality was investigated by Bickel (1974), Chan and Wierman (1977), Callaert and Janssen (1978) and Helmers and van Zwet (1982) who showed in increasing generality that

$$\sup_x \left| P(\sigma_N^{-1}U_N \le x) - \Phi(x) \right| = O(N^{-1/2}), \tag{1.8}$$

where Φ denotes the standard normal distribution function (d.f.). If (1.2) and (1.7) are satisfied, so that asymptotic normality is ensured, then $E|g(X_1)|^3 < \infty$ suffices to establish (1.8). Moreover, the assumption $Eh^2(X_1, X_2) < \infty$ may be relaxed, provided σ_N is replaced by $\sigma(\hat{U}_N)$ in (1.8).

The next step in the asymptotic analysis of $\sigma_N^{-1}U_N$, is to obtain an Edgeworth expansion for its d.f., and for statistical purposes one typically needs such an expansion up to a remainder term which is $o(N^{-1})$. To be specific, let

$$\kappa_3 = \sigma_g^{-3}\{Eg^3(X_1) + 3Eg(X_1)g(X_2)\psi(X_1, X_2)\}, \tag{1.9}$$

$$\begin{aligned}\kappa_4 = \sigma_g^{-4}\{&Eg^4(X_1) - 3\sigma_g^4 + 12Eg^2(X_1)g(X_2)\psi(X_1, X_2) \\ &+ 12Eg(X_1)g(X_2)\psi(X_1, X_3)\psi(X_2, X_3)\}.\end{aligned} \tag{1.10}$$

Straightforward calculation shows that if $Eh^4(X_1, X_2) < \infty$—which we shall *not* generally require in this paper—then $\kappa_3 N^{-1/2}$ and $\kappa_4 N^{-1}$ are asymptotic expressions with error $o(N^{-1})$ for the third and fourth cumulants of $\sigma_N^{-1}U_N$, respectively. Define

$$\begin{aligned}F_N(x) = \Phi(x) - \phi(x)\bigg\{&\frac{\kappa_3}{6}N^{-1/2}(x^2 - 1) + \frac{\kappa_4}{24}N^{-1}(x^3 - 3x) \\ &+ \frac{\kappa_3^2}{72}N^{-1}(x^5 - 10x^3 + 15x)\bigg\},\end{aligned} \tag{1.11}$$

where ϕ denotes the standard normal density. We wish to show that

$$\sup_x \left| P(\sigma_N^{-1}U_N \le x) - F_N(x) \right| = o(N^{-1}) \tag{1.12}$$

as $N \to \infty$.

The validity of the Edgeworth expansion (1.11)–(1.12) was established by Janssen (1978) and by Callaert, Janssen and Veraverbeke (1980) under a complicated condition which these authors were able to verify only for certain cases where the distribution of $h(X_1, X_2)$ possesses an absolutely continuous part. An inspection of special cases, however, quickly reveals that the expansion may be valid even when h assumes only two values. In this respect the situation appears to be more favorable than it is for sums of i.i.d. random variables, where the lattice case has to be excluded. The explanation of this phenomenon is simple: the left-hand side of (1.12) cannot be smaller than the largest jump of the d.f. of U_N and in the lattice case the jumps are of the order $N^{-1/2}$ for sums, but $N^{-3/2}$ for most U-statistics. An exception is, of course, the U-statistic $\{\Sigma 1_B(X_i)\}^2$ which is distributed like the square of a binomial random variable, so that the jumps are of the order $N^{-1/2}$.

The aim of the present paper is to establish the Edgeworth expansion under very mild assumptions that are easy to verify and do not involve smoothness of the distribution of $h(X_1, X_2)$. Suppose that there exist positive numbers $\delta, \delta_1, \delta_2, \delta_3, C$ and positive and continuous functions $\chi_j: (0, \infty) \to (0, \infty)$, $j = 1, 2$, satisfying

$$\lim_{t\to\infty} \chi_1(t) = 0, \tag{1.13}$$

$$\lim_{t\to\infty} \chi_2(t) \geq \delta_1 > 0, \tag{1.14}$$

as well as a real number r such that

$$r \geq 2 + \delta > 2, \tag{1.15}$$

$$E|\psi(X_1, X_2)|^r \leq C, \tag{1.16}$$

$$Eg^4(X_1)1_{[t,\infty)}(|g(X_1)|) \leq \chi_1(t) \quad \text{for all } t > 0, \tag{1.17}$$

$$|Ee^{itg(X_1)}| \leq 1 - \chi_2(t) < 1 \quad \text{for all } t > 0. \tag{1.18}$$

Let $\lambda_1, \lambda_2, \ldots$ denote the eigenvalues of the kernel ψ with respect to P_X, ranked according to descending absolute values and with multiple eigenvalues repeated. Thus, for some orthonormal sequence of eigenfunctions $\omega_1, \omega_2, \ldots$,

$$\int \psi(x, y)\omega_j(x)\, dP_X(x) = \lambda_j \omega_j(y), \qquad |\lambda_1| \geq |\lambda_2| \geq \cdots. \tag{1.19}$$

Assume, in addition to (1.13)–(1.18), that there exists a natural number k such that

$$|\lambda_k| \geq \delta_2 > 0. \tag{1.20}$$

Finally, assumptions (1.15), (1.16) and (1.20) are linked by requiring that

$$(r-2)(k-4) \geq 8 + \delta_3 > 8. \tag{1.21}$$

We note that (1.18) implies the existence of a positive number δ_4 depending only on χ_2 and such that

$$Eg^2(X_1) \geq \delta_4 > 0, \tag{1.22}$$

so that the conditions for asymptotic normality of $\sigma_N^{-1}U_N$ are satisfied. We shall prove

THEOREM 1.1. *Suppose that positive numbers $\delta, \delta_1, \delta_2, \delta_3, C$ and positive continuous functions χ_1 and χ_2 exist such that* (1.13)–(1.21) *are satisfied. Then there exists a sequence $\varepsilon_N \downarrow 0$ depending only on $\delta, \delta_1, \delta_2, \delta_3, C, \chi_1$ and χ_2 such that for $N = 2, 3, \dots$,*

$$\sup_x \left| P\left(\sigma_N^{-1} U_N \le x\right) - F_N(x) \right| \le \varepsilon_N N^{-1}, \tag{1.23}$$

where σ_N^2 and F_N are given by (1.6) *and* (1.9)–(1.11).

The laborious way in which we have phrased the assumptions as well as the conclusion of the theorem is caused by our insistence to define uniformity classes: for any class of pairs (h, P_X) for which the assumptions are satisfied for fixed δ, δ_j, C and χ_j, (1.12) holds uniformly. It will therefore continue to hold if we let h and P_X vary with N, provided $(h_N, P_{X,N})$, $N = 1, 2, \dots,$ are all in such a class. If we do not insist on uniformity and simply consider a fixed pair (h, P_X), then the result is much easier to state:

COROLLARY 1.1. *Suppose that there exist a number $r > 2$ and an integer k such that $(r-2)(k-4) > 8$ and that the following assumptions are satisfied*

$$E\left|\psi(X_1, X_2)\right|^r < \infty, \tag{1.24}$$

$$E\left|g(X_1)\right|^4 < \infty, \tag{1.25}$$

$$\limsup_{|t| \to \infty} \left| E e^{itg(X_1)} \right| < 1, \tag{1.26}$$

(1.27) ψ *possesses k nonzero eigenvalues with respect to P_X.*

Then (1.12) *holds.*

In the theorem as well as in the corollary, the role of all but one of the conditions is immediately clear. Since $Eg^2(X_1) > 0$ [cf. (1.22)] and $E\psi^2(X_1, X_2) < \infty$, $\hat{U}_N$ is the dominating term on the right in (1.5) and the conditions on $g(X_1)$ establish an Edgeworth expansion for $\hat{U}_N$. The moment assumption $E|\psi(X_1, X_2)|^r < \infty$ for some $r > 2$ allows us to correct the expansion for the remainder term Δ_N in (1.5). The existence of k nonzero eigenvalues of ψ, however, plays a much more subtle part which we shall discuss after the proof of the theorem has been given. We note that this kind of assumption first occurs in this context in Götze (1979).

If we are content to have an Edgeworth expansion with remainder $o(N^{-1/2})$ instead of $o(N^{-1})$, then we can do without the eigenvalue assumption. At the same time we may, of course, replace 4 by 3 in (1.17) and delete (1.14) so that

(1.18) becomes a nonlattice condition. Define

$$\tilde{F}_n(x) = \Phi(x) - \tfrac{1}{6}\kappa_3 N^{-1/2}\phi(x)(x^2 - 1), \tag{1.28}$$

where κ_3 is given by (1.9).

Theorem 1.2. *Suppose that positive numbers δ, C and positive continuous functions χ_1 and χ_2 exist such that* (1.13) *and* (1.15)–(1.18) *are satisfied, with g^4 replaced by $|g|^3$ in* (1.17). *Then there exists a sequence $\varepsilon_N \downarrow 0$ depending only on δ, C, χ_1 and χ_2 such that for $N = 2, 3, \ldots,$*

$$\sup_x \left| P\left(\sigma_N^{-1} U_N \le x\right) - \tilde{F}_N(x)\right| \le \varepsilon_N N^{-1/2}. \tag{1.29}$$

To prove Theorem 1.1 we shall have to study the characteristic function (c.f.) of $\sigma_N^{-1}U_N$. This is done separately for small (and intermediate) and for large values of the argument in Sections 2 and 3, respectively. After the extensive previous work on the asymptotics of U-statistics, the arguments in the first part are almost standard; the essential difficulties arise in the second part. Combination of the results of Sections 2 and 3 immediately yields Theorem 1.1. Theorem 1.2 follows from an analysis closely resembling that of Section 2, the only difference being that the use of the fourth moment of $g(X_1)$ should now be avoided. The proof that this can be done is easy and we omit it.

In Section 4 we discuss various aspects of assumption (1.20) and in Section 5 we give an application of Theorem 1.1. Two technical results—a moment inequality and a concentration inequality—which are needed in Section 3 but which may be of wider interest, are dealt with in the Appendix.

2. The c.f. for small values of the argument. Let ϕ_N denote the c.f. of $\sigma_N^{-1}U_N$,

$$\phi_N(t) = E\exp\{it\sigma_N^{-1}U_N\} \tag{2.1}$$

and, for κ_3 and κ_4 as in (1.9)–(1.10), let

$$\phi_N^*(t) = e^{-t^2/2}\left[1 - \frac{i\kappa_3}{6}N^{-1/2}t^3 + \frac{\kappa_4}{24}N^{-1}t^4 - \frac{\kappa_3^2}{72}N^{-1}t^6\right] \tag{2.2}$$

be the Fourier–Stieltjes transform $\int \exp(itx)\, dF_N(x)$ of F_N in (1.11). By Esseen's smoothing lemma [cf. Feller (1971), page 538] we have proved (1.23) if we construct sequences $\{T_N\}$ and $\{\varepsilon_N'\}$ depending only on δ, δ_1, δ_2, δ_3, C, χ_1 and χ_2 such that $N^{-1}T_N \to \infty$, $\varepsilon_N' \to 0$, and

$$\int_{-T_N}^{T_N}\left|\frac{\phi_N(t) - \phi_N^*(t)}{t}\right| dt \le \varepsilon_N' N^{-1}. \tag{2.3}$$

We begin by studying $\phi_N(t)$ for small $|t|$ and prove

LEMMA 2.1. *Suppose that* (1.13)–(1.18) *are satisfied. Then there exists a sequence* $\varepsilon_N'' \downarrow 0$ *depending only on* δ, δ_1, C, χ_1 *and* χ_2 *such that for*

$$t_N = N^{(r-1)/r}(\log N)^{-1}, \tag{2.4}$$

$$\int_{-t_N}^{t_N} \left| \frac{\phi_N(t) - \phi_N^*(t)}{t} \right| dt \le \varepsilon_N'' N^{-1}. \tag{2.5}$$

PROOF. To prevent the laborious formulation of our results from occurring throughout the proofs also, we shall make extensive use of o and O symbols rather than explicit error bounds. It will be tacitly understood that every statement involving o and O holds uniformly for all h and P_X satisfying the assumptions of the lemma to be proved for a fixed choice of the δ, δ_j, C and χ_j involved, and also uniformly for the values of t being considered.

Assume without loss of generality that $\delta \in (0, 1]$ and define

$$\varepsilon = \frac{\delta}{3(2+\delta)} \in (0, 1/9]. \tag{2.6}$$

Combining (2.1), (1.5) and

$$\left| e^{ix} - \sum_{\nu=0}^{m} \frac{(ix)^\nu}{\nu!} \right| \le \frac{2}{m!}|x|^{m+\theta} \quad \text{for every } \theta \in [0,1], \tag{2.7}$$

we can write

$$\phi_N(t) = E\exp\{it\sigma_N^{-1}\hat{U}_N\}\left(1 + it\sigma_N^{-1}\Delta_N - \tfrac{1}{2}t^2\sigma_N^{-2}\Delta_N^2\right) + O\left(E|t\sigma_N^{-1}\Delta_N|^{2+\delta}\right). \tag{2.8}$$

Let

$$\gamma_N(t) = E\exp\{it\sigma_N^{-1}(N-1)g(X_1)\}$$

denote the c.f. of $\sigma_N^{-1}(N-1)g(X_1)$. In view of (1.6), (1.22) and the fact that $E|\Delta_N|^{2+\delta} = O(N^{2+\delta})$ [cf. Callaert and Janssen (1978)] we may rewrite (2.8) as

$$\begin{aligned} \phi_N(t) &= \gamma_N^N(t) + \gamma_N^{N-2}(t) it\sigma_N^{-1}\binom{N}{2} E\exp\left\{it\sigma_N^{-1}(N-1)\sum_{j=1}^{2} g(X_j)\right\}\psi(X_1, X_2) \\ &\quad - \frac{1}{2}\gamma_N^{N-2}(t)t^2\sigma_N^{-2}\binom{N}{2} E\exp\left\{it\sigma_N^{-1}(N-1)\sum_{j=1}^{2} g(X_j)\right\}\psi^2(X_1, X_2) \\ &\quad - 3\gamma_N^{N-3}(t)t^2\sigma_N^{-2}\binom{N}{3} E\exp\left\{it\sigma_N^{-1}(N-1)\sum_{j=1}^{3} g(X_j)\right\} \\ &\quad \times \psi(X_1, X_3)\psi(X_2, X_3) \\ &\quad - 3\gamma_N^{N-4}(t)t^2\sigma_N^{-2}\binom{N}{4}\left[E\exp\left\{it\sigma_N^{-1}(N-1)\sum_{j=1}^{2} g(X_j)\right\}\psi(X_1, X_2)\right]^2 \\ &\quad + O(|N^{-1/2}t|^{2+\delta}). \end{aligned} \tag{2.9}$$

Next we expand the exponentials and find, e.g.,

$$E\exp\left\{it\sigma_N^{-1}(N-1)\sum_{j=1}^{2}g(X_j)\right\}\psi(X_1,X_2)$$

$$= E\Bigg[\prod_{j=1}^{2}\left(\exp\{it\sigma_N^{-1}(N-1)g(X_j)\}-1-it\sigma_N^{-1}(N-1)g(X_j)\right)$$

$$+2it\sigma_N^{-1}(N-1)\Bigg(\exp\{it\sigma_N^{-1}(N-1)g(X_1)\}$$

$$-\sum_{\nu=0}^{2}\{it\sigma_N^{-1}(N-1)g(X_1)\}^{\nu}/\nu!\Bigg)g(X_2)$$

$$-t^2\sigma_N^{-2}(N-1)^2g(X_1)g(X_2)$$

$$-it^3\sigma_N^{-3}(N-1)^3g^2(X_1)g(X_2)\Bigg]\psi(X_1,X_2)$$

$$= -t^2\sigma_N^{-2}(N-1)^2Eg(X_1)g(X_2)\psi(X_1,X_2)$$

$$-it^3\sigma_N^{-3}(N-1)^3Eg^2(X_1)g(X_2)\psi(X_1,X_2)$$

$$+O(N^{-2}t^4+|N^{-1/2}t|^{3(1+2\varepsilon)}),$$

with ε as in (2.6). To see this, use (2.7), (1.15)–(1.17), (1.22), (1.6) and

$$Eg^2(X_1)g^2(X_2)|\psi(X_1,X_2)|\le Eg^4(X_1)\{E\psi^2(X_1,X_2)\}^{1/2},$$

$$E|g(X_1)|^{2+6\varepsilon}|g(X_2)\psi(X_1,X_2)|\le\left[Eg^4(X_1)E|g(X_1)|^{(2+\delta)/(1+\delta)}\right]^{(1+\delta)/(2+\delta)}$$

$$\times\left[E|\psi(X_1,X_2)|^{2+\delta}\right]^{1/(2+\delta)}.$$

The other exponentials in (2.9) may be expanded in a similar fashion and after some further simplification (2.9) reduces to

$$\phi_N(t)=\gamma_N^N(t)+\gamma_N^{N-2}(t)\left(-\tfrac{1}{2}it^3\sigma_N^{-3}N^4Eg(X_1)g(X_2)\psi(X_1,X_2)\right.$$

$$\left.+\tfrac{1}{2}t^4\sigma_N^{-4}N^5Eg^2(X_1)g(X_2)\psi(X_1,X_2)-\tfrac{1}{4}t^2\sigma_N^{-2}N^2E\psi^2(X_1,X_2)\right)$$

$$+\tfrac{1}{2}\gamma_N^{N-3}(t)t^4\sigma_N^{-4}N^5Eg(X_1)g(X_2)\psi(X_1,X_3)\psi(X_2,X_3) \tag{2.10}$$

$$-\tfrac{1}{8}\gamma_N^{N-4}(t)t^6\sigma_N^{-6}N^8[Eg(X_1)g(X_2)\psi(X_1,X_2)]^2$$

$$+O\left(|\gamma_N(t)|^{N-4}|t|P(|t|)N^{-1-3\varepsilon}+|N^{-1/2}t|^{2+\delta}\right),$$

where P is a fixed polynomial.

For $\sigma_g^2 = Eg^2(X_1)$ as in (1.7), let

$$\gamma(t) = E\exp\{it\sigma_g^{-1}g(X_1)\}$$

denote the c.f. of $\sigma_g^{-1}g(X_1)$. From the classical theory of Edgeworth expansions for sums of i.i.d. random variables we know that (1.17) and (1.22) imply that for sufficiently small $\varepsilon' > 0$ and for $|t| \le \varepsilon' N^{1/2}$,

$$\gamma^N(N^{-1/2}t) = e^{-t^2/2}\left[1 - \frac{i\tilde{\kappa}_3}{6}N^{-1/2}t^3 + \frac{\tilde{\kappa}_4}{24}N^{-1}t^4 - \frac{\tilde{\kappa}_3^2}{72}N^{-1}t^6\right] + o\left(N^{-1}|t|e^{-t^2/4}\right), \tag{2.11}$$

where

$$\tilde{\kappa}_3 = \sigma_g^{-3}Eg^3(X_1), \qquad \tilde{\kappa}_4 = \sigma_g^{-4}Eg^4(X_1) - 3$$

are the third and fourth cumulants of $\sigma_g^{-1}g(X_1)$. Since $\gamma_N(t) = \gamma(\sigma_g\sigma_N^{-1}(N-1)t)$, an easy calculation shows that for $m = 0, 2, 3, 4$,

$$\gamma_N^{N-m}(t) = \gamma^N(N^{-1/2}t) + e^{-t^2/2}\left[\frac{1}{4}\sigma_g^{-2}E\psi^2(X_1, X_2) + \frac{m}{2}\right]N^{-1}t^2 + o\left(N^{-1}|t|e^{-t^2/4}\right) \tag{2.12}$$

for $|t| \le \varepsilon' N^{1/2}$. Substitution of (2.11), (2.12) and (1.6) in (2.10) shows after some rearrangement that for $|t| \le \varepsilon' N^{1/2}$,

$$\phi_N(t) = \phi_N^*(t) + o\left(N^{-1}|t|P(|t|)e^{-t^2/4}\right) + O(N^{-1-\delta/2}|t|^{2+\delta}), \tag{2.13}$$

where ϕ_N^* is given by (2.2), (1.9) and (1.10) and P is a fixed polynomial. It follows that for ε as given by (2.6),

$$\int_{-N^\varepsilon}^{N^\varepsilon}\left|\frac{\phi_N(t) - \phi_N^*(t)}{t}\right|dt = o(N^{-1}). \tag{2.14}$$

Obviously,

$$\int_{|t|\ge N^\varepsilon}\left|\frac{\phi_N^*(t)}{t}\right|dt = o(N^{-1})$$

and it therefore remains to be shown that for t_N as in (2.4),

$$\int_{N^\varepsilon \le |t| \le t_N}\left|\frac{\phi_N(t)}{t}\right|dt = o(N^{-1}). \tag{2.15}$$

Define, for $m = 1, \dots, N-1$,

$$\Delta_N(m) = \sum_{i=1}^{m}\sum_{j=i+1}^{N}\psi(X_i, X_j). \tag{2.16}$$

As $E|\Delta_N(m)|^r = O((mN)^{r/2})$ [cf. Callaert and Janssen (1978) for $r = 3$], we obtain

$$(2.17)\qquad \begin{aligned}|\phi_N(t)| \le &\left| E\exp\{it\sigma_N^{-1}(U_N - \Delta_N(m))\} \sum_{\nu=0}^{[r]} \frac{(it\sigma_N^{-1}\Delta_N(m))^\nu}{\nu!}\right| \\ &+ O\left(\left|\frac{m^{1/2}t}{N}\right|^r\right),\end{aligned}$$

where $[r]$ denotes the integer part of r. Since $(N-1)\sum_{i=1}^m g(X_i)$ are the only terms in $(U_N - \Delta_N(m))$ involving $X_1, \ldots, X_m$, we find that for $m \ge 2\nu$,

$$(2.18)\qquad \begin{aligned}&|E\exp\{it\sigma_N^{-1}(U_N - \Delta_N(m))\}\Delta_N^\nu(m)| \\ &\le |\gamma_N^{m-2\nu}(t)|(mN)^\nu E|\psi(X_1, X_2)|^\nu.\end{aligned}$$

Also, for sufficiently small $\tilde{\varepsilon} > 0$ and $|t| \le \tilde{\varepsilon}N^{1/2}$, we have

$$(2.19)\qquad |\gamma_N(t)| \le 1 - \frac{t^2}{3N} \le \exp\left\{-\frac{t^2}{3N}\right\}.$$

First take $N^\varepsilon \le |t| \le \tilde{\varepsilon}N^{1/2}$ and $m = m(t) = [3rN\log N/t^2] + 1$. For sufficiently large N, we see that indeed $1 \le m \le N - 1$ and (2.17)–(2.19) yield

$$(2.20)\qquad |\phi_N(t)| = O\left(\left(\frac{\log N}{N}\right)^{r/2}\right)$$

for $N^\varepsilon \le |t| \le \tilde{\varepsilon}N^{1/2}$.

Next we take $\tilde{\varepsilon}N^{1/2} \le |t| \le t_N$. In view of (1.14), (1.18) and the continuity of χ_2, there exists $\eta > 0$ such that for sufficiently large N,

$$(2.21)\qquad |\gamma_N(t)| \le 1 - \eta.$$

Choose $m = -r\log N/\log(1-\eta)$. For sufficiently large N, (2.17), (2.18) and (2.21) imply that

$$(2.22)\qquad |\phi_N(t)| = O\left((\log N)^{r/2}N^{-r}|t|^r\right)$$

for $\tilde{\varepsilon}N^{1/2} \le |t| \le t_N$. Since (2.20) and (2.22) hold uniformly not only for fixed δ, δ_1, C, χ_1 and χ_2 but also for the values of t being considered, (2.15) follows and the proof of Lemma 2.1 is complete. □

3. The c.f. for large values of the argument. In this section we prove

LEMMA 3.1. *Suppose that* (1.13), (1.15)–(1.17) *and* (1.19)–(1.22) *are satisfied. Then there exists a sequence* $\varepsilon_N''' \downarrow 0$ *depending only on* δ, δ_2, δ_3, δ_4, C *and* χ_1 *such that for* t_N *as in* (2.4) *and*

$$(3.1)\qquad T_N = N\log N,$$

$$(3.2)\qquad \int_{t_N \le |t| \le T_N} \left|\frac{\phi_N(t)}{t}\right| dt \le \varepsilon_N''' N^{-1}.$$

PROOF. We begin by noting that (1.21) implies that $k \geq 5$ and in view of (1.15) we may assume without loss of generality that

$$2 + \delta \leq r \leq 10 + \delta_3, \qquad 5 \leq k \leq 5 + \frac{8 + \delta_3}{\delta}. \tag{3.3}$$

Though of course not essential, these restrictions make it easier to go from error bounds in terms of r and k to bounds in terms of δ and δ_3 as required in the statement of the lemma.

In Section 2, the proof that $|\phi_N(t)|$ is sufficiently small for $N^\varepsilon \leq |t| \leq t_N$ was based on the fact that for these values of t the behaviour of $|\phi_N(t)|$ is still determined to some extent by that of the c.f. of $\hat{U}_N$, and hence by the c.f. of $g(X_1)$. For larger values of $|t|$, however, the influence of the remainder term Δ_N may completely destroy that of $\hat{U}_N$. It seems that we have no more use for the $g(X_i)$ and we shall remove them by a conditioning argument.

Define random variables $Y_1, \ldots, Y_N$ such that $X_1, \ldots, X_N, Y_1, \ldots, Y_N$ are i.i.d. and let $V_i = (X_i, Y_i)$, $i = 1, \ldots, N$. Let n be an integer with $1 \leq n \leq (N-1)/4$. Then

$$\begin{aligned}
|\phi_N(t)|^2 &\leq E\left|E\left(\exp\{it\sigma_N^{-1}U_N\} | X_1, \ldots, X_{4n}\right)\right|^2 \\
&\leq E\left|E\left(\exp\left\{it\sigma_N^{-1}\sum_{j=1}^{l-1}\sum_{l=4n+1}^{N} h(X_j, X_l)\right\}\middle| X_1, \ldots, X_{4n}\right)\right|^2 \\
&= E\exp\left\{it\sigma_N^{-1}\left[\sum_{j=1}^{4n}\sum_{l=4n+1}^{N}\left(h(X_j, X_l) - h(X_j, Y_l)\right)\right.\right. \\
&\qquad \left.\left. + \sum_{j=4n+1}^{l-1}\sum_{l=4n+1}^{N}\left(h(X_j, X_l) - h(Y_j, Y_l)\right)\right]\right\} \\
&\leq E\left|E\left(\exp\left\{it\sigma_N^{-1}\sum_{j=1}^{4n}\sum_{l=4n+1}^{N}\left(h(X_j, X_l) - h(X_j, Y_l)\right)\right\}\middle| V_{4n+1}, \ldots, V_N\right)\right| \\
&= E\exp\left\{it\sigma_N^{-1}\sum_{j=1}^{2n}\sum_{l=4n+1}^{N}\left(h(X_j, X_l)\right.\right. \\
&\qquad \left.\left. - h(Y_j, X_l) - h(X_j, Y_l) + h(Y_j, Y_l)\right)\right\} \\
&= E\exp\left\{it\sigma_N^{-1}\sum_{j=1}^{2n}\sum_{l=4n+1}^{N}\left(\psi(X_j, X_l)\right.\right. \\
&\qquad \left.\left. - \psi(Y_j, X_l) - \psi(X_j, Y_l) + \psi(Y_j, Y_l)\right)\right\} \\
&= E\exp\left\{it\sigma_N^{-1}\sum_{j=1}^{2n}\sum_{l=4n+1}^{N}\Psi(V_j, V_l)\right\},
\end{aligned} \tag{3.4}$$

where for $v_j = (x_j, y_j)$, $j = 1, \dots, N$, we have defined

$$\Psi(v_j, v_l) = \psi(x_j, x_l) - \psi(y_j, x_l) - \psi(x_j, y_l) + \psi(y_j, y_l). \tag{3.5}$$

Our next step is to truncate the random variables $X_1, \dots, X_{2n}, Y_1, \dots, Y_{2n}$, while losing half of them in the process. Consider a measurable set $B \in \mathscr{B}$ with

$$\alpha = P(V_1 \in B \times B) = P_X^2(B). \tag{3.6}$$

For every $\alpha \in [0,1]$, $x \in [0,1]$ and $\rho \in (0,1)$ we have

$$\alpha x + (1-\alpha) \le x^{\rho} \vee \left(\frac{\rho}{\alpha}\right)^{\rho/(1-\rho)}, \tag{3.7}$$

where $(x \vee y)$ denotes the larger of x and y. It follows from (3.4)–(3.7) that

$$|\phi_N(t)|^2 \le E\left[E\left(\exp\left\{it\sigma_N^{-1} \sum_{l=4n+1}^{N} \Psi(V_1, V_l)\right\} \middle| V_{4n+1}, \dots, V_N\right)\right]^{2n}$$

$$\le E\left[\alpha E\left(\exp\left\{it\sigma_N^{-1} \sum_{l=4n+1}^{N} \Psi(V_1, V_l)\right\} \middle| V_{4n+1}, \dots, V_N; V_1 \in B \times B\right)\right.$$

$$\left. + (1-\alpha)\right]^{2n} \tag{3.8}$$

$$\le E\left[E\left(\exp\left\{it\sigma_N^{-1} \sum_{l=4n+1}^{N} \Psi(V_1, V_l)\right\} \middle| V_{4n+1}, \dots, V_N; V_1 \in B \times B\right)\right]^{2\rho n}$$

$$+ \left(\frac{\rho}{\alpha}\right)^{2\rho n/(1-\rho)}$$

for every $\rho \in (0,1)$. Take $\rho = \frac{1}{2}$ and define $\tilde{V}_j = (\tilde{X}_j, \tilde{Y}_j)$, $j = 1, \dots, n$, in such a way that $\tilde{X}_1, \dots, \tilde{X}_n, \tilde{Y}_1, \dots, \tilde{Y}_n$ are i.i.d. with common distribution

$$P(\tilde{X}_j \in A) = P(\tilde{Y}_j \in A) = \frac{P_X(A \cap B)}{P_X(B)} \tag{3.9}$$

and independent of $V_{4n+1}, \dots, V_N$. Then (3.8) may be rewritten as

$$|\phi_N(t)|^2 \le E \exp\left\{it\sigma_N^{-1} \sum_{j=1}^{n} \sum_{l=4n+1}^{N} \Psi(\tilde{V}_j, V_l)\right\} + (2\alpha)^{-2n} \tag{3.10}$$

$$= E\left[E\left(\exp\{it\sigma_N^{-1} Z_n\} | \tilde{V}_1, \dots, \tilde{V}_n\right)\right]^{N-4n} + (2\alpha)^{-2n},$$

where

$$
\begin{aligned}
Z_n &= \sum_{j=1}^{n} \Psi(\tilde{V}_j, V_N) \\
&= \sum_{j=1}^{n} \left[\psi(\tilde{X}_j, X_N) - \psi(\tilde{Y}_j, X_N) - \psi(\tilde{X}_j, Y_N) + \psi(\tilde{Y}_j, Y_N) \right].
\end{aligned} \tag{3.11}
$$

It remains to choose the set B and we take

$$B = \left\{ x \colon \int |\psi(x, y)|^r \, dP_X(y) \le C_1 \tau \right\}, \tag{3.12}$$

for a large but fixed $\tau > 0$ to be specified later.

Let us now consider the conditional expectation in (3.10). Since $|\exp\{ix\} - 1 - ix + \frac{1}{2}x^2| \le x^2/6 + |x|^r$ for $r > 2$, we have

$$
\begin{aligned}
0 \le E\left(\exp\{it\sigma_N^{-1} Z_n\} | \tilde{V}_1, \ldots, \tilde{V}_n\right) &\le 1 - \tfrac{1}{3} t^2 \sigma_N^{-2} E(Z_n^2 | \tilde{V}_1, \ldots, \tilde{V}_n) \\
&\quad + |t|^r \sigma_N^{-r} E(|Z_n|^r | \tilde{V}_1, \ldots, \tilde{V}_n).
\end{aligned} \tag{3.13}
$$

By (3.11) and (3.12)

$$E\left(\Psi(\tilde{V}_1, V_N) | \tilde{V}_1\right) = E\left(\Psi(\tilde{V}_1, V_N) | V_N\right) = 0 \quad \text{a.s.}, \tag{3.14}$$

$$E\left(|\Psi(\tilde{V}_1, V_N)|^r | \tilde{V}_1\right) \le 4^r C_1 \tau \quad \text{a.s.} \tag{3.15}$$

It follows from Lemma A.1 in the Appendix together with (3.3) that for every integer $m \ge 1$

$$E\left[E(|Z_n|^r | \tilde{V}_1, \ldots, \tilde{V}_n)\right]^m = O(n^{rm/2}).$$

Taking $m = 10k/\delta_3$, we find by (3.3) and Markov's inequality that

$$P\left(E(|Z_n|^r | \tilde{V}_1, \ldots, \tilde{V}_n) \ge n^{r/2} N^{\delta_3/(4k)}\right) = O(N^{-5/2}). \tag{3.16}$$

Next we turn to the quadratic term in (3.13). Let $\lambda_1, \lambda_2, \ldots$ be the eigenvalues of ψ with respect to P_X with $|\lambda_1| \ge |\lambda_2| \ge \cdots$ and let $\omega_1, \omega_2, \ldots$ be an orthonormal sequence of eigenfunctions corresponding to $\lambda_1, \lambda_2, \ldots$, i.e., (1.19) holds and for all ν and ν',

$$\int \omega_\nu(x) \, dP_X(x) = 0, \qquad \int \omega_\nu(x) \omega_{\nu'}(x) \, dP_X(x) = \delta_{\nu, \nu'}, \tag{3.17}$$

where $\delta_{\nu,\nu'} = 0$ or 1 according as $\nu \ne \nu'$ or $\nu = \nu'$. Assume (1.20) is satisfied. We have

$$\psi(x, y) = \sum_{\nu=1}^{k} \lambda_\nu \omega_\nu(x) \omega_\nu(y) + R(x, y), \tag{3.18}$$

where R is a symmetric function of its two variables satisfying

$$\int R(x, y) \omega_\nu(y) \, dP_X(y) = 0 \quad \text{for } \nu = 1, \ldots, k. \tag{3.19}$$

As a consequence we find

$$E\left(Z_n^2 \mid \tilde{V}_1, \ldots, \tilde{V}_n\right) = 2\int\left[\sum_{j=1}^{n}\left\{\psi(\tilde{X}_j, y) - \psi(\tilde{Y}_j, y)\right\}\right]^2 dP_X(y)$$

$$= 2\int\left[\sum_{j=1}^{n}\left\{\sum_{\nu=1}^{k}\lambda_\nu\omega_\nu(y)\left(\omega_\nu(\tilde{X}_j) - \omega_\nu(\tilde{Y}_j)\right)\right.\right.$$

$$\left.\left. + R(\tilde{X}_j, y) - R(\tilde{Y}_j, y)\right\}\right]^2 dP_X(y)$$

$$(3.20)\qquad \begin{aligned} &= 2\sum_{j=1}^{n}\sum_{j'=1}^{n}\sum_{\nu=1}^{k}\sum_{\nu'=1}^{k}\int\left[\lambda_\nu\omega_\nu(y)\left(\omega_\nu(\tilde{X}_j) - \omega_\nu(\tilde{Y}_j)\right)\right. \\ &\qquad\qquad \left. + R(\tilde{X}_j, y) - R(\tilde{Y}_j, y)\right] \\ &\qquad\qquad \times\left[\lambda_{\nu'}\omega_{\nu'}(y)\left(\omega_{\nu'}(\tilde{X}_{j'}) - \omega_{\nu'}(\tilde{Y}_{j'})\right)\right. \\ &\qquad\qquad \left. + R(\tilde{X}_{j'}, y) - R(\tilde{Y}_{j'}, y)\right] dP_X(y) \\ &= 2\sum_{\nu=1}^{k}\lambda_\nu^2\left[\sum_{j=1}^{n}\left\{\omega_\nu(\tilde{X}_j) - \omega_\nu(\tilde{Y}_j)\right\}\right]^2 \\ &\quad + 2k^2\int\left[\sum_{j=1}^{n}\left\{R(\tilde{X}_j, y) - R(\tilde{Y}_j, y)\right\}\right]^2 dP_X(y) \\ &\geq 2\delta_2^2\sum_{\nu=1}^{k}\left[\sum_{j=1}^{n}\left\{\omega_\nu(\tilde{X}_j) - \omega_\nu(\tilde{Y}_j)\right\}\right]^2. \end{aligned}$$

We shall have to investigate the covariance matrix Σ of the random vector $(\omega_1(\tilde{X}_1) - \omega_1(\tilde{Y}_1), \ldots, \omega_k(\tilde{X}_1) - \omega_k(\tilde{Y}_1))$. First note that (1.16) and (1.20) imply that for $\nu = 1, \ldots, k$

$$(3.21)\qquad \begin{aligned} E|\omega_\nu(X_1)|^r &\leq \lambda_\nu^{-r}E\left|\int\psi(X_1, y)\omega_\nu(y)\,dP_X(y)\right|^r \\ &\leq \delta_2^{-r}E\left\{\int\psi^2(X_1, y)\,dP_X(y)\right\}^{r/2} \leq \delta_2^{-r}C. \end{aligned}$$

Let $\sigma_{\nu,\nu'} = E(\omega_\nu(\tilde{X}_1) - \omega_\nu(\tilde{Y}_1))(\omega_{\nu'}(\tilde{X}_1) - \omega_{\nu'}(\tilde{Y}_1))$, $\nu, \nu' = 1, \ldots, k$, denote the elements of Σ. For $\nu \neq \nu'$, (3.21) and Hölder's inequality ensure that

$$(3.22)\qquad \begin{aligned} |\sigma_{\nu,\nu'}| &\leq \left|\frac{1}{\alpha}E(\omega_\nu(X_1) - \omega_\nu(Y_1))(\omega_{\nu'}(X_1) - \omega_{\nu'}(Y_1))1_{(B\times B)^c}(X_1, Y_1)\right| \\ &\leq 4\delta_2^{-2}C^{2/r}\alpha^{-1}(1-\alpha)^{(r-2)/r}, \end{aligned}$$

whereas for $\nu = \nu'$ we find similarly

$$-4\delta_2^{-2}C^{2/r}\alpha^{-1}(1-\alpha)^{(r-2)/r} \le \sigma_{\nu,\nu} - 2 \le 2\alpha^{-1}(1-\alpha). \tag{3.23}$$

Now we may still choose τ in (3.12) and since

$$1 - \alpha^{1/2} = P_X(B^c) \le \tau^{-1}, \tag{3.24}$$

by (3.12) and Markov's inequality, we can force α to be arbitrarily close to 1 by taking τ large. In view of (3.22)–(3.24) and (3.3), we can choose $\tau = \tau(\delta, \delta_2, \delta_3, C)$ in such a way that

$$2\alpha \ge e^{1/2}, \tag{3.25}$$

$$|\sigma_{\nu,\nu'} - 2\delta_{\nu,\nu'}| \le k^{-1} \quad \text{for all } \nu, \nu' = 1,\dots,k. \tag{3.26}$$

If ρ_k denotes the smallest eigenvalue of Σ, then (3.26) yields

$$\rho_k \ge 2 - \left\{ \sum_{\nu=1}^{k} \sum_{\nu'=1}^{k} (\sigma_{\nu,\nu'} - 2\delta_{\nu,\nu'})^2 \right\}^{1/2} \ge 1. \tag{3.27}$$

Also (3.21), (3.25) and (3.3) imply that $E|\omega_\nu(\tilde{X}_1) - \omega_\nu(\tilde{Y}_1)|^{2+\delta}$, $\nu = 1,\dots,k$, as well as k are bounded. It follows that we may apply Lemma A.2 in the Appendix to the right-hand side of (3.20) to obtain

$$\begin{aligned} &P\Big(E\big(Z_n^2|\tilde{V}_1,\dots,\tilde{V}_n\big) \le nN^{-4/k}(\log N)^{-6/k}\Big) \\ &\qquad = O\big(N^{-2}(\log N)^{-3} + n^{-k/2}\big). \end{aligned} \tag{3.28}$$

Let us now combine the results obtained in (3.10), (3.13), (3.16) and (3.28). First we note that (3.25) ensures that the term $(2\alpha)^{-2n}$ in (3.10) is $O(e^{-n})$ and that σ_N^2 is of exact order N^3 by (1.16), (1.17) and (1.22). Take t_N and T_N as in (2.4) and (3.1), choose any t such that $t_N \le |t| \le T_N$ and then define

$$n = n(t) = \left[\frac{\sigma_N^2 N^{(4/k)-1}(\log N)^{2+(6/k)}}{t^2} \right], \tag{3.29}$$

where $[x]$ denotes the integer part of x. As $t_N \le |t| \le T_N$, it follows from (2.4), (3.1) and (1.21) that

$$\begin{aligned} \sigma_N^2 N^{(4/k)-3}(\log N)^{6/k} - 1 \le n &= O\big(N^{(4/k)+(2/r)}(\log N)^{4+(6/k)}\big) \\ &= O\big(N^{1-\delta_3/(kr)}(\log N)^6\big), \end{aligned} \tag{3.30}$$

and in view of (3.3) this means that $1 \le n \le (N-1)/4$ for sufficiently large N, so that (3.29) is indeed a possible choice of n. Similarly, one easily checks that (3.29), (1.21) and (3.3) imply that for sufficiently large N,

$$|t|^r \sigma_N^{-r} n^{r/2} N^{\delta_3/(4k)} \le \tfrac{1}{6} t^2 \sigma_N^{-2} n N^{-4/k} (\log N)^{-6/k}. \tag{3.31}$$

Together (3.10), (3.13), (3.16), (3.28), (3.31), (3.29) and (3.30) show that for sufficiently large N

$$|\phi_N(t)|^2 \le \left[1 - \tfrac{1}{6}t^2 n\sigma_N^{-2}N^{-4/k}(\log N)^{-6/k}\right]^{N-4n} + O\left(N^{-2}(\log N)^{-3} + n^{-k/2}\right)$$
$$\le \exp\left\{-\tfrac{1}{6}N^{-1}(N-4n)(\log N)^2\right\} + O\left(N^{-2}(\log N)^{-3} + n^{-k/2}\right)$$
$$= O\left(N^{-2}(\log N)^{-3}\right),$$

so that

$$|\phi_N(t)| = O\left(N^{-1}(\log N)^{-3/2}\right) \tag{3.32}$$

uniformly for $t_N \le |t| \le T_N$. This proves Lemma 3.1 and Theorem 1.1 at the same time. □

4. The eigenvalue assumption. In Section 1 we noted that the meaning of assumption (1.20) concerning the eigenvalues of ψ, is not intuitively clear. From the analysis in Sections 2 and 3, however, we can at least see the part that it plays in the proof of Theorem 1.1. As we pointed out at the beginning of the proof of Lemma 3.1, the analysis of $|\phi_N(t)|$ for $|t| \le N^{(r-1)/r}(\log N)^{-1}$ proceeds by showing that up to that point, the properties of $\hat{U}_N$ determine the behaviour of $|\phi_N(t)|$, because the influence of $|t|\sigma_N^{-1}\Delta_N$ is still small. For larger values of $|t|$, $\hat{U}_N$ does not play a role any longer and we have to show that $|t|\sigma_N^{-1}\Delta_N$ is large enough to take over the task of making $|\phi_N(t)|$ small. Since, in general, sums of independent random variables can be unpleasantly close to zero with probabilities that are nonnegligible for our purposes, assumption (1.20) is there to prevent this.

Still, we are unable to show that without assumption (1.20), the theorem would indeed fail. Our search for a counterexample, however, has convinced us that such an example would have to be extremely pathological.

To compute the eigenvalues $\lambda_1, \ldots, \lambda_k$ of ψ can of course be laborious, but fortunately this is not necessary in order to verify assumption (1.20). Consider functions $f_1, \ldots, f_k$ with

$$\int f_j^2(x)\, dP_X(x) \le 1, \qquad j = 1, \ldots, k, \tag{4.1}$$

and define random variables

$$W_j = \int \psi(X_1, y) f_j(y)\, dP_X(y). \tag{4.2}$$

Let Σ_W denote the covariance matrix of the random vector $W = (W_1, \ldots, W_k)$ and suppose that it has a smallest eigenvalue $\tilde{\lambda}_k$ satisfying

$$\tilde{\lambda}_k \ge \delta_2 > 0. \tag{4.3}$$

LEMMA 4.1. *Suppose that in the set of conditions of Theorem* 1.1 *we replace* (1.19)–(1.20) *by the assumption that* $f_1, \ldots, f_k$ *exist such that* (4.1)–(4.3) *are satisfied. Then the set of conditions obtained is equivalent to the original set.*

PROOF. If (1.19)–(1.20) hold, we may choose $f_j = \omega_j$, $W_j = \lambda_j \omega_j(X_1)$ and hence $\tilde{\lambda}_k = \lambda_k^2 \geq \delta_2^2$. Replacing δ_2 by $\delta_2^{1/2}$ yields (4.3).

Conversely, suppose that (4.1) and (4.3) hold for certain $f_1, \ldots, f_k$. Let $\mathscr{F}$ denote the linear space spanned by $f_1, \ldots, f_k$ and define $\| f \|$ and Tf by

$$\| f \|^2 = \int f^2 \, dP_X, \qquad (Tf)(x) = \int \psi(x, y) f(y) \, dP_X(y).$$

For $f = \sum_{j=1}^k c_j f_j$, we have

$$\|Tf\|^2 = E\left(\sum_{j=1}^k c_j W_j \right)^2 = c' \Sigma_W c \geq \tilde{\delta}_2 \sum_{j=1}^k c_j^2,$$

$$\| f \|^2 \leq \sum_{j=1}^k c_j^2 \sum_{j=1}^k \| f_j \|^2 \leq k \sum_{j=1}^k c_j^2$$

in view of (4.2), (4.3) and (4.1). Together this yields

$$\|Tf\|^2 \geq \frac{\tilde{\delta}_2}{k} \| f \|^2 \quad \text{for every } f \in \mathscr{F}. \tag{4.4}$$

On the other hand, (4.3) ensures that $f_1, \ldots, f_k$ are linearly independent in $L_2(P_X)$ and hence $\mathscr{F}$ must contain functions orthogonal to $\omega_1, \omega_2, \ldots, \omega_{k-1}$ defined in (1.19). But this implies that

$$\inf_{f \in \mathscr{F}} \frac{\|Tf\|^2}{\| f \|^2} \leq \lambda_k^2, \tag{4.5}$$

where $\lambda_1, \lambda_2, \ldots,$ are given by (1.19). Combining (4.4) and (4.5) we find

$$|\lambda_k| \geq \left(\frac{\tilde{\delta}_2}{k} \right)^{1/2}. \tag{4.6}$$

Because of (1.15) and (1.21) we may assume k to be bounded [cf. (3.3)] and the proof is complete. □

Of course (4.3) will usually be easier to verify than (1.20). The situation is even simpler in Corollary 1.1 or, more generally, in all cases where ψ is fixed. Assumption (1.27) may then be replaced by the nonsingularity of Σ_W, i.e., by the fact that $W_1, \ldots, W_k$ are not almost surely linearly dependent. A simple sufficient condition for the existence of such $W_1, \ldots, W_k$ is that there exist points $y_1, \ldots, y_k$ in the support of F such that the functions $\psi(\cdot, y_1), \ldots, \psi(\cdot, y_k)$ are linearly independent.

5. An example. Let $X_1, \ldots, X_N$ be i.i.d. random variables with a common continuous d.f. F on $\mathbb{R}$. Let

$$R_i^+ = \sum_{j=1}^N 1_{\{|X_j| \leq |X_i|\}} \tag{5.1}$$

and let W_N^+ denote Wilcoxon's one-sample signed rank statistic for testing the hypothesis that the distribution of X_1 is symmetric about zero, thus

$$W_N^+ = \sum_{i=1}^{N} 1_{\{X_i \geq 0\}} R_i^+. \tag{5.2}$$

If we define

$$\begin{aligned} U_N &= W_N^+ - EW_N^+ \\ &= W_N^+ - N(1 - F(0)) \\ &\quad - N(N-1)\int_0^\infty (F(x) - F(-x))\, dF(x), \end{aligned} \tag{5.3}$$

then U_N is clearly a U-statistic. An easy computation yields

$$U_N = (N-1)\sum_{i=1}^{N} g_N(X_i) + \sum_{i=1}^{N}\sum_{j=i+1}^{N} \psi(X_i, X_j), \tag{5.4}$$

where

$$\begin{aligned} g_N(x) &= 1 - F(-x) - \int (1 - F(-x))\, dF(x) \\ &\quad + \frac{1}{N-1}\left\{1_{[0,\infty)}(x) - 1 + F(0)\right\}, \end{aligned} \tag{5.5}$$

$$\begin{aligned} \psi(x, y) &= 1_{[0,\infty)}(x + y) - (1 - F(-x)) - (1 - F(-y)) \\ &\quad + \int (1 - F(-x))\, dF(x). \end{aligned} \tag{5.6}$$

Note that $Eg_N(X_1) = 0$ and $E(\psi(X_1, X_2)|X_1) = 0$ a.s.

Having decomposed U_N in the manner of Section 1, we check the conditions of Theorem 1.1. Since both g_N and ψ are bounded, (1.13) and (1.15)–(1.17) are satisfied for every r. Next, (1.14) and (1.18) will hold if the distribution of $F(-X_1)$ has an absolutely continuous component. It remains to verify (1.20) for some $k \geq 5$. In view of Lemma 4.1 and the fact that ψ does not depend on N it suffices to find functions $f_1, \ldots, f_k$ with $\int f_j^2\, dF \leq 1$ such that the random variables,

$$W_j = \int \psi(X_1, y) f_j(y)\, dF(y), \qquad j = 1, \ldots, k, \tag{5.7}$$

are not almost surely linearly dependent. Take

$$f_j(x) = F^j(x), \qquad j = 1, \ldots, k, \tag{5.8}$$

so that

$$W_j = \frac{1}{j+1}\left\{F(-X_1) - F^{j+1}(-X_1) - \int (F(-x) - F^{j+1}(-x))\, dF(x)\right\}. \tag{5.9}$$

Then

$$\sum_{j=1}^{k} c_j W_j = \sum_{i=0}^{k+1} a_i F^i(-X_1), \tag{5.10}$$

with

$$a_0 = -\sum_{j=1}^{k} c_j \int \left(F(-x) - F^{j+1}(-x)\right) dF(x),$$

$$a_1 = \sum_{j=1}^{k} \frac{c_j}{j+1}, \qquad a_i = -\frac{c_{j-1}}{j} \quad \text{for } i = 2, \dots, k+1.$$

Since the distribution of $F(-X_1)$ is supposed to have an absolutely continuous part, (5.10) can vanish almost surely only if $a_0 = a_1 = \cdots = a_{k+1} = 0$ which implies $c_1 = \cdots = c_k = 0$. It follows that assumption (1.20) holds every k.

Thus we have established the validity of the Edgeworth expansion with remainder $o(N^{-1})$ for Wilcoxon's one-sample rank statistic under the assumptions that F is continuous and that the distribution of $F(-X_1)$ has an absolutely continuous component. We stress the fact that previous results on Edgeworth expansions for U-statistics would fail in this case because U_N has a pure lattice distribution. Edgeworth expansions for one-sample rank statistics were obtained in Albers, Bickel and van Zwet (1976) by a completely different method.

APPENDIX

In this appendix we prove a moment inequality and a concentration inequality which are needed in Section 3 of the present paper, but which may also be of independent interest.

LEMMA A.1. *Let P and Q be probability measures on arbitrary sample spaces $\mathscr{X}$ and $\mathscr{Y}$ and let $X_1, \dots, X_n$ be i.i.d. with common distribution P. Let ψ: $\mathscr{X} \times \mathscr{Y} \to \mathbb{R}$ satisfy $\int \psi(x, y)\, dP(x) = 0$ for Q—almost all $y \in \mathscr{Y}$, and $\int \psi(x, y)\, dQ(y) = 0$ for P—almost all $x \in \mathscr{X}$. Then, for every real $p \geq 2$ and integer $k \geq 1$, there exists a positive number $A = A(p, k)$ which is bounded for bounded p and k and such that*

$$E\left\{\int \left|\sum_{i=1}^{n} \psi(X_i, y)\right|^p dQ(y)\right\}^k \leq A n^{kp/2} E\left\{\int |\psi(X_1, y)|^p\, dQ(y)\right\}^k.$$

PROOF. If the expectation on the right equals $+\infty$, then there is nothing to prove. Assume therefore that

$$C = E\left\{\int |\psi(X_1, y)|^p\, dQ(y)\right\}^k < \infty.$$

Let $Y_1, \dots, Y_k$ be i.i.d. with common distribution Q and independent of

$X_1, \ldots, X_n$. Then

$$B = E\left\{\int \left|\sum_{i=1}^{n} \psi(X_i, y)\right|^p dQ(y)\right\}^k = E\left|\prod_{j=1}^{k} \sum_{i=1}^{n} \psi(X_i, Y_j)\right|^p$$

$$= E\left|\sum_{i_1=1}^{n} \cdots \sum_{i_k=1}^{n} \prod_{j=1}^{k} \psi(X_{i_j}, Y_j)\right|^p.$$

Let $m_1, \ldots, m_r$ be integers ≥ 2 with $\sum_{\nu=1}^{r} m_\nu = k - l$, $l \geq 0$, and let $I(m_1, \ldots, m_r)$ denote the collection of sequences $i_1, \ldots, i_k \in \{1, 2, \ldots, n\}$ which contain $(l + r)$ distinct values, out of which l occur with multiplicity 1 and r with multiplicities $m_1, \ldots, m_r$, respectively. Define

$$Z(m_1, \ldots, m_r) = \sum_{(i_1, \ldots, i_k) \in I(m_1, \ldots, m_r)} \cdots \sum \prod_{j=1}^{k} \psi(X_{i_j}, Y_j), \tag{A.1}$$

and note that each term in this sum has the same distribution. There are at most n^r different ways of choosing the indices with multiplicities $m_1, \ldots, m_r$ and at most $k!$ different ways of permuting $i_1, \ldots, i_k$. It follows that

$$E|Z(m_1, \ldots, m_r)|^p \leq (k! n^r)^p E\left|\sum_{1 \leq i_1 < \cdots < i_l \leq n-r} \cdots \sum \prod_{j=1}^{l} \psi(X_{i_j}, Y_j)\right.$$

$$\left. \times \prod_{j=1}^{m_1} \psi(X_{n-r+1}, Y_{l+j}) \times \cdots \times \prod_{j=1}^{m_r} \psi(X_n, Y_{l+m_1+\cdots+m_{r-1}+j})\right|^p$$

$$\leq (k! n^r)^p \prod_{\nu=1}^{r} E\left\{\int |\psi(X_1, y)|^p dQ(y)\right\}^{m_\nu} \tag{A.2}$$

$$\times E\left|\sum_{1 \leq i_1 < \cdots < i_l \leq n-r} \cdots \sum \prod_{j=1}^{l} \psi(X_{i_j}, Y_j)\right|^p$$

$$\leq (k! n^r)^p C^{(k-l)/k} E|W_l(n-r)|^p,$$

where, for $l = 1, 2, \ldots, k$ and $t = l, l+1, \ldots, n$,

$$W_l(t) = \sum_{1 \leq i_1 < i_2 < \cdots < i_l \leq t} \cdots \sum \prod_{j=1}^{l} \psi(X_{i_j}, Y_j)$$

and we define $W_l(l-1) = 0$ for $l = 1, 2, \ldots, k$ and $W_0(t) = 1$ for $t = 0, 1, \ldots, n$.

For fixed $l \geq 1$, $W_l(t)$, $t = l-1, l, \ldots, n$, is a martingale with $W_l(l-1) = 0$. It follows from an inequality of Dharmadhikari, Fabian and Jogdeo (1968) that for $l \geq 1$ and $t = l, l+1, \dots, n$

$$
\begin{aligned}
E|W_l(t)|^p &\leq a(p)(t-l+1)^{p/2-1} \sum_{s=l}^{t} E|W_l(s) - W_l(s-1)|^p \\
&= a(p)(t-l+1)^{p/2-1} E|\psi(X_1, Y_1)|^p \sum_{s=l}^{t} E|W_{l-1}(s-1)|^p \\
&\leq \left\{a(p)t^{p/2-1} E|\psi(X_1, Y_1)|^p\right\}^l \sum_{s_1=l}^{t} \sum_{s_2=l-1}^{s_1-1} \cdots \sum_{s_l=1}^{s_{l-1}-1} 1 \\
&\leq \left\{a(p)E|\psi(X_1, Y_1)|^p\right\}^l t^{lp/2}
\end{aligned}
\tag{A.3}
$$

for $a(p) = 2^{2p^2}$. Clearly (A.3) will continue to hold for $l = 0$ and $t = 0, \dots, n$ provided we define $0^0 = 1$. Combining this with (A.2) we find

$$E|Z(m_1, \dots, m_r)|^p \leq (k!n^r)^p C^{(k-l)/k} a^l(p) C^{l/k} n^{lp/2}$$

and as $2r + l \leq k$,

$$B \leq (k!)^{p+1} 2^{2kp^2} n^{kp/2} C.$$

The lemma is proved. □

Lemma A.2. *Let $X_1, \dots, X_n$ be i.i.d. k-dimensional random vectors with common distribution P with a positive definite covariance matrix Σ with smallest eigenvalue ρ_k. Define $S_n = n^{-1/2}\Sigma_{i=1}^n X_i$. Then there exists a positive number B depending only on k and P such that for every $\varepsilon > 0$ and $n = 1, 2, \dots$,*

$$P(\|S_n\| \leq \varepsilon) \leq B(\varepsilon^k + n^{-k/2}).$$

B is constant over any class of distributions with k bounded, ρ_k bounded away from zero and $E\|X\|^{2+\delta}$ bounded for a fixed $\delta > 0$.

Proof. Let $\tilde{P}$ be the distribution of $(X_1 - X_2)$ and for $t \in \mathbb{R}^k$ let

$$\psi(t) = E\exp\{it'(X_1 - EX_1)\}.$$

It follows that

$$|\psi(t)|^2 = \int_{\mathbb{R}^k} e^{it'x}\, d\tilde{P}(x) = \int_{\mathbb{R}^k} \cos(t'x)\, d\tilde{P}(x),$$

$$\frac{1 - |\psi(t)|^2}{\|t\|^2} \geq \int_{\|x\| \leq \theta^{-1}} \frac{1 - \cos(t'x)}{\|t\|^2}\, d\tilde{P}(x)$$

for every $\theta > 0$. For $\|t\| \leq \theta$, we have $|t'x| \leq 1$ for $\|x\| \leq \theta^{-1}$ and $1 - \cos(t'x) \geq \frac{1}{2}\cos(1)(t'x)^2$. Hence for $\|t\| \leq \theta$ we see that

$$
\begin{aligned}
\frac{1 - |\psi(t)|^2}{\|t\|^2} &\geq \frac{1}{2}\cos(1) \int_{\|x\| \leq \theta^{-1}} \frac{(t'x)^2}{\|t\|^2}\, d\tilde{P}(x) \\
&= \frac{1}{2}\cos(1) \int_{\|x\| \leq \theta^{-1}} (\tau'x)^2\, d\tilde{P}(x),
\end{aligned}
\tag{A.4}
$$

where $\tau = t/\|t\|$, so that $\|\tau\| = 1$. By the dominated convergence theorem we obtain

$$\lim_{\theta \downarrow 0} \int_{\|x\| \le \theta^{-1}} (\tau' x)^2 \, d\tilde{P}(x) = E\{\tau'(X_1 - X_2)\}^2 = 2\sigma^2(\tau' X_1) \ge 2\rho_k,$$

and hence for sufficiently small $\theta_0 > 0$ and $\|t\| \le \theta_0$, we find

$$|\psi(t)|^2 \le 1 - \tfrac{1}{2}\rho_k \cos(1)\|t\|^2 \le \exp\{-\tfrac{1}{2}\rho_k \cos(1)\|t\|^2\}. \tag{A.5}$$

Let $U = (U_1, \ldots, U_k)$ be a random vector which is independent of $X_1, \ldots, X_N$ and which has i.i.d. components $U_1, \ldots, U_k$ with a common density $g(u) = (1 - \cos u)/(\pi u^2)$ and corresponding c.f.

$$\gamma(t) = E e^{itU_1} = (1 - |t|) 1_{[0,1]}(|t|).$$

Choose a_k such that $P(|U_1| \le a_k) = 2^{-1/k}$ and $\varepsilon \ge a_k k^{1/2}/(\theta_0 n^{1/2})$. It is clear that

$$\begin{aligned} P\big(\|S_n + n^{-1/2}\theta_0^{-1}U\| \le 2\varepsilon\big) &\ge P\big(\|U\| \le \varepsilon\theta_0 n^{1/2}\big) P\big(\|S_n\| \le \varepsilon\big) \\ &\ge P\big(\|U\| \le a_k k^{1/2}\big) P\big(\|S_n\| \le \varepsilon\big) \ge \tfrac{1}{2} P\big(\|S_n\| \le \varepsilon\big), \end{aligned}$$

and using (A.5) we arrive at

$$\begin{aligned} P\big(\|S_n\| \le \varepsilon\big) &\le 2P\big(\|S_n + n^{-1/2}\theta_0^{-1}U\| \le 2\varepsilon\big) \\ &\le \frac{2}{\pi^k} \int_{\mathbb{R}^k} \prod_{j=1}^k \left| \frac{\sin(2\varepsilon t_j)}{t_j} \right| |\psi(n^{-1/2}t)|^n \prod_{j=1}^k \gamma\left(\frac{t_j}{\theta_0 n^{1/2}} \right) dt \\ &\le \frac{2}{\pi^k} \int_{-\theta_0 n^{1/2}}^{\theta_0 n^{1/2}} \cdots \int_{-\theta_0 n^{1/2}}^{\theta_0 n^{1/2}} (2\varepsilon)^k \exp\left\{ -\frac{1}{4}\rho_k \cos(1)\|t\|^2 \right\} dt \\ &\le 2(2\varepsilon/\pi)^k \int_{\mathbb{R}^k} \exp\left\{ -\frac{1}{4}\rho_k \cos(1)\|t\|^2 \right\} dt = 2^{2k+1}\big(\pi\cos(1)\rho_k\big)^{-k/2}\varepsilon^k \end{aligned} \tag{A.6}$$

for all $|\varepsilon| \ge a_k k^{1/2}/(\theta_0 n^{1/2})$. For $|\varepsilon| < a_k k^{1/2}/(\theta_0 n^{1/2})$ (A.6) yields the trivial bound

$$\begin{aligned} P\big(\|S_n\| \le \varepsilon\big) &\le P\big(\|S_n\| \le a_k k^{1/2}/\big(\theta_0 n^{1/2}\big)\big) \\ &\le 2^{2k+1} \left[\frac{k a_k^2}{\pi \cos(1)\rho_k \theta_0^2} \right]^{k/2} n^{-k/2}. \end{aligned} \tag{A.7}$$

Addition of (A.6) and (A.7) proves the lemma for fixed P.

If we assume that $E\|X_1\|^{2+\delta} \le C$, then this implies that for every $\theta > 0$ and $\|\tau\| = 1$,

$$\int_{\|x\| > \theta^{-1}} (\tau' x)^2 \, d\tilde{P}(x) \le \int_{\|x\| > \theta^{-1}} \|x\|^2 \, d\tilde{P}(x) \le 2^{(2+\delta)} C \theta^{\delta}.$$

Returning to (A.4)–(A.5) we now see that we can choose

$$\theta_0 = 2^{-(2+\delta)/\delta} (\rho_k / C)^{1/\delta} \tag{A.8}$$

and ensure the validity of (A.5) for $\|t\| \le \theta_0$. Substituting (A.8) in (A.7) and assuming in addition that k and ρ_k^{-1} are bounded, we conclude that B in Lemma A.2 is also bounded and the proof of the lemma is complete. □

Acknowledgments. It is a pleasure to acknowledge the help of J. Bretagnolle which very much improved our proof of the concentration inequality of Lemma A.2. The authors are indebted to the Associate Editor and the referees for their constructive criticism.

REFERENCES

ALBERS, W., BICKEL, P. J. and VAN ZWET, W. R. (1976). Asymptotic expansions for the power of distribution-free tests in the one-sample problem. *Ann. Statist.* 4 108–156.

BICKEL, P. J. (1974). Edgeworth expansions in nonparametric statistics. *Ann. Statist.* **2** 1–20.

CALLAERT, H. and JANSSEN, P. (1978). The Berry–Esseen theorem for *U*-statistics. *Ann. Statist.* **6** 417–421.

CALLAERT, H., JANSSEN, P. and VERAVERBEKE, N. (1980). An Edgeworth expansion for *U*-statistics. *Ann. Statist.* **8** 299–312.

CHAN, Y.-K. and WIERMAN, J. (1977). On the Berry–Esseen theorem for *U*-statistics. *Ann. Probab.* **5** 136–139.

DHARMADHIKARI, S. W., FABIAN, V. and JOGDEO, K. (1968). Bounds on the moments of martingales. *Ann. Math. Statist.* **39** 1719–1723.

FELLER, W. (1971). *An Introduction to Probability Theory and Its Applications* **2**, 2nd ed. Wiley, New York.

GÖTZE, F. (1979). Asymptotic expansions for bivariate von Mises functionals. *Z. Wahrsch. verw. Gebiete* **50** 333–355.

HELMERS, R. and VAN ZWET, W. R. (1982). The Berry–Esseen bound for *U*-statistics. *Statistical Decision Theory and Related Topics, III* (S. S. Gupta and J. O. Berger, eds.) **1** 497–512. Academic, New York.

HOEFFDING, W. (1948). A class of statistics with asymptotically normal distributions. *Ann. Math. Statist.* **19** 293–325.

JANSSEN, P. L. (1978). De Berry–Esseen Stelling en een Asymptotische Ontwikkeling voor *U*-statistieken. Ph.D. thesis, Dept. of Mathematics, Catholic Univ. of Leuven.

P. J. BICKEL
DEPARTMENT OF STATISTICS
UNIVERSITY OF CALIFORNIA
BERKELEY, CALIFORNIA 94720

F. GÖTZE
DEPARTMENT OF MATHEMATICS
UNIVERSITY OF BIELEFELD
POSTFACH 8640
4800 BIELEFELD 12
FEDERAL REPUBLIC OF GERMANY

W. R. VAN ZWET
DEPARTMENT OF MATHEMATICS
UNIVERSITY OF LEIDEN
POSTBUS 9512
2300 RA LEIDEN
THE NETHERLANDS

Chapter 15
Entropic instability of Cramer's characterization of the normal law

S.G. Bobkov and G.P. Chistyakov and F. Götze

Abstract We establish instability of the characterization of the normal law in Cramer's theorem with respect to the total variation norm and the entropic distance. Two constructions of counter-examples are provided.

15.1 Introduction

A well-known theorem of Cramer (1936, [Cr]) indicates that, if the sum $X+Y$ of two independent random variables X and Y has a normal distribution, then necessarily both X and Y are normal. Soon after Cramér had proved his theorem (which answered a question raised by P. Lévy in 1931), P. Lévy established stability of this characterization property of normal distributions. In a qualitative form it states that, for independent random variables X and Y,

if $X+Y$ is nearly normal then both X and Y are nearly normal.

Here "nearly" is understood in the sense of the topology of weak convergence of probability distributions on the real line. For example, with respect to the Lévy distance, Lévy's theorem is formulated as follows. Given $\varepsilon > 0$ and distribution functions F_1, F_2,

Sergey G. Bobkov
School of Mathematics, University of Minnesota, 127 Vincent Hall, 206 Church St. S.E., Minneapolis, MN 55455, USA
e-mail: bobkov@math.umn.edu

Gennadiy P. Chistyakov
Fakultät für Mathematik, Universität Bielefeld, Postfach 100131, 33501 Bielefeld, Germany
e-mail: chistyak@math.uni-bielefeld.de

Friedrich Götze
Fakultät für Mathematik, Universität Bielefeld, Postfach 100131, 33501 Bielefeld, Germany
e-mail: goetze@math.uni-bielefeld.de

S. van de Geer and M. Wegkamp (eds.), *Selected Works of Willem van Zwet*, Selected Works in Probability and Statistics, DOI 10.1007/978-1-4614-1314-1_15, © Springer Science+Business Media, LLC 2012

$$L(F_1 * F_2, \Phi) < \varepsilon \;\Rightarrow\; L(F_1, \Phi_{a_1,\sigma_1}) < \delta_\varepsilon,\; L(F_2, \Phi_{a_2,\sigma_2}) < \delta_\varepsilon,$$

for some $a_1, a_2 \in \mathbf{R}$ and $\sigma_1, \sigma_2 > 0$, where δ_ε only depends on ε, and in a such way that $\delta_\varepsilon \to 0$, as $\varepsilon \to 0$.

Here $\Phi_{a,\sigma}$ stands for the distribution functions of the normal law $N(a, \sigma^2)$ with mean a and variance σ^2, and we omit indices in the standard case $a = 0$, $\sigma = 1$. As usual, $F_1 * F_2$ denotes the convolution of the distribution functions.

In 1950s Linnik [L2] extended this result to arbitrary probability distributions on the real line: If the convolution $F_1 * F_2$ is close to F, then both F_1 and F_2 have to be close to the class of all components of F. Linnik noted as well that Cramér's theorem may be viewed as a particular case of Darmois–Skitovich's theorem on the independence of independent linear statistics (cf. [L1]).

Another important issue which attracted many researchers is the problem of quantitative versions of the stability property of the normal law. This problem has been studied for a long time, starting with results by Sapogov in the 1950s [S1-2] (who considered the Kolmogorov distance and was apparently unaware of the work of P. Lévy) and ending with results by Chistyakov and Golinskii [C-G] in the 1990s, who found the correct asymptotics of the best possible error function $\varepsilon \to \delta_\varepsilon$ for the Lévy distance. See also [Z], [Se].

In this note we address the following natural question in connection with Lévy's theorem. Given independent random variables X and Y, assume that the distribution of $X + Y$ is known to be nearly normal in a stronger sense. What does this imply for X and Y in terms of closeness to the normal? When saying "stronger", we mean classical distances between distributions such as the total variation norm $\|F - G\|_{\mathrm{TV}}$, or the entropic distance $D(X)$ from a given distribution F of X to the associated normal law. Thus, we wonder whether or not X and Y need to be nearly normal with respect to these distances. In case of the entropic distance, this question was raised in the mid 1960's by Kac and McKean ([MC], pp. 365–366; cf. also [C-S] for some related aspects of the problem).

As it turns out, in general the answer is negative in both cases.

Theorem 1. *For any $\varepsilon > 0$, there exist independent random variables X and Y with absolutely continuous symmetric distributions F_1, F_2, and with* $\mathrm{Var}(X) = \mathrm{Var}(Y) = 1$, *such that*

a) $\|F_1 * F_2 - \Phi * \Phi\|_{\mathrm{TV}} < \varepsilon$;

b) $\|F_1 - \Phi_{a,\sigma}\|_{\mathrm{TV}} > c$ *and* $\|F_2 - \Phi_{a,\sigma}\|_{\mathrm{TV}} > c$, *for all* $a \in \mathbf{R}$ *and* $\sigma > 0$,

where $c > 0$ denotes an absolute constant.

As we will see, Theorem 1 holds for any number $c \in (0, 1/2)$.

The statement of the theorem may be strengthened in terms of the entropic distance. Recall that, if a random variable X with finite second moment has a density $p(x)$, its entropy

$$h(X) = -\int_{-\infty}^{+\infty} p(x) \log p(x)\, dx$$

is well-defined and, what is classical, it is bounded from above by the entropy of the normal random variable Z, having the same variance $\sigma^2 = \mathrm{Var}(Z) = \mathrm{Var}(X)$. The entropic distance to the normal is given by the formula

$$D(X) = h(Z) - h(X) = \int_{-\infty}^{+\infty} p(x) \log \frac{p(x)}{\varphi_{a,\sigma}(x)}\, dx,$$

where $\varphi_{a,\sigma}$ stands for the density of the normal law $N(a,\sigma^2)$ with parameters $a = \mathbb{E}X$, $\sigma^2 = \mathrm{Var}(X)$. Alternatively, it may be described as the shortest distance from the distribution F of X to the family of all normal laws on the line in the sense of the Kullback-Leibler distance.

Similarly to the total variation, the quantity $D(X)$ is homogeneous of order zero with respect to X, that is, $D(\lambda X) = D(X)$, for all $\lambda > 0$. In particular, it does not depend on the variance of X. The two distances are related by virtue of the Pinsker-Csiszár-Kullback inequality ([P], [Cs], [K]), which gives

$$D(X) \geq \frac{1}{2} \|F - \Phi_{a,\sigma}\|_{\mathrm{TV}}^2.$$

In this sense the entropic distance is stronger than the total variation. Therefore, one may wonder whether or not the stability property in Cramer's theorem still holds when replacing the Lévy distance with the entropic distance. If so, this could also be viewed as the inverse to the concavity of the entropy functional (or to the so-called entropy power inequality, cf. [D-C-T]), which implies that

$$D(X+Y) \leq \frac{D(X) + D(Y)}{2},$$

whenever X and Y are independent and have equal variances.

It turns out however, this is not the case.

Theorem 2. *For any $\varepsilon > 0$, there exist independent random variables X and Y with absolutely continuous symmetric distributions F_1, F_2, and with* $\mathrm{Var}(X) = \mathrm{Var}(Y) = 1$, *such that*

a) $D(X+Y) < \varepsilon$;

b) $\|F_1 - \Phi_{a,\sigma}\|_{\mathrm{TV}} > c$ *and* $\|F_2 - \Phi_{a,\sigma}\|_{\mathrm{TV}} > c$, *for all $a \in \mathbf{R}$ and $\sigma > 0$,*

where $c > 0$ denotes an absolute constant. In particular, both $D(X)$ and $D(Y)$ are separated from zero.

In the next section we describe how such random variables may be constructed. In fact, our (counter-)examples for Theorem 1 still work for Theorem 2. We consider two constructions. The first one explicitly specifies densities for X and Y, while the other one deals with their distribution functions, which are explicitly provided, as well.

In Section 3 we show that the distributions of X and Y are separated from the normal law, thus proving claim $b)$ of Theorem 1. Finally, in Section 4 we provide computations for the convolutions, which will justify claim $a)$ of Theorem 1.

15.2 Constructions of examples

In this section we describe two types of the construction of random variables. We use the standard notations

$$\varphi(x) = \frac{1}{\sqrt{2\pi}}\, e^{-x^2/2}, \quad \Phi(x) = \int_{-\infty}^{x} \varphi(y)\, dy \qquad (x \in \mathbf{R})$$

for the density and the distribution function of the standard normal law.

Construction I (by an explicit formula for densities).
Given $T > 0$, let X_T be a random variable with density function

$$p_T(x) = c_T \sin^2(Tx)\, \varphi(x), \qquad x \in \mathbf{R},$$

where $c_T = 2/(1 - e^{-2T^2})$ is the normalizing constant. Introduce a further random variable, X_{2T}, independent of X_T, with density p_{2T}.

Clearly, X_T has a symmetric distribution with

$$\mathbb{E}X_T^2 = \frac{c_T}{2}\left(1 - (1 - 4T^2)e^{-2T^2}\right) \to 1, \quad \text{as } T \to +\infty.$$

Based on this choices, in the proof of Theorems 1–2 we consider

$$X = \frac{X_T}{\sqrt{\mathbb{E}X_T^2}}, \quad Y = \frac{X_{2T}}{\sqrt{\mathbb{E}X_{2T}^2}}$$

for large values of T.

Note that we may rewrite our densities as

$$p_T(x) = \frac{c_T}{2}\, (\varphi(x) - \cos(2Tx)\, \varphi(x)).$$

As another variant one may also consider densities of the form

$$p(x) = \varphi(x) + \sin(Tx)\, \varphi(x),$$

which are somewhat simpler. However, they are not symmetric about the origin.

Construction II (by an explicit formula for distribution functions).
Given $T > 0$, let X_T be a random variable with the distribution function

$$F_T(x) = \Phi(x) + \frac{1}{2T}\sin(Tx)\,\varphi(x)\,1_{\{|x|<T\}}.$$

Their densities are given by

$$p_T(x) = \varphi(x) + \frac{1}{2}\left(\cos(Tx) - \frac{x}{T}\sin(Tx)\right)\varphi(x)\,1_{\{|x|<T\}}.$$

Clearly, $p_T(x) > 0$ everywhere (perhaps except for $|x| = T$), so F_T is increasing. Since also $F_T(-\infty) = 0$, $F_T(+\infty) = 1$, F_T is indeed a distribution function. Note that p_T is even, so the distribution of X_T is symmetric about the origin.

Again introduce a second independent random variable X_{2T} with the distribution function F_{2T}.

To see that $\mathrm{Var}(X_T) \to 1$, as $T \to +\infty$, we may apply well-known identities which can be obtained by the successive differentiation of the identity $\int_{-\infty}^{+\infty}\cos(Tx)\,\varphi(x)\,dx = e^{-T^2/2}$ with respect to the variable T:

1) $\int_{-\infty}^{+\infty} x\sin(Tx)\,\varphi(x)\,dx = T\,e^{-T^2/2}$,

2) $\int_{-\infty}^{+\infty} x^2\cos(Tx)\,\varphi(x)\,dx = (1-T^2)\,e^{-T^2/2}$,

3) $\int_{-\infty}^{+\infty} x^3\sin(Tx)\,\varphi(x)\,dx = (3T-T^3)\,e^{-T^2/2}$.

Write

$$\mathbb{E}X_T^2 = 1 + \frac{1}{2}\int_{-T}^{T} x^2\left(\cos(Tx) - \frac{x}{T}\sin(Tx)\right)\varphi(x)\,dx.$$

By 2)–3), extending integration to the whole line, we get that

$$\mathbb{E}X_T^2 = 1 - e^{-T^2/2} - \frac{1}{2}\int_{\{|x|>T\}} x^2\left(\cos(Tx) - \frac{x}{T}\sin(Tx)\right)\varphi(x)\,dx.$$

Clearly, the last integral tends to zero.

Based on this choices, for the proof of Theorems 1–2 one may similarly take

$$X = \frac{X_T}{\sqrt{\mathbb{E}X_T^2}}, \qquad Y = \frac{X_{2T}}{\sqrt{\mathbb{E}X_{2T}^2}}$$

for large values of T.

Although seemingly more complicated, the second construction is more convenient, when measuring the distance to the normal for metrics, such as Lévy and Kantorovich-Rubinshtein, which explicitly involve distribution functions (rather than densities).

15.3 Separation from the normal

The distributions F_T of X_T, constructed in the previous section, are close to the standard normal in the sense of the topology of weak convergence. To see this, let us look at the characteristic functions for the distributions from Construction I:

$$\begin{aligned} f_T(t) = \mathbb{E}\, e^{itX_T} &= \int_{-\infty}^{+\infty} e^{itx} p_T(x)\,dx \\ &= \frac{c_T}{2}\int_{-\infty}^{+\infty} \cos(tx)\,(1-\cos(2Tx))\,\varphi(x)\,dx \\ &= \frac{c_T}{2}\int_{-\infty}^{+\infty} \Big(\cos(tx) - \frac{\cos((t+2T)x)+\cos((t-2T)x)}{2}\Big)\,\varphi(x)\,dx \\ &= \frac{c_T}{2}\Big(e^{-t^2/2} - \frac{e^{-(t+2T)^2/2}+e^{-(t-2T)^2/2}}{2}\Big). \end{aligned}$$

Hence, for any fixed real t,

$$f_T(t) = \frac{1}{1-e^{-2T^2/2}}\Big(e^{-t^2/2} - \frac{e^{-(t+2T)^2/2}+e^{-(t-2T)^2/2}}{2}\Big) \to e^{-t^2/2},$$

and thus weakly in distribution

$$X_T \Rightarrow N(0,1), \quad \text{as} \quad T \to +\infty.$$

By a compactness argument, it is easy to see that $\rho(F_T,\Phi) \to 0$, for any metric metrizing the weak convergence in the space of all probability distributions on the line. If the second moments of distributions are known to be bounded, one may use, for example, the Kantorovich-Rubinshtein distance, which in our case is given by

$$W_1(F_T,\Phi) = \int_{-\infty}^{+\infty} |F_T(x) - \Phi(x)|\,dx.$$

By the very definition of the distributions from Construction II, we obtain immediately that $W_1(F_T,\Phi) < \frac{1}{2T}$.

As a consequence, the normalized random variables X and Y are also close to the standard normal law for the Kantorovich-Rubinshtein metric.

On the other hand, let us look at the total variation distance. One may apply the general elementary estimate

$$\sup_{t\in\mathbf{R}} |f(t) - g(t)| \le \|F-G\|_{\mathrm{TV}},$$

holding for arbitrary probability distributions F and G on the real line with characteristic functions f and g, respectively. In particular, for the distributions from the first construction (choosing $t = 2T$), we have

$$\|F_T - \Phi\|_{\rm TV} \geq \sup_{t \in \mathbf{R}} |f_T(t) - e^{-t^2/2}| \geq |f_T(2T) - e^{-2T^2}| \to \frac{1}{2},$$

as $T \to +\infty$. Hence,

$$\liminf_{T \to +\infty} \|F_T - \Phi\|_{\rm TV} \geq \frac{1}{2}.$$

This observation can be strengthened by considering the shortest total variation distance from F_T to the class of all normal laws on the line.

Lemma 1. *We have*

$$\liminf_{T \to +\infty} \inf_{a,\sigma} \|F_T - \Phi_{a,\sigma}\|_{\rm TV} \geq \frac{1}{2}.$$

Proof. As was discussed above, we may use the bounds

$$\begin{aligned} \|F_T - \Phi_{a,\sigma}\|_{\rm TV} &\geq \sup_{t \in \mathbf{R}} \left| f_T(t) - e^{iat - \sigma^2 t^2/2} \right| \\ &\geq \sup_{t \in \mathbf{R}} \left| |f_T(t)| - e^{-\sigma^2 t^2/2} \right|. \end{aligned}$$

It follows from the formula for $f_T(t)$ that uniformly over all $t \geq 0$,

$$\left| f_T(t) - e^{-\sigma^2 t^2/2} \right| \geq \left| |e^{-t^2/2} - \frac{1}{2}\, e^{-(t-2T)^2/2}| - e^{-\sigma^2 t^2/2} \right| - o(T),$$

as $T \to +\infty$, so

$$\liminf_{T \to +\infty} \inf_{a,\sigma} \|F_T - \Phi_{a,\sigma}\|_{\rm TV} \geq \liminf_{T \to +\infty} \inf_{\sigma > 0} \sup_{t > 0} \left| |e^{-t^2/2} - \frac{1}{2}\, e^{-(t-2T)^2/2}| - e^{-\sigma^2 t^2/2} \right|.$$

Here and in the sequel, $o(T)$ denotes a quantity which tends to zero, as $T \to +\infty$, uniformly over all t from the indicated range.

To estimate the supremum on the right-hand side uniformly over all $\sigma > 0$, fix a (large) number N. In case $\sigma \geq N/T$, choose $t = 2T$, which gives

$$\left| |e^{-t^2/2} - \frac{1}{2}\, e^{-(t-2T)^2/2}| - e^{-\sigma^2 t^2/2} \right| = \frac{1}{2} - e^{-2\sigma^2 T^2} + o(T) \geq \frac{1}{2} - e^{-2N^2} + o(T).$$

In case $\sigma < N/T$, choose $t = 2\sqrt{T}$, which gives

$$\left| |e^{-t^2/2} - \frac{1}{2}\, e^{-(t-2T)^2/2}| - e^{-\sigma^2 t^2/2} \right| = e^{-2\sigma^2 T} + o(T) \geq e^{-2N^2/T} + o(T),$$

where the right-hand side tends to 1, as $T \to +\infty$. Altogether this yields

$$\liminf_{T\to+\infty}\inf_{a,\sigma}\|F_T-\Phi_{a,\sigma}\|_{\rm TV}\geq\frac{1}{2}-e^{-2N^2}.$$

Since the left-hand side does not depend on N, we may let $N\to+\infty$, and the lemma follows.

As we mentioned in the previous section, the random variables X and Y in Theorems 1–2 are obtained from X_T and X_{2T} by normalizing, so that $\mathrm{Var}(X)=\mathrm{Var}(Y)$. Since the total variation norm is invariant under rescaling of the coordinates, Lemma 1 also implies that,

$$\liminf_{T\to+\infty}\inf_{a,\sigma}\|F-\Phi_{a,\sigma}\|_{\rm TV}\geq\frac{1}{2},\qquad\liminf_{T\to+\infty}\inf_{a,\sigma}\|G-\Phi_{a,\sigma}\|_{\rm TV}\geq\frac{1}{2},$$

where F and G denote distributions of X and Y (which also depend on T).

Recalling also Pinsker-Csiszár-Kullback's inequality, we may conclude the property $b)$ in these theorems.

Conclusion 1. For random variables X and Y from Construction I, we have

$$\|F-\Phi_{a,\sigma}\|_{\rm TV}>c,\qquad\|G-\Phi_{a,\sigma}\|_{\rm TV}>c,$$

for all T large enough, where c is any prescribed number in $(0,1/2)$. In particular, $D(X)>c^2/4$ and $D(Y)>c^2/4$.

A similar approach may be used to study the distributions F_T from the second construction. The corresponding characteristic functions are given by

$$\begin{aligned}f_T(t)&=e^{-t^2/2}+\frac{1}{2}\int_{-T}^{T}e^{itx}\left(\cos(Tx)-\frac{x}{T}\sin(Tx)\right)\varphi(x)\,dx\\&=e^{-t^2/2}+\frac{e^{-(t+T)^2/2}+e^{-(t-T)^2/2}}{4}\\&\quad-\frac{1}{2}\int_{|x|>T}e^{itx}\cos(Tx)\,\varphi(x)\,dx-\frac{1}{2T}\int_{-T}^{T}e^{itx}\,x\sin(Tx)\,\varphi(x)\,dx.\end{aligned}$$

Clearly, the first integral is bounded in absolute value by $2(1-\Phi(T))<e^{-T^2/2}$, while the absolute value of the second integral is smaller than $\int|x|\,\varphi(x)\,dx<1$. Hence, uniformly over all $t\in\mathbf{R}$

$$f_T(t)=e^{-t^2/2}+\frac{e^{-(t+T)^2/2}+e^{-(t-T)^2/2}}{4}+o(T),$$

as $T\to+\infty$. Next one can repeat the line of arguments from the proof of Lemma 1.

Conclusion 2. For the random variables X and Y of Construction II, Conclusion 1 holds with level $1/4$ replacing $1/2$ (for constants c).

15.4 Convolutions of distributions from Construction I

Write the density of random variables X_T from Construction I in the form

$$p_T(x) = \frac{c_T}{2}\left(\varphi(x) - \cos(2Tx)\,\varphi(x)\right),$$

where $c_T = 2/(1-e^{-2T^2})$ is the normalizing constant. Note that $\frac{c_T}{2} \to 1$, as $T \to +\infty$.

Instead of the sum $X+Y$ (which is a bit more complicated), we consider the sum $X_T + X_{2T}$ of two independent random variables, assuming that X_T has density p_T and X_{2T} has density p_{2T}. The density of this sum represents the convolution $p_T * p_{2T}$.

In analogy with notations for distribution functions, for integrable functions $p(x)$ and $q(x)$ we write $(p*q)(x) = p(x)*q(x) = \int_{-\infty}^{+\infty} p(x-y)q(y)\,dy$.

To simplify the computations, introduce

$$q_T(x) = \varphi(x) - \cos(2Tx)\,\varphi(x)$$

and write

$$\begin{aligned}(q_T * q_{2T})(x) - (\varphi*\varphi)(x) = &-\varphi(x) * \left[(\cos(2Tx)+\cos(4Tx))\,\varphi(x)\right] \\ &+ \left[\cos(2Tx)\,\varphi(x)\right] * \left[\cos(4Tx))\,\varphi(x)\right].\end{aligned}$$

Note that

$$(\varphi*\varphi)(x) = \frac{1}{2\sqrt{\pi}}\, e^{-x^2/4}.$$

To compute convolutions, we need one simple relation. Given a complex variable α, consider the integral

$$\frac{1}{\sqrt{\pi}} \int_{-\infty}^{+\infty} e^{-\frac{(x-y)^2+y^2}{2}}\, e^{\alpha y}\,dy.$$

Changing the variable $y = \frac{x}{2} - \frac{t}{\sqrt{2}}$, we obtain $\frac{(x-y)^2+y^2}{2} = \frac{x^2}{4} + \frac{t^2}{2}$, and the integral becomes

$$\frac{1}{\sqrt{\pi}}\,\frac{1}{\sqrt{2}}\, e^{-x^2/4}\, e^{\alpha x/2} \int_{-\infty}^{+\infty} e^{-\alpha t/\sqrt{2}}\, e^{-t^2/2}\,dt = e^{-x^2/4}\, e^{\alpha x/2+\alpha^2/4}.$$

Therefore,

$$\int_{-\infty}^{+\infty} \varphi(x-y)\,\varphi(y)\,e^{\alpha y}\,dy = (\varphi*\varphi)(x)\, e^{\alpha x/2+\alpha^2/4}.$$

Taking $\alpha = iT$, we get

$$A_T \equiv \int_{-\infty}^{+\infty} \varphi(x-y)\,\varphi(y)\,\cos(Ty)\,dy = (\varphi*\varphi)(x)\,e^{-T^2/4}\,\cos(Tx/2),$$

$$B_T \equiv \int_{-\infty}^{+\infty} \varphi(x-y)\,\varphi(y)\,\sin(Ty)\,dy = (\varphi*\varphi)(x)\,e^{-T^2/4}\,\sin(Tx/2).$$

Hence, the convolution $\varphi(x) * [(\cos(2Tx)+\cos(4Tx))\,\varphi(x)]$ is given by

$$\int_{-\infty}^{+\infty} \varphi(x-y)\,\varphi(y)\,(\cos(2Ty)+\cos(4Ty))\,dy = A_{2T}+A_{4T}.$$

Similarly, the convolution $[\cos(2Tx)\,\varphi(x)] * [\cos(4Tx))\,\varphi(x)]$ is

$$\begin{aligned}&\textstyle\int_{-\infty}^{+\infty} \varphi(x-y)\,\varphi(y)\cos(2T(x-y))\cos(4Ty))\,dy\\ &\quad= \textstyle\int_{-\infty}^{+\infty} \varphi(x-y)\,\varphi(y)\,\frac{\cos(2Tx-6Ty)+\cos(2Tx+2Ty)}{2}\,dy\\ &\qquad= \tfrac{1}{2}\,(A_{6T}\cos(2Tx)+A_{2T}\cos(2Tx)+B_{6T}\sin(2Tx)-B_{2T}\sin(2Tx)).\end{aligned}$$

Collecting the two convolutions together, we obtain for $(q_T * q_{2T})(x)$ the representation

$$(\varphi*\varphi)(x) - A_{2T} - A_{4T} + \frac{A_{6T}+A_{2T}}{2}\cos(2Tx) + \frac{B_{6T}-B_{2T}}{2}\sin(2Tx).$$

Now, using the obvious bound $|A_{2T}| \le (\varphi*\varphi)(x)\,e^{-T^2}$ and similarly for B_{2T}, we arrive at

$$\left|\frac{(q_T*q_{2T})(x)}{(\varphi*\varphi)(x)} - 1\right| \le 4e^{-T^2}.$$

But $p_T * p_{2T} = (1+\varepsilon_T)\,q_T * q_{2T}$, where $\varepsilon_T = \frac{1}{4}\,c_T c_{2T} - 1 \to 0$, as $T \to +\infty$, and moreover $|\varepsilon_T| \le Ce^{-2T^2}$, whenever $T \ge 1$. Hence, we get:

Lemma 2. *For all* $T \ge 1$ *and* $x \in \mathbf{R}$,

$$\left|\frac{(p_T*p_{2T})(x)}{(\varphi*\varphi)(x)} - 1\right| \le Ce^{-T^2}$$

with some absolute constant C.

This estimate is quite sufficient to see that

$$\|F_T * F_{2T} - \Phi*\Phi\|_{\mathrm{TV}} \le Ce^{-T^2}$$

and also for the Kullback-Leibler's distance

$$D(X_T+X_{2T}||Z) = \int_{-\infty}^{+\infty} (p_T*p_{2T})(x)\log\frac{(p_T*p_{2T})(x)}{(\varphi*\varphi)(x)}\,dx \to 0,$$

as $T \to +\infty$, where $Z \sim N(0,2)$. So the (closest) entropic distance to the normal

$$D(X_T + X_{2T}) \to 0.$$

A similar property, $D(X+Y) \to 0$, as $T \to +\infty$, also holds for normalized random variables, since $\mathrm{Var}(X_T) \to 1$, although this conclusion requires a certain justification. What is needed is the property

$$D(\alpha_T X_T + X_{2T}) \to 0$$

where $\alpha_T \to 1$. This may be done, for example, by a slight modification of the arguments used in the proof of Lemma 2. With this in mind Theorems 1–2 are proved.

We leave it to the reader to check that the same conclusion is true for probability distributions from Construction II.

References

[C-S] Carlen, E. A., Soffer, A. Entropy production by block variable summation and central limit theorems. Comm. Math. Phys. 140 (1991), no. 2, 339–371.

[C-G] Chistyakov, G. P., Golinskii, L. B. Order-sharp estimates for the stability of decompositions of the normal distribution in the Levy metric. (Russian) Translated in J. Math. Sci. 72 (1994), no. 1, 2848–2871. Stability problems for stochastic models (Russian) (Moscow, 1991), 16–40, Vsesoyuz. Nauchno-Issled. Inst. Sistem. Issled., Moscow, 1991.

[Cr] Cramér, H. Ueber eine Eigenschaft der Normalen Verteilungsfunktion. Math. Zeitschrift, Bd. 41 (1936), 405–414.

[Cs] Csiszár, I. Information-type measures of difference of probability distributions and indirect observations. Studia Sci. Math. Hungar., 2 (1967), 299–318.

[D-C-T] Dembo, A., Cover, T. M., Thomas, J. A. Information-theoretic inequalities. IEEE Trans. Inform. Theory, 37 (1991), no. 6, 1501–1518.

[K] Kullback, S. A lower bound for discrimination in terms of variation. IEEE Trans. Inform. Theory, T-13, 1967, 126–127.

[L1] Linnik, Yu. V. A remark on Cramer's theorem on the decomposition of the normal law. (Russian) Teor. Veroyatnost. i Primenen. 1 (1956), 479–480.

[L2] Linnik, Yu. V. General theorems on the factorization of infinitely divisible laws. III. Sufficient conditions (countable bounded Poisson spectrum; unbounded spectrum; "stability"). Theor. Probability Appl. 4 (1959), 142–163.

[MK] McKean, H. P., Jr. Speed of approach to equilibrium for Kac's caricature of a Maxwellian gas. Arch. Rational Mech. Anal. 21 (1966), 343–367.

[P] Pinsker, M. S. Information and information stability of random variables and processes. Translated and edited by Amiel Feinstein Holden-Day, Inc., San Francisco, Calif.-London-Amsterdam, 1964, xii+243 pp.

[S1] Sapogov, N. A. The stability problem for a theorem of Cramér. (Russian) Izvestiya Akad. Nauk SSSR. Ser. Mat. 15 (1951), 205–218.

[S2] Sapogov, N. A. The problem of stability for a theorem of Cramér. (Russian) Vestnik Leningrad. Univ. 10 (1955), no. 11, 61–64.

[Se] Senatov, V. V. Refinement of estimates of stability for a theorem of H. Cramér. (Russian) Continuity and stability in problems of probability theory and mathematical statistics. Zap. Naucn. Sem. Leningrad. Otdel. Mat. Inst. Steklov. (LOMI) 61 (1976), 125–134.

[Z] Zolotarev, V. M. On the problem of the stability of the decomposition of the normal law into components. (Russian) Teor. Verojatnost. i Primenen., 13 (1968), 738–742.

Part IV
Resampling

The Annals of Statistics
1996, Vol. 24, No. 6, 2297–2318

RESAMPLING: CONSISTENCY OF SUBSTITUTION ESTIMATORS[1]

By Hein Putter and Willem R. van Zwet
University of Leiden and University of North Carolina, Chapel Hill

On the basis of N i.i.d. random variables with a common unknown distribution P we wish to estimate a functional $\tau_N(P)$. An obvious and very general approach to this problem is to find an estimator $\hat{P}_N$ of P first, and then construct a so-called substitution estimator $\tau_N(\hat{P}_N)$ of $\tau_N(P)$. In this paper we investigate how to choose the estimator $\hat{P}_N$ so that the substitution estimator $\tau_N(\hat{P}_N)$ will be consistent.

Although our setup covers a broad class of estimation problems, the main substitution estimator we have in mind is a general version of the bootstrap where resampling is done from an estimated distribution $\hat{P}_N$. We do not focus in advance on a particular estimator $\hat{P}_N$, such as, for example, the empirical distribution, but try to indicate which resampling distribution should be used in a particular situation. The conclusion that we draw from the results and the examples in this paper is that the bootstrap is an exceptionally flexible method which comes into its own when full use is made of its flexibility. However, the choice of a good bootstrap method in a particular case requires rather precise information about the structure of the problem at hand. Unfortunately, this may not always be available.

1. Substitution estimators. Let $(\mathscr{X}, \mathscr{A})$ be a measurable space and let $\mathscr{P}$ be a collection of probability measures on $(\mathscr{X}, \mathscr{A})$. Let Π be a topology on $\mathscr{P}$, so that $(\mathscr{P}, \Pi)$ is a topological space. Finally, let $X_1, X_2, \dots$ denote a sequence of i.i.d. random variables with values in $\mathscr{X}$ and (unknown) common distribution $P \in \mathscr{P}$.

For $N = 1, 2, \dots,$ we consider a map $\tau_N: (\mathscr{P}, \Pi) \to (\mathscr{R}, r)$, where $(\mathscr{R}, r)$ is a metric space. Both spaces $(\mathscr{P}, \Pi)$ and $(\mathscr{R}, r)$ are equipped with the σ-algebra of Borel sets $\mathscr{B}(\mathscr{P}, \Pi)$ and $\mathscr{B}(\mathscr{R}, r)$, which are generated by the open sets in $(\mathscr{P}, \Pi)$ and $(\mathscr{R}, r)$, respectively. Probability distributions on these spaces are probability measures on the Borel sets and are induced by measurable maps from $(\mathscr{X}^\infty, \mathscr{A}^\infty, P^\infty)$ to $(\mathscr{P}, \mathscr{B}(\mathscr{P}, \Pi))$ or $(\mathscr{R}, \mathscr{B}(\mathscr{R}, r))$. We assume throughout that each τ_N is measurable.

Having observed the i.i.d. sample $X_1, \dots, X_N$ with common distribution $P \in \mathscr{P}$, our aim is to estimate the somewhat abstract $\mathscr{R}$-valued "parameter"

Received October 1994; revised December 1995.

[1]This paper presents the second part of the 1992 Wald Memorial Lectures. The first part of these lectures is the subject of a companion paper [van Zwet (1996)]. Research was supported by the Netherlands Organization for Scientific Research (NWO) and the Sonderforschungsbereich 343 "Diskrete Strukturen in der Mathematik" at the University of Bielefeld, Germany.

AMS 1991 *subject classifications.* Primary 62G09; secondary 62F12.
Key words and phrases. Resampling, bootstrap, substitution estimator, consistency, set of first category, equicontinuity, local uniform convergence.

S. van de Geer and M. Wegkamp (eds.), *Selected Works of Willem van Zwet*, Selected Works in Probability and Statistics, DOI 10.1007/978-1-4614-1314-1_16, © Springer Science+Business Media, LLC 2012

$\tau_N(P)$. For a measurable map $t_N\colon \mathscr{X}^N \to \mathscr{R}$, let $T_N = t_N(X_1, \ldots, X_N)$ be an estimator of $\tau_N(P)$ based on $X_1, \ldots, X_N$. We shall say that T_N is a *consistent estimator* of $\tau_N(P)$ for $P \in \mathscr{P}$ if

$$(1.1) \qquad r(T_N, \tau_N(P)) \to_P 0 \quad \text{for every } P \in \mathscr{P},$$

where $\to_P$ indicates convergence in probability under P as $N \to \infty$. The more formally inclined reader should view this expression as shorthand for the correct but laborious statement that the sequence $\{T_N\}_{N=1}^{\infty}$ is a consistent sequence of estimators of the sequence $\{\tau_N(P)\}_{N=1}^{\infty}$. If we wish to stress the role of the metric r in (1.1), we call T_N *r-consistent*.

In the absence of any special structural properties of $\tau_N(P)$, a popular estimator of $\tau_N(P)$ is the *substitution estimator* $\tau_N(\hat{P}_N)$. (This is commonly called a "plug-in estimator," but this expression is of the same sad grammatical level as "see-through clothes.") It is obtained by first estimating P by $\hat{P}_N = p_N(X_1, \ldots, X_N)$ for a measurable map $p_N\colon \mathscr{X}^N \to \mathscr{P}$ and then substituting this estimator in τ_N. We shall call the estimator $\hat{P}_N$ consistent with respect to the topology $\boldsymbol{\Pi}$ (*$\boldsymbol{\Pi}$-consistent*) if for every $P \in \mathscr{P}$ and every neighborhood U of P,

$$(1.2) \qquad P^N\big(\hat{P}_N \in U\big) \to 1 \quad \text{as } N \to \infty.$$

In the particular applications we have in mind, the topology $\boldsymbol{\Pi}$ on $\mathscr{P}$ will often be metrized by a metric ρ, so that the topological space $(\mathscr{P}, \boldsymbol{\Pi})$ is a metric space $(\mathscr{P}, \rho)$. Consistency of $\hat{P}_N$ will then be *ρ-consistency*, defined by

$$(1.3) \qquad \rho\big(\hat{P}_N, P\big) \to_P 0 \quad \text{for every } P \in \mathscr{P}.$$

We shall study the consistency of $\tau_N(\hat{P}_N)$ as an estimator of $\tau_N(P)$, assuming that $\hat{P}_N$ is a consistent estimator of P.

The metric ρ in (1.3) will often be the Hellinger metric H on $\mathscr{P}$. Recall that for P, $Q \in \mathscr{P}$ with densities f and g with respect to a common σ-finite measure μ on $(\mathscr{X}, \mathscr{A})$, the Hellinger distance H of P and Q is defined by

$$(1.4) \qquad H(P, Q) = \left\{ \int_{\mathscr{X}} \left(f^{1/2} - g^{1/2}\right)^2 d\mu \right\}^{1/2}.$$

Note that this definition does not depend on the choice of the dominating measure μ and that H is indeed a metric on $\mathscr{P}$. If $\rho = H$, (1.3) becomes $H(\hat{P}_N, P) \to_P 0$ for every $P \in \mathscr{P}$ and we say that $\hat{P}_N$ is *Hellinger-consistent*. We call $\hat{P}_N$ a *$\sqrt{N}$-Hellinger-consistent estimator* of P when

$$(1.5) \qquad H\big(\hat{P}_N, P\big) = \mathscr{O}_P(N^{-1/2}) \quad \text{for every } P \in \mathscr{P},$$

which means that for every $P \in \mathscr{P}$ and $\varepsilon > 0$ there exists a $C > 0$ such that

$$P^N\Big(H\big(\hat{P}_N, P\big) \geq CN^{-1/2}\Big) \leq \varepsilon \quad \text{for all } N.$$

Many results in asymptotic statistics do not hold for all underlying distributions $P \in \mathscr{P}$, but only for $P \in \mathscr{P} \setminus D$, where the exceptional set D is in some sense small compared to $\mathscr{P}$. For a finite dimensional parametric family $\mathscr{P} = \{P_\theta\colon \theta \in \Theta\}$ with $\Theta \subset \mathbb{R}^k$, we may identify $\mathscr{P}$ with Θ, and the exceptional subset of Θ will typically be small in the sense that it has Lebesgue measure zero. On the more general spaces of distributions $\mathscr{P}$ that we consider in this paper, there is no obvious analogue of Lebesgue measure for which "small" sets can naturally be described as sets of measure zero. Moreover, our formulation of the consistency problem as well as our proofs of the results are largely topological rather than measure theoretic. It is therefore hardly surprising that the exceptional set D in our results will be small in a topological sense: D will be a *set of the first category* in $(\mathscr{P}, \boldsymbol{\Pi})$. We recall that a set of the first category is a countable union of nowhere dense sets, and that a set is *nowhere dense* in $(\mathscr{P}, \boldsymbol{\Pi})$ if its closure does not contain an open set in $(\mathscr{P}, \boldsymbol{\Pi})$.

We begin our study of the consistency of substitution estimators with an elementary observation. Suppose that the sequence $\tau_N\colon (\mathscr{P}, \boldsymbol{\Pi}) \to (\mathscr{R}, r)$ is *equicontinuous* on $\mathscr{P}$, that is, for every $P \in \mathscr{P}$ and $\varepsilon > 0$ there exists a neighborhood U_ε of P such that $r(\tau_N(P), \tau_N(Q)) < \varepsilon$ for all $Q \in U_\varepsilon$ and $N = 1, 2, \ldots$. Then consistency of $\hat{P}_N$ clearly implies consistency of $\tau_N(\hat{P}_N)$, since for every $P \in \mathscr{P}$ and $\varepsilon > 0$,

$$(1.6)\qquad P^N\Big(r\big(\tau_N\big(\hat{P}_N\big), \tau_N(P)\big) \geq \varepsilon\Big) \leq P^N\big(\hat{P}_N \notin U_\varepsilon\big) \to 0$$

as $N \to \infty$. Trivial though this observation may be, we shall dignify it by including it among the four theorems in this section.

THEOREM 1.1. *Suppose the following statements hold*:

(i) *The sequence of maps* $\tau_N\colon (\mathscr{P}, \boldsymbol{\Pi}) \to (\mathscr{R}, r)$ *is equicontinuous on* $\mathscr{P}$.

(ii) *There exists an estimator* $\hat{P}_N = p_N(X_1, \ldots, X_N)$ *of* P *with values in* $\mathscr{P}$, *which is* $\boldsymbol{\Pi}$*-consistent for* $P \in \mathscr{P}$.

Then $\tau_N(\hat{P}_N)$ *is an* r*-consistent estimator of* $\tau_N(P)$; *thus,*

$$(1.7)\qquad r\big(\tau_N\big(\hat{P}_N\big), \tau_N(P)\big) \to_P 0 \quad \textit{for every } P \in \mathscr{P}.$$

We can push this argument a little bit further by assuming that $\mathscr{P} = \bigcup_{i \in I} \mathscr{P}_i$ for an arbitrary index set I and disjoint measurable $\mathscr{P}_i$, and that the assumptions of Theorem 1.1 hold on each $\mathscr{P}_i$ separately. If $\boldsymbol{\Pi}_i = \{U \cap \mathscr{P}_i\colon U \in \boldsymbol{\Pi}\}$ denotes the relative topology on $\mathscr{P}_i$, we have the following corollary:

COROLLARY 1.1. *Suppose that* $\mathscr{P} = \bigcup_{i \in I} \mathscr{P}_i$ *and that the following statements hold*:

(i) *For each* $i \in I$, *the sequence of maps* $\tau_N\colon (\mathscr{P}_i, \boldsymbol{\Pi}_i) \to (\mathscr{R}, r)$ *is equicontinuous on* $\mathscr{P}_i$.

(ii) *There exists a* $\boldsymbol{\Pi}$*-consistent estimator* $\hat{P}_N = p_N(X_1, \dots, X_N)$ *of* $P \in \mathscr{P}$ *with the additional property that for each* $i \in I$, $P^N(\hat{P}_N \in \mathscr{P}_i) \to 1$ *for every* $P \in \mathscr{P}_i$.

Then $\tau_N(\hat{P}_N)$ *is an r-consistent estimator of* $\tau_N(P)$.

This result also follows directly from Theorem 1.1 by replacing the topology $\boldsymbol{\Pi}$ by the smallest topology containing $\boldsymbol{\Pi}_i$ for all $i \in I$. This has the effect of isolating the $\mathscr{P}_i$ from one another by making each $\mathscr{P}_i$ both open and closed. Note that assumption (ii) of Corollary 1.1 implies that $\hat{P}_N$ can serve as a test statistic for testing the hypothesis $P = P_i$ versus $P = P_j$, whenever $P_i \in \mathscr{P}_i$, $P_j \in \mathscr{P}_j$ and $i \neq j$. This test is asymptotically perfect in the sense that both error probabilities tend to zero as $N \to \infty$.

It is clear that the equicontinuity assumption for τ_N cannot be weakened much further unless one is willing to impose even more severe restrictions on the estimator $\hat{P}_N$. However, the equicontinuity of τ_N does merit further attention. It is often the case that τ_N converges pointwise to a function τ: $(\mathscr{P}, \boldsymbol{\Pi}) \to (\mathscr{R}, r)$, that is,

$$r(\tau_N(P), \tau(P)) \to 0 \quad \text{for every } P \in \mathscr{P}. \tag{1.8}$$

We shall show that in this case continuity of each τ_N ensures equicontinuity of τ_N outside of a set of the first category. As a result we have the following theorem:

THEOREM 1.2. *Suppose the following statements hold*:

(i) *For every* N, *the map* $\tau_N: (\mathscr{P}, \boldsymbol{\Pi}) \to (\mathscr{R}, r)$ *is continuous.*

(ii) *For every* $P \in \mathscr{P}$, $\tau_N(P)$ *converges to a limit* $\tau(P)$ *in* $(\mathscr{R}, r)$.

(iii) *There exists an estimator* $\hat{P}_N = p_N(X_1, \dots, X_N)$ *of* P *with values in* $\mathscr{P}$, *which is* $\boldsymbol{\Pi}$*-consistent for* $P \in \mathscr{P}$.

Then there exists a set D *of the first category in* $(\mathscr{P}, \boldsymbol{\Pi})$ *such that the sequence* τ_N *is equicontinuous at every point* $P \in \mathscr{P} \setminus D$, *and hence*

$$r\big(\tau_N\big(\hat{P}_N\big), \tau(P)\big) \to_P 0 \quad \textit{for every } P \in \mathscr{P} \setminus D. \tag{1.9}$$

Since $\tau_N \to \tau$ in Theorem 1.2, we have replaced $\tau_N(P)$ by $\tau(P)$ in (1.9): consistent estimation of $\tau_N(P)$ and $\tau(P)$ amounts to the same thing in this case. We do insist, however, that the substitution estimator be of the form $\tau_N(\hat{P}_N)$, rather than $\tau(\hat{P}_N)$. This is because in applications such as the bootstrap one often has no way of knowing the functional form of τ. Nevertheless, the reader should note that under the assumptions of Theorem 1.2, $\tau(\hat{P}_N)$ is indeed a consistent estimator of $\tau(P)$ for $P \in \mathscr{P} \setminus D$, because equicontinuity of τ_N implies continuity of τ.

Let us briefly discuss the results stated so far. Theorem 1.1 makes it clear that, as long as we make no assumptions about the speed of convergence of $\hat{P}_N$ to P, the equicontinuity of the sequence τ_N is the key to consistency of

$\tau_N(\hat{P}_N)$. Of course one can reduce the severity of the equicontinuity assumption somewhat by placing restrictions in probability on the possible values of $\hat{P}_N$. Corollary 1.1 is an example of this. If $\tau_N \to \tau$, Theorem 1.2 provides a worst case scenario: the substitution estimator can only fail to be consistent on a set D of exceptional points, which is at most a set of the first category. Without further investigation, however, such a statement is of only limited practical value. After all, the true underlying distribution P may be one of the exceptional points. Also the convergence may of course be slow near these points. Thus Theorem 1.2 merely indicates the structure of the consistency problem rather than providing a complete solution. In any particular case one will have to investigate whether such exceptional points actually exist, and if so, where they are located. It often turns out that with a judicious choice of the topology $\boldsymbol{\Pi}$ and the estimator $\hat{P}_N$, there are no exceptional points and the substitution estimator will be consistent for all $P \in \mathscr{P}$.

This last remark may need further clarification. In applications, the choice of the metric r on $\mathscr{R}$ will usually be determined in advance by the type of consistency that one would like the substitution estimator to possess. On the other hand, the choice of the topology $\boldsymbol{\Pi}$ on $\mathscr{P}$, or of the metric ρ inducing it, is completely open to us. If $\boldsymbol{\Pi}$ is a coarse topology, it will be relatively easy to find a consistent estimator $\hat{P}_N$ of P, but relatively many sequences of maps τ_N will possess only limited continuity properties and the set of exceptional P for which $\tau_N(\hat{P}_N)$ is not consistent will be relatively large. Conversely, if $\boldsymbol{\Pi}$ is a fine topology, there will be few, if any, consistent $\hat{P}_N$, but having found one, it will produce substitution estimators $\tau_N(\hat{P}_N)$ which are consistent for relatively many sequences τ_N, except on relatively small sets of exceptional P. If the sequence τ_N is given in a particular application, the trick will be to find a topology which is fine enough to provide τ_N with sufficient continuity properties, yet coarse enough to admit a consistent estimate $\hat{P}_N$ of P. In Section 3 we illustrate this search for an appropriate topology and for an estimator $\hat{P}_N$ which is consistent in this topology by a number of examples.

Another point worth noting concerns our interpretation of a set D of the first category as a "small" set. In a certain sense, this is indeed correct if $(\mathscr{P}, \boldsymbol{\Pi})$ is topologically complete. In this case the category theorem asserts that $\mathscr{P} \setminus D$ is at least dense in $\mathscr{P}$ [cf. Dudley (1989), page 44]. In more general cases, however, D may be quite large. In fact, the entire space $\mathscr{P}$ may be of the first category in $(\mathscr{P}, \boldsymbol{\Pi})$ and we may have $D = \mathscr{P}$, so that Theorem 1.2 is vacuous. We discuss an example of this phenomenon in Section 3. Fortunately it turns out that the pathological character of this example is due to an unfortunate choice of the topology $\boldsymbol{\Pi}$. A different choice of topology leads to an estimator $\hat{P}_N$ for which the substitution estimator $\tau_N(\hat{P}_N)$ is consistent for all $P \in \mathscr{P}$.

In the preceding paragraphs we have stressed the constructive aspects of our results so far by explaining how these results may be used to arrive at an estimator $\hat{P}_N$ which makes the substitution estimator $\tau_N(\hat{P}_N)$ consistent. However, one may also approach the consistency problem from a different angle and investigate the existence of a consistent substitution estimator without worrying about its construction. For a result of this type, a logical assumption is the existence of a consistent estimator $T_N = t_N(X_1, \dots, X_N)$ of $\tau_N(P)$. If no such estimator exists, there is no hope of finding a consistent substitution estimator.

THEOREM 1.3. *Suppose the following statements hold*:

(i) *For every* N, *the map* $\tau_N: (\mathscr{P}, \boldsymbol{\Pi}) \to (\mathscr{R}, r)$ *is measurable.*

(ii) *The metric space* $(\mathscr{R}, r)$ *is separable.*

(iii) *There exist an* r-*consistent estimator* $T_N = t_N(X_1, \dots, X_N)$ *of* $\tau_N(P)$ *for* $P \in \mathscr{P}$.

Then there exists an estimator $\hat{P}_N = p_N(X_1, \dots, X_N)$ *with values in* $\mathscr{P}$ *such that* $\tau_N(\hat{P}_N)$ *is an* r-*consistent estimator of* $\tau_N(P)$ *for every* $P \in \mathscr{P}$.

In the proof of Theorem 1.3 we construct the estimator $\hat{P}_N$ explicitly on the basis of T_N. Hence, if T_N is not known to us, we cannot construct $\hat{P}_N$, and if it is known, it may not make much sense to construct $\hat{P}_N$ and $\tau_N(\hat{P}_N)$ since we already have a consistent estimator T_N of $\tau_N(P)$. Thus Theorem 1.3 should indeed be viewed purely as an existence statement to the effect that anything that can be estimated consistently at all, can be estimated consistently by a substitution estimator. The problem is of course to find an appropriate $\hat{P}_N$.

The final result of this section allows us to construct a substitution estimator in some cases where T_N is not known, but its existence is. We consider the case where $(\mathscr{P}, \boldsymbol{\Pi})$ is a metric space $(\mathscr{P}, H)$, the τ_N are assumed to be continuous but not necessarily convergent and $\hat{P}_N$ is $\sqrt{N}$-Hellinger-consistent. If $\tau_N(P)$ can be estimated consistently at all, we show that $r(\tau_N(P_N), \tau_N(P)) \to 0$ for every sequence P_N with $H(P_N, P) = \mathscr{O}(N^{-1/2})$ and for all P outside of a set of the first category. Substituting $\hat{P}_N$ for P_N we find the following theorem:

THEOREM 1.4. *Let the topology* $\boldsymbol{\Pi}$ *be metrized by the Hellinger metric* H *and suppose the following statements hold*:

(i) *For every* N, *the map* $\tau_N: (\mathscr{P}, \boldsymbol{\Pi}) \to (\mathscr{R}, r)$ *is continuous.*

(ii) *There exists an* r-*consistent estimator* $T_N = t_N(X_1, \dots, X_N)$ *of* $\tau_N(P)$ *for* $P \in \mathscr{P}$.

(iii) *There exists an estimator* $\hat{P}_N = p_N(X_1, \dots, X_N)$ *with values in* $\mathscr{P}$, *which is* $\sqrt{N}$-*Hellinger-consistent for* $P \in \mathscr{P}$.

Then there exists a set D *of the first category in* $(\mathscr{P}, H)$ *such that* $\tau_N(\hat{P}_N)$ *is an* r-*consistent estimator of* $\tau_N(P)$ *for* $P \in \mathscr{P} \setminus D$, *that is,*

$$r\Big(\tau_N\big(\hat{P}_N\big), \tau_N(P)\Big) \to_P 0 \quad \textit{for every } P \in \mathscr{P} \setminus D. \tag{1.10}$$

Because the requirement that $\hat{P}_N$ is $\sqrt{N}$-Hellinger-consistent may be somewhat unexpected, we shall show by means of a counterexample (Example 3.4 in Section 3) that this assumption is really needed. Ordinary Hellinger consistency is not sufficient. It is clear from the work of Le Cam (1973, 1986) and Birgé (1983, 1986) that $\sqrt{N}$-Hellinger-consistent estimators will generally exist for finite dimensional families $\mathscr{P}$ where dimension is defined in terms of metric entropy. Typical examples of such families are parametric families $\mathscr{P} = \{P_\theta \colon \theta \in \Theta\}$ with $\Theta \subset \mathbb{R}^k$, provided that Hellinger distance in $\mathscr{P}$ and Euclidean distance in Θ are compatible in some sense. For these families Theorem 1.4 enables us to find a substitution estimator that will work for "most" P if anything does. For infinite dimensional families $\mathscr{P}$, $\sqrt{N}$-Hellinger-consistent estimators of P will generally not exist, and without further assumptions on τ_N, one will generally not be able to construct a satisfactory estimator of $\tau_N(P)$ either.

We also note that for parametric families $\mathscr{P} = \{P_\theta \colon \theta \in \Theta\}$ with $\Theta \subset \mathbb{R}^k$, it is possible to prove results similar to Theorem 1.4, where the exceptional set equals $D = \{P_\theta \colon \theta \in \Theta_0\}$ and Θ_0 has Lebesgue measure 0 [cf. Putter (1994)].

The two main results in this section are concerned with the interplay between conditions on τ_N and conditions on $\hat{P}_N$, needed to obtain reasonable substitution estimators $\tau_N(\hat{P}_N)$ of $\tau_N(P)$. Theorem 1.2 discusses what is needed for τ_N under the weakest possible condition (consistency) on $\hat{P}_N$. Theorem 1.4, on the other hand, operates under the weakest possible condition [estimability of $\tau_N(P)$] on τ_N.

In the remainder of the paper we proceed as follows. In Section 2 we apply the results of this section to the bootstrap and discuss the significance of our results in this context. Section 3 provides a number of examples that clarify the relationship between our results and standard bootstrap theory. Proofs of Theorems 1.2, 1.4 and 1.3 are given in Sections 4, 5 and 6, respectively.

2. The bootstrap. In the setup of the previous section, consider a sequence of random variables $Y_N = y_N(X_1, \dots, X_N; P)$, where y_N is a measurable map from $\mathscr{X}^N \times \mathscr{P}$ to a *separable* metric space $(\mathscr{S}, s)$. Let $\mathscr{R}$ be the space of all probability distributions on $(\mathscr{S}, s)$ equipped with a metric r, which metrizes weak convergence. An obvious choice for r is Prohorov's metric ϱ. For distributions $R_1, R_2 \in \mathscr{R}$ this is defined by

$$\varrho(R_1, R_2) = \inf\{\varepsilon > 0 \colon R_1(A) \leq R_2(A^\varepsilon) + \varepsilon, \text{ for all } A \in \mathscr{B}(\mathscr{S}, s)\}, \tag{2.1}$$

where A^ε is an ε-neighborhood of A. Since $(\mathscr{S}, s)$ is separable, ϱ does indeed metrize weak convergence of probability measures in $\mathscr{R}$ [cf. Dudley (1989), Section 11.3], but of course other choices of r are also possible. Note that the separability of $(\mathscr{S}, s)$ also implies that $(\mathscr{R}, r)$ is separable [cf. Billingsley (1968), page 239]. Our aim is to estimate the law $\tau_N(P)$ of Y_N under P. Obviously $\tau_N(P) \in \mathscr{R}$.

As before, let us estimate P by $\hat{P}_N = p_N(X_1, \dots, X_N)$ for a measurable map $p_N \colon \mathscr{X}^N \to \mathscr{P}$. With $\hat{P}_N$ as the resampling distribution, the bootstrap estimator of $\tau_N(P)$ is simply a substitution estimator $\tau_N(\hat{P}_N)$. To see this, note that if the resampling distribution is $\hat{P}_N$, the bootstrap estimates the distribution of Y_N by that of

$$Y_N^* = y_N\big(X_1^*, \dots, X_N^*; \hat{P}_N\big),$$

where $X_1^*, \ldots, X_N^*$ are i.i.d. with distribution $\hat{P}_N$. However, this is just a description of $\tau_N(\hat{P}_N)$. The bootstrap estimate $\tau_N(\hat{P}_N)$ can be computed either analytically or by Monte Carlo simulation, but we shall not be concerned with that question here.

Note that the resampling distribution $\hat{P}_N$ is not necessarily the empirical distribution of $X_1, \ldots, X_N$, as is customary for the nonparametric bootstrap. In fact, our requirement that $\hat{P}_N$ takes its values in $\mathscr{P}$ prohibits this in many cases. If, for example, $\mathscr{P}$ is a parametric family $\{P_\theta : \theta \in \Theta\}$, our estimate $\hat{P}_N = P_{\hat{\theta}_N}$ will typically be based on an estimate $\hat{\theta}_N$ of the parameter θ and our bootstrap procedure will be the so-called *parametric bootstrap*. As we have indicated in Section 1, the purpose of this paper is to emphasize the importance of a judicious choice of the resampling distribution $\hat{P}_N$ so as to satisfy the requirements of our theorems. All but one of the examples in Section 3 will concern cases where the nonparametric bootstrap fails, but a proper choice of $\hat{P}_N$ will make the bootstrap work. On the one hand, this illustrates the great flexibility of the bootstrap method. On the other hand, it also shows that precise information about the behavior of the distribution $\tau_N(P)$ of Y_N as a function of P is needed to arrive at the correct resampling distribution $\hat{P}_N$. Unfortunately, such information may often not be available.

With the present choice of $(\mathscr{R}, r)$ and τ_N, Theorems 1.1–1.4 and Corollary 1.1 become results on the consistency of the bootstrap. For the sake of brevity, we shall not reformulate these results in this particular context. All the reader has to remember is that $\tau_N(P)$ is now the distribution of the random variable $Y_N = y_N(X_1, \ldots, X_N; P)$ taking values in a separable metric space, r is a metric metrizing weak convergence of probability distributions on this space and $\tau_N(\hat{P}_N)$ is the bootstrap estimate of $\tau_N(P)$ with $\hat{P}_N$ as the resampling distribution. As we pointed out above, assumption (ii) of Theorem 1.3 is automatically satisfied.

We begin by noting that the equicontinuity condition of Theorem 1.1 has been used to prove consistency of the bootstrap estimator ever since the beginning of research on bootstrap asymptotics [cf., e.g., Bickel and Freedman (1981) and Beran (1984)].

In Theorem 1.2, the choice of the metric r is irrelevant as long as it metrizes weak convergence of probability distributions on the separable metric space $(\mathscr{S}, s)$. Assumption (ii) of Theorem 1.2 now means that the distributions $\tau_N(P)$ of Y_N converge weakly to a limit distribution $\tau(P)$. The conclusion of Theorem 1.2 is that for $P \in \mathscr{P} \setminus D$, the bootstrap estimator $\tau_N(\hat{P}_N)$ converges weakly to $\tau(P)$ in probability.

The situation is more complicated in the remaining results where we do not require convergence of $\tau_N(P)$. The conclusions of these theorems refer to sequences $\tau_N(\hat{P}_N)$ and $\tau_N(P)$ of distributions for which the distance tends to zero in r-metric. This is not a matter which depends only on the topology of weak convergence which is induced by r. For different choices of the metric r metrizing weak convergence, $r(P_N, Q_N) \to 0$ may mean the same or different things [cf. Dudley (1989), Theorem 11.7.1 and problem 8 on page 313]. This problem disappears if the sequence $\tau_N(P)$ is uniformly tight.

In the context of the bootstrap, the assumption that the distributions $\tau_N(P)$ of Y_N converge weakly to a limit distribution $\tau(P)$ is important for another reason as well. It allows us to use the $(M\text{–}N)$-bootstrap $\tau_M(\hat{P}_N)$, where $M = M_N$ tends to infinity with N, but at a slower rate $M_N = o(N)$. Since consistent estimation of $\tau_M(P)$, $\tau_N(P)$ or $\tau(P)$ amounts to the same thing in this case, the $(M\text{–}N)$-bootstrap may be viewed as an attempt to estimate $\tau_M(P)$ with the advantage of having at our disposal a resampling distribution $\hat{P}_N$ which is much closer to the underlying P than $\hat{P}_M$. As a result the $(M\text{–}N)$-bootstrap is consistent much more generally than the traditional $(N\text{–}N)$-bootstrap [cf. Politis and Romano (1994)]. Viewed in this light, we may weaken condition (iii) of Theorem 1.4 to Hellinger consistency at an arbitrarily slower rate $H(\hat{P}_N, P) = \mathcal{O}_P(a_N)$ with $a_N \to 0$, but $Na_N^2 \to \infty$, provided that we replace the bootstrap $\tau_N(\hat{P}_N)$ by the $(M\text{–}N)$-bootstrap $\tau_M(\hat{P}_N)$ with $M = a_N^{-2}$ and that $\tau_N(P) \to \tau(P)$. However, with these modifications, Theorem 1.4 is simply contained in Theorem 1.2 and it follows that we have nothing new to say about this method of improving the bootstrap by employing a smaller resample size. We are solely concerned with an appropriate choice of the resampling distribution $\hat{P}_N$.

The assumption (i) in Theorems 1.2 and 1.4 that τ_N is continuous for each N will generally not cause any problems. For most reasonable choices of the topology $\boldsymbol{\Pi}$ and the metric r, the distribution τ_N of Y_N for a fixed value of N would be continuous if Y_N did not depend on P. The direct dependence of Y_N on P is not likely to make matters worse, and the assumption that τ_N is continuous for every fixed N will be satisfied in all reasonable cases. It is the equicontinuity that may be lacking for certain P.

It was mentioned in Section 1 that for a parametric model $\mathcal{P} = \{P_\theta\colon \theta \in \Theta\}$ with $\Theta \subset \mathbb{R}^k$, a $\sqrt{N}$-Hellinger-consistent parametric estimator $\hat{P}_N = P_{\hat{\theta}_N}$ will typically exist, provided that Hellinger distance in $\mathcal{P}$ and Euclidean distance in Θ are compatible in an appropriate sense. In this case, Theorem 1.4 asserts that for continuous τ_N, the parametric bootstrap with resampling distribution $P_{\hat{\theta}_N}$ will work for "most" P if anything does, even if $\tau_N(P)$ does not converge to a limit distribution $\tau(P)$.

As we pointed out at the end of Section 1, Theorems 1.2 and 1.4 deal with two extreme cases with minimal conditions on $\hat{P}_N$ and τ_N, respectively. In the context of the bootstrap this distinction attains an added significance. Before applying the bootstrap one should answer two questions:

1. What should one bootstrap?
2. How should one bootstrap?

The first of these questions refers in particular to choosing the proper dependence of Y_N—and hence of $\tau_N(P)$—on P. As a general rule one should do this in such a way that τ_N depends on P as little as possible. Theorem 1.2 suggests that whenever possible one should normalize Y_N so that its distribution $\tau_N(P)$ tends to a limit distribution $\tau(P)$. Any consistent choice of a resampling distribution $\hat{P}_N$ will then produce a bootstrap $\tau_N(\hat{P}_N)$ that works outside a set D of the first category. To get rid of this set D, one may search for a topology Π on $\mathscr{P}$ which is fine enough to make τ_N equicontinuous on $\mathscr{P}$, and then for a Π-consistent estimator $\hat{P}_N$. Unfortunately, this step will often be impossible for lack of the necessary knowledge of τ_N.

The second question refers to the choice of the resampling distribution $\hat{P}_N$. Theorem 1.4 asserts that even if we do not know how to normalize Y_N properly, the parametric bootstrap will generally still work on $\mathscr{P}\setminus D$. In nonparametric models, however, we had better make sure that τ_N converges.

In the extensive literature on the bootstrap it is usually shown that the bootstrap is strongly consistent, in the sense that $r(\tau_N(\hat{P}_N), \tau_N(P)) \to 0$ P-almost surely. For perfectly good reasons, strong consistency has not played an important role in the development of statistics so far, and hence we have been content to formulate our results in terms of ordinary (weak) consistency $r(\tau_N(\hat{P}_N), \tau_N(P)) \to_P 0$.

3. Examples. In this section we shall give some examples that illustrate the importance of choosing an appropriate resampling distribution $\hat{P}_N$ in applying the bootstrap. The first example exhibits a function $\tau_N(P)$ that is continuous in P with respect to Hellinger distance for every fixed N, but where the pointwise limit $\tau(P) = \lim_{N\to\infty} \tau_N(P)$ has a single discontinuity at a point P_0. It is shown that a parametric bootstrap fails in the point of discontinuity of τ. With a suitable metric that isolates that point, the equicontinuity is recaptured.

EXAMPLE 3.1. Let $\mathscr{P} = \{P_\alpha\colon 0 \le \alpha < 1/2\}$, where P_α is the probability distribution on $\mathbb{R}$ with distribution function F_α, defined for $0 < \alpha < 1/2$ by

$$F_\alpha(x) = \begin{cases} 0, & \text{if } x \le 0,\\ 1-(1+\alpha x)^{-1/\alpha}, & \text{if } x > 0, \end{cases} \tag{3.1}$$

and for $\alpha = 0$ by

$$F_0(x) = \lim_{\alpha\to 0} F_\alpha(x) = \begin{cases} 0, & \text{if } x \le 0,\\ 1-e^{-x}, & \text{if } x > 0. \end{cases} \tag{3.2}$$

Let $\mathscr{P}$ be equipped with Hellinger distance H and let the metric r on $\mathscr{R}$ be Lévy's metric. The Hellinger distance on $\mathscr{P}$ is related to Euclidean distance on the parameter space $[0, 1/2)$ by the relation

$$H(P_\alpha, P_\beta) = \frac{|\alpha-\beta|}{\sqrt{2(1+\alpha)(1+2\alpha)}} + o(|\alpha-\beta|) \quad \text{for } \alpha, \beta \in (0, 1/2) \tag{3.3}$$

and

$$H(P_0, P_\alpha) = \frac{\alpha}{\sqrt{2}} + o(\alpha). \tag{3.4}$$

We are interested in the distribution of the random variables

$$Y_N = N^{-\alpha}(M_N - \log N) \quad \text{for } 0 \le \alpha < 1/2, \tag{3.5}$$

where M_N stands for $\max_{i=1,\dots,N} X_i$. Note that Y_N depends on the underlying distribution through $N^{-\alpha}$.

Let $\tau_N(P_\alpha)$ be the law of Y_N when $X_1,\dots,X_N$ are i.i.d. with distribution P_α and let $G_{N,\alpha}$ denote the distribution function of $\tau_N(P_\alpha)$. Then for $0<\alpha<1/2$,

$$G_{N,\alpha}(x) = P(N^{-\alpha}(M_N - \log N) \le x) = \left[F_\alpha(N^\alpha x + \log N)\right]^N$$
$$= \left[1 - (1 + \alpha N^\alpha x + \alpha \log N)^{-1/\alpha}\right]^N, \qquad x > -N^{-\alpha}\log N,$$

and for $\alpha = 0$,

$$G_{N,0}(x) = P(M_N - \log N \le x) = (1 - \exp(-x - \log N))^N, \quad x > -\log N.$$

For fixed N, it is easily seen that τ_N is continuous at P_α for $\alpha > 0$. Furthermore, $G_{N,\alpha}(x) \to G_{N,0}(x)$ as $\alpha \to 0$ for every $x > -\log N$ and hence $r(\tau_N(P_\alpha), \tau_N(P_0)) \to 0$ as $\alpha \to 0$, so that $\tau_N(P)$ is continuous at P_0. Hence, for each N, τ_N is clearly continuous on $\mathscr{P}$.

Now let N tend to infinity and let $\tau(P_\alpha)$ be the pointwise limit of $\tau_N(P_\alpha)$. The distribution function of $\tau(P_\alpha)$ will be denoted by G_α. Then for $0<\alpha<1/2$,

$$G_\alpha(x) = \lim_{N\to\infty} G_{N,\alpha}(x) = \exp(-(\alpha x)^{-1/\alpha}), \qquad x > 0,$$

and for $\alpha = 0$,

$$G_0(x) = \lim_{N\to\infty} G_{N,0}(x) = \exp(-e^{-x}), \qquad x \in \mathbb{R}.$$

We find that $\tau(P)$ is not continuous at P_0, since $\lim_{\alpha\to 0} G_\alpha(x) = 0$ for all x.

Application of Theorem 1.2 yields the existence of a set D of the first category in $(\mathscr{P}, H)$ such that the sequence τ_N is equicontinuous at P for all $P \in \mathscr{P}\setminus D$. Consequently if $\hat{P}_N$ is a Hellinger-consistent sequence of estimators of P, the bootstrap with resampling distribution $\hat{P}_N$ is consistent for all $P \in \mathscr{P}\setminus D$. Note that P_0 belongs to the exceptional set D since the limit τ is not continuous at P_0. A closer analysis reveals that τ_N is equicontinuous at P_α for $\alpha > 0$.

It appears therefore that P_0 is the only trouble spot in the model $\mathscr{P}$, so the problem can be resolved by choosing a metric on $\mathscr{P}$ that isolates P_0. Take, for instance,

$$\text{(3.6)} \qquad \pi(P,Q) = \begin{cases} H(P,Q), & \text{if } P, Q \ne P_0, \\ \sqrt{2}, & \text{if } P = P_0 \ne Q \text{ or } P \ne P_0 = Q. \end{cases}$$

Clearly π defines a metric on $\mathscr{P}$ and the sequence τ_N is trivially equicontinuous with respect to π at P_0, and hence on $\mathscr{P}$. A π-consistent estimator of P will have to satisfy $P_0^N(\hat{P}_N = P_0) \to 1$ and for all $P \in \mathscr{P}\setminus\{P_0\}$ both $P^N(\hat{P}_N = P_0) \to 0$ and $H(\hat{P}_N, P) \to_P 0$. If we set $\hat{P}_N = P_{\hat{\alpha}_N}$, then this implies that $\hat{\alpha}_N$ has to be a consistent estimate of α satisfying

$$P_0^N(\hat{\alpha}_N = 0) \to 1 \quad \text{and} \quad P_\alpha^N(\hat{\alpha}_N = 0) \to 0 \quad \text{for } 0 < \alpha < 1/2.$$

It is indeed possible to detect the isolated point P_0 in $\mathscr{P}$ with probability tending to 1 by choosing, for instance,

$$\hat{\alpha}_N = \begin{cases} 0, & \text{if } M_N \le 2\log N, \\ \left[1 - \left(\overline{X}_N\right)^{-1}\right]^+, & \text{otherwise.} \end{cases} \tag{3.7}$$

Since τ_N is equicontinuous with respect to π and $\pi(P_{\hat{\alpha}_N}, P) \to_P 0$, for every $P \in \mathscr{P}$, the bootstrap with resampling distribution $P_{\hat{\alpha}_N}$ is consistent for every P in $\mathscr{P}$. Note that this result may also be obtained directly by applying Corollary 1.1 combined with (3.7), but it seemed instructive to exhibit a metric π that separates $\{P_0\}$ and $\mathscr{P} \setminus \{P_0\}$ explicitly.

EXAMPLE 3.2. If the class of all possible distributions $\mathscr{P}$ is complete, the exceptional set of the first category, appearing in the result of Theorems 1.2 and 1.4, is small in the sense that its complement is dense in $\mathscr{P}$. If $\mathscr{P}$ is not complete, however, these sets can be quite large. In this example we discuss a particular statistical model $\mathscr{P}$ equipped with Hellinger metric H, such that $\mathscr{P}$ is of the first category in $(\mathscr{P}, H)$. This model is not an artificial construct, but it is the natural model for a statistical situation of interest.

Let $\mathscr{P}$ be the class of probability distributions P on $(0, \infty)$ with distribution functions F satisfying

$$\lim_{x \searrow 0} \frac{F(x)}{x} = a(P) \in (0, \infty). \tag{3.8}$$

Let $X_1, X_2, \ldots$ be i.i.d. random variables taking values in $(0, \infty)$ with unknown common distribution P in $\mathscr{P}$ and distribution function F. Consider the random variable

$$Y_N = N \min\{X_1, \ldots, X_N\} \tag{3.9}$$

and let $\tau_N(P)$ be the distribution of Y_N under P. Note that $\mathscr{P}$ is precisely the class of underlying distributions P for which $\tau_N(P)$ converges to a nondegenerate limit $\tau(P)$, which is an exponential distribution with parameter $a(P)$. Let $\mathscr{P}$ be equipped with Hellinger distance H. Then assumptions (i) and (ii) of Theorem 1.2 are satisfied. It is shown in Putter and van Zwet (1994) that application of Theorem 1.2 does not yield any positive information in the sense that the exceptional set D appearing in the conclusion of the theorem equals the entire space $\mathscr{P}$. The aforementioned paper also contains a direct proof that $\mathscr{P}$ is indeed a set of the first category in $(\mathscr{P}, H)$. Luckily, all this trouble is caused only by a wrong choice of the metric ρ. If we define

$$\pi(P, Q) = \sup_{x>0} \frac{|F(x) - G(x)|}{x}, \tag{3.10}$$

where F and G denote the distribution functions corresponding to P and Q, then it is shown in Putter and van Zwet (1994) that the sequence τ_N is equicontinuous with respect to π for all $P \in \mathscr{P}$. A π-consistent estimator $\hat{P}_N$

of P is also provided. Its distribution function $\hat{F}_N$ is given by

$$(3.11) \qquad \hat{F}_N(x) = \begin{cases} \dfrac{F_N(\xi_N)}{\xi_N} x, & \text{if } 0 < x < \xi_N, \\ F_N(x), & \text{if } x \geq \xi_N, \end{cases}$$

where F_N denotes the empirical distribution function and ξ_N is a sequence of positive numbers converging to zero with $N\xi_N \to \infty$. It follows that the bootstrap with $\hat{P}_N$ as resampling distribution works for all P in $\mathscr{P}$.

In practice, this example occurs in a slightly modified form. Instead of $X_1, \ldots, X_N$, one observes $Z_1 = \theta + X_1, \ldots, Z_N = \theta + X_N$ for the purpose of estimating the parameter $\theta \in \mathbb{R}$ which is the lower endpoint of the support of the distribution of the Z_i. When using $\min(Z_1, \ldots, Z_N)$ as an estimator of θ, one is indeed interested in the distribution of $Y_N = N(\min(Z_1, \ldots, Z_N) - \theta) = N \min(X_1, \ldots, X_N)$. Obviously, $\hat{P}_N$ as defined by (3.11) cannot be used for the resampling distribution of $X_1^*, \ldots, X_N^*$, since the empirical distribution function F_N of $X_1, \ldots, X_N$ is now unknown. However, a slight modification will work. If G_N denotes the empirical distribution function of $Z_1, \ldots, Z_N$, one can estimate the distribution P of X_1 by a distribution $\overline{P}_N$ with distribution function.

$$\overline{F}_N(x) = \begin{cases} \dfrac{G_N(\min(Z_1, \ldots, Z_N) + \xi_N)}{\xi_N} x, & \text{if } 0 < x < \xi_N, \\ G_N(\min(Z_1, \ldots, Z_N) + x), & \text{if } x \geq \xi_N, \end{cases}$$

where $\xi_N \to 0$ and $N\xi_N \to \infty$. It is easy to see that $\pi(\overline{P}_N, \hat{P}_N) \to_P 0$, so that $\pi(\overline{P}_N, P) \to_P 0$ and the resampling distribution $\overline{P}_N$ will produce a consistent bootstrap for all $P \in \mathscr{P}$.

EXAMPLE 3.3 (Superefficiency). Another example, related to Example 3.1, is provided by Beran (1982). Consider an i.i.d. sequence $X_1, \ldots, X_N$ with a common normal distribution P_θ with unknown mean θ and unit variance. The Hodges estimator of θ is given by

$$(3.12) \qquad T_N = \begin{cases} \overline{X}_N, & \text{if } |\overline{X}_N| > N^{-1/4}, \\ b\overline{X}_N, & \text{if } |\overline{X}_N| \leq N^{-1/4}, \end{cases}$$

where $\overline{X}_N = (1/N)\Sigma_{i=1}^N X_i$ and $b \in (0, 1)$. We wish to find a bootstrap estimate for the distribution $\tau_N(P_\theta)$ of

$$Y_N = N^{1/2}(T_N - \theta)$$

under P_θ. We equip the class $\mathscr{P} = \{P_\theta \colon \theta \in \mathbb{R}\}$ with the Euclidean metric d in the parameter space, that is, $d(P_\theta, P_{\theta'}) = |\theta - \theta'|$. Thus, in effect we are identifying $\mathscr{P}$ and its parameter space $\mathbb{R}$. Let $(\mathscr{R}, l)$ denote the class of all distributions on $\mathbb{R}$ equipped with the Lévy metric l. Consider τ_N as a map from $(\mathscr{P}, d)$ to $(\mathscr{R}, l)$.

If we denote the distribution function corresponding to $\tau_N(P_\theta)$ by $G_{N,\theta}$, we find

$$G_{N,\theta}(x) = \begin{cases} \Phi(x), & \text{if } N^{-1/4}|x + \theta N^{1/2}| \geq 1, \\ \Phi\left(\dfrac{x + (1-b)\theta N^{1/2}}{b}\right), & \text{if } N^{-1/4}|x + \theta N^{1/2}| \leq b, \\ \Phi(-N^{1/4} - \theta N^{1/2}), & \text{if } -1 < N^{-1/4}(x + \theta N^{1/2}) < -b, \\ \Phi(N^{1/4} - \theta N^{1/2}), & \text{if } b < N^{-1/4}(x + \theta N^{1/2}) < 1, \end{cases}$$

where Φ denotes the standard normal distribution function. It follows that the pointwise limit G_θ of $G_{N,\theta}$ is given by

$$G_\theta(x) = \begin{cases} \Phi(x/b), & \text{if } \theta = 0, \\ \Phi(x), & \text{otherwise.} \end{cases} \tag{3.13}$$

This implies that the limit $\tau(P_\theta)$ of $\tau_N(P_\theta)$ is a normal distribution with variance b^2 if $\theta = 0$ and unity otherwise. Since $0 < b < 1$, the Hodges estimator is superefficient at $\theta = 0$.

Obviously, τ_N is continuous at every P_θ and since $\tau_N \to \tau$, Theorem 1.2 applies, and hence the sequence τ_N is equicontinuous on $\mathscr{P} \setminus D$, where D is of the first category. As τ has a discontinuity at P_0, this distribution clearly belongs to D. Observing that $G_{N,\theta}(x) = G_{N,\theta'}(x) = \Phi(x)$ if both $|x + \theta N^{1/2}| \geq N^{1/4}$ and $|x + \theta' N^{1/2}| \geq N^{1/4}$, we see that $l(\tau_N(P_\theta), \tau_N(P_{\theta'}))$ can be made arbitrarily small for θ' in a small neighborhood of a fixed $\theta \neq 0$ and large N, so that τ_N is equicontinuous at every P_θ with $\theta \neq 0$. Hence, $D = \{P_0\}$, the single point of discontinuity of the limit distribution τ.

Of course this does not imply that

$$\liminf_N l\left(\tau_N(P_0), \tau_N(P_{\theta_N})\right) > 0 \tag{3.14}$$

for sequences $\theta_N \to 0$. However, as Beran (1982) points out, (3.14) does hold for sequences θ_N converging at rate $N^{-1/2}$. To see this, notice that for such sequences

$$G_{N,\theta_N}(x) = \Phi\left(\frac{x + (1-b)\theta_N N^{1/2}}{b}\right) + o(1).$$

In fact, for every $\varepsilon > 0$, (3.14) holds uniformly for $|\theta_N| \geq \varepsilon N^{-1/2}$. This shows that for $\hat{\theta}_N = \overline{X}_N$, for instance, the parametric bootstrap with resampling distribution $P_{\hat{\theta}_N}$ will work for $\theta \neq 0$, but fails for $\theta = 0$.

In Example 3.1 we have shown how to deal with a situation like this. By applying Corollary 1.1, we find that all we have to do to make the parametric bootstrap work is to modify the estimator $\hat{\theta}_N = \overline{X}_N$ to ensure that

$$P_0^N(\hat{\theta}_N = 0) \to 1, \qquad P_\theta^N(\hat{\theta}_N = 0) \to 0 \quad \text{for } \theta \neq 0.$$

Choosing

$$\hat{\theta}_N = \begin{cases} \overline{X}_N, & \text{if } |\overline{X}_N| > N^{-1/4}, \\ 0, & \text{if } |\overline{X}_N| \le N^{-1/4}, \end{cases}$$

which is the Hodges estimator for $b = 0$, will accomplish this and the corresponding parametric bootstrap $\tau_N(P_{\hat{\theta}_N})$ will work for all θ.

The reason we discuss the Hodges estimator T_N is that it is perhaps the best known example of an estimator which is superefficient for a single parameter value $\theta = 0$. Le Cam (1953) has pointed out that one can modify T_N in an obvious way to construct an estimator which is superefficient for all θ belonging to a countable closed set in $\mathbb{R}$. Moreover, Le Cam showed that an estimator of θ can only be superefficient on a set of the first category in $\mathbb{R}$ equipped with the Euclidean metric. Since superefficiency can only occur at points where τ_N is not equicontinuous, this may be viewed as a consequence of Theorem 1.2.

EXAMPLE 3.4. Our last example concerns the question of existence of a consistent bootstrap estimator. It may also clarify why a $\sqrt{N}$-Hellinger-consistent estimator $\hat{P}_N$ is needed in Theorem 1.4.

Let $X_1, X_2, \ldots$ be i.i.d. random variables with a common normal distribution with expectation $\theta \in \mathbb{R}$ and variance 1, which we shall indicate as P_θ or $\mathscr{N}(\theta, 1)$. Define $\overline{X}_N = (1/N)\Sigma_{i=1}^N X_i$ and

$$Y_N = N^{1/2}\big(\overline{X}_N - a_N\theta\big).$$

We distinguish three different cases.

Case (i): $a_N \equiv 1$. The distribution $\tau_N(P_\theta)$ of Y_N is $\mathscr{N}(0,1)$ independent of θ, which can obviously be estimated consistently for any metric r on $\mathscr{R}$. Also the sequence τ_N is equicontinuous for any topology Π on $\mathscr{P} = \{P_\theta\colon \theta \in \mathbb{R}\}$ and any metric r on $\mathscr{R}$. As we can choose any Π and r, Theorem 1.1 ensures that the bootstrap $\tau_N(P_{\hat{\theta}_N})$ equals the true distribution $\tau_N(P_\theta)$ for any "estimator" $\hat{\theta}_N$ of θ, consistent or not.

Case (ii): $a_N \equiv 0$. Now $\tau_N(P_\theta)$ is $\mathscr{N}(N^{1/2}\theta, 1)$, which cannot be estimated consistently in Prohorov or Lévy metric. The reason for this is that any estimator of θ has an error which is at least of order $N^{-1/2}$ in probability and as a result $N^{1/2}\theta$ cannot be estimated consistently. It follows that there is no consistent bootstrap estimator of $\tau_N(P_\theta)$ either.

Case (iii): $a_N = 1 - \varepsilon_N$, $\varepsilon_N \searrow 0$. Now $\tau_N(P_\theta)$ is $\mathscr{N}(\varepsilon_N N^{1/2}\theta, 1)$ which can be estimated consistently by $\mathscr{N}(\varepsilon_N N^{1/2}\overline{X}_N, 1)$ in Prohorov or Lévy metric. If $\hat{\theta}_N$ is an estimator of θ, the bootstrap $\tau_N(P_{\hat{\theta}_N})$ will work if and only if

$$\varepsilon_N N^{1/2}\big(\hat{\theta}_N - \theta\big) \to_{P_\theta} 0$$

for every $\theta \in \mathbb{R}$. This is true for every sequence $\varepsilon_N \searrow 0$ if and only if $\hat{\theta}_N - \theta = \mathscr{O}_{P_\theta}(N^{-1/2})$ or $H(P_{\hat{\theta}_N}, P_\theta) = \mathscr{O}_{P_\theta}(N^{-1/2})$. It follows that the assumption in Theorem 1.4 that $\hat{P}_N$ is $\sqrt{N}$-Hellinger-consistent cannot be relaxed. The bootstrap $\tau_N(P_{\overline{X}_N})$, which incidentally is the same as $\mathscr{N}(\varepsilon_N N^{1/2}\overline{X}_N, 1)$, is obviously consistent.

4. Proof of Theorem 1.2. In this section $(\mathscr{P}, \boldsymbol{\Pi})$ will be a topological space, $(\mathscr{R}, r)$ a metric space and τ_N a sequence of continuous maps from $(\mathscr{P}, \boldsymbol{\Pi})$ to $(\mathscr{R}, r)$, converging to a limit $\tau: (\mathscr{P}, \boldsymbol{\Pi}) \to (\mathscr{R}, r)$.

DEFINITION. τ_N is *locally uniformly convergent* at P_0 if for every $\varepsilon > 0$ there exists a neighborhood U_ε of P_0 and a number N_ε such that $r(\tau_N(P), \tau(P)) \le \varepsilon$ for all $N \ge N_\varepsilon$ and for all $P \in U_\varepsilon$.

DEFINITION. τ_N is *equicontinuous* at P_0 if for every $\varepsilon > 0$ there exists a neighborhood U_ε of P_0 such that $P \in U_\varepsilon$ implies

$$\sup_N r(\tau_N(P), \tau_N(P_0)) \le \varepsilon.$$

For the sequence of continuous maps τ_N, define

$$E_1 = \{P \in \mathscr{P}: \tau_N \text{ is locally uniformly convergent at } P\}, \tag{4.1}$$

$$E_2 = \{P \in \mathscr{P}: \tau_N \text{ is equicontinuous at } P\}. \tag{4.2}$$

The following lemma asserts that E_1 and E_2 are equal.

LEMMA 4.1. *Suppose that τ_N is continuous for $N = 1, 2, \dots$ and that τ_N converges pointwise to a limit τ. Then τ_N is equicontinuous at $P_0 \in \mathscr{P}$ if and only if τ_N is locally uniformly convergent at P_0.*

PROOF. Suppose that τ_N is locally uniformly convergent at P_0 and fix $\varepsilon > 0$. Then there exists a neighborhood U_ε of P_0 and an integer N_ε such that

$$r(\tau_N(P), \tau(P)) \le \varepsilon \quad \text{for } P \in U_\varepsilon \text{ and } N \ge N_\varepsilon.$$

Hence, for $P \in U_\varepsilon$ and $N \ge N_\varepsilon$,

$$\begin{aligned} r(\tau_N(P), \tau_N(P_0)) &\le r(\tau_N(P), \tau_{N_\varepsilon}(P)) + r(\tau_N(P_0), \tau_{N_\varepsilon}(P_0)) \\ &\quad + r(\tau_{N\varepsilon}(P), \tau_{N_\varepsilon}(P_0)) \\ &\le 4\varepsilon + r(\tau_{N_\varepsilon}(P), \tau_{N_\varepsilon}(P_0)). \end{aligned}$$

Since τ_N is continuous for every N, there exists a neighborhood U'_ε of P_0 such that for $P \in U'_\varepsilon$, $r(\tau_N(P), \tau_N(P_0)) \le \varepsilon$ for $N \le N_\varepsilon$, so

$$r(\tau_N(P), \tau_N(P_0)) \le 5\varepsilon$$

for $P \in U_\varepsilon \cap U'_\varepsilon$ and all N. Hence τ_N is equicontinuous at P_0.

Conversely, suppose that τ_N is equicontinuous at P_0. Fix $\varepsilon > 0$. Then there exists a neighborhood U_ε of P_0 such that $r(\tau_N(P), \tau_N(P_0)) \le \varepsilon$ for $P \in U_\varepsilon$ and all N. Since $\tau_N \to \tau$, this implies that $r(\tau(P), \tau(P_0)) \le \varepsilon$ for $P \in U_\varepsilon$. Hence, for $P \in U_\varepsilon$ and all N,

$$\begin{aligned} r(\tau_N(P), \tau(P)) &\le r(\tau_N(P), \tau_N(P_0)) + r(\tau(P), \tau(P_0)) + r(\tau_N(P_0), \tau(P_0)) \\ &\le 2\varepsilon + r(\tau_N(P_0), \tau(P_0)). \end{aligned}$$

As $\tau_N \to \tau$, there exists N_ε such that for all $P \in U_\varepsilon$ and $N \ge N_\varepsilon$,

$$r(\tau_N(P), \tau(P)) \le 3\varepsilon,$$

so τ_N is locally uniformly convergent at P_0. □

LEMMA 4.2. *Suppose that τ_N is continuous for each N and that τ_N converges pointwise to a limit τ. Let E_1 and E_2 be defined as in* (4.1) *and* (4.2). *Then E_1^c (and hence also E_2^c) is a set of the first category in* $(\mathscr{P}, \Pi)$.

PROOF. The sequence of maps τ_N is locally uniformly convergent at P_0 iff for every m there exists $M = M(m)$ such that P_0 is an interior point of

$$G_{mM} = \left\{P: r(\tau_N(P), \tau(P)) \leq \frac{1}{2m} \text{ for all } N \geq M\right\},$$

and hence an interior point of the larger set

$$F_{mM} = \{P: r(\tau_N(P), \tau_{N'}(P)) \leq 1/m \text{ for all } N, N' \geq M\}.$$

As τ_N is continuous, F_{mM} is the intersection of closed sets and is therefore closed. Since $\tau_N \to \tau$, clearly $\bigcup_{M=1}^{\infty} F_{mM} = \mathscr{P}$ for every m.

Let $\mathring{F}_{mM}$ denote the interior of F_{mM}. The sequence τ_N is not locally uniformly convergent at P iff there exists m such that $P \notin \bigcup_M \mathring{F}_{mM}$, that is, iff for some m,

$$P \in \bigcup_{M=1}^{\infty} F_{mM} \setminus \bigcup_{M'=1}^{\infty} \mathring{F}_{mM'}.$$

Hence

$$E_1^c = \bigcup_{m=1}^{\infty}\left(\bigcup_{M=1}^{\infty} F_{mM} \setminus \bigcup_{M'=1}^{\infty} \mathring{F}_{mM'}\right) \subset \bigcup_{m=1}^{\infty}\bigcup_{M=1}^{\infty}\left(F_{mM} \setminus \mathring{F}_{mM}\right).$$

Since F_{mM} is closed, $F_{mM} \setminus \mathring{F}_{mM}$ is a closed set with empty interior and therefore nowhere dense. It follows that the set $\bigcup_m \bigcup_M (F_{mM} \setminus \mathring{F}_{mM})$ is a countable union of nowhere dense sets, and hence of the first category and a fortiori so is E_1^c. The lemma is proved. □

PROOF OF THEOREM 1.2. When τ_N is equicontinuous at P and $\hat{P}_N$ is a consistent estimator of P, it is clear from the argument leading to Theorem 1.1 that $r(\tau_N(\hat{P}_N), \tau_N(P)) \to_P 0$. It follows from Lemmas 4.1 and 4.2 that there exists a set D of the first category in $\mathscr{P}$ such that τ_N is equicontinuous at all $P \in \mathscr{P} \setminus D$. The theorem is proved. □

5. Proof of Theorem 1.4. In this section we shall assume that $\mathscr{P}$ is a metric space, equipped with Hellinger distance H and that τ_N: $(\mathscr{P}, H) \to (\mathscr{R}, r)$ is continuous for each N. We shall omit the assumption that τ_N converges pointwise. For the proof of Theorem 1.4 we follow an entirely different path. Let us first collect some results about Hellinger distance. Suppose that for each $N = 1, 2, \ldots$, $(\mathscr{X}_N, \mathscr{A}_N)$ is a measurable space and Q_{1N} and Q_{2N} are probability measures on $(\mathscr{X}_N, \mathscr{A}_N)$ with densities q_{1N} and q_{2N} with respect to a σ-finite dominating measure μ_N. Let us define an *asymptotically perfect test* for distinguishing between $\{Q_{1N}\}$ and $\{Q_{2N}\}$ as a sequence of tests for Q_{1N} against Q_{2N} for which the probabilities of errors of both type I and type II tend to zero as N tends to infinity. Existence of such an asymptotically perfect test between $\{Q_{1N}\}$ and $\{Q_{2N}\}$ is related to the Hellinger distance between Q_{1N} and Q_{2N}. In particular, an asymptotically perfect test cannot exist if $\limsup_{N\to\infty} H^2(Q_{1N}, Q_{2N}) < 2$.

LEMMA 5.1. *Suppose that* $\limsup_{N\to\infty} H^2(Q_{1N}, Q_{2N}) < 2$. *Then for any sequence* $A_N \in \mathscr{A}_N$,

$$\liminf_{N\to\infty}(Q_{1N}(A_N) + Q_{2N}(A_N^c)) > 0.$$

PROOF. We can write

$$H^2(Q_{1N}, Q_{2N}) = \int \left(q_{1N}^{1/2} - q_{2N}^{1/2}\right)^2 d\mu_N = 2 - 2\int (q_{1N}q_{2N})^{1/2}\, d\mu_N, \tag{5.1}$$

and hence $\liminf \int (q_{1N}q_{2N})^{1/2}\, d\mu_N > 0$. The Cauchy–Schwarz inequality ensures that for every N and for any set $A_N \in \mathscr{A}_N$,

$$\begin{aligned}\int (q_{1N}q_{2N})^{1/2}\, d\mu_N &= \int_{A_N} (q_{1N}q_{2N})^{1/2}\, d\mu_N + \int_{A_N^c} (q_{1N}q_{2N})^{1/2}\, d\mu_N \\ &\le \left(\int_{A_N} q_{1N}\, d\mu_N \int_{A_N} q_{2N}\, d\mu_N\right)^{1/2} \\ &\quad + \left(\int_{A_N^c} q_{1N}\, d\mu_N \int_{A_N^c} q_{2N}\, d\mu_N\right)^{1/2} \\ &\le (Q_{1N}(A_N))^{1/2} + (Q_{2N}(A_N^c))^{1/2}.\end{aligned}$$

The lemma follows. □

Let P and P_N be probability measures on a measurable space $(\mathscr{X}, \mathscr{A})$ with densities p and p_N with respect to a σ-finite measure μ on $(\mathscr{X}, \mathscr{A})$. On the product measurable space $(\mathscr{X}^N, \mathscr{A}^N)$ we define the product measures $Q_{1N} = P^N$ and $Q_{2N} = P_N^N$ with densities $\prod_{i=1}^N p(x_i)$ and $\prod_{i=1}^N p_N(x_i)$ with respect to $\mu_N = \mu^N$. By (5.1) we have

$$\begin{aligned}1 - \tfrac{1}{2}H^2(Q_{1N}, Q_{2N}) &= \int \cdots \int \prod_{i=1}^N \{p(x_i)p_N(x_i)\}^{1/2}\, d\mu_N \\ &= \left[\int \{p(x)p_N(x)\}^{1/2}\, d\mu\right]^N \qquad (5.2) \\ &= \left[1 - \tfrac{1}{2}H^2(P, P_N)\right]^N.\end{aligned}$$

It follows that $H(P, P_N) = \mathscr{O}(N^{-1/2})$ implies that $\limsup_N H^2(Q_{1N}, Q_{2N}) < 2$, and Lemma 5.1 yields the following corollary:

COROLLARY 5.1. *Suppose* $H(P, P_N) = \mathscr{O}(N^{-1/2})$. *Then for any sequence* $A_N \in \mathscr{A}^N$,

$$\liminf_{N\to\infty}(P^N(A_N) + P_N^N(A_N^c)) > 0.$$

LEMMA 5.2. *Suppose that* $\tau_N\colon (\mathscr{P}, H) \to (\mathscr{R}, r)$ *is continuous for every* N *and that* $T_N = t_N(X_1, \dots, X_N)$ *is a consistent estimator of* $\tau_N(P)$, *that is*,

$$r(T_N, \tau_N(P)) \to_P 0 \quad \textit{for every } P \in \mathscr{P}.$$

Then there exists a set D *of the first category in* $(\mathscr{P}, H)$ *such that for every* $P_0 \in \mathscr{P} \setminus D$, *every* $\varepsilon > 0$ *and every sequence* $\delta_N \searrow 0$,

$$\lim_{N \to \infty} \sup_{\{P:\, H(P, P_0) \le \delta_N\}} P^N(r(T_N, \tau_N(P)) \ge \varepsilon) = 0. \tag{5.3}$$

PROOF. Fix an integer $k > 0$ and define $\psi_N^{(k)}\colon (\mathscr{P}, H) \to \mathbb{R}$ with Euclidean distance by

$$\psi_N^{(k)}(P) = P^N(r(T_N, \tau_N(P)) \ge 1/k), \qquad N = 1, 2, \dots .$$

Clearly, $\psi_N^{(k)}(P) \to 0$ as $N \to \infty$ for every $P \in \mathscr{P}$. Since we would like to apply Lemma 4.2 to $\psi_N^{(k)}$, we would also need continuity and hence we modify $\psi_N^{(k)}$ slightly: choose $\delta_N \searrow 0$ and define

$$\tilde{\psi}_N^{(k)}(P) = \frac{1}{\delta_N} \int_{1/k}^{1/k + \delta_N} P^N(r(T_N, \tau_N(P)) \ge u)\, du.$$

We have

$$0 \le \tilde{\psi}_N^{(k)}(P) \le \psi_N^{(k)}(P) \le \tilde{\psi}_N^{(k+1)}(P), \tag{5.4}$$

the last inequality for $N \ge N_0 = N_0(k)$, such that $\delta_{N_0} \le 1/(k(k+1))$. Therefore, $\tilde{\psi}_N^{(k)} \to 0$ on $\mathscr{P}$, but $\tilde{\psi}_N^{(k)}$ is also continuous on $\mathscr{P}$. To see this, note that for any $P_1, P_2 \in \mathscr{P}$ and $A \in \mathscr{A}^N$, $|P_1^N(A) - P_2^N(A)| \le H(P_1^N, P_2^N)$, and for fixed N, we can make this arbitrarily small by taking $H(P_1, P_2)$ small [cf. (5.2)]. Hence, for every fixed N,

$$\left| \tilde{\psi}_N^{(k)}(P_1) - \frac{1}{\delta_N} \int_{1/k}^{1/k + \delta_N} P_2^N(r(T_N, \tau_N(P_1)) \ge u)\, du \right|$$

can be made as small as we wish to taking $H(P_1, P_2)$ small. Since τ_N is continuous and the integral defining $\tilde{\psi}_N^{(k)}$ depends continuously on the upper and lower bound of the range of integration, the same is true for

$$\left| \frac{1}{\delta_N} \int_{1/k}^{1/k + \delta_N} P_2^N(r(T_N, \tau_N(P_1)) \ge u)\, du - \tilde{\psi}_N^{(k)}(P_2) \right|,$$

which proves the continuity of each of the functions $\tilde{\psi}_N^{(k)}$.

Application of Lemma 4.2 yields the existence of a set $D^{(k)}$ of the first category in $(\mathscr{P}, H)$ such that $\{\tilde{\psi}_N^{(k)}\}$ is locally uniformly convergent at P_0 for every $P_0 \in \mathscr{P} \setminus D^{(k)}$, so a fortiori

$$\sup_{\{P:\, H(P, P_0) \le \delta_N\}} \tilde{\psi}_N^{(k)}(P) \to 0 \tag{5.5}$$

for every $P_0 \in \mathscr{P} \setminus D^{(k)}$ and every sequence $\delta_N \searrow 0$. Taking $D = \bigcup_{k=1}^{\infty} D^{(k)}$ and noting that D is also of the first category, we find that (5.5) holds for all $k = 1, 2, \dots,$ provided that $P_0 \in \mathscr{P} \setminus D$. Because of (5.4), this implies the same for $\psi_N^{(k)}$ itself, and hence we obtain (5.3) and the lemma. □

LEMMA 5.3. *Suppose that* $\tau_N: (\mathscr{P}, H) \to (\mathscr{R}, r)$ *is continuous for every* N *and that* $T_N = t_N(X_1, \dots, X_N)$ *is a consistent estimator of* $\tau_N(P)$. *Then there exists a set* D *of the first category in* $(\mathscr{P}, H)$ *such that for every* $P_0 \in \mathscr{P} \setminus D$ *and every* $C > 0$,

$$\sup_{\{P:\, H(P, P_0) \le CN^{-1/2}\}} r(\tau_N(P), \tau_N(P_0)) \to 0. \tag{5.6}$$

PROOF. According to Lemma 5.2 we can choose a set D of the first category in $(\mathscr{P}, H)$ such that for every $P_0 \in \mathscr{P} \setminus D$, every $C > 0$ and every $\varepsilon > 0$,

$$\sup_{\{P:\, H(P, P_0) \le CN^{-1/2}\}} P^N(r(T_N, \tau_N(P)) \ge \varepsilon) \to 0. \tag{5.7}$$

Fix $P_0 \in \mathscr{P} \setminus D$ and take any sequence $P_N \in \mathscr{P}$ with $H(P_N, P_0) \le CN^{-1/2}$. For $N = 1, 2, \dots,$ define

$$A_N = \{r(T_N, \tau_N(P_0)) \ge r(T_N, \tau_N(P_N))\} \in \mathscr{A}^N.$$

On A_N we have

$$\begin{aligned} r(T_N, \tau_N(P_0)) &\ge 1/2\{r(T_N, \tau_N(P_N)) + r(T_N, \tau_N(P_0))\} \\ &\ge 1/2 r(\tau_N(P_N), \tau_N(P_0)), \end{aligned}$$

and similarly on A_N^c,

$$r(T_N, \tau_N(P_N)) > 1/2 r(\tau_N(P_N), \tau_N(P_0)).$$

It follows that

$$\begin{aligned} \left[P_0^N(A_N) + P_N^N(A_N^c)\right] \le\ & 2\mathbf{I}_{(0, 2\varepsilon)}(r(\tau_N(P_N), \tau_N(P_0))) \\ & + 2 \sup_{\{P:\, H(P, P_0) \le CN^{-1/2}\}} P^N(r(T_N, \tau_N(P)) \ge \varepsilon). \end{aligned}$$

Because $H(P_N, P_0) \le CN^{-1/2}$, we can combine Corollary 5.1 and (5.7) to conclude that

$$\limsup_N r(\tau_N(P_N), \tau_N(P_0)) \le 2\varepsilon.$$

Since ε is an arbitrary positive number and P_N is an arbitrary sequence with $H(P_N, P_0) \le CN^{-1/2}$ the proof is complete. □

PROOF OF THEOREM 1.4. Take D as in Lemma 5.3, fix $P \in \mathscr{P} \setminus D$ and take $\varepsilon > 0$. By assumption (iii), we can find $C > 0$ such that for every N,

$$P^N\left(H(\hat{P}_N, P) \le CN^{-1/2}\right) \ge 1 - \varepsilon.$$

Application of Lemma 5.3 yields that for every positive δ and for every $P \in \mathscr{P} \setminus D$,

$$\limsup_N P^N\left(r\left(\tau_N(\hat{P}_N), \tau_N(P)\right) > \delta\right) \le \varepsilon,$$

and as ε is positive but otherwise arbitrary, this proves the theorem. □

6. Proof of Theorem 1.3. In this section we only assume measurability of the maps τ_N, separability of $(\mathscr{R}, r)$ and the existence of an r-consistent estimator T_N of $\tau_N(P)$. To construct an r-consistent substitution estimator $\tau_N(\hat{P}_N)$, we begin by choosing $\hat{P}_N = p_N(X_1, \dots, X_N)$ to be an approximate minimum distance estimator, that is, an estimator satisfying

$$r\big(T_N, \tau_N(\hat{P}_N)\big) \le \inf_{P \in \mathscr{P}} r(T_N, \tau_N(P)) + \varepsilon_N \tag{6.1}$$

for some sequence $\varepsilon_N \searrow 0$. If this can be done in such a way that $p_N: \mathscr{X}^N \to \mathscr{P}$ is measurable so that $\hat{P}_N$ is a proper estimator, then the consistency of $\tau_N(\hat{P}_N)$ will follow, since for every $P_0 \in \mathscr{P}$,

$$\begin{aligned} r\big(\tau_N(\hat{P}_N), \tau_N(P_0)\big) &\le r\big(T_N, \tau_N(\hat{P}_N)\big) + r(T_N, \tau_N(P_0)) \\ &\le \inf_{P \in \mathscr{P}} r(T_N, \tau_N(P)) + \varepsilon_N + r(T_N, \tau_N(P_0)) \\ &\le 2r(T_N, \tau_N(P_0)) + \varepsilon_N \to_{P_0} 0 \end{aligned}$$

because of the consistency of T_N.

It remains to be shown that (6.1) can be satisfied for a measurable p_N. Define $\mathscr{R}^0 = \{\tau_N(P): P \in \mathscr{P}\}$, let $\overline{\mathscr{R}}^0$ denote the closure of $\mathscr{R}^0$ in $(\mathscr{R}, r)$ and let T_N^0 denote the projection of $T_N \in \mathscr{R}$ on $\overline{\mathscr{R}}^0$. Inequality (6.1) asserts that $\hat{P}_N$ must be chosen in such a way that $\tau_N(\hat{P}_N)$ lies in a ball with center T_N and radius $r(T_N, T_N^0) + \varepsilon_N$. One easily convinces oneself that this implies that for any ball $B \subset \mathscr{R}$ with radius $\varepsilon_N/3$, we can choose a fixed $\hat{P}_N$ satisfying (6.1) for all $T_N \in B$. As $(\mathscr{R}, r)$ is separable, we can cover $\mathscr{R}$ with a countable number of balls B_k with radius $\varepsilon_N/3$. For every m, we can therefore define $\hat{P}_N$ as a fixed point of $\mathscr{P}$ for all T_N in the measurable set $B_m \cap \bigcap_{k=1}^{m-1} B_k^c$, so that $\hat{P}_N$ is an elementary measurable function of T_N. As T_N is a measurable function of $X_1, \dots, X_N$, the proof is complete. □

Acknowledgments. The idea that contiguity of $\hat{P}_N$ with respect to P or, as it turned out, $\sqrt{N}$-Hellinger consistency of $\hat{P}_N$, would be relevant in this context grew out of conversations with Dimitri Chibisov. Jaap Fabius patiently explained measure-theoretic difficulties to us. An Associate Editor and three referees provided constructive comments. Our sincere thanks go to all of them.

REFERENCES

BERAN, R. J. (1982). Estimated sampling distributions: the bootstrap and competitors. *Ann. Statist.* **10** 212–225.

BERAN, R. J. (1984). Bootstrap methods in statistics. *Jber. Deutsch. Math.-Verein.* **86** 14–30.

BICKEL, P. J. and FREEDMAN, D. A. (1981). Some asymptotic theory for the bootstrap. *Ann. Statist.* **9** 1196–1217.

BILLINGSLEY, P. (1986). *Convergence of Probability Measures*. Wiley, New York.

BIRGÉ, L. (1983). Approximation dans les espaces métriques et théorie de l'estimation. *Z. Wahrsch. Verw. Gebiete* **65** 181–237.

BIRGÉ, L. (1986). On estimating a density using Hellinger distance and some other strange facts. *Probab. Theory Related Fields* **71** 271–291.

DUDLEY, R. M. (1989). *Real Analysis and Probability*. Wadsworth, Belmont, CA.

LE CAM, L. M. (1953). On some asymptotic properties of maximum likelihood estimates and related Bayes' estimates. *University of California Publications in Statistics* **1** 277–330.

LE CAM, L. M. (1973). Convergence of estimates under dimensionality restrictions. *Ann. Statist.* **1** 38–53.

LE CAM, L. M. (1986). *Asymptotic Methods in Statistical Decision Theory*. Springer, New York.

LE CAM, L. M. and YANG, G. L. (1990). *Asymptotics in Statistics: Some Basic Concepts*. Springer, New York.

POLITIS, D. and ROMANO, J. (1994). Large sample confidence regions based on subsamples under minimal assumptions. *Ann. Statist.* **22** 2031–2050.

PUTTER, H. (1994). Consistency of resampling methods. Ph.D. thesis, Univ. Leiden.

PUTTER, H. and VAN ZWET, W. R. (1996). On a set of the first category. In *Festschrift for Lucien Le Cam*. Springer, New York. To appear.

VAN ZWET, W. R. (1996). Resampling: the jackknife and the naïve bootstrap. Unpublished manuscript.

DEPARTMENT OF MATHEMATICS
FREE UNIVERSITY OF AMSTERDAM
1081 DE BOELELAAN
1081 HV AMSTERDAM
THE NETHERLANDS

DEPARTMENT OF MATHEMATICS
AND COMPUTER SCIENCE
UNIVERSITY OF LEIDEN
P.O. BOX 9512
2300 RA LEIDEN
THE NETHERLANDS

Statistica Sinica **7**(1997), 1-31

RESAMPLING FEWER THAN n OBSERVATIONS: GAINS, LOSSES, AND REMEDIES FOR LOSSES

P. J. Bickel, F. Götze and W. R. van Zwet

University of California, Berkeley,
University of Bielefeld and University of Leiden

Abstract: We discuss a number of resampling schemes in which $m = o(n)$ observations are resampled. We review nonparametric bootstrap failure and give results old and new on how the m out of n with replacement and without replacement bootstraps work. We extend work of Bickel and Yahav (1988) to show that m out of n bootstraps can be made second order correct, if the usual nonparametric bootstrap is correct and study how these extrapolation techniques work when the nonparametric bootstrap does not.

Key words and phrases: Asymptotic, bootstrap, nonparametric, parametric, testing.

1. Introduction

Over the last 10-15 years Efron's nonparametric bootstrap has become a general tool for setting confidence regions, prediction, estimating misclassification probabilities, and other standard exercises of inference when the methodology is complex. Its theoretical justification is based largely on asymptotic arguments for its consistency or optimality. A number of examples have been addressed over the years in which the bootstrap fails asymptotically. Practical anecdotal experience seems to support theory in the sense that the bootstrap generally gives reasonable answers but can bomb.

In a recent paper Politis and Romano (1994), following Wu (1990), and independently Götze (1993) showed that what we call the m out of n without replacement bootstrap with $m = o(n)$ typically works to first order both in the situations where the bootstrap works and where it does not.

The m out of n with replacement bootstrap with $m = o(n)$ has been known to work in all known realistic examples of bootstrap failure. In this paper,

- We show the large extent to which the Politis, Romano, Götze property is shared by the m out of n with replacement bootstrap and show that the latter has advantages.
- If the usual bootstrap works the m out of n bootstraps pay a price in efficiency. We show how, by the use of extrapolation the price can be avoided.

S. van de Geer and M. Wegkamp (eds.), *Selected Works of Willem van Zwet*, Selected Works in Probability and Statistics, DOI 10.1007/978-1-4614-1314-1_17,

- We support some of our theory with simulations.

The structure of our paper is as follows. In Section 2 we review a series of examples of success and failure to first order (consistency) of (Efron's) nonparametric bootstrap (nonparametric). We try to isolate at least heuristically some causes of nonparametric bootstrap failure. Our framework here is somewhat novel. In Section 3 we formally introduce the m out of n with and without replacement bootstrap as well as what we call "sample splitting", and establish their first order properties restating the Politis-Romano-Götze result. We relate these approaches to smoothing methods. Section 4 establishes the deficiency of the m out of n bootstrap to higher order if the nonparametric bootstrap works to first order and Section 5 shows how to remedy this deficiency to second order by extrapolation. In Section 6 we study how the improvements of Section 5 behave when the nonparametric bootstrap doesn't work to first order. We present simulations in Section 7 and proofs of our new results in Section 8. The critical issue of choice of m and applications to testing will be addressed elsewhere.

2. Successes and Failure of the Bootstrap

We will limit our work to the i.i.d. case because the issues we discuss are clearest in this context. Extension to the stationary mixing case, as done for the m out of n without replacement bootstrap in Politis and Romano (1994), are possible but the study of higher order properties as in Sections 4 and 5 of our paper is more complicated.

We suppose throughout that we observe $X_1, \ldots, X_n$ taking values in $X = R^p$ (or more generally a separable metric space). i.i.d. according to $F \in \mathcal{F}_0$. We stress that $\mathcal{F}_0$ need not be and usually isn't the set of all possible distributions. In hypothesis testing applications, $\mathcal{F}_0$ is the hypothesized set, in looking at the distributions of extremes, $\mathcal{F}_0$ is the set of populations for which extremes have limiting distributions. We are interested in the distribution of a symmetric function of $X_1, \ldots, X_n$; $T_n(X_1, \ldots, X_n, F) \equiv T_n(\hat{F}_n, F)$ where $\hat{F}_n$ is defined to be the empirical distribution of the data. More specifically we wish to estimate a parameter which we denote $\theta_n(F)$, of the distribution of $T_n(\hat{F}_n, F)$, which we denote by $\mathcal{L}_n(F)$. We will usually think of θ_n as real valued, for instance, the variance of $\sqrt{n}$ median $(X_1, \ldots, X_n)$ or the 95% quantile of the distribution of $\sqrt{n}(\bar{X} - E_F(X_1))$.

Suppose $T_n(\cdot, F)$ and hence θ_n is defined naturally not just on $\mathcal{F}_0$ but on $\mathcal{F}$ which is large enough to contain all discrete distributions. It is then natural to estimate F by the nonparametric maximum likelihood estimate, (NPMLE), $\hat{F}_n$, and hence $\theta_n(F)$ by the plug in $\theta_n(\hat{F}_n)$. This is Efron's (ideal) nonparametric bootstrap. Since $\theta_n(F) \equiv \gamma(\mathcal{L}_n(F))$ and, in the cases we consider, computation of γ is straightforward the real issue is estimation of $\mathcal{L}_n(F)$. Efron's (ideal) bootstrap is to estimate $\mathcal{L}_n(F)$ by the distribution of $T_n(X_1^*, \ldots, X_n^*, \hat{F}_n)$ where, given $X_1, \ldots, X_n$ the X_i^* are i.i.d. $\hat{F}_n$, i.e. the bootstrap distribution of T_n. In practice, the bootstrap distribution is itself estimated by Monte Carlo or more sophisticated resampling schemes, (see DeCiccio and Romano (1989) and Hikley (1988)). We will not enter into this question further.

Theoretical analyses of the bootstrap and its properties necessarily rely on asymptotic theory, as $n \to \infty$ coupled with simulations. We restrict analysis to $T_n(\hat{F}_n, F)$ which are asymptotically stable and nondegenerate on $\mathcal{F}_0$. That is, for all $F \in \mathcal{F}_0$, at least weakly

$$\mathcal{L}_n(F) \to \mathcal{L}(F) \text{ non degenerate}$$

$$\theta_n(F) \to \theta(F) \tag{2.1}$$

as $n \to \infty$.

Using m out of n bootstraps or sample splitting implicitly changes our goal from estimating features of $\mathcal{L}_n(F)$ to features of $\mathcal{L}_m(F)$. This is obviously nonsensical without assuming that the laws converge.

Requiring non degeneracy of the limit law means that we have stabilized the scale of $T_n(\hat{F}_n, F)$. Any functional of $\mathcal{L}_n(F)$ is also a functional of the distribution of $\sigma_n T_n(\hat{F}_n, F)$ where $\sigma_n \to 0$ which also converges in law to point mass at 0. Yet this degenerate limit has no functional $\theta(F)$ of interest.

Finally, requiring that stability need occur only on $\mathcal{F}_0$ is also critical since failure to converge off $\mathcal{F}_0$ in a reasonable way is the first indicator of potential bootstrap failure.

2.1. When does the nonparametric bootstrap fail?

If θ_n does not depend on n, the bootstrap works, (is consistent on $\mathcal{F}_0$), if θ is continuous at all points of $\mathcal{F}_0$ with respect to weak convergence on $\mathcal{F}$. Conversely, the nonparametric bootstrap can fail if,

1. θ is not continuous on $\mathcal{F}_0$.
   An example we explore later is $\theta_n(F) = 1(F \text{ discrete})$ for which $\theta_n(\hat{F}_n)$ obviously fails if F is continuous.
   Dependence on n introduces new phenomena. In particular, here are two other reasons for failure we explore below.
2. θ_n is well defined on all of $\mathcal{F}$ but θ is defined on $\mathcal{F}_0$ only or exhibits wild discontinuities when viewed as a function on $\mathcal{F}$. This is the main point of examples 3-6.
3. $T_n(\hat{F}_n, F)$ is not expressible as or approximable on $\mathcal{F}_0$ by a continuous function of $\sqrt{n}(\hat{F}_n - F)$ viewed as an object weakly converging to a Gaussian limit in a suitable function space. (See Giné and Zinn (1989).) Example 7 illustrate this failure. Again this condition is a diagnostic and not necessary for failure as Example 6 shows.

We illustrate our framework and discuss prototypical examples of bootstrap success and failure.

2.2. Examples of bootstrap success

Example 1. Confidence intervals: Suppose $\sigma^2(F) \equiv \text{Var}_F(X_1) < \infty$ for all $F \in \mathcal{F}_0$.
(a) Let $T_n(\hat{F}_n, F) \equiv \sqrt{n}(\bar{X} - E_F X_1)$. For the percentile bootstrap we are interested in $\theta_n(F) \equiv P_F[T_n(\hat{F}_n, F) \le t]$. Evidently $\theta(F) = \Phi(\frac{t}{\sigma(F)})$. In fact, we want to estimate the quantiles of the distribution of $T_n(\hat{F}_n, F)$. If $\theta_n(F)$ is the $1-\alpha$ quantile then $\theta(F) = \sigma(F) z_{1-\alpha}$ where z is the Gaussian quantile.
(b) Let $T_n(\hat{F}_n, F) = \sqrt{n}(\bar{X} - E_F X_1)/s$ where $s^2 = \frac{1}{n-1}\sum_{i=1}^n (X_i - \bar{X})^2$. If $\theta_n(F) \equiv P_F(T_n(\hat{F}_n, F) \le t]$ then, $\theta(F) = \Phi(t)$, independent of F. It seems silly to be estimating a parameter whose value is known but, of course, interest now centers on $\theta'(F)$ the next higher order term in $\theta_n(F) = \Phi(t) + \frac{\theta'(F)}{\sqrt{n}} + O(n^{-1})$.

Example 2. Estimation of variance: Suppose F has unique median $m(F)$, continuous density $f(m(F)) > 0$, $E_F|X|^\delta < \infty$, some $\delta > 0$ for all $F \in \mathcal{F}_0$ and $\theta_n(F) = \text{Var}_F(\sqrt{n}\text{ median }(X_1, \ldots, X_n))$. Then $\theta(F) = [4f^2(m(F))]^{-1}$ on $\mathcal{F}_0$.

Note that, whereas θ_n is defined for all empirical distributions F in both examples the limit $\theta(F)$ is 0 or ∞ for such distributions in the second. Nevertheless, it is well known (see Efron (1979)) that the nonparametric bootstrap is consistent in both examples in the sense that $\theta_n(\hat{F}_n) \xrightarrow{P} \theta(F)$ for $F \in \mathcal{F}_0$.

2.3. Examples of bootstrap failure

Example 3. Confidence bounds for an extremum: This is a variation on Bickel Freedman (1981). Suppose that all $F \in \mathcal{F}_0$ have a density f continuous and positive at $F^{-1}(0) > -\infty$. It is natural to base confidence bounds for $F^{-1}(0)$ on the bootstrap distribution of

$$T_n(\hat{F}_n, F) = n(\min_i X_i - F^{-1}(0)).$$

Let

$$\theta_n(F) = P_F[T_n(\hat{F}_n, F) > t] = (1 - F(\frac{t}{n} + F^{-1}(0))^n.$$

Evidently $\theta_n(F) \to \theta(F) = \exp(-f(F^{-1}(0))t)$ on $\mathcal{F}_0$.

The nonparametric bootstrap fails. Let

$$N_n^*(t) = \sum_{i=1}^n 1(X_i^* \le \frac{t}{n} + X_{(1)}), t > 0,$$

where $X_{(1)} \equiv \min_i X_i$ and $1(A)$ is the indicator of A. Given $X_{(1)}$, $n\hat{F}_n(\frac{t}{n} + X_{(1)})$ is distributed as 1+ binomial $(n-1, \frac{F(\frac{t}{n}+X_{(1)})-F(X_{(1)})}{(1-F(X_{(1)}))})$ which converges weakly to a Poisson $(f(F^{-1}(0))t)$ variable. More generally, $n\hat{F}_n(\frac{\cdot}{n}+X_{(1)})$ converges weakly conditionally to $1+N(\cdot)$, where N is a homogeneous Poisson process with parameter $f(F^{-1}(0))$. It follows that $N_n^*(\cdot)$ converges weakly (marginally) to a process $M(1+N(\cdot))$ where M is a standard Poisson process independent of $N(\cdot)$. Thus if, in Efron's notation, we use P^* to denote conditional probability given $\hat{F}_n$ and let $\hat{F}_n^*$, be the empirical d.f. of $X_1^*,\ldots,X_n^*$ then $P^*[T_n(\hat{F}_n^*)>t]=P^*[N_n^*(t)=0]$ converges weakly to the random variable $P[M(1+N(t))=0|N]=e^{-(N(t)+1)}$ rather than to the desired $\theta(F)$.

Example 4. Extrema for unbounded distributions: (Athreya and Fukuchi (1994), Deheuvels, Mason, Shorack (1993))

Suppose $F\in\mathcal{F}_0$ are in the domain of attraction of an extreme value distribution. That is: for some constants $A_n(F)$, $B_n(F)$,

$$n(1-F)(A_n(F)+B_n(F)x)\to H(x,F),$$

where H is necessarily one of the classical three types (David (1981), p.259): $e^{-\beta x}1(\beta x\geq 0)$, $\alpha x^{-\beta}1(x\geq 0)$, $\alpha(-x)^{\beta}1(x\leq 0)$, for $\alpha,\beta\neq 0$. Let,

$$\theta_n(F)\equiv P[(\max(X_1,\ldots,X_n)-A_n(F))/B_n(F)\leq t]\to e^{-H(t,F)}\equiv\theta(F). \quad (2.2)$$

Particular choices of $A_n(F)$, for example, $F^{-1}(1-\frac{1}{n})$ and $B_n(F)$ are of interest in inference. However, the bootstrap does not work. It is easy to see that

$$n(1-\hat{F}_n(A_n(F)+tB_n(F)))\xrightarrow{w}N(t), \quad (2.3)$$

where N is an inhomogeneous Poisson process with parameter $H(t,F)$ and $\xrightarrow{w}$ denotes weak convergence. Hence if $T_n(\hat{F}_n,F)=(\max(X_1,\ldots,X_n)-A_n(F))/B_n(F)$ then

$$P^*[T_n(\hat{F}_n^*,F)\leq t]\xRightarrow{w}e^{-N(t)}. \quad (2.4)$$

It follows that the nonparametric bootstrap is inconsistent for this choice of A_n, B_n. If it were consistent, then

$$P^*[T_n(\hat{F}_n^*,\hat{F}_n)\leq t]\xrightarrow{P}e^{-H(t,F)} \quad (2.5)$$

for all t and (2.5) would imply that it is possible to find random A real and $B\neq 0$ such that $N(Bt+A)=H(t,F)$ with probability 1. But $H(t,F)$ is continuous except at 1 point. So (2.4) and (2.5) contradict each other. Again, $\theta(F)$ is well defined for $F\in\mathcal{F}_0$ but not otherwise. Furthermore, small perturbations in F can lead to drastic changes in the nature of H, so that θ is not continuous if $\mathcal{F}_0$ is as large as possible.

Essentially the same bootstrap failure arises when we consider estimating the mean of distributions in the domain of attraction of stable laws of index $1<\alpha\leq 2$. (See Athreya (1987))

Example 5. Testing and improperly centered U and V statistics: (Bretagnolle (1983))

Let $\mathcal{F}_0 = \{F : F[-c, c] = 1, E_F X_1 = 0\}$ and let $T_n(\hat{F}_n) = n\bar{X}^2 = n \int xy d\hat{F}_n(x) d\hat{F}_n(y)$. This is a natural test statistic for $H : F \in \mathcal{F}_0$. Can one use the nonparametric bootstrap to find the critical value for this test statistic? Intuitively, $\hat{F}_n \notin \mathcal{F}_0$ and this procedure is rightly suspect. Nevertheless, in more complicated contexts, it is a mistake made in practice. David Freedman pointed us to Freedman et al. (1994) where the Bureau of the Census appears to have fallen into such a trap. (see Hall and Wilson (1991) for other examples.) The nonparametric bootstrap may, in general, not be used for testing as will be shown in a forthcoming paper.

In this example, due to Bretagnolle (1983), we focus on $\mathcal{F}_0$ for which a general U or V statistic T is degenerate and show that the nonparametric bootstrap doesn't work. More generally, suppose $\psi : R^2 \to R$ is bounded and symmetric and let $\mathcal{F}_0 = \{F : \int \psi(x, y) dF(x) = 0 \text{ for all } y\}$.

Then, it is easy to see that

$$T_n(\hat{F}_n) = \int \psi(x, y) dW_n^0(x) dW_n^0(y), \tag{2.6}$$

where $W_n^0(x) \equiv \sqrt{n}(\hat{F}_n(x) - F(x))$ and well known that

$$\theta_n(F) \equiv P_F[T_n(\hat{F}_n) \le t] \to P\Big[\int \psi(xy) dW^0(F(x)) dW^0(F(y)) \le t\Big] \equiv \theta(F),$$

where W^0 is a Brownian Bridge. On the other hand it is clear that,

$$\begin{aligned} T_n(\hat{F}_n^*) &= n \int \psi(x, y) d\hat{F}_n^*(x) d\hat{F}_n(y) \\ &= \int \psi(x, y) dW_n^*(x) dW_n^{0*}(y) + 2 \int \psi(x, y) dW_n^0(x) dW_n^{0*}(y) \\ &\quad + \int \psi(x, y) dW_n^0(x) dW_n^0(y), \end{aligned} \tag{2.7}$$

where $W_n^{0*}(x) \equiv \sqrt{n}(\hat{F}_n^*(x) - \hat{F}_n(x))$. It readily follows that,

$$\begin{aligned} P^*[T_n(\hat{F}_n^*) \le t] \overset{w}{\Rightarrow} P\Big[\int \psi(x, y) &dW^0(F(x)) dW^0(F(y)) \\ +2 \int \psi(x, y) &dW^0(F(x)) d\tilde{W}^0(F(y)) \\ + \int \psi(x, y) &d\tilde{W}^0(F(x)) d\tilde{W}^0(F(y)) \le t | \tilde{W}^0\Big], \end{aligned} \tag{2.8}$$

where $\tilde{W}^0, W^0$ are independent Brownian Bridges.

This is again an instance where $\theta(F)$ is well defined for $F \in \mathcal{F}$ but $\theta_n(F)$ does not converge for $F \notin \mathcal{F}_0$

Example 6. Nondifferentiable functions of the empirical: (Beran and Srivastava (1985) and Dümbgen (1993))

Let $\mathcal{F}_0 = \{F : E_F X_1^2 < \infty\}$ and

$$T_n(\hat{F}_n, F) = \sqrt{n}(h(\bar{X}) - h(\mu(F)))$$

when $\mu(F) = E_F X_1$. If h is differentiable the bootstrap distribution of T_n is, of course, consistent. But take $h(x) = |x|$, differentiable everywhere except at 0. It is easy to see then that if $\mu(F) \neq 0$, $\mathcal{L}_n(F) \to \mathcal{N}(0, \mathrm{Var}_F(X_1))$ but if $\mu(F) = 0$, $\mathcal{L}_n(F) \to |\mathcal{N}(0, \mathrm{Var}_F(X_1))|$.

The bootstrap is consistent if $\mu \neq 0$ but not if $\mu = 0$. We can argue as follows. Under $\mu = 0$, $\sqrt{n}(\bar{X}^* - \bar{X})$, $\sqrt{n}\bar{X}$ are asymptotically independent $\mathcal{N}(0, \sigma^2(F))$. Call these variables Z and Z'. Then, $\sqrt{n}(|\bar{X}^*| - |\bar{X}|) \stackrel{w}{\Rightarrow} |Z + Z'| - |Z'|$, a variable whose distribution is not the same as that of $|Z|$. The bootstrap distribution, as usual, converges (weakly) to the (random) conditional distribution of $|Z + Z'| - |Z'|$ given Z'. This phenomenon was first observed in a more realistic context by Beran and Srivastava (1985). Dümbgen (1993) constructs similar reasonable though more complicated examples where the bootstrap distribution never converges. If we represent $T_n(\hat{F}_n, F) = \sqrt{n}(T(\hat{F}_n) - T(F))$ in these cases then there is no linear $\dot{T}(F)$ such that $\sqrt{n}(T(\hat{F}_n) - T(F)) \approx \sqrt{n}\dot{T}(F)(\hat{F}_n - F)$ which permits the argument of Bickel-Freedman (1981).

2.4. Possible remedies

Putter and van Zwet (1993) show that if $\theta_n(F)$ is continuous for every n on $\mathcal{F}$ and there is a consistent estimate $\tilde{F}_n$ of F then bootstrapping from $\tilde{F}_n$ will work, i.e. $\theta_n(\tilde{F}_n)$ will be consistent except possibly for F in a "thin" set.

If we review our examples of bootstrap failure, we can see that constructing suitable $\tilde{F}_n \in \mathcal{F}_0$ and consistent is often a remedy that works for all $F \in \mathcal{F}_0$ not simply the complement of a set of the second category. Thus in Example 3 taking $\tilde{F}_n$ to be $\hat{F}_n$ kernel smoothed with bandwidth $h_n \to 0$ if $nh_n^2 \to 0$ works. In the first and simplest case of Example 4 it is easy to see, Freedman (1981), that taking $\tilde{F}_n$ as the empirical distribution of $X_i - \bar{X}$, $1 \leq i \leq n$ which has mean 0 and thus belongs to $\mathcal{F}_0$ will work. The appropriate choice of $\tilde{F}_n$ in the other examples of bootstrap failure is less clear. For instance, Example 4 calls for $\tilde{F}_n$ with estimated tails of the right order but how to achieve this is not immediate.

A general approach which we believe is worth investigating is to approximate $\mathcal{F}_0$ by a nested sequence of parametric models, (a sieve), $\{\mathcal{F}_{0,m}\}$, and use the M.L.E. $\tilde{\mathcal{F}}_{m(n)}$ for $\mathcal{F}_{0,m(n)}$, for a suitable sequence $m(n) \to \infty$. See Shen and Wong (1994) for example.

The alternative approach we study is to change θ_n itself as well as possibly its argument. The changes we consider are the m out of n with replacement bootstrap, the $(n-m)$ out of n jackknife or $\binom{n}{m}$ bootstrap discussed by Wu (1990) and Politis and Romano (1994), and what we call sample splitting.

3. The m Out of n Bootstraps

Let h be a bounded real valued function defined on the range of T_n, for instance, $t \to 1(t \leq t_0)$.

We view as our goal estimation of $\theta_n(F) \equiv E_F(h(T_n(\hat{F}_n, F)))$. More complicated functionals such as quantiles are governed by the same heuristics and results as those we detail below. Here are the procedures we discuss.

(i) *The n/n bootstrap (The nonparametric bootstrap)*

Let,

$$B_n(F) = E^* h(T_n(\hat{F}_n^*, F)) = n^{-n} \sum_{(i_1,\ldots,i_n)} h(T_n(X_{i_1},\ldots,X_{i_n},F)).$$

Then, $B_n \equiv B_n(\hat{F}_n) = \theta_n(\hat{F})$ is the n/n bootstrap.

(ii) *The m/n bootstrap*

Let

$$B_{m,n}(F) \equiv n^{-m} \sum_{(i_1,\ldots,i_m)} h(T_m(X_{i_1},\ldots,X_{i_m},F)).$$

Then, $B_{m,n} \equiv B_{m,n}(\hat{F}_n) = \theta_m(\hat{F}_n)$ is the m/n bootstrap.

(iii) *The $\binom{n}{m}$ bootstrap*

Let

$$J_{m,n}(F) = \binom{n}{m}^{-1} \sum_{i_1<\cdots<i_m} h(T_m(X_{i_1},\ldots,X_{i_m},F)).$$

Then, $J_{m,n} \equiv J_{m,n}(\hat{F}_n)$ is the $\binom{n}{m}$ bootstrap.

(iv) *Sample splitting*

Suppose $n = mk$. Define,

$$N_{m,n}(F) \equiv k^{-1} \sum_{j=0}^{k-1} h(T_m(X_{jm+1},\ldots,X_{(j+1)m},F))$$

and $N_{m,n} \equiv N_{m,n}(\hat{F}_n)$ as the sample splitting estimates. For safety in practice one should start with a random permutation of the X_i.

The motivation behind $B_{m(n),n}$ for $m(n) \to \infty$ is clear. Since, by (2.1), $\theta_{m(n)}(F) \to \theta(F)$, $\theta_{m(n)}(\hat{F}_n)$ has as good a rationale as $\theta_n(\hat{F}_n)$. To justify $J_{m,n}$ note that we can write $\theta_m(F) = \theta_m(\underbrace{F \times \cdots \times F}_{m})$ since it is a parameter of the law of $T_m(X_1, \ldots, X_m, F)$. We now approximate $F \times \cdots \times F$ not by the m dimensional product measure $\underbrace{\hat{F}_n \times \cdots \times \hat{F}_n}_{m}$ but by sampling without replacement. Thus sample splitting is just k fold cross validation and represents a crude approximation to $\underbrace{F \times \cdots \times F}_{m}$.

The sample splitting method requires the least computation of any of the lot. Its obvious disadvantages are that it relies on an arbitrary partition of the sample and that since both m and k should be reasonably large, n has to be really substantial. This method and compromises between it and the $\binom{n}{m}$ bootstrap are studied in Blom (1976) for instance. The $\binom{n}{m}$ bootstrap differs from the m/n by $o_P(1)$ if $m = o(n^{1/2})$. Its advantage is that it never presents us with the ties which make resampling not look like sampling. As a consequence, as we note in Theorem 1, it is consistent under really minimal conditions. On the other hand it is somewhat harder to implement by simulation. We shall study both of these methods further, below, in terms of their accuracy.

A simple and remarkable result on $J_{m(n),n}$ has been obtained by Politis and Romano (1994), generalizing Wu (1990). This result was also independently noted and generalized by Götze (1993). Here is a version of the Götze result and its easy proof. Write J_m for $J_{m,n}$, B_m for $B_{m,n}$, N_m for $N_{m,n}$.

Theorem 1. *Suppose* $\frac{m}{n} \to 0$, $m \to \infty$.
Then,

$$J_m(F) = \theta_m(F) + O_P((\frac{m}{n})^{\frac{1}{2}}). \tag{3.1}$$

If h is continuous and

$$T_m(X_1, \ldots, X_m, F) = T_m(X_1, \ldots, X_m, \hat{F}_n) + o_p(1) \tag{3.2}$$

then

$$J_m = \theta_m(F) + o_p(1). \tag{3.3}$$

Proof. Suppose T_m does not depend on F. Then, J_m is a U statistic with kernel $h(T_m(x_1, \ldots, x_m))$ and $E_F J_m = \theta_m(F)$ and (3.1) follows immediately. For (3.2) note that

$$\begin{aligned} &E_F \Big| J_m - \binom{n}{m}^{-1} \sum_{i_1 < \cdots < i_m} h(T_m(X_{i_1}, \ldots, X_{i_m}, F)) \Big| \\ &\leq E_F | h(T_m(X_1, \ldots, X_m, \hat{F}_n)) - h(T_m(X_1, \ldots, X_m, F))| \end{aligned} \tag{3.4}$$

and (3.2) follows by bounded convergence. These results follows in the same way and even more easily for N_m. Note that if T_m does not depend on F, $E_F N_m = \theta_m(F)$ and,

$$\operatorname{Var}_F(N_m) = \frac{m}{n} \operatorname{Var}_F(h(T_m(X_1, \ldots, X_m))) > \operatorname{Var}_F(J_m). \tag{3.5}$$

Note. It may be shown, more generally under (3.2), that, for example, distances between the $\binom{n}{m}$ bootstrap distributions of $T_m(\hat{F}_m, F)$ and $\mathcal{L}_m(F)$ are also $O_P(m/n)^{1/2}$.

Let $X_j^{(i)} = (X_j, \ldots, X_j)_{1\times i}$

$$h_{i_1,\ldots,i_r}(X_1,\ldots,X_r) = \frac{1}{r!}\sum_{1\le j_1\ne\cdots\ne j_r\le r} h(T_m(X_{j_1}^{(i_1)},\ldots,X_{j_r}^{(i_r)},F)), \tag{3.6}$$

for vectors $\boldsymbol{i} = (i_1, \ldots, i_r)$ in the index set

$$\Lambda_{r,m} = \{(i_1,\ldots,i_r) : 1 \le i_1 \le \cdots \le i_r \le m, i_1 + \cdots + i_r = m\}.$$

Then

$$B_{m,n}(F) = \sum_{r=1}^{m}\sum_{\boldsymbol{i}\in\Lambda_{r,m}} \omega_{m,n}(\boldsymbol{i})\frac{1}{\binom{n}{r}}\sum_{1\le j_1\le\cdots\le j_r\le m} h_{\boldsymbol{i}}(X_{j_1},\ldots,X_{j_r},F), \tag{3.7}$$

where

$$\omega_{m,n}(\boldsymbol{i}) = \binom{n}{r}\binom{m}{i_1,\ldots,i_r}/n^m.$$

Let

$$\theta_{m,n}(F) = E_F B_{m,n}(F) = \sum_{r=1}^{m}\sum_{\boldsymbol{i}\in\Lambda_{r,m}} \omega_{m,n}(\boldsymbol{i}) E_F h_{\boldsymbol{i}}(X_1,\ldots,X_r). \tag{3.8}$$

Finally, let

$$\delta_m(\frac{r}{m}) \equiv \max\{|E_F h_{\boldsymbol{i}}(X_1,\ldots,X_r) - \theta_m(F)| : \boldsymbol{i}\in\Lambda_{r,m}\} \tag{3.9}$$

and define $\delta_m(x)$ by extrapolation on $[0,1]$. Note that $\delta_m(1) = 0$.

Theorem 2. *Under the conditions of Theorem 1*

$$B_{m,n}(F) = \theta_{m,n}(F) + O_P(\frac{m}{n})^{\frac{1}{2}}. \tag{3.10}$$

If further,

$$\delta_m(1 - xm^{-1/2}) \to 0 \tag{3.11}$$

uniformly for $0 \le x \le M$, *all* $M < \infty$, *and* $m = o(n)$, *then*

$$\theta_{m,n}(F) = \theta_m(F) + o(1). \tag{3.12}$$

Finally if,

$$T_m(X_1^{(i_n)},\ldots,X_r^{(i_r)},F) = T_m(X_1^{(i_1)},\ldots,X_r^{(i_r)},\hat{F}_n) + o_P(1) \tag{3.13}$$

whenever $i \in \Lambda_{r,m}, m \to \infty$ *and* $\max\{i_1, \ldots, i_r\} = O(m^{1/2})$ *then, if* $m \to \infty, m = o(n)$,

$$B_m = \theta_m(F) + o_p(1). \tag{3.14}$$

The proof of Theorem 2 will be given in the Appendix. There too we will show briefly that, in the examples we have discussed and some others, $J_{m(n)}$, $B_{m(n)}$, $N_{m(n)}$ are consistent for $m(n) \to \infty$, $\frac{m}{n} \to 0$.

According to Theorem 2, if T_n does not depend om F the m/n bootstrap works as well as the $\binom{n}{m}$ bootstrap if the value of T_m is not greatly affected by a number on the order of $\sqrt{m}$ ties in its argument. Some condition is needed. Consider $T_n(X_1, \ldots, X_n) = 1(X_i = X_j \text{ for some } i \neq j)$ and suppose F is continuous. The $\binom{n}{m}$ bootstrap gives $T_m = 0$ as it should. If $m \neq o(\sqrt{n})$ so that the $\binom{n}{m}$ and m/n bootstraps do not coincide asymptotically the m/n bootstrap gives $T_m = 1$ with positive probability. Finally, (3.13) is the natural extension of (3.2) and is as easy to verify in all our examples.

A number of other results are available for m out of n bootstraps.

Giné and Zinn (1989) have shown quite generally that when $\sqrt{n}(\hat{F}_n - F)$ is viewed as a member of a suitable Banach space $\mathcal{F}$ and,
(a) $T_n(X_1, \ldots, X_n, F) = t(\sqrt{n}(\hat{F}_n - F))$ for t continuous
(b) $\mathcal{F}$ is not too big
then B_n and $B_{m(n)}$ are consistent.

Praestgaard and Wellner (1993) extended these results to $J_{m(n)}$ with $m = o(n)$. Finally, under the Giné-Zinn conditions,

$$||\sqrt{m}(\hat{F}_n - F)|| = (\frac{m}{n})||\sqrt{n}(\hat{F}_n - F)|| = O_P(\frac{m}{n})^{1/2} \tag{3.15}$$

if $m = o(n)$. Therefore,

$$t(\sqrt{m}(\hat{F}_m - \hat{F}_n)) = t(\sqrt{m}(\hat{F}_m - F)) + o_p(1) \tag{3.16}$$

and consistency of N_m if $m = o(n)$ follows from the original Giné-Zinn result.

We close with a theorem on the parametric version of the m/n bootstrap which gives a stronger property than that of Theorem 1.

Let $\mathcal{F}_0 = \{F_\theta : \theta \in \Theta \subset R^p\}$ where Θ is open and the model is regular. That is, θ is identifiable, the F_θ have densities f_θ with respect to a σ finite μ and the map $\theta \to \sqrt{f_\theta}$ is continuously Hellinger differentiable with nonsingular derivative. By a result of LeCam (see Bickel, Klaassen, Ritov, Wellner (1993) for instance), there exists an estimate $\hat{\theta}_n$ such that, for all θ,

$$\int (f_{\hat{\theta}_n}^{1/2}(x) - f_\theta^{1/2}(x))^2 d\mu(x) = O_{P_\theta}(\frac{1}{n}). \tag{3.17}$$

Theorem 3. *Suppose $\mathcal{F}_0$ is as above. Let $F_\theta^m \equiv \underbrace{F_\theta \times \cdots \times F_\theta}_{m}$ and $\|\cdot\|$ denote the variational norm. Then*

$$\|F_{\hat{\theta}_n}^m - F_\theta^m\| = O_p((\frac{m}{n})^{1/2}). \tag{3.18}$$

Proof. This is consequence of the relations (LeCam (1986)).

$$\|F_{\theta_0}^m - F_{\theta_1}^m)\| \le H(F_{\theta_0}^m, F_{\theta_1}^m)[(2 - H^2(F_{\theta_0}^m, F_{\theta_1}^m)], \tag{3.19}$$

where

$$H^2(F,G) = \frac{1}{2}\int(\sqrt{dF} - \sqrt{dG})^2 \tag{3.20}$$

and

$$H^2(F_{\theta_0}^m, F_{\theta_1}^m) = 1 - (\int\sqrt{f_{\theta_0} f_{\theta_1} d\mu})^m = 1 - (1 - H^2(F_{\theta_0}, F))^m. \tag{3.21}$$

Substituting (3.21) into (3.20) and using (3.17) we obtain

$$\|F_{\hat{\theta}_n}^m - F_\theta^m\| = O_{P_\theta}(1 - \exp O_{P_\theta}(\frac{m}{n}))^{\frac{1}{2}}(1 + \exp O_{P_\theta}(\frac{m}{n})^{\frac{1}{2}}) = O_{P_\theta}(\frac{m}{n})^{\frac{1}{2}}. \tag{3.22}$$

This result is weaker than Theorem 1 since it refers only to the parametric bootstrap. It is stronger since even for $m = 1$, when sampling with and without replacement coincide, $\|\hat{F}_n - F_\theta\| = 1$ for all n if F_θ is continous.

4. Performance of B_m, J_m, and N_m as Estimates of $\theta_n(F)$

As we have noted, if we take $m(n) = o(n)$ then in all examples considered in which B_n is inconsistent, $J_{m(n)}$, $B_{m(n)}$, $N_{m(n)}$ are consistent. Two obvious questions are,

(1) How do we choose $m(n)$?

(2) Is there a price to be paid for using $J_{m(n)}$, $B_{m(n)}$, or $N_{m(n)}$ when B_n is consistent?

We shall turn to the first very difficult question in a forthcoming paper on diagnostics. The answer to the second is, in general, yes. To make this precise we take the point of view of Beran (1982) and assume that at least on $\mathcal{F}_0$,

$$\theta_n(F) = \theta(F) + \theta'(F)n^{-1/2} + O(n^{-1}), \tag{4.1}$$

where $\theta(F)$ and $\theta'(F)$ are regularly estimable on $\mathcal{F}_0$ in the sense of Bickel, Klaassen, Ritov and Wellner (1993) and $O(n^{-1})$ is uniform on Hellinger compacts. There are a number of general theorems which lead to such expansions. See, for example, Bentkus, Götze and van Zwet (1994).

Somewhat more generally than Beran, we exhibit conditions under which $B_n = \theta_n(\hat{F}_n)$ is fully efficient as an estimate of $\theta_n(F)$ and show that the m out n bootstrap with $\frac{m}{n} \to 0$ has typically relative efficiency 0.

We formally state a theorem which applies to fairly general parameters θ_n. Suppose ρ is a metric on $\mathcal{F}_0$ such that

$$\rho(\hat{F}_n, F_0) = O_{P_{F_0}}(n^{-1/2}) \text{ for all } F_0 \in \mathcal{F}_0. \tag{4.2}$$

Further suppose

A. $\theta(F), \theta'(F)$ are ρ Fréchet differentiable in $\mathcal{F}$ at $F_0 \in \mathcal{F}_0$. That is,

$$\theta(F) = \theta(F_0) + \int \psi(x, F_0)dF(x) + o(\rho(F, F_0)) \tag{4.3}$$

for $\psi \in L_2^0(F_0) \equiv \{h : \int h^2(x)dF_0(x) < \infty, \int h(x)dF_0(x) = 0\}$ and θ' obeys a similar identity with ψ replaced by another function $\psi' \in L_2^0(F_0)$. Suppose further

B. The tangent space of $\mathcal{F}_0$ at F_0 as defined in Bickel et al. (1993) is $L_2^0(F_0)$ so that ψ and ψ' are the efficient influence functions of θ, θ'. Essentially, we require that in estimating F there is no advantage in knowing $F \in \mathcal{F}_0$.

Finally, we assume,

C. For all $M < \infty$,

$$\sup\{|\theta_m(F) - \theta(F) - \theta'(F)m^{-1/2}| : \rho(F, F_0) \leq M_n^{-1/2}, F \in \mathcal{F}\} = O(m^{-1}) \tag{4.4}$$

a strengthened form of (4.1). Then,

Theorem 4. *Under regularity of θ, θ' and A and C at F_0,*

$$\begin{aligned}\theta_m(\hat{F}_n) &\equiv \theta(F_0) + \theta'(F_0)m^{-1/2} + \frac{1}{n}\sum_{i=1}^{n}(\psi(X_i, F_0) + \psi'(X_i, F_0)m^{-1/2}) \\ &\quad + O(m^{-1}) + o_p(n^{-1/2}).\end{aligned} \tag{4.5}$$

If B also holds, $\theta_n(\hat{F}_n)$ is efficient. If in addition, $\theta'(F_0) \neq 0$, and $\frac{m}{n} \to 0$ the efficiency of $\theta_m(\hat{F}_n)$ is 0.

Proof. The expansions of $\theta(\hat{F}_n)\theta'(\hat{F}_n)$ are immediate by Fréchet differentiability and (4.5) follows by plugging these into (4.1). Since θ, θ' are assumed regular, ψ and ψ' are their efficient influence functions. Full efficiency of $\theta_n(\hat{F}_n)$ follows by general theory as given in Beran (1983) for special cases or by extending Theorem 2, p.63 of Bickel et al. (1993) in an obvious way. On the other hand, if $\theta'(F_0) \neq 0$, $\sqrt{n}(\theta_m(\hat{F}_n) - \theta_n(F_0))$ has asymptotic bias $(\sqrt{\frac{n}{m}} - 1)\theta'(F_0) + O(\frac{\sqrt{n}}{m}) = \sqrt{\frac{n}{m}}(1 + o(1))\theta'(F_0) \to \pm\infty$ and inefficiency follows.

Inefficiency results of the same type or worse may be proved about J_m and N_m but require going back to $T_m(X_1, \ldots, X_m, F)$ since J_m and B_n are not related in a simple way. We pursue this only by way of Example 1. If $\theta_n(F) = \text{Var}_F(\sqrt{n}(\bar{X} - \mu(F)) = \theta(F)$, $B_m = B_n$ but,

$$J_m = \sigma^2(\hat{F}_n)(1 - \frac{m-1}{n-1}). \tag{4.6}$$

Thus, since $\theta'(F) = 0$ here, B_m is efficient but J_m has efficiency 0 if $\frac{m}{\sqrt{n}} \to \infty$. N_m evidently behaves in the same way.

It is true that the bootstrap is often used not for estimation but for setting confidence bounds. This is clearly the case for Example (1b), the bootstrap of t where $\theta(F)$ is known in advance. For example, Efron's percentile bootstrap uses the $(1-\alpha)$th quantile of the bootstrap distribution of $\bar{X}$ as a level $(1-\alpha)$ approximate upper confidence bound for μ. As is well known by now (see Hall (1992)), for example, this estimate although, when suitably normalized, efficiently estimating the $(1-\alpha)$th quantile of the distribution of $\sqrt{n}(\bar{X}-\mu)$ does not improve to order $n^{-1/2}$ over the coverage probability of the usual Gaussian based $\bar{X} + z_{1-\alpha}\frac{s}{\sqrt{n}}$. However, the confidence bounds based on the bootstrap distribution of the t statistic $\sqrt{n}(\bar{X}-\mu(F))/s$ get the coverage probability correct to order $n^{-1/2}$. Unfortunately, this advantage is lost if one were to use the $1-\alpha$ quantile of the bootstrap distribution of $T_m(\hat{F}_m, F) = \sqrt{m}(\bar{X}_m - \mu(F))/s_m$ where $\bar{X}_m$ and s_m^2 are the mean and usual estimate of variance bsed on a sample of size m. The reason is that, in this case, the bootstrap distribution function is

$$\Phi(t) - m^{-1/2}c(\hat{F}_n)\varphi(t)H_2(t) + O_P(m^{-1}) \tag{4.7}$$

rather than the needed,

$$\Phi(t) - n^{-1/2}c(\hat{F}_n)\varphi(t)H_2(t) + O_P(n{-}1).$$

The error committed is of order $m^{-1/2}$. More general formal results can be stated but we do not pursue this.

The situation for $J_{m(n)}$ and $N_{m(n)}$ which function under minimal conditions, is even worse as we discuss in the next section.

5. Remedying the Deficiencies of $B_{m(n)}$ when B_n is Correct: Extrapolation

In Bickel and Yahav (1988), motivated by considerations of computational economy, situations were considered in which θ_n has an expansion of the form (4.1) and it was proposed using B_m at $m = n_0$ and $m = n_1$, $n_0 < n_1 << n$ to produce estimates of θ_n which behave like B_n. We sketch the argument for a special case.

Suppose that, as can be shown for a wide range of situations, if $m \to \infty$,

$$B_m = \theta_m(\hat{F}_n) = \theta(\hat{F}_n) + \theta'(\hat{F}_n)m^{-1/2} + O_P(m^{-1}). \tag{5.1}$$

Then, if $n_1 > n_0 \to \infty$

$$\theta'(\hat{F}_n) = (B_{n_0} - B_{n_1})(n_0^{-1/2} - n_1^{-1/2})^{-1} + O_P(n_0^{-1/2}) \tag{5.2}$$

$$\theta(\hat{F}_n) = \frac{n_0^{-1/2} B_{n_1} - n_1^{-1/2} B_{n_0}}{n_0^{-1/2} - n_1^{-1/2}} + O_P(n_0^{-1}) \tag{5.3}$$

and hence a reasonable estimate of B_n is,

$$B_{n_0,n_1} \equiv \frac{n_0^{-1/2} B_{n_1} - n_1^{-1/2} B_{n_0}}{n_0^{-1/2} - n_1^{-1/2}} + \frac{(B_{n_0} - B_{n_1})}{n_0^{-1/2} - n_1^{-1/2}} n^{-1/2}.$$

More formally,

Proposition. *Suppose $\{\theta_m\}$ obey C of Section 4 and $n_0 n^{-1/2} \to \infty$. Then,*

$$B_{n_0,n_1} = B_n + o_p(n^{-1/2}). \tag{5.4}$$

Hence, under the conditions of Theorem 3 B_{n_0,n_1} is efficient for estimating $\theta_n(F)$.

Proof. Under C, (5.4) holds. By construction,

$$\begin{aligned} B_{n_0,n_1} &= \theta(\hat{F}_n) + \theta'(\hat{F}_n)n^{-1/2} + O_P(n_0^{-1}) + O_P(n_0^{-1/2} n^{-1/2}) \\ &= \theta_n(\hat{F}_n) + O_P(n_0^{-1}) + O_P(n_0^{-1/2} n^{-1/2}) + O_P(n^{-1}) \\ &= \theta_n(\hat{F}_n) + O_P(n_0^{-1}) \end{aligned} \tag{5.5}$$

and (5.4) follows.

Assorted variations can be played on this theme depending on what we know or assume about θ_n. If, as in the case where T_n is a t statistic, the leading term $\theta(F)$ in (4.1) is $\equiv \theta_0$ independent of F, estimation of $\theta(F)$ is unnecessary and we need only one value of $m = n_0$. We are led to a simple form of estimate, since ψ of Theorem 4 is 0,

$$\hat{\theta}_{n_0} = (1 - (\frac{n_0}{n})^{1/2})\theta_0 + (\frac{n_0}{n})^{1/2} B_{n_0}. \tag{5.6}$$

This kind of interpolation is used to improve theoretically the behaviour of B_{m_0} as an estimate of a parameter of a stable distribution by Hall and Jing (1993) though we argue below that the improvement is somewhat illusory.

If we apply (5.4) to construct a bootstrap confidence bound we expect the coverage probability to be correct to order $n^{-1/2}$ but the error is $O_P((n_0 n)^{-1/2})$ rather than $O_P(n^{-1})$ as with B_n. We do not pursue a formal statement.

5.1. Extrapolation of J_m and N_m

We discuss extrapolation for J_m and N_m only in the context of the simplest Example 1, where the essential difficulties become apparent and we omit general theorems.

In work in progress, Götze and coworkers are developing expansions for general symmetric statistics under sampling from a finite population. These results will permit general statements of the same qualitative nature as in our discussion of Example 1. Consider $\theta_m(F) = P_F[\sqrt{m}(\bar{X}_m - \mu(F)) \le t]$. If $EX_1^4 < \infty$ and the X_i obey Cramér's condition, then

$$\theta_m(F) = \Phi(\frac{t}{\sigma(F)}) - K_3(F)\frac{\varphi}{6\sqrt{m}}(\frac{t}{\sigma(F)})H_2(\frac{t}{\sigma(F)}) + O(m^{-1}), \tag{5.7}$$

where $\sigma^2(F)$ and $K_3(F)$ are the second and third cumulants of F and $H_k(t) = \frac{(-1)^k}{\varphi(t)}\frac{d\varphi^k(t)}{dt^k}$. By Singh (1981), $B_m = \theta_m(\hat{F}_n)$ has the same expansion with F replaced by $\hat{F}_n$. However, by an easy extension of results of Robinson (1978) and Babu and Singh (1985),

$$J_m = \Phi(\frac{t}{\hat{K}_{2m}}) - \varphi(\frac{t}{\hat{K}_{2m}^{1/2}})\frac{\hat{K}_{3m}}{6m^{1/2}}H_2(\frac{t}{\hat{K}_{2m}^{1/2}}) + O_P(m^{-1}), \tag{5.8}$$

where

$$\hat{K}_{2m} = \sigma^2(\hat{F}_n)(1 - \frac{m-1}{n-1}) \tag{5.9}$$

$$\hat{K}_{3m} = K_3(\hat{F}_n)(1 - \frac{m-1}{n-1})(1 - \frac{2(m-1)}{n-2}). \tag{5.10}$$

The essential character of expansion (5.8), if $m/n = o(1)$, is

$$J_m = \theta(\hat{F}_n) + m^{-1/2}\theta'(\hat{F}_n) + \frac{m}{n}\gamma_n + O_P(m^{-1} + (\frac{m}{n})^2 + \frac{m^{\frac{1}{2}}}{n}), \tag{5.11}$$

where γ_n is $O_P(1)$ and independent of m. The m/n terms essentially come from the finite population correction to the variance and highter order cumulants of means of samples from a finite population. They reflect the obvious fact that if $m/n \to \lambda > 0$, J_m is, in general, incorrect even to first order. For instance, the variance of the $\binom{n}{m}$ bootstrap distribution corresponding to $\sqrt{m}(\bar{X} - \mu(F))$ is $1/n\sum(X_i - \bar{X})^2(1 - \frac{m-1}{n-1}))$ which converges to $\sigma^2(F)(1-\lambda)$ if $m/n \to \lambda > 0$. What this means is that if expansions (4.1), (5.1) and (5.11) are valid, then using $J_{m(n)}$ again gives efficiency 0 compared to B_n. Worse is that (5.2) with J_{n_0}, J_{n_1} replacing B_{n_0}, B_{n_1} will not work since the n_1/n terms remain and make a contribution larger than $n^{-1/2}$ if $n_1/n^{1/2} \to \infty$. Essentially it is necessary to estimate the coefficient of m/n and remove the contribution of this term at the same time while keeping the three required values of m: $n_0 < n_1 < n_2$ such that the error $O(\frac{1}{n_0} + (\frac{n_2}{n})^2)$ is $o(n^{-1/2})$. This essentially means that n_0, n_1, n_2 have order larger than $n^{1/2}$ and smaller that $n^{3/4}$.

This effect persists if we seek to use an extrapolation of J_m for the t statistic. The coefficient of m/n as well as $m^{-1/2}$ needs to be estimated. An alternative here and perhaps more generally is to modify the t statistic being bootstrapped and extrapolated. Thus $T_m(X_1, \ldots, X_m, F) \equiv \sqrt{m}\frac{(\bar{X}_m - \mu(F))}{\hat{\sigma}(1-\frac{m-1}{n-1})^{1/2}}$ leads to an expansion for J_m of the form,

$$J_m = \Phi(t) + \theta'(\hat{F}_n)m^{-1/2} + O_P(m^{-1} + m/n), \tag{5.12}$$

and we again get correct coverage to order $n^{-1/2}$ by fitting the $m^{-1/2}$ term's coefficient, weighting it by $n^{-1/2} - m^{-1/2}$ and adding it to J_m.

If we know, as we sometimes at least suspect in symmetric cases, that $\theta(F) = 0$, we should appropriately extrapolate linearly in m^{-1} rather than $m^{-1/2}$.

The sample splitting situation is less satisfactory in the same example. Under (5.1), the coefficient of $1/\sqrt{m}$ is asymptotically constant. Put another way, the asymptotic correlation of B_m, $B_{\lambda m}$ as $m, n \to \infty$ for fixed $\lambda > 0$ is 1. This is also true for J_m under (5.11). However, consider N_m and N_{2m} (say) if $T_m = \sqrt{m}(\bar{X}_m - \mu(F))$. Let h be continuously boundedly differentiable, $n = 2km$. Then

$$\operatorname{Cov}(N_m, N_{2m}) = \frac{1}{k}\operatorname{Cov}\Big(h(m^{-1/2}(\sum_{j=1}^{m}(X_j - \bar{X}))), h((2m)^{-1/2}\sum_{j=1}^{2m}(X_j - \bar{X}))\Big). \tag{5.13}$$

Thus, by the central limit theorem,

$$\operatorname{Corr}(N_m, N_{2m}) \to \frac{1}{2}\frac{\operatorname{Cov}}{\operatorname{Var}(Z_1)}\Big(h(Z_1), h\frac{(Z_1 + Z_2)}{\sqrt{2}}\Big), \tag{5.14}$$

where Z_1, Z_2 are independent Gaussian $\mathcal{N}(0, \sigma^2(F))$ and $\sigma^2(F) = \operatorname{Var}_F(X_1)$. More generally, viewed as a process in m for fixed n, N_m centered and normalized is converging weakly to a non degenerate process. Thus, extrapolation does not make sense for N_m.

Two questions naturally present themselves.

(a) How do these games play out in practice rather than theory?

(b) If the expansions (5.1) and (5.11) are invalid beyond the 0th order, the usual situation when the nonparametric bootstrap is inconsistent, what price do we pay theoretically for extrapolation?

Simulations giving limited encouragement in response to question (a) are given in Bickel and Yahav (1988). We give some further evidence in Section 7. We now turn to question (b) in the next section.

6. Behaviour of the Smaller Resample Schemes When B_n is Inconsistent, and Presentation of Alternatives

The class of situations in which B_n does not work is too poorly defined for us to come to definitive conclusions. But consideration of the examples suggests the following,

A. When, as in Example 6, $\theta(F)$, $\theta'(F)$ are well defined and regularly estimable on $\mathcal{F}_0$ we should still be able to use extrapolation (suitably applied) to B_m and possibly to J_m to produce better estimates of $\theta_n(F)$.

B. When, as in all our other examples of inconsistency, $\theta(F)$ is not regularly estimable on $\mathcal{F}_0$ extrapolation should not improve over the behaviour of B_{n_0}, B_{n_1}.

C. If n_0, n_1 are comparable extrapolation should not do particularly worse either.

D. A closer analysis of T_n and the goals of the bootstrap may, in these "irregular" cases, be used to obtain procedures which should do better than the m/n or $\binom{n}{m}$ or extrapolation bootstraps.

The only one of these claims which can be made general is C.

Proposition 1. *Suppose*

$$B_{n_1} - \theta_n(F) \asymp B_{n_0} - \theta_n(F), \tag{6.1}$$

where $\asymp$ *indicates that the ratio tends to* 1. *Then, if* $n_0/n_1 \not\to 1$

$$B_{n_0,n_1} - \theta_n(F) \asymp B_{n_0} - \theta_n(F). \tag{6.2}$$

Proof. Evidently, $\frac{B_{n_0}+B_{n_1}}{2} = \theta_n(F) + \Omega(\epsilon_n)$ where $\Omega(\epsilon_n)$ means that the exact order of the remainder is ϵ_n. On the other hand,

$$\frac{B_{n_0} - B_{n_1}}{n_0^{-1/2} - n_1^{-1/2}}\Big(\frac{1}{\sqrt{n}} - \frac{1}{2}\big(\frac{1}{\sqrt{n_0}} + \frac{1}{\sqrt{n_1}}\big)\Big) = \Omega(\epsilon_n)\Big(\sqrt{\frac{n_0}{n}} + \Omega(1)\Big)$$

and the proposition follows.

We illustrate the other three claims in going through the examples.

Example 3. Here, $F^{-1}(0) = 0$,

$$\theta_n(F) = e^{f(0)t}\Big(1 + n^{-1} f'(0)\frac{t^2}{2}\Big) + O(n^{-2}) \tag{6.3}$$

which is of the form (5.1). But the functional $\theta(F)$ is not regular and only estimable at rate $n^{-1/3}$ if one puts a first order Lipschitz condition on $F \in \mathcal{F}_0$. On the other hand,

$$\begin{aligned}\log B_m &= m\log(1-\hat{F}_n(\frac{t}{m})) = m\log(1-(\hat{F}_n(\frac{t}{m})-\hat{F}_n(0))) \\ &= -m(F(\frac{t}{m})-F(0)) - \frac{m}{\sqrt{n}}\sqrt{n}(\hat{F}_n(\frac{t}{m})-F(\frac{t}{m})) + O_P(m(\hat{F}_n(\frac{t}{m})-F(\frac{t}{m}))^2) \\ &= tf(0) + \Omega(\frac{1}{m}) + \Omega_P(\sqrt{\frac{m}{n}}) + O_P(\frac{1}{n}), \qquad (6.4)\end{aligned}$$

where as before Ω, Ω_p indicate exact order. As Politis and Romano (1994) point out, $m = \Omega(n^{1/3})$ yields the optimal rate $n^{-1/3}$ (under f Lipschitz). Extrapolation does not help because the $\sqrt{\frac{m}{n}}$ term is not of the form $\gamma_n\sqrt{\frac{m}{n}}$ where γ_n is independent of m. On the contrary, as a process in m, $\sqrt{mn}(\hat{F}_n(\frac{t}{m}) - F(\frac{t}{m}))$ behaves like the sample path of a stationary Gaussian process. So conclusion B holds in this case.

Example 4. A major difficulty here is defining $\mathcal{F}_0$ narrowly enough so that it is meaningful to talk about expansions of $\theta_n(F)$, $B_n(F)$ etc. If $\mathcal{F}_0$ in these examples is in the domain of attraction of stable laws or extreme value distributions it is easy to see that $\theta_n(F)$ can converge to $\theta(F)$ arbitrarily slowly. This is even true in Example 1 if we remove the Lipschitz condition on f. By putting on conditions as in Example 1, it is possible to obtain rates. Hall and Jing (1993) specify a possible family for the stable law attraction domain estimation of the mean mentioned in Example 4 in which $B_n = \Omega(n^{-\frac{1}{\alpha}})$ where α is the index of the stable law and α and the scales of the (assumed symmetric) stable distribution are not regularly estimable but for which rates such as $n^{-2/5}$ or a little better are possible. The expansions for $\theta_n(F)$ are not in powers of $n^{-1/2}$ and the expansion for B_n is even more complex. It seems evident that extrapolation does not help. Hall and Jing's (1993) theoretical results and simulations show that $B_{m(n)}$ though consistent, if $m(n)/n \to 0$, is a very poor estimate of $\theta_n(F)$. They obtain at least theoretically superior results by using interpolation between B_m and the, "known up to the value of the stable law index α", value of $\theta(F)$. However, the conditions defining $\mathcal{F}_0$ which permit them to deduce the order of B_n are uncheckable so that this improvement appears illusory.

Example 6. The discontinuity of $\theta(F)$ at $\mu(F) = 0$ under any reasonable specification of $\mathcal{F}_0$ makes it clear that extrapolation cannot succeed. The discontinuity in $\theta(F)$ persists even if we assume $\mathcal{F}_0 = \{\mathcal{N}(\mu, 1) : \mu \in R\}$ and use the parametric bootstrap. In the parametric case it is possible to obtain constant level confidence bounds by inverting the tests for $H : |\mu| = |\mu_0|$ vs $K : |\mu| > |\mu_0|$ using the noncentral χ_1^2 distribution of $(\sqrt{n}\bar{X})^2$. Asymptotically conservative confidence bounds can be constructed in the nonparametric case by forming a bootstrap confidence interval for $\mu(F)$ using $\bar{X}$ and then taking the image of this interval into $\mu \to |\mu|$. So this example illustrates points B and D.

We shall discuss claims A and D in the context of Example 5 or rather its simplest case with $T_n(\hat{F}_n, F) = n\bar{X}^2$. We begin with,

Proposition 2. *Suppose $E_F X_1^4 < \infty$, $E_F X_1 = 0$, and F satisfies Cramer's condition. Then,*

$$B_m \equiv P^*[|\sqrt{m}\bar{X}^*|^2 \leq t^2] = 2\Phi(\frac{t}{\hat{\sigma}}) - 1 \frac{m\bar{X}^2}{\hat{\sigma}^3} t\varphi(\frac{t}{\hat{\sigma}}) - \frac{\hat{K}_3\bar{X}}{3\hat{\sigma}^4}\varphi H_3(\frac{t}{\hat{\sigma}}) + O_P(\frac{m}{n})^{3/2} + O_P(m^{-1}). \quad (6.5)$$

If $m = \Omega(n^{1/2})$ then

$$P^*[|\sqrt{m}\bar{X}^*|^2 \leq t^2] = P_F[n\bar{X}^2 \leq t] + O_P(n^{-1/4}) \quad (6.6)$$

and no better choice of $\{m(n)\}$ is possible. If $n_0 < n_1$, $n_0 n^{-1/2} \to \infty$, $n_1 = o(n^{3/4})$,

$$B^{n_0,n_1} \equiv B_{n_0} - n_0\{(B_{n_1} - B_{n_0})/(n_1 - n_0)\} = P_F[n\bar{X}^2 \leq t] + O_P(n^{-1/2}). \quad (6.7)$$

Proof. We make a standard application of Singh (1981). If $\hat{\sigma}^2 \equiv \frac{1}{n}\sum(X_i - \bar{X})^2$, $\hat{K}_3 \equiv \frac{1}{n}\sum(X_i - \bar{X})^3$ we get, after some algebra and Edgeworth expansion,

$$P^*[\sqrt{m}\bar{X}^* \leq t] = \Phi\Big(\frac{t - \sqrt{m}\bar{X}}{\hat{\sigma}}\Big) - \frac{1}{\sqrt{m}}\varphi\Big(\frac{t - \sqrt{m}\bar{X}}{\hat{\sigma}}\Big)\frac{\hat{K}_3}{6}H_2\Big(\frac{t - \sqrt{m}\bar{X}}{\hat{\sigma}}\Big) + O_p(m^{-1}).$$

After Taylor expansion in $\sqrt{m}\frac{\bar{X}}{\hat{\sigma}}$ we conclude,

$$P^*[m\bar{X}_m^{*2} \leq t^2] = 2\Phi(\frac{t}{\hat{\sigma}}) - 1 + \frac{\varphi'}{2}(\frac{t}{\hat{\sigma}})m\bar{X}^2 - \frac{\hat{K}_3}{3\hat{\sigma}^4}[\varphi H_3](\frac{t}{\hat{\sigma}})\bar{X} + O_P(\frac{m}{n})^{3/2} + O_P(m^{-1}) \quad (6.8)$$

and (6.5) follows. Since $m\bar{X}^2 = \Omega_P(m/n)$, (6.6) follows. Finally, from (6.5), if $n_0 n^{-1/2}$, $n_1 n^{-1/2} \to \infty$

$$B_{n_0} - n_0\{(B_{n_1} - B_{n_0})/(n_1 - n_0)\} = 2\Phi(\frac{t}{\hat{\sigma}}) - 1 - \frac{K_3}{6}\varphi H_2(\frac{t}{\hat{\sigma}})\bar{X} + O_P(n^{-3/4}) + O_P(n^{-1/2}) + O_P(n^{-1/2}). \quad (6.9)$$

Since $\bar{X} = O_P(n^{-1/2})$, (6.7) follows.

Example 5. As we noted, the case $T_n(\hat{F}_n, F) = n\bar{X}^2$ is the prototype of the use of the m/n bootstrap for testing discussed in Bickel and Ren (1995). From (6.7) of proposition 2 it is clear that extrapolation helps. However, it is not true that B^{n_0,n_1} is efficient since it has an unnecessary component of variance $(\hat{K}_3/6)[\varphi H_2](\frac{t}{\bar{\sigma}})\bar{X}$ which is negligible only if $K_3(F) = 0$. On the other hand it is easy to see that efficient estimation can be achieved by resampling not the X_i but the residuals $X_i - \bar{X}$, that is, a consistent estimate of F belonging to $\mathcal{F}_0$. So this example illustrates both A and D. Or in the general U or V statistic case, bootstrapping not $T_m(\hat{F}_n, F) \equiv n \int \psi(x,y) d\hat{F}_n(x) d\hat{F}_n(y)$ but rather $n \int \psi(x,y) d(\hat{F}_n - F)(x) d(\hat{F}_n - F)(y)$ is the right thing to do.

7. Simulations and Conclusions

The simulation algorithms were written and carried out by Adele Cutler and Jiming Jiang. Two situations were simulated, one already studied in Bickel and Yahav (1988) where the bootstrap is consistent (essentially Example 1) the other (essentially Example 3) where the bootstrap is inconsistent.

Sample size: $n = 50, 100, 400$
Bootstrap sample size: $B = 500$
Simulation size: $N = 2000$
Distributions: Example 1: $F = \chi_1^2$; Example 3: $F = \chi_2^3$
Statistics:
Example 1(a) modified: $T_m^{(a)} = \sqrt{m}(\sqrt{\bar{X}_m} - \sqrt{\mu(F)})$
Example 1(b): $T_m^{(b)} = \sqrt{m}\frac{(\bar{X} - \mu(F))}{s_m}$ where $s_m^2 = \frac{1}{m-1}\sum_{i=1}^m (X_i - \bar{X}_m)^2$.
Example 3. $T_m^{(c)} = m(\min(X_1, \ldots, X_m) - F^{-1}(0))$
Parameters of resampling distributions: $G_m^{-1}(.1)$, $G_m^{-1}(.9)$ where G_m is the distribution of T_m under the appropriate resampling scheme. We use B, J, N to distinguish the schemes m/n, $\binom{n}{m}$ and sample splitting respectively.

In Example 1 the G_m^{-1} parameters were used to form upper and lower "90%" confidence bounds for $\theta \equiv \sqrt{\mu(F)}$. Thus, from $T_m^{(a)}$,

$$\bar{\theta}_{mB} = \sqrt{\bar{X}_n - \frac{1}{\sqrt{n}} G_{mB}^{-1}(.1))} \tag{7.1}$$

for the "90%" upper confidence bound based on the m/n bootstrap and, from $T_m^{(b)}$,

$$\bar{\theta}_{mB} = ((\bar{X}_n - \frac{s_n}{\sqrt{n}} G_{mB}^{-1}(.1))_+)^{1/2}, \tag{7.2}$$

where G_{mB} now corresponds to the t statistic. $\underline{\theta}_{mB}$, is defined similarly. The $\bar{\underline{\theta}}_{mJ}$ bounds are defined with G_{mJ} replacing G_{mB}. The $\bar{\underline{\theta}}_{mN}$ bounds are considered only for the unambiguous case m divides n and α an integer multiple of m/n.

Thus if $m = n/10$, $G^{-1}_{mN}(.1)$ is simply the smallest of the 10 possible values $\{T_m(X_{jm+1}, \ldots, X_{(j+1)m}, \hat{F}_n), 0 \le j \le 9\}$.

We also specify 2 subsample sizes $n_0 < n_1$ for the extrapolation bounds, $\underline{\theta}_{n_0,n_1}\bar{\theta}_{n_0,n_1}$. These are defined for $T_m^{(a)}$, for example, by.

$$\bar{\theta}_{n_0,n_1} = \sqrt{\bar{X}_n} - \frac{1}{\sqrt{n}}\Big\{\frac{(G^{-1}_{n_0B}(.1) + G^{-1}_{n_1B}(.1))}{2} + (n^{-1/2} - \frac{1}{2}(n_0^{-1/2} + n_1^{-1/2}))(G^{-1}_{n_0B}(.1) - G^{-1}_{n_1B}(.1))/(n_0^{-1/2} - n_1^{-1/2})\Big\}. \quad (7.3)$$

We consider roughly, $n_0 = 2\sqrt{n}, n_1 = 4\sqrt{n}$ and specifically, the triples (n, n_0, n_1): $(50, 15, 30), (100, 20, 40)$ and $(400, 40, 80)$.

In Example 3, we similarly study the lower confidence bound on $\theta = F^{-1}(0)$ given by,

$$\bar{\theta}_m = \max(X_1, \ldots, X_n) - \frac{1}{n}G^{-1}_{mB}(.9). \quad (7.4)$$

and the extrapolation lower confidence bound

$$\underline{\theta}_{n_0,n_1} = \min(X_1, \ldots, X_n) - \frac{1}{n}\frac{(G^{-1}_{n_0B}(.9) + G^{-1}_{n_1B}(.9))}{2} + (n^{-1} - \frac{(n_0^{-1} + n_1^{-1})}{2})(G^{-1}_{n_0B}(.9) - G^{-1}_{n_1B}(.9))(n_0^{-1} - n_1^{-1}). \quad (7.5)$$

Note that we are using $1/m$ rather than $1/\sqrt{m}$ for extrapolation.

Measures of performance:

$CP \equiv$ Coverage probability, the actual probability under the situation simulated that the region prescribed by the confidence bound covers the true value of the parameter being estimated.

$$RMSE = \sqrt{E(\text{Bound}-\text{Actual quantile bound})^2}\ .$$

Here the actual quantile bound refers to what we would use if we knew the distribution of $T_n(X_1, \ldots, X_n, F)$. For example for $T_m^{(a)}$ we would replace $G^{-1}_{mB}(.1)$ in (7.1) for $F = \chi^2_1$ by the .1 quantile of the distribution of $\sqrt{n}(\sqrt{\frac{S_m}{m}} - 1)$ where S_m has a χ^2_m distribution, call it $G^{*-1}_m(.1)$. Thus, here,

$$MSE = \frac{1}{n}E(G^{-1}_{mB}(.1) - G^{*-1}_m(.1))^2.$$

We give in Table 1 results for the B_{n_1}, B_n and B_{n_0,n_1} bounds, based on $T_m^{(b)}$. The $T_m^{(a)}$ bootstrap, as in Bickel and Yahav (1988), has CP and $RMSE$ for

B_n, B_{n_0,n_1} and B_{n_1} agreeing to the accuracy of the Monte Carlo and we omit these tables.

We give the corresponding results for lower confidence bounds based on $T_m^{(c)}$ in Table 2. Table 3 presents results for sample splitting for $T_m^{(a)}$. Table 4 presents $T_m^{(a)}$ results for the $\binom{n}{m}$ bootstrap.

Table 1. The t bootstrap: Example 1(b) at 90% nominal level

n		Coverage probabilities (CP) B	B1	BR	$RMSE$ B	B1	BR
50							
	UB	.88	.90	.88	.19	.21	.19
	LB	.90	.90	.90	.15	.15	.15
100							
	UB	.90	.93	.89	.13	.14	.12
	LB	.91	.90	.91	.11	.10	.11
400							
	UB	.91	.94	.90	.06	.07	.06
	LB	.91	.90	.91	.05	.05	.05

Notes: (a) B1 corresponds to (6.2) or its LCB analogue for $m=n_1(n)=30$, $40, 80$. Similarly B corresponds to $m = n$.

(b) BR corresponds to (6.3) or its LCB analogue with $(n_0, n_1) = (15, 30), (20, 40), (40, 80)$.

Table 2. The min statistic bootstrap: Example 3 at the nominal 90% level

n		CP	$RMSE$
50			
	B	.75	.01
	B1	.78	.07
	BR	.70	.07
	B1S	.82	.07
	BRS	.80	.07

n		CP	$RMSE$
100			
	B	.75	.04
	B1	.82	.03
	BR	.76	.04
	B1S	.87	.03
	BRS	.86	.03
400			
	B	.75	.09
	B1	.86	.01
	BR	.83	.01

Notes: (a) B corresponds to (6.4) with $m = n$, B1 with $m=n_1=30, 40, 80$, B1S with $m=n_1=16$.

(b) BR corresponds to (6.5) with $(n_0, n_1)=(15, 30), (20, 40), (40, 80)$, BRS with $(n_0, n_1)=(4, 16)$.

Table 3. Sample splitting in Example 1(a)

n		CP N	$B_{m(n)}$	$RMSE$ N	$B_{m(n)}$
50					
	UB	.82	.86	.32	.18
	LB	.86	.91	.28	.16
100					
	UB	.86	.89	.30	.14
	LB	.84	.90	.26	.12
400					
	UB	.85	.89	.28	.08
	LB	.86	.91	.27	.09

Note: N here refers to $m = .1n$ and $\alpha = .1$.

Table 4. The $\binom{n}{m}$ bootstrap and the m/n bootstrap in Example 1(a)

n	CP m	J	B	E(Length) J	B
50	16	.82	.88	.07	.09
100	16	.86	.88	.04	.05
400	40	.88	.90	.01	.01

Note: These figures are for simulation sizes of $N = 500$ and for 90% confidence intervals. Thus, the end points of the intervals are given by (7.1) and its UCB counterpart for B and J but with .1 replaced by .05. Similarly, $[E(\text{Bound}-\text{Actual quantile bound})^2]^{1/2}$ is replaced by the expected length of the confidence interval.

Conclusions. The conclusions we draw are limited by the range of our simulations. We opted for realistic sample sizes, of $50, 100$ and a less realistic 400. For $n = 50, 100$ the subsample sizes $n_1 = 30$ (for $n = 50$) and 40 (for $n = 100$) are of the order $n/2$ rather than $o(n)$. For all sample sizes $n_0 = 2\sqrt{n}$ is not really "of larger order than $\sqrt{n}$". The simulations in fact show the asymptotics as very good when the bootstrap works even for relatively small sample sizes. The story when the bootstrap doesn't work is less clear.

When the bootstrap works (Example 1)

- BR and B are very close both in terms of CP, and $RMSE$ even for $n = 50$ from Table 1.
- B1's CP though sometimes better than B's consistently differs more from B's and its $RMSE$ follows suit In particular, for UB in Table 1, the $RMSE$ of B1 is generally larger. LB exhibits less differences but this reflects that UB is

governed by the behaviour of χ_1^2 at 0. In simulations we do not present we get similar sharper differences for LB when F is a heavy tailed distribution such as Pareto with $EX^5 = \infty$

- The effects, however, are much smaller than we expected. This reflects that these are corrections to the coefficient of the $n^{-1/2}$ term in the expansion. Perhaps the most surprising aspect of these tables is how well B1 performs.
- From Table 3 we see that because the m we are forced to by the level considered is small, CP for the sample splitting bounds differs from the nominal level. If $n \to \infty$, $m/n \to .1$ the coverage probability doesn't tend to .1 since the estimated quantile doesn't tend to the actual quantile and both CP and $RMSE$ behave badly compared to B_m. This naive method can be fixed up (see Blom (1976) for instance). However, its simplicity is lost and the $\binom{n}{m}$ or m/n bootstrap seem preferable.
- The $\binom{n}{m}$ bounds are inferior as Table 4 shows. This reflects the presence of the finite population correction m/n, even though these bounds were considered for the more favorable sample size $m = 16$ for $n = 50, 100$ rather than $m = 30, 40$. Corrections such as those of Bertail (1994) or simply applying the finite population correction to s would probably bring performance up to that of B_{n_1}. But the added complication doesn't seem worthwhile.

When the bootstrap doesn't work (Example 3)

- From Table 2, as expected, the CP of the n/n bootstrap for the lower confidence bound was poor for all n. For $n_0 = 2\sqrt{n}, n_1 = 4\sqrt{n}$, CP for B1 was constantly better than B for all n. BR is worse than B1 but improves with n and was nearly as good as B1 for $n = 400$. For small n_0, n_1 both B1 and BR do much better. However, it is clear that the smaller m of B1S is better than all other choices.

We did not give results for the upper confidence bound because the granularity of the bootstrap distribution of $\min_i X_i$ for these values of m and n made $CP = 1$ in all cases.

Evidently, n_0, n_1 play a critical role here. What apparently is happening is that for n_0, n_1 not sufficiently small compared with n extrapolation picks up the wrong slope and moves the not so good B1 bound even further towards the poor B bound.

A message of these simulations to us is that extrapolation of the B_m plot may carry risks not fully revealed by the asymptotics. On the other hand, if n_0 and n_1 are chosen in a reasonable fashion extrapolation on the $\sqrt{n}$ scale works well when the bootstrap does. Two notes, based on simulations we do not present, should be added to the optimism of Bickel, Yahav (1988) however. There may be risk if n_0 is really small compared to $\sqrt{n}$. We obtained poor results for BR for the t statistics for $n_0 = 4$ and 2. Thus $n_0 = 4$, $n_1 = 16$ gave the wrong slope to the extrapolation which tended to overshoot badly. Also, taking n_1 and n_0 close to each other, as the theory of the 1988 paper suggests is appropriate for statistics possessing high order expansions when the expansion coefficients are deterministic, gives poor results. It can also be seen theoretically that the sampling variability of the bootstrap for m of the order $\sqrt{n}$ makes this prescription unreasonable.

The principal message we draw is that it is necessary to develop data driven methods of selection of m which lead to reasonable results over situations where both the bootstrap works and where it doesn't. Such methods are being pursued.

Acknowledgement

We are grateful to Jiming Jiang and Adele Cutler for essential programming, to John Rice for editorial comments, and to Kjell Doksum for the Blom reference. This research was supported by NATO Grant CRG 920650, Sonderforschungsbereich 343 Diskrete Strukturen der Mathematik, Bielefeld and NSA Grant MDA 904-94-H-2020.

Appendix

Proof of Theorem 2. For $\boldsymbol{i} = (i_1, \ldots, i_r) \in \Lambda_{r,m}$ let $U(\boldsymbol{i}) = \frac{1}{\binom{n}{r}} \sum \{h_{\boldsymbol{i}}(X_{j_1}, \ldots, X_{j_r}, F) : 1 \leq j_1 < \cdots < j_r \leq n\}$. Then, since $h_{\boldsymbol{i}}$ as defined is symmetric in its arguments it is a U statistic and $\|h\|_\infty$ is an upper bound to its kernel. Hence

(a) $$\operatorname{Var}_F U(\boldsymbol{i}) \leq \|h\|_\infty^2 \frac{r}{n}.$$ On the other hand,

(b) $$EU(\boldsymbol{i}) = E_F h_{\boldsymbol{i}}(X_1, \ldots, X_r, F) \quad \text{and}$$

(c) $$B_{m,n}(F) = \sum_{r=1}^{m} \sum \{w_{m,n}(\boldsymbol{i}) U(\boldsymbol{i}) : \boldsymbol{i} \in \Lambda_{r,m}\}$$ by (3.7). Thus, by (c),

(d) $$\operatorname{Var}_F^{1/2} B_{m,n}(F) \leq \sum_{r=1}^{m} \sum \{w_{m,n}(\boldsymbol{i}) \operatorname{Var}_F^{1/2} U(\boldsymbol{i}) : \boldsymbol{i} \in \Lambda_{r,m}\}$$

$$\leq \max \operatorname{Var}_F^{1/2} U(\boldsymbol{i}) \leq \|h\|_\infty \left(\frac{m}{n}\right)^{1/2}$$

by (a). This completes the proof of (3.10).

The proof of (3.11) is more involved. By (3.8)

(e) $$|\theta_{m,n}(F) - \theta(F)| \leq \sum_{r=1}^{m} \sum \{|E_F h_{\boldsymbol{i}}(X_1, \ldots, X_r) - \theta_m(F)| w_{m,n}(\boldsymbol{i}) : \boldsymbol{i} \in \Lambda_{r,m}\}.$$

Let,

(f) $$P_{m,n}[R_m = r] = \sum\{w_{m,n}(\boldsymbol{i}) : \boldsymbol{i} \in \Lambda_{r,m}\}.$$

Expression (f) is easily recognized as the probability of getting $n - r$ empty cells when throwing n balls independently into m boxes without restrictions (see Feller (1968), p.19). Then it is well known or easily seen that

(g) $$E_{m,n}(R_m) = n(1 - (1 - \frac{1}{n})^m)$$

(h) $$\text{Var}_{m,n}(R_m) = n\{(1 - \frac{1}{n})^m - (1 - \frac{2}{n})^m\} + n^2\{(1 - \frac{2}{n})^m - (1 - \frac{1}{n})^{2m}\}.$$

It is easy to check that, if $m = o(n)$

(i) $$E_{m,n}(R_m) = m(1 + O(\frac{m}{n}))$$

(j) $$\text{Var}_{m,n}(R_m) = O(m)$$

so that,

(k) $$\frac{R_m}{m} = 1 + O_P(m^{-1/2}).$$

From (e),

(l) $$|\theta_{m,n}(F) - \theta(F)| \leq \sum_{r=1}^{m} \delta_m(\frac{r}{m}) P_{m,n}[R_m = r].$$

By (k), (l) and the dominated convergence theorem (3.12) follows from (3.11) and (k).

Finally, as in Theorem 1, we bound, as in (3.4),

(m) $$|B_{m,n}(F) - B_m(F)| \leq \sum_{r=1}^{m} \sum\{E_F|h_{\boldsymbol{i}}(X_1, \ldots, X_r) - h_{\boldsymbol{i}}(X_1, \ldots, X_r, \hat{F}_n)| : \boldsymbol{i} \in \Lambda_{r,m}\} w_{m,n}(\boldsymbol{i}),$$

where

(n) $$h_{\boldsymbol{i}}(X_1, \ldots, X_r, \hat{F}_n) = \frac{1}{r!} \sum_{1 \leq j_1 \neq \cdots \neq j_r \leq r} h(T_m(X_{j_1}^{(i_1)}, \ldots, X_{j_r}^{(i_r)}, \hat{F}_n)).$$

Let R_m be distributed according to (f) and given $R_m = r$, let $(I_1, \ldots, I_r)$ be uniformly distributed on the set of partitions of m into r ordered integers, $I_1 \leq I_2 \leq \cdots \leq I_r$. Then, from (m) we can write

$$\text{(o)} \qquad |B_{m,n}(F) - B_m(F)| \leq E\Delta(I_1, \ldots, I_{R_m}),$$

where $\|\Delta\|_\infty \leq \|h\|_\infty$. Further, by the continuity of h and (3.13), since $I_1 \leq \cdots \leq I_{R_m}$,

$$\text{(p)} \qquad \Delta(I_1, \ldots, I_{R_m})1(I_{R_m} \leq \epsilon_m m) \xrightarrow{P} 0$$

whenever $\epsilon_m = O(m^{-1/2})$. Now, $I_{R_m} > \epsilon_m m$,

$$\text{(q)} \qquad m = \sum_{j=1}^{R_m} I_j$$

and $I_j \geq 1$ imply that,

$$\text{(r)} \qquad m(1 - \epsilon_m) \geq \sum_{j=1}^{R_m - 1} I_j \geq (R_m - 1).$$

Thus,

$$\text{(s)} \qquad P_{m,n}(I_{R_m} > \epsilon_m m) \leq P_{m,n}(\frac{R_m}{m} - 1 \leq -\epsilon_m + O(m^{-1})) \rightarrow 0$$

if $\epsilon_m m^{1/2} \rightarrow \infty$. Combining (s), (k) and (p) we conclude that

$$\text{(t)} \qquad E\Delta(I_1, \ldots, I_{R_m}) \rightarrow 0$$

and hence (o) implies (3.14).

The corollary follows from (e) and (f).

Note that this implies that the m/n bootstrap works if about $\sqrt{m}$ ties do not affect the value of T_m much.

Checking that J_m, B_m, N_m $m = o(n)$ works

The arguments we give for B_m also work for J_m only more easily since Theorem 1 can be verified. It is easier to directly verify that, in all our examples, the m/n bootstrap distribution of $T_n(\hat{F}_n, F)$ converges weakly (in probability) to its limit $\mathcal{L}(F)$ and conclude that Theorem 2 holds for all h continuous and bounded than to check the conditions of Theorem 2. Such verifications can be found in the papers we cite. We sketch in what follows how the conditions of Theorem 1 and 2 can be applied.

Example 1. (a) We sketch heuristically how one would argue for functionals considered in Section 2 rather than quantiles. For J_m we need only check that (2.6) holds since $\sqrt{m}(\bar{X} - \mu(F)) = o_p(1)$. For B_m note that the distribution of $m^{-1/2}(i_1 X_1 + \cdots + i_r X_r)$ differs from that of $m^{-1/2}(X_1 + \cdots + X_m)$ by $O(\sum_{j=1}^{r} \frac{(i_j^2 - 1)}{m})$. If we maximize $\sum_{j=1}^{r}(i_j^2 - 1)$ subject to $\sum_{j=1}^{r} i_j = m$, $i_j \geq 1$ we obtain $\frac{2(m-r)}{m} + \frac{(m-r)^2}{m}$. Thus for suitable h, $\delta_m(x) = 2(1-x) + \frac{1}{\sqrt{m}}(1-x)^2$ and the hypotheses of Theorem 2 hold.

(b) Note that,

$$P\Big[\sqrt{n}\frac{(\bar{X} - \mu(F))}{s} \leq t\Big] = P[\sqrt{n}(\bar{X} - \mu(F)) - st \leq 0]$$

and apply the previous arguments to $T_n(\hat{F}_n, F) \equiv \sqrt{n}(\bar{X} - \mu(F)) - st$.

Example 2. In Example 2 the variance corresponds to $h(x) = x^2$ if $T_m(\hat{F}_m, F) = m^{1/2}(\mathrm{med}(X_1, \ldots, X_m) - F^{-1}(\frac{1}{2}))$. An argument parallel to that in Efron (1979) works. Here is a direct argument for h bounded.

$$\text{(a)} \qquad P[\mathrm{med}(X_1^{(i_1)}, \ldots, X_r^{(i_r)}) \neq \mathrm{med}(X_1^{(i_1)}, \ldots, X_r^{(i_r - 1)}, X_{r+1})] \leq \frac{1}{r+1}.$$

Thus,

$$\text{(b)} \qquad P[\mathrm{med}(X_1^{(i_1)}, \ldots, X_r^{(i_r)}) \neq \mathrm{med}(X_1, \ldots, X_m)] \leq \sum_{j=r+1}^{m} \frac{1}{j} \leq \log(\frac{m}{r}).$$

Hence for h bounded,

$$\delta_m(x) \leq \|h\|_\infty \log(\frac{1}{x})$$

and we can apply Theorem 2.

Example 3. Follows by checking (3.2) in Theorem 1 and that Theorem 2 applies for J_m by arguing as above for B_m. Alternatively, argue as in Athreya and Fukushi (1994).

Arguments similar to those given so far can be applied to the other examples.

References

Athreya, K. B. (1987). Bootstrap of the mean in the infinite variance case. *Ann. Statist.* **15**, 724-731.

Athreya, K. B. and Fukuchi, J. (1994). Bootstrapping extremes of I.I.D. random variables. Proceedings of Conference on Extreme Value Theory (NIST).

Babu, G. J. and Singh, K. (1985). Edgeworth expansions for sampling without replacement from finite populations. *J. Multivariate Anal.* **17**, 261-278.

Bentkus, V., Götze, F. and van Zwet, W. R. (1994). An Edgeworth expansion for symmetric statistics. Tech Report Univ. of Bielefeld.

Beran, R. (1982). Estimated sampling distributions: The bootstrap and competitors. *Ann. Statist.* **10**, 212-225.

Beran, R. and Srivastava, M. S. (1985). Bootstrap tests and confidence regions for functions of a covariance matrix. *Ann. Statist.* **13**, 95-115.

Bertail, P. (1994). Second order properties of an extrapolated bootstrap without replacement. Submitted to *Bernoulli.*

Bhattacharya, R. and Ghosh, J. K. (1978). On the validity of the formal Edgeworth expansion. *Ann. Statist.* **6**, 434-451.

Bickel, P. J. and Freedman, D. A. (1981). Some asymptotic theory for the bootstrap. *Ann. Statist.* **9**, 1196-1217.

Bickel, P. J. and Ren, J. J. (1995). The m out of n bootstrap and goodness of fit tests with double censored data. Robust Statistics, Data Analysis and Computer Intensive Methods Ed. H. Rieder Lecture Notes in Statistics, Springer-Verlag.

Bickel, P. J., Klaassen, C. K., Ritov, Y. and Wellner, J. (1993). *Efficient and Adaptive Estimation in Semiparametric Models.* Johns Hopkins University Press, Baltimore.

Bickel, P. J. and Yahav, J. A. (1988). Richardson extrapolation and the bootstrap. J. *Amer. Statist. Assoc.* **83**, 387-393.

Blom, G. (1976). Some properties of incomplete U statistics. *Biometrika* **63**, 573-580.

Bretagnolle, J. (1981). Lois limites du bootstrap de certaines fonctionelles. *Ann. Inst. H. Poincaré, Ser.B* **19**, 281-296.

David, H. A. (1981). *Order Statistics.* 2nd edition, John Wiley, New York.

Deheuvels, P., Mason, D. and Shorack, G. (1993). Some results on the influence of extremes on the bootstrap. *Ann. Inst. H. Poincaré* **29**, 83-103.

DeCiccio T. J. and Romano, J. P. (1989). The automatic percentile method: Accurate confidence limits in parametric models, *Canad. J. Statist.* **17**, 155-169.

Dümbgen, L. (1993). On nondifferentiable functions and the bootstrap. *Probab. Theory Related Fields* **95**, 125-140.

Efron, B. (1979). Bootstrap methods: another look at the jackknife. *Ann. Statist.* **7**, 1-26.

Efron, B. and Tibshirani, R. J. (1993). *An Introduction to the Bootstrap.* Chapman & Hall, London, New York.

Feller W. (1968). *Probability Theory* v1. John Wiley, New York.

Freedman D. A. (1981). Bootstrapping regression models. *Ann. Statist.* **9**, 1218-1228.

Giné. E. and Zinn, J. (1989). Necessary conditions for the bootstrap of the mean. *Ann. Statist.* **17**, 684-691.

Götze, F. (1993). *Bulletin* I. M. S.

Hall, P. (1992). *The Bootstrap and Edgeworth Expansion.* Springer Verlag, New York.

Hall, P. and Wilson, S. (1991). Two guidelines for bootstrap hypothesis testing. *Biometrics* **47**, 757-762.

Hall, P. and Jing B. Y. (1993). Performance of boostrap for heavy tailed distributions. Tech. Report A. N. U. Canberra.

Hinkley, D. V. (1988). Bootstrap methods (with discussion). *J. Roy. Statist. Soc. Ser.B* **50**, 321-337.

Mammen, E. (1992). When does bootstrap work? Springer Verlag, New York.

Politis, D. N. and Romano, J. P. (1994). A general theory for large sample confidence regions based on subsamples under minimal assumptions. *Ann. Statist.* **22**, 2031-2050.

Praestgaard, J. and Wellner, J. (1993). Exchangeably weighted bootstraps of the general empirical process. *Ann. Probab.* **21**, 2053-2086.

Putter, H. and van Zwet, W. R. (1993). Consistency of plug in estimators with applications to the bootstrap. Submitted to *Ann. Statist.*

Robinson, J. (1978). An asymptotic expansion for samples from a finite population. *Ann. Statist.* **6**, 1005-1011.

Shen, X. and Wong, W. (1994). Convergence rates of sieve estimates. *Ann. Statist.* **22**, 580-615.

Singh, K. (1981). On the asymptotic accuracy of Efron's bootstrap. *Ann. Statist.* **9**, 1187-1195.

Wu, C. F. J. (1990). On the asymptotic properties of the jackknife histogram. *Ann. Statist.* **18**, 1438-1452.

Department of Statistics, University of California, Berkeley, 367 Evans Hall, #3860, Berkeley, CA 94720-3860, U. S. A.

Department of Mathematics, University of Bielefeld, Universitatsstrasse 4800, Bielefeld, Germany.

Department of Mathematics, University of Leiden, PO Box 9512 2300RA, Leiden, Netherlands.

(Received August 1995; accepted June 1996)

20

On a Set of the First Category

Hein Putter[1]
Willem R. van Zwet[2]

ABSTRACT In an analysis of the bootstrap Putter & van Zwet (1993) showed that under quite general circumstances, the bootstrap will work for "most" underlying distributions. In fact, the set of exceptional distributions for which the bootstrap does not work was shown to be a set D of the first category in the space $\mathcal{P}$ of all possible underlying distributions, equipped with a topology Π. Such a set of the first category is usually "small" in a topological sense. However, it is known that this concept of smallness may sometimes be deceptive and in unpleasant cases such "small" sets may in fact be quite large.

Here we present a striking and hopefully amusing example of this phenomenon, where the "small" subset D equals all of $\mathcal{P}$. We show that as a result, a particular version of the bootstrap for the sample minimum will never work, even though our earlier results tell us that it can only fail for a "small" subset of underlying distributions. We also show that when we change the topology on $\mathcal{P}$—and as a consequence employ a different resampling distribution—this paradox vanishes and a satisfactory version of the bootstrap is obtained. This demonstrates the importance of a proper choice of the resampling distribution when using the bootstrap.

20.1 Introduction

Many of the results of asymptotic statistics cannot be established in complete generality. One often has to allow the possibility that the result will not hold if the underlying probability distribution belongs to a small subset D of the collection of all possible underlying probability distributions $\mathcal{P}$. In many concrete examples, D will turn out to be empty, but in general one has to take the existence of such an exceptional set into account.

If $\mathcal{P}$ is a parametric model, the exceptional set D will typically be small in the sense that it is indexed by a set of Lebesgue measure zero in the Euclidean parameter space. From a technical point of view, its occurrence is caused by an application of a result like Egorov's or Lusin's theorem where exceptional sets of arbitrarily small Lebesgue measure occur. In more general models one could conceivably use similar tools for more general

[1]University of Leiden
[2]University of Leiden and University of North Carolina.

S. van de Geer and M. Wegkamp (eds.), *Selected Works of Willem van Zwet*, Selected Works in Probability and Statistics, DOI 10.1007/978-1-4614-1314-1_18, © Springer Science+Business Media, LLC 2012

measures, but it is difficult to think of a measure on $\mathcal{P}$ which is such that we can agree that a set of measure zero is indeed small in a relevant sense.

In a recent study of resampling, we have followed a different path and established asymptotic results where the exceptional set is small in a topological rather than a measure-theoretic sense (Putter & van Zwet 1993). If we equip the set $\mathcal{P}$ with a metric ρ, the exceptional set D in these results is a set of the first category in the metric space $(\mathcal{P}, \rho)$. We recall that a set of the first category is a countable union of nowhere dense sets, and that a set is nowhere dense if its closure has empty interior. Equivalently, a set is of the first category if it can be covered by a countable union of closed sets, each of which has empty interior.

This concept of a small set was used by Le Cam as early as Le Cam (1953), where it is shown that superefficiency can only occur on a set of the first category. In a parametric setting, Le Cam was careful to point out that under the right conditions the exceptional set also corresponds to a set of Lebesgue measure zero in the parameter space. The same is true for the results in Putter & van Zwet (1993), as shown by Putter (1994).

Of course the question remains whether a set of the first category is indeed small in any accepted sense. If $(\mathcal{P}, \rho)$ is complete, we know that a set of the first category is small, for example in the sense that it has a dense complement (cf. Dudley 1989, pp. 43–44). If $(\mathcal{P}, \rho)$ is not complete, then a set of the first category can be uncomfortably large: in fact we shall see that the entire space $\mathcal{P}$ may be of the first category itself.

In this note we discuss a particular statistical model $\mathcal{P}_0$ equipped with Hellinger metric H, such that $(\mathcal{P}_0, H)$ is not complete and $\mathcal{P}_0$ is of the first category in $(\mathcal{P}_0, H)$. An application of our results on resampling shows that a particular version of the bootstrap will work except if the underlying distribution belongs to a set D of the first category. Unfortunately, it turns out that $D = \mathcal{P}_0$ so that we have no guarantee that this version of the bootstrap will ever work, and indeed it may never do. Luckily, our analysis also shows that we need not despair. It turns out that our problems are not caused by any inherent pathology of the model $\mathcal{P}_0$, but by a wrong choice of metric on $\mathcal{P}_0$. If we replace H by a different, complete, metric and modify the construction of the bootstrap accordingly, the pathology disappears and we obtain a version of the bootstrap that will work for any $P \in \mathcal{P}_0$. In fact the example may serve to clarify the importance of a correct choice of the resampling distribution when using the bootstrap.

In Section 2 we exhibit the particular class of distributions $\mathcal{P}_0$ which is of the first category in $(\mathcal{P}_0, H)$. In section 3 we show that this class is not an artificial construct, but that it is the natural model for a statistical situation of interest. We then proceed to make the connection with a result on the bootstrap in Putter & van Zwet (1993) and show that this result doesn't produce a satisfactory version of the bootstrap for this model. Finally we show that a different choice of metric on $\mathcal{P}_0$ will resolve our problems.

20.2 A set of the first category

Let us consider the class $\mathcal{P}_0$ of probability distributions P on $(0,\infty)$ which have distribution functions F satisfying

(1) $$\lim_{x\downarrow 0} \frac{F(x)}{x} = a(P) \in (0,\infty).$$

We equip $\mathcal{P}_0$ with Hellinger metric H. For distributions $P, Q \in \mathcal{P}_0$, with densities f and g with respect to a common σ-finite measure ν, this is defined by

$$H(P,Q)^2 = \int \left(\sqrt{f} - \sqrt{g}\right)^2 d\nu.$$

(2) **Proposition** *The set $\mathcal{P}_0$ is of the first category in $(\mathcal{P}_0, H)$.*

Proof. For $k = 1, 2, \ldots$, let $\delta_k = 1/k$ and

$$B_k = \{P \in \mathcal{P}_0 : \left|\frac{F(x)}{x} - \frac{F(\delta_k)}{\delta_k}\right| \leq 1 \quad \text{for } 0 < x \leq \delta_k\}.$$

Clearly, $\mathcal{P}_0 \subset \bigcup_{k=1}^{\infty} B_k$, and since convergence in Hellinger metric implies pointwise convergence of distribution functions, we see that each B_k is closed in $(\mathcal{P}_0, H)$. It remains to be shown that no B_k contains an open set.

Fix k and choose a distribution $P \in B_k$ with distribution function F and with $a(P) = \alpha$. Define $G_n(x) = \min\left(n^{-1}, (3+\alpha)x\right)$, $F_n = \max(G_n, F)$, and let P_n be the distribution with distribution function F_n. Then $a(P_n) = 3 + \alpha$ but, for n large enough, $F_n(\delta_k)/\delta_k = F(\delta_k)/\delta_k \leq 1 + \alpha$ because $P \in B_k$. It follows that $P_n \notin B_k$ for large n, even though P_n converges to P in Hellinger metric. □

20.3 A bootstrap fiasco

Let $\mathcal{P}$ be a class of probability distributions on $\mathbb{R}$. We equip $\mathcal{P}$ with a metric ρ. Let $X_1, X_2, \ldots$ be independent and identically distributed (i.i.d.) random variables with (unknown) common distribution $P \in \mathcal{P}$. We are interested in the large sample behavior of a random variable

(3) $$Y_N = y_N(X_1, \ldots, X_N; P).$$

Let $\tau_N(P)$ denote the distribution of Y_N under $P \in \mathcal{P}$, and suppose that, for every $P \in \mathcal{P}$, $\tau_N(P)$ converges weakly to a limit distribution $\tau(P)$. If $\widehat{P}_N = P_N(X_1, \ldots, X_N)$ is an estimator of P taking values in $\mathcal{P}$, then $\tau_N(\widehat{P}_N)$ is called a bootstrap estimator of $\tau_N(P)$, or of $\tau(P)$, with resampling distribution $\widehat{P}_N$. For all P and $\widehat{P}_N$, the distributions $\tau_N(P)$, $\tau(P)$, and $\tau_N(\widehat{P}_N)$ are elements of the class $\mathcal{R}$ of all probability measures on $\mathbb{R}$. We equip this class with Lévy distance ℓ, or any other metric which metrizes weak convergence. The bootstrap is said to work for a particular $P \in \mathcal{P}$ if it is an ℓ-consistent estimator of $\tau_N(P)$, i.e. if $\ell(\tau_N(\widehat{P}_N), \tau_N(P))$ converges to zero in probability under P. As $\ell(\tau_N(P), \tau(P)) \to 0$, this is the same as ℓ-consistency for estimating the limit distribution $\tau(P)$.

The following two propositions are taken from Putter & van Zwet (1993).

(4) **Proposition** *Suppose that*

(i) The sequence of maps $\tau_N : (\mathcal{P}, \rho) \to (\mathcal{R}, \ell)$ is equicontinuous on $\mathcal{P}$;

(ii) $\widehat{P}_N$ takes values in $\mathcal{P}$ and is a ρ-consistent estimator of P, i.e. $\rho(\widehat{P}_N, P) \xrightarrow{P} 0$ for every $P \in \mathcal{P}$.

Then the bootstrap $\tau_N(\widehat{P}_N)$ works for every $P \in \mathcal{P}$.

(5) **Proposition** *Suppose that*

(i) $\tau_N : (\mathcal{P}, \rho) \to (\mathcal{R}, \ell)$ is continuous for every N;

(ii) For every $P \in \mathcal{P}$, $\tau_N(P)$ converges weakly to a limit $\tau(P)$;

(iii) $\widehat{P}_N$ takes values in $\mathcal{P}$ and is a ρ-consistent estimator of P, i.e. $\rho(\widehat{P}_N, P) \xrightarrow{P} 0$ for every $P \in \mathcal{P}$.

Then there exists a set D of the first category in $(\mathcal{P}, \rho)$ such that the sequence τ_N is equicontinuous at every $P \in \mathcal{P} \setminus D$ and hence the bootstrap $\tau_N(\widehat{P}_N)$ works for every $P \in \mathcal{P} \setminus D$.

Usually these results are used with Hellinger distance H for ρ, and on closer inspection it often turns out that the exceptional set D may be taken to be empty.

In the remainder of this paper we shall consider a specific example of this situation. We choose $\mathcal{P} = \mathcal{P}_0$, the class of distributions defined in (1). For i.i.d. random variables $X_1, \ldots, X_N$ taking values in $(0, \infty)$ with common distribution $P \in \mathcal{P}_0$, we define

$$Y_N^0 = N \min\{X_1, \ldots, X_N\}. \tag{6}$$

Note that $\mathcal{P}_0$ is a natural model for studying the large sample behavior of Y_N^0, since it is precisely the class of underlying distributions for which the distributions $\tau_N(P)$ of Y_N^0 under P converge weakly to a non-degenerate limit, which is an exponential distribution with parameter $a(P)$.

Bootstrapping the sample minimum is a problem of some notoriety as it is an early example where the usual choice of the empirical distribution P_N for the resampling distribution $\widehat{P}_N$ does not work. To check whether the bootstrap with a different choice of $\widehat{P}_N$ will work for "most" $P \in \mathcal{P}_0$, we may appeal to Proposition 5. In doing so, we are still free to choose a metric ρ on $\mathcal{P}_0$ and we shall make the usual choice by taking ρ to be Hellinger distance H. Since Y_N^0 is a function of $X_1, \ldots, X_N$ only, and not of P, it is easy to see that $\tau_N : (\mathcal{P}_0, H) \to (\mathcal{R}, \ell)$ is continuous for each N. As $\tau_N(P)$ converges weakly to a limit $\tau(P)$ for every $P \in \mathcal{P}_0$, Proposition 5 asserts that if $\widehat{P}_N$ is a Hellinger consistent estimator with values in $\mathcal{P}_0$, then the bootstrap $\tau_N(\widehat{P}_N)$ will work except for P in a set

D of the first category in $(\mathcal{P}_0, H)$. The content of Proposition 2 having made us somewhat suspicious, we may want to investigate the nature of the exceptional set D where the functions $\tau_N : (\mathcal{P}_0, H) \to (\mathcal{R}, \ell)$ are not equicontinuous. Since $\tau(P)$ depends on P only through $a(P)$, and any $P \in \mathcal{P}_0$ may be approximated arbitrarily well in Hellinger distance by a sequence $P_r \in \mathcal{P}_0$ with a constant value of $a(P_r)$ different from $a(P)$, we know that the limit distribution τ is nowhere continuous in P. This implies that the functions τ_N are not equicontinuous at any $P \in \mathcal{P}_0$, so that our worst suspicions are confirmed: the exceptional set D equals the entire set of possible distributions in this case. Our application of Proposition 5 with $\rho = H$ has therefore produced no positive information concerning this example at all.

Even though Proposition 5 is vacuous in this case, it might still by a stroke of luck be true that the bootstrap estimate $\tau_N(\widehat{P}_N)$ would work for most reasonable Hellinger-consistent estimators $\widehat{P}_N$ of P. First of all we note that it is indeed possible to construct an estimator of P which is Hellinger-consistent for every distribution P on $\mathbb{R}$ which has no singular part (cf. Devroye & Győrfi 1990, p. 1497). All we have to do is to assign probability k/N to all values which were observed $k > 1$ times, and add a kernel density estimator based on the remaining values which have only been observed once. Using the normal kernel we arrive at an estimator F_N^* for the distribution function F of P which is given by

$$F_N^*(x) = \frac{1}{N}\sum_{i=1}^{N}\left(\delta_i 1_{(0,x]}(X_i) + (1-\delta_i)\Phi\left(\frac{x-X_i}{h_N}\right)\right) \tag{7}$$

where Φ is the standard normal distribution function,

$$\delta_i = \begin{cases} 0 & \text{if } X_j \neq X_i \text{ for } j \neq i, \\ 1 & \text{otherwise,} \end{cases} \tag{8}$$

and $h_N \to 0$ but $Nh_N \to \infty$. Admittedly, F_N^* does not satisfy (1) and hence the corresponding estimator P_N^* of P does not take its values in $\mathcal{P}_0$ as is required in Proposition 5. However this defect is easily cured by considering the following slight modification of F_N^*,

$$\widehat{F}_N(x) = \begin{cases} xF_N^*(M_N)/M_N & \text{for } 0 \le x < M_N, \\ F_N^*(x) & \text{for } x \ge M_N, \end{cases} \tag{9}$$

where $M_N = \min(X_1, \ldots, X_N)$. Clearly $\widehat{F}_N$ satisfies (1), and hence the corresponding estimator $\widehat{P}_N$ of P takes its values in $\mathcal{P}_0$ and is Hellinger consistent for every $P \in \mathcal{P}_0$ which has no singular part. Nevertheless we shall see that the bootstrap $\tau_N(\widehat{P}_N)$ based on this estimator does not work for any $P \in \mathcal{P}_0$.

The bootstrap $\tau_N(\widehat{P}_N)$ has distribution function

$$\widehat{H}_N(y) = 1 - \left(1 - \widehat{F}_N(y/N)\right)^N .$$

For $P \in \mathcal{P}_0$, the limit distribution $\tau(P)$ is exponential with parameter $a(P) \in (0, \infty)$, and hence the bootstrap $\tau_N(\widehat{P}_N)$ will work for a particular $P \in \mathcal{P}_0$ if and only if

$$\sup_{y>0} |\widehat{H}_N(y) - [1 - \exp\{-a(P)y\}]| \xrightarrow{P} 0.$$

This is easily seen to be equivalent to

$$N\widehat{F}_N(\frac{y}{N}) - a(P)y \xrightarrow{P} 0, \tag{10}$$

for every $y > 0$.

However, (10) cannot hold. If F_N denotes the empirical distribution function, (7) implies that for all x,

$$F_N^*(x) \geq \frac{1}{N} \sum_{i=1}^{N} 1_{(0,x]}(X_i)[\delta_i + \frac{1-\delta_i}{2}] \geq 1/2F_N(x).$$

As $F_N(x) = 0$ for $x < M_N$, we also find that for all x,

$$\widehat{F}_N(x) \geq 1/2F_N(x).$$

Hence, for every $y > 0$, the definition (1) ensures that as $N \to \infty$,

$$P\Big(|N\widehat{F}_N(\frac{y}{N}) - a(P)y| \geq a(P)y\Big) \geq P\Big(N\widehat{F}_N(\frac{y}{N}) \geq 2a(P)y\Big)$$
$$\geq P\Big(NF_N(\frac{y}{N}) \geq 4a(P)y\Big) = P\Big(Z \geq 4a(P)y\Big) + o(1) \not\to 0,$$

where Z has a Poisson distribution with expectation $a(P)y > 0$. This shows that (10) is false, and as a consequence, the bootstrap based on $\widehat{P}_N$ does not work for any $P \in \mathcal{P}_0$, and the fiasco is indeed complete.

20.4 A bootstrap success

Luckily, the disastrous results of the previous section also indicate quite clearly how the damage may be repaired. Our problems in the previous section originate from the fact that the parameter of the exponential limit distribution $a(P)$ is not a continuous function of the underlying distribution $P \in \mathcal{P}_0$ with respect to Hellinger distance on $\mathcal{P}_0$. Hence we should look for a different metric on $\mathcal{P}_0$, and in view of the definition of $a(P)$ in (1), one obvious candidate is a metric π defined by

$$\pi(P, Q) = \sup_{x>0} \frac{|F(x) - G(x)|}{x} \tag{11}$$

where F and G denote the distribution functions corresponding to P and Q.

With this new metric π, things immediately fall into place. The metric space $(\mathcal{P}_0, \pi)$ is easily seen to be complete and hence sets of the first category have dense complements. Clearly the exponential limit distribution

$\tau(P)$ is continuous when viewed as a map $\tau : (\mathcal{P}_0, \pi) \to (\mathcal{R}, \ell)$. Also, the sequence of distributions $\tau_N(P)$ of Y_N is equicontinuous on $\mathcal{P}_0$. To see this, note that for underlying distributions P and Q with distribution functions F and G, Y_N has distribution functions

$$H_{N,P}(y) = 1 - (1 - F(y/N))^N$$

and

$$H_{N,Q}(y) = 1 - (1 - G(y/N))^N .$$

Fix $P \in \mathcal{P}_0$ and $0 < \epsilon < 1$. Choose positive numbers y_0 and z_0 such that

$$y_0 = \frac{4\log(4/\epsilon)}{a(P)} \quad \text{and} \quad F(z) \geq 1/2\, a(P)\, z \quad \text{for } 0 \leq z \leq z_0$$

and note that $|a^N - b^N| \leq N|a - b|$ if $0 \leq a, b \leq 1$. If $N \geq y_0/z_0$ we choose $\pi(P, Q) \leq \frac{\epsilon}{2y_0} \leq \frac{a(P)}{4}$ and find

$$\begin{aligned}
\sup_y |H_{N,P}(y) - H_{N,Q}(y)| &= \sup_y \left|(1 - F(y/N))^N - (1 - G(y/N))^N\right| \\
&\leq N \sup_{y \leq y_0} |F(y/N) - G(y/N)| + (1 - F(y_0/N))^N + (1 - G(y_0/N))^N \\
&\leq y_0 \pi(P, Q) + \exp\{-NF(y_0/N)\} + \exp\{-NG(y_0/N)\} \\
&\leq \frac{\epsilon}{2} + \exp\{-1/2\, a(P)\, y_0\} + \exp\{-1/2\, a(P)\, y_0 + y_0 \pi(P, Q)\} \\
&\leq \frac{\epsilon}{2} + \left(\frac{\epsilon}{4}\right)^2 + \exp\{-1/4\, a(P) y_0\} = \frac{\epsilon}{2} + \frac{\epsilon^2}{16} + \frac{\epsilon}{4} < \epsilon .
\end{aligned}$$

On the other hand, if $1 \leq N < y_0/z_0$, we choose y_1 such that

$$1 - F\left(\frac{z_0 y_1}{y_0}\right) \leq \frac{\epsilon}{4}.$$

For $\pi(P, Q) \leq \frac{\epsilon}{4y_1}$ we find

$$\begin{aligned}
&\sup_y |H_{N,P}(y) - H_{N,Q}(y)| \\
&\qquad \leq y_1 \pi(P, Q) + (1 - F(y_1/N))^N + (1 - G(y_1/N))^N \\
&\qquad \leq \frac{\epsilon}{4} + (1 - F(y_1/N)) + \left(1 - F(y_1/N) + \frac{y_1}{N}\pi(P, Q)\right) \\
&\qquad \leq \frac{\epsilon}{4} + \left(1 - F\left(\frac{z_0 y_1}{y_0}\right)\right) + \left(1 - F\left(\frac{z_0 y_1}{y_0}\right) + \frac{\epsilon}{4}\right) \leq \epsilon .
\end{aligned}$$

Hence for every $0 < \epsilon < 1$ there exists $\delta > 0$ depending on P but not on Q, such that $\pi(P, Q) \leq \delta$ implies $\sup_y |H_{N,P}(y) - H_{N,Q}(y)| \leq \epsilon$ for all N, which establishes the equicontinuity of $\{\tau_N\}$ on $\mathcal{P}_0$.

By Proposition 4 the equicontinuity of $\tau_N : (\mathcal{P}_0, \pi) \to (\mathcal{R}, \ell)$ implies that the bootstrap $\tau_N(\widetilde{P}_N)$ will work for all $P \in \mathcal{P}_0$ if $\widetilde{P}_N$ is a π-consistent estimator of P. An example of such an estimator is the random distribution

$\widetilde{P}_N$ with distribution function

$$\widetilde{F}_N(x) = \begin{cases} xF_N(\xi_N)/\xi_N & \text{for } 0 \le x < \xi_N \\ F_N(x) & \text{for } x \ge \xi_N , \end{cases} \tag{12}$$

where F_N denotes the empirical distribution function, $\xi_N \to 0$ but $N\xi_N \to \infty$. To see this, let F denote the distribution function corresponding to the underlying distribution $P \in \mathcal{P}_0$. Then $F_N(\xi_N)/\xi_N \xrightarrow{P} a(P)$ if $\xi_N \to 0$ and $N\xi_N \to \infty$, since

$$E\frac{F_N(\xi_N)}{\xi_N} = \frac{F(\xi_N)}{\xi_N} \to a(P)$$

and

$$\sigma^2\left(\frac{F_N(\xi_N)}{\xi_N}\right) = \frac{F(\xi_N)(1-F(\xi_N))}{N\xi_N^2} \to 0.$$

It follows that, if $\xi_N \to 0$ and $N\xi_N \to \infty$ as $N \to \infty$, then

$$\sup_{x \le \xi_N} \frac{|\widetilde{F}_N(x) - F(x)|}{x} \le \left|\frac{F_N(\xi_N)}{\xi_N} - a(P)\right| + \sup_{x \le \xi_N}\left|\frac{F(x)}{x} - a(P)\right| \xrightarrow{P} 0.$$

Also, for every sequence $\eta_N \to 0$ with $\eta_N > \xi_N$,

$$\sup_{\xi_N \le x \le \eta_N} \frac{|\widetilde{F}_N(x) - F(x)|}{x} = \sup_{\xi_N \le x \le \eta_N} \frac{|F_N(x) - F(x)|}{F(x)/a(P)} + o(1)$$

and this tends to zero in probability if $N\xi_N \to \infty$ (cf. Chang 1955, Theorem 1; see also Shorack & Wellner 1986, p. 424). Finally, taking η_N such that $N^{1/2}\eta_N \to \infty$, we have

$$\sup_{x > \eta_N} \frac{|F_N(x) - F(x)|}{x} \le \sup_x \frac{|F_N(x) - F(x)|}{\eta_N} = O_P(N^{-1/2}\eta_N^{-1}).$$

Thus the π-consistency of $\widetilde{F}_N$ follows and we have shown

(13) **Proposition** *If $\widetilde{P}_N$ is an estimator of P with distribution function $\widetilde{F}_N$ given by (12), then the bootstrap estimator $\tau_N(\widetilde{P}_N)$ of the distribution $\tau_N(P)$ of Y_N^0 is consistent for all $P \in \mathcal{P}_0$.*

Acknowledgments: This research was supported by the Netherlands' Organization for Scientific Research (NWO) and by the Sonderforschungsbereich 343 "Diskrete Strukturen in der Mathematik" at the University of Bielefeld, German Federal Republic.

We are grateful to David Pollard for his careful reading of this paper which led to many improvements.

20.5 References

Chang, L.-C. (1955), 'On the ratio of the empirical distribution to the theoretical distribution function', *Acta Math. Sinica* **5**, 347–368. (English Translation in: *Selected Translations in Mathematical Statististics and Probability*, **4**, 17–38 (1964).).

Devroye, L. & Győrfi, L. (1990), 'No empirical probability measure can converge in the total variation sense for all distributions', *Annals of Statistics* **18**, 1496–1499.

Dudley, R. M. (1989), *Real Analysis and Probability*, Wadsworth, Belmont, California.

Le Cam, L. (1953), 'On some asymptotic properties of maximum likelihood estimates and related Bayes estimates', *University of California Publications in Statistics* **1**, 277–330.

Putter, H. (1994), Consistency of Resampling Methods, PhD thesis, University of Leiden.

Putter, H. & van Zwet, W. R. (1993), Consistency of plug-in estimators with application to the bootstrap, Technical report, University of Leiden.

Shorack, G. R. & Wellner, J. A. (1986), *Empirical Processes with Applications to Statistics*, Wiley, New York.

Chapter 19
Discussion of three resampling papers

Peter J. Bickel

Abstract
Discussion of:
Putter H., and van Zwet W.R. (1996). Resampling: Consistency of substitution estimators, *Annals of Statistics* **24** 2297–2318.
Putter H., and van Zwet W.R. (1997). On a set of the first category. *Festschrift for Lucien Le Cam, Springer Verlag*, 315–324.
Bickel P., Götze F. and van Zwet W.R. (1997). Resampling fewer than n observations: Gains, Losses and Remedies for Losses, *Statistica Sinica* **1** 1-31.

It is a pleasure to return to these three papers of van Zwet's on the bootstrap, two coauthored with Hein Putter and the other with Friedrich Götze and myself.

They marked van Zwet's attempt to understand the behaviour of bootstrap estimates of parameters, when observations, $X_1, \ldots, X_n$ were i.i.d. for $P \in \mathscr{P}$. This was done for clarity of conception only. It was evident that generalizations to weakly dependent data should hold.

In the first two papers Putter and van Zwet considered a sequence $\tau_N(P)$ of parameters which were themselves probability distributions, and endowed with a suitable metric (e.g. Prohorov-Levy). $\mathscr{P}$ was endowed with a metric ρ. If P_N is a ρ consistent estimate of P and

$$d\big(\tau_N(\hat{P}_N), \tau_N(P)\big) \xrightarrow{P} 0\,, \qquad (19.1)$$

the bootstrap $\tau_N(\hat{P}_N)$ was defined as successfully estimating $\tau_N(P)$.

Their main emphasis in the first paper was to show the general feasibility of constructions satisfying (refequation1) for continuous τ_N except on sets of the first category in $\mathscr{P}$. This led then, on the one hand, to the parametric bootstrap and, on the other, to the view that bootstraps and $\hat{P}_N$ had to be tailored to the specific problem, for interesting τ_N. Essentially this is the case unless ρ is the Hellinger

Peter J. Bickel
Department of Statistics, 367 Evans Hall. Berkeley, CA 94710-3860
e-mail: bickel@stat.berkeley.edu

S. van de Geer and M. Wegkamp (eds.), *Selected Works of Willem van Zwet*, Selected Works in Probability and Statistics, DOI 10.1007/978-1-4614-1314-1_19, © Springer Science+Business Media, LLC 2012

metric in which case, ρ consistency of any $\hat{P}_N$ typically fails unless $\mathscr{P}$ is regular parametric – but see Golubev et al [1].

The second paper gave interesting examples of situations where such $\hat{P}_N$ could be constructed but, unless a great deal of attention to the structure of $\mathscr{P}$ was given, would have the set of the first category mentioned be all of $\mathscr{P}$!

The third paper jointly with Götze and myself took a different track, concentrating on sequences, $\tau_N(P,\hat{P}_N)$ such as those that usually arise in setting confidence bounds and other statistical questions, e.g. $\tau_N(P_N,P) = N^{-\frac{1}{2}}\left(\int x d\hat{P}_N - \int x d\hat{P}\right)$ for $\mathscr{P} = \{P : \int x^2 dP < \infty\}$.

Here, and throughout the sequel, $\hat{P}_N$ is the empirical distribution. As in the above example, the τ_N are assumed to have a basic property quite different from the τ_N considered by Putter and van Zwet:

(I) $\tau_N(\hat{P}_N,P)$ converges weakly to a limiting probability distribution τ_P .

In the example above $\tau_P = \mathscr{N}(0, \mathrm{Var}_P(X_1))$. Our focus was on estimating the probability distribution of $\tau_N(\hat{P}_N,P)$.

Condition I trivially implies that, if $m_N \to \infty$, $\tau_{m_N}(\hat{P}_{m_N},P)$ converge weakly to τ_P as well. The condition suggests that we use implicit scaling as in the example and estimate the distribution of $\tau_N(\hat{P}_N,P)$ by that of $\tau_{m_N}(\hat{P}^*_{m_N},\hat{P}_N)$ where $m_N \to \infty$ slowly, $\{\hat{P}^*_m\}$ depends on $\hat{P}_N$ and converges to P in an appropriate metric ρ.

Independently, Politis and Romano (1994) and Götze (1993) considered the basic but statistically less interesting case that $\tau_N(\hat{P}_N,P) \equiv \tau_N(P)$, so that τ_p is degenerate. They showed that if $\hat{P}^*_m$ is the empirical distribution of a sample drawn *without replacement* from $X_1,\ldots,X_N$, $m_N \to \infty$ and $\frac{m_N}{N} \to 0$, then the conditional distributions of $\tau(\hat{P}^*_m,\hat{P}_N)$ given the data converge weakly, with probability 1, to τ_P without any further conditions. Our (1997) paper goes on to investigate when $\tau_{m_N}(\hat{P}^*_{m_N},\hat{P}_N)$ converges to τ_P generally, both when $\hat{P}^*_{m_N}$ corresponds to sampling without replacement and with replacement. A number of other issues are also studied. Not surprisingly when $\tau_N(\hat{P}^*_N,\hat{P}_N)$ converges weakly to τ_P, it typically does so faster than $\tau_{m_N}(P^*_{m_N},\hat{P}_N)$ with $\frac{m_N}{N} \to 0$. We discuss ways of removing this disability and propose a crude rule for data determined selection of m_N.

This approach and general applications to situations where the Efron bootstrap fails are analyzed further in Götze and Rakauskas (2001), Bickel and Sakov (2002a) and Bickel and Sakov (2002b).

The papers with Putter exhibit van Zwet's typical approach to research: A general question is sharply posed followed by a definitive, technically subtle answer, in this case, I think, not as satisfactory as van Zwet originally hoped. The 1997 paper though considerably less elegant and definitive than the work with Putter would seem to have the general applicability that van Zwet initially hoped for – but that's obviously a biased opinion.

References

1. Golubev G.K., Levit B., Tsybakov, A.B. (1996). Asymptotically efficient estimation of analytic functions in Gaussian noise, *Bernoulli, 2*, 167–181.
2. Götze F. and Rakauskas, A. (2001). Adaptive choice of bootstrap sample sizes, *State of the Art in Probability and Statistics*, 286–309.
3. Bickel P. and Sakov A. (2002). Extrapolation and the bootstrap, *Sankhya, Ser. A, 3*, 640–652.
4. Bickel, P. and Sakov, A. (2002). On the choice of *m* in the *m* out of *n* bootstrap and confidence bounds for extrema, *Statist. Sinica, 18*, 967–985.
5. Politis, D. and Romano, J. (1994). Large sample confidence regions based on subsamples under minimal assumptions, *Ann. Statist. 22*, 2031–2050.
6. Götze F. (1993). Asymptotic approximations and the bootstrap, *JMS Bulletin*, 305.

Part V
Applications

The Annals of Applied Probability
1993, Vol. 3, No. 4, 1112–1144

A NON-MARKOVIAN MODEL FOR CELL POPULATION GROWTH: TAIL BEHAVIOR AND DURATION OF THE GROWTH PROCESS

BY MATHISCA C. M. DE GUNST[1] AND WILLEM R. VAN ZWET

Free University of Amsterdam, and University of Leiden and University of North Carolina

De Gunst has formulated a stochastic model for the growth of a certain type of plant cell population that initially consists of n cells. The total cell number $N_n(t)$ as predicted by the model is a non-Markovian counting process. The relative growth of the population, $n^{-1}(N_n(t) - n)$, converges almost surely uniformly to a nonrandom function X. In the present paper we investigate the behavior of the limit process $X(t)$ as t tends to infinity and determine the order of magnitude of the duration of the process $N_n(t)$. There are two possible causes for the process N_n to stop growing, and correspondingly, the limit process $X(t)$ has a derivative $X'(t)$ that is the product of two factors, one or both of which may tend to zero as t tends to infinity. It turns out that there is a remarkable discontinuity in the tail behavior of the processes. We find that if only one factor of $X'(t)$ tends to zero, then the rate at which the limit process reaches its final limit is much faster and the order of magnitude of the duration of the process N_n is much smaller than when both occur approximately at the same time.

1. Biological background. Much of the research in plant cell biotechnology is directed at biosynthesis of secondary metabolites in plant cell cultures [Morris, Scragg, Stafford and Fowler (1986)]. Control of the productivity of these cell cultures in multiliter vessels in industry requires detailed knowledge of the kinetics of growth, division, differentiation and product formation of cells grown under different environmental conditions. However, our understanding of these kinetics is still very incomplete. In collaboration with K. R. Libbenga of the Department of Plant Molecular Biology at the University of Leiden, we have developed a mathematical model for the division, differentiation and population growth of plant cells in a liquid medium. This model is based on the presently available experimental knowledge of the behavior of individual cells, and takes into account the influence

Received September 1992; revised March 1993.

[1]Part of this research was done while visiting the Department of Statistics, Stanford University. This work was partially supported by a NATO Science Fellowship from the Netherlands Organization for Scientific Research (NWO).

AMS 1991 *subject classifications.* Primary 60G55, 60F99; secondary 62P10.
Key words and phrases. Stochastic model, population growth, non-Markovian counting process, tail behavior, duration.

S. van de Geer and M. Wegkamp (eds.), *Selected Works of Willem van Zwet*, Selected Works in Probability and Statistics, DOI 10.1007/978-1-4614-1314-1_20, © Springer Science+Business Media, LLC 2012

of the depletion of two components of the medium that are indispensable for the growth and division of the cells.

Before we formulate the model in Section 2, let us first describe the biological background. We start at the level of a single plant cell, which is transferred at time $t = 0$ to a fresh liquid medium containing the substances needed for the growth, division and survival of the cell. The cell will go through a sequence of events called the cell cycle, which starts at time 0 and ends with the division of the cell. The cell cycle is illustrated in the diagram of Figure 1. The cycle starts with the G_1-phase (G for gap) during which the biosynthetic activity of the cell proceeds at a high rate. The S-phase (synthesis) that follows starts when DNA synthesis begins and ends when the DNA content of the cell nucleus has doubled and the chromosomes have replicated. The cell then enters the G_2-phase (another gap), which continues until the final M-phase (mitosis), which is the brief period of actual cell division. During the M-phase, the biosynthetic activity of the cell proceeds very slowly and increases again rapidly after division as the two new cells enter the G_1-phase of their cell cycles. Together, the G_1-, S- and G_2-phases are also called the interphase.

It is a well verified fact that the duration of the cell cycle varies considerably, even among cells of the same type under the same external conditions. Most of the variability is observed in the length of the G_1-phase; the remainder of the cycle time shows far less variation. Moreover, it is known that the G_1-phase tends to last longer if the supply of certain nutrients is reduced; the

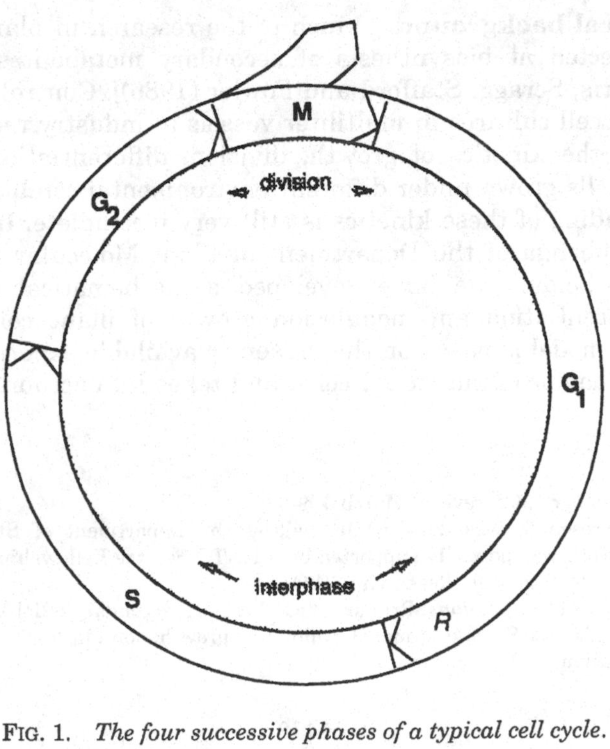

FIG. 1. *The four successive phases of a typical cell cycle.*

duration of the other parts of the cycle is scarcely affected [cf. Alberts, Bray, Lewis, Raff, Roberts and Watson (1989), page 745].

Another well established fact is that once the cell has left the G_1-phase, it is committed to complete the cell cycle regardless of environmental conditions [Alberts, Bray, Lewis, Raff, Roberts and Watson (1989), page 745]. This indicates the existence of a point of no return—often called the restriction point $\mathscr{R}$—somewhere late in the G_1-phase. It is thought that, as the cell approaches the point $\mathscr{R}$, it must "wait" for some endogenous trigger or stimulus that moves it past $\mathscr{R}$, and that under fixed external conditions the probability per unit time of passing $\mathscr{R}$ is roughly constant. In a mathematical model this would correspond to an exponentially distributed waiting time for a stimulus to occur, and if the remainder of the length of the G_1-phase—like that of the other phases—is almost constant, then the total duration of the cell cycle would be the sum of a constant time and an exponentially distributed one. This hypothesis was advanced in a seminal paper by Smith and Martin (1973) and verified on the basis of experimental data [see also Shields (1977)]. Later authors have criticized this so-called transition probability model [cf. Nelson and Green (1981)], but at present the existence of the restriction point seems to be firmly established. Others [Brooks, Bennett and Smith (1980), Castor (1980) and Cooper (1982)] have proposed alternative probability distributions for the duration of the cell cycle, incorporating more than one waiting time, for instance, but it seems difficult to distinguish between these models on the basis of the existing experimental data. At this time the transition probability model appears to be the accepted theory in the biological literature [Alberts, Bray, Lewis, Raff, Roberts and Watson (1989), pages 733 and 746].

We have already noted that the average duration of the G_1-phase increases as the supply of nutrients is reduced, and when no nutrients are present, cells cannot pass the restriction point at all. Hence we shall assume in our model that the parameter of the exponential waiting time for a stimulus is an increasing function of the concentration of nutrients, which tends to zero as the concentration does. The remaining part of the cycle is not affected by the concentration of nutrients. Of course, a cell also consumes nutrients, especially during the first part of the G_1-phase leading up to the restriction point $\mathscr{R}$. In our model we shall telescope this process and assume for simplicity that a cell only consumes a fixed amount of nutrient at the time it receives the stimulus to pass $\mathscr{R}$.

A plant cell that takes part in the cycling process is usually small and spherical, with its nucleus positioned at the center. However, if one watches a population of plant cells grow by cell division, one also notices after some time the presence of larger, more stretched out cells, with nuclei close to the cell wall. These cells are in an early stage of differentiation and do not divide. Such a differentiating cell most probably resides in the G_1-phase, before the restriction point $\mathscr{R}$, in a so-called quiescent, or G_0-state [Alberts, Bray, Lewis, Raff, Roberts and Watson (1989), page 750]. It is possible for such a cell to restart its cycle, but it needs a much more powerful trigger to do so than the one needed for passing $\mathscr{R}$: it has to dedifferentiate first before it resumes its cell cycle. Transfer of a cell to a fresh liquid medium is apparently such a trigger and will make the cell start its cycle almost at once.

Although the underlying biochemical mechanisms are far from known, there are strong indications [Bayliss (1985); Trewavas (1985)] that plant hormones play a crucial role in determining differentiation: the higher their concentration in the medium, the larger the proportion of cells that are actively cycling, and the smaller the proportion of cells that will never divide but will differentiate instead. In the absence of hormones, there will be no cycling cells. We shall model this phenomenon by assuming that at the time of cell division, the two new cells independently become cycling cells (type A cells) with probability P or differentiating cells (type B cells) with probability $(1 - P)$. Here P is assumed to be an increasing function of the hormone concentration in the medium at the time of division, which vanishes as the hormone concentration does. A cell also takes up hormones, and we shall assume that a fixed amount of hormone is used up by each cell at the time of its division.

Having described the behavior of a single cell, we now turn to the behavior of a population of plant cells in a liquid medium. Such populations can occur either as batch cultures or as continuous cultures [Street (1973)]. In either case, the culture consists of isolated cells—or very small cell aggregates—that remain dispersed as they grow in the liquid medium. This is achieved by continuous stirring of the fermentor in which the cells grow. A culture is started by the transfer of a certain number of cells to a fresh medium containing known quantities of nutrients and hormones. In contrast to a continuous culture, a batch culture does not have any inflow of fresh medium or outflow of culture. As such, the batch culture is the appropriate system to study the growth of the number of cells of a population and to investigate the influence of the different components of the medium on the population growth. In what follows we shall, therefore, restrict our attention to plant cells in batch culture.

The transfer of the cells to a fresh medium at time 0 triggers all cells to start their cycles almost at once, and we shall, therefore, assume that at time 0 all cells are of type A and that their cell cycles have been synchronized. If the amount of nutrient were kept constant or varied over time in a nonrandom fashion, it would be reasonable to assume that the duration of the cycles of different cells would be independent. However, in batch culture, the concentration of nutrient decreases at the random times when stimuli arrive, and cells compete for the available nutrient. This creates a complicated type of dependence between the division times for different cells.

Similarly, the hormone concentration decreases at the times of cell division. It follows that the cell population will ultimately stop growing, either because the nutrient is exhausted and no more stimuli can occur, or because the hormone concentration has become so low that the population of cycling cells can no longer be sustained. In the model this occurs when the probability P of becoming an A cell has fallen below 0.5.

In batch culture, cell death is observed only at the beginning when cells are transferred to a fresh medium, and at a time when the cell density has been very high for a considerable period. The effect of the former is removed by simply never counting these dead cells, whereas the latter occurs in practice only after the growth of the population has stopped. We shall, therefore, assume that cell death does not occur.

On the basis of this biological description we shall build a mathematical model for the growth of a plant cell population in batch culture in Section 2. In this model we shall study the duration of the growth process (i.e., the time until the population stops growing) when the initial population size n is large. We shall find that the duration is usually proportional to $\log n$. However, if the nutrient is exhausted at approximately the same time as the hormone concentration becomes too low to sustain the process, the duration is proportional to $n^{1/2} \log n$. Similar results are proved for an appropriately defined limit process.

2. The mathematical model. Let us turn this biological description into a mathematical model for the growth of a population of plant cells in batch culture. We start at time $t = 0$ with n cells. Because we intend to consider the growth of this population as n tends to infinity, we use n as an index throughout. The Smith–Martin model tells us that the duration of a cell cycle is of the form $(W + c)$, where W is a random waiting time for a stimulus to arrive and $c > 0$ is a constant. In a constant environment W has an exponential distribution with parameter λ (i.e., with expectation $1/\lambda$) and stimuli arrive independently for different cells. To fix thoughts, we assume that the cell cycle starts with the exponential waiting time for the stimulus and that the cell divides a constant time c after receiving the stimulus.

At time t, there will be $N_n(t)$ cells, of which $N_{An}(t)$ are A cells (i.e., cycling cells). Of these $N_{An}(t)$ A cells, $N_{An}^0(t)$ cells are at time t waiting for a stimulus to arrive, whereas the remaining $N_{An}(t) - N_{An}^0(t)$ A cells are somewhere in the time span of length c between arrival of a stimulus and division, and will, therefore, divide before or at time $(t + c)$. Thus

$$(2.1)\qquad \begin{aligned} N_{An}(t) - N_{An}^0(t) &= N_n(t+c) - N_n(t) \quad \text{or} \\ N_{An}^0(t) &= N_{An}(t) - (N_n(t+c) - N_n(t)). \end{aligned}$$

At time $t = 0$, all cells are of type A and at the beginning of their cycle, so

$$(2.2)\qquad N_n(0) = N_{An}(0) = N_{An}^0(0) = n.$$

We shall also need normalized versions of the three processes defined so far, and we write

$$(2.3)\qquad \begin{aligned} X_n(t) &= n^{-1}(N_n(t) - n), \\ X_{An}(t) &= n^{-1}N_{An}(t), \\ X_{An}^0(t) &= n^{-1}N_{An}^0(t). \end{aligned}$$

At the time of a cell division, the two new cells independently become A cells with probability P and B cells with probability $(1 - P)$. This probability P is an increasing function of the hormone concentration immediately before the division, and in batch culture this concentration decreases as time goes on. Suppose that the amount of hormone at time $t = 0$ equals $[nb_h]$, with b_h a positive constant, and that an amount 1 is used up at each division. Here $[x]$ denotes the largest integer less than or equal to x. Immediately before the ith division, the amount of hormone is $([nb_h] - (i - 1))$, and hence the probability of a cell becoming an A cell at the ith division equals

$$(2.4)\qquad P_{in} = P\left(\frac{[nb_h] - i + 1}{n}\right), \qquad i = 1, 2, \dots,$$

where P is increasing on $[0, \infty)$ and $P(u) = 0$ for $u \le 0$. According to Monod kinetics, which is the standard model for these biochemical processes [see, for instance, Roels (1983)], P is given by

$$(2.5)\qquad P(u) = \begin{cases} 0, & u \le 0, \\ \dfrac{u}{a_h + u} = 1 - \dfrac{a_h}{a_h + u}, & u > 0, \end{cases}$$

where a_h denotes a positive constant. Note that $P_{in} = 0$ for $i \ge [nb_h] + 1$. Because P is nonnegative, nondecreasing and concave, one easily verifies that for $m = 1, 2, \dots, [nb_h]$,

$$(2.6)\qquad \left| \sum_{i=1}^{m} P_{in} - n \int_0^{m/n} P(b_h - u)\, du \right| \le P(b_h) = \frac{b_h}{a_h + b_h} \le 1.$$

Let $Z_n = (Z_{1n}, Z_{2n}, \dots)$ denote a random sequence, where $Z_{1n}, Z_{2n}, \dots$ are independent and Z_{in} has a binomial distribution with parameters 2 and P_{in}. Here Z_{in} models the number of A cells created at the ith division, and hence

$$(2.7)\qquad \begin{aligned} N_{An}(t) &= 2n - N_n(t) + \sum_{i=1}^{N_n(t)-n} Z_{in} \\ &= n + \sum_{i=1}^{N_n(t)-n} (Z_{in} - 1). \end{aligned}$$

In view of (2.1), for $t > c$,

$$(2.8)\qquad N_{An}^0(t - c) = 2n - N_n(t) + \sum_{i=1}^{N_n(t-c)-n} Z_{in}.$$

Note that, conditional on Z_n, $\{N_{An}(s)\colon s \le t\}$ depends only on $\{N_n(s)\colon s \le t\}$, but $\{N_{An}^0(s)\colon s \le t\}$ depends on $\{N_n(s)\colon s \le (t + c)\}$.

The parameter λ of the exponential waiting time for a stimulus is an increasing function of the amount of substrate (or nutrients) present, and in batch culture this concentration decreases over time. Suppose that the amount of substrate at time $t = 0$ equals $[nb_s]$, for a positive constant b_s, and that an amount 1 is used up for each stimulus. At time t, $(N_n(t) - n)$ divisions have taken place, and this is also the number of stimuli that have arrived before or at time $(t - c)$. Hence, the amount of substrate at time $(t - c)$ equals

$$[nb_s] - (N_n(t) - n) = n\left(\frac{[nb_s]}{n} - X_n(t)\right).$$

Thus, the time dependent rate at which a stimulus arrives at time $(t - c)$ is an increasing function of $\{[nb_s]/n\} - X_n(t)$, say $Q(\{[nb_s]/n\} - X_n(t))$. According to Monod kinetics once more, Q is defined by

$$(2.9) \qquad Q(u) = \begin{cases} 0, & u \le 0, \\ \dfrac{u}{d(a_s + u)} = \dfrac{1}{d} - \dfrac{a_s}{d(a_s + u)}, & u > 0, \end{cases}$$

where d and a_s are positive constants. Note that the rate becomes zero after $[nb_s]$ stimuli have arrived.

The amount of substrate $n(\{[nb_s]/n\} - X_n(t))$ that is present at time $(t - c)$ will remain unchanged until the random time $(\tau - c)$ when the first stimulus after $(t - c)$ arrives. Thus, $(\tau - c)$ is distributed as the minimum of the times of the first event in independent Poisson processes with intensity $Q(\{[nb_s]/n\} - X_n(t))$. For the $N^0_{An}(t - c)$ waiting A cells that are already present at time $(t - c)$, the corresponding Poisson processes start at time $(t - c)$, and for A cells created after time $(t - c)$ the processes start at the time of their creation. Thus, given Z_n, the conditional intensity of the stimulus process at time $(t - c)$ equals the left-continuous version of $N^0_{An}(t - c)Q(\{[nb_s]/n\} - X_n(t))$. Because a stimulus at time $(t - c)$ corresponds to a cell division at time t, it follows that, *conditional on* Z_n, the process $\{N_n(t) - n: t \ge 0\}$ is a counting process with the left-continuous version of

$$(2.10) \qquad \Lambda_{Zn}(t) = \begin{cases} 0, & 0 \le t < c, \\ N^0_{An}(t - c)Q\left(\dfrac{[nb_s]}{n} - X_n(t)\right), & t \ge c, \end{cases}$$

as its *conditional intensity*. Together with the distribution of Z_n given before, this determines our mathematical model for the growth of a plant cell population in batch culture.

The process N_n stops growing for one of two entirely different reasons: either A cells become extinct or the rate at which the stimuli arrive becomes zero. Thus the process $N_n(t)$ reaches its final value at the first time t when either $N_{An}(t) = 0$ or $N_n(t) = n + [nb_s]$. Note that $N^0_{An}(t - c) = 0$ is not sufficient for N_n to stop growing at time t, because new A cells may be born between time $(t - c)$ and t. If T_n denotes the random time of the final cell division, then by (2.7),

$$T_n = \inf\left\{t: \sum_{i=1}^{N_n(t)-n} (Z_{in} - 1) = -n\right\} \wedge \inf\{t: N_n(t) = n + [nb_s]\},$$

where $(a \wedge b)$ means the smaller of a and b [similarly, the larger of a and b will be denoted by $(a \vee b)$]. The random level that N_n has reached by then equals

$$N_n(T_n) = n + M_n \wedge [nb_s], \tag{2.11}$$

where

$$M_n = \inf\left\{m\colon \sum_{i=1}^{m} (Z_{in} - 1) = -n\right\} \geq n. \tag{2.12}$$

Clearly this final level $N_n(T_n)$ depends only on Z_n and b_s. The process stops because $N(T_n) = n + [nb_s]$ or $N_{An}(T_n) = 0$, or both, depending on whether $M_n > [nb_s]$ or $M_n < [nb_s]$ or $M_n = [nb_s]$, respectively. An alternative expression for T_n is

$$T_n = \inf\{t\colon N_n(t) = n + M_n \wedge [nb_s]\}, \tag{2.13}$$

which shows that given Z_n, T_n is the time at which the counting process N_n reaches a fixed level.

Suppose that, for large n, the process $X_n(t) = (N_n(t) - n)/n$ is close to a deterministic function $X(t)$ in $D[0,\infty)$, the space of right-continuous, $\mathbb{R}$-valued functions on $[0,\infty)$ with left-hand limits everywhere. Then by (2.8), (2.10) and the fact that

$$n^{-1}\sum_1^m Z_{in} \sim n^{-1}\sum_1^m 2P_{in} \sim 2\int_0^{m/n} P(b_h - u)\,du$$

by (2.6), we find that $\Lambda_{Zn}(t)/n$ will be close to $F(t, X)$, where $F\colon [0,\infty) \times D[0,\infty) \to \mathbb{R}$ is defined by

$$F(t,x) = \left\{1 - x(t) + 2\int_0^{x(t-c)} P(b_h - u)\,du\right\} \times Q(b_s - x(t))1_{[c,\infty)}(t). \tag{2.14}$$

Here $a_n \sim b_n$ means that the quotient of a_n and b_n tends to 1 (in probability) as n tends to infinity. Thus it seems plausible that, if a deterministic limit X of the processes X_n exists, it should satisfy the equation

$$x(t) = \begin{cases} 0, & 0 \leq t < c, \\ \int_c^t F(s,x)\,ds, & t \geq c. \end{cases} \tag{2.15}$$

It is shown in De Gunst (1989) that (2.15) has a unique solution X in $D[0,\infty)$. This function X is continuous, nonnegative, nondecreasing and bounded on $[0,\infty)$, and differentiable on (c,∞) with a continuous, positive and bounded derivative. Hence $X(t)$ tends to a finite limit $X(\infty)$ as t tends to infinity, and in view of (2.14) and (2.15), it follows that $X'(t)$ also tends to a limit, which must necessarily be zero:

$$X(\infty) = \lim_{t\to\infty} X(t) < \infty, \qquad \lim_{t\to\infty} X'(t) = 0. \tag{2.16}$$

For biologically plausible values of the parameters, the graph of X exhibits alternating intervals of slow and rapid increase, which level off as time progresses. This reflects the synchronism of cells at time $t = 0$, which is gradually destroyed by the variability of the cycle times. Figure 2 shows a graph of X that was obtained by fitting a numerical solution of (2.15) to the only experimental data that are available so far, and that are also shown in this figure. Figure 2, as well as a statistical analysis of the experimental data, shows that for appropriate parameter values, the function X describes actual batch culture growth quite well. A detailed description of the experimental procedures, the statistical analysis of the data, and a further discussion of the relevance of the results can be found in De Gunst, Harkes, Val, Van Zwet and Libbenga (1990).

Having defined X as the nonrandom counterpart of X_n, we proceed to define the counterparts of the other processes in (2.3) by

$$X_A(t) = 1 - X(t) + 2\int_0^{X(t)} P(b_h - u)\,du, \tag{2.17}$$
$$X_A^0(t-c) = 1 - X(t) + 2\int_0^{X(t-c)} P(b_h - u)\,du,$$

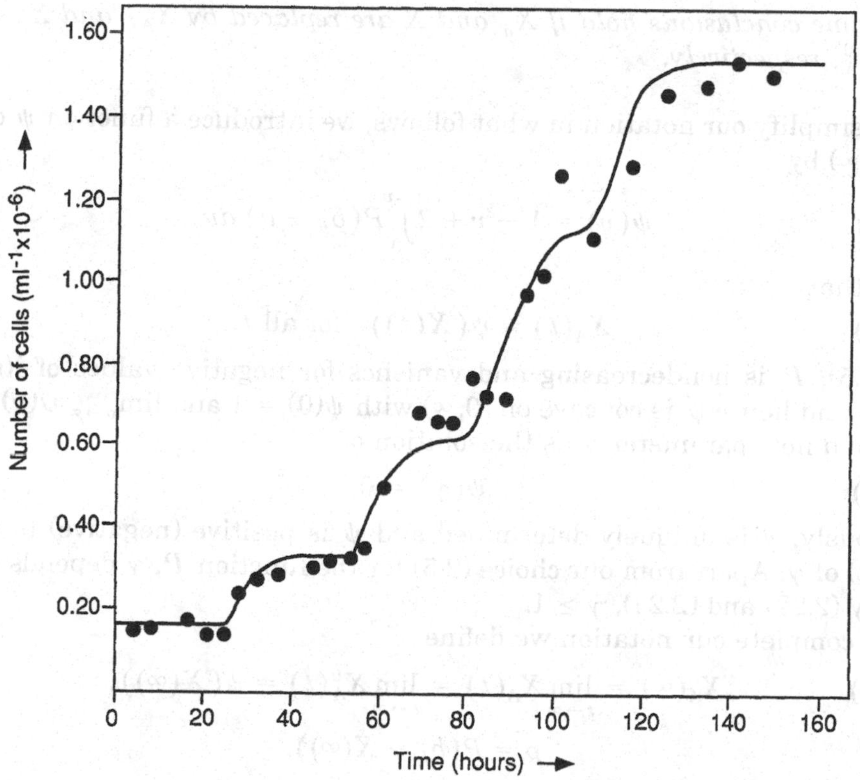

FIG. 2. *Experimental data of the growth of a batch culture of tobacco cells (dots). The curve through the data was fitted using numerical solutions of* (2.10). *The parameter values are* $n = 1.625 \times 10^8 l^{-1}$, $c = 26$h, $d = 4$ h, $b_s = 29.9$, $k_s = 2.2 \times 10^{-11}$ mol, $y_s = 5.4 \times 10^{10}$ mol^{-1}, $b_h = 4.7$, $k_h = 9.7 \times 10^{-17}$ mol, $y_h = 3.4 \times 10^{15}$ mol^{-1}.

in analogy with (2.7) and (2.8). Thus, by (2.10), (2.14) and (2.15) we have the two corresponding expressions

$$\Lambda_{Zn}(t) = nX^0_{An}(t-c)Q(\{[nb_s]/n\} - X_n(t)) \quad \text{for } t \geq c, \tag{2.18}$$

$$X'(t) = X^0_A(t-c)Q(b_s - X(t)) \quad \text{for } t \geq c, \tag{2.19}$$

for the conditional intensity of $(N_n - n)$ and the derivative of X.

In De Gunst and Van Zwet (1992) it is shown that indeed X_n converges in probability to X at a rate of $n^{-1/2}$ uniformly on $[0, \infty)$. Moreover, there is an exponential bound for the tail probability. The same holds for X_{An} and X_A, as well as for X^0_{An} and X^0_A. Theorem 2.1 summarizes these results.

THEOREM 2.1. *Let X be the solution of* (2.15). *Then there exist positive numbers A and α, such that for $n = 1, 2, \dots$ and $x \geq 0$,*

$$\mathscr{P}\Big(\sup_{t \geq 0} |X_n(t) - X(t)| \geq x\Big) \leq A \exp\{-\alpha x^2 n\}, \tag{2.20}$$

and hence

$$\sup_{t \geq 0} |X_n(t) - X(t)| = \mathscr{O}_{\mathscr{P}}(n^{-1/2}). \tag{2.21}$$

The same conclusions hold if X_n and X are replaced by X_{An} and X_A or X^0_{An} and X^0_A, respectively.

To simplify our notation in what follows, we introduce a function ψ defined on $[0, \infty)$ by

$$\psi(v) = 1 - v + 2\int_0^v P(b_h - u)\, du. \tag{2.22}$$

Note that

$$X_A(t) = \psi(X(t)) \quad \text{for all } t. \tag{2.23}$$

By (2.5), P is nondecreasing and vanishes for negative values of its argument, and hence ψ is concave on $[0, \infty)$ with $\psi(0) = 1$ and $\lim_{v \to \infty} \psi(v) = -\infty$. Define a new parameter γ as the solution of

$$\psi(\gamma) = 0. \tag{2.24}$$

Obviously, γ is uniquely determined and ψ is positive (negative) to the left (right) of γ. Apart from our choice (2.5) for the function P, γ depends only on b_h. By (2.22) and (2.24), $\gamma \geq 1$.

To complete our notation we define

$$X_A(\infty) = \lim_{t \to \infty} X_A(t) = \lim_{t \to \infty} X^0_A(t) = \psi(X(\infty)), \tag{2.25}$$

$$\begin{aligned} p &= P(b_h - X(\infty)), \\ q &= Q(b_s - X(\infty)), \\ q' &= Q'(b_s - X(\infty)), \end{aligned} \tag{2.26}$$

where Q' denotes the derivative of Q.

We have already noted that the process N_n will stop because one or both of the factors in (2.18) vanish. Similarly, for the deterministic limit function X, (2.16), (2.19) and (2.25) imply that

$$(2.27) \qquad X_A(\infty)Q(b_s - X(\infty)) = \psi(X(\infty))Q(b_s - X(\infty)) = 0,$$

so that we have three different cases: either $X(\infty) = b_s$ and hence $q = Q(b_s - X(\infty)) = 0$ or $X_A(\infty) = \psi(X(\infty)) = 0$ or both. An alternative way to express this is to write

$$(2.28) \qquad X(\infty) = \inf\{v: \psi(v) \le 0\} \wedge b_s = \gamma \wedge b_s,$$

which is the analogue of (2.11) and (2.12).

In the present paper, we shall investigate two closely related issues: the behavior of $X(t)$ for large values of t and the duration of the process N_n, that is, the time T_n to the final cell division. These issues concern the tail behavior of the processes X and X_n, respectively, the former being an (easier) deterministic version of the latter. We shall show that, depending on the values of the biological parameters, there is a remarkable discontinuity in this tail behavior of X and X_n.

Let us consider more closely the three cases that (2.27) allows and classify their occurrence in terms of the parameters γ and b_s.

$\gamma > b_s$:

(2.29)(i) By (2.28), $X(\infty) = b_s$ and $q = Q(b_s - X(\infty)) = 0$.
On the other hand, $X_A(\infty) = \psi(X(\infty)) = \psi(b_s) > 0$;

$\gamma < b_s$:

By (2.28), $X(\infty) = \gamma$ and $q = Q(b_s - X(\infty)) = Q(b_s - \gamma) > 0$.
(2.29)(ii) On the other hand, $X_A(\infty) = \psi(X(\infty)) = \psi(\gamma) = 0$.
Note that in this case $p = P(b_h - X(\infty)) = P(b_h - \gamma) < \frac{1}{2}$,
because (2.22) implies $0 = \psi(\gamma) \ge 1 + \gamma\{2P(b_h - \gamma) - 1\}$;

$\gamma = b_s$:

By (2.28), $X(\infty) = b_s = \gamma$ and $q = Q(b_s - X(\infty)) = 0$.
(2.29)(iii) However, in this case, $X_A(\infty) = \psi(X(\infty)) = \psi(\gamma) = 0$.
As in the previous case, $p = P(b_h - b_s) = P(b_h - \gamma) < \frac{1}{2}$.

We shall show that if $\gamma \ne b_s$,

$$X(\infty) - X(t) \sim Ae^{-at} \quad \text{as } t \to \infty,$$

$$\frac{T_n}{\log n} \to \frac{1}{a} \quad \text{in probability as } n \to \infty$$

and give expressions for a in case $(\gamma - b_s)$ is positive or negative. If $\gamma = b_s$, however, we find a very different behavior:

$$X(\infty) - X(t) \sim \frac{1}{at} \quad \text{as } t \to \infty,$$

$$\frac{T_n}{n^{1/2} \log n} \quad \text{is of exact order 1 in probability as } n \to \infty$$

and, again, we give an expression for a in this case.

The biological interest of these results lies in the fact that for a certain balance ($\gamma = b_s$) between the initial amounts of substrate $[nb_s]$ and hormone $[nb_h]$, the model predicts a much longer duration of the growth process than for other parameter values. This phenomenon, which is indeed a very essential aspect of the model, lends itself to experimental verification. Because the experiments considered here are very expensive and time consuming, this verification has not yet been carried out. Another aspect of these results that is of some practical importance is that some of the secondary metabolites are known to be synthesized only at the end of the cell culture's growth process, and one would, therefore, like to avoid values of γ close to b_s.

From a mathematical point of view, the results of this paper are rather more delicate than those of Theorem 2.1. Though the main feature of Theorem 2.1 is the uniformity in t, the theorem still provides very little information about the behavior of $X_n(t)$ and $X(t)$ for very large t, which is needed here. There seem to be few results on the duration of processes similar to the one we study. Kurtz (1982) discusses a case where the limit process X reaches its ultimate value in finite time; Barbour (1975) and Nagaev and Mukhomor (1975) study the duration of an epidemic. The problems that these authors face are very different from ours.

In Section 3, we prove the results on the behavior of $X(\infty) - X(t)$ for large t. A result on a class of differential equations that plays a key role in this analysis is given in Appendix A. In Section 4 we tackle the estimation of T_n. A maximal inequality and a fluctuation inequality that are needed in Section 4 are given in Appendix B.

3. Tail behavior of X. In this section we investigate the behavior of $X(t)$ for large values of t. Our starting point will be expression (2.19): For $t \geq c$,

$$X'(t) = X_A^0(t-c)Q(b_s - X(t)) = \big(X_A(\infty) + \big[X_A^0(t-c) - X_A(\infty)\big]\big) \times\big(q + \big[Q(b_s - X(t)) - q\big]\big),$$

and Taylor expansion of the terms in square brackets. By (2.17) and because both P and Q have bounded derivatives of every order, we find

$$\begin{aligned} X_A^0(t-c) - X_A(\infty) &= X(\infty) - X(t) - 2\int_{X(t-c)}^{X(\infty)} P(b_h - u)\,du \\ &= (X(\infty) - X(t)) - 2p(X(\infty) - X(t-c)) \\ &\quad + \mathscr{O}\big((X(\infty) - X(t-c))^2\big), \end{aligned} \tag{3.1}$$

$$Q(b_s - X(t)) - q = q'(X(\infty) - X(t)) + \mathscr{O}\big((X(\infty) - X(t))^2\big). \tag{3.2}$$

Here p, q and q' are defined in (2.26). By (2.27),

$$X_A(\infty)q = 0. \tag{3.3}$$

It follows that

$$\begin{aligned} X'(t) = {} & (q + q'X_A(\infty))(X(\infty) - X(t)) \\ & - 2pq(X(\infty) - X(t-c)) \\ & + q'(X(\infty) - X(t))^2 \\ & - 2pq'(X(\infty) - X(t))(X(\infty) - X(t-c)) \\ & + \mathscr{O}\Big(q(X(\infty) - X(t-c))^2 + X_A(\infty)(X(\infty) - X(t))^2 \\ & \qquad + (X(\infty) - X(t))(X(\infty) - X(t-c))^2\Big). \end{aligned} \tag{3.4}$$

LEMMA 3.1.

$$X(\infty) - X(t-c) = \mathscr{O}(X(\infty) - X(t)) \quad \text{as } t \to \infty.$$

If $\gamma = b_s$, *then*

$$X(t) - X(t-c) = \mathscr{O}\big((X(\infty) - X(t))^2\big) \quad \text{as } t \to \infty.$$

PROOF. By (3.3), $X_A(\infty)q = 0$. If $q = 0$, then $X(\infty) = b_s$. By (3.1), $X_A^0(t-c) \le X(\infty) + X_A(\infty) \le 2b_s$, and as Q' decreases, $Q(b_s - X(t)) \le Q'(0)(X(\infty) - X(t))$. Hence $X'(t) \le 2b_s Q'(0)(X(\infty) - X(t))$ for $t > c$ and

$$\int_{t-c}^{t} \frac{X'(s)}{X(\infty) - X(s)}\, ds = \log\left(\frac{X(\infty) - X(t-c)}{X(\infty) - X(t)}\right) \le 2cb_s Q'(0),$$

so that $X(\infty) - X(t-c) \le e^{2cb_s Q'(0)}(X(\infty) - X(t))$ for $t \ge 2c$.

If $X_A(\infty) = 0$, then (3.1) implies that $0 \le X_A^0(t-c) \le X(\infty) - X(t)$, whereas $Q(b_s - X(t)) \le Q(b_s)$ as Q is increasing. Repeating the foregoing argument, we see that $X(\infty) - X(t-c) \le e^{cQ(b_s)}(X(\infty) - X(t))$ for $t \ge 2c$.

Finally, if $\gamma = b_s$, then we are in case (2.29)(iii) and $X_A(\infty) = q = 0$. Now (3.1) and (3.2) imply that for $t \ge c$,

$$X'(t) \le q'(X(\infty) - X(t))^2 + \mathscr{O}\big((X(\infty) - X(t))^3\big).$$

Integrating over $(t-c, t)$, we find

$$X(t) - X(t-c) = \mathscr{O}\big((X(\infty) - X(t-c))^2\big) = \mathscr{O}\big((X(\infty) - X(t))^2\big). \quad \square$$

Theorems 3.1, 3.2 and 3.3 deal with the three essentially different cases (2.29)(i)–(iii) that we discussed in Section 2. We note that ψ and γ are defined in (2.22) and (2.24).

THEOREM 3.1. *Let* $\gamma > b_s$ *and define* $a = \psi(b_s)/(da_s) > 0$. *Then* $X(\infty) = b_s$ *and there exists a positive number* A, *such that*

$$\lim_{t\to\infty} e^{at}(X(\infty) - X(t)) = A. \tag{3.5}$$

PROOF. This is case (2.29)(i), so that $X(\infty) = b_s$, $X_A(\infty) = \psi(b_s) > 0$, $q = 0$ and $q' = Q'(0) = 1/(da_s)$. Hence (3.4) and Lemma 3.1 imply that

$$X'(t) = q'X_A(\infty)(X(\infty) - X(t)) + \mathcal{O}((X(\infty) - X(t))^2). \tag{3.6}$$

The theorem follows from Lemma A.1 in Appendix A with $v = 0$, $w = a = q'X_A(\infty) = \psi(b_s)/(da_s)$ and $f(t) = X(\infty) - X(t)$. □

THEOREM 3.2. *Let $\gamma < b_s$ and define $a \in (0, Q(b_s - \gamma)]$ as the unique solution of $Q(b_s - \gamma) - a = 2P(b_h - \gamma)Q(b_s - \gamma)e^{ac}$. Then $X(\infty) = \gamma$ and there exists a positive number A such that*

$$\lim_{t\to\infty} e^{at}(X(\infty) - X(t)) = A. \tag{3.7}$$

PROOF. This is case (2.29)(ii), so that $X(\infty) = \gamma$, $X_A(\infty) = 0$, $q = Q(b_s - \gamma) > 0$ and $p = P(b_h - \gamma) < \frac{1}{2}$. Together with Lemma 3.1, this implies that (3.4) reduces to

$$\begin{aligned} X'(t) &= q(X(\infty) - X(t)) - 2pq(X(\infty) - X(t-c)) \\ &\quad + \mathcal{O}((X(\infty) - X(t))^2). \end{aligned} \tag{3.8}$$

The theorem follows from Lemma A.1 in Appendix A with $v = 2pq = 2P(b_h - \gamma)Q(b_s - \gamma)$, $w = q = Q(b_s - \gamma)$ and $f(t) = X(\infty) - X(t)$. Note that, because $p = P(b_h - \gamma) < \frac{1}{2}$, we have indeed $v < w$. □

THEOREM 3.3. *Let $\gamma = b_s$ and define $a = (1 - 2P(b_h - b_s))/(da_s) > 0$. Then $X(\infty) = b_s$ and*

$$\lim_{t\to\infty} at(X(\infty) - X(t)) = 1. \tag{3.9}$$

PROOF. We are now in case (2.29)(iii), so that $X(\infty) = b_s = \gamma$, $X_A(\infty) = \psi(\gamma) = 0$, $q = Q(0) = 0$, $q' = Q'(0) = 1/(da_s) > 0$ and $p = P(b_h - b_s) < \frac{1}{2}$. Hence $a = q'(1 - 2p)$ is positive. Together with Lemma 3.1, this implies that (3.4) reduces to

$$X'(t) = a(X(\infty) - X(t))^2 + \mathcal{O}((X(\infty) - X(t))^3). \tag{3.10}$$

Dividing by $(X(\infty) - X(t))^2$ and integrating, we find for $c < s < t$,

$$\frac{1}{X(\infty) - X(t)} - \frac{1}{X(\infty) - X(s)} = a(t-s) + \mathcal{O}((t-s)(X(\infty) - X(s))).$$

Dividing by at and then letting first t and then s tend to infinity, we find that $(at(X(\infty) - X(t)))^{-1}$ tends to 1. □

4. Duration of the growth process. We now turn to the duration T_n of the growth process N_n. To simplify our notation we shall write

$$b_{sn} = \frac{[nb_s]}{n} \in \left(b_s - \frac{1}{n}, b_s\right], \qquad b_{hn} = \frac{[nb_h]}{n} \in \left(b_h - \frac{1}{n}, b_h\right] \tag{4.1}$$

throughout this section. In view of (2.28) and the fact that $\gamma \geq 1$ on the one hand and (2.11) and (2.12) on the other, we know that $X(\infty) \geq 1 \wedge b_s$ and $X_n(T_n) \geq 1 \wedge b_{sn}$. For $\varepsilon \in (0, 1 \wedge b_s)$ we may, therefore, define $\tau(\varepsilon)$ and $\tau_n(\varepsilon)$ by

$$X(\infty) - X(\tau(\varepsilon)) = \varepsilon, \tag{4.2}$$

$$\tau_n(\varepsilon) = \inf\{t \geq 0\colon X_n(T_n) - X_n(t) \leq \varepsilon\}, \tag{4.3}$$

with the convention that $\tau_n(\varepsilon) = 0$ if $X_n(T_n) < \varepsilon$.

LEMMA 4.1. *There exist positive numbers A and α such that for $0 < \varepsilon < 1 \wedge b_s$ and $n = 1, 2, \ldots$,*

$$\mathscr{P}\left(\tau_n(\varepsilon) > \tau\left(\frac{\varepsilon}{2}\right)\right) \leq A \exp\{-\alpha\varepsilon^2 n\}. \tag{4.4}$$

PROOF.

$$\begin{aligned}\mathscr{P}\left(\tau_n(\varepsilon) > \tau\left(\frac{\varepsilon}{2}\right)\right) &\leq \mathscr{P}\left(X_n(T_n) - X_n\left(\tau\left(\frac{\varepsilon}{2}\right)\right) > \varepsilon\right)\\ &\leq \mathscr{P}\left(\sup_{t\geq 0}|X_n(t) - X(t)| > \frac{\varepsilon}{4}\right)\end{aligned}$$

because $X(\infty) = X(\tau(\varepsilon/2)) + \varepsilon/2$ and $X_n(T_n) = X_n(\infty)$. Hence (4.4) follows from Theorem 2.1. □

Lemma 4.1 ensures that $\tau_n(\varepsilon)$ is bounded except on a set of exponentially small probability for every fixed $\varepsilon > 0$. Of course, T_n tends to infinity in probability as $n \to \infty$ in view of Theorem 2.1 and the results of Section 3. For our study of the first order asymptotic behavior of T_n, any bounded contribution to T_n will be irrelevant, and we may, therefore, study $T_n - \tau_n(\varepsilon)$ instead. This implies that we need only take the times needed for the final $[\varepsilon n]$ cell divisions into account, for arbitrarily small positive ε.

As in Section 3, the cases $\gamma > b_s$, $\gamma < b_s$ and $\gamma = b_s$ are essentially different and we shall discuss these cases in three separate subsections.

4.1. *The case $\gamma > b_s$.* If $\gamma > b_s$ we are in case (2.29)(i), so $X(\infty) = b_s$, $X_A(\infty) = \psi(b_s) > 0$, $q = Q(0) = 0$ and $q' = Q'(0) = 1/(da_s)$. Theorem 2.1 implies that, except on a set Ω_n^c of negligible probability for large n, $X_{An}^0(T_n) = X_{An}(T_n) = X_{An}(\infty)$ will be close to $X_A^0(\infty) = X_A(\infty) = \psi(b_s) > 0$, and the same is true for $X_{An}^0(t - c)$ for sufficiently large t. In particular, $X_{An}(T_n) > 0$ on Ω_n. Because the process X_n stops when either $X_{An} = 0$ or $X_n = b_{sn}$, we must have $X_n(T_n) = b_{sn}$ and hence $Q(b_{sn} - X_n(t))$ is approximately equal to $q'(X_n(T_n) - X_n(t))$ for large t. It follows from (2.10) that on Ω_n, $\Lambda_{Zn}(t)$ is close to $na(X_n(T_n) - X_n(t))$ for large t, where $a = \psi(b_s)/(da_s) > 0$. But this means that, going back in time from T_n, the times between the last $[\varepsilon n]$ consecutive cell divisions are approximately independent and exponentially distributed random variables with means $1/a, 1/(2a), 1/(3a), \ldots, 1/([\varepsilon n]a)$, provided that $\varepsilon > 0$ is sufficiently small. This implies that $(T_n - \tau_n(\varepsilon))$, which is the sum of these variables, is asymptotic to $(\log n)/a$, and by Lemma 4.1 this remains true for T_n.

In the remainder of this subsection we make this argument rigorous. In Lemma 4.2 we study the appropriate sum of independent exponential random variables. Theorem 4.1 provides a precise statement of the result and in its proof we fill in the gaps in the heuristic argument given before.

LEMMA 4.2. *Let $V_1, V_2, \dots$ be i.i.d. random variables that are exponentially distributed with mean* 1. *For $\varepsilon \in (0,1]$ and $m \geq 1/\varepsilon$, define*

$$S_m(\varepsilon) = \sum_{j=1}^{[\varepsilon m]} \frac{V_j}{j}. \tag{4.5}$$

Then for $\varepsilon \in (0,1]$, $m \geq 1/\varepsilon$ and $x \geq 0$,

$$\mathscr{P}(|S_m(\varepsilon) - \log m| \geq x) \leq \frac{3}{\varepsilon}\exp\left\{-\frac{x}{2}\right\}.$$

PROOF. Writing $S_m(\varepsilon) = S$ we have

$$Ee^{S/2} = \prod_{j=1}^{[\varepsilon m]} \frac{j}{j-\frac{1}{2}} \leq 2\prod_{j=2}^{[\varepsilon m]}\left(\frac{j}{j-1}\right)^{1/2} \leq 2(\varepsilon m)^{1/2},$$

$$\mathscr{P}(S \geq \log m + x) \leq \frac{Ee^{S/2}}{\exp\{\frac{1}{2}\log m + \frac{1}{2}x\}} \leq 2\varepsilon^{1/2}e^{-x/2},$$

$$Ee^{-S} = \prod_{j=1}^{[\varepsilon m]} \frac{j}{j+1} \leq \frac{1}{\varepsilon m},$$

$$\mathscr{P}(S \leq \log m - x) \leq \frac{Ee^{-S}}{\exp\{-\log m + x\}} \leq \frac{1}{\varepsilon}e^{-x}.$$

As $\varepsilon^{1/2} \leq \varepsilon^{-1}$, the lemma is proved. □

THEOREM 4.1. *Let $\gamma > b_s$ and define $a = \psi(b_s)/(da_s) > 0$ as in Theorem* 3.1. *Then, for every $\delta > 0$ there exist positive numbers A and α such that for $n = 1, 2, \dots$ and $0 \leq x \leq n$,*

$$\mathscr{P}\left(\left|T_n - \frac{\log n}{a}\right| \geq \delta \log n + x\right) \leq A\exp\{-\alpha x\}, \tag{4.6}$$

and hence

$$\frac{T_n}{\log n} \to \frac{1}{a} \quad \textit{in probability}. \tag{4.7}$$

PROOF. Choose $\varepsilon \in (0, 1 \wedge b_s \wedge X_A(\infty))$ and consider the event

$$\Omega_n = \{|X_{An}(T_n) - X_A(\infty)| \leq \varepsilon\}.$$

In view of (2.3), (2.7), (2.11) and (2.12) this event is measurable with respect to the σ-algebra $\mathscr{F}_{Zn}$ generated by $Z_n = (Z_{1n}, Z_{2n}, \ldots)$. Because $X_{An}(T_n) = X_{An}(\infty)$, Theorem 2.1 ensures that

$$\mathscr{P}(\Omega_n^c) \le A_1 \exp\{-\alpha_1 \varepsilon^2 n\} \tag{4.8}$$

for positive A_1 and α_1.

As $\varepsilon < X_A(\infty)$, we have $X_{An}(T_n) > 0$ and hence $X_n(T_n) = b_{sn}$ on Ω_n. Also, by (2.7), (2.8) and (4.3),

$$\begin{aligned}\sup_{t \ge \tau_n(\varepsilon)} X_{An}^0(t-c) &\le \sup_{t \ge \tau_n(\varepsilon)} X_{An}(t)\\ &= X_{An}(T_n) + \sup_{t \ge \tau_n(\varepsilon)} (X_{An}(t) - X_{An}(T_n))\\ &\le X_{An}(T_n) + X_n(T_n) - X_n(\tau_n(\varepsilon)) \le X_{An}(T_n) + \varepsilon\end{aligned}$$

and

$$\begin{aligned}\inf_{t \ge \tau_n(\varepsilon)+c} X_{An}^0(t-c) &= X_{An}(T_n) + \inf_{t \ge \tau_n(\varepsilon)} (X_{An}^0(t) - X_{An}^0(T_n))\\ &\ge X_{An}(T_n) - 2(X_n(T_n) - X_n(\tau_n(\varepsilon))) \ge X_{An}(T_n) - 2\varepsilon.\end{aligned}$$

It follows that on Ω_n,

$$\begin{aligned}\sup_{t \ge \tau_n(\varepsilon)} X_{An}^0(t-c) &\le X_A(\infty) + 2\varepsilon,\\ \inf_{t \ge \tau_n(\varepsilon)+c} X_{An}^0(t-c) &\ge X_A(\infty) - 3\varepsilon.\end{aligned}$$

On Ω_n, $X_n(T_n) = b_{sn}$, and hence (2.9) and (4.3) imply that for $t \ge \tau_n(\varepsilon)$,

$$\frac{1}{d(a_s + \varepsilon)}(b_{sn} - X_n(t)) \le Q(b_{sn} - X_n(t)) \le \frac{1}{da_s}(b_{sn} - X_n(t)).$$

Define $a = X_A(\infty)/(da_s) = \psi(b_s)/(da_s)$ as in the statement of the theorem and choose $\delta \in (0, a^{-1})$. Combining (2.10) and the inequalities derived so far, we find that by taking $\varepsilon > 0$ sufficiently small, we can make sure that on Ω_n,

$$\Lambda_{Zn}(t) \le \frac{a}{1 - \delta a} n(b_{sn} - X_n(t)) \quad \text{for } t \ge \tau_n(\varepsilon), \tag{4.9}$$

$$\Lambda_{Zn}(t) \ge \frac{a}{1 + \delta a} n(b_{sn} - X_n(t)) \quad \text{for } t \ge \tau_n(\varepsilon) + c. \tag{4.10}$$

Conditionally on $Z_n = z$ with $\{Z_n = z\} \subset \Omega_n$, the process $(N_n(t) - n)$ is a counting process with intensity $\Lambda_{Zn}(t)$, which is bounded above and below by (4.9) and (4.10). In view of the argument in the first paragraph of this subsection, this implies that conditionally on $Z_n = z$ with $\{Z_n = z\} \subset \Omega_n$,

$(T_n - \tau_n(\varepsilon))$ is stochastically larger than a sum of $[\varepsilon n]$ independent exponential random variables with means $(1-\delta a)/(aj)$, $j = 1, 2, \dots, [\varepsilon n]$, or equivalently, than $((1-\delta a)/a)S_n(\varepsilon)$ with $S_n(\varepsilon)$ as in Lemma 4.2 with $m = n$. Hence by (4.8),

$$\begin{aligned}
&\mathscr{P}\left(T_n - \frac{\log n}{a} \le -\delta \log n - x\right)\\
&\quad\le \mathscr{P}\left(T_n - \tau_n(\varepsilon) \le \frac{1-\delta a}{a}\log n - x\right)\\
&\quad\le \mathscr{P}\left(S_n(\varepsilon) \le \log n - \frac{ax}{1-\delta a}\right) + A_1 \exp\{-\alpha_1 \varepsilon^2 n\}.
\end{aligned}$$

Similarly, (4.10) implies that conditionally on $Z_n = z$ with $\{Z_n = z\} \subset \Omega_n$, $(T_n - \tau_n(\varepsilon))$ is stochastically smaller than $((1+\delta a)/a)S_n(\varepsilon) + c$, and hence by Lemma 4.1 and (4.8),

$$\begin{aligned}
&\mathscr{P}\left(T_n - \frac{\log n}{a} \ge \delta \log n + x\right)\\
&\quad\le \mathscr{P}\left(T_n - \tau_n(\varepsilon) \ge \frac{1+\delta a}{a}\log n + x - \tau\left(\frac{\varepsilon}{2}\right)\right)\\
&\qquad + A_2 \exp\{-\alpha_2 \varepsilon^2 n\}\\
&\quad\le \mathscr{P}\left(S_n(\varepsilon) \ge \log n + \frac{a}{1+\delta a}\left(x - \tau\left(\frac{\varepsilon}{2}\right) - c\right)\right)\\
&\qquad + A_3 \exp\{-\alpha_3 \varepsilon^2 n\}
\end{aligned}$$

for appropriate positive A_2, A_3, α_2 and α_3. Combining these results with Lemma 4.2, we find that for $n = 1, 2, \dots$ and $\tau(\varepsilon/2) + c \le x \le n$,

$$\begin{aligned}
&\mathscr{P}\left(\left|T_n - \frac{\log n}{a}\right| \ge \delta \log n + x\right)\\
&\quad\le \mathscr{P}\left(|S_n(\varepsilon) - \log n| \ge \frac{a}{1+\delta a}\left(x - \tau\left(\frac{\varepsilon}{2}\right) - c\right)\right) + \frac{A}{2}\exp\{-\alpha n\}\\
&\quad\le \frac{3}{\varepsilon}\exp\left\{\frac{a}{1+\delta a}\left(\tau\left(\frac{\varepsilon}{2}\right) + c - x\right)\right\} + \frac{A}{2}\exp\{-\alpha n\}\\
&\quad\le A \exp\{-\alpha x\}
\end{aligned}$$

for appropriately chosen positive A and α. If $A \ge \exp\{\alpha(\tau(\varepsilon/2) + c)\}$, then this bound remains valid for $0 \le x < \tau(\varepsilon/2) + c$ and the proof of (4.6) is complete. For $x = \delta \log n$, (4.6) yields (4.7). □

4.2. *The case $\gamma < b_s$.* If $\gamma < b_s$ we are in case (2.29)(ii), so $X(\infty) = \gamma < b_s$, $X_A(\infty) = 0$, $q = Q(b_s - \gamma) > 0$ and $p = P(b_h - \gamma) < \frac{1}{2}$. Theorem 2.1 implies that, except on a set $\tilde{\Omega}_n^c$ of negligible probability for large n, $X_n(T_n) = X_n(\infty)$ will be close to $X(\infty) = \gamma$, and the same is true for $X_n(t)$ for sufficiently large t. In particular, $X_n(T_n) < b_{sn}$ on $\tilde{\Omega}_n$, and hence the process X_n must stop because $X_{An}(T_n) = 0$. Moreover, on $\tilde{\Omega}_n$, $Q(b_{sn} - X_n(t))$ is close to q for large t and the probabilities P_{in} of A cells are approximately equal to p for the final $[\varepsilon n]$ cell divisions if $\varepsilon > 0$ is small [cf. (2.4)]. It follows that on $\tilde{\Omega}_n$ and for small $\varepsilon > 0$, $(T_n - \tau_n(\varepsilon))$ is approximately equal to the duration of a process that starts at time $\tau_n(\varepsilon)$ with $N_{An}(\tau_n(\varepsilon))$ A cells—of which $N_{An}^0(\tau_n(\varepsilon))$ are waiting for a stimulus—and has a fixed stimulus rate $q > 0$ and a fixed probability of A cells $p < \frac{1}{2}$. Both $N_{An}(\tau_n(\varepsilon))$ and $N_{An}^0(\tau_n(\varepsilon))$ are of exact order n with probability close to 1, because $X_{An}(\tau_n(\varepsilon))$ and $X_{An}^0(\tau_n(\varepsilon))$ are close to $X_A(\tau(\varepsilon))$ and $X_A^0(\tau(\varepsilon))$, which are positive.

Let $S(p, q, c)$ be the duration of a process that starts with a single waiting A cell and has a fixed stimulus rate $q > 0$ and a fixed probability of A cells $p < \frac{1}{2}$. In Lemma 4.3 we show that the right tail of the distribution of $S(p, q, c)$ behaves like that of an exponential distribution with mean a^{-1}, where a is the solution of $q - a = 2pqe^{ac}$. Hence $(T_n - \tau_n(\varepsilon))$ is approximately distributed as the maximum of a (random) number M of independent exponentially distributed random variables with mean a^{-1}, and this number M is of exact order n. But this means that $(T_n - \tau_n(\varepsilon)) \sim (\log n)/a$, and by Lemma 4.1 we also have $T_n \sim (\log n)/a$.

In the remainder of this subsection we first prove Lemma 4.3 concerning the distribution of $S(p, q, c)$. Theorem 4.2 provides a precise formulation of the result for T_n and some additional details will be found in the proof of this theorem.

Thus in Lemma 4.3 we consider the following situation. At time $t = 0$ there is a single A cell waiting for a stimulus. A cells independently receive a stimulus after an exponential waiting time with mean $(\tilde{q})^{-1}$ and divide a constant time c later. With each division the new cells independently become A cells with probability $\tilde{p}$ and B cells with probability $(1 - \tilde{p})$. B cells do not divide. Let $S(\tilde{p}, \tilde{q}, c)$ denote the time until the final division.

LEMMA 4.3. *Suppose that $\tilde{q} > 0$ and $0 \le \tilde{p} < \frac{1}{2}$ and define $\tilde{a} \in (0, \tilde{q}]$ as the unique solution of $\tilde{q} - \tilde{a} = 2\tilde{p}\tilde{q}e^{\tilde{a}c}$. Then there exists a positive number $\tilde{A}$ such that*

$$\lim_{t\to\infty} e^{\tilde{a}t}\mathscr{P}(S(\tilde{p}, \tilde{q}, c) > t) = \tilde{A}. \tag{4.11}$$

PROOF. Writing $f(t) = \mathscr{P}(S(\tilde{p}, \tilde{q}, c) > t)$ we find for $t > c$,

$$f(t) = e^{-\tilde{q}(t-c)} + 2\tilde{p}(1 - \tilde{p})\tilde{q}\int_0^{t-c} e^{-\tilde{q}s} f(t - c - s)\, ds$$

$$+ \tilde{p}^2\tilde{q}\int_0^{t-c} e^{-\tilde{q}s}\left[2f(t - c - s) - (f(t - c - s))^2\right] ds$$

$$= e^{-\tilde{q}(t-c)} + 2\tilde{p}\tilde{q}\int_0^{t-c} e^{-\tilde{q}s} f(t-c-s)\,ds$$
$$-\tilde{p}^2\tilde{q}\int_0^{t-c} e^{-\tilde{q}s}\big(f(t-c-s)\big)^2\,ds$$
$$= e^{-\tilde{q}(t-c)}\left[1 + 2\tilde{p}\tilde{q}\int_0^{t-c} e^{\tilde{q}s} f(s)\,ds - \tilde{p}^2\tilde{q}\int_0^{t-c} e^{\tilde{q}s}\big(f(s)\big)^2\,ds\right].$$

Multiplying by $e^{\tilde{q}(t-c)}$, differentiating and dividing again by $e^{\tilde{q}(t-c)}$, we obtain for $t > c$,

$$f'(t) + \tilde{q}f(t) = 2\tilde{p}\tilde{q}f(t-c) - \tilde{p}^2\tilde{q}\big(f(t-c)\big)^2$$

or

$$f'(t) = 2\tilde{p}\tilde{q}f(t-c)\left(1 - \frac{\tilde{p}}{2}f(t-c)\right) - \tilde{q}f(t).$$

Obviously, f is strictly decreasing and $f' < 0$ on $[c,\infty)$. Because $0 \le p < \frac{1}{2}$, we have $\lim_{t\to\infty} f(t) = 0$ and we may apply Lemma A.1 in Appendix A with $v = 2\tilde{p}\tilde{q}$ and $w = \tilde{q}$ to complete the proof. □

THEOREM 4.2. *Let $\gamma < b_s$ and define $a \in (0, Q(b_s - \gamma)]$ as the unique solution of $Q(b_s - \gamma) - a = 2P(b_h - \gamma)Q(b_s - \gamma)e^{ac}$ as in Theorem 3.2. Then, for every $\delta > 0$ there exist positive numbers A and α such that for $n = 1, 2, \ldots$ and $0 \le x \le n$,*

$$\mathscr{P}\left(\left|T_n - \frac{\log n}{a}\right| \ge \delta \log n + x\right) \le A\exp\{-\alpha x\}, \tag{4.12}$$

and hence

$$\frac{T_n}{\log n} \to \frac{1}{a} \quad \textit{in probability}. \tag{4.13}$$

PROOF. Choose $\varepsilon \in (0, 1 \wedge (b_s - \gamma))$ and define the event

$$\tilde{\Omega}_n = \{|X_n(T_n) - X(\infty)| \le \varepsilon\}.$$

By (2.3), (2.11) and (2.12, $\tilde{\Omega}_n$ is measurable with respect to the σ-algebra $\mathscr{F}_{Zn}$ generated by $Z_n = (Z_{1n}, Z_{2n}, \ldots)$, and by Theorem 2.1,

$$\mathscr{P}\big(\tilde{\Omega}_n^c\big) \le A_1 \exp\{-\alpha_1\varepsilon^2 n\} \tag{4.14}$$

for positive A_1 and α_1. On $\tilde{\Omega}_n$, $X_n(T_n) \le X(\infty) + \varepsilon = \gamma + \varepsilon < b_s$ and hence $X_n(T_n) < b_{sn}$ for $n \ge n_0$. It follows that for $n \ge n_0$, $X_{An}(T_n) = 0$ on $\tilde{\Omega}_n$.

As $q = Q(b_s - \gamma) > 0$ and $p = P(b_h - \gamma) < \frac{1}{2}$, we can choose $\varepsilon' \in (0, q \wedge (\frac{1}{2} - p))$ and define

$$\begin{aligned} p_1 &= (p - \varepsilon') \vee 0 \in \left[0, \tfrac{1}{2}\right), & p_2 &= p + \varepsilon' \in \left(0, \tfrac{1}{2}\right), \\ q_1 &= q - \varepsilon' > 0, & q_2 &= q + \varepsilon' > 0. \end{aligned} \tag{4.15}$$

By taking $\varepsilon > 0$ sufficiently small we can obviously ensure that on $\tilde{\Omega}_n$ and for $n \geq n_1$ and $t \geq \tau_n(\varepsilon)$,

$$p_1 \leq P\left(b_{hn} - X_n(t) + \frac{1}{n}\right) \leq p_2, \tag{4.16}$$
$$q_1 \leq Q(b_{sn} - X_n(t)) \leq q_2.$$

Consider a process $X_n^+(t)$ obtained by modifying $X_n(t)$ as follows: For $t > \tau_n(\varepsilon)$ the rate $Q(b_{sn} - X_n(t))$ and the probability $P(b_{hn} - X_n(t) + 1/n)$ are replaced by q_1 and p_2, respectively, and all A cells present at time $\tau_n(\varepsilon)$ are replaced by A cells waiting for a stimulus. On $\tilde{\Omega}_n$ and for $n \geq n_1$, (4.16) implies that X_n^+ is obtained from X_n by adding a random number of A cells and increasing the length of the cell cycle of a number of cells by a random amount, and as a result the duration T_n^+ of X_n^+ is stochastically larger than T_n on $\tilde{\Omega}_n$ for $n \geq n_1$. Moreover, $(T_n^+ - \tau_n(\varepsilon))$ is distributed as the maximum of $N_{An}(\tau_n(\varepsilon))$ independent and identically distributed random variables, each distributed as $S(p_2, q_1, c)$ discussed in Lemma 4.3. As $N_{An}(\tau_n(\varepsilon)) \leq n(1 + b_s)$, (4.14) and Lemma 4.1 yield for $n \geq n_1$,

$$\begin{aligned}\mathscr{P}(T_n \geq t) &\leq \mathscr{P}(T_n^+ \geq t) + \mathscr{P}(\tilde{\Omega}_n^c)\\ &\leq \mathscr{P}(T_n^+ - \tau_n(\varepsilon) \geq t - \tau(\varepsilon/2)) + A_2 \exp\{-\alpha_2 \varepsilon^2 n\}\\ &\leq n(1 + b_s)\mathscr{P}(S(p_2, q_1, c) \geq t - \tau(\varepsilon/2)) + A_2 \exp\{-\alpha_2 \varepsilon^2 n\}\end{aligned}$$

for positive A_2 and α_2, and all t.

Define a as in the statement of the theorem as the solution of $q - a = 2pqe^{ac}$. Similarly, suppose that $\bar{a}$ satisfies $q_1 - \bar{a} = 2p_2q_1e^{\bar{a}c}$. As $p_2 > p$ and $q_1 < q$, we have $\bar{a} < a$ and $\bar{a} \uparrow a$ as ε' in (4.15) tends to zero. Hence, for every $\delta > 0$, we can choose $\varepsilon' > 0$ sufficiently small to ensure that $a^{-1} \leq \bar{a}^{-1} \leq a^{-1} + \delta$. By Lemma 4.3 we find that for $n \geq n_1$ and $2\tau(\varepsilon/2) \leq x \leq n$,

$$\begin{aligned}&\mathscr{P}\left(T_n - \frac{\log n}{a} \geq \delta \log n + x\right)\\ &\quad\leq n(1 + b_s)\mathscr{P}\left(S(p_2, q_1, c) \geq \left(\frac{1}{a} + \delta\right)\log n + x - \tau(\varepsilon/2)\right)\\ &\quad\quad + A_2 \exp\{-\alpha_2 \varepsilon^2 n\} \leq n(1 + b_s)\bar{A}\exp\left\{-\log n - \frac{\bar{a}x}{2}\right\}\\ &\quad\quad + A_2 \exp\{-\alpha_2 \varepsilon^2 n\} \leq A_3 \exp\{-\alpha_3 x\}\end{aligned} \tag{4.17}$$

for positive $\bar{A}$, A_3 and α_3. Obviously an appropriate choice of A_3 will guarantee the validity of this bound for all n and $0 \leq x \leq n$.

We may also modify the process $X_n(t)$ by replacing $Q(b_{sn} - X_n(t))$ and $P(b_{hn} - X_n(t) + 1/n)$ by q_2 and p_1 for $t > \tau_n(\varepsilon)$, and simply removing all A cells that have received a stimulus before or at time $\tau_n(\varepsilon)$. On $\tilde{\Omega}_n$ and for $n \geq n_1$, (4.16) obviously implies that the duration T_n^- of this new process $X_n^-(t)$ is stochastically smaller than T_n. Moreover, $(T_n^- - \tau_n(\varepsilon))$ is distributed as the maximum of $N^0_{An}(\tau_n(\varepsilon))$ independent copies of $S(p_1, q_2, c)$. As $X'(t) > 0$ for all $t > c$, (2.14), (2.15) and (2.17) imply that for every $\varepsilon > 0$ there exists $\eta > 0$ such that $X^0_A(t) \geq 2\eta$ for $0 \leq t \leq \tau(\varepsilon/2)$. It follows from Theorem 2.1 and Lemma 4.1 that

$$\mathscr{P}\big(N^0_{An}(\tau_n(\varepsilon)) \leq \eta n\big) \leq A_4 \exp\{-\alpha_4 n\}$$

for positive A_4 and α_4. Hence, for $n \geq n_1$, (4.14) yields

$$\begin{aligned}\mathscr{P}(T_n \leq t) &\leq \mathscr{P}(T_n^- \leq t) + \mathscr{P}(\tilde{\Omega}_n^c)\\ &\leq \big[\mathscr{P}(S(p_1, q_2, c) \leq t)\big]^{\eta n} + A_5 \exp\{-\alpha_5 n\}\end{aligned}$$

for positive A_5 and α_5.

Define $\tilde{a}$ as the solution of $q_2 - \tilde{a} = 2p_1 q_2 e^{\tilde{a}c}$. As $p_1 \leq p$ and $q_2 > q$, we have $\tilde{a} > 0$ and $\tilde{a} \downarrow a$ as ε' in (4.15) tends to zero. Hence, for every $\delta > 0$ we can choose $\varepsilon' > 0$ sufficiently small to ensure that $a^{-1} - \delta \leq \tilde{a}^{-1} \leq a^{-1}$. By Lemma 4.3 we find that for $n \geq n_1$ and $0 \leq x \leq n$,

$$\begin{aligned}&\mathscr{P}\left(T_n - \frac{\log n}{a} \leq -\delta \log n - x\right)\\ &\leq \left[\mathscr{P}\left(S(p_1, q_2, c) \leq \left(\frac{1}{a} - \delta\right)\log n - x\right)\right]^{\eta n} + A_5 \exp\{-\alpha_5 n\} \qquad (4.18)\\ &\leq \big[1 - \tilde{A}\exp\{-\log n + \tilde{a}x\}\big]^{\eta n} + A_5 \exp\{-\alpha_5 \eta\}\\ &\leq \exp\{-\tilde{A}\eta e^{\tilde{a}x}\} + A_5 \exp\{-\alpha_5 n\} \leq A_6 \exp\{-\alpha_6 x\}\end{aligned}$$

for positive $\tilde{A}$, A_6, and α_6. Obviously the bound will hold for all n and $0 \leq x \leq n$ for an appropriate choice of A_6. Together, (4.17) and (4.18) prove (4.12). Taking $x = \delta \log n$ in (4.12) we complete the proof of the theorem. □

4.3. *The case* $\gamma = b_s$. If $\gamma = b_s$ we are in case (2.29)(iii), so $X(\infty) = b_s = \gamma$, $X_A(\infty) = 0$, $q = Q(0) = 0$ and $p = P(b_h - b_s) < \frac{1}{2}$. The process X_n may stop because $X_n(T_n) = b_{sn}$ or $X_{An}(T_n) = 0$, and in contrast to the two previous cases, neither of these possibilities can be ruled out with large probability. We shall, therefore, have to deal with both possibilities and our approach will combine the main elements of the proofs in the two previous subsections.

In the cases $\gamma > b_s$ and $\gamma < b_s$, either $(b_{sn} - X_n(T_n))$ or $X_{An}(T_n)$ equals zero and the other one of these two quantities is of exact order 1 with high probability. Because now both $b_s - X(\infty) = 0$ and $X_A(\infty) = 0$, the latter part of this statement is no longer true and we shall have to assess the exact order of magnitude of the nonzero quantity among $(b_{sn} - X_n(T_n))$ and $X_{An}(T_n)$. In view of the complicated dependence of these two random variables, some care is needed here. We shall proceed by bounding the nonzero variable in terms of $Z_n = (Z_{1n}, Z_{2n}, \ldots)$ in Lemma 4.4, and then showing that this implies that it is of exact order $n^{-1/2}$ in probability in Corollary 4.1.

LEMMA 4.4. *Let* $\gamma = b_s$. *If* $X_n(T_n) = b_{sn}$, *then*

$$\frac{1}{n}\left|\sum_{i=1}^{[nb_s]}(Z_{in} - 2P_{in})\right| - \frac{3}{n} \le X_{An}(T_n) \le \frac{1}{n}\left|\sum_{i=1}^{[nb_s]}(Z_{in} - 2P_{in})\right| + \frac{3}{n}. \tag{4.19}$$

If $X_{An}(T_n) = 0$, *then*

$$\begin{aligned}\frac{1}{n}\left|\sum_{i=1}^{[nb_s]}(Z_{in} - 2P_{in})\right| - \frac{3}{n} &\le b_{sn} - X_n(T_n)\\ &\le b_s\left\{\frac{1}{n}\max_{1\le m\le [nb_s]}\left|\sum_{i=1}^{m}(Z_{in} - 2P_{in})\right| + \frac{2}{n}\right\}.\end{aligned} \tag{4.20}$$

PROOF. Because $\gamma = b_s$, we have by (2.22) and (2.24),

$$2\int_0^{b_s} P(b_h - u)\,du = b_s - 1. \tag{4.21}$$

Because P is nondecreasing, this implies for $0 \le v \le b_s$,

$$2\int_{b_s - v}^{b_s} P(b_h - u)\,du \le \frac{(b_s - 1)v}{b_s}. \tag{4.22}$$

If $X_n(T_n) = b_{sn}$ we use (2.7), (4.21) and straightforward algebra to obtain

$$\begin{aligned}X_{An}(T_n) &= \frac{1}{n}\sum_{i=1}^{[nb_s]}(Z_{in} - 2P_{in}) + 2\left(\frac{1}{n}\sum_{i=1}^{nb_{sn}} P_{in} - \int_0^{b_{sn}} P(b_h - u)\,du\right)\\ &\quad + \int_{b_{sn}}^{b_s}(1 - 2P(b_h - u))\,du\\ &= \frac{1}{n}\sum_{i=1}^{[nb_s]}(Z_{in} - 2P_{in}) + \mathscr{R},\end{aligned} \tag{4.23}$$

where $|\mathscr{R}| \le 3/n$ by (2.6) and because $0 \le P(u) \le 1$ for all u and $0 \le b_s - b_{sn} < 1/n$. Because $X_{An}(T_n)$ is nonnegative, it also equals the absolute value of the expression on the right in (4.23), and (4.19) follows.

If $X_{An}(T_n) = 0$, we again use (2.7) and (4.21) to obtain

$$\begin{aligned}b_{sn} - X_n(T_n) &= 2\int_0^{b_s} P(b_h - u)\,du\\ &\quad - \frac{1}{n}\sum_{i=1}^{nX_n(T_n)} Z_{in} - (b_s - b_{sn}).\end{aligned} \tag{4.24}$$

Hence by (4.22),

$$b_{sn} - X_n(T_n) \le \frac{b_s - 1}{b_s}(b_s - X_n(T_n)) + 2\int_0^{X_n(T_n)} P(b_h - u)\,du$$
$$- \frac{1}{n}\sum_{i=1}^{nX_n(T_n)} Z_{in} - (b_s - b_{sn})$$
$$\le \frac{b_s - 1}{b_s}(b_{sn} - X_n(T_n)) - \frac{1}{n}\sum_{i=1}^{nX_n(T_n)} (Z_{in} - 2P_{in})$$
$$- 2\left(\frac{1}{n}\sum_{i=1}^{nX_n(T_n)} P_{in} - \int_0^{X_n(T_n)} P(b_h - u)\,du\right)$$

or

$$b_{sn} - X_n(T_n) \le b_s\left\{-\frac{1}{n}\sum_{i=1}^{nX_n(T_n)} (Z_{in} - 2P_{in})\right.$$
$$\left. -2\left(\frac{1}{n}\sum_{i=1}^{nX_n(T_n)} P_{in} - \int_0^{X_n(T_n)} P(b_h - u)\,du\right)\right\},$$

and the inequality on the right in (4.20) follows from (2.6).

To prove the lower bound in (4.20) we start once more with (4.24) and write

$$b_{sn} - X_n(T_n) - \frac{1}{n}\sum_{i=nX_n(T_n)+1}^{[nb_s]} Z_{in}$$
$$= 2\int_0^{b_s} P(b_h - u)\,du - \frac{1}{n}\sum_{i=1}^{[nb_s]} Z_{in} - (b_s - b_{sn})$$
$$= -\frac{1}{n}\sum_{i=1}^{[nb_s]} (Z_{in} - 2P_{in}) - \mathscr{R}$$

with $\mathscr{R}$ as in (4.23) so that $|\mathscr{R}| \le 3/n$. Because $0 \le Z_{in} \le 2$ for all i and n, this yields

$$\frac{1}{n}\left|\sum_{i=1}^{[nb_s]} (Z_{in} - 2P_{in})\right| - \frac{3}{n} \le \left|b_{sn} - X_n(T_n) - \frac{1}{n}\sum_{i=nX_n(T_n)+1}^{[nb_s]} Z_{in}\right|$$
$$\le b_{sn} - X_n(T_n),$$

which completes the proof of the lemma. □

As before, let $\mathscr{F}_{Zn}$ denote the σ-algebra generated by $Z_n = (Z_{1n}, Z_{2n}, \ldots)$.

COROLLARY 4.1. *Let $\gamma = b_s$. For $0 < b < B$, define the events*

(4.25) $$\Omega_{1n} = \{b \le n^{1/2} X_{An}(T_n) \le B\},$$

(4.26) $$\Omega_{2n} = \{b \le n^{1/2}(b_{sn} - X_n(T_n)) \le B\}.$$

These events are $\mathscr{F}_{Zn}$-measurable and for every $\varepsilon > 0$ there exist $0 < b < B$ such that for sufficiently large n,

$$\mathscr{P}(\Omega_{1n} \cup \Omega_{2n}) \geq 1 - \varepsilon. \tag{4.27}$$

PROOF. $\mathscr{F}_{Zn}$-measurability follows from (2.11), (2.12) and (2.7). Obviously, $b_{sn} - X_n(T_n) = 0$ on Ω_{1n} and $X_{An}(T_n) = 0$ on Ω_{2n}. Hence (4.27) follows from Lemma 4.4, the central limit theorem and Lemma B.1 in Appendix B. □

So we have shown that both $(b_{sn} - X_n(T_n))$ and $X_{An}(T_n) = X_{An}^0(T_n)$ are either equal to zero or of exact order $n^{-1/2}$ in probability. However, to analyze the process $N_n(t)$ for large t we shall have to determine the exact order of magnitude of both factors $Q(b_{sn} - X_n(t))$ and $N_{An}^0(t-c) = nX_{An}^0(t-c)$ of the conditional intensity $\Lambda_{Zn}(t)$ of the process [cf. (2.10)]. The factor $Q(b_{sn} - X_n(t))$ is monotone in t and our knowledge concerning $(b_{sn} - X_n(t))$ will suffice. Determining the exact order of $X_{An}^0(t-c)$ for large t is a more delicate matter. In Lemma 4.5 we establish an asymptotic expression for $X_{An}^0(t-c)$ for large n and t in terms of $X_{An}(T_n)$ and $(X_n(T_n) - X_n(t))$. A key step in obtaining this expression is to show that $(X_n(t) - X_n(t-c))$—and hence the difference between $X_{An}^0(t-c)$ and $X_{An}(t)$—is negligible for our purposes.

LEMMA 4.5. *Let $\gamma = b_s$, so $p = P(b_h - b_s) < \frac{1}{2}$. Then for every $D > 0$,*

$$\sup_{t \geq \tau_n(Dn^{-1/2})+c} (X_n(t) - X_n(t-c)) = \mathscr{O}_{\mathscr{P}}\left(\frac{\log n}{n}\right), \tag{4.28}$$

$$\sup_{t \geq \tau_n(Dn^{-1/2})+c} \Big| X_{An}^0(t-c) - \{X_{An}(T_n) + (1-2p)(X_n(T_n) - X_n(t))\}\Big| = \mathscr{O}_{\mathscr{P}}(n^{-3/4}). \tag{4.29}$$

PROOF. Take $\varepsilon > 0$. In (4.25) and (4.26) we choose $0 < b < B$ so that (4.27) holds for sufficiently large n. On $\Omega_n^* = \Omega_{1n} \cup \Omega_{2n}$ we have for $t \geq \tau_n(Dn^{-1/2})$,

$$X_{An}^0(t-c) = X_{An}(T_n) + (X_n(T_n) - X_n(t)) - \sum_{i=nX_n(t-c)+1}^{nX_n(T_n)} Z_{in}$$

$$\leq (B + D)n^{-1/2},$$

$$Q(b_{sn} - X_n(t)) \leq Q'(0)(b_{sn} - X_n(t)) \leq \frac{(B+D)}{da_s} n^{-1/2},$$

because $Q'(0) = 1/(da_s)$. By (2.10) this implies that on Ω_n^*,

$$\sup_{t \geq \tau_n(Dn^{-1/2})} \Lambda_{Zn}(t) \leq \lambda = \frac{(B+D)^2}{da_s}.$$

Conditionally on $Z_n = z$ with $\{Z_n = z\} \subset \Omega_n^*$, N_n is a counting process with intensity $\Lambda_{Z_n}(t) \leq \lambda$. Hence, if $\Pi(t)$ denotes a unit Poisson process and k a positive integer,

$$\mathscr{P}\left(\sup_{t \geq \tau_n(Dn^{-1/2})+c} (X_n(t) - X_n(t-c)) \geq \frac{2k}{n} \right)$$

$$\leq \mathscr{P}\left(\sup_{\substack{t \geq c \\ \Pi(\lambda t) \leq Dn^{1/2}}} (\Pi(\lambda t) - \Pi(\lambda(t-c))) \geq 2k \right) + \varepsilon$$

$$\leq \mathscr{P}\left(\sup_{\substack{t \geq \lambda c \\ \Pi(t) \leq Dn^{1/2}}} (\Pi(t) - \Pi(t - \lambda c)) \geq 2k \right) + \varepsilon$$

$$\leq \frac{Dn^{1/2}}{k} \exp\{2\lambda c - k\} + \varepsilon,$$

by Lemma B.2 in Appendix B. Because $\lambda = \lambda(\varepsilon)$ is finite for every $\varepsilon > 0$, this proves (4.28).

By (2.7) and (2.8),

$$\begin{aligned} &X_{An}^0(t-c) - \{X_{An}(T_n) + (1-2p)(X_n(T_n) - X_n(t))\} \\ &\quad = -\frac{1}{n} \sum_{i=nX_n(t-c)+1}^{nX_n(T_n)} Z_{in} - \frac{1}{n} \sum_{i=nX_n(t)+1}^{nX_n(T_n)} (Z_{in} - 2P_{in}) \\ &\quad\quad - 2\left(\frac{1}{n} \sum_{i=nX_n(t)+1}^{nX_n(T_n)} (P_{in} - p) \right). \end{aligned} \tag{4.30}$$

Because $0 \leq Z_{in} \leq 2$, the first term on the right in (4.30) is bounded in absolute value by $2(X_n(t) - X_n(t-c)) = \mathscr{O}_{\mathscr{P}}((\log n)/n)$ uniformly for $t \geq \tau_n(Dn^{-1/2}) + c$ by (4.28). To deal with the next term we note that $nX_n(T_n) \leq [nb_s]$ and that for $t \geq \tau_n(Dn^{-1/2}) + c$, $nX_n(t) \geq [nb_s] - ([nb_s] - nX_n(T_n)) - [Dn^{1/2}] = [nb_s] - \mathscr{O}_{\mathscr{P}}(n^{1/2})$ by Corollary 4.1. Application of Lemma B.1 in Appendix B for $M = \mathscr{O}(n^{1/2})$ yields that the second term on the right in (4.30) is $\mathscr{O}_{\mathscr{P}}(n^{-3/4})$ uniformly for $t \geq \tau_n(Dn^{-1/2}) + c$. Finally, (2.4), (2.5) and Corollary 4.1 imply that uniformly for $t \geq \tau_n(Dn^{-1/2}) + c$,

$$\begin{aligned} 0 &\leq \frac{1}{n} \sum_{i=nX_n(t)+1}^{nX_n(T_n)} (P_{in} - p) \\ &\leq (X_n(T_n) - X_n(t))(P(b_{hn} - X_n(t)) - P(b_h - b_s)) \\ &\leq Dn^{-1/2} P'(b_h - p_s)\left(\frac{1}{n} + b_{sn} - X_n(T_n) + Dn^{-1/2} \right) = \mathscr{O}_{\mathscr{P}}(n^{-1}). \end{aligned}$$

Together with (4.30), these estimates establish (4.29) and the lemma. □

We are now in a position to determine the exact order of magnitude of T_n. Roughly speaking, we shall argue that if $X_n(T_n) = b_{sn}$, the situation is similar to the one in Theorem 4.1, the main difference being that in the present case $X^0_{An}(t-c)$, and hence $\Lambda_{Zn}(t)$, is smaller by a factor $n^{1/2}$ for large t. This implies that T_n is of order $n^{1/2}\log n$ rather than $\log n$ as is the case in Theorem 4.1. Similarly, if $X_{An}(T_n) = 0$, the situation is comparable to that
in Theorem 4.2, but now $Q(b_{sn} - X_n(t))$ is smaller by a factor $n^{1/2}$ for large t. Again the conclusion is that T_n is of order $n^{1/2}\log n$ instead of $\log n$ as in Theorem 4.2. The basic reason underlying all of this is that in the case $\gamma = b_s$ we have $X_A(\infty) = b_s - X(\infty) = 0$, and hence the nonzero quantity among $X_{An}(T_n)$ and $(b_{sn} - X_n(T_n))$ is of order $n^{-1/2}$ by Corollary 4.1. As we already noted, this is essentially different from what happens if $\gamma > b_s$ or $\gamma < b_s$, when either $X_A(\infty)$ or $(b_s - X(\infty))$ is positive and hence either $X_{An}(T_n)$ or $(b_{sn} - X_n(T_n))$ is of exact order 1.

THEOREM 4.3. *Let $\gamma = b_s$. Then for every $\varepsilon > 0$ there exist positive numbers a and A such that for $n = 2, 3, \dots,$*

$$\mathscr{P}\left(an^{1/2}\log n \le T_n \le An^{1/2}\log n\right) \ge 1 - \varepsilon, \tag{4.31}$$

and hence

$$\frac{T_n}{n^{1/2}\log n} \text{ is of exact order 1 in probability.} \tag{4.32}$$

PROOF. Take $\varepsilon > 0$ and define Ω_{1n} and Ω_{2n} as in (4.25) and (4.26) with $0 < b < B$ such that (4.27) holds for sufficiently large n. By Lemma 4.1, Theorem 3.3 and Lemma 4.5, we can also choose positive numbers C and D, and an event $\overline{\Omega}_n$ with $\mathscr{P}(\overline{\Omega}_n) \ge 1 - \varepsilon$ and such that on $\overline{\Omega}_n$,

$$\tau_n(Dn^{-1/2}) \le \tau\left(\frac{D}{2}n^{-1/2}\right) \le Cn^{1/2}, \tag{4.33}$$

$$\begin{aligned}\sup_{t \ge \tau_n(Dn^{-1/2})+c} &\Big|X^0_{An}(t-c) \\ &- \{X_{An}(T_n) + (1-2p)(X_n(T_n) - X_n(t))\}\Big| \\ &\le Cn^{-3/4}.\end{aligned} \tag{4.34}$$

Note that on $\Omega_{1n} \cup \Omega_{2n}$ we also have the trivial inequality

$$\begin{aligned}X^0_{An}(t-c) &\le X_{An}(t) \le X_{An}(T_n) + (X(T_n) - X_n(t)) \\ &\le (B+D)n^{-1/2}\end{aligned} \tag{4.35}$$

for $t \ge \tau_n(Dn^{-1/2})$.

Given $Z_n = z$ with $\{Z_n = z\} \subset \Omega_{1n}$, $(N_n(t) - n)$ is a counting process with conditional intensity $\Lambda_{Z_n}(t)$. On $\Omega_{1n} \cap \overline{\Omega}_n$, (4.25), (4.34) and (4.35) imply that for sufficiently large n,

$$\Lambda_{Z_n}(t) \geq (bn^{1/2} - Cn^{1/4})Q'(b_s)(b_{sn} - X_n(t)) \quad \text{for } t \geq \tau_n(Dn^{-1/2}) + c,$$

$$\Lambda_{Z_n}(t) \leq (B + D)n^{1/2}Q'(0)(b_{sn} - X_n(t)) \quad \text{for } t \geq \tau_n(Dn^{-1/2}). \tag{4.36}$$

By redefining Λ_{Z_n} on the subset of Ω_{1n}, where these inequalities do not hold, we can ensure that (4.36) is satisfied on Ω_{1n} while changing the process only on a subset of $\Omega_{1n} \cap \overline{\Omega}_n^c$. In the proof of Theorem 4.1 we now replace $\tau_n(\varepsilon)$ by $\tau_n(Dn^{-1/2})$ and (4.9) and (4.10) by (4.36), and repeat the argument following (4.10) to conclude that $(T_n - \tau_n(Dn^{-1/2}))$ is stochastically bounded above and below by two constant multiples of $n^{1/2}S_m(\varepsilon)$, where $S_m(\varepsilon)$ is as in Lemma 4.2 with $\varepsilon m = Dn^{1/2}$. Application of Lemma 4.2 yields the existence of positive numbers a_1 and A_1 such that

$$\mathscr{P}\left(a_1 n^{1/2} \log n \leq T_n - \tau_n(Dn^{-1/2}) \leq A_1 n^{1/2} n^{1/2} \log n | \Omega_{1n}\right) \geq 1 - \varepsilon - \mathscr{P}\left(\overline{\Omega}_n^c | \Omega_{1n}\right) \tag{4.37}$$

for sufficiently large n.

On Ω_{2n}, (4.26) implies that for $t \geq \tau_n(Dn^{-1/2})$,

$$q_1 n^{-1/2} \leq Q(b_{sn} - X_n(t)) \leq q_2 n^{-1/2} \tag{4.38}$$

for positive $q_1 = Q'(b_s)b$ and $q_2 = Q'(0)(B + D)$. Arguing as in the proof of Theorem 4.2, we find that (4.35) and (4.38) imply that on Ω_{2n}, $(T_n - \tau_n(Dn^{-1/2}))$ is stochastically smaller than the maximum of $(B + D)n^{1/2}$ independent copies of $S(p_2, q_1 n^{-1/2}, c)$, with $S(\tilde{p}, \tilde{q}, c)$ as in Lemma 4.3. Moreover, $S(p_2, q_1 n^{-1/2}, c)$ is distributed as $n^{1/2}S(p_2, q_1, cn^{-1/2})$, which is stochastically smaller than $n^{1/2}S(p_2, q_1, c)$.

On the other hand, on Ω_{2n}, $(T_n - \tau_n(Dn^{-1/2}))$ is stochastically larger than the maximum of $N_{An}^0(\tau_n(Dn^{-1/2}))$ independent copies of $S(p_1, q_2 n^{-1/2}, c)$. Also $S(p_1, q_2 n^{-1/2}, c)$ is stochastically larger than $n^{1/2}S(p_1, q_2, 0)$ and on $\Omega_{2n} \cap \overline{\Omega}_n$, $N_{An}^0(\tau_n(Dn^{-1/2})) \geq bn^{1/2} - Cn^{1/4}$ by (4.34). As in the proof of Theorem 4.2, we apply Lemma 4.3 to these upper and lower bounds and find that there exist positive numbers a_2 and A_2 such that

$$\mathscr{P}\left(a_2 n^{1/2} \log n \leq T_n - \tau_n(Dn^{-1/2}) \leq A_2 n^{1/2} \log n | \Omega_{2n}\right) \geq 1 - \varepsilon - \mathscr{P}\left(\overline{\Omega}_n^c | \Omega_{2n}\right) \tag{4.39}$$

for sufficiently large n.

Because $\mathscr{P}(\overline{\Omega}_n^c) \leq \varepsilon$ and $\mathscr{P}(\Omega_{1n} \cup \Omega_{2n}) \geq 1 - \varepsilon$, (4.33), (4.37) and (4.39) ensure the validity of (4.31) for large n and, therefore, trivially for all $n \geq 2$. Because (4.32) is merely a restatement of (4.31), this completes the proof of the theorem. □

APPENDIX A

LEMMA A.1. *For real numbers $c > 0$ and $0 \le v < w$, let f: $[0, \infty) \to (0, \infty)$ be continuously differentiable on (c, ∞) with derivative $f' < 0$ and $\lim_{t\to\infty} f(t) = 0$, and suppose that as $t \to \infty$,*

$$
\begin{aligned}
\text{(A.1)} \qquad f'(t) &= vf(t-c)(1 + \mathscr{O}(f(t-c))) \\
&\quad - wf(t)(1 + \mathscr{O}(f(t-c))).
\end{aligned}
$$

Then the equation $w - a = ve^{ac}$ has a unique solution $a \in (0, w]$ and there exists a positive number A such that

$$\text{(A.2)} \qquad \lim_{t\to\infty} e^{at} f(t) = A.$$

PROOF. Because $w - x > ve^{cx}$ for $x = 0$, ve^{cx} is nonnegative and nondecreasing in x and $(w - x)$ decreases strictly to 0 as $x \uparrow w$, the equation $w - a = ve^{ac}$ does indeed have a unique solution $a \in (0, w]$. Note that $a = w$ if $v = 0$.

Take $\varepsilon = (w - v)/4$. As $f(t) \to 0$ for $t \to \infty$, there exists $t_0 > c$ such that for $t \ge t_0$,

$$f'(t) \le (v + \varepsilon)f(t-c) - (w - \varepsilon)f(t),$$

and hence, for $t \ge t_0$,

$$
\begin{aligned}
f(t) &= -\int_t^\infty f'(u)\,du \ge -(v+\varepsilon)\int_{t-c}^\infty f(u)\,du + (w-\varepsilon)\int_t^\infty f(u)\,du \\
&= (w - v - 2\varepsilon)\int_t^\infty f(u)\,du - (v+\varepsilon)\int_{t-c}^t f(u)\,du \\
&\ge \frac{w-v}{2}\int_t^\infty f(u)\,du - wcf(t-c).
\end{aligned}
$$

As a result,

$$\text{(A.3)} \qquad \int_t^\infty f(u)\,du \le \frac{2}{w-v}\{f(t) + wcf(t-c)\} \to 0 \quad \text{as } t \to \infty$$

and we have shown that f is integrable.

The lemma is now trivial for $v = 0$. We have

$$\frac{f'(t)}{f(t)} = -w + \mathscr{O}(f(t-c)),$$

and for $s, t \to \infty$,

$$\log\left(\frac{e^{wt}f(t)}{e^{ws}f(s)}\right) = \int_s^t \left(\frac{f'(u)}{f(u)} + w\right) du = \mathscr{O}\left(\int_{s-c}^{t-c} f(u)\,du\right) = o(1).$$

Because $a = w$ in this case, this proves (A.2).

We may therefore assume that $0 < v < w$ and $0 < a < w$, and that

$$\text{(A.4)} \qquad f'(t) = vf(t-c) - wf(t) + \mathscr{O}\big((f(t-c))^2\big).$$

Define

$$g(t) = e^{at}f(t), \qquad t \geq 0. \tag{A.5}$$

Rewriting (A.4) in terms of g with the aid of the equation $w - a = ve^{ac}$, we find for $t > c$,

$$g'(t) = (w-a)(g(t-c) - g(t)) + \mathcal{O}(f(t-c)g(t-c)). \tag{A.6}$$

Hence, for an appropriate constant $C > 0$ and $t > c$,

$$\begin{aligned}(w-a)&[(1-\varepsilon(t))g(t-c) - g(t)]\\ &\leq g'(t) \leq (w-a)[(1+\varepsilon(t))g(t-c) - g(t)],\end{aligned} \tag{A.7}$$

where $\varepsilon(t) = Cf(t-c)$.

For $k = 1, 2, \ldots,$ define

$$m_k = \min_{(k-1)c \leq t \leq kc} g(t), \qquad M_k = \max_{(k-1)c \leq t \leq kc} g(t), \qquad \varepsilon_k = \varepsilon(kc).$$

Choose k_0 so that $\varepsilon_{k_0} \leq \frac{1}{2}$. For $k \geq k_0$ and $kc \leq t \leq (k+1)c$, we have $\varepsilon_k \leq \frac{1}{2}$ and by (A.7),

$$\begin{aligned}-(w-a)&[g(t) - (1-\varepsilon_k)m_k]\\ &\leq g'(t) \leq (w-a)[(1+\varepsilon_k)M_k - g(t)].\end{aligned} \tag{A.8}$$

For $t = kc$, both $[g(t) - (1-\varepsilon_k)m_k]$ and $[(1+\varepsilon_k)M_k - g(t)]$ are positive and the inequalities (A.8) for $g'(t)$ ensure that both remain so throughout the interval $kc \leq t \leq (k+1)c$. However, this implies that for $k \geq k_0$,

$$m_{k+1} \geq (1-\varepsilon_k)m_k, \qquad M_{k+1} \leq (1+\varepsilon_k)M_k. \tag{A.9}$$

Because $\varepsilon_k \leq \frac{1}{2}$ for $k \geq k_0$, we find that for every $k \geq k_0$,

$$\begin{aligned}\prod_{r=k}^{\infty}(1-\varepsilon_r) &\geq \exp\left\{-2\sum_{r=k}^{\infty}\varepsilon_r\right\} \geq \exp\left\{-\frac{2C}{c}\int_{(k-2)c}^{\infty} f(u)\,du\right\},\\ \prod_{r=k}^{\infty}(1+\varepsilon_r) &\leq \exp\left\{\sum_{r=k}^{\infty}\varepsilon_r\right\} \leq \exp\left\{\frac{C}{c}\int_{(k-2)c}^{\infty} f(u)\,du\right\}.\end{aligned} \tag{A.10}$$

In view of (A.3), this yields the existence of a sequence $\delta_k \downarrow 0$ and $M > m > 0$ such that for every $k \geq k_0$,

$$\inf_{r \geq k} m_r \geq m_k \prod_{r=k}^{\infty}(1-\varepsilon_r) \geq m_k(1-\delta_k) \geq m > 0,$$

$$\sup_{r \geq k} M_r \leq M_k \prod_{r=k}^{\infty}(1+\varepsilon_r) \leq M_k(1+\delta_k) \leq M < \infty.$$

It follows that for every $k \geq k_0$,

$$\begin{aligned}0 < m \leq m_k(1-\delta_k) &\leq \liminf_{t\to\infty} g(t) \leq \limsup_{t\to\infty} g(t)\\ &\leq M_k(1+\delta_k) \leq M < \infty.\end{aligned} \tag{A.11}$$

Let us look at (A.8) somewhat more carefully. Because $g(t) - (1 - \varepsilon_k)m_k > 0$ and $(1 + \varepsilon_k)M_k - g(t) > 0$ for $t \in [kc, (k+1)c]$, we find that for $kc \le t \le (k+1)c$ and $k \ge k_0$,

$$\log\left(\frac{g(t) - (1 - \varepsilon_k)m_k}{g(kc) - (1 - \varepsilon_k)m_k}\right) = \int_{kc}^{t} \frac{g'(u)}{g(u) - (1 - \varepsilon_k)m_k}\,du \ge -(w - a)c,$$

and after replacing $g(t)$ by m_{k+1},

$$\text{(A.12)} \qquad m_{k+1} \ge (1 - \varepsilon_k)m_k + e^{-(w-a)c}[g(kc) - (1 - \varepsilon_k)m_k].$$

Similarly, the right-hand inequality in (A.8) ensures that for $k \ge k_0$,

$$\text{(A.13)} \qquad M_{k+1} \le (1 + \varepsilon_k)M_k - e^{-(w-a)c}[(1 + \varepsilon_k)M_k - g(kc)],$$

and subtracting (A.12) from (A.13) we see that for $k \ge k_0$,

$$\begin{aligned} M_{k+1} - m_{k+1} &\le (1 - e^{-(w-a)c})[(1 + \varepsilon_k)M_k - (1 - \varepsilon_k)m_k] \\ &\le \beta(M_k - m_k) + 2M\varepsilon_k, \end{aligned}$$

where $\beta = 1 - \exp\{-(w - a)c\} \in (0, 1)$. By iterating this inequality, we find that for $r \ge 1$,

$$\begin{aligned} M_{k_0+r} - m_{k_0+r} &\le \beta^r(M_{k_0} - m_{k_0}) + 2M\sum_{\nu=1}^{r} \beta^{r-\nu}\varepsilon_{k_0+\nu-1} \\ &\le \beta^r M + M \sum_{j=[(r+1)/2]}^{\infty} \beta^j + 2M \sum_{j=k_0+[r/2]}^{\infty} \varepsilon_j \\ &\le \beta^r M + \frac{M}{1-\beta}\beta^{r/2} + \frac{2MC}{c}\int_{1/2(2k_0+r-5)c}^{\infty} f(u)\,du. \end{aligned}$$

As $0 < \beta < 1$, (A.3) ensures that

$$\text{(A.14)} \qquad \lim_{k\to\infty}(M_k - m_k) = 0,$$

and because $\delta_k \downarrow 0$, (A.11) and (A.14) show that $g(t)$ tends to a positive and finite limit as $t \to \infty$. In view of (A.5), the proof is complete. □

APPENDIX B

LEMMA B.1. *If $X_1, X_2, \ldots$ are independent bounded random variables, $0 \le X_j \le a$ for $j = 1, 2, \ldots$, then for all $M \in \mathbb{N}$ and every $x \ge 0$,*

$$\text{(B.1)} \qquad \mathscr{P}\left(\max_{1\le m\le M}\left|\sum_{j=1}^{m}(X_j - EX_j)\right| \ge x\right) \le 4\exp\left\{-\frac{2x^2}{9a^2M}\right\}.$$

PROOF. The lemma follows from Theorem 2 in Hoeffding (1963) combined with Lévy's inequality [Shorack and Wellner (1986), page 844]. □

LEMMA B.2. *If $\Pi(t)$ is a unit Poisson process, a and b are positive numbers and $m \in \mathbb{N}$, then*

$$\text{(B.2)} \qquad \mathscr{P}\left(\sup_{\substack{t \geq b \\ \Pi(t) \leq a}} (\Pi(t) - \Pi(t-b)) \geq 2m \right) \leq \frac{a}{m} \exp\{2b - m\}.$$

PROOF. Let $0 = Y_0 < Y_1 < Y_2 < \cdots$ be the consecutive jump times of Π. Then

$$\begin{aligned} &\mathscr{P}\left(\sup_{\substack{t \geq b \\ \Pi(t) \leq a}} (\Pi(t) - \Pi(t-b)) \leq 2m - 1 \right) \\ &\quad = \mathscr{P}\left(\min_{2m-1 \leq k \leq a} (Y_k - Y_{k-(2m-1)}) \geq b \right) \\ &\quad \geq \mathscr{P}\left(\min_{1 \leq r \leq [a/m]} (Y_{rm} - Y_{(r-1)m}) \geq b \right) \\ &\quad = \{\mathscr{P}(\Pi(b) \leq m-1)\}^{[a/m]}. \end{aligned}$$

Because $\Pi(b)$ has a Poisson distribution with mean b, we have $\mathscr{P}(\Pi(b) \geq m) \leq \exp\{2b - m\}$ and hence for $a \geq m$ and $m \geq 2b$,

$$\begin{aligned} &\mathscr{P}\left(\sup_{\substack{t \geq b \\ \Pi(t) \leq a}} (\Pi(t) - \Pi(t-b)) \leq 2m - 1 \right) \\ &\quad \geq (1 - \exp\{2b - m\})^{a/m} \geq 1 - \frac{a}{m}\exp\{2b - m\}, \end{aligned}$$

which proves (B.2) for $a \geq m$ and $m \geq 2b$. For $a < m$ or $m < 2b$, (B.2) is trivially true. □

Acknowledgements. We thank the Editor, an Associate Editor and the referees for their constructive criticism that very much improved the readability of the paper.

REFERENCES

ALBERTS, B., BRAY, D., LEWIS, J., RAFF, M., ROBERTS, K. and WATSON, J. D. (1989). *Molecular Biology of the Cell*, 2nd ed. Garland, New York.

BARBOUR, A. D. (1975). The duration of a closed stochastic epidemic. *Biometrika* **62** 477–482.

BAYLISS, M. W. (1985). The regulation of the cell division cycle in cultured plant cells. In *The Cell Division Cycle in Plants* (J. A. Bryant and D. Francis, eds.). Cambridge Univ. Press.

BROOKS, R. F., BENNETT, D. C. and SMITH, J. A. (1980). Mammalian cells need two random transitions. *Cell* **19** 493–504.

CASTOR, F. A. L. (1980). A G_1 rate model accounts for cell-cycle kinetics attributed to 'transition probability.' *Nature* **287** 857–859.

COOPER, S. (1982). The continuum model: statistical implications. *J. Theoret. Biol.* **94** 783–800.

DE GUNST, M. C. M. (1989). *A Random Model for Plant Cell Population Growth. CWI Tract* **58**. Math. Centrum, Amsterdam.

DE GUNST, M. C. M., HARKES, P. A. A., VAL, J., VAN ZWET, W. R. and LIBBENGA, K. R. (1990). Modelling the growth of a batch culture of plant cells: a corpuscular approach. *Enzyme Microb. Technol.* **12** 61–71.

DE GUNST, M. C. M. and VAN ZWET, W. R. (1992). A non-Markovian model for cell population growth: speed of convergence and central limit theorem. *Stochastic Process. Appl.* **41** 297–324.

HOEFFDING, W. (1963). Probability inequalities for sums of bounded random variables. *J. Amer. Statist. Assoc.* **58** 13–30.

KURTZ, T. G. (1982). Representation and approximation of counting processes. In *Advances in Filtering and Optimal Stochastic Control Lecture Notes in Control and Inform. Sci.* **42** 177–191. Springer, New York.

MORRIS, P., SCRAGG, A. H., STAFFORD, A. and FOWLER, M. W., EDS. (1986). *Secondary Metabolism in Plant Cell Cultures*. Cambridge Univ. Press.

NAGAEV, A. V. and MUKHOMOR, T. P. (1975). A limit distribution of the duration of an epidemic. *Theory Probab. Appl.* **20** 805–818.

NELSON, S. and GREEN, P. J. (1981). The random transition model of the cell cycle. A critical review. *Cancer Chemother. Pharmacol.* **6** 11–18.

ROELS, J. A. (1983). *Energetics and Kinetics in Biotechnology*. North-Holland, Amsterdam.

SHIELDS, R. (1977). Transition probability and the origin of variation in the cell cycle. *Nature* **267** 704–707.

SHORACK, G. R. and WELLNER, J. A. (1986). *Empirical Processes with Applications to Statistics*. Wiley, New York.

SMITH, J. A. and MARTIN, L. (1973). Do cells cycle? *Proc. Nat. Acad. Sci. U.S.A.* **70** 1263–1267.

STREET, H. E. (1973). *Plant Tissue and Cell Culture. Botanical Monographs* **11**. Blackwell, Oxford.

TREWAVAS, A. J. (1985). Growth substances, calcium and the regulation of cell division. In *The Cell Division Cycle in Plants*. (J. A. Bryant and D. Francis, eds.). Cambridge Univ. Press.

DEPARTMENT OF MATHEMATICS
FREE UNIVERSITY
DE BOELELAAN 1081 A
1081 HV AMSTERDAM
THE NETHERLANDS

DEPARTMENT OF MATHEMATICS
UNIVERSITY OF LEIDEN
P.O. BOX 9512
2300 RA LEIDEN
THE NETHERLANDS
AND
DEPARTMENT OF STATISTICS
UNIVERSITY OF NORTH CAROLINA
CHAPEL HILL, NORTH CAROLINA 27599-3260

Bernoulli **9**(6), 2003, 1071–1092

Parameter estimation for the supercritical contact process

MARTA FIOCCO[1] and WILLEM R. VAN ZWET[2]

[1]*Department of Medical Statistics, Leiden University Medical Centre, P.O. Box 9604, 2300 RC Leiden, The Netherlands. E-mail: m.fiocco@lumc.nl*
[2]*Mathematical Institute, University of Leiden, P.O. Box 9512, 2300 RA Leiden, The Netherlands. E-mail: vanzwet@math.leidenuniv.nl*

Contact processes – and, more generally, interacting particle processes – can serve as models for a large variety of statistical problems, especially if we allow some simple modifications that do not essentially complicate the mathematical treatment of these processes. We begin a statistical study of the supercritical contact process that starts with a single infected site at the origin and is conditioned on survival of the infection. We consider the statistical problem of estimating the parameter λ of the process on the basis of an observation of the process at a single time t. We propose an estimator of λ and show that it is consistent and asymptotically normal as $t \to \infty$.

Keywords: contact process; parameter estimation; random mask; shrinking; supercritical

1. Introduction

A d-dimensional contact process is a simplified model for the spread of a biological organism or an infection on the lattice $\mathbb{Z}^d$. At each time $t \geqslant 0$, every point of the lattice (or site) is either infected or healthy. As time passes, a healthy site is infected at Poisson rate λ by each of its $2d$ immediate neighbours which is itself infected; an infected site recovers and becomes healthy at Poisson rate 1. Given the set of infected sites ξ_t at time t, the processes involved are independent until a change occurs. If the process starts with a set $A \subset \mathbb{Z}^d$ of infected sites at time $t = 0$, then ξ_t^A will denote the set of infected sites at time $t \geqslant 0$ and $\{\xi_t^A : t \geqslant 0\}$ will denote the contact process. For example, $\{\xi_t^{\mathbb{Z}^d} : t \geqslant 0\}$ or $\{\xi_t^{\{0\}} : t \geqslant 0\}$ will denote the processes starting with every site infected, or with a single infected site at the origin. If the starting set is chosen at random according to a probability distribution α, then the process will be written as $\{\xi_t^\alpha : t \geqslant 0\}$. If we do not want to specify the initial state of the process at all, we simply write $\{\xi_t : t \geqslant 0\}$.

We also need a compact notation for the state of a single site $x \in \mathbb{Z}^d$ at time t. For any contact process ξ_t, we write

$$\xi_t(x) = 1_{\xi_t}(x) = \begin{cases} 1 & \text{if } x \text{ is infected at time } t, \\ 0 & \text{if } x \text{ is healthy at time } t, \end{cases} \tag{1.1}$$

thus using the same symbol ξ_t for both the set of infected points and its indicator function. Of course $\xi_t^A(x)$ and $\xi_t^\alpha(x)$ will refer to the processes ξ_t^A and ξ_t^α in the same manner.

The first thing to note about the contact process is that for all non-empty $A \subset \mathbb{Z}^d$, the infection will continue forever with positive probability if and only if λ exceeds a certain *critical value* λ_d. Such a process is called *supercritical*. Thus, if we define the random hitting time

$$\tau^A = \inf\{t : \xi_t^A = \varnothing\}, \qquad A \subset \mathbb{Z}^d, \tag{1.2}$$

with the convention that $\tau^A = \infty$ if $\xi_t^A \neq \varnothing$ for all $t \geqslant 0$, then for the supercritical contact process

$$\mathbb{P}(\tau^A = \infty) > 0 \tag{1.3}$$

for every non-empty $A \subset \mathbb{Z}^d$. Moreover, if A has infinite cardinality $|A| = \infty$, then

$$\mathbb{P}(\tau^A = \infty) = 1. \tag{1.4}$$

In the supercritical case, the process $\xi_t^{\mathbb{Z}^d}$ that starts with all sites infected converges in distribution to the so-called upper invariant measure $\nu = \nu_\lambda$. Here convergence in distribution means convergence of probabilities of events defined by the behaviour of the process on finite subsets of $\mathbb{Z}^d$, and 'invariant' refers to the fact that the process $\{\xi_t^\nu : t \geqslant 0\}$ is stationary. In particular, the distribution of ξ_t^ν is equal to ν for all t. Obviously, ν is also invariant under integer-valued translations of $\mathbb{Z}^d$. The long-range behaviour of the supercritical contact process $\{\xi_t^A : t \geqslant 0\}$ for arbitrary non-empty $A \subset \mathbb{Z}^d$ is described by the *complete convergence theorem*. Let μ_t^A denote the probability distribution of ξ_t^A and $\delta_\varnothing$ the distribution that assigns probability 1 to the empty set.

Theorem 1.1. *Let $A \subset \mathbb{Z}^d$ and $\lambda > \lambda_d$. Then, as $t \to \infty$,*

$$\mu_t^A \xrightarrow{w} \mathbb{P}(\tau^A < \infty)\delta_\varnothing + \mathbb{P}(\tau^A = \infty)\nu_\lambda. \tag{1.5}$$

For a proof see Liggett (1999, p. 55).

If $\lambda > \lambda_d$ and $A = \mathbb{Z}^d$, the process $\xi_t^{\mathbb{Z}^d}$ survives forever with probability 1 by (1.4) and converges exponentially to the limit process, that is, for positive C and γ and all $t \geqslant 0$,

$$0 \leqslant \mathbb{P}\left(\xi_t^{\mathbb{Z}^d}(x) = 1\right) - \mathbb{P}(\xi^\nu(x) = 1) \leqslant C\mathrm{e}^{-\gamma t} \tag{1.6}$$

(Liggett 1999, p. 57).

Another major result concerning the contact process is the *shape theorem*. To formulate this result we first have to describe the graphical representation of contact processes due to Harris (1978). This is a particular coupling of all contact processes of a given dimension d and with a given value of λ, but with every possible initial state A or initial distribution α. Consider space-time $\mathbb{Z}^d \times [0, \infty)$. For every site $x \in \mathbb{Z}^d$ we define on the line $x \times [0, \infty)$ a Poisson process with rate 1; for every ordered pair (x, y) of neighbouring sites in $\mathbb{Z}^d$ we define a Poisson process with rate λ. All of these Poisson processes are independent.

We now draw a picture of $\mathbb{Z}^d \times [0, \infty)$ where, for each site $x \in \mathbb{Z}^d$, we remove the points of the corresponding Poisson process with rate 1 from the line $x \times [0, \infty)$; for each ordered pair of neighbouring sites (x, y) we draw an arrow going perpendicularly from the line $x \times [0, \infty)$ to the line $y \times [0, \infty)$ at the points of the Poisson processes with rate λ corresponding to the pair (x, y).

For any set $A \subset \mathbb{Z}^d$, define ξ_t^A to be the set of sites that can be reached by starting at time 0 at some site in A and travelling until time t along unbroken segments of lines $x \times [0, \infty)$ in the direction of increasing time, as well as along arrows. Clearly, $\{\xi_t^A : t \geqslant 0\}$ is distributed as a contact process with initial state A. By choosing the initial set at random with distribution α, we define $\{\xi_t^\alpha : t \geqslant 0\}$. The obvious beauty of this coupling is that for two initial sets of infected sites $A \subset B$, we have $\xi_t^A \subset \xi_t^B$ for all $t \geqslant 0$.

Unless indicated otherwise, we shall assume that all contact processes are defined according to this graphical construction. We shall also restrict attention to the supercritical case and assume that $\lambda > \lambda_d$ throughout.

Before formulating the shape theorem we need to introduce some notation. Let $\|\cdot\|$ denote the L^∞ norm on $\mathbb{R}^d$, that is,

$$\|x\| = \max_{1 \leqslant i \leqslant d} |x_i|$$

for $x = (x_1, \ldots, x_d) \in \mathbb{R}^d$, and let $Q = (x \in \mathbb{R}^d : \|x\| \leqslant \frac{1}{2}\}$ denote the unit hypercube centred at the origin. For $A, B \subset \mathbb{R}^d$, $A \oplus B = \{x + y : x \in A, y \in B\}$ will denote the direct sum of A and B, and for real r, $rA = \{rx : x \in A\}$. Define

$$H_t = \bigcup_{s \leqslant t} \xi_s^{\{0\}} \oplus Q, \tag{1.7}$$

$$K_t = \{x \in \mathbb{Z}^d : \xi_t^{\{0\}}(x) = \xi_t^{\mathbb{Z}^d}(x)\} \oplus Q. \tag{1.8}$$

Thus for the process $\{\xi_t^{\{0\}} : t \geqslant 0\}$ that starts with a single infected site at the origin, H_t is obtained by taking the union of the sites that have been infected up to or at time t, and replacing these sites by unit hypercubes centred at these sites in order to fill in the space between neighbouring sites. Similarly, K_t is the filled-in version of the set of sites where $\xi_t^{\{0\}}$ and $\xi_t^{\mathbb{Z}^d}$ coincide. We are now in a position to formulate the shape theorem (cf. Durrett 1991; Bezuidenhout and Grimmett 1990).

Theorem 1.2. *There exists a bounded convex subset U of $\mathbb{R}^d$ with the origin as an interior point and such that, for any $\epsilon \in (0, 1)$,*

$$(1 - \epsilon)tU \subset H_t \cap K_t \subset H_t \subset (1 + \epsilon)tU, \tag{1.9}$$

eventually almost surely on the event $\{\tau^{\{0\}} = \infty\}$ where $\xi_t^{\{0\}}$ survives forever.

The shape theorem describes the growth of the set of infected sites if the process $\xi_t^{\{0\}}$ survives forever. Roughly speaking, the convex hull of the set of infected sites will grow linearly in time as $t \to \infty$ and acquire an asymptotic shape tU, where U is a fixed convex set with the origin as an interior point. Inside this set, say in $(1 - \epsilon)tU$, the smallest and the largest possible process $\xi_t^{\{0\}}$ and $\xi_t^{\mathbb{Z}^d}$ are equal eventually a.s., and this must mean that, for large t, their distribution is close to the equilibrium distribution ν. Together, the complete convergence theorem and the shape theorem describe the peculiar type of convergence of the supercritical contact process to its limiting distribution. The infection spreads at a constant speed and, relatively soon after it has reached a site x, equilibrium will set in at that site.

A third important property of the contact process is its self-duality. If, in the graphical representation, time is run backwards and all arrows representing infection of one site by another are reversed, then the new graphical representation has precisely the same probabilistic structure as the original one. In particular,

$$\mathbb{P}(\xi_t^A \cap B \neq \emptyset) = \mathbb{P}(\xi_t^B \cap A \neq \emptyset), \text{ for all } A, B \subset \mathbb{Z}^d \text{ and } t \geqslant 0. \tag{1.10}$$

With $A = \{0\}$ and $B = \mathbb{Z}^d$ this yields

$$\mathbb{P}\left(\tau^{\{0\}} > t\right) = \mathbb{P}\left(\xi_t^{\mathbb{Z}^d}(0) = 1\right)$$

which, letting $t \to \infty$ in the supercritical case, reduces to

$$\mathbb{P}\left(\tau^{\{0\}} = \infty\right) = \mathbb{P}\left(\xi_t^{\nu}(0) = 1\right).$$

Combining this with (1.6), we see that if $\lambda > \lambda_d$, then

$$\mathbb{P}(t < \tau^{\{0\}} < \infty) \leqslant C\mathrm{e}^{-\gamma t} \tag{1.11}$$

(cf. Liggett 1999, p. 57).

In this paper we shall study the estimation problem for the parameter λ of the supercritical contact process $\xi_t^{\{0\}}$, given that it does not die out. Based on an observation of $\xi_t^{\{0\}}$ at a single time t, we derive an estimator $\hat{\lambda}_t^{\{0\}}$ and show that it is consistent and asymptotically normal as $t \to \infty$.

The informal description of the convergence of the contact process immediately suggests a way to derive an estimator of the parameter λ. If $\xi_t^{\{0\}}$ survives forever, then observing $\xi_t^{\{0\}}(x)$ for all sites x contained in $(1-\epsilon)tU$ is asymptotically the same as observing the limit process $\xi_t^{\nu}(x)$ on this set. This asymptotic 'equivalence' of ξ_t^{ν} and $\xi_t^{\{0\}}$ on $(1-\epsilon)tU$ should allow us to derive an estimator of λ based on the limit process $\xi_t^{\nu}(x)$ for sites $x \in (1-\epsilon)tU$, and hope that this estimator will also work for the process $\xi_t^{\{0\}}$. The advantage of deriving the estimator under ξ_t^{ν} is that we can use the stationarity of this process to set up the estimating equation.

For $D \subset \mathbb{Z}^d$, define the total number of infected sites in the set D at time t as

$$n_t(D) = \sum_{x \in D} \xi_t(x), \tag{1.12}$$

and the total number of pairs of neighbouring sites for which one site is healthy and lies in D and the other is infected as

$$k_t(D) = \sum_{x \in D} k_t(x), \tag{1.13}$$

where

$$k_t(x) = (1 - \xi_t(x)) \sum_{|x-y|=1} \xi_t(y). \tag{1.14}$$

Here $|x-y| = \sum |x_i - y_i|$ denotes the L^1 distance between sites x and y. When we need to specify the initial state of the process we shall use an appropriate notation. For example, $n_t^{\{0\}}$ and $k_t^{\{0\}}$ will indicate that we are referring to the process $\xi_t^{\{0\}}$. Similarly, for the process ξ_t^ν, we write $n_t^\nu(x)$ and k_t^ν.

For the ξ_t^ν process, $\xi_t^\nu(x)$ increases by 1 at rate $\lambda k_t^\nu(x)$ and decreases by 1 at rate $\xi_t^\nu(x)$. As ξ_t^ν is stationary, this implies that $\lambda \mathbb{E} k_t^\nu(x) = \mathbb{E}\xi_t^\nu(x)$ and, since ξ_t^ν is spatially translation-invariant, we have

$$\lambda = \frac{\mathbb{E}\xi_t^\nu(x)}{\mathbb{E}k_t^\nu(x)} = \frac{\mathbb{E}\xi_t^\nu(0)}{\mathbb{E}k_t^\nu(0)}. \tag{1.15}$$

Notice that these expectations are independent of t because of the stationarity of ξ_t^ν. For $t \geqslant 0$, let $A_t \subset \mathbb{Z}^d$ be finite sets of cardinality $|A_t| \to \infty$ as $t \to \infty$. It seems reasonable to expect that some form of the law of large numbers will ensure that, as $t \to \infty$,

$$\frac{n_t^\nu(A_t)}{|A_t|} = \frac{\sum_{x \in A_t} \xi_t^\nu(x)}{|A_t|} \sim \mathbb{E}\xi_t^\nu(0)$$

and

$$\frac{k_t^\nu(A_t)}{|A_t|} = \frac{\sum_{x \in A_t} k_t^\nu(x)}{|A_t|} \sim \mathbb{E}k_t^\nu(0).$$

This would imply that $n_t^\nu(A_t)/k_t^\nu(A_t)$ is a plausible estimator of λ on the basis of an observation of the process ξ_t^ν at a single time t. If, in addition to $|A_t| \to \infty$, we also require that $A_t \subset (1-\epsilon)tU$ for some $\epsilon > 0$, then the shape theorem suggests that, conditional on $\xi_t^{\{0\}}$ surviving forever, the probabilistic behaviour of $\xi_t^{\{0\}}$ and ξ_t^ν should be asymptotically the same on the set $A_t \subset \mathbb{Z}^d$. But this indicates that if we observe the process $\xi_t^{\{0\}}$ instead of ξ_t^ν, then $n_t^{\{0\}}(A_t)/k_t^{\{0\}}(A_t)$ would be a plausible estimator of λ based on $\xi_t^{\{0\}}$, provided that $\xi_t^{\{0\}}$ survives. Unfortunately, the set U is unknown – as is t in many applications – and hence we cannot implement this estimation procedure directly. However, the shape theorem also suggests that if $\xi_t^{\{0\}}$ survives forever, the convex hull $\mathcal{C}(\xi_t^{\{0\}})$ of the set $\xi_t^{\{0\}}$ of infected sites behaves asymptotically like tU. Hence we may expect that if we define a mask

$$C_t = (1-\delta)\mathcal{C}\left(\xi_t^{\{0\}}\right),$$

for some $\delta > 0$, and $\xi_t^{\{0\}}$ survives, then $|C_t \cap \mathbb{Z}^d| \to \infty$ and $C_t \subset (1-\epsilon)tU$ for some $\epsilon > 0$. Combining these ideas, we arrive at

$$\hat{\lambda}_t^{\{0\}} = \hat{\lambda}_t^{\{0\}}(C_t) = \frac{n_t^{\{0\}}(C_t)}{k_t^{\{0\}}(C_t)} \tag{1.16}$$

as a plausible estimator of λ on the basis of an observation of $\xi_t^{\{0\}}$ at a single time t. In fact we shall use masks C_t which are obtained by shrinking the set $\mathcal{C}(\xi_t^{\{0\}})$ in a more general manner than through multiplication by $1-\delta$ (cf. Section 3).

The aim of this paper is to prove that $\hat{\lambda}_t^{\{0\}}$ is a consistent and asymptotically normal estimator of λ on the event where $\xi_t^{\{0\}}$ survives forever. To do this we not only have the considerable problem of making the above heuristic argument precise, but in order to prove the asymptotic normality, we also have to show that, for the $\xi_t^{\{0\}}$ process conditional on survival, distant sites evolve almost independently. The technical tools for dealing with these problems are provided in Fiocco and van Zwet (2003).

We should stress at this point that shrinking $C(\xi_t^{\{0\}})$ to obtain the mask C_t is absolutely essential to obtain an estimator that works well in practice. Without shrinking, the mask will contain the boundary area of the set of infected points where equilibrium has not yet set in and the infected points are therefore less dense. This has the effect of lowering the estimator of λ. Simulation shows that the resulting negative bias is considerable and that 20–40% of the sites have to be removed by shrinking to eliminate this bias (cf. Fiocco 1997). From a theoretical point of view we shall find that without shrinking – i.e. if $\delta = 0$ and hence $C_t = C(\xi_t^{\{0\}})$ – we can still show consistency of the estimator $\hat{\lambda}_t^{\{0\}}$, but not its asymptotic normality.

2. Technical tools

In this section we provide the reader with a number of tools that will be used in this paper for establishing the properties of $\hat{\lambda}_t^{\{0\}}$. These results may be found in Fiocco and van Zwet (2003). Let $C(\xi_t^{\{0\}})$ be the convex hull of the set of infected sites. Theorems 1.3–1.5 in Fiocco and van Zwet (2003) provide eventually almost sure bounds on this set, and probability bounds for the lower inclusion for $H_t \cap K_t$ as well as $C(\xi_t^{\{0\}})$ in (1.9) and (2.1).

Theorem 2.1. *For every $\epsilon \in (0, 1)$,*

$$(1-\epsilon)tU \subset C(\xi_t^{\{0\}}) \subset (1+\epsilon)tU \tag{2.1}$$

eventually a.s. on the set $\{\tau^{\{0\}} = \infty\}$. Moreover, for every $\epsilon \in (0, 1)$ and $r > 0$, there exists a positive number $A_{r,\epsilon}$ such that, for every $t > 0$,

$$\mathbb{P}((1-\epsilon)tU \subset H_t \cap K_t | \tau^{\{0\}} = \infty) \geqslant 1 - A_{r,\epsilon} t^{-r}$$
$$\mathbb{P}((1-\epsilon)tU \subset C(\xi_t^{\{0\}}) | \tau^{\{0\}} = \infty) \geqslant 1 - A_{r,\epsilon} t^{-r}.$$

Before formulating the next result we need to introduce some notation. Let $H = \{0, 1\}^{\mathbb{Z}^d}$ denote the state space for the contact process. For $f : H \to \mathbb{R}$ and $x \in \mathbb{Z}^d$, define

$$\Delta_f(x) = \sup\{|f(\eta) - f(\zeta)| : \eta, \zeta \in H \text{ and } \eta(y) = \zeta(y) \text{ for all } y \neq x\}, \tag{2.2}$$

$$|||f||| = \sum_{x \in \mathbb{Z}^d} \Delta_f(x).$$

For $R_1, F_2 \subset \mathbb{Z}^d$, let $d(R_1, R_2)$ denote the L^1 distance of R_1 and R_2:

$$d(R_1, R_2) = \inf_{x \in R_1, y \in R_2} |x - y| = \inf_{x \in R_1, y \in R_2} \sum_{i=1}^{d} |x_i - y_i|.$$

Let

$$D_R = \{f : H \to \mathbb{R},\ |||f||| < \infty,\ f(\eta) \text{ depends on } \eta \text{ only through } \eta \cap R\}, \tag{2.3}$$

that is, D_R is the class of functions f with $|||f||| < \infty$ such that $f(\eta)$ depends on η only through $\eta(x)$ with $x \in R$.

Theorem 2.2. *There exist positive numbers γ and C such that for every R_1, $R_2 \subset \mathbb{Z}^d$, $f \in D_{R_1}$ $g \in_{R_2}$, and $t \geqslant 0$,*

$$\left|\operatorname{cov}\left(f(\xi_t^{\mathbb{Z}^d}),\ g(\xi_t^{\mathbb{Z}^d})\right)\right| \leqslant C|||f||| \cdot g|||e^{-\gamma d(R_1,R_2)}. \tag{2.4}$$

In particular, there exist positive numbers γ and C such that, for all $t \geqslant 0$, and $x, y \in \mathbb{Z}^d$,

$$\left|\operatorname{cov}\left(\xi_t^{\mathbb{Z}^d}(x),\ \xi_t^{\mathbb{Z}^d}(y)\right)\right| \leqslant Ce^{-\gamma|x-y|}, \tag{2.5}$$

and

$$\left|\operatorname{cov}\left(k_t^{\mathbb{Z}^d}(x),\ k_t^{\mathbb{Z}^d}(y)\right)\right| \leqslant Ce^{-\gamma|x-y|}. \tag{2.6}$$

Proof. The first part of the theorem is Theorem 1.7 in Fiocco and van Zwet (2003), and is proved in Section 3 of that paper. Inequalities (2.5) and (2.6) follow because $|||f||| = |||g||| = 1$ and 8, respectively. □

Obviously (2.5) and (2.6) imply that $\sigma^2(n_t^{\mathbb{Z}^d}(D))$ and $\sigma^2(k_t^{\mathbb{Z}^d}(D))$ are of order $|D|$ for large D. The following theorem extends this results to all moments of even order.

Theorem 2.3. *For any $k = 1, 2, \ldots,$ there exists a number $C_k > 0$ such that for every $D \subset \mathbb{Z}^d$ and $t \geqslant 0$,*

$$\mu_{2k} = \mathbb{E}\left(n_t^{\mathbb{Z}^d}(D) - \mathbb{E}n_t^{\mathbb{Z}^d}(D)\right)^{2k} \leqslant C_k|D|^k. \tag{2.7}$$

and

$$\nu_{2k} = \mathbb{E}\left(k_t^{\mathbb{Z}^d}(D) - \mathbb{E}k_t^{\mathbb{Z}^d}(D)\right)^{2k} \leqslant C_k|D|^k. \tag{2.8}$$

Proof. The proof follows from Theorem 4.1 in Fiocco and van Zwet (2003). □

Let $\bar{\xi}_t^{\mathbb{Z}^d}$ denote a process distributed as $\xi_t^{\mathbb{Z}^d}$ conditioned on $\{\tau^{\{0\}} = \infty\}$. Theorem 1.6 in Fiocco and van Zwet (2003) asserts that we can couple the processes $\bar{\xi}_t^{\mathbb{Z}^d}$ and $\xi_t^{\mathbb{Z}^d}$ in such a way that they coincide on tU except on a set of exponentially small probability. We shall not explicitly describe this coupling, other than to note that it is not in accordance with the graphical representation since the two processes are defined on essentially different subsets of the sample space. We repeat the theorem for the reader's convenience:

Theorem 2.4. *There exist a coupling $({}_c\xi_t^{\mathbb{Z}^d}, {}_c\bar{\xi}_t^{\mathbb{Z}^d})$ of $(\xi_t^{\mathbb{Z}^d}, \bar{\xi}_t^{\mathbb{Z}^d})$ and positive constants C and γ such that for all $t > 0$,*

$$\mathbb{P}\left({}_c\xi_t^{\mathbb{Z}^d} \cap tU = {}_c\bar{\xi}_t^{\mathbb{Z}^d} \cap tU\right) > 1 - Ce^{-\gamma t}.$$

Let $\bar{\xi}_t^{\{0\}}$ denote a process which is distributed as $\xi_t^{\{0\}}$ conditioned on $\{\tau^{\{0\}} = \infty\}$. The final result in this section is a restatement of Theorem 1.8 in Fiocco and van Zwet (2003) and asserts that for this process, distant sites evolve almost independently for large t.

Theorem 2.5. *For every $\epsilon \in (0, 1)$ and $r > 0$ there exist a positive number $A_{r,\epsilon}$, as well as positive constants C and γ, such that, for all $t > 0$ and all f and g satisfying $f \in D_{R_1}$ with $R_1 \subset (1-\epsilon)tU \cap \mathbb{Z}^d$, and $g \in D_{R_2}$ with $R_2 \subset \mathbb{Z}^d$,*

$$\left|\mathrm{cov}\left(f(\bar{\xi}_t^{\{0\}}),\, g(\bar{\xi}_t^{\{0\}})\right)\right| \leqslant |||f||| \cdot |||g|||\ \left(Ce^{-\gamma d(R_2, R_1)} + A_{r,\epsilon} t^{-r}\right). \tag{2.9}$$

3. Shrinking

As we have argued in the Introduction, we choose the mask C_t for computing the estimator $\hat{\lambda}_t^{\{0\}}$ as a shrunken version of the convex hull $\mathcal{C}(\xi_t^{\{0\}})$ that is guaranteed to lie in $(1-\epsilon)tU$ with large probability. As an example we discussed the choice $C_t = (1-\delta)\mathcal{C}(\xi_t^{\{0\}})$, about which we shall have more to say later in this section (see Example 3.2). However, we also noted that it is possible to consider more general methods of shrinking, and this is the topic of the present section.

For a set $A \subset \mathbb{R}^d$ the interior of A is denoted by $\mathring{A}$ and the discrete cardinality of A as $|A|_D = |A \cap \mathbb{Z}^d|$. Define a shrinking operation as follows.

Definition 3.1. *Suppose that to any convex set $V \subset \mathbb{R}^d$ there corresponds a convex set $V^- \subset \mathbb{R}^d$. Then the map $V \to V^-$ is called a shrinking if, for every convex V and W with $0 \in \mathring{V}$,*

$$V^- \subset V, \tag{3.1}$$

$$V \subset W \Rightarrow V^- \subset W^-, \tag{3.2}$$

$$|(tV)^-|_D \to \infty \text{ as } t \to \infty, \tag{3.3}$$

and

$$\text{if } s, t \to \infty \text{ with } t/s \to 1, \text{ then } \frac{|(tV)^-|_D}{|(sV)^-|_D} \to 1. \tag{3.4}$$

Property (3.3) guarantees that if V contains a ball centred at the origin and hence tV grows linearly in t in any direction, then the number of lattice points in $(tV)^-$ tends to infinity. By a standard argument one finds that (3.4) is equivalent to the following condition: if $0 \in \mathring{V}$, then for every $\delta > 0$ there exist $\epsilon > 0$ and $t_0 > 0$ such that

$$\left|\frac{|[(1+\epsilon)tV]^-|_D}{|[(1-\epsilon)tV]^-|_D} - 1\right| \leqslant \delta \qquad \text{for all } t \geqslant t_0. \tag{3.5}$$

We shall base the estimator of λ on a shrunken version C_t of $\mathcal{C}(\xi_t^{\{0\}})$, that is,

$$C_t = [C(\xi_t^{\{0\}})]^- \tag{3.6}$$

and

$$\hat{\lambda}_t^{\{0\}} = \hat{\lambda}_t^{\{0\}}(C_t) = \frac{n_t^{\{0\}}(C_t)}{k_t^{\{0\}}(C_t)} \tag{3.7}$$

The set defined in (3.6) is called the random mask or window. Notice that 0 is an interior point of U and hence of $C(\xi_t^{\{0\}})$ eventually a.s., so that C_t satisfies (3.1)–(3.4) eventually a.s. Since we are concerned with limit behaviour of $\xi_t^{\{0\}}$ as $t \to \infty$, this is sufficient for our purpose.

Together (3.6), (3.7) and Definition 3.1 will allow us to prove consistency of $\hat{\lambda}_t^{\{0\}}$ on the set where $\xi_t^{\{0\}}$ survives forever. However, in order to prove strong consistency of $\hat{\lambda}_t^{\{0\}}$, we need to strengthen assumption (3.3) and require that if $0 \in \mathring{V}$, then

$$\text{for some } \delta > 0, \qquad \liminf_{t\to\infty} \frac{|(tV)^-|_D}{t^\delta} > 0. \tag{3.8}$$

To prove asymptotic normality of our estimator given $\{\tau^{\{0\}} = \infty\}$ we need to assume that if $0 \in \mathring{V}$, then

$$V^- \subset (1-\delta)V, \tag{3.9}$$

while at the same time strengthening (3.3) in a different direction and requiring that

$$(tV)^- \to \mathbb{R}^d \qquad \text{as } t \to \infty. \tag{3.10}$$

We end this section by presenting various ways of shrinking that one may wish to apply to the convex hull of the set of infected sites $C(\xi_t^{\{0\}})$ in order to obtain the mask C_t.

Example 3.1 $V^- = V$. This satisfies Definition 3.1 as well as (3.8) and (3.10), but not (3.9). In this case we do not shrink but simply choose $C_t = C(\xi_t^{\{0\}})$ for computing $\hat{\lambda}_t^{\{0\}}$.

Example 3.2 $V^- = (1-\delta)V$, $0 < \delta < 1$. Obviously Definition 3.1 as well as (3.8)–(3.10) are satisfied. In determining the mask $C_t = (1-\delta)C(\xi_t^{\{0\}})$ we have to face the problem that we observe the set $\xi_t^{\{0\}}$, but not necessarily the location of the origin. As C_t is determined by shrinking $C(\xi_t^{\{0\}})$ towards the origin, we have to estimate the origin and shrink towards this estimated origin instead. An obvious estimate of the origin is the coordinatewise average of all sites in $C(\xi_t^{\{0\}})$, that is, the centre of gravity of this set of sites. In view of Theorem 2.1 and the fact that the set U is obviously symmetric with respect to the origin, it is easy to see that the estimate of the origin has error $o_P(t)$ on the set where $\xi_t^{\{0\}}$ survives forever. But this implies that shrinking $C(\xi_t^{\{0\}})$ towards the estimated rather than the true origin will not affect the consistency of $\hat{\lambda}_t^{\{0\}}$ in the conclusion of Theorem 4.1. The asymptotic normality of $\hat{\lambda}_t^{\{0\}}$ in Theorem 5.1 will not be affected either by a slightly more complicated argument.

Example 3.3 $V^- = \text{peeling}(V)$. This type of shrinking avoids the estimation of the origin of the picture. For an arbitrary convex set $V \subset \mathbb{R}^d$, the peeling procedure starts with the set $V_0 = C(V \cap \mathbb{Z}^d)$, the convex hull of the lattice points of V. Notice that, in the particular case we are considering, $V = C(\xi_t^{\{0\}})$ and hence $V_0 = V$. The peeling of V is now obtained by removing all lattice points in the L^1 contour of V_0, constructing the convex hull of the remaining lattice points of V_0, and repeating this procedure k times until a fraction α of the lattice points in V_0 has been removed. This amounts to stripping away the k outermost layers of the blob. Obviously peeling satisfies Definition 3.1 as well as (3.8)–(3.10). In view of the problems encountered in Example 3.2, we prefer peeling over multiplication by $1-\delta$ as a shrinking operation. For more details on peeling, see Fiocco (1997).

Example 3.4 $V^- = B_{\{c,r\}}$. The mask is computed by taking a Euclidean ball inside the set of infected sites with centre c and radius r, where the centre is estimated by taking the coordinatewise average of all sites in $C(\xi_t^{\{0\}})$ and the radius r is computed by averaging the L^1 distances between the estimated centre and the sites in $(C)\xi_t^{\{0\}})$.

It should be clear from these four examples that we have a great deal of freedom in choosing our mask as a shrunken version of $C(\xi_t^{\{0\}})$. In order to satisfy (3.1)–(3.4), we mainly have to watch out that we do not remove all but a bounded number of lattice points of $C\,(\xi_t^{\{0\}})$, and that for large sets the fraction α of lattice points deleted depends on the size of the set in a smooth manner. Conditions (3.8) and (3.10) are not likely to be violated for any sensible procedure either. Assumption (3.9) asserts that the shrinking is non-trivial.

Simulation of the estimator for dimension $d = 2$ indicates that for best results, the optimal fraction α of sites to be deleted by shrinking should generally be between 0.2 and 0.4, and should decrease for increasing t. For $\alpha = 0$, i.e. without shrinking, the performance of the estimator is generally disastrous. On theoretical grounds one can argue that α should be chosen proportional to t^{-1}.

4. The estimation problem: Consistency

In the proof of the consistency of $\hat{\lambda}_t^{\{0\}}$ we shall not follow the same route as we did in Section 1 to arrive at the estimator $\hat{\lambda}_t^{\{0\}}(C_t)$. Rather than introducing a new coupling to compare $\xi_t^{\{0\}}$ on $\{\tau^{\{0\}} = \infty\}$ with ξ_t^{ν}, we shall simply employ the standard graphical representation for comparison with $\xi_t^{\mathbb{Z}^d}$ instead. In Theorem 2.1 we showed that on $\{\tau^{\{0\}} = \infty\}$, $C(\xi_t^{\{0\}})$ can be bracketed between two non-random convex sets. By applying the shape theorem (Theorem 1.2) we reduce the problem to one concerning the $\xi_t^{\mathbb{Z}^d}$ process on a non-random convex set and then show that the difference between the random and the non-random masks is negligible.

Let $A_t \subset \mathbb{Z}^d$ be a finite non-random set with $|A_t| \to \infty$ as $t \to \infty$. By analogy with (1.12) and (1.13), define

$$n_t^{\mathbb{Z}^d}(A_t) = \sum_{x \in A_t} \xi_t^{\mathbb{Z}^d}(x) \tag{4.1}$$

$$k_t^{\mathbb{Z}^d}(A_t) = \sum_{x \in A_t} k_t^{\mathbb{Z}^d}(x), \qquad k_t^{\mathbb{Z}^d}(x) = (1 - \xi_t^{\mathbb{Z}^d}(x)) \sum_{|x-y|=1} \xi_t^{\mathbb{Z}^d}(y). \tag{4.2}$$

Lemma 4.1. *Suppose that for $t \geqslant 0$, the sets $A_t \subset \mathbb{Z}^d$ satisfy $A_t \subset A_{t'}$ if $t < t'$, $|A_t| < \infty$ and $|A_t| \to \infty$ for $t \to \infty$. Then, as $t \to \infty$,*

$$\frac{n_t^{\mathbb{Z}^d}(A_t)}{|A_t|} \xrightarrow{P} \mathbb{E}\xi^\nu(0), \tag{4.3}$$

$$\frac{k_t^{\mathbb{Z}^d}(A_t)}{|A_t|} \xrightarrow{P} \mathbb{E}k^\nu(0). \tag{4.4}$$

Moreover, if, for some $\delta > 0$,

$$\liminf_{t \to \infty} \frac{|A_t|}{t^\delta} > 0, \tag{4.5}$$

then, as $t \to \infty$,

$$\frac{n_t^{\mathbb{Z}^d}(A_t)}{|A_t|} \to \mathbb{E}\xi^\nu(0) \text{ a.s.}, \tag{4.6}$$

$$\frac{k_t^{\mathbb{Z}^d}(A_t)}{|A_t|} \to \mathbb{E}k^\nu(0) \text{ a.s.} \tag{4.7}$$

Proof. We shall only prove (4.3) and (4.6). The proof of (4.4) and (4.7) is almost exactly the same.

By Theorem 2.3 and the Markov inequality,

$$\mathbb{P}\left(\left| \frac{n_t^{\mathbb{Z}^d}(A_t)}{|A_t|} - \frac{\mathbb{E}n_t^{\mathbb{Z}^d}(A_t)}{|A_t|} \right| \geqslant \epsilon \right) \leqslant C_{k,\epsilon} |A_t|^{-k} \tag{4.8}$$

for every $k = 1, 2, \dots$ and appropriate $C_{k,\epsilon} > 0$. By (1.6),

$$\frac{\mathbb{E}n_t^{\mathbb{Z}^d}(A_t)}{|A_t|} = \mathbb{E}\xi_t^{\mathbb{Z}^d}(0) \to \mathbb{E}\xi^\nu(0) \tag{4.9}$$

as $t \to \infty$. Since $|A_t| \to \infty$, this proves (4.3).

For every $\epsilon > 0$ and $A \subset \mathbb{Z}^d$, we have

$$\mathbb{P}\left(\sup_{0 \leqslant s \leqslant h} \left| n_{t+s}^{\mathbb{Z}^d}(A) - n_t^{\mathbb{Z}^d}(A) \right| \geqslant \epsilon |A| \right) \leqslant \mathbb{P}(Z \geqslant \epsilon |A|), \tag{4.10}$$

where Z has a Poisson distribution with $\mathbb{E}Z = \mu = c \cdot h \cdot |A|$, where $c = 1 \vee 2d\lambda$. To see this, note that between time t and $t + h$ a change at any particular site in A occurs at rate at most c. As

$$\mathbb{E}e^Z = e^{(e-1)\mu} \leqslant e^{2\mu},$$

we find that if $h \leqslant \epsilon/(4c)$, then

$$\mathbb{P}(Z \geqslant \epsilon|A|) \leqslant e^{2\mu-\epsilon|A|} \leqslant e^{-\epsilon|A|/2}. \tag{4.11}$$

Take $t_0 = 0$ and define $t_0 < t_1 < t_2 < \ldots$ recursively by

$$t_{m+1} = (t_m + \epsilon/(4c)) \wedge \inf\{t > t_m : A_{t-} \neq A_{t+}\},$$

where

$$A_{t-} = \lim_{s \uparrow t} A_s = \bigcup_{s<t} A_s, \qquad A_{t+} = \lim_{s \downarrow t} A_s = \bigcap_{s>t} A_s.$$

Hence t_{m+1} is obtained by adding to t_m until one either arrives at $t_m + \epsilon/(4c)$ or encounters a change in A_t. Because A_t is non-decreasing, this implies that by passing from t_m to t_{m+1}, one either increases t by $\epsilon/(4c)$ or $|A_t|$ by at least 1. It follows that $t_m \to \infty$ as $m \to \infty$. To see this, note that either $t_m \to \infty$ or $|A_{t_m}| \to \infty$. Since $|A_t| < \infty$ for all t, we must have $t_m \to \infty$ in both cases. Obviously there exists $0 \leqslant k \leqslant m-1$ such that $t_m \geqslant k\epsilon/(4c)$ and $|A_{t_{m+}}| \geqslant |A_{t_m}| \geqslant |A_{t_m-}| \geqslant m-k-1$. By (4.5) this implies that

$$\liminf_m \frac{|A_{t_m-}|}{m^{\delta'}} > 0$$

for $\delta' = \delta \wedge 1$. It follows from (4.8) that, for every $k = 1, 2 \ldots,$

$$\mathbb{P}\left(\left|\frac{n_{t_m}^{\mathbb{Z}^d}(A_{t_m})}{|A_{t_m}|} - \mathbb{E}\xi_{t_m}^{\mathbb{Z}^d}(0)\right| \geqslant \epsilon\right) \leqslant C'_{k,\epsilon} m^{-\delta' k}, \tag{4.12}$$

and the same is true with A_{t_m} replaced by A_{t_m-} or A_{t_m+}.

As $t_{m+1} - t_m \leqslant \epsilon/(4c)$ and $A_t = A_{t_m+}$ for $t_m < t < t_{m+1}$, (4.10) and (4.11) yield

$$\begin{aligned}\mathbb{P}\left(\sup_{t_m<t<t_{m+1}} \left|n_t^{\mathbb{Z}^d}(A_t) - n_{t_m}^{\mathbb{Z}^d}(A_{t_m+})\right| \geqslant \epsilon|A_{t_m+}|\right) &\leqslant e^{-\epsilon|A_{t_m+}|/2} \\ &\leqslant e^{-\epsilon C m^{\delta'}/2},\end{aligned} \tag{4.13}$$

for some $C > 0$ and $m > m_0$. By (4.12) with $k > 1/\delta'$, (4.13) and the Borel–Cantelli lemma we find

$$\frac{n_t^{\mathbb{Z}^d}(A_t)}{|A_t|} - \mathbb{E}\xi_t^{\mathbb{Z}^d}(0) \to 0 \text{ a.s.},$$

and, together with (4.9), this proves (4.6). □

Lemma 4.1 allows us to prove both the consistency and the strong consistency of $\hat{\lambda}_t^{\{0\}}$ as $t \to \infty$.

Theorem 4.1. *Let $\hat{\lambda}_t^{\{0\}}(C_t)$ be the estimator of λ for the process $\xi_t^{\{0\}}$ defined in (3.6)–(3.7) and Definition 3.1. Then on the set where $\xi_t^{\{0\}}$ survives forever,*

$$\hat{\lambda}_t^{\{0\}}(C_t) \xrightarrow{P} \lambda \quad \textit{as } t \to \infty. \tag{4.14}$$

If, in addition, (3.8) *holds, then*

$$\hat{\lambda}_t^{\{0\}}(C_t) \to \lambda \quad \textit{as } t \to \infty \tag{4.15}$$

a.s. on the set where $\xi_t^{\{0\}}$ *survives forever.*

Proof. Choose $\delta > 0$ and $\epsilon > 0$ such that (3.5) is satisfied for some $t_0 > 0$. Define $A_t = [(1-\epsilon)tU]^- \cap \mathbb{Z}^d$ and $B_t = [(1+\epsilon)tU]^- \cap \mathbb{Z}^d$. By Theorem 2.1, (3.2) and (3.6), $A_t \subset C_t \cap \mathbb{Z}^d \subset B_t$ eventually a.s. on $\{\tau^{\{0\}} = \infty\}$, and then (3.5) ensures that

$$n_t^{\{0\}}(A_t) \leqslant n_t^{\{0\}}(C_t) \leqslant \{n_t^{\{0\}}(A_t) + |B_t \backslash A_t|\} \leqslant n_t^{\{0\}}(A_t) + \delta|A_t|.$$

Again by (3.5), it follows that

$$\begin{aligned}(1+\delta)^{-1} \liminf_{t\to\infty} n_t^{\{0\}}(A_t)/|A_t| &\leqslant \liminf_{t\to\infty} n_t^{\{0\}}(C_t)/|C_t| \\ &\leqslant \limsup_{t\to\infty} n_t^{\{0\}}(C_t)/|C_t| \leqslant \limsup_{t\to\infty} n_t^{\{0\}}(A_t)/|A_t| + \delta.\end{aligned}$$

By (3.1)–(3.3), A_t satisfies the assumptions for (4.3) to hold and as $A_t \subset (1-\epsilon)tU$, Theorem 1.2 implies that $n_t^{\{0\}}(A_t) = n_t^{\mathbb{Z}^d}(A_t)$ eventually a.s. on $\{\tau^{\{0\}} = \infty\}$. Letting $t \to \infty$ and then $\delta \to 0$, we find that $n_t^{\{0\}}(C_t)/|C_t| \xrightarrow{P} \mathbb{E}\xi^\nu(0)$ on $\{\tau^{\{0\}} = \infty\}$. In exactly the same way one may use (4.4) to prove that $k_t^{\{0\}}(C_t)/|C_t| \xrightarrow{P} \mathbb{E}k_t^\nu(0)$ on $\{\tau^{\{0\}} \to \infty\}$, and (4.14) follows by combining these results and using (1.15). By using (4.6) and (4.7) instead of (4.3) and (4.4), one establishes (4.15) under the additional condition (3.8). □

Remark 4.1. By (4.14),

$$\mathbb{P}\{|\hat{\lambda}_t^{\{0\}} - \lambda| \geqslant \epsilon | \tau^{\{0\}} = \infty\} \to 0 \qquad \text{as } t \to \infty, \tag{4.16}$$

for every $\epsilon > 0$. From a statistical point of view this appears unsatisfactory since we shall never know whether the process will survive forever and hence whether $\hat{\lambda}_t^{\{0\}}$ will be close to λ even for very large t. However, for the supercritical contact process (4.16) is obviously equivalent to

$$\mathbb{P}\{|\hat{\lambda}_t^{\{0\}} - \lambda| \geqslant \epsilon | \xi_t^{\{0\}} \neq \emptyset\} \to 0, \tag{4.17}$$

for every $\epsilon > 0$, and this statement does have statistical relevance. Of course our result does not provide any information in the subcritical case ($\lambda \leqslant \lambda_d$).

5. The estimation problem: Asymptotic normality

This section is devoted to the proof of a conditional central limit theorem for the estimator $\hat{\lambda}_t^{\{0\}} = \hat{\lambda}_t^{\{0\}}(C_t)$ based on the random mask C_t. First, we establish the joint asymptotic normality of

$$|A_t|^{-1/2}\Big(n_t^{\mathbb{Z}^d}(A_t) - |A_t|\mathbb{E}\xi^\nu(0),\ k_t^{\mathbb{Z}^d}(A_t) - |A_t|\mathbb{E}k^\nu(0)\Big)$$

for a non-random mask $A_t \subset \mathbb{Z}^d$, with $|A_t| < \infty$ for all $t \geq 0$ but $|A_t| \to \infty$ as $t \to \infty$. Next we show that this result carries over to the $\bar{\xi}_t^{\{0\}}$ process, that is, the $\xi_t^{\{0\}}$ process conditioned on $\{\tau^{\{0\}} = \infty\}$. This proves the asymptotic normality of the estimator $\hat{\lambda}_t^{\{0\}}(A_t)$ given $\{\tau^{\{0\}} = \infty\}$ for a non-random mask A_t. Then we show that the contribution to the standardized estimator which is due to the randomness of the mask $C_t = [\mathcal{C}(\xi_t^{\{0\}})]^-$ vanishes as $t \to \infty$. The asymptotic normality of

$$|C_t|^{-1/2}(\hat{\lambda}_t^{\{0\}}(C_t) - \lambda)$$

given $\{\tau^{\{0\}} = \infty\}$ then follows.

A very general central limit theorem for a translation-invariant random field was proved by Bolthausen (1982) under mixing conditions. Let $\zeta(x)$, $x \in \mathbb{Z}^d$, denote a real-valued translation-invariant random field, that is, $\{\zeta(x) : x \in \mathbb{Z}^d\}$ is a collection of random variables and the joint law of the $\zeta(x)$ is invariant under integer-valued shifts in $\mathbb{Z}^d$. It is assumed that $\mathbb{E}\zeta^2(x) < \infty$. For $x = (x_1, \ldots, x_d)$, $y = (y_1, \ldots, y_d) \in \mathbb{Z}^d$, define the L^∞ distance of x and y as

$$\rho(x, y) = \max_{1 \leq i \leq d} |x_i - y_i|.$$

Let $A_n \subset \mathbb{Z}^d$, $n = 1, 2, \ldots$, with $|A_n| < \infty$ for all n, $|A_n| \to \infty$ as $n \to \infty$ and

$$\frac{|\partial A_n|}{|A_n|} \to 0 \quad \text{as } n \to \infty. \tag{5.1}$$

Here

$$\partial A_n = \{x \in A_n : \exists\ y \in \mathbb{Z}^d \backslash A_n \text{ with } \rho(x, y) = 1\} \tag{5.2}$$

denotes the L^∞ contour of A_n in $\mathbb{Z}^d$. Consider

$$S_n = \sum_{x \in A_n} (\zeta(x) - \mathbb{E}\zeta(0)).$$

If $C \subset \mathbb{Z}^d$, let $\mathcal{B}_C$ be the σ-algebra generated by $\{\zeta(x), x \in C\}$. For $C_1, C_2 \subset \mathbb{Z}^d$, let

$$\rho(C_1, C_2) = \inf\{\rho(x, y) : x \in C_1, y \in C_2\}.$$

For $m \in \mathbb{N}$, $k, l \in \mathbb{N} \cup \{\infty\}$, define the mixing coefficients

$$\alpha_{k,l}(m) = \sup\{|\mathbb{P}(B_1 \cap B_2) - \mathbb{P}(B_1)\mathbb{P}(B_2)| : B_i \in \mathcal{B}_{C_i}, |C_1| \leq k, \\ |C_2| \leq l, \rho(C_1, C_2) \geq m\}. \tag{5.3}$$

Let $N(\mu, \sigma^2)$ denote the univariate normal distribution with expectation μ and variance σ^2 and $N(\mu, \Sigma)$ the bivariate normal distribution with expectation vector μ and covariance matrix Σ. Part of Bolthausen's theorem reads as follows.

Lemma 5.1. *Suppose that, as $m \to \infty$,*

$$\sum_{m=1}^{\infty} m^{d-1}\alpha_{k,l}(m) < \infty, \qquad \text{for } k+l \leqslant 4, \tag{5.4}$$

$$\alpha_{1,\infty}(m) = o(m^{-d}), \tag{5.5}$$

and that, for some $\delta > 0$,

$$\mathbb{E}|\zeta(x)|^{2+\delta} < \infty \quad \text{and} \quad \sum_{m=1}^{\infty} m^{d-1}\alpha_{1,1}(m)^{\delta/(2+\delta)} < \infty. \tag{5.6}$$

Then $\sum_{x\in\mathbb{Z}^d}|\text{cov}(\zeta(0), \zeta(x))| < \infty$. *If, in addition,* $\sigma^2 = \sum_{x\in\mathbb{Z}^d}\text{cov}(\zeta(0), \zeta(x)) > 0$ *and* (5.1) *holds, then* $|A_n|^{-1/2}S_n/\sigma$ *converges in distribution to* $N(0, 1)$.

For our purposes we have to modify this result slightly. First of all, we allow a different stationary random field $\zeta_n(x)$ for each n, so that S_n becomes

$$\tilde{S}_n = \sum_{x\in A_n}(\zeta_n(x) - \mathbb{E}\zeta_n(0)).$$

As a result, we also have to replace the assumptions of the lemma by versions which are uniform in n. This means that in the assumptions of the lemma we replace $\alpha_{k,l}(m)$ by the supremum over n of expression (5.3) for $\zeta_n(x)$. Similarly, the integrability of $|\zeta_n(x)|^{2+\delta}$ in (5.6) is replaced by the uniform integrability of $|\zeta_n(x)|^{2+\delta}$. Then Bolthausen's proof goes through to show that $\sup_n\sum_{x\in\mathbb{Z}^d}|\text{cov}(\zeta_n(0), \zeta_n(x))| < \infty$ and that $|A_n|^{-1/2}\tilde{S}_n/\sigma_n \xrightarrow{\mathcal{D}} N(0, 1)$, provided that $\liminf \sigma_n^2 > 0$, where $\sigma_n^2 = \sum_{x\in\mathbb{Z}^d}\text{cov}(\zeta_n(0), \zeta_n(x))$.

A second modification of Lemma 5.1 concerns assumption (5.5). It is clear from Bolthausen's proof that (5.5) may be replaced by

$$\alpha_{1,l}(l^{1/(2d+1)}) = o(l^{-1/2}) \qquad \text{as } l \to \infty. \tag{5.7}$$

With these modifications, Lemma 5.1 allows us to prove:

Lemma 5.2. *Choose* $\epsilon \in (0, 1)$ *and* $A_t \subset \mathbb{Z}^d$ *for* $t \geqslant 0$ *such that*

$$A_t \subset (1-\epsilon)tU, \quad |A_t| \to \infty, \quad \text{and} \quad |\partial A_t|/|A_t| \to 0 \quad \text{as } t \to \infty. \tag{5.8}$$

As $t \to \infty$, *the conditional distribution of the random vector*

$$|A_t|^{-1/2}\left(\sum_{x\in A_t}(\xi_t^{\{0\}}(x) - \mathbb{E}\xi^\nu(0)), \sum_{x\in A_t}(k_t^{\{0\}}(x) - \mathbb{E}k^\nu(0))\right) \tag{5.9}$$

given $\{\tau^{\{0\}} = \infty\}$ *converges weakly to* $N(0, \Sigma)$, *where*

$$\Sigma = \begin{pmatrix} \sigma_1^2 & \sigma_{1,2} \\ \sigma_{1,2} & \sigma_2^2 \end{pmatrix} \tag{5.10}$$

and

$$\sigma_1^2 = \sum_{x\in\mathbb{Z}^d} \operatorname{cov}(\xi^\nu(0), \xi^\nu(x)), \qquad \sigma_2^2 = \sum_{x\in\mathbb{Z}^d} \operatorname{cov}(k^\nu(0), k^\nu(x)), \qquad \sigma_{1,2} = \sum_{x\in\mathbb{Z}^d} \operatorname{cov}(k^\nu(0), \xi^\nu(x)). \tag{5.11}$$

Proof. The lemma concerns the process $\bar{\xi}_t^{\{0\}}$ which is distributed as $\xi_t^{\{0\}}$ conditioned on $\{\tau^{\{0\}} = \infty\}$, restricted to the set $(1-\epsilon)tU$. By Theorems 1.2 and 2.4 we may first replace this process by the conditional process $\bar{\xi}_t^{\mathbb{Z}^d}$ and then by the unconditional process $\xi_t^{\mathbb{Z}^d}$. Similarly, we may replace $\mathbb{E}\xi^\nu(0)$ by $\mathbb{E}\xi_t^{\mathbb{Z}^d}(0)$ since $|A_t|^{1/2}|\mathbb{E}\xi_t^{\mathbb{Z}^d}(0) - \mathbb{E}\xi^\nu(0)| = \mathcal{O}(t^{d/2}\mathrm{e}^{-\gamma t}) \to 0$ by (1.6). The same holds for $\mathbb{E}k^\nu(0)$ and $\mathbb{E}k_t^{\mathbb{Z}^d}(0)$. Hence, it suffices to prove that

$$|A_t|^{-1/2}\left(\sum_{x\in At}\left(\xi_t^{\mathbb{Z}^d}(x) - \mathbb{E}\xi_t^{\mathbb{Z}^d}(0)\right), \sum_{x\in At}\left(k_t^{\mathbb{Z}^d}(x) - \mathbb{E}k_t^{\mathbb{Z}^d}(0)\right)\right).$$

is asymptotically $N(0, \Sigma)$.

Let u and v be real numbers and define

$$\zeta_t(x) = u\xi_t^{\mathbb{Z}^d}(x) + vk_t^{\mathbb{Z}^d}(x).$$

Clearly $\{\zeta_t(x), x \in \mathbb{Z}^d\}$ is a real-valued, translation-invariant random field for each t. Consider

$$\tilde{S}_t = \sum_{x\in A_t}(\zeta_t(x) - \mathbb{E}\zeta_t(0)).$$

The fact that $\tilde{S}_t$ depends on a real-valued index $t \to \infty$, instead of an integer $n \to \infty$ as in our version of Bolthausen's result, is of course immaterial in what follows. Note that $|\zeta_t(x) \leqslant |u| + 4|v|$ so that all moments of $|\zeta_t(x)|$ are bounded independent of t.

Let us write $\alpha_{klt}(m)$ for the quantity defined in (5.3) computed for ζ_t. By Theorem 2.2 and because $\rho(x, y) \leqslant d(x, y) = \sum_{i=1}^d |x_i - y_i|$, there exist positive C and γ such that

$$\alpha_{klt}(m) \leqslant Ckl\mathrm{e}^{-\gamma m},$$

independent of t. This means that assumptions (5.4), (5.6) and (5.7) are satisfied uniformly in t. Note that (5.5) is not satisfied since we cannot allow $l = \infty$, but, as we have indicated, (5.7) serves just as well. Hence, we have proved that

$$|A_t|^{-1/2}\sigma_t^{-1}\sum_{x\in A_t}\left(u\left(\xi_t^{\mathbb{Z}^d}(x) - \mathbb{E}\xi^{\mathbb{Z}^d}(0)\right) + v\left(k_t^{\mathbb{Z}^d}(x) - \mathbb{E}k_t^{\mathbb{Z}^d}(0)\right)\right) \tag{5.12}$$

has a standard normal limit distribution provided that $\liminf \sigma_t^2 > 0$. Here

$$\sigma_t^2 = \sum_{x\in\mathbb{Z}^d} \operatorname{cov}\left(u\xi_t^{\mathbb{Z}^d}(0) + vk_t^{\mathbb{Z}^d}(0), u\xi_t^{\mathbb{Z}^d}(x) + vk_t^{\mathbb{Z}^d}(x)\right). \tag{5.13}$$

By (1.6) the terms in (5.13) converge to $\operatorname{cov}(u\xi^\nu(0) + vk_t^\nu(0), u\xi^\nu(x) + vk_t^\nu(x))$ as $t \to \infty$, and by Theorem 2.2 the terms are bounded by $C'\exp\{-\gamma\sum_{1\leqslant i\leqslant d}|x_i|\}$, independent of t. It follows that the sum also converges, so σ_t^2 tends to

$$\sigma^2(u, v) = \sum_{x\in\mathbb{Z}^d} \operatorname{cov}\big(u\xi^\nu(0) + vk_t^\nu(0),\, u\xi^\nu(x) + vk_t^\nu(x)\big). \tag{5.14}$$

Hence $\tilde{S}_t$ is asymptotically $N(0, \sigma^2(u, v))$ if $\sigma^2(u, v) > 0$ and asymptotically degenerate at 0 if $\sigma^2(u, v) = 0$. The lemma is proved by the Cramér–Wold device. □

To prove the joint asymptotic normality of $n_t^{\{0\}}(C_t)$ and $k_t^{\{0\}}(C_t)$ – and hence of $\hat{\lambda}_t^{\{0\}}(C_t)$ – conditional on $\{\tau^{\{0\}} = \infty\}$, we have to consider the difference between these quantities computed for the random mask C_t and a non-random mask which is close to C_t. For $\epsilon > 0$ and $t > 0$, define

$$A_t = [(1-\epsilon)tU]^- \cap \mathbb{Z}^d, \qquad B_t = [(1+\epsilon)tU]^- \cap \mathbb{Z}^d, \tag{5.15}$$

that is, A_t and B_t consist of the sites in the shrunken versions of the sets $(1-\epsilon)tU$ and $(1+\epsilon)tU$ respectively, where the shrinking operation $V \to V^-$ is defined in Definition 3.1.

Lemma 5.3. *For $\epsilon \in (0, 1)$ define A_t and B_t as in (5.15) and let $D_t = (B_t \backslash A_t) \cap C_t$, with $C_t = [C(\xi_t^{\{0\}})]^-$ as given by (3.6) and Definition 3.1. If the shrinking operation $V \to V^-$ satisfies (3.9) for some $\delta \in (0, 1)$, then, for every $z > 0$,*

$$\lim_{\epsilon\to 0} \limsup_{t\to\infty} \mathbb{P}\left(|A_t|^{-1/2} \left| \sum_{x\in D_t} (\xi_t^{\{0\}}(x) - \mathbb{E}\xi^\nu(0)) \right| \geqslant z \,|\, \tau^{\{0\}} = \infty \right) = 0, \tag{5.16}$$

$$\lim_{\epsilon\to 0} \limsup_{t\to\infty} \mathbb{P}\left(|A_t|^{-1/2} \left| \sum_{x\in D_t} (k_t^{\{0\}}(x) - \mathbb{E}k^\nu(0)) \right| \geqslant z \,|\, \tau^{\{0\}} = \infty \right) = 0. \tag{5.17}$$

Proof. We shall only prove (5.16) as the proof of (5.17) is almost the same. As before, we write $\bar{\xi}_t^{\{0\}}$ for the conditional process $(\xi_t^{\{0\}} | \tau^{\{0\}} = \infty\}$; $\overline{\mathbb{P}}$ will denote the conditional probability $\mathbb{P}(\cdot | \tau^{\{0\}} = \infty)$.

Without loss of generality we assume that $\epsilon \leqslant \delta/4$ so that $(1-\delta)(1+\epsilon) \leqslant 1 - 3\delta/4$ and, by (3.9),

$$B_t = [(1+\epsilon)tU]^- \cap \mathbb{Z}^d \subset (1-\delta)(1+\epsilon)tU \subset \left(1 - \frac{3\delta}{4}\right)tU. \tag{5.18}$$

As $|D_t| \leqslant |B_t| = \mathcal{O}(t^d)$ and $|A_t| \to \infty$, we note that in (5.16) we may replace $\mathbb{E}\xi^\nu(0)$ first by $\mathbb{E}\xi_t^{\mathbb{Z}^d}(0)$ because of (1.6) and then by $\mathbb{E}\bar{\xi}_t^{\mathbb{Z}^d}(0)$ because of Theorem 2.4, and finally by $\mathbb{E}\bar{\xi}_t^{\{0\}}(0)$ in view of Theorem 2.1. Hence, in order to prove (5.23), it is enough to show that

$$\lim_{\epsilon\to 0} \limsup_{t\to\infty} \mathbb{P}\left(|A_t|^{-1/2} \left| \sum_{x\in D_t} \left(\bar{\xi}_t^{\{0\}}(x) - \mathbb{E}\bar{\xi}_t^{\{0\}}(x)\right) \right| \geqslant z \right) = 0. \tag{5.19}$$

Define

$$C_t^* = \left[C(\{\bar{\xi}_t^{\{0\}} \cup (1-\epsilon)tU\} \cap (1+\epsilon)tU \right]^-. \tag{5.20}$$

By (2.1), $(1-\epsilon)tU \subset C(\bar{\xi}_t^{\{0\}}) \subset (1+\epsilon)tU$, and hence

$$\mathcal{C}\Big(\{\bar{\xi}_t^{\{0\}} \cup (1-\epsilon)tU\} \cap (1+\epsilon)tU\Big) = \mathcal{C}\Big(\bar{\xi}_t^{\{0\}} \cup (1-\epsilon)tU\Big)$$

$$= \mathcal{C}\Big(\mathcal{C}(\bar{\xi}_t^{\{0\}} \cup (1-\epsilon)tU\Big) = \mathcal{C}\Big(\bar{\xi}_t^{\{0\}}\Big)$$

eventually a.s. ($\overline{\mathbb{P}}$). It follows that

$$C_t^* = [\mathcal{C}(\bar{\xi}_t^{\{0\}})]^- = C_t \tag{5.21}$$

eventually a.s. ($\overline{\mathbb{P}}$). Obviously this implies that

$$\sum_{x \in D_t} \Big(\bar{\xi}_t^{\{0\}}(x) - \mathbb{E}\bar{\xi}_t^{\{0\}}(x)\Big) = \sum_{x \in B_t \setminus A_t} \Big(\bar{\xi}_t^{\{0\}}(x) - \mathbb{E}\bar{\xi}_t^{\{0\}}(x)\Big) I_{C_t}(x)$$

$$= \sum_{x \in B_t \setminus A_t} \Big(\bar{\xi}_t^{\{0\}}(x) - \mathbb{E}\bar{\xi}_t^{\{0\}}(x)\Big) I_{C_t^*}(x)$$

eventually a.s. ($\overline{\mathbb{P}}$). Instead of (5.19), it is therefore sufficient to show that

$$\lim_{\epsilon \to 0} \limsup_{t \to \infty} \mathbb{P}\left(|A_t|^{-1/2} \left| \sum_{x \in B_t \setminus A_t} (\bar{\xi}_t^{\{0\}}(x) - \mathbb{E}\bar{\xi}_t^{\{0\}}(x)) I_{C_t^*}(x) \right| \geqslant z \right) = 0.$$

Clearly this will follow if we prove that

$$\lim_{\epsilon \to 0} \limsup_{t \to \infty} |A_t|^{-1} \mathbb{E}\left[\sum_{x \in B_t \setminus A_t} (\bar{\xi}_t^{\{0\}}(x) - \mathbb{E}\bar{\xi}_t^{\{0\}}(x)) I_{C_t^*}(x) \right]^2 = 0. \tag{5.22}$$

By (5.20) the random set C_t^* is determined by the random set $\{\bar{\xi}_t^{\{0\}} \cup (1-\epsilon)tU\} \cap (1+\epsilon)tU$ which is bracketed by the non-random convex sets $(1-\epsilon)tU$ and $(1+\epsilon)tU$. It follows that C_t^* is determined by the values of $\bar{\xi}_t^{\{0\}}(y)$ for sites $y \in (1+\epsilon)tU \setminus (1-\epsilon)tU$. Put differently, for every $x \in \mathbb{Z}^d$, the function $g_x : H \to \{0, 1\}$ defined by

$$g_x(\bar{\xi}_t^{\{0\}}) = I_{C_t^*}(x) \tag{5.23}$$

satisfies

$$g_x \in D_R, \qquad \text{with } R = \{(1+\epsilon)tU \setminus (1-\epsilon)tU\} \cap \mathbb{Z}^d \tag{5.24}$$

and D_R defined by (2.3).

The expected value in (5.22) can be written as

$$\mathbb{E}\left[\sum_{x\in B_t\backslash A_t}\left(\bar{\xi}_t^{\{0\}}(x)-\mathbb{E}\bar{\xi}_t^{\{0\}}(x)\right)I_{C_t^*}(x)\right]^2$$

$$= \sum_{x,x'\in B_t\backslash A_t}\mathbb{E}\left(\bar{\xi}_t^{\{0\}}(x)-\mathbb{E}\bar{\xi}_t^{\{0\}}(x)\right)\left(\bar{\xi}_t^{\{0\}}(x')-\mathbb{E}\bar{\xi}_t^{\{0\}}(x')\right)I_{C_t^*}(x)I_{C_t^*}(x') \tag{5.25}$$

$$= \sum_{x,x'\in B_t\backslash A_t}\mathbb{E}f_x(\bar{\xi}_t^{\{0\}})f_{x'}(\bar{\xi}_t^{\{0\}})g_x(\bar{\xi}_t^{\{0\}})g_{x'}(\bar{\xi}_t^{\{0\}}),$$

with $f_x(\bar{\xi}_t^{\{0\}})=\bar{\xi}_t^{\{0\}}(x)-\mathbb{E}\bar{\xi}_t^{\{0\}}(x)$ and g_x defined by (5.23). Obviously

$$f_x\cdot f_{x'}\in D_{\{x,x'\}},\qquad g_x\cdot g_{x'}\in D_R, \tag{5.26}$$

in view of (5.24). If $x, x'\in B_t\backslash A_t$, then (5.18) ensures that $\{x,x'\}\subset(1-3\delta/4)tU$ and, because $\epsilon\leqslant\delta/4$, (5.24) implies that $R\subset\{(1-\delta/4)tU\}^c$. Hence, if $d(\cdot,\cdot)$ denotes L^1 distance, then

$$d(\{x,x'\},R)\geqslant b_\delta'''t\qquad \textit{for all } x,x'\in B_t\backslash A_t, \tag{5.27}$$

where b_δ''' is a positive number depending only on δ. Finally, we use (2.2) to compute

$$|||f_x\cdot f_x'|||=2,\qquad |||g_x\cdot g_x'|||\leqslant|R|\leqslant a\epsilon t^d\leqslant a\delta t^d, \tag{5.28}$$

for an appropriate constant $a>0$. Combining (5.25)–(5.28) and invoking Theorem 2.5 with $r=3d$, we obtain

$$|A_t|^{-1}\mathbb{E}\left[\sum_{x\in B_t\backslash A_t}(\bar{\xi}_t^{\{0\}}(x)-\mathbb{E}\bar{\xi}_t^{\{0\}}(x))I_{C_t^*}(x)\right]^2$$

$$\leqslant |A_t|^{-1}\sum_{x,x'\in B_t\backslash A_t}\mathbb{E}f_x(\bar{\xi}_t^{\{0\}})f_{x'}(\bar{\xi}_t^{\{0\}})\mathbb{E}g_x(\bar{\xi}_t^{\{0\}})g_{x'}(\bar{\xi}_t^{\{0\}})+M_t$$

$$\leqslant |A_t|^{-1}\sum_{x,x'\in B_t\backslash A_t}|\mathrm{cov}(\bar{\xi}_t^{\{0\}}(x),\bar{\xi}_t^{\{0\}}(x'))|+M_t,$$

where the remainder term M_t satisfies, for appropriate positive c_δ and c_δ',

$$|M_t|\leqslant|A_t|^{-1}|B_t\backslash A_t|^2c_\delta|||f_x\cdot f_x'|||\cdot|||g_x\cdot g_x'|||t^{-3d}$$
$$\leqslant c_\delta'|A_t|^{-1}\to 0\qquad \text{as } t\to\infty,$$

since $|B_t\backslash A_t|\leqslant|B_t|\leqslant|(1+\epsilon)tU|_D\leqslant|(1+\delta/4)tU|_D=\mathcal{O}(t^d)$ and $|A_t|\to\infty$ by (3.3).

To prove (5.22), it therefore remains to be shown that

$$\lim_{\epsilon\to 0}\limsup_{t\to\infty}|A_t|^{-1}\sum_{x,x'\in B_t\backslash A_t}\left|\mathrm{cov}\left(\bar{\xi}_t^{\{0\}}(x),\bar{\xi}_t^{\{0\}}(x')\right)\right|=0. \tag{5.29}$$

Invoking Theorem 2.5 once more, this time with $r=d+1$, we find that, for $x, x'\in B_t\backslash A_t$, $x\neq x'$, and appropriate $c_\delta''>0$,

$$\left|\mathrm{cov}\left(\bar{\xi}_t^{\{0\}}(x), \bar{\xi}_t^{\{0\}}(x')\right)\right| \leqslant c_\delta''|x-x'|^{-(d+1)},$$

since $x, x' \in B_t \backslash A_t$ implies $|x-x'| = \mathcal{O}(t)$. It follows that

$$\sum_{x,x' \in B_t\backslash A_t} \left|\mathrm{cov}\left(\bar{\xi}_t^{\{0\}}(x), \bar{\xi}_t^{\{0\}}(x')\right)\right| \leqslant |B_t\backslash A_t| \left(1 + c_\delta'' \sum_{x\in\mathbb{Z}^d, x\neq 0} |x|^{-(d+1)}\right)$$
$$\leqslant c_\delta'''|B_t\backslash A_t|$$

for some $c_\delta''' > 0$, as $\sum_{x\in\mathbb{Z}^d\backslash\{0\}} |x|^{-(d+1)}$ converges. Hence, (5.29) holds if

$$\lim_{\epsilon\to 0} \limsup_{t\to\infty} \frac{|B_t\backslash A_t|}{|A_t|} = 0.$$

But since $A_t = [(1-\epsilon)tU]^- \cap \mathbb{Z}^d$ and $B_t = [(1+\epsilon)tU]^- \cap \mathbb{Z}^d$, this is a consequence of (3.5). This proves (5.29) and the lemma. □

We are now in a position to prove the main result of this paper.

Theorem 5.1. *Let $\hat{\lambda}_t^{\{0\}}(C_t)$ be the estimator of λ for the process $\xi_t^{\{0\}}$ defined in* (3.6)–(3.7) *and Definition 3.1. If the shrinking operation $V \to V^-$ satisfies* (3.9) *for some $\delta \in (0, 1)$ as well as* (3.10), *then, as $t \to \infty$, the conditional distribution of*

$$|C_t|_D^{1/2}[\hat{\lambda}_t^{\{0\}}(C_t) - \lambda], \tag{5.30}$$

given that $\{\tau^{\{0\}} = \infty\}$, converges weakly to $N(0, \sigma^2)$. Here

$$\sigma^2 = \lambda^2 \left[\frac{\sigma_1^2}{\{\mathbb{E}\xi^\nu(0)\}^2} + \frac{\sigma_2^2}{\{\mathbb{E}k^\nu(0)\}^2} - \frac{2\sigma_{1,2}}{\{\mathbb{E}\xi^\nu(0)\mathbb{E}k^\nu(0)\}}\right], \tag{5.31}$$

where σ_1, σ_2 and $\sigma_{1,2}$ are given by (5.11).

Proof. In the proof we write $\bar{\xi}_t^{\{0\}}$ for the conditional process $(\xi_t^{\{0\}}|\tau^{\{0\}} = \infty\}$. For $t \geqslant 0$, define A_t and B_t by (5.15). Since $0 \in \mathring{U}$, we have $A_t \subset (1-\epsilon)tU$ by (3.1) and $[(1-\epsilon)tU]^- \to \mathbb{R}^d$ by (3.10). Because $[(1-\epsilon)tU]^-$ is bounded and convex, it follows that $|\partial A_t|/|A_t| \to 0$ as $t \to \infty$ by an easy argument. Hence A_t satisfies condition (5.8) of Lemma 5.2 and we find that, for every $\epsilon \in (0, 1)$, the random vector

$$\left(|A_t|^{-1/2}\left[\sum_{x\in A_t}(\bar{\xi}_t^{\{0\}}(x) - \mathbb{E}\xi^\nu(0))\right], |A_t|^{-1/2}\left[\sum_{x\in A_t}(\bar{k}_t^{\{0\}}(x) - \mathbb{E}k^\nu(0))\right]\right), \tag{5.32}$$

has a limiting $N(0, \Sigma)$ distribution with Σ given by (5.10)–(5.11).

In view of (3.9), we may apply Lemma 5.3 to obtain

$$\Psi_1(\epsilon) = \limsup_{t\to\infty} \mathbb{P}\left(|A_t|^{-1/2} \left| \sum_{x\in D_t} (\xi_t^{\{0\}}(x) - \mathbb{E}\xi^\nu(0)) \right| \geqslant z \right) \to 0,$$

$$\Psi_1(\epsilon) = \limsup_{t\to\infty} \mathbb{P}\left(|A_t|^{-1/2} \left| \sum_{x\in D_t} (\bar{k}_t^{\{0\}}(x) - \mathbb{E}k^\nu(0)) \right| \geqslant z \right) \to 0, \tag{5.33}$$

as $\epsilon \to 0$ for every $z > 0$. Here $D_t = (B_t \backslash A_t) \cap C_t$. Notice that by (2.1) and (3.2) we have $A_t \subset C_t \cap \mathbb{Z}^d \subset B_t$ and hence

$$C_t \cap \mathbb{Z}^d = A_t \cup D_t, \qquad A_t \cap D_t = \varnothing, \tag{5.34}$$

eventually a.s. on $\{\tau^{\{0\}} = \infty\}$.

Next we note that $|B_t|/|A_t| \to 1$ by (3.5) and hence $|C_t|/|A_t| \to 1$ eventually a.s. on $\{\tau^{\{0\}} = \infty\}$ as $t \to \infty$. It now follows by a standard argument that the limit distribution of (5.32) will remain unchanged if A_t is replaced by C_t and $|A_t|$ by $|C_t|$. Finally, (3.7) and another standard argument establish the theorem. □

6. The asymptotic variance of $\hat{\lambda}_t^{\{0\}}(C_t)$

If the variance σ^2 of the normal limit distribution in Theorem 5.1 were known, then this would allow us to assess the accuracy of the estimator or to set up asymptotic confidence intervals for λ of the form

$$\hat{\lambda}_t^{\{0\}}(C_t) - u_{\alpha/2}|C_t|_D^{-1/2}\sigma < \lambda < \hat{\lambda}_t^{\{0\}}(C_t) + u_{\alpha/2}|C_t|_D^{-1/2}\sigma, \tag{6.1}$$

where u_α is the upper α-point of the standard normal distribution. This asymptotic confidence interval would be valid provided that $\xi_t^{\{0\}}$ survives forever, but, as we pointed out in Remark 4.1, it is enough that $\xi_t^{\{0\}} \neq \varnothing$, that is, that the process has survived up to time t.

Since σ^2 is unknown we have to find an estimator of σ^2. One way to achieve this would be to estimate $\sigma^2 = \sigma^2(\lambda)$ as a function of λ by simulating $\xi_t^{\{0\}}$ a large number of times for each λ, each time computing the value of $\hat{\lambda}_t^{\{0\}}(C_t)$ and using $|C_t|_D$ times the sample variance of these values as an estimate of $\sigma^2(\lambda)$. One could then use $\sigma^2(\hat{\lambda}_t^{\{0\}}(C_t))$ as an estimate of σ^2. Of course in any particular instance it would be enough to carry out these simulations only for $\lambda = \hat{\lambda}_t^{\{0\}}(C_t)$.

An alternative way to estimate σ^2 would be to use the observed process $\xi_t^{\{0\}}$ itself. First, we subdivide the mask C_t into k subsets $C_{t,1}, \ldots, C_{t,k}$ of (approximately) equal size and compute the values $\hat{\lambda}_t^{\{0\}}(C_{t,i})$ for $i = 1, \ldots, k$. We then use $k^{-1}|C_t|_D$ times the sample variance of these values as an estimate of σ^2.

An obvious advantage of the second method is that it is not as dependent on the model as the first. It is quite conceivable that the estimator $\hat{\lambda}_t^{\{0\}}(C_t)$ is a useful statistic in a much broader class of models than the contact process. In this case the second method is more likely to produce a sensible result than the first.

Acknowledgements

The authors are indebted to an associate editor and two referees for constructive criticism. They also wish to thank the past editor, Ole Barndorff-Nielsen, for his willingness to continue handling this paper well after the end of his tenure.

References

Bezuidenhout, C. and Grimmett, G. (1990) The critical contact process dies out. *Ann. Probab.*, **18**, 1462–1482.

Bolthausen, E. (1982) On the central limit theorem for stationary mixing random fields. *Ann. Probab.*, **10**, 1047–1050.

Durrett, R. (1991) The contact process, 1974–1989. In W.E. Kohler and B.S. White (eds), *Mathematics of Random Media*, Lectures in Appl. Math. 27, pp. 1–18. Providence, RI: American Mathematical Society.

Fiocco, M. (1997) Statistical estimation for the supercritical contact process. Doctoral thesis, University of Leiden, The Netherlands.

Fiocco, M. and van Zwet, W.R. (2003) Decaying correlations for the supercritical contact process conditioned on survival. *Bernoulli*, **9**, 763–781.

Harris, T.E. (1978) Additive set-valued Markov processes and graphical methods. *Ann. Probab.*, **6**, 355–378.

Liggett, T. (1999) *Stochastic Interacting Systems: Contact, Voter and Exclusion Processes*. New York: Springer-Verlag.

Received April 1999 and revised April 2003

The Annals of Applied Probability
2004, Vol. 14, No. 2, 881–902

ON THE MINIMAL TRAVEL TIME NEEDED TO COLLECT n ITEMS ON A CIRCLE

BY NELLY LITVAK AND WILLEM R. VAN ZWET

University of Twente and University of Leiden

Consider n items located randomly on a circle of length 1. The locations of the items are assumed to be independent and uniformly distributed on $[0, 1)$. A picker starts at point 0 and has to collect all n items by moving along the circle at unit speed in either direction. In this paper we study the minimal travel time of the picker. We obtain upper bounds and analyze the exact travel time distribution. Further, we derive closed-form limiting results when n tends to infinity. We determine the behavior of the limiting distribution in a positive neighborhood of zero. The limiting random variable is closely related to exponential functionals associated with a Poisson process. These functionals occur in many areas and have been intensively studied in recent literature.

1. Introduction. This paper is devoted to the properties of the optimal route of the picker who has to collect n items independently and uniformly distributed on a circle. By *optimal* route we mean the route providing the minimal travel time (see Figure 1). The problem has applications in performance analysis of carousel systems. A carousel is an automated storage and retrieval system which is widely used in modern warehouses. The system consists of a large number of shelves or drawers rotating in a closed loop in either direction. Orders are represented by a list of items. The list specifies the type and retrieval quantity of each item. The picker has a fixed position in front of the carousel, which rotates the required items to the picker. In this paper we study the minimal travel (rotation) time of the carousel while picking one order of n items, the locations of which are assumed to be independent and uniformly distributed on the carousel.

Let $U_0 = 0$ be the picker's starting point and, for $i = 1, 2, \dots, n$, let the random variable U_i denote the position of the ith item. The random variables $U_1, U_2, \dots, U_n$ are independent and uniformly distributed on $[0, 1)$. Set $U_{n+1} = 1$. Let

$$0 = U_{0:n} < U_{1:n} < \cdots < U_{n:n} < U_{n+1:n} = 1$$

denote the ordered $U_0, U_1, \dots, U_{n+1}$. Then the picker's starting point and the positions of the n items partition the circle into $n + 1$ uniform spacings

$$D_{i,n} = U_{i:n} - U_{i-1:n}, \qquad 1 \le i \le n + 1.$$

Received December 2002; revised May 2003.
AMS 2000 subject classifications. Primary 90B05; secondary 62E15, 60F05, 60G51.
Key words and phrases. Uniform spacings, carousel systems, exact distributions, asymptotics, exponential functionals.

S. van de Geer and M. Wegkamp (eds.), *Selected Works of Willem van Zwet*, Selected Works in Probability and Statistics, DOI 10.1007/978-1-4614-1314-1_22, © Springer Science+Business Media, LLC 2012

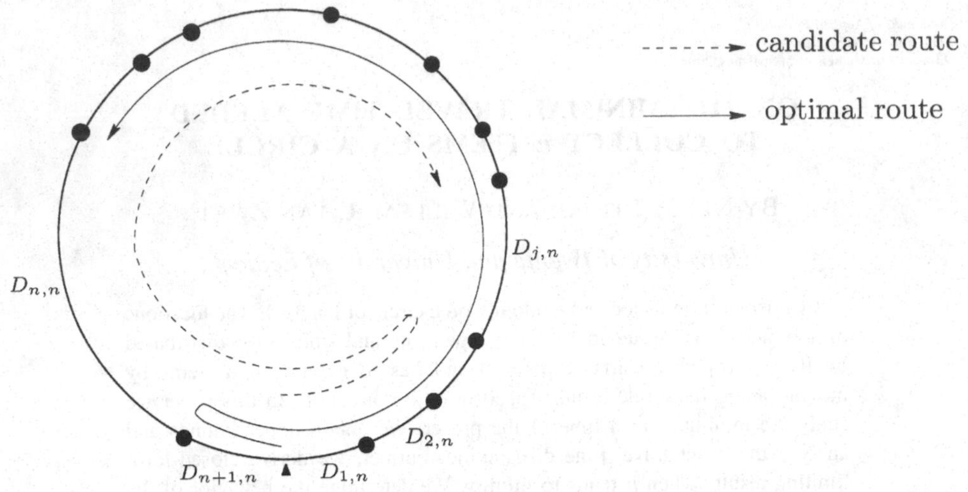

FIG. 1. *Minimal travel time on a circle.*

Let $X_1, X_2, \ldots$ be independent exponential random variables with mean 1 and write

$$S_0 = 0, \qquad S_i = \sum_{j=1}^{i} X_j, \qquad i \geq 1.$$

It is well known that [cf. Pyke (1965)]

$$(1.1) \quad (D_{1,n}, D_{2,n}, \ldots, D_{n+1,n}) \stackrel{d}{=} (X_1/S_{n+1}, X_2/S_{n+1}, \ldots, X_{n+1}/S_{n+1}),$$

that is, the spacings are distributed as normalized exponentials. According to Pyke (1965), this construction is useful "to show that an ordering of uniform spacings may be considered as an ordering of the exponential random variables."

Now, let T_n be the minimal travel time. We explore T_n in terms of the uniform $(n+1)$-spacings $D_{1,n}, D_{2,n}, \ldots, D_{n+1,n}$. For $n = 1$, the problem is trivial. The picker just chooses the shorter distance from the starting point to the item, and thus the travel time T_1 is distributed as $(1/2)D_{1,1}$ (a normalized minimum of two exponentials). For $n = 2$, one can easily verify that the optimal route is guaranteed by the nearest item heuristic where the next item to be picked is always the nearest one. The travel time distribution for this heuristic was obtained by Litvak and Adan (2001). It follows from their result that T_2 is distributed as $(1/2)D_{1,2} + (3/4)D_{2,2}$. For $n \geq 3$, the problem becomes much more difficult.

A crucial and simple observation made by many authors [see, e.g., Bartholdi and Platzman (1986)] is that the optimal route admits at most one turn. Obviously, it is never optimal to cover the same segment of the circle more than twice. Thus, in general, T_n can be expressed as

$$T_n = 1 - \max\left\{ \max_{1\le j\le n} \{D_{j,n} - U_{j-1:n}\}, \max_{1\le j\le n} \{D_{n+2-j,n} - (1 - U_{n+2-j:n})\}\right\}. \tag{1.2}$$

This formula is easy to understand by means of Figure 1. Clearly, for $j = 1, 2, \dots, n$, the term $D_{j,n} - U_{j-1:n}$ is the gain in travel time (compared to one full rotation) obtained by skipping the spacing $D_{j,n}$ and going back instead (ending in a clockwise direction). The same can be said about $D_{n+2-j,n} - (1 - U_{n+2-j:n})$, but here the picker ends in a counterclockwise direction. Under the optimal strategy, the picker chooses the largest possible gain.

Let $T_n^{(m)}$ be the travel time under so-called m-step strategies: the picker chooses the shortest route among $2(m+1)$ candidate routes that change direction at most once (as does the optimal route) and only do so after collecting no more than m items. It was proved by Litvak and Adan (2002) that for $2m < n$,

$$\begin{aligned} T_n^{(m)} &= 1 - \max\left\{ \max_{1\le j\le m+1} \{D_{j,n} - U_{j-1:n}\}, \max_{1\le j\le m+1} \{D_{n+2-j,n} - (1 - U_{n+2-j:n})\}\right\} \\ &\stackrel{d}{=} 1 - \frac{1}{S_{n+1}} \max\left\{ \max_{1\le j\le m+1} \{X_j - S_{j-1}\}, \max_{1\le j\le m+1} \{X_{n+2-j} - (S_{n+1} - S_{n+2-j})\}\right\} \\ &\stackrel{d}{=} 1 - \max\left\{ \sum_{j=1}^{m+1} \frac{1}{2^j - 1} D_{j,n}, \sum_{j=1}^{m+1} \frac{1}{2^j - 1} D_{n+2-j,n}\right\}. \end{aligned} \tag{1.3}$$

Formula (1.3) follows from the following curious property of exponential random variables obtained by Litvak (2001).

LEMMA 1.1. *For any $m = 0, 1, \dots$ and $0 < q < 1$,*

$$\max_{1\le j\le m+1} \{X_j - (q^{-1} - 1)S_{j-1}\} \stackrel{d}{=} (q^{-1} - 1) \sum_{j=1}^{m+1} q^j (1 - q^j)^{-1} X_j. \tag{1.4}$$

The proof also implies that for any $m = 0, 1, \dots, n$,

$$\max_{1\le j\le m+1} \{D_{j,n} - (q^{-1} - 1)U_{j-1:n}\} \stackrel{d}{=} (q^{-1} - 1) \sum_{j=1}^{m+1} q^j (1 - q^j)^{-1} D_{j,n}. \tag{1.5}$$

If $2m < n$, then the two internal maxima in the third expression of (1.3) are independent and (1.4) can be used to rewrite each of them separately. Moreover, the same argument applies for the normalized exponentials yielding (1.3). If $2m \geq n$, then the two internal maxima become dependent and the argument fails.

In fact, the optimal strategy is the m-step strategy with $m = n - 1$. Intuitively, it is clear, however, that with high probability the optimal route has only a few steps before a turn. That is, the m-step strategy often prescribes the optimal picking sequence even when m is relatively small. It was shown by Litvak and Adan (2002) that already for $m = 2$, the m-step strategy is quite close to optimal and, on average, outperforms the nearest item heuristic.

Let $K_n^{(m)}$ and K_n denote a number of items collected before a turn under the m-step strategy and the optimal strategy, respectively. If there is no turn, these numbers are set equal to zero. It was proved by Litvak and Adan (2002) that: (i) $T_n^{(m)}$ and $K_n^{(m)}$ are independent random variables; (ii) for any $k = 0, 1, \ldots, m$ and $2m < n$,

$$\begin{aligned} \mathbb{P}(K_n^{(m)} = k) &= \mathbb{P}\left(\left[\arg \max_{1 \leq j \leq m+1} \{D_{j,n} - U_{j-1:n}\} = k + 1\right]\right) \\ &= \mathbb{P}\left(\left[\arg \max_{1 \leq j \leq m+1} \{X_j - S_{j-1}\} = k + 1\right]\right) \\ &= \frac{1}{2^{k+1} - 2^{k-m}}; \end{aligned} \tag{1.6}$$

(iii) for any $k = 0, 1, \ldots, n - 2$,

$$\mathbb{P}(K_n > k) < 1/2^k. \tag{1.7}$$

The last estimate is helpful in the analysis of the limiting properties of the optimal route. For example, it was proved by Litvak and Adan (2002) that for any fixed $k = 0, 1, \ldots,$

$$\lim_{n \to \infty} \mathbb{P}(K_n = k) = 1/2^{k+1}. \tag{1.8}$$

Indeed, observe that for $k = 0, 1, \ldots, m$,

$$\mathbb{P}(K_n^{(m)} = k) - \mathbb{P}(K_n > m) \leq \mathbb{P}(K_n = k) \leq \mathbb{P}(K_n^{(m)} = k).$$

Now, let m and n go to infinity in such a way that the inequality $2m < n$ is always satisfied. Then (1.8) follows readily from (1.6) and (1.7).

In this paper we first derive simple upper bounds for the minimal travel time. Then we analyse the distribution of T_n. Further, we obtain the limiting behavior of T_n when n tends to infinity.

2. Upper bounds. Let T_n be the minimal travel time needed to collect n items independently and uniformly distributed on a circle of length 1. The following lemma gives an upper bound that holds for *any* realization of the random items' locations.

LEMMA 2.1. *For any $n \geq 1$, the travel time T_n never exceeds $1 - \alpha_{n+1}$, where*

$$\alpha_{n+1} = \frac{1}{2^{m+1} + 2^m - 2}, \qquad n = 2m;$$

$$\alpha_{n+1} = \frac{1}{2 \cdot 2^{m+1} - 2}, \qquad n = 2m + 1.$$

This upper bound is tight.

PROOF. Assume that $n = 2m + 1$. For $n = 2m$ the proof is similar. The positions of the items plus the picker's starting point partition the circle into $n + 1$ spacings with lengths $d_1, d_2, \ldots, d_{n+1}$. Note that for any collection $d_1, d_2, \ldots, d_{n+1} \geq 0$ there exists a number $j = 1, 2, \ldots, m + 1$ such that either (i) $d_j \geq 2^{j-1}\alpha_{n+1}$, $d_l < 2^{l-1}\alpha_{n+1}$, $l = 1, 2, \ldots, j - 1$, or (ii) $d_{n+2-j} \geq 2^{j-1}\alpha_{n+1}$, $d_{n+2-l} < 2^{l-1}\alpha_{n+1}$, $l = 1, 2, \ldots, j - 1$. This follows since

$$2 \sum_{j=1}^{m+1} 2^{j-1}\alpha_{n+1} = d_1 + d_2 + \cdots + d_{n+1} = 1.$$

Without loss of generality assume (i). Then the route that skips the spacing d_j and goes back instead has length

$$1 - d_j + d_1 + d_2 + \cdots + d_{j-1} \leq 1 - \alpha_{n+1},$$

and its length must be greater or equal than T_n. This proves the upper bound.

To show the tightness we just put $d_j = d_{n+2-j} = 2^{j-1}\alpha_{n+1}$, $j = 1, 2, \ldots, m + 1$. In this case the travel time under the optimal strategy equals $1 - \alpha_{n+1}$. □

Let us now consider the following approximation of T_n in (1.2),

$$T_n^0 \stackrel{d}{=} 1 - \frac{1}{S_{n+1}} \max\left\{ \max_{1 \leq j \leq m+1} \{X_j - S_{j-1}\}, \max_{1 \leq j \leq m'+1} \{X_{n+2-j} - (S_{n+1} - S_{n+2-j})\} \right\},$$

where $m = m' = (n - 1)/2$ if n is odd and $m = m' + 1 = n/2$ if n is even. In both cases we have $m + m' = n - 1$ so that the X_j's from the first internal maximum are not involved in the second internal maximum. That is, the two internal maxima are independent, and we can apply Lemma 1.1 to each of these separately to arrive at

$$T_n^0 \stackrel{d}{=} 1 - \max\left\{ \sum_{j=1}^{m+1} \frac{1}{2^j - 1} D_{j,n}, \sum_{j=1}^{m'+1} \frac{1}{2^j - 1} D_{n+2-j,n} \right\}. \tag{2.1}$$

Clearly, T_n^0 gives a tight stochastic upper bound for T_n. In fact, T_n^0 and T_n differ with probability of order $2^{-n/2}$ according to (1.7). It was shown by Litvak and Adan (2002) that T_n^0 is stochastically larger than the weighted sum

$$T_n^* = \sum_{j=2}^{n+1} (1-\alpha_j) D_{j,n}.$$

Straightforward estimation of the expected difference between T_n^* and T_n^0 yields

$$\mathbb{E}(T_n^0 - T_n^*) < 0.09\mathbb{E}(D_{1,n}) = \frac{0.09}{n+1}.$$

Thus,

$$\mathbb{E}(T_n) < \mathbb{E}(T_n^0) < \frac{1}{n+1}\sum_{j=2}^{n+1}(1-\alpha_j) + \frac{0.09}{n+1}. \tag{2.2}$$

In Table 1 (see Section 4), we compare the mean travel time $\mathbb{E}(T_n)$ obtained by simulation with upper estimate (2.2) and approximation (4.8), which follows from the limiting results in Section 4. The results prove that the bound (2.2) is quite sharp. For larger n, however, (4.8) gives a slightly better approximation.

3. The minimal travel time distribution. In this section we produce an explicit expression for $\mathbb{P}(T_n \geq 1-t)$. First, note that it is never optimal to turn after covering half of a circle. Now, consider the events

$$A_{n,k}(u,v) = [U_{k:n} = u < 1/2 < 1 - v = U_{k+1:n}],$$
$$0 \leq u, v < 1/2, k = 0, 1, \dots, n.$$

For $k = 2, 3, \dots, n-2$, the joint distribution of $U_{1:n}, \dots, U_{k-1:n}, 1-U_{k+2:n}, \dots, 1-U_{n:n}$ given $A_{n,k}(u,v)$ is that of

$$uU_{1:k-1}, \dots, uU_{k-1:k-1}, \qquad vV_{n-k-1:n-k-1}, \dots, vV_{1:n-k-1},$$

where $\mathbf{U}$ and $\mathbf{V}$ are independent vectors of uniform order statistics. As the event $[T_n \geq 1-t]$ implies $1 - v - u - u \wedge v \leq t$, we have for $k = 2, 3, \dots, n-2$,

$$\begin{aligned}
&\mathbb{P}(T_n \geq 1-t | A_{n,k}(u,v)) \\
&\quad = \mathbb{P}\Big(\max_{1\leq j\leq k}\{(U_{j:k-1} - U_{j-1:k-1}) - U_{j-1:k-1}\} \leq t/u, \\
&\qquad\qquad \max_{1\leq j\leq n-k}\{(V_{j:n-k-1} - V_{j-1:n-k-1}) - V_{j-1:n-k-1}\} \leq t/v\Big) \qquad (3.1)\\
&\qquad \times \mathbf{1}_{[1-v-u-u\wedge v\leq t]} \\
&\quad = P_{k-1}(t/u) P_{n-k-1}(t/v) \mathbf{1}_{[1-v-u-u\wedge v\leq t]}.
\end{aligned}$$

Here $u \wedge v = \min\{u, v\}$ denotes the smaller of u and v and

$$P_m(t) = \mathbb{P}\left(\max_{1\le j\le m+1}\{D_{j,m} - U_{j-1:m}\} \le t\right), \qquad m = 1, 2, \ldots; t \ge 0. \tag{3.2}$$

One readily verifies that the final expression in (3.1) continues to hold for $k = 1$ and $k = n - 1$, provided we define

$$P_0(t) = \mathbf{1}_{[t>1]}, \qquad t \ge 0. \tag{3.3}$$

For $k = 0$ and $k = n$, we find

$$\mathbb{P}(T_n \ge 1 - t \mid A_{n,0}(0, v)) = P_{n-1}(t/v)\mathbf{1}_{[1-v\le t]},$$
$$\mathbb{P}(T_n \ge 1 - t \mid A_{n,n}(u, 0)) = P_{n-1}(t/u)\mathbf{1}_{[1-u\le t]}.$$

It follows that

$$\begin{aligned} &\mathbb{P}(T_n \ge 1 - t) \\ &= \int_0^{1/2}\int_0^{1/2}\sum_{k=1}^{n-1}\binom{n}{k} k u^{k-1}(n-k)v^{n-k-1} \\ &\qquad \times P_{k-1}(t/u)P_{n-k-1}(t/v)\mathbf{1}_{[1-v-u-u\wedge v\le t]}\,du\,dv \\ &\quad + 2\cdot\mathbf{1}_{[t>1/2]}\int_{u=1-t}^{1/2} nu^{n-1}P_{n-1}(t/u)\,du. \end{aligned} \tag{3.4}$$

Formula (1.5) and Theorem 2 of Ali and Obaidullah (1982) imply an expression for $P_m(t)$. Writing

$$c_j = (2^j - 1)^{-1}, \qquad j = 1, 2, \ldots,$$

and $x_+ = \max\{x, 0\}$ for the positive part of a number x, we find that for $m = 1, 2, \ldots,$

$$P_m(t) = \mathbb{P}\left(\sum_{j=1}^{m+1} c_j D_{j,m} \le t\right) = \sum_{j=1}^{m+1}\{(t - c_j)_+\}^m \prod_{\substack{l=1,\\ l\ne j}}^{m+1}(c_l - c_j)^{-1}. \tag{3.5}$$

The last expression is also valid for $m = 0$. Of course, for $t > 1$, the terms in (3.5) sum to 1.

Alternatively, one can determine $P_m(t)$, recursively. Conditioning on $U_{1:m}$, we find for $m = 2, 3, \ldots,$

$$\begin{aligned} &P\left(\max_{1\le i\le m+1}\{D_{i,m} - U_{i-1:m}\} < t \,\middle|\, U_{1:m} = u\right) \\ &= P\left((1-u)\max_{1\le i\le m}\{D_{i,m-1} - U_{i-1:m-1}\} - u < t\right)\mathbf{1}_{[u\le t]} \\ &= P_{m-1}\left(\frac{t+u}{1-u}\right)\mathbf{1}_{[u\le t]}. \end{aligned}$$

This yields the recursive equation

$$(3.6)\qquad P_m(t) = \int_0^t m(1-u)^{m-1} P_{m-1}\left(\frac{t+u}{1-u}\right) du,$$

which is valid for $m = 1, 2, \ldots$.

We can now find the distribution of T_n by substituting (3.3) and either (3.5) or (3.6) in (3.4) and integrating. One obtains, for example, for $t \geq 0$,

$$P_1(t) = \tfrac{1}{2}(3t-1)_+ - \tfrac{3}{2}(t-1)_+,$$
$$P_2(t) = \tfrac{1}{8}\{(7t-1)_+\}^2 - \tfrac{7}{8}\{(3t-1)_+\}^2 + \tfrac{7}{4}\{(t-1)_+\}^2$$

and for $0 \leq t \leq 1$,

$$\mathbb{P}(T_1 \geq 1-t) = (2t-1)_+,$$
$$\mathbb{P}(T_2 \geq 1-t) = \tfrac{1}{3}\{(4t-1)_+\}^2 - 2\{(2t-1)_+\}^2,$$
$$\mathbb{P}(T_3 \geq 1-t) = \tfrac{1}{4}\{(6t-1)_+\}^3 - \tfrac{41}{36}\{(4t-1)_+\}^3 - \tfrac{1}{4}\{(4t-1)_+\}^2 + \tfrac{11}{4}\{(2t-1)_+\}^3.$$

Although the general structure of these functions is fairly easy to understand, it seems quite useless to provide explicit expressions for $\mathbb{P}(T_n \geq 1-t)$ for much larger values of n. Instead, we study their asymptotic behavior in Section 4.

4. Limiting results. In this section we shall obtain the limiting distribution of $(n+1)(1-T_n)$. First of all, let us consider the limiting behavior of

$$P_m(t/(m+1)) = \mathbb{P}\left((m+1)\sum_{j=1}^{m+1} \frac{1}{2^j-1} D_{j,m} < t\right).$$

THEOREM 4.1. *Let $X_1, X_2, \ldots$ be independent exponential random variables with mean 1. Then*

$$(4.1)\qquad (m+1)\sum_{j=1}^{m+1} \frac{1}{2^j-1} D_{j,m} \xrightarrow{d} \sum_{j=1}^{\infty} \frac{1}{2^j-1} X_j,$$

and the limiting distribution is given by

$$P(t) = \lim_{m\to\infty} P_m(t/(m+1))$$
$$(4.2)\qquad = 1 - \sum_{j=1}^{\infty} (-1)^{j-1} 2^j \exp\{-(2^j-1)t\} \prod_{l=1}^{j} \frac{1}{2^l-1}.$$

The distribution function P satisfies the integral equation

$$(4.3)\qquad e^{-t}P(t) = \int_t^{2t} e^{-u} P(u)\, du.$$

PROOF. The argument essentially repeats the proof of Theorem 4 of Litvak and Adan (2001). Define

$$J_m = \sum_{j=1}^{m+1} \frac{1}{2^j - 1} X_j, \qquad J = \sum_{j=1}^{\infty} \frac{1}{2^j - 1} X_j. \tag{4.4}$$

By the monotone convergence theorem, $\mathbb{E}(J) = \lim_{m\to\infty} \mathbb{E}(J_m) < \infty$. In particular, it implies $\mathbb{P}(J < \infty) = 1$.

Now, using (1.1), we write

$$(m+1) \sum_{j=1}^{m+1} \frac{1}{2^j - 1} D_{j,m} \overset{d}{=} \frac{(m+1) J_m}{S_{m+1}}.$$

By definition, the sequence $\{J_m\}$ converges a.s. to J. The strong law of large numbers implies that the sequence $\{(m+1)/S_{m+1}\}$ converges a.s. to 1. Thus, $\{(m+1)J_m/S_{m+1}\}$ converges a.s. to J which immediately gives (4.1).

The distribution P of J can be obtained via inversion of its Laplace–Stieltjes transform

$$\varphi(s) = \mathbb{E}(\exp(-sJ)) = \prod_{j=1}^{\infty} \frac{2^j - 1}{2^j - 1 + s}.$$

One can expand $\varphi(s)$ in rational fractions of s and obtain

$$\varphi(s) = \sum_{j=1}^{\infty} \frac{(-1)^{j-1} 2^j}{2^j - 1 + s} \prod_{l=1}^{j-1} \frac{1}{2^l - 1}. \tag{4.5}$$

Here, in order to write the formula for the residues of $\varphi(s)$, one can apply well-known expressions from so-called q-calculus [see, e.g., Gasper and Rahman (1990)], but in our case it is not difficult to verify this formula directly. Inversion of (4.5) yields (4.2).

Finally, we use (3.6) and the dominated convergence theorem to obtain

$$\begin{aligned} P(t) &= \lim_{m\to\infty} P_m(t/(m+1)) \\ &= \lim_{m\to\infty} \int_0^{t/(m+1)} m(1-u)^{m-1} P_{m-1}\left(\frac{t/(m+1)+u}{1-u}\right) du \\ &= \int_0^t e^{-u} P(t+u)\, du \\ &= e^t \int_t^{2t} e^{-u} P(u)\, du, \end{aligned}$$

which proves (4.3). □

Obviously, we also have convergence of moments. For the kth moment of P, (4.2) yields

$$\mathbb{E}(J^k) = \int_0^\infty t^k \, dP(t) = k! \sum_{j=1}^{\infty} (-1)^{j-1} \frac{2^j}{(2^j-1)^k} \prod_{l=1}^{j} \frac{1}{2^l - 1}.$$

Alternatively, one can directly use (4.4) to find a simple expression for cumulants κ_ν, $\nu \geq 1$, of P. It is immediate that

$$\mathbb{E}(J) = \kappa_1 = \sum_{j=1}^{\infty} (2^j - 1)^{-1},$$

$$\mathrm{Var}(J) = \kappa_2 = \sum_{j=1}^{\infty} (2^j - 1)^{-2}.$$

Furthermore,

$$\log(\mathbb{E}\exp(itJ)) = -\sum_{j=1}^{\infty} \log(1 - (2^j-1)^{-1} it) = \sum_{j=1}^{\infty}\sum_{\nu=1}^{\infty} \frac{(it)^\nu}{\nu(2^j-1)^\nu},$$

where i is the imaginary unit. Since $\log(\mathbb{E}\exp(itJ)) = \sum_{\nu=1}^{\infty} \kappa_\nu (it)^\nu (\nu!)^{-1}$, it follows that

$$\kappa_\nu = (\nu - 1)! \sum_{j=1}^{\infty} (2^j - 1)^{-\nu}, \qquad \nu \geq 1.$$

The distribution function P on $[0, \infty)$ has the remarkable property that it is infinitely often differentiable and that all of its derivatives $P^{(k)}$ vanish at the origin. This is most easily seen by differentiating (4.3), but one may also use (4.2) to show analytically that $P^{(k)}(0) = 0$ for all $k = 1, 2, \ldots$. It follows that P is not analytic at the origin. The series (4.2) diverges for all $t < 0$ and, hence, P cannot be represented by its Taylor series around $t = 0$.

Now repeating the argument from the proof of Theorem 4.1, one can show that

$$(n+1)(1 - T_n^0) \xrightarrow{d} \max\left\{\sum_{j=1}^{\infty} \frac{1}{2^j - 1} X_j, \sum_{j=1}^{\infty} \frac{1}{2^j - 1} X_j'\right\},$$

where $X_1, X_2, \ldots, X_1', X_2', \ldots$ are independent exponentials with mean 1. Since the two sums in the maximum are independent and (1.7) ensures that

$$\mathbb{P}(T_n \neq T_n^0) < 2^{-(n-2)/2},$$

we have proved the following statement.

THEOREM 4.2. *Let $X_1, X_2, \ldots, X'_1, X'_2, \ldots$ be independent exponential random variables with mean 1. Then*

$$(n+1)(1-T_n) \xrightarrow{d} \max\left\{\sum_{j=1}^{\infty}\frac{1}{2^j-1}X_j, \sum_{j=1}^{\infty}\frac{1}{2^j-1}X'_j\right\}, \tag{4.6}$$

and the limiting distribution is

$$\lim_{n\to\infty}\mathbb{P}(T_n > 1 - t/(n+1)) = [P(t)]^2, \tag{4.7}$$

where $P(t)$ is given by (4.2).

Again we have moment convergence and for the kth moment we find

$$\lim_{n\to\infty}\mathbb{E}[(n+1)(1-T_n)]^k$$

$$=2k!\sum_{j=1}^{\infty}(-1)^{j-1}\frac{2^j}{(2^j-1)^k}\prod_{l=1}^{j}\frac{1}{2^l-1}$$

$$-2k!\sum_{j=1}^{\infty}\sum_{i=1}^{\infty}(-1)^{i+j}\frac{2^{i+j}}{(2^i+2^j-2)^{k+1}}\prod_{l=1}^{j}\frac{1}{2^l-1}\prod_{r=1}^{i-1}\frac{1}{2^r-1}.$$

An equivalent expression for the expectation can be obtained as

$$\lim_{n\to\infty}\mathbb{E}[(n+1)(1-T_n)]$$

$$=\int_0^\infty(1-[P(t)]^2)\,dt$$

$$=2\sum_{j=1}^{\infty}\frac{1}{2^j-1}-\sum_{i=1}^{\infty}\sum_{j=1}^{\infty}(-1)^{i+j}\frac{2^{i+j}}{2^i+2^j-2}\prod_{l=1}^{j}\frac{1}{2^l-1}\prod_{r=1}^{i}\frac{1}{2^r-1}$$

$$\approx 2.1578.$$

For large n we therefore have the estimate

$$\mathbb{E}(T_n)\approx 1-\frac{2.1578}{n+1}. \tag{4.8}$$

In Table 1 we compare the mean travel time obtained by simulation with upper estimate (2.2) (see Section 2) and approximation (4.8). We see that both approximations are quite sharp, but (4.8) performs somewhat better. It is no surprise that both (2.2) and (4.8) are close to $\mathbb{E}(T_n)$ for large n since all three quantities converge to 1 as $n\to\infty$. What is encouraging is that, already for $n=30$, both approximations of $(n+1)(1-\mathbb{E}(T_n))$ are very good. That (4.8) yields a better approximation of $\mathbb{E}(T_n)$ than (2.2) is to be expected since it is asymptotically correct up to and including order n^{-1}, whereas (2.2) has a slight asymptotic error of about $+0.006/(n+1)$. After all, (2.2) was derived as an upper bound.

TABLE 1
Estimation of the mean travel time under the optimal strategy

n	3	5	10	15	20	30
$\mathbb{E}(T_n)$	0.5262	0.6591	0.8052	0.8653	0.8972	0.9304
$\mathbb{E}[(n+1)(1-T_n)]$	1.8952	2.0454	2.1423	2.1548	2.1592	2.1572
Upper estimate (2.2)	0.5433	0.6670	0.8068	0.8658	0.8976	0.9306
$(n+1)$[1-upper estimate(2.2)]	1.8268	1.9980	2.1252	2.1472	2.1504	2.1514
Approximation (4.8)	0.4605	0.6404	0.8038	0.8651	0.8972	0.9304
$(n+1)$[1-approximation (4.8)]	2.1578	2.1578	2.1578	2.1578	2.1578	2.1578

5. Asymptotic behavior in the neighborhood of zero. In this section we study the behavior of $P(t)$ as $t \to +0$. So far we have found only that P has vanishing derivatives at the origin and can not be expanded in a Taylor expansion around $t = 0$. We shall, therefore, have to attack this problem in a different manner.

Let $X_1, X_2, \dots$ be independent exponential random variables with mean 1, let

$$c_j = (2^j - 1)^{-1}, \qquad j = 1, 2, \dots,$$

and define

$$J = \sum_{j=1}^{\infty} c_j X_j.$$

We want to determine the behavior of

$$P(t) = \mathbb{P}(J \leq t)$$

for small positive values of t. In principle this problem is solved in Theorem 3.2 of Davis and Resnick (1991), but we need to do a substantial amount of analysis to make their result explicit, even in our relatively simple case.

In our case, the distribution function $F(x) = P(X_1 < x) = 1 - \exp\{-x\}$ and the density $f(x) = \exp\{-x\}$ are regularly varying at 0 with index $\alpha = 1$ and $\alpha - 1 = 0$, respectively. The c_j's are positive and nonincreasing, their sum converges and for every $\theta \in (0, 1)$,

$$\theta^n \sum_{j=1}^{\infty} \{c_j^2 / c_n^2\} \mathbf{1}_{[j \geq \theta^{-n}]} = \theta^n \sum_{j=1}^{\infty} \{(2^n - 1)/(2^j - 1)\}^2 \mathbf{1}_{[j \geq \theta^{-n}]} \to 0$$

as $n \to \infty$. The density f satisfies

$$\int_0^{\infty} e^{-2\lambda x} f^2(x)\, dx = 1/\{2(1+\lambda)\} \qquad \text{for } \lambda > 0.$$

Hence, we have verified the assumptions of Theorem 3.2 of Davis and Resnick (1991) in our case. The theorem states that

$$P(m_\lambda) \sim \exp\{\lambda m_\lambda\} \varphi_J(\lambda) / (\lambda S_\lambda \sqrt{2\pi}) \qquad \text{as } \lambda \to \infty. \tag{5.1}$$

Here

$$m_\lambda = \sum_{j=1}^{\infty} \frac{c_j}{1+\lambda c_j} = \sum_{j=1}^{\infty} \frac{1}{2^j - 1 + \lambda},$$

$$\varphi_J(\lambda) = \prod_{j=1}^{\infty} \frac{1}{1+\lambda c_j} = \prod_{j=1}^{\infty} \frac{2^j - 1}{2^j - 1 + \lambda}$$

and

$$S_\lambda^2 = \sum_{j=1}^{\infty} \frac{c_j^2}{(1+\lambda c_j)^2} = \sum_{j=1}^{\infty} \frac{1}{(2^j - 1 + \lambda)^2}.$$

We obviously have to study the behavior of these quantities as $\lambda \to \infty$ and, hence, $m_\lambda \to 0$. It is easier to deal with integrals than sums. For $k = 1, 2, \ldots$ and $\lambda \to \infty$, we have

$$\begin{aligned} 0 &\le \int_0^\infty (2^x - 1 + \lambda)^{-k}\, dx - \sum_{j=1}^{\infty} (2^j - 1 + \lambda)^{-k} \\ &\le \sum_{j=0}^{\infty} (2^j - 1 + \lambda)^{-k} - \sum_{j=1}^{\infty} (2^j - 1 + \lambda)^{-k} \\ &= \lambda^{-k} \end{aligned}$$

and, hence,

$$\sum_{j=1}^{\infty} (2^j - 1 + \lambda)^{-k} = \int_0^\infty (2^x - 1 + \lambda)^{-k}\, dx + O(\lambda^{-k}).$$

For $k = 2$, this yields

$$\begin{aligned} S_\lambda^2 &= (\log 2)^{-1} \int_0^\infty (y+\lambda)^{-2} (y+1)^{-1}\, dy + O(\lambda^{-2}) \\ &= (\log \lambda)/(\lambda^2 \log 2) + O(\lambda^{-2}), \end{aligned} \tag{5.2}$$

as $\lambda \to \infty$. If we apply the same approach to λm_λ and $\log \varphi_J(\lambda)$, however, then the error caused by approximating these sums by integrals is of the order $O(1)$ and $O(\log \lambda)$, respectively, which yields a multiplicative error factor $(1 + O(a\lambda^b))$ in (5.1) for some positive a and b. Of course this is not good enough so we shall have to expand the series representing λm_λ and $\log \varphi_J(\lambda)$ directly with remainder $o(1)$ in both cases.

Let k be a natural number and $\theta \in [0, 1)$ be such that

$$\lambda = 2^{k+\theta},$$

and thus

$$k = (\log \lambda)/\log 2 - \theta = \lfloor (\log \lambda)/\log 2 \rfloor,$$
$$\theta = (\log \lambda)/\log 2 - k = \operatorname{frac}((\log \lambda)/\log 2).$$

Here $\lfloor x \rfloor$ and $\operatorname{frac}(x)$ are the integer and the fractional part of x, respectively. In order for $\lambda \to \infty$, it is necessary and sufficient that $k \to \infty$, while θ may vary arbitrarily in [0, 1) with k. Using (5.2) we find

$$\begin{aligned}
\lambda m_\lambda &= \sum_{j=1}^{\infty} \frac{\lambda}{2^j - 1 + \lambda} \\
&= \sum_{j=1}^{\infty} \frac{2^{k+\theta}}{2^j - 1 + 2^{k+\theta}} \\
&= \sum_{j=1}^{\infty} \frac{2^{k+\theta}}{2^j + 2^{k+\theta}} + O(\lambda^{-1} \log \lambda) \\
&= \sum_{j=1}^{k} \frac{1}{2^{j-k-\theta} + 1} + \sum_{j=k+1}^{\infty} \frac{1}{2^{j-k-\theta} + 1} + O(\lambda^{-1} \log \lambda) \\
&= \sum_{j=0}^{k-1} \frac{1}{2^{-j-\theta} + 1} + \sum_{j=1}^{\infty} \frac{1}{2^{j-\theta} + 1} + O(\lambda^{-1} \log \lambda) \\
&= \sum_{j=1}^{k} \frac{2^j}{2^j + 2^{1-\theta}} + \sum_{j=1}^{\infty} \frac{1}{2^{j-\theta} + 1} + O(\lambda^{-1} \log \lambda) \\
&= k - \sum_{j=1}^{k} \frac{2^{1-\theta}}{2^j + 2^{1-\theta}} + \sum_{j=1}^{\infty} \frac{2^\theta}{2^j + 2^\theta} + O(\lambda^{-1} \log \lambda) \\
&= \frac{\log \lambda}{\log 2} - \sum_{j=1}^{\infty} \frac{2^{1-\theta}}{2^j + 2^{1-\theta}} + \sum_{j=1}^{\infty} \frac{2^\theta}{2^j + 2^\theta} - \theta + O(\lambda^{-1} \log \lambda).
\end{aligned}$$

Hence,

$$\lambda m_\lambda = (\log \lambda)/\log 2 + A(\theta) + O(\lambda^{-1} \log \lambda), \tag{5.3}$$

with

$$A(\theta) = -\sum_{j=1}^{\infty} \frac{2^{1-\theta}}{2^j + 2^{1-\theta}} + \sum_{j=1}^{\infty} \frac{2^\theta}{2^j + 2^\theta} - \theta.$$

Notice that the term $A(\theta)$ of order 1 is not constant but depends on $\theta \in [0, 1)$. The expansion (5.3) is obviously uniform in θ.

Similarly, by (5.3),

$$\begin{aligned}\log\varphi_J(\lambda) &= \sum_{j=1}^{\infty}\log\{(2^j-1)/(2^j-1+\lambda)\}\\ &= \sum_{j=1}^{\infty}\log\{2^j/(2^j+\lambda)\}+\sum_{j=1}^{\infty}\log(1-2^{-j})\\ &\quad+\sum_{j=1}^{\infty}\log\{1+1/(2^j-1+\lambda)\}\\ &= \sum_{j=1}^{\infty}\log\{2^j/(2^j+2^{k+\theta})\}+\sum_{j=1}^{\infty}\log(1-2^{-j})\\ &\quad+O(\lambda^{-1}\log\lambda).\end{aligned}$$

Furthermore,

$$\begin{aligned}&\sum_{j=1}^{\infty}\log\{2^j/(2^j+2^{k+\theta})\}\\ &\quad= \sum_{j=1}^{k}\log\{2^{1-\theta}/(2^j+2^{1-\theta})\}+\sum_{j=1}^{\infty}\log\{2^j/(2^j+2^{\theta})\}\\ &\quad= k(1-\theta)\log 2-(1/2)k(k+1)\log 2\\ &\quad\quad-\sum_{j=1}^{k}\log(1+2^{1-\theta-j})-\sum_{j=1}^{\infty}\log(1+2^{\theta-j}).\end{aligned}$$

Substituting $k=(\log\lambda)/\log 2-\theta$ and using

$$\sum_{j=k+1}^{\infty}\log(1+2^{1-\theta-j})\le\sum_{j=k+1}^{\infty}2^{1-\theta-j}=O(\lambda^{-1}),$$

we finally find

$$\log\varphi_J(\lambda)=-\frac{(\log\lambda)^2}{2\log 2}+\frac{\log\lambda}{2}+B(\theta)+O(\lambda^{-1}\log\lambda), \tag{5.4}$$

where

$$\begin{aligned}B(\theta) &= \sum_{j=1}^{\infty}\log\{(1-2^{-j})/[(1+2^{\theta-j})(1+2^{1-\theta-j})]\}\\ &\quad-(1/2)\theta(1-\theta)\log 2.\end{aligned} \tag{5.5}$$

Again the term $B(\theta)$ of order 1 depends on θ and the expansion is uniform in $\theta \in [0, 1)$.

Substituting (5.3), (5.4) and (5.2) in (5.1), we obtain, for $\lambda \to \infty$,

$$\begin{aligned} P(m_\lambda) \sim \sqrt{\frac{\log 2}{2\pi \log \lambda}} \exp\biggl\{ &-\frac{1}{2\log 2}(\log \lambda)^2 \\ &+ \left(\frac{1}{2} + \frac{1}{\log 2}\right)\log \lambda + A(\theta) + B(\theta)\biggr\}. \end{aligned} \tag{5.6}$$

It remains to find approximations of $\log \lambda$ and $(\log \lambda)^2$ as functions of

$$t = m_\lambda = (\log \lambda)/(\lambda \log 2) + A(\theta)/\lambda + O(\lambda^{-2} \log \lambda).$$

We find

$$\begin{aligned} \log(1/t) &= \log \lambda - \log\log \lambda + \log\log 2 - A(\theta)(\log 2)/\log \lambda \\ &\quad + O\bigl((\log \lambda)^{-2}\bigr), \\ \log\log(1/t) &= \log\log \lambda - (\log\log \lambda)/\log \lambda + (\log\log 2)/\log \lambda \\ &\quad + O\bigl((\log\log \lambda)^2/(\log \lambda)^2\bigr) \end{aligned}$$

and, hence,

$$\begin{aligned} \log \lambda &= \log(1/t) + \log\log(1/t) - \log\log 2 + \bigl(\log\log(1/t)\bigr)/\log(1/t) \\ &\quad - (\log\log 2)/\log(1/t) + A(\theta)(\log 2)/\log(1/t) \\ &\quad + O\bigl((\log\log(1/t))^2/(\log(1/t))^2\bigr), \\ (\log \lambda)^2 &= [\log(1/t) + \log\log(1/t) - \log\log 2]^2 + 2\log\log(1/t) \\ &\quad - 2\log\log 2 + 2A(\theta)\log 2 \\ &\quad + O\bigl((\log\log(1/t))^2/\log(1/t)\bigr). \end{aligned}$$

Together with (5.5) and (5.6), this yields that, for $t \to 0$,

$$\begin{aligned} P(t) \sim C(\theta) &\exp\{-(2\log 2)^{-1}[\log(1/t) + \log\log(1/t) - \log\log 2]^2\} \\ &\times t^{-(1/2 + 1/\log 2)} \end{aligned}$$

with

$$C(\theta) = 2^{-\theta(1-\theta)/2} \frac{1}{\sqrt{2\pi}} \prod_{j=1}^{\infty} \frac{1 - 2^{-j}}{(1 + 2^{\theta - j})(1 + 2^{1-\theta-j})}. \tag{5.7}$$

The factor $C(\theta)$ depends on $\theta = \operatorname{frac}((\log \lambda)/\log 2)$.

It remains to express θ in terms of t. Define

$$\psi(t) = (\log 2)^{-1}[\log(1/t) + \log\log(1/t) - \log\log 2].$$

We have $k+\theta = (\log\lambda)/\log 2 = \psi(t) + o(1)$, and as C is positive and bounded, the derivative of C is positive and bounded and $C(\theta) = C(1-\theta)$. This implies that $C(\mathrm{frac}\{\psi(t)\}) = C(\theta)(1+o(1))$. It follows that, as $t \to +0$,

$$P(t) \sim C(\mathrm{frac}\{\psi(t)\}) \exp\left\{-\frac{\log 2}{2}[\psi(t)]^2\right\} t^{-(1/2+1/\log 2)}, \tag{5.8}$$

with C defined in (5.7). This is an exact asymptotic expression for $P(t)$ as $t \to +0$.

The dependence on $\mathrm{frac}(\psi(t))$ in (5.8) is a most unusual feature. In fact, preliminary numerical calculations make one wonder whether there is any dependence at all, since one finds that $C(\theta)$ equals a constant (≈ 0.01013) throughout the interval $0 \le \theta < 1$ to any reasonable degree of accuracy. Thus, in order to properly understand the asymptotic expression (5.8), we have to analyze $C(\theta)$ in more detail. Proposition 5.1 states that $C(\theta)$ does indeed depend on θ, but in a very peculiar way. In fact, for any real θ,

$$C(\theta) = \left[\frac{\sqrt{\log 2}}{2^{1/8} 2\pi} \prod_{j=1}^{\infty}(1-2^{-j})^2\right](\tilde{\vartheta}_3(\theta))^{-1} \approx 0.01013(\tilde{\vartheta}_3(\theta))^{-1},$$

where

$$\begin{aligned}\tilde{\vartheta}_3(\theta) &= 1 + 2\sum_{k=1}^{\infty} \exp\{-2k^2\pi^2/\log 2\}\cos\{2k\pi(1/2-\theta)\} \\ &= \vartheta_3(\pi(1/2-\theta), \exp\{-2\pi^2/\log 2\}).\end{aligned} \tag{5.9}$$

Here ϑ_3 is a theta function

$$\vartheta_3(z,q) = 1 + 2\sum_{k=1}^{\infty} q^{k^2}\cos(2kz).$$

Note that for all θ,

$$|\tilde{\vartheta}_3(\theta) - 1| < 10^{-12}$$

is a quantity which is difficult to reveal numerically!

PROPOSITION 5.1. *For any real* θ,

$$\begin{aligned}&\prod_{j=1}^{\infty}(1+2^{\theta-j})(1+2^{1-\theta-j}) \\ &= 2^{-\theta(1-\theta)/2}\,\tilde{\vartheta}_3(\theta)\frac{2^{1/8}\sqrt{2\pi}}{\sqrt{\log 2}}\prod_{j=1}^{\infty}(1-2^{-j})^{-1},\end{aligned} \tag{5.10}$$

where $\tilde{\vartheta}_3(\theta)$ *is given by* (5.9).

PROOF. We first apply Jacobi's triple product identity [see, e.g., Askey (1980) and Gasper and Rahman (1990)]. For any $q \in (0, 1)$,

$$\prod_{j=0}^{\infty}(1-xq^j)(1-x^{-1}q^{j+1})(1-q^{j+1}) = \sum_{n=-\infty}^{\infty}(-1)^n q^{\binom{n}{2}} x^n. \tag{5.11}$$

Take $x = -2^{-\theta}$, $q = 1/2$. Then (5.11) becomes

$$\prod_{j=1}^{\infty}(1+2^{\theta-j})(1+2^{1-\theta-j})(1-2^{-j}) = \sum_{n=-\infty}^{\infty} 2^{-n(n-1)/2}2^{-\theta n}. \tag{5.12}$$

The right-hand side of (5.12) is of the form

$$c(\theta)\sum_{n=-\infty}^{\infty} g(n),$$

where

$$c(\theta) = \frac{2^{1/8}\sqrt{2\pi}}{\sqrt{\log 2}}2^{-\theta(1-\theta)/2},$$

and

$$g(x) = \frac{\sqrt{\log 2}}{\sqrt{2\pi}}\exp\{-(1/2)(\log 2)(x-1/2+\theta)^2\}$$

is a normal density with mean $\mu = 1/2 - \theta$ and standard deviation $\sigma = 1/\sqrt{\log 2}$. The characteristic function of g is given by

$$\gamma(t) = \exp\{-t^2/(2\log 2) + it(1/2-\theta)\},$$

where i is the imaginary unit. For each fixed λ and for each real ξ, the Poisson summation formula [see Feller (1970)] gives

$$\sum_{k=-\infty}^{+\infty}\gamma(\xi+2k\lambda) = \frac{\pi}{\lambda}\sum_{n=-\infty}^{+\infty} g(n\pi/\lambda)\exp\{in(\pi/\lambda)\xi\}. \tag{5.13}$$

Put $\lambda = \pi$, $\xi = 0$. Then the right-hand side of (5.13) becomes $\sum_{n=-\infty}^{\infty} g(n)$ and on the left-hand side we have

$$\begin{aligned}\sum_{k=-\infty}^{\infty}\gamma(2k\pi) &= \gamma(0) + \sum_{\substack{k=-\infty\\ k\neq 0}}^{\infty}\exp\{-2k^2\pi^2/\log 2\}\exp\{i(1/2-\theta)2k\pi\}\\ &= 1 + 2\sum_{k=1}^{\infty}\exp\{-2k^2\pi^2/\log 2\}\cos\{2k\pi(1/2-\theta)\}\\ &= \tilde{\vartheta}_3(\theta).\end{aligned}$$

Hence, (5.13) reduces to

$$\sum_{n=-\infty}^{+\infty} g(n) = \tilde{\vartheta}_3(\theta),$$

implying that the right-hand side of (5.12) equals $c(\theta)\tilde{\vartheta}_3(\theta)$. This immediately yields (5.10). The proposition is proved. □

We summarize our findings in the following theorem.

THEOREM 5.2. *Let $X_1, X_2, \dots$ be independent exponential random variables with mean* 1, *and let*

$$J = \sum_{j=1}^{\infty} (2^j - 1)^{-1} X_j.$$

Then

$$\mathbb{P}(J \le t) \sim \frac{\sqrt{\log 2}\prod_{j=1}^{\infty}(1-2^{-j})^2}{2^{1/8} 2\pi \tilde{\vartheta}_3(\mathrm{frac}\{\psi(t)\})} \times \exp\left\{-\frac{\log 2}{2}[\psi(t)]^2\right\} t^{-(1/2+1/\log 2)} \qquad \text{as } t \to +0,$$

where

$$\psi(t) = (\log 2)^{-1}[\log(1/t) + \log\log(1/t) - \log\log 2]$$

and $\tilde{\vartheta}_3$ is defined in (5.9).

6. Related results. In a similar fashion we can also analyze more general linear combinations of i.i.d. exponential random variables than J. For any $q \in (0, 1)$, define

$$J^{(q)} = (q^{-1} - 1) \sum_{j=1}^{\infty} (q^{-j} - 1)^{-1} X_j.$$

Clearly, $J \equiv J^{(1/2)}$. One can show that

$$(m+1)(q^{-1} - 1) \sum_{j=1}^{m+1} \frac{1}{q^{-j} - 1} D_{j,m} \xrightarrow{d} J^{(q)} \qquad \text{as } m \to \infty,$$

where the expression on the left occurs in the right-hand side of (1.5) for $n = m$. The random variable $J^{(q)}$ can be written in the following way. Let $N(t)$ be a standard Poisson process. Then

$$(q^{-1}-1)^{-1}J^{(q)} = \int_0^\infty q^{N(t)+1}(1-q^{N(t)+1})^{-1}\,dt$$
$$= \sum_{j=1}^\infty q^j \int_0^\infty q^{jN(t)}\,dt$$
$$= \sum_{j=1}^\infty q^j I^{(q^j)},$$

where

$$I^{(q)} = \int_0^\infty q^{N(t)}\,dt = \sum_{j=1}^\infty q^{j-1}X_j$$

is an exponential functional associated with a Poisson process. The functional $I^{(q)}$ has been intensively studied in recent literature. Its density was obtained independently by Dumas, Guillemin and Robert (2002), Bertoin, Biane and Yor (2002) and Litvak and Adan (2001), for $q = 1/2$. Carmona, Petit and Yor (1997) derived a density of $\int_0^\infty h(N(t))\,dt$ for a large class of functions $h:\mathbb{N}\to\mathbb{R}_+$, in particular, for $h(n) = q^n$. Bertoin, Biane and Yor (2002) found the fractional moments of $I^{(q)}$. If $i^{(q)}(t)$ is a density of $I^{(q)}$, then $i^{(q)}(t)$ and all its derivatives equal 0 at the point $t = 0$. This implies, by the way, that all moments of $1/I^{(q)}$ are finite. However, for $q = 1/e$, it was proved by Bertoin and Yor (2002a) that $1/I^{(1/e)}$ is not determined by its moments.

The functional $I^{(q)}$ appears in a number of applications. Let T_n^{NI} be the travel time needed to collect n items independently and uniformly distributed on a circle of length 1 operating under the nearest item heuristic (the picker always travels to the nearest item to be retrieved). Then it was shown by Litvak and Adan (2001) that $(n+1)(1-T_n^{NI})$ converges in distribution to $I^{(1/2)}$. Dumas, Guillemin and Robert (2002) showed that the distribution of $I^{(q)}$ plays a key role in the analysis of limiting behavior of a Transmission Control Protocol connection. These results were extended by Guillemin, Robert and Zwart (2002), who found the distribution and the fractional moments of the exponential functional

$$I(\xi) = \int_0^\infty e^{-\xi(t)}\,dt, \tag{6.1}$$

where $(\xi(t), t \ge 0)$ is a compound Poisson process. An exponential functional (6.1) associated with a Levy process $\xi(t)$ appears in mathematical finance and many other fields. It has been studied recently by Bertoin and Yor (2001, 2002a, b), Bertoin, Biane and Yor (2002), Carmona, Petit and Yor (1997) and Yor (2001).

Along the same lines as in Section 5, one can prove theorems similar to Theorem 5.2 for $I^{(q)}$ and $J^{(q)}$. In fact, it is straightforward to repeat the calculations for

$$qI^{(q)} = \sum_{j=1}^{\infty} q^j X_j \quad \text{and} \quad \frac{q}{1-q} J^{(q)} = \sum_{j=1}^{\infty} \frac{1}{q^{-j}-1} X_j.$$

We obtain, for $q \in (0, 1)$ as $t \to +0$,

$$\mathbb{P}(qI^{(q)} \le t) \sim \frac{1}{2\pi} q^{1/8} \sqrt{\log(1/q)} \left[\prod_{j=1}^{\infty} (1-q^j)\right] t^{-(1/2+1/\log(1/q))}$$
$$\times \exp\left\{-\frac{\log(1/q)}{2} [\psi^{(q)}(t)]^2\right\} [\tilde{\vartheta}_3^{(q)}(\mathrm{frac}\{\psi^{(q)}(t)\})]^{-1},$$

$$\mathbb{P}\left(\frac{q}{1-q} J^{(q)} \le t\right) \sim \mathbb{P}(qI^{(q)} \le t) \prod_{j=1}^{\infty} (1-q^j),$$

where

$$\psi^{(q)}(t) = (\log(1/q))^{-1} [\log(1/t) + \log\log(1/t) - \log(\log(1/q))],$$

$$\tilde{\vartheta}_3^{(q)}(\theta) = 1 + 2 \sum_{k=1}^{\infty} \exp\{-2k^2\pi^2/\log(1/q)\} \cos\{2k\pi(1/2-\theta)\}.$$

This agrees with the result of Bertoin and Yor (2002a) that

$$\log i(t) \sim -\tfrac{1}{2} (\log(1/t))^2 \qquad \text{as } t \to +0,$$

where $i(t)$ is a density of

$$I = \int_0^{\infty} e^{-N(t)}\, dt = \sum_{j=1}^{\infty} e^{-j} X_j.$$

For the functional $I^{(1/2)}$, which describes the limiting behavior of the travel time under the nearest item heuristic, we find

$$\mathbb{P}(I^{(1/2)} \le 2t) \sim \mathbb{P}(J \le t) \prod_{j=1}^{\infty} (1-2^{-j})^{-1}, \qquad t \to +0.$$

Acknowledgements. We are grateful to Bert Zwart for drawing our attention to relevant literature and particularly to the paper of Davis and Resnick (1991), which inspired the results in Section 5. Discussions with Fred Steutel also proved very helpful.

REFERENCES

ALI, M. M. and OBAIDULLAH, M. (1982). Distribution of linear combination of exponential variates. *Comm. Statist. Theory Methods* **11** 1453–1463.

ASKEY, R. (1980). Ramanujan's extensions of the gamma and beta functions. *Amer. Math. Monthly* **87** 346–359.

BARTHOLDI, J. J. III and PLATZMAN, L. K. (1986). Retrieval strategies for a carousel conveyor. *IIE Transactions* **18** 166–173.

BERTOIN, J. and YOR, M. (2001). On subordinators, self-similar Markov processes and some factorizations of the exponential variable. *Electron. Comm. Probab.* **6** 95–106.

BERTOIN, J. and YOR, M. (2002a). On the entire moments of self-similar Markov processes and exponential functionals of Lévy processes. *Ann. Fac. Sci. Toulouse Math.* (6) **11** 33–45.

BERTOIN, J. and YOR, M. (2002b). The entrance laws of self-similar Markov processes and exponential functionals of Lévy processes. *Potential Anal.* **17** 389–400.

BERTOIN, J., BIANE, P. and YOR, M. (2002). Poissonian exponential functionals, q-series, q-integrals, and the moment problem for log-normal distributions. Technical Report PMA-705, Univ. Paris 6, Laboratoire de Probabilités.

CARMONA, P., PETIT, F. and YOR, M. (1997). On the distribution and asymptotic results for exponential functionals of Lévy processes. In *Exponential Functionals and Principal Values Related to Brownian Motion* (M. Yor, ed.) 73–130. Iberoamericana, Madrid.

DAVIS, R. A. and RESNICK, S. I. (1991). Extremes of moving averages of random variables with finite endpoint. *Ann. Probab.* **19** 312–328.

DUMAS, V., GUILLEMIN, F. and ROBERT, PH. (2002). A Markovian analysis of additive-increase multiplicative-decrease algorithms. *Adv. in Appl. Probab.* **34** 85–111.

FELLER, W. (1970). *An Introduction to Probability Theory and Its Applications* **II**. Wiley, London.

GASPER, G. and RAHMAN, M. (1990). *Basic Hypergeometric Series*. Cambridge Univ. Press.

GUILLEMIN, F., ROBERT, PH. and ZWART, B. (2002). AIMD algorithms and exponential functionals. Technical Report 4447, INRIA.

LITVAK, N. and ADAN, I. (2001). The travel time in carousel systems under the nearest item heuristic. *J. Appl. Probab.* **38** 45–54.

LITVAK, N. and ADAN, I. (2002). On a class of order pick strategies in paternosters. *Oper. Res. Lett.* **30** 377–386.

LITVAK, N. (2001). Some peculiarities of exponential random variables. *J. Appl. Probab.* **38** 787–792.

PYKE, R. (1965). Spacings. *J. R. Stat. Soc. Ser. B Stat. Methodol.* **27** 395–449.

YOR, M. (2001). *Exponential Functionals of Brownian Motion and Related Processes*. Springer, Berlin.

FACULTY OF ELECTRICAL ENGINEERING
MATHEMATICS AND COMPUTER SCIENCE
UNIVERSITY OF TWENTE
P.O. BOX 217
7500 AE ENSCHEDE
THE NETHERLANDS
E-MAIL: n.litvak@math.utwente.nl

DEPARTMENT OF MATHEMATICS
UNIVERSITY OF LEIDEN
P.O. BOX 9512
2300 RA LEIDEN
THE NETHERLANDS
E-MAIL: vanzwet@math.leidenuniv.nl

Chapter 23
Applications: simple models and difficult theorems

Nelly Litvak

Abstract In this short article I will discuss three papers written by Willem van Zwet with three different co-authors: Mathisca de Gunst, Marta Fiocco, and myself. Each of the papers focuses on one particular application: growth of the number of biological cells [3], spreading of an infection [7], and the optimal travel time in warehousing carousel systems [8].

23.1 Introduction

In this short article I will discuss three papers written by Willem van Zwet with three different co-authors: Mathisca de Gunst, Marta Fiocco, and myself. Each of the papers focuses on one particular application: growth of the number of biological cells [3], spreading of an infection [7], and the optimal travel time in warehousing carousel systems [8]. To my opinion, each of these papers displays the attitude that I personally value a lot in mathematics. An application is the strong starting point for each of the papers. Further, the model is simple and transparent. Yet, the analysis involves advanced mathematics and brings to the results that not only give new insights into the applications but also are of a pure mathematical interest. The present volume contains [7] and [8], and the follow-up paper [4] of [3] which I will also briefly discuss.

The papers are written in a clear language and do not try to look more fancy than they are. In fact, I remember Willem laughing at my attempts to make the paper more general by replacing 1/2 with $b \in (0,1)$: 'What have you done? Please, bring the $1/2$ back! It is more natural and makes the whole thing much easier to read'. And on my sceptical remark about the number of people who are actually going to

Nelly Litvak
Department of Applied Mathematics, University of Twente, P.O. Box 217, 7500 AE Enschede, The Netherlands
e-mail: n.litvak@ewi.utwente.nl

S. van de Geer and M. Wegkamp (eds.), *Selected Works of Willem van Zwet*, Selected Works in Probability and Statistics, DOI 10.1007/978-1-4614-1314-1_23, © Springer Science+Business Media, LLC 2012

read this text he smiled again: 'Well, you have to assume people will read it.' Now, assuming that people will read the introduction to this chapter, I will try, to the best of my own understanding, to describe the essence of the models and the results for each of the papers, what I think was difficult and why it worked. I will try to stick to common sense and intuition, so please forgive me if I am not very precise and go ahead, read the papers for correct formulations and exact results.

23.2 A non-Markovian model for cell population growth

Biostatistics is an extremely important topic, popularity of which has grown hugely in the last years. The paper [3] describes a model for a cell population growth. Initially, we have n plant cells transferred to a medium of a known composition at time $t = 0$. The cells can divide, and we are interested in the number $N_n(t)$ of cells at time $t > 0$. Specifically, we want to obtain a law of large numbers and a central limit theorem for the process $N_n(t)$ as n grows large. The motivation for this problem formulation is that in reality the number of cells is quite large.

The division happens as follows. From the medium, the cells receive a stimulus at a random time, and after that it takes a cell exactly c time units before it divides. The time it takes to receive a stimulus depends on the concentration of a substrate (sugar) in the medium. Clearly, with time, the substrate is being used up and thus it takes longer before a cell receives a stimulus. As described so far, the model already contains two non-trivial features. First, the rate at which the cells receive a stimulus is variable (non-increasing). Second, the cells' 'pregnancy' of length c obviously makes the process $N_n(t)$ non-Markovian. There is also a third interesting feature of the model, namely, the authors distinguish between A-cells and B-cells, where only A-cells are able to divide. As a result of a division, two cells are produced, each of which can be an A-cell with a probability that depends on the concentration of a hormone in the medium. Again, with time the hormone is being used up and thus the probability of producing an A-cell is decreasing.

Altogether, the model description is not hard and very natural but each of the model assumptions brings essential new features in the analysis. Then, what makes this model solvable? One helping feature is the 'boundedness' of the process. First of all, at most two cells can be born at each division. This makes the number of born A-cells bounded, and we can apply the inequalities of the type presented in Lemma 4.2, which resembles the Azuma's inequality for martingales (see e.g. [13, p. 307]). Second, the authors assume that the amount of the substrate and the hormone is proportional to the original number of cells. This is a natural scaling, which ensures that, on average, each cell can potentially receive a certain fixed amount of both ingredients. For each cell, this makes the whole process bounded. Therefore, intuitively, it is clear that after a random finite time T_n no division will happen for one of the two possible reasons: either the substrate is finished and thus no cell can receive a stimulus, or the hormone is finished and thus no more A-cell is born. Moreover, the total amount of cells remains of the order n at any time, which ensures that the usual scalings for the large deviation result (Theorem 4.2) and the central limit theorem (Theorem 5.1) work in this setting. Another feature that makes the model tractable is that, despite the process being non-Markovian, the time it takes a new cell to obtain a stimulus is exponential, which allows to talk about the rates and use the bounds developed for Markov processes (e.g. Lemma 4.1).

The large deviation result established in Theorem 4.2 implies that that $N_n(t)/n$ converges to a function $X(t)$ in probability, uniformly in t at exponential rate when n grows to infinity. Here the function $X(t)$ is the averaged integrated intensity of the process. To obtain $X(t)$, the authors need to make several steps of conditioning and averaging, where the first important step is the conditioning on the number of A-cells produced at each division. Obviously, in this model, the intensity at time t depends on the aggregated intensity before time t because this aggregated intensity defines how much substrate and hormone has been used before t. Hence, it is natural that $X(t)$ is defined as a solution of an integral equation. Technically, the uniform convergence result is very difficult and requires a lot of preliminary work. Totally different argument is used to prove the convergence for a bounded t (Theorem 4.1) and for $t \to \infty$ (Lemma 4.6). Finally, the proof of the main theorem combines all the preliminary results plus uses a very elegant argument to control the deviation of the integrated intensity process from $X(t)$.

The central limit theorem in Section 5 describes in detail the convergence of the process $V_n(t) = n^{1/2}(N_n(t)/n - X(t))$ to its limit $V(t)$ in distribution, where the convergence is in the sense of the Skorohod metric. The process $V(t)$ involves two independent Wiener processes: one of them, W_0, is responsible for the random deviation of $N_n(t)$ from the integrated intensity process, and another one, W_1, reflects the randomness due to a random number of A-cells produced at each division. Clearly, $V(t)$ is again a solution of an integral equation that involves both W_0 and W_1 in a non-trivial way. The form of $V(t)$ and its covariance structure are really complicated, and, as noticed by the authors, 'almost impossible to guess without going into the special structure of the underlying process...'.

Last section contains numerical examples, which show that the scaling results are in good agreement with experimental data.

The above summarises paper [3], which is the first part of the analysis of the non-markovian model of the population growth. This volume contains the second part of this work, paper [4], where the duration of the growth is analysed. Here the authors obtain a remarkable discontinuity result. It turns out that with a certain balance between the initial amount of hormone and substrate the number of divisions and the duration of the process is much larger than for other values of the parameters. Another example of surprising properties of this deep interesting model.

23.3 Parameter estimation for the supercritical contact process

The paper studies a contact process on a d-dimensional grid. The model description is typical for processes of this sort. Each site in $\mathbb{Z}^d$ is either infected or healthy.

A healthy site gets infected at rate λ by any of its infected neighbors. An infected site becomes healthy at rate 1. The process is supercritical, that is, with positive probability, an infection started by one infected site, will last forever. This is ensured by the inequality $\lambda > \lambda_d$, where λ_d is a critical value.

The goal of the paper is to estimate the parameter λ. Intuitively, it is not hard to imagine what the estimator should be. The authors follow a most natural path. Provided that the process started by a single infected site in 0 and survives forever, we take some set D where a stationary regime has been established. Then the estimator for λ at time t is simply

$$\hat{\lambda} = \frac{\#\text{ infected sites in } D \text{ at } t}{\#\text{ sites in } D \text{ that are healthy but have infected neighbors}}. \tag{23.1}$$

The fraction above is a result of a balance equation in stationarity: the denominator multiplied by λ is the rate at which new sites get infected, and the nominator (multiplied by one) is the rate at which infected sites get healthy. In stationarity, both rates should be equal.

The description of the proposed estimator will be complete once we decide on how to choose D, and this is where the main difficulty lies because each of the requirements to D is quite tricky: how do we know whether the stationary regime has been established? how do we know which site started the infection? and what if infection has been started by a set of sites? The authors resolve this questions by employing the shape theorem (Theorem 1.2 in the paper,e.g. [5]). The meaning of this theorem is very well described in the paper right after its statement on p. 1073. In summary, the shape theorem has two consequences. First, the set of infected sites grows with time t roughly as tU where U is a non-random set. Second, inside U, the processes started with one infected site and with all infected sites are equal eventually almost surely. Both consequences are extremely important for establishing the results of the paper. In Theorem 2.1, the authors prove that for the process started with one infected site, the convex hull of all infected sites is squeezed between $(1-\varepsilon)Ut$ and $(1+\varepsilon)Ut$ eventually a.s. provided that the process survives forever. Thus, the convex hull of infected sites becomes a starting point for creating a suitable set D. Next, the similarity of the process started with one and all infected sites provides the tool for proving the consistency of the estimator as $t \to \infty$, see Section 4.

Two other important elements of the model and the approach must be mentioned: shrinking and bounded correlations. Throughout the paper authors work not directly with the convex hull of infected sites $\mathscr{C}_t$ but rather with a so-called shrinking of this set, C_t. Shrinking is defied in Section 3 in a very general sense, and several possible procedures are suggested to obtain a shrinking. Essentially, shrinking means that the 'border' sites have to be removed. The reason is that the equilibrium has not yet set on these sites, and this may (and will!) distort the estimator. The consistency of the estimator (23.1) with $D = C_t$ as $t \to \infty$ holds under very mild shrinking conditions. However, for the asymptotic normality to hold, a certain fraction of nodes from $\mathscr{C}_t$ has to be removed. The authors notice that in fact, to obtain a good estimator, one should remove 20% to 40% of sites. Further, for the asymptotic normality it is crucial that correlations between any two sites decrease exponentially with the distance between these sites. These short-range dependencies, that ensure that some sort of central limit theorem must hold, are stated in Theorem 2.2 and further in Lemma 5.1. The asymptotic normality of the estimator is established in Theorem 5.1. This is not however the end of the story because the asymptotic variance of the estimator involves unknown parameters. In Section 6 the authors discuss this difficulty and provide a possible plausible solution.

I would like to add that a quantitative analysis of infection spread is definitely a very important topic, for example, in social and computer networks. Such networks however are usually not a grid. On the contrary, they exhibit power law degree distributions and the well known small-world phenomenon. These fascinating properties of real-life networks motivated an emergence of a new research area, devoted to the studies of complex systems, that has boosted in the last ten years. We refer to e.g. [1] for a survey of the field and its relation to statistical mechanics and interacting particle systems. The problem of infection spread in complex networks is for sure one of the key topics in this new area (see e.g. [2, 6, 10, 11]). Rigorous mathematical studies in this direction have just started. Obviously, the problem of parameter estimation for existing computer viruses and pandemics is highly relevant and offers an endless number of new mathematical challenges.

23.4 Collecting n items on a circle

Finally, my own paper [8]. This work was a continuation of my PhD thesis that I did at EURANDOM, in Eindhoven. I was lucky to have a PhD project that I could explain to anyone even without the famous back-side of an envelop. Imagine a circle and suppose that n items are distributed randomly at its circumference, which we assume to have a length 1. We start at point zero and move at a constant (unit) speed with the goal to collect all n items. We may move in one direction or turn, following any strategy we like. For instance we may choose to never change a direction, or always collect an item nearest to our current position, or pick the shortest route. The problem is to find the distribution of the travel time under different strategies. The question arises in automated storage and retrieval systems known as warehousing carousels. A circle represents a carousel that consists of a large number of shelves or drawers moving in a closed loop in either direction, and the items are locations of the products to be picked. The objective is to evaluate the rotation time, which is an important part of the response time of the system.

Clearly, if we just move in one direction, the problem is trivial: the probability to collect all items within time $t \in [0, 1]$ is just t^n. However, already for the nearest-item strategy a straightforward approach results in hopelessly messy calculations, which do not lead to any meaningful outcomes. Nevertheless, the problem has an elegant solution, and the distribution of the travel time often can be written in a very simple form. The fruitful idea is to recall that the intervals between adjacent items are uniform spacings that are distributed as i.i.d. exponential random variables, divided by their sum. Then the travel time can be written as a function of exponential random variables. In order to find the distribution of this function, the memory-less property can be used yielding surprisingly simple outcomes like in Lemma 1.1 in the paper. This way, in the papers with Ivo Adan, we derived elegant formulas for the travel time distribution under the nearest item heuristic and some other close-to-optimal strategies. For the optimal route, the problem however remained open.

It may take at most one-two minutes to guess what the optimal route on a circle should be. Clearly, it is not optimal to turn more than once. Thus, we just have to choose the shortest out of the $2n$ routes with no turn or one turn. The distribution of the optimal travel time however remains tricky even if we employ the spacings. The difficulty arises from the theoretical possibility that we may have to collect more than a half of the items before the turn. Although this scenario is all but irrelevant in practice, it has to be taken into account in the analysis, messing up the calculations. In the thesis I could not solve the problem and presented only some preliminary results on the upper bounds for the optimal route (Section 2 of the paper). Willem liked the problem from the very beginning and always believed that the distribution of the optimal route can be obtained. This paper started with obtaining the recursive equation for the optimal route (Section 3). Although the equations are not explicit, we do provide a recursion, which makes it possible to find the minimal travel time distribution for any n.

The results became much cleaner and the focus of the paper actually shifted when we turned to the asymptotic behavior as n goes to infinity. Theorem 4.2 states that in this case the difference between the shortest travel time and one complete rotation behaves as $1/(n+1)$ multiplied by the maximum between two independent random variables of the form $J = \sum_{i=1}^{\infty}(2^i - 1)^{-1}X_i$, where X_i's are independent standard exponential random variables.

Interestingly, at that time such weighted sums of exponentials attracted a lot of attention as a special case of an exponential functional of a Poisson process (see Section 6). In particular, Fabrice Guillemin, Philippe Robert and Bert Zwart encountered such functionals in the analysis of a transmission control protocols on the Internet. One intriguing and unresolved question about such random variables was their lower-tail behavior, that is, the asymptotic expression of $P(J < t)$ as $t \to 0$. To this end, only the asymptotics of $\log P(J < t)$ was known. The article by Davis and Resnick that Bert Zwart pointed to us was highly relevant but the results could not be applied directly because they were given in the form of transforms. After long calculations we arrived to the formula (5.8) that provided the exact asymptotic behavior in a closed-form. Compared to the logarithmic asymptotics, this formula contained several additional terms that were not known before. However, we were not completely satisfied because one of the factors (the function C in (5.8)) was defined by an infinite product. When plotted, this function looked liked a constant. Was it a yet another weird way to write a constant? It was tempting to prove it. We were delighted when a more detailed analysis (Proposition 5.1) revealed that our function C had an unexpected oscillating behavior involving theta-functions.

The oscillations were so small that they simply could not be seen in the plots, the analysis was needed to find them!

The explanation of why the oscillations appear seems to lie in the sort of a 'binary tree structure' of our functional J, whose coefficients are negative powers of two. Later on, Philippe Robert found that such oscillating asymptotic behavior is a typical feature of algorithms with a tree structure. For further reading I recommend his very interesting papers [9] and [12]. I think that the oscillating asymptotic behavior of algorithms is a highly compelling phenomenon, and I am very happy that our paper contributed in its study.

23.5 Acknowledgement

I would like to thank Sara van de Geer and Marten Wegkamp for creating this volume and for inviting me to contribute in it.

References

1. Réka Albert and Albert-László Barabási (2002). Statistical mechanics of complex networks.*Rev. Mod. Phys.* **74** 47–97.
2. N. Berger, C. Borgs, J.T. Chayes, and A. Saberi (2005). On the spread of viruses on the Internet. In:*Proceedings of the sixteenth annual ACM-SIAM symposium on Discrete algorithms*, 301310. Society for Industrial and Applied Mathematics Philadelphia, PA, USA, 301–310.
3. Mathisca C. M. de Gunst and Willem R. van Zwet (1992). A non-Markovian model for cell population growth: speed of convergence and central limit theorem. *Stochastic Process. Appl.* **41**, 297–324.
4. Mathisca C. M. de Gunst and Willem R. van Zwet (1993). A non-Markovian model for cell population growth: tail behavior and duration of the growth process. *Ann. Appl. Probab.* **3** 1112–1144.
5. Rick Durrett. The contact process, 19741989 (1991). In: *Mathematics of Random Media*, (Blacksburg, VA, 1989), volume 27 of Lectures in Appl. Math., pages 1–18. Amer. Math. Soc., Providence, RI.
6. Rick Durrett and Paul Jung (2007). Two phase transitions for the contact process on small worlds. *Stochast. Process. Appl.* **117**, 1910–1927.
7. Marta Fiocco and Willem R. van Zwet (2003). Parameter estimation for the supercritical contact process. *Bernoulli* **9** 1071–1092.
8. N. Litvak and W. R. van Zwet (2004). On the minimal travel time needed to collect n items on a circle. *Ann. Appl. Probab.* **14** 881–902.
9. Hanène Mohamed and Philippe Robert (2005). A probabilistic analysis of some tree algorithms. *Ann. Appl. Probab.* **15** 2445–2471.
10. M. E. J. Newman. Spread of epidemic disease on networks (2002). *Physical Review E* **66** 16–28.
11. R. Pastor-Satorras and A. Vespignani. Epidemic spreading in scale-free networks (2001). *Physical review letters* **86** 3200–3203.
12. Philippe Robert. On the asymptotic behavior of some algorithms (2005). *Random Structures Algorithms* **27** 235–250.
13. Sheldon M. Ross (1996). *Stochastic processes*. Wiley Series in Probability and Statistics: Probability and Statistics. John Wiley & Sons Inc., New York, second edition.

Part VI
Probability

The Annals of Probability
1978, Vol. 6, No. 1, 133-137

A PROOF OF KAKUTANI'S CONJECTURE ON RANDOM SUBDIVISION OF LONGEST INTERVALS

BY W. R. VAN ZWET

University of Leiden

Choose a point at random, i.e., according to the uniform distribution, in the interval (0, 1). Next, choose a second point at random in the largest of the two subintervals into which (0, 1) is divided by the first point. Continue in this way, at the nth step choosing a point at random in the largest of the n subintervals into which the first $(n-1)$ points subdivide (0, 1). Let F_n be the empirical distribution function of the first n points chosen. Kakutani conjectured that with probability 1, F_n converges uniformly to the uniform distribution function on (0, 1) as n tends to infinity. It is shown in this note that this conjecture is correct.

1. Introduction. Let X_1 be uniformly distributed on (0, 1). For $n = 2, 3, \cdots$, the conditional distribution of X_n given $X_1, \cdots, X_{n-1}$ is uniform on the largest of the n subintervals into which $X_1, \cdots, X_{n-1}$ subdivide (0, 1). Let F_n denote the empirical distribution function (df) of $X_1, \cdots, X_n$, thus $F_n(x) = n^{-1} \sum_{i=1}^{n} 1_{\{X_i \leq x\}}$.

THEOREM. *With probability* 1

$$\lim_{n\to\infty} \sup_{x\in(0,1)} |F_n(x) - x| = 0 . \tag{1.1}$$

At first sight the truth of this statement seems intuitively obvious. The Glivenko-Cantelli theorem tells us that (1.1) holds with probability 1 if X_1, $X_2, \cdots$ are independent and identically distributed (i.i.d.) according to the uniform distribution on (0, 1). Compared with this case, one feels that F_n should converge to the uniform df even faster in the present situation, because at each step one is putting a point where it is needed most, i.e., in the largest subinterval. At the same time, however, it is clear that the procedure by which the points are chosen makes their joint distribution extremely complicated. To be convinced of this, one only has to try and write down what happens in just the first few steps.

The main idea of the proof is the introduction of a stopping rule for which the stopped sequence has an essentially simpler character than the original one. For $t \in (0, 1)$, let N_t be the smallest natural number n for which $X_1, \cdots, X_n$ subdivide (0, 1) into $(n + 1)$ subintervals of length $\leq t$. Correspondingly, define $N_t = 0$ for $t \geq 1$. The basic property of the stopped sequence $X_1, \cdots, X_{N_t}$ is that any (sub-) interval appearing during its construction will receive another random point before the sequence is stopped, if and only if its length exceeds t. It follows that the joint distribution of N_t and the set $\{X_1, \cdots, X_{N_t}\}$ remains unchanged if at each step the next point is chosen at random in any one of the

Received February 28, 1977.
AMS 1970 *subject classifications.* Primary 60F15; Secondary 60K99.
Key words and phrases. Glivenko-Cantelli type theorem.

S. van de Geer and M. Wegkamp (eds.), *Selected Works of Willem van Zwet*, Selected Works in Probability and Statistics, DOI 10.1007/978-1-4614-1314-1_24, © Springer Science+Business Media, LLC 2012

existing subintervals of length $> t$ rather than in the largest subinterval as prescribed by the original procedure. In the first place this implies that for $t \in (0, 1)$, the conditional distribution of $(N_t - 1)$ given $X_1 = x$ is that of the sum of the numbers of random points needed to subdivide the intervals $(0, x)$ and $(x, 1)$ independently and in the prescribed way into subintervals of length $\leqq t$. By blowing up these intervals to length 1 and replacing t by t/x and $t/(1 - x)$ respectively one sees that

$$\mathscr{L}(N_t \mid X_1 = x) = \mathscr{L}(N_{t/x} + N_{t/(1-x)} + 1), \qquad 0 < t < 1, \tag{1.2}$$

where for $N_{t/x}$ and $N_{t/(1-x)}$ independent copies are chosen.

Another consequence of the abovementioned property of the stopped sequence is the following. Take $x \in (0, 1)$ and let $N_t(x)$ denote the number of values in $(0, x]$ among $X_1, \cdots, X_{N_t}$, thus $N_t(x) = N_t F_{N_t}(x)$. Suppose that $0 < t < x$ and let ξ be the first value in the interval $[x - t, x]$ occurring in the sequence $X_1, \cdots, X_{N_t}$. If from the $N_t(x)$ values in $(0, x]$ we delete all values in $(\xi, x]$, the number remaining is distributed as the number of random points needed to subdivide $(0, x]$ into subintervals of length $\leqq t$ in the prescribed way, i.e., as $N_{t/x}$. If also $t < 1 - x$, the same argument applied to the interval $(x, 1)$ shows that there exist copies of $N_{t/x}$ and $N_{t/(1-x)}$ such that

$$N_{t/x} \leqq N_t(x) \leqq N_t - N_{t/(1-x)} \tag{1.3}$$

with probability 1. This clearly holds for all t since $N_{t/x} = 0$ for $t \geqq x$ and $N_{t/(1-x)} = 0$ for $t \geqq 1 - x$.

2. Proof of the theorem. For $t \in [\frac{1}{2}, 1)$, the stopped sequence $X_1, \cdots, X_{N_t}$ never returns to a subinterval it has left. Hence the Markov inequality yields

$$P(N_t > k) = P(\textstyle\prod_{i=1}^k \{U_i \vee (1 - U_i)\} > t) \leqq (\frac{3}{4})^k t^{-1}, \qquad \frac{1}{2} \leqq t < 1, \tag{2.1}$$

where $U_1, U_2, \cdots$ are i.i.d. with a uniform distribution on $(0, 1)$, so that $E\{U_i \vee (1 - U_i)\} = \frac{3}{4}$. It follows that $EN_t^m < \infty$ for every $m \geqq 0$ and $\frac{1}{2} \leqq t < 1$. For $s, t \in (0, 1)$, N_{st} is stochastically smaller than a sum of $(N_t + 1)$ copies of N_s and hence $EN_t^m < \infty$ for $m \geqq 0$ and $0 < t < 1$. Since EN_t^m is nonincreasing in t,

$$\sup_{t_0 \leqq t < 1} EN_t^m < \infty \qquad \text{for} \quad 0 < t_0 < 1 \quad \text{and} \quad m \geqq 0. \tag{2.2}$$

Clearly $EN_t^m = 0$ for $t \geqq 1$ and $m \geqq 0$ because $N_t = 0$ for $t \geqq 1$. Another consequence of (2.1) is that for $\frac{1}{2} < t < 1$

$$P(N_t > k) \leqq \textstyle\prod_{i=1}^k P(\{U_i \vee (1 - U_i)\} > t) \leqq \{2(1 - t)\}^k,$$

$$EN_t = \textstyle\sum_{k=0}^\infty P(N_t > k) \leqq \dfrac{1}{2t - 1}.$$

Since $N_t \geqq 1$ a.s. for $t < 1$, it follows that

$$\lim_{t \uparrow 1} EN_t = 1. \tag{2.3}$$

Define $\mu(t) = EN_t$. For $t \geqq 1$, $\mu(t) = 0$ and in view of (1.2) one finds that for $0 < t < 1$,

$$\mu(t) = \int_0^1 \left\{\mu\left(\frac{t}{x}\right) + \mu\left(\frac{t}{1-x}\right) + 1\right\} dx = 2\int_0^1 \mu\left(\frac{t}{x}\right) dx + 1 \tag{2.4}$$

$$= 2\int_t^1 \mu\left(\frac{t}{x}\right) dx + 1 = 2t\int_t^1 \frac{\mu(y)}{y^2}\, dy + 1\,.$$

Now $\sup_{y\geq t}\mu(y) < \infty$ for $t > 0$ because of (2.2) and hence (2.4) implies that μ is continuous and even differentiable on (0, 1) with

$$\left(\frac{\mu(t)-1}{t}\right)' = \frac{\mu'(t)}{t} - \frac{\mu(t)-1}{t^2} = -2\frac{\mu(t)}{t^2},$$

or

$$\frac{\mu'(t)}{\mu(t)+1} = -\frac{1}{t}\,.$$

Together with (2.3) this yields

$$\mu(t) = \frac{2}{t} - 1 \qquad \text{for} \quad 0 < t < 1\,. \tag{2.5}$$

Let $v(t)$ denote the variance of N_t and apply (1.2) again, this time also using the independence of $N_{t/x}$ and $N_{t/(1-x)}$ in (1.2). In view of (2.5) one obtains for $0 < t \leqq \frac{1}{2}$,

$$v(t) = E\left(N_t - \frac{2}{t} + 1\right)^2$$

$$= \int_0^1 E\left(N_{t/x} + N_{t/(1-x)} - \frac{2x}{t} - \frac{2(1-x)}{t} + 2\right)^2 dx$$

$$= \int_0^1 E\left(N_{t/x} - \frac{2x}{t} + 1\right)^2 dx + \int_0^1 E\left(N_{t/(1-x)} - \frac{2(1-x)}{t} + 1\right)^2 dx$$

$$+ 2\int_0^1 E\left(N_{t/x} - \frac{2x}{t} + 1\right) E\left(N_{t/(1-x)} - \frac{2(1-x)}{t} + 1\right) dx$$

$$= 2\int_0^1 E\left(N_{t/x} - \frac{2x}{t} + 1\right)^2 dx\,,$$

where the cross-product term vanishes because of (2.5) and because either $t/x < 1$ or $t/(1-x) < 1$ for $t \in (0, \frac{1}{2}]$ and $x \in (0, 1)$, $x \neq \frac{1}{2}$. So for $t \in (0, \frac{1}{2}]$,

$$v(t) = 2\int_t^1 v(t/x)\, dx + 2\int_0^t \left(\frac{2x}{t} - 1\right)^2 dx = 2t\int_t^1 \frac{v(y)}{y^2} dy + \frac{2t}{3}\,. \tag{2.6}$$

Because of (2.2), $\sup_{y\geq t} v(y) < \infty$ for $t > 0$, and together with (2.6) this ensures that v is continuous on $(0, \frac{1}{2}]$ and differentiable on $(0, \frac{1}{2})$ with

$$\left(\frac{v(t)}{t}\right)' = \frac{v'(t)}{t} - \frac{v(t)}{t^2} = -2\frac{v(t)}{t^2}$$

or

$$\frac{v'(t)}{v(t)} = -\frac{1}{t}\,.$$

Hence, if $c = \frac{1}{2}v(\frac{1}{2})$,

$$v(t) = \frac{c}{t} \quad \text{for} \quad 0 < t < \tfrac{1}{2}. \tag{2.7}$$

For $m = 2, 3, \cdots$, define $M_m = N_{m-2}$ and $M_m(x) = N_{m-2}(x)$ for $x \in (0, 1)$. Then (2.5), (2.7) and the Bienaymé–Chebyshev inequality imply that

$$P(|M_m - 2m^2 + 1| \geqq m^{\frac{5}{3}}) \leqq \frac{\sigma^2(M_m)}{m^{\frac{10}{3}}} = cm^{-\frac{4}{3}}.$$

By the Borel–Cantelli lemma

$$\limsup_m m^{-\frac{5}{3}}|M_m - 2m^2| \leqq 1 \quad \text{a.s.}$$

so that

$$\lim_{m\to\infty} \frac{M_m}{2m^2} = 1 \quad \text{a.s.}, \tag{2.8}$$

$$\lim_{m\to\infty} \frac{M_{m+1}}{M_m} = 1 \quad \text{a.s.} \tag{2.9}$$

For fixed $x \in (0, 1)$ and $t = m^{-2}$, the reasoning leading to (2.8) may also be applied to each of the three terms on the left- and right-hand sides of (1.3). Since the argument does not involve joint distributions for different values of m, it follows without further specification of the copies chosen in (1.3) that for any fixed $x \in (0, 1)$

$$\lim_{m\to\infty} \frac{M_m(x)}{2m^2} = x \quad \text{a.s.},$$

or, in view of (2.8),

$$\lim_{m\to\infty} F_{M_m}(x) = x \quad \text{a.s.} \tag{2.10}$$

For $M_m \leqq n \leqq M_{m+1}$,

$$\begin{aligned} |F_n(x) - x| &\leqq \frac{M_m}{n}|F_{M_m}(x) - x| + \frac{n - M_m}{n}\{x \vee (1 - x)\} \\ &\leqq |F_{M_m}(x) - x| + \left(1 - \frac{M_m}{M_{m+1}}\right) \end{aligned}$$

and together with (2.8), (2.9) and (2.10) this implies that for every fixed $x \in (0, 1)$,

$$\lim_{n\to\infty} F_n(x) = x \quad \text{a.s.} \tag{2.11}$$

By a standard argument this yields (1.1) and the theorem is proved.

Acknowledgement. The author recalls with pleasure the 1976 stochastics meeting at Oberwolfach where R. M. Dudley introduced the participants to Kakutani's conjecture and proceeded to shoot down our combined attempts at solving the problem.

Note added in proof. After this paper was submitted it has come to the author's attention that J. Komlós and G. Tusnády had also arrived at the conclusion that Kakutani's conjecture can be proved by the method employed in this paper. More recently essentially the same proof was given again independently in Lootgieter (1977a); an outline of this paper is given in Lootgieter (1977b). For the solution of a related nonrandom problem the reader is referred to Kakutani (1975), Adler and Flatto (1977) and Lootgieter (loc. cit.).

REFERENCES

ADLER, R. L. and FLATTO, L. (1977). Uniform distribution of Kakutani's interval splitting procedure. *Z. Wahrscheinlichkeitstheorie und Verw. Gebiete* **38** 253–259.

KAKUTANI, S. (1975). A problem in equidistribution. *Lecture Notes in Math*. **541** 369–376. Springer, Berlin.

LOOTGIETER, J. C. (1977a). Sur la répartition des suites de Kakutani. Technical Report, Université Paris VI; submitted to *Ann. Henri Poincaré*.

LOOTGIETER, J. C. (1977b). Sur la répartition des suites de Kakutani. *C.R. Acad. Sci. Paris* **285A** 403–406.

INSTITUTE OF APPLIED MATHEMATICS
AND COMPUTER SCIENCE
UNIVERSITY OF LEIDEN
WASSENAARSEWEG 80, LEIDEN
THE NETHERLANDS

The Annals of Probability
1980, Vol. 8, No. 5, 986–990

A STRONG LAW FOR LINEAR FUNCTIONS OF ORDER STATISTICS

BY W. R. VAN ZWET

University of Leiden

A strong law of large numbers for linear combinations of order statistics is proved under integrability conditions only. Together with some straightforward extensions, the theorem generalizes previous results of Wellner, Helmers and Sen.

1. Introduction. Let $U_1, U_2, \cdots$ be random variables defined on a single probability space $(\Omega, \mathcal{Q}, P)$ and suppose that $U_1, U_2, \cdots$ are independent and identically distributed (i.i.d.) according to the uniform distribution on (0, 1). For $N = 1, 2, \cdots$, $U_{1:N} < U_{2:N} < \cdots < U_{N:N}$ denote the ordered $U_1, \cdots, U_N$. Introduce Lebesgue measurable functions $J_N : (0, 1) \to \mathbb{R}$, $N = 1, 2, \cdots$, a Borel measurable function $g : (0, 1) \to \mathbb{R}$ and define $g_N : (0, 1) \to \mathbb{R}$, $N = 1, 2, \cdots$, by

$$g_N(t) = g(U_{[Nt]+1:N}), \tag{1.1}$$

where $[x]$ denotes the integer part of x. We adopt the convention that when integration is with respect to Lebesgue measure λ on (0, 1), we shall write $\int f$ for $\int f d\lambda$. The range of integration will be (0, 1) unless explicitly indicated otherwise. For $1 \leqslant p \leqslant \infty$, L_p is the Lebesgue space of measurable functions $f : (0, 1) \to \mathbb{R}$ with finite norm $\| f \|_p = \{\int | f |^p\}^{1/p}$ for $1 \leqslant p < \infty$ and $\| f \|_\infty = \text{ess sup}| f |$ for $p = \infty$.

The purpose of this note is to show that under integrability assumptions on J_N and g,

$$M_N = \int J_N(g_N - g) = \Sigma_{i=1}^N g(U_{i:N}) \int_{(i-1)/N}^{i/N} J_N - \int J_N g \tag{1.2}$$

converges to zero for $N \to \infty$ with probability 1 (w.p. 1). If, moreover, J_N converges in an appropriate sense to a function J which shares the integrability properties of J_N, we prove that

$$\tilde{M}_N = \int J_N g_N - \int J g \tag{1.3}$$

also converges to zero w.p. 1.

If $J_N(t) = c_{N,i}$ for $(i - 1)/N < t \leqslant i/N$, $i = 1, \cdots, N$, and $g = h \circ F^{-1}$ for a probability distribution function (df) F on $\mathbb{R}$ and a Borel measurable function $h : \mathbb{R} \to \mathbb{R}$, then the joint distribution of $\int J_N g_N$, $N = 1, 2, \cdots$, is that of $N^{-1} \Sigma c_{N,i} h(X_{i:N})$, $N = 1, 2, \cdots$, where the $X_{i:N}$ are order statistics of a sequence of i.i.d. random variables with common df F. We are therefore concerned with the almost sure convergence of suitably standardized linear combinations of a function of order statistics.

Received September 25, 1978; revised May 22, 1979.
AMS 1970 subject classifications. Primary 60F15; secondary 62G30.
Key words and phrases. Strong law, order statistics.

S. van de Geer and M. Wegkamp (eds.), *Selected Works of Willem van Zwet*, Selected Works in Probability and Statistics, DOI 10.1007/978-1-4614-1314-1_25, © Springer Science+Business Media, LLC 2012

Previous results in this direction may be found in Wellner (1977), Helmers (1977) and Sen (1978). Wellner restricts attention to the case where $J_N(t) = c_{N,i}$ for $(i-1)/N < t \leq i/N$, $i = 1, \cdots, N$, and assumes that g is left continuous and of bounded variation on closed subintervals of (0, 1). He proves that $M_N \to 0$ w.p. 1 if numbers b_1, b_2 and C, as well as $\delta > 0$ exist such that, for all N and $t \in (0, 1)$,

$$|g(t)| \leq Ct^{-1+b_1+\delta}(1-t)^{-1+b_2+\delta}, \tag{1.4}$$

$$|J_N(t)| \leq Ct^{-b_1}(1-t)^{-b_2}, \tag{1.5}$$

$$\int t^{1-b_1-\frac{1}{2}\delta}(1-t)^{1-b_2-\frac{1}{2}\delta}d|g| < \infty, \tag{1.6}$$

where $d|g|$ denotes integration with respect to the total variation measure induced by g. He shows that $\tilde{M}_N \to 0$ w.p. 1 under the additional assumption that J_N converges to J pointwise.

It is clear that Wellner's result will cover most cases that one is likely to come across in practice, the main flaw being that it just fails to contain the strong law for the sample mean, i.e., the case where $J_N \equiv 1$ and $g \in L_1$. This gap is closed in Helmers (1977) where it is shown that $\tilde{M}_N \to 0$ w.p. 1 for $c_{N,i} = J(i/(N+1))$, J piecewise continuous and bounded and $g = F^{-1} \in L_1$.

For $b_1, b_2 \in [0, 1)$, Wellner's conditions (1.4) and (1.5) imply integrability of g and J_N and for this case a mathematically more satisfactory result was obtained in Sen (1978, Theorem 4.1). Sen also takes $J_N(t) = c_{N,i}$ for $(i-1)/N < t \leq i/N$, $i = 1, \cdots, N$, and assumes that J_N converges pointwise to J, but now J is required to be continuous and of bounded variation on closed subintervals of (0, 1). This switching of the smoothness condition from g (Wellner) to J (Helmers and Sen) is quite common in problems concerning linear functions of order statistics, where one can use both kinds of smoothness almost interchangeably. The improvement, however, is that instead of (1.4)–(1.6), Sen requires that $g \in L_q$ and $\sup_N \| J_N \|_p < \infty$ for some $p, q \in (1, \infty)$ with $p^{-1} + q^{-1} = 1$, to prove that $\tilde{M}_N \to 0$ w.p. 1. Note that $J_N \to J$ pointwise and $\sup_N \| J_N \|_p < \infty$ imply $J \in L_p$ by Fatou's lemma and together with $g \in L_q$ this ensures that Sen's assumption that $Jg \in L_1$ is automatically satisfied. Apparently unaware of Wellner (1977), Sen also proves another result (Theorem 4.2) which is strictly contained in Wellner's.

The present note constitutes an attempt to provide a mathematically cleaner version of the above results. Roughly speaking we shall show that all smoothness conditions on g and J, including (1.6), are superfluous and that the pointwise convergence of J_N can be relaxed. We do not assume that J_N is a step function.

2. A strong law. Let $g: (0, 1) \to \mathbb{R}$ be Borel measurable and let g_N be defined by (1.1). We begin by proving

LEMMA 2.1. *With probability* 1, g_N *converges to* g *in Lebesgue measure*, i.e., $\lim_{N\to\infty} \lambda\{t: |g_N(t) - g(t)| \geq \delta\} = 0$ *for every* $\delta > 0$.

PROOF. Choose $\varepsilon > 0$. By Lusin's theorem there exists a Borel set $B \subset (0,1)$ and a continuous function $\tilde{g}: (0,1) \to \mathbb{R}$ such that $\lambda(B) \leqslant \varepsilon$ and $g = \tilde{g}$ on $(0,1) \cap B^c$. Define $\tilde{g}_N(t) = \tilde{g}(U_{[Nt]+1:N})$ and $B_N = \{t: U_{[Nt]+1:N} \in B\}$, so that $g_N = \tilde{g}_N$ on $(0,1) \cap B_N^c$. Since $\lambda(B_N) = P_N(B)$, where P_N denotes the empirical distribution of $U_1, \cdots, U_N$, it follows from the strong law that $\limsup_N \lambda(B_N) \leqslant \varepsilon$ w.p. 1. In view of the Glivenko-Cantelli theorem and the continuity of $\tilde{g}$, this implies that w.p. 1 we have for every $\delta > 0$

$$\limsup_N \lambda\{t: |g_N(t) - g(t)| \geqslant \delta\} \leqslant \lambda(B) + \limsup_N \lambda(B_N)$$
$$+ \limsup_N \lambda\{t: |\tilde{g}_N(t) - \tilde{g}(t)| \geqslant \delta\} \leqslant 2\varepsilon.$$

Since $\varepsilon > 0$ is arbitrary the lemma is proved.

THEOREM 2.1. *Let* $1 \leqslant p \leqslant \infty$, $p^{-1} + q^{-1} = 1$, *and suppose that* $J_N \in L_p$ *for* $N = 1, 2, \cdots$ *and* $g \in L_q$. *If either*

(i) $1 < p \leqslant \infty$ *and* $\sup_N \| J_N \|_p < \infty$, *or*

(ii) $p = 1$ *and* $\{J_N : N = 1, 2, \cdots\}$ *is uniformly integrable,*

then $\lim_{N\to\infty} M_N = 0$ *with probability* 1.

PROOF. Suppose first that $1 < p \leqslant \infty$, so $q < \infty$; w.p. 1, $g_N \to g$ in Lebesgue measure and $\int |g_N|^q = N^{-1}\Sigma|g(U_i)|^q \to \int |g|^q$ by the strong law. By Vitali's theorem this implies that $\int |g_N - g|^q \to 0$, and Hölder's inequality yields $|M_N| \leqslant \| J_N \|_p \| g_N - g \|_q \to 0$ w.p. 1.

Suppose now that $p = 1$, so $q = \infty$. Because of the uniform integrability of J_N and Lemma 2.1, we have w.p. 1

$$\limsup_N |M_N| \leqslant \delta \limsup_N \| J_N \|_1 + 2\| g \|_\infty \limsup_N \int_{\{|g_n - g| > \delta\}} |J_N|$$
$$= \delta \limsup_N \| J_N \|_1$$

for every $\delta > 0$. Since $\sup_N \| J_N \|_1 < \infty$, the proof is complete.

For $J_N \in L_p$, $N = 1, 2, \cdots$, consider the type of convergence to $J \in L_p$ defined by $\lim_{N\to\infty} \int J_N f = \int J f$ for every $f \in L_q$. For $1 \leqslant p < \infty$ this is weak convergence in L_p and for $p = \infty$ it is weak* convergence in L_∞. Necessary and sufficient conditions for a set $\{J_N, N = 1, 2, \cdots\} \subset L_p$ to be sequentially relatively compact in the topology of this convergence are precisely conditions (i) and (ii) in Theorem 2.1 (for $1 \leqslant p < \infty$ see Dunford and Schwartz (1958), IV.8.4 and IV.8.11; for $p = \infty$ see Banach (1932), page 131, for the sufficiency of (i); the necessity is easy). To ensure that J_N converges to $J \in L_p$ in the above sense one only has to add to conditions (i) and (ii) the further assumption that $\int_0^t J_N \to \int_0^t J$ for every $t \in (0,1)$ (see Dunford and Schwartz (1958), IV. 13.23, 25, 27, and Banach (1932), page 135–136). Under this additional assumption we may therefore replace $\int J_N g$ by $\int J g$ in Theorem 2.1 to obtain

COROLLARY 2.1. *Suppose that the conditions of Theorem* 2.1 *are satisfied and that there exists a function* $J \in L_p$ *such that* $\lim_{N\to\infty} \int_0^t J_N = \int_0^t J$ *for every* $t \in (0,1)$. *Then* $\lim_{N\to\infty} \tilde{M}_N = 0$ *with probability* 1.

Note that the remarks preceding Corollary 2.1 also imply that the conditions on J_N and J in the corollary are necessary as well as sufficient to ensure that $\tilde{M}_N \to 0$ w.p. 1 for every $g \in L_q$.

3. Variations on a theme. Theorem 2.1 and its corollary clearly contain Wellner's result for $0 \leq b_1 = b_2 < 1$ as well as those of Helmers and Sen (cf. Section 1). In this section we extend our results to cover the other cases discussed by Wellner, which enlarges the range of applications considerably. Though these extensions are straightforward, the conditions inevitably become more cumbersome to state.

For different b_1 and b_2 in $[0, 1)$, (1.4) and (1.5) allow a different balance between the rates of growth of g and J_N near 0 and 1. Correspondingly, we shall show that in Theorem 2.1 and Corollary 2.1 one may allow different values of p and q on different subintervals of $(0, 1)$, provided these subintervals overlap; the existence of such overlapping subintervals is easily seen to be equivalent to the assumptions of Theorem 3.1. The indicator function of a set A is denoted by $\chi(A)$ or $\chi(A,.)$.

THEOREM 3.1. *Let* $0 = t_0 < t_1 < \cdots < t_k = 1$ *and* $\varepsilon > 0$. *For* $j = 1, \cdots, k$, *let* $1 \leq p_j \leq \infty$, $p_j^{-1} + q_j^{-1} = 1$ *and define intervals* $A_j = (t_{j-1}, t_j)$ *and* $B_j = (t_{j-1} - \varepsilon, t_j + \varepsilon) \cap (0, 1)$. *Suppose that, for* $j = 1, \cdots, k$, $J_N\chi(A_j) \in L_{p_j}$ *for* $N = 1, 2, \cdots$, $g\chi(B_j) \in L_{q_j}$ *and either*

(i) $1 < p_j \leq \infty$ *and* $\sup_N \|J_N\chi(A_j)\|_{p_j} < \infty$, *or*

(ii) $p_j = 1$ *and* $\{J_N\chi(A_j) : N = 1, 2, \cdots\}$ *is uniformly integrable.*

Then $\lim_{N\to\infty} M_N = 0$ *with probability* 1. *If, moreover, there exists a function* J *with* $J\chi(A_j) \in L_{p_j}$ *for* $j = 1, \cdots, k$, *such that* $\lim_{N\to\infty} \int_0^t J_N = \int_0^t J$ *for every* $t \in (0, 1)$, *then also* $\lim_{N\to\infty} \tilde{M}_N = 0$ *with probability* 1.

PROOF. Consider an index j with $1 < p_j \leq \infty$, so $q_j < \infty$. Choose $\delta \in (0, \varepsilon]$ and define $C_j = (t_{j-1} - \delta, t_j + \delta) \cap (0, 1)$. The Glivenko-Cantelli theorem and the strong law ensure that w.p. 1

$$\lim_N \sup \int_{A_j} |g_N|^{q_j} \leq \lim_N \sup \frac{1}{N} \Sigma |g(U_i)|^{q_j} \chi(C_j, U_i) = \int_{C_j} |g|^{q_j} < \infty.$$

Since $\delta \in (0, \varepsilon]$ is arbitrary, this implies that $\int |g_N|^{q_j}\chi(A_j) \to \int |g|^{q_j}\chi(A_j)$ w.p. 1 by Fatou's lemma. Arguing as in the proof of Theorem 2.1, we conclude that $\int J_N(g_N - g)\chi(A_j) \to 0$ w.p. 1.

For an index j with $p_j = 1$ and $q_j = \infty$, the Glivenko-Cantelli theorem ensures that $\lim \sup_N \|g_N\chi(A_j)\|_\infty \leq \|g\chi(B_j)\|_\infty < \infty$ w.p. 1, and again copying the proof of Theorem 2.1, we find that $\int J_N(g_N - g)\chi(A_j) \to 0$ w.p. 1. This proves the first statement of the theorem. The second statement is obvious because the assumptions of the theorem imply that $\int J_N g\chi(A_j) \to \int Jg\chi(A_j)$ for $j = 1, \cdots, k$.

A second extension of our results concerns, e.g., the case where near a point $t_0 \in [0, 1]$, $|g|$ (or $|J_N|$) grows faster than (uniform) integrability would allow, but where the effect of this is cancelled by the fact that J_N (or g) tends to zero at t_0 at an appropriate rate. Since we are concerned with the product of J_N at the point t

and g at the point $U_{[Nt]+1:N}$, this cancellation will work best if we can pin down the order statistics near t_0 quite close to their expected values. This means that the best results are to be obtained for $t_0 = 0$ and/or 1, which corresponds to (1.4) and (1.5) for the case where one or both of the b_i are outside the interval [0, 1). As this is also the most common situation in applications, we shall restrict the few remarks we make to this case. The reader can easily formulate a similar result for arbitrary t_0 for himself.

Take any $\delta > 0$ and define intervals $K_{N,i}$ for $i = 1, \cdots, [(N+1)/2]$, $N = 1, 2, \cdots$, by

$$K_{N,i} = \left[\frac{i}{N}\left\{\log\left(\frac{N^2}{i(N-i+1)}\right)\right\}^{-1-\delta},\right.$$

$$\left.(1+\delta)\frac{i}{N}\log\log\left(\frac{4N^2}{i(N-i+1)}\right)\right] \cap (0, 1).$$

For $P-$ almost every $\omega \in \Omega$, there exists $N(\omega)$ such that for $N \geqslant N(\omega)$ and $i = 1, \cdots, [(N+1)/2]$ we have $U_{i:N} \in K_{N,i}$. This follows easily from Theorems 2 and 3 in Shorack and Wellner (1978) together with Bernstein's inequality for binomial tails. For $N = 1, 2, \cdots$, define $\bar{g}_N : (0, 1) \to [0, \infty]$ by

$$\bar{g}_N(t) = \sup\{|g(s)| : s \in K_{N,[Nt]+1}\} \qquad \text{for } t \in \left(0, \tfrac{1}{2}\right),$$

$$= \sup\{|g(1-s)| : s \in K_{N,N-[Nt]}\} \qquad \text{for } t \in \left[\tfrac{1}{2}, 1\right).$$

Then, w.p. 1, $|g_N| \leqslant \bar{g}_N$ on (0, 1) for sufficiently large N.

For $\eta \in (0, \frac{1}{2})$, let D_η denote the interval $(\eta, 1-\eta)$. The following result is now an immediate consequence of Theorem 3.1.

THEOREM 3.2. *Suppose that, for every* $\eta \in (0, \frac{1}{2})$, $g\chi(D_\eta)$, $J_N\chi(D_\eta)$ *and* $J\chi(D_\eta)$ *satisfy the conditions on* g, J_N *and* J *in Theorem* 3.1. *If also*

$$\lim_{\eta\to 0}\lim_N \sup \int_{D_\eta^c} |J_N|(\bar{g}_N + |g|) = \lim_{\eta\to 0}\int_{D_\eta^c}|Jg| = 0,$$

then the conclusions of Theorem 3.1 *continue to hold.*

REFERENCES

[1] BANACH, S. (1932). *Théorie des Opérations Linéaires*. Chelsea, New York.
[2] DUNFORD, N. and SCHWARTZ, J. T. (1958). *Linear Operators, Vol. I*. Interscience, New York.
[3] HELMERS, R. (1977). A strong law of large numbers for linear combinations of order statistics. Report SW 50/77, Mathematisch Centrum, Amsterdam.
[4] SEN, P. K. (1978). An invariance principle for linear combinations of order statistics. *Z. Wahrscheinlichkeitstheorie und Verw. Gebiete* **42** 327–340.
[5] SHORACK, G. R. and WELLNER, J. A. (1978). Linear bounds on the empirical distribution function. *Ann. Probability* **6** 349–353.
[6] WELLNER, J. A. (1977). A Glivenko-Cantelli theorem and strong laws of large numbers for functions of order statistics. *Ann. Statist.* **5** 473–480.

DEPARTMENT OF MATHEMATICS
UNIVERSITY OF LEIDEN
WASSENAARSEWEG 80
POSTBUS 9512 2300 RA LEIDEN
THE NETHERLANDS

The Annals of Probability
1987, Vol. 15, No. 3, 871–884

A REFINEMENT OF THE KMT INEQUALITY FOR THE UNIFORM EMPIRICAL PROCESS

BY DAVID M. MASON[1] AND WILLEM R. VAN ZWET

University of Munich and University of Leiden

A refinement of the Komlós, Major and Tusnády (1975) inequality for the supremum distance between the uniform empirical process and a constructed sequence of Brownian bridges is obtained. This inequality leads to a weighted approximation of the uniform empirical and quantile processes by a sequence of Brownian bridges dual to that recently given by M. Csörgő, S. Csörgő, Horváth and Mason (1986). The present theory approximates the uniform empirical process more closely than the uniform quantile process, whereas the former theory more closely approximates the uniform quantile process.

1. Introduction. Let $U_1, U_2, \ldots,$ be a sequence of independent uniform $(0,1)$ random variables, and for each $n \geq 1$, let G_n denote the uniform empirical distribution function and $U_{1,n} \leq \cdots \leq U_{n,n}$ the order statistics based on the first n of these uniform $(0,1)$ random variables. Define the uniform empirical quantile function to be, for each $n \geq 1$,

$$U_n(s) = U_{k,n}, \qquad (k-1)/n < s \leq k/n,\ k = 1, \ldots, n,$$

where $U_n(0) = U_{1,n}$, and the uniform quantile process

$$\beta_n(s) = n^{1/2}(s - U_n(s)), \qquad 0 \leq s \leq 1.$$

Also let

$$\alpha_n(s) = n^{1/2}(G_n(s) - s), \qquad 0 \leq s \leq 1$$

denote the uniform empirical process.

M. Csörgő, S. Csörgő, Horváth and Mason (Cs-Cs-H-M) (1986) recently constructed a probability space on which sit a sequence $U_1, U_2, \ldots,$ of independent uniform $(0,1)$ random variables and a sequence $B_1, B_2, \ldots,$ of Brownian bridges such that for universal positive constants a, b, c and n_0

$$P\Big(\sup_{0 \leq s \leq d/n} n^{1/2}|\beta_n(s) - B_n(s)| > a \log d + x\Big) < be^{-cx}, \tag{1}$$

for all $n_0 \leq d \leq n$, $0 \leq x \leq d^{1/2}$, with the same inequality holding for the supremum taken over $1 - d/n \leq s \leq 1$. Setting $d = n$ in (1) yields the M. Csörgő and Révész (1978) inequality for the Brownian bridge approximation to the uniform quantile process.

Received July 1985; revised May 1986.

[1]Research supported by the Alexander von Humboldt Foundation while visiting the University of Munich on leave from the University of Delaware.

AMS 1980 *subject classifications*. Primary 60F99, 60F17.

Key words and phrases. Weighted empirical and quantile processes, Brownian bridge approximation.

S. van de Geer and M. Wegkamp (eds.), *Selected Works of Willem van Zwet*, Selected Works in Probability and Statistics, DOI 10.1007/978-1-4614-1314-1_26, © Springer Science+Business Media, LLC 2012

From inequality (1) we obtain immediately that on the Cs-Cs-H-M (1986) probability space

$$\text{(2)} \qquad \sup_{0 \le s \le 1} n^{1/2} |\beta_n(s) - B_n(s)| = O(\log n) \quad \text{a.s.}$$

We note that this particular sequence B_n does not approximate the empirical process α_n as closely as it does β_n, since by Kiefer (1970)

$$\text{(3)} \quad \limsup_{n \to \infty} \sup_{0 \le s \le 1} (2n)^{1/4} |\alpha_n(s) - \beta_n(s)| / \left((\log\log n)^{1/4} (\log n)^{1/2} \right) = 1 \quad \text{a.s.},$$

which in combination with (2) yields

$$\text{(4)} \quad \limsup_{n \to \infty} \sup_{0 \le s \le 1} (2n)^{1/4} |\alpha_n(s) - B_n(s)| / \left((\log\log n)^{1/4} (\log n)^{1/2} \right) = 1 \quad \text{a.s.}$$

Inequality (1) leads to the following important weighted approximation statements [cf. Cs-Cs-H-M (1986)]:

On the Cs-Cs-H-M (1986) probability space we have, for any $0 \le \nu_1 < \frac{1}{2}$,

$$\text{(5)} \qquad \sup_{1/(n+1) \le s \le n/(n+1)} |\beta_n(s) - B_n(s)| / (s(1-s))^{1/2 - \nu_1} = O_P(n^{-\nu_1}),$$

and for any $0 \le \nu_2 < \frac{1}{4}$,

$$\text{(6)} \qquad \sup_{0 \le s \le 1} |\alpha_n(s) - \overline{B}_n(s)| / (s(1-s))^{1/2 - \nu_2} = O_P(n^{-\nu_2}),$$

where for $n \ge 2$, $\overline{B}_n(s) = B_n(s)$ when $1/n \le s \le 1 - 1/n$ and zero elsewhere. It can be shown that statements (5) and (6) do not hold for $\nu_1 \ge \frac{1}{2}$ and $\nu_2 \ge \frac{1}{4}$.

The construction of the Cs-Cs-H-M (1986) probability space is based on the Komlós, Major and Tusnády (KMT) (1976) strong approximation to the partial sums of independent random variables. In Cs-Cs-H-M (1986) it was remarked that an analogous theory should be feasible starting out instead from the KMT (1975) strong approximation to the uniform empirical process. The purpose of this paper is to present this alternative theory.

Just as the key result in the Cs-Cs-H-M (1986) theory is inequality (1), a refinement of the M. Csörgő and Révész (1978) inequality, the key result in the present alternative theory is a refinement of the KMT (1975) inequality for the Brownian bridge approximation to the uniform empirical process.

THEOREM 1. *There exist a sequence of independent uniform* (0, 1) *random variables* $U_1, U_2, \ldots$, *and a sequence of Brownian bridges* $B_1, B_2, \ldots$, *sitting on the same probability space* $(\Omega, \mathscr{A}, P)$ *such that for universal positive constants* C, K *and* λ,

$$\text{(7)} \qquad P\left(\sup_{0 \le s \le d/n} n^{1/2} |\alpha_n(s) - B_n(s)| > C \log d + x \right) < K e^{-\lambda x},$$

for all $-\infty < x < \infty$ *and* $1 \le d \le n$, *with the same inequality holding for the supremum taken over* $1 - d/n \le s \le 1$.

Setting $d = n$ in (7) yields the original KMT (1975) inequality.

From (7) we have immediately that on the probability space of Theorem 1

$$(8)\qquad \sup_{0\le s\le 1} n^{1/2}|\alpha_n(s) - B_n(s)| = O(\log n) \quad \text{a.s.},$$

whereas now by the Kiefer result quoted in (3)

$$(9)\quad \limsup_{n\to\infty} \sup_{0\le s\le 1} (2n)^{1/4}|\beta_n(s) - B_n(s)|/\big((\log\log n)^{1/4}(\log n)^{1/2}\big) = 1 \quad \text{a.s.}$$

By essentially copying the proofs of Theorems 2.1 and 2.2 of Cs-Cs-H-M (1986), we obtain the following versions of the above weighted approximation statements:

THEOREM 2. *On the probability space of Theorem* 1, *statement* (5) *holds for all* $0 \le \nu_1 < \frac{1}{4}$ *and statement* (6) *holds for all* $0 \le \nu_2 < \frac{1}{2}$.

We see that not only are the almost sure approximation statements reversed, but so are the weighted approximation statements. Hence, we have a theory completely dual to that given in Cs-Cs-H-M (1986). In applications of this approximation methodology in probability and statistics, one now has the choice of working on the Cs-Cs-H-M (1986) probability space or on the probability space of Theorem 1 depending on whether in the particular problem in question one needs to approximate more closely the uniform empirical or the uniform quantile process by a sequence of Brownian bridges. For some of the wide-ranging applications of this weighted approximation theory the reader is referred to Cs-Cs-H-M (1986).

The remainder of this paper is devoted to a proof of Theorem 1. This proof resembles that of the KMT (1975) inequality for the empirical process and it would have been convenient if we could merely have pointed out the modifications needed to produce the refinement of Theorem 1. Unfortunately, the proof in KMT (1975) contains few details and we shall have to provide these in the present paper. The inequality for the tail of a multinomial distribution that is given in Lemma 3, may be of independent interest.

2. Outline of the proof of Theorem 1. Let B denote a fixed Brownian bridge. For each integer $n \ge 1$ we construct n independent uniform (0, 1) random variables $U_1^{(n)},\dots,U_n^{(n)}$ as random functions of increments of the Brownian bridge B exactly as in KMT (1975), pages 123–124. Let $\tilde{G}_n$ and $\tilde{\alpha}_n$ denote the empirical distribution function and empirical process based on $U_1^{(n)},\dots,U_n^{(n)}$. For any nonnegative integers i and k such that $0 < (k+1)2^{-i} \le 1$, write

$$\Delta_{i,k}^{(n)} = n\big(\tilde{G}_n((k+1)2^{-i}) - \tilde{G}_n(k2^{-i})\big)$$

and

$$D_{i,k}^{(n)} = n^{1/2}\big(B((k+1)2^{-i}) - B(k2^{-i})\big).$$

Also let

$$\tilde{\Delta}_{i,k}^{(n)} = \Delta_{i+1,2k}^{(n)} - \Delta_{i+1,2k+1}^{(n)}$$

and

$$\tilde{D}_{i,k}^{(n)} = D_{i+1,2k}^{(n)} - D_{i+1,2k+1}^{(n)}.$$

For the sequence of random vectors $(U_1^{(n)}, \dots, U_n^{(n)}, B)$, $n = 1, 2, \dots,$ the following fundamental inequality holds:

LEMMA 1 [Lemma 2 of KMT (1975)]. *There exist positive constants C_1, C_2 and η such that*

$$|\tilde{\Delta}_{i,k}^{(n)} - \tilde{D}_{i,k}^{(n)}| \le C_1 2^i n^{-1}\Big\{\big(\tilde{\Delta}_{i,k}^{(n)}\big)^2 + \big(\Delta_{i,k}^{(n)} - n2^{-i}\big)^2\Big\} + C_2,$$

whenever

$$|\tilde{\Delta}_{i,k}^{(n)}| \le \eta n 2^{-i} \quad \textit{and} \quad |\Delta_{i,k}^{(n)} - n2^{-i}| \le \eta n 2^{-i}.$$

To prove Theorem 1 it will be enough to show that the following inequality is valid:

INEQUALITY 1. *There exist universal positive constants K, C and λ such that for all $-\infty < x < \infty$, $n \ge 1$ and $1 \le d \le n$,*

$$P\Big(\sup_{0 \le s \le d/n} n^{1/2}|B(s) - \tilde{\alpha}_n(s)| > C \log d + x\Big) < Ke^{-\lambda x}, \tag{10}$$

with the same inequality holding for the supremum taken over $1 - d/n \le s \le 1$.

The fact that the second part of Inequality 1 is true follows from the first part and the underlying symmetry of the KMT construction, i.e.,

$$\{(\tilde{\alpha}_n(s), B(s)) : 0 \le s \le 1\} =_{\mathscr{D}} \{(\tilde{\alpha}_n(1-s), B(1-s)) : 0 \le s \le 1\}.$$

Having established the inequality for $(\tilde{\alpha}_n, B)$, $n = 1, 2, \dots,$ one can then construct a sequence of independent uniform $(0,1)$ random variables $U_1, U_2, \dots,$ and a sequence of Brownian bridges $B_1, B_2, \dots,$ sitting on the same probability space $(\Omega, \mathscr{A}, P)$, say, such that Inequality 1 holds with $\tilde{\alpha}_n$ replaced by α_n and B by B_n. The general technique of constructing such a probability space is described in Lemma 3.1.2 in M. Csörgő (1983).

Inequality 1 is almost a direct consequence of the following inequality:

INEQUALITY 2. *There exist universal positive constants a, b and λ such that for any $n \ge 1$ and $1 \le z \le n$,*

$$P\Big(\sup_{0 \le s \le 2^{-j}} n^{1/2}|B(s) - \tilde{\alpha}_n(s)| > z\Big) < a \exp\{b(p - j) - \lambda z\},$$

where p is a nonnegative integer such that

$$n2^{-(p+1)} < z/32 \le n2^{-p}$$

and j is any integer $0 \le j \le p$.

To see that Inequality 2 implies Inequality 1, we choose any $n \geq 1$ and $1 \leq d \leq n$. Select an x such that $1 \leq C \log d + x \leq n$, where $C = b/(\lambda \log 2)$ with b and λ as in Inequality 2. Define integers p and j by

$$n2^{-(p+1)} < (x + C \log d)/32 \leq n2^{-p},$$
$$j = \min([\log_2(n/d)], p),$$

where $[y]$ denotes the integer part of y.

Now $d/n \leq 2^{-j}$ and by Inequality 2

$$\begin{aligned} &P\Big(\sup_{0 \leq s \leq d/n} n^{1/2}|B(s) - \tilde{\alpha}_n(s)| > C \log d + x\Big) \\ &\quad \leq P\Big(\sup_{0 \leq s \leq 2^{-j}} n^{1/2}|B(s) - \tilde{\alpha}_n(s)| > C \log d + x\Big) \\ &\quad \leq a \exp\{b(p - j) - \lambda C \log d - \lambda x\}. \end{aligned} \tag{11}$$

Since $p \leq \log_2 n + 5$, our choice of C and j implies that

$$b(p - j) - \lambda C \log d \leq b \max(\log_2 d + 6, 0) - b \log_2 d \leq 6b$$

and hence (10) holds for $1 \leq C \log d + x \leq n$ with $K = a \exp\{6b\}$. If $C \log d + x < 1$, then necessarily $x < 1$ and (10) holds with $K = \exp\{\lambda\}$.

Finally, let $C \log d + x > n$. There exists a positive K_0 such that

$$P\Big(\sup_{0 \leq s \leq 1} |B(s)| > r\Big) + P\Big(\sup_{0 \leq s \leq 1} |\tilde{\alpha}_n(s)| > r\Big) < K_0 \exp\{-2r^2\},$$

for all $n \geq 1$ and $r \geq 0$ [cf. M. Csörgő and Révész (1981) and Dvoretzky, Kiefer and Wolfowitz (1956)]. Since now $(C \log d + x)^2/n > x$, it follows that (10) holds with $K = K_0$ and $\lambda = \frac{1}{2}$. Combining these results we find that Inequality 1 holds, if we assume the validity of Inequality 2.

The proof of Theorem 1 will be complete once we establish Inequality 2. This will be done in Section 3.

3. Proof of Inequality 2. The proof of Inequality 2 will consist of a number of lemmas. Repeated use will be made of the following special case of Bernstein's inequality: Let X have a binomial distribution with parameters $n \geq 1$ and $0 < p < 1$. Then for any $r \geq 0$ [cf. Bennett (1962)],

$$P(|X - np| > (np)^{1/2} r) < 2 \exp\left\{-\frac{r^2}{2\left(1 + r/\left(3(np)^{1/2}\right)\right)}\right\}. \tag{12}$$

For each $i = 1, 2, \ldots,$ set

$$\xi_{i,n} = \left(n\tilde{G}_n(2^{-i}) - n2^{-i}\right)^2$$

and for any choice of integers $0 \leq j \leq p$ and $l \geq 1$, define

$$S_{j,p} = \sum_{i=j+1}^{p+1} 2^i \xi_{i,n}, \qquad T_l = \sum_{i=1}^{l} 2^{2i-l} \xi_{i,n}.$$

We shall first be concerned with establishing bounds for the tails of the distributions of $S_{j,p}$ and T_l.

LEMMA 2. *For every $A > 0$ there exists a positive number λ_1 such that for all $n \ge 1$, $z \ge 0$ and p so that $z \le An\,2^{-p}$, and $1 \le l \le p+1$,*

$$P\left(n^{-1}T_l > z\right) < 2\exp\{-\lambda_1 z\}. \tag{13}$$

PROOF. Introduce the independent and identically distributed random vectors

$$Y_j = \left(1_{\{U_j \le 2^{-1}\}} - 2^{-1}, \dots, 1_{\{U_j \le 2^{-l}\}} - 2^{-l}\right), \qquad j = 1, \dots, n,$$

and the inner product and norm on $\mathbb{R}^l$ given by

$$\langle x, y\rangle = 2^{-l}\sum_{i=1}^{l} 2^{2i}x_i y_i, \qquad \|x\| = \langle x, x\rangle^{1/2}.$$

Notice that

$$\sigma^2 = E\|Y_1\|^2 = \sum_{i=1}^{l} 2^{i-l}(1-2^{-i}) = 2 - (l+2)2^{-l},$$

so that $\frac{1}{2} \le \sigma^2 < 2$. By Hölder's inequality we have, for $m \ge 2$,

$$E\|Y_1\|^m \le \sigma^2\left(\sum_{i=1}^{l} 2^{2i-l}\right)^{(m-2)/2} < \sigma^2\left(2^{(l+1)/2}\right)^{m-2}.$$

Applying an exponential bound given by Yurinskiĭ (1976), page 491, we obtain

$$P\left(n^{-1}T_l > z\right) = P\left(\left\|\sum_{j=1}^{n} Y_j\right\| > (nz)^{1/2}\right)$$

$$\le 2\exp\left\{-\frac{z}{2\sigma^2}\left[1 + \frac{1.62}{\sigma^2}\left(\frac{2^{l+1}z}{n}\right)^{1/2}\right]^{-1}\right\}.$$

Since $2^{l+1} \le 2^{p+2} \le 4An/z$ and $\frac{1}{2} \le \sigma^2 \le 2$, the lemma follows. □

In order to bound the tail of the distribution of $S_{j,p}$ we require a technical lemma which is likely to be of separate interest. Let $(X_1, \dots, X_{k+1})$ have a multinomial distribution with parameters $n, p_1, \dots, p_{k+1}$. Assume that $p_i > 0$ for $i = 1, \dots, k$ and define

$$s = \sum_{i=1}^{k} p_i \in (0, 1].$$

We shall prove

LEMMA 3. *For every $C > 0$ and $\delta > 0$, there exist positive numbers a, b and λ such that for all $n \ge 1$, $k \ge 1$ and positive $z, p_1, \dots, p_k$ satisfying $z \le$*

$Cn \min\{p_i\colon 1 \le i \le k\}$ *and* $s \le 1 - \delta$,

$$P\left(\sum_{i=1}^{k} \frac{(X_i - np_i)^2}{np_i} > z \right) < a \exp\{bk - \lambda z\}. \tag{14}$$

PROOF. For every $\varepsilon > 0$, (12) and the upper bound on z ensure that

$$P\left(\max_{1 \le i \le k} \frac{|X_i - np_i|}{np_i} > \varepsilon \right) \le 2 \sum_{i=1}^{k} \exp\left\{ - \frac{\varepsilon^2 np_i}{2(1 + \varepsilon/3)} \right\}$$
$$\le 2k \exp\{-\lambda(\varepsilon) z\} \le \exp\{k - \lambda(\varepsilon) z\},$$

with $\lambda(\varepsilon) = \varepsilon^2 \{2C(1 + \varepsilon/3)\}^{-1}$. Hence, it suffices to show that

$$P\left(\sum_{i=1}^{k} Y_i > z \right) < a \exp\{bk - \lambda z\}, \tag{15}$$

where

$$Y_i = \begin{cases} \dfrac{(X_i - np_i)^2}{np_i}, & \text{if } \dfrac{(X_i - np_i)^2}{np_i} \le \varepsilon^2 np_i, \\ 0, & \text{otherwise,} \end{cases}$$

for some constant $\varepsilon > 0$ to be chosen below.

Let $\tilde{X}_1, \dots, \tilde{X}_{k+1}$ be independent with $\tilde{X}_i$ having a Poisson distribution with parameter np_i, and define

$$\tilde{Y}_i = \begin{cases} \dfrac{(\tilde{X}_i - np_i)^2}{np_i}, & \text{if } \dfrac{(\tilde{X}_i - np_i)^2}{np_i} \le \varepsilon^2 np_i, \\ 0, & \text{otherwise.} \end{cases}$$

Clearly there exists $\varepsilon > 0$, independent of $n, p_1, \dots, p_k$, such that for $i = 1, \dots, k$ and $|h| \le \frac{1}{2}\varepsilon (np_i)^{1/2}$,

$$R_i(h) = E \exp\left\{ h \frac{\tilde{X}_i - np_i}{(np_i)^{1/2}} \right\} = \exp\left\{ np_i \exp\left(\frac{h}{(np_i)^{1/2}} \right) - np_i - h(np_i)^{1/2} \right\}$$
$$\le \exp\{h^2\},$$

and this determines our choice of ε. Thus, for $0 < y \le \varepsilon^2 np_i$ and $h = y^{1/2}/2$,

$$P(\tilde{Y}_i > y) \le P\left(\frac{\tilde{X}_i - np_i}{(np_i)^{1/2}} > y^{1/2} \right) + P\left(\frac{\tilde{X}_i - np_i}{(np_i)^{1/2}} < -y^{1/2} \right)$$
$$\le R_i(h) \exp\{-hy^{1/2}\} + R_i(-h) \exp\{-hy^{1/2}\}$$
$$\le 2 \exp\{-y/4\}.$$

It follows that

$$E\exp\{\tilde{Y}_i/6\} = \int_0^\infty P(\tilde{Y}_i > 6\log x)\,dx$$
$$= 1 + \tfrac{1}{6}\int_0^{\varepsilon^2 np_i} P(\tilde{Y}_i > y)e^{y/6}\,dy$$
$$\le 1 + \tfrac{1}{3}\int_0^\infty e^{-y/12}\,dy = 5$$

and hence

$$(16)\qquad P\left(\sum_{i=1}^k \tilde{Y}_i > z\right) \le E\exp\left\{\tfrac{1}{6}\left(\sum_{i=1}^k \tilde{Y}_j - z\right)\right\}$$
$$\le \exp\{k\log 5 - z/6\}.$$

The transition from (16) to (15) is achieved by conditioning. We have

$$(17)\qquad P\left(\sum_{i=1}^k Y_i > z\right) = P\left(\sum_{i=1}^k \tilde{Y}_i > z \,\middle|\, \sum_{i=1}^{k+1} \tilde{X}_i = n\right)$$
$$= \sum_{m=0}^n A_m P\left(\sum_{i=1}^k \tilde{Y}_i > z,\ \sum_{i=1}^k \tilde{X}_i = m\right),$$

where

$$A_m = \frac{P\left(\sum_{i=1}^k \tilde{X}_i = m \,|\, \sum_{i=1}^{k+1} \tilde{X}_i = n\right)}{P\left(\sum_{i=1}^k \tilde{X}_i = m\right)} = \frac{n!}{(n-m)!\,n^m}(1-s)^{n-m}e^{ns}.$$

Application of Stirling's formula in the form

$$\exp\left\{\frac{1}{12k+1}\right\} \le \frac{k!\,e^k}{(2\pi k)^{1/2}k^k} \le \exp\left\{\frac{1}{12k}\right\}$$

to the cases $1 \le m \le n-1$ and $m = n$ separately, yields

$$A_m \le 3\{n/(n-m+1)\}^{1/2} \quad \text{for all } 0 \le m \le n.$$

By considering the ratio A_{m+1}/A_m one sees that A_m attains its maximum value for $m = [ns] + 1$. Hence, for all $0 \le m \le n$,

$$A_m \le 3\left(\frac{n}{n-[ns]}\right)^{1/2} \le 3(1-s)^{-1/2} \le 3\delta^{-1/2}.$$

Together with (17) and (16) this implies

$$P\left(\sum_{i=1}^k Y_i > z\right) \le 3\delta^{-1/2}P\left(\sum_{i=1}^k \tilde{Y}_i > z\right) \le 3\delta^{-1/2}\exp\{k\log 5 - z/6\}.$$

The proof is complete. □

LEMMA 4. *For every $A > 0$, there exist positive numbers a_2, b_2 and λ_2 such that for all $n \geq 1$, $z \geq 0$ and p so that $z \leq An2^{-p}$, and $0 \leq j \leq p$,*

$$P\left(n^{-1}S_{j,p} > z\right) < a_2 \exp\{b_2(p-j) - \lambda_2 z\}. \tag{18}$$

PROOF. Define

$$X_i = n\{\tilde{G}_n(2^{-i}) - \tilde{G}_n(2^{-(i+1)}) - 2^{-(i+1)}\}, \qquad i = j+1,\dots,p,$$

$$X_{p+1} = n\{\tilde{G}_n(2^{-(p+1)}) - 2^{-(p+1)}\}.$$

For $\nu = j+1,\dots,p+1$, we have

$$\begin{aligned}\xi_\nu = \left(n\tilde{G}_n(2^{-\nu}) - n2^{-\nu}\right)^2 &= \left(\sum_{i=\nu}^{p+1} X_i\right)^2\\ &\leq \sum_{i=\nu}^{p+1} 2^{-i/2} \sum_{i=\nu}^{p+1} 2^{i/2}X_i^2\\ &\leq 4\left\{2^{-\nu/2} \sum_{i=\nu}^{p+1} 2^{i/2}X_i^2\right\},\end{aligned}$$

and hence

$$\begin{aligned}n^{-1}S_{j,p} &\leq 4n^{-1} \sum_{\nu=j+1}^{p+1} 2^{\nu/2} \sum_{i=\nu}^{p+1} 2^{i/2}X_i^2\\ &\leq 16n^{-1} \sum_{i=j+1}^{p+1} 2^i X_i^2\\ &\leq 16n^{-1}\left\{\sum_{i=j+1}^{p} 2^{i+1}X_i^2 + 2^{p+1}X_{p+1}^2\right\}.\end{aligned}$$

Now $(X_{j+1},\dots,X_{p+1}, n - X_{j+1} - \cdots - X_{p+1})$ has a multinomial distribution with parameters $n, 2^{-(j+2)},\dots,2^{-(p+1)}, 2^{-(p+1)}, 1 - 2^{-(j+1)}$. Since $z \leq 2An2^{-(p+1)}$, application of Lemma 3 for $C = A/8$, $k = p - j + 1$ and $\delta = \frac{1}{2}$ yields

$$P\left(n^{-1}S_{j,p} > z\right) < a \exp\{b(p - j + 1) - \lambda z/16\}$$

and the assertion of the lemma follows. □

Choose $\varepsilon > 0$. For any integers $p \geq 0$, $0 \leq l \leq i - 1$ and $j \geq 0$, define the events

$$C_p^{(n)} = \left\{\max_{1 \leq i \leq p+1} 2^i|\Delta_{i,0}^{(n)} - n2^{-i}| > \varepsilon n\right\},$$

$$C_{i,l}^{(n)} = \left\{\max_{2^{-l+i} \leq m \leq 2^{-l+i+1}-1} 2^{i+1}|\Delta_{i+1,m}^{(n)} - n2^{-(i+1)}| > \varepsilon n\right\},$$

$$F_{j,p}^{(n)} = \bigcup_{l=j}^{p-1} \bigcup_{i=l+1}^{p} C_{i,l}^{(n)},$$

where $F_{j,p}^{(n)} = \varnothing$ if $j \geq p$ by convention. Finally, for $0 \leq j \leq p$, set

$$E_{j,p}^{(n)} = C_p^{(n)} \cup F_{j,p}^{(n)}.$$

LEMMA 5. *For every $A > 0$ and $\varepsilon > 0$, there exist positive constants a_3, b_3 and λ_3 such that for all $n \geq 1$, $z \geq 0$ and p so that $z \leq An2^{-p}$, and $0 \leq j \leq p$,*

$$P\left(E_{j,p}^{(n)}\right) < a_3 \exp\{b_3(p-j) - \lambda_3 z\}.$$

PROOF. Take $\lambda = \varepsilon^2/\{4A(1+\varepsilon/3)\}$. As $n2^{-p} \geq z/A$, inequality (12) yields

$$P\left(C_p^{(n)}\right) < 2\sum_{i=1}^{p+1} \exp\{-\lambda z 2^{p-i+1}\} < \frac{2\exp\{-\lambda z\}}{1-\exp\{-\lambda z\}},$$

which is bounded by $2(1-\exp\{-\lambda\})^{-1}\exp\{-\lambda z\}$ if $z \geq 1$. For $0 \leq z < 1$, $P(C_p^{(n)}) < \exp\{\lambda\}\exp\{-\lambda z\}$, so that for $z \geq 0$

$$P\left(C_p^{(n)}\right) < a\exp\{-\lambda z\},$$

with positive a depending only on A and ε.

For $j = p$, $F_{j,p}^{(n)} = \varnothing$. For $0 \leq j \leq p-1$, we have

$$P\left(C_{i,l}^{(n)}\right) \leq 2^{-l+i+1}\exp\{-\lambda z\},$$

for each $j \leq l \leq p-1$ and $l+1 \leq i \leq p$. Hence, for $0 \leq j \leq p-1$,

$$P\left(F_{j,p}^{(n)}\right) \leq \sum_{l=j}^{p-1} \sum_{i=l+1}^{p} 2^{-l+i+1}\exp\{-\lambda z\} \leq 2^{p-j+3}\exp\{-\lambda z\},$$

which completes the proof. □

For the proof of our next lemma we need the following combinatorial identity that can be inferred from a similar identity given on page 118 of KMT (1975): Let f be any function on $[0,1]$. For nonnegative integers i and m such that $0 < (2m+1)/2^{i+1} < 1$ define the second differences

$$\phi(i,m,f) = 2f((2m+1)2^{-(i+1)}) - f(m2^{-i}) - f((m+1)2^{-i}).$$

Then for any choice of nonnegative integers k, p and l such that

$$2^{-(l+1)} < (2k+1)2^{-(p+1)} \leq 2^{-l},$$

we have

$$\begin{aligned} f((2k+1)2^{-(p+1)}) &= (2-(2k+1)2^{l-p})f(2^{-(l+1)}) \\ &\quad + ((2k+1)2^{l-p}-1)f(2^{-l}) \\ &\quad + \sum_{i=l+1}^{p} c(i,p,k)\phi(i,k(i),f), \end{aligned} \tag{19}$$

where the sum is defined to be zero if $l \geq p$ and for $i = l+1, \dots, p$,

$$k(i) = \left[(2k+1)/2^{p+1-i}\right], \qquad 0 \leq c(i,p,k) \leq 1. \tag{20}$$

In addition, we shall use the elementary identity

$$f(2^{-r}) = \sum_{i=0}^{r-1} \phi(i,0,f)2^{-r+i}, \quad \text{for } r \geq 1, \tag{21}$$

valid whenever $f(0) = f(1) = 0$.

For any two nonnegative integers $j \geq p$, let

$$\mathscr{I}_{j,p} = \{k \geq 0: (2k+1)2^{-(p+1)} \leq 2^{-j}\}.$$

LEMMA 6. *For every $A > 0$ there exist positive numbers a_4, b_4 and λ_4 such that for all $n \geq 1$, $z \geq 0$ and p so that $z \leq An2^{-p}$, and $0 \leq j \leq p$,*

$$P\Big(\max_{k \in \mathscr{I}_{j,p}} n^{1/2}\big|\tilde{\alpha}_n\big((2k+1)2^{-(p+1)}\big) - B\big((2k+1)2^{-(p+1)}\big)\big| > z\Big)$$
$$< a_4 \exp\{b_4(p-j) - \lambda_4 z\}.$$

PROOF. Choose positive C_1, C_2 and η for which the assertion of Lemma 1 holds and take $\varepsilon = \eta/2$ in the definition of the event $E_{j,p}^{(n)}$ in Lemma 5. We shall write E for $E_{j,p}^{(n)}$ and E^c for its complement, and we define

$$Z_k = n^{1/2}\big|\tilde{\alpha}_n\big((2k+1)2^{-(p+1)}\big) - B\big((2k+1)2^{-(p+1)}\big)\big|.$$

In view of Lemma 5 it suffices to find positive a, b and λ such that

$$\sum_{k \in \mathscr{I}_{j,p}} P(Z_k > z, E^c) < a\exp\{b(p-j) - \lambda z\}. \tag{22}$$

Obviously, $0 \in \mathscr{I}_{j,p}$; if $k \in \mathscr{I}_{j,p}$, $k \neq 0$, then $(2k+1)2^{-(p+1)}$ cannot be equal to 2^{-l} for any l. It follows that

$$\mathscr{I}_{j,p} = \{0\} \cup \bigcup_{l=j}^{p-1} \mathscr{I}_l, \qquad \mathscr{I}_l = \{k: 2^{-(l+1)} < (2k+1)2^{-(p+1)} < 2^{-l}\}. \tag{23}$$

We begin by studying Z_0. Identity (21) yields

$$Z_0 \leq \tfrac{1}{2} \sum_{i=0}^{p} |\tilde{D}_{i,0}^{(n)} - \tilde{\Delta}_{i,0}^{(n)}|2^{-(p-i)},$$

and on the set E^c we have

$$|\Delta_{i,0}^{(n)} - n2^{-i}| \leq \tfrac{1}{2}\eta n 2^{-i}, \quad \text{for } i = 0, \dots, p+1,$$

the conclusion for $i = 0$ being trivial since $\Delta_{0,0}^{(n)} = n$. This also ensures that on E^c

$$\begin{aligned} |\tilde{\Delta}_{i,0}^{(n)}| &= |2\Delta_{i+1,0}^{(n)} - \Delta_{i,0}^{(n)}| \\ &\leq |2\Delta_{i+1,0}^{(n)} - n2^{-i}| + |\Delta_{i,0}^{(n)} - n2^{-i}| \leq \eta n 2^{-i}, \end{aligned} \tag{24}$$

for $i = 0, \dots, p$ and hence Lemma 1 implies that on E^c

$$\begin{aligned} Z_0 &\leq C_1 n^{-1} 2^{-(p+1)} \sum_{i=0}^{p} 2^{2i}\Big\{\big(\tilde{\Delta}_{i,0}^{(n)}\big)^2 + \big(\Delta_{i,0}^{(n)} - n2^{-i}\big)^2\Big\} + C_2 \\ &\leq 5C_1 n^{-1} 2^{-(p+1)} \sum_{i=1}^{p+1} 2^{2i}\big(\Delta_{i,0}^{(n)} - n2^{-i}\big)^2 + C_2 \\ &= 5C_1 n^{-1} T_{p+1} + C_2. \end{aligned} \tag{25}$$

Application of Lemma 2 shows that

$$(26)\qquad P(Z_0 > z, E^c) \le 2\exp\{-\lambda_1(z - C_2)/(5C_1)\}.$$

Next we consider the case $k \in \mathscr{I}_l$ for some l satisfying $j \le l \le p - 1$. Since $2^{-(l+1)} < (2k+1)2^{-(p+1)} < 2^{-l}$, we have $1 < (2k+1)2^{l-p} < 2$ and identities (19) and (21) ensure that

$$\begin{aligned} Z_k &\le 2\sum_{i=0}^{l} |\tilde{D}_{i,0}^{(n)} - \tilde{\Delta}_{i,0}^{(n)}| 2^{-(l-i)} + \sum_{i=l+1}^{p} |\tilde{D}_{i,k(i)}^{(n)} - \tilde{\Delta}_{i,k(i)}^{(n)}| \\ &\le 20C_1 n^{-1} T_{l+1} + 4C_2 + \sum_{i=l+1}^{p} |\tilde{D}_{i,k(i)}^{(n)} - \tilde{\Delta}_{i,k(i)}^{(n)}|, \end{aligned}$$

where the second inequality follows by the same argument that led to (25). Notice that if $k \in \mathscr{I}_l$, then necessarily $2^{p-l} + 1 \le 2k + 1 \le 2^{p-l+1} - 1$ and hence for each $i = l+1, \dots, p$, we find

$$2^{-(l+1)} \le 2k(i)2^{-(i+1)} < (2k(i)+1)2^{-(i+1)} < 2(k(i)+1)2^{-(i+1)} \le 2^{-l}.$$

But this implies that for $i = l+1, \dots, p$ and on the set E^c,

$$\begin{aligned} |\tilde{\Delta}_{i,k(i)}^{(n)}| &\le |\Delta_{i+1,2k(i)}^{(n)} - n2^{-(i+1)}| + |\Delta_{i+1,2k(i)+1}^{(n)} - n2^{-(i+1)}| \\ &\le \tfrac{1}{2}\eta n 2^{-i}, \\ |\Delta_{i,k(i)}^{(n)} - n2^{-i}| &\le |\Delta_{i+1,2k(i)}^{(n)} - n2^{-(i+1)}| + |\Delta_{i+1,2k(i)+1}^{(n)} - n2^{-(i+1)}| \\ &\le \tfrac{1}{2}\eta n 2^{-i}, \end{aligned}$$

and Lemma 2 yields

$$\begin{aligned} Z_k &\le 20C_1 n^{-1} T_{l+1} + 4C_2 \\ &\quad + C_1 n^{-1} \sum_{i=l+1}^{p} 2^i \Big\{ \big(\tilde{\Delta}_{i,k(i)}^{(n)}\big)^2 + \big(\Delta_{i,k(i)}^{(n)} - n2^{-i}\big)^2 \Big\} + C_2(p-l). \end{aligned}$$

Arguing as on page 120 of KMT (1975) it can be shown that

$$\Big\{ \Big(\Delta_{i,k(i)}^{(n)}, \big(\tilde{\Delta}_{i,k(i)}^{(n)}\big)^2 \Big) : i = l+1, \dots, p \Big\} =_{\mathscr{D}} \Big\{ \Big(\Delta_{i,0}^{(n)}, \big(\tilde{\Delta}_{i,0}^{n}\big)^2 \Big) : i = l+1, \dots, p \Big\},$$

and because

$$\sum_{i=l+1}^{p} 2^i \Big\{ \big(\tilde{\Delta}_{i,0}^{(n)}\big)^2 + \big(\Delta_{i,0}^{(n)} - n2^{-i}\big)^2 \Big\} \le 7 \sum_{i=l+1}^{p+1} 2^i \big(\Delta_{i,0}^{(n)} - n2^{-i}\big)^2 = 7S_{l,p}$$

by (24), we obtain for $k \in \mathscr{I}_l$,

$$\begin{aligned} &P(Z_k > z, E^c) \\ &\quad \le P\big(20C_1 n^{-1} T_{l+1} + 4C_2 > z/2\big) + P\big(7C_1 n^{-1} S_{l,p} + C_2(p-l) > z/2\big) \\ &\quad \le 2\exp\{-\lambda_1(z - 8C_2)/(40C_1)\} \\ &\qquad + a_2 \exp\{b_2(p-j) - \lambda_2(z - 2C_2(p-l))/(14C_1)\} \\ &\quad \le \tilde{a}\exp\{\tilde{b}(p-j) - \tilde{\lambda} z\}, \end{aligned}$$

for positive $\tilde{a}$, $\tilde{b}$ and $\tilde{\lambda}$. Notice that $\mathscr{J}_l$ has 2^{p-l-1} elements, so

$$\sum_{l=j}^{p-1} \sum_{k \in \mathscr{J}_l} P(Z_k > z, E^c) \le \tilde{a} \exp\{(\tilde{b} + \log 2)(p-j) - \tilde{\lambda} z\}.$$

Together with (23) and (26) this yields (22) and the lemma. □

For any $0 \le j \le p$ and $n \ge 1$, let

$$\alpha_n(j,p) = \sup_{0 \le h \le 2^{-p}} \sup_{h \le s \le 2^{-j}} n^{1/2} |\tilde{\alpha}_n(s) - \tilde{\alpha}_n(s-h)|,$$

$$B_n(j,p) = \sup_{0 \le h \le 2^{-p}} \sup_{h \le s \le 2^{-j}} n^{1/2} |B(s) - B(s-h)|.$$

Lemma 7. *There exist positive constants a_5, b_5 and λ_5 such that for all $n \ge 1$, $z > 0$ and p so that $n2^{-(p+1)} < z/16 \le n2^{-p}$, and $0 \le j \le p$,*

$$P(\alpha_n(j,p) + B_n(j,p) > z) < a_5 \exp\{b_5(p-j) - \lambda_5 z\}.$$

Proof. Whenever $(k-1)2^{-p} \le s - h \le k2^{-p}$ for some $k = 1, \dots, 2^{p-j}$, then

$$\begin{aligned} -n2^{-p} &\le n^{1/2}(\tilde{\alpha}_n(s) - \tilde{\alpha}_n(s-h)) \\ &\le n^{1/2}(\tilde{\alpha}_n((k+1)2^{-p}) - \tilde{\alpha}_n((k-1)2^{-p})) + 2n2^{-p}. \end{aligned}$$

Because $z/2 > 4n2^{-p}$ we find by inequality (12)

$$\begin{aligned} P(\alpha_n(j,p) > z/2) &\le P(\alpha_n(j,p) > 4n2^{-p}) \\ &\le 2^{p-j} P\left(n^{1/2} |\tilde{\alpha}_n(2^{-(p-1)})| > n2^{-(p-1)}\right) \\ &< 2\exp\left\{(p-j)\log 2 - \frac{3n}{4} 2^{-p}\right\} \qquad (27) \\ &\le 2\exp\left\{(p-j)\log 2 - \frac{3}{64} z\right\}. \end{aligned}$$

If W denotes a standard Wiener process, then

$$\{B(s): 0 \le s \le 1\} =_{\mathscr{D}} \{W(s) - sW(1): 0 \le s \le 1\},$$

and hence

$$P(B_n(j,p) > z/2) \le P\left(\sup_{0 \le h \le 2^{-p}} \sup_{h \le s \le 2^{-j}} n^{1/2} |W(s) - W(s-h)| > z/4\right) + P(n^{1/2} 2^{-p} W(1) > z/4).$$

It now follows from Lemma 1.2.1 on page 29 of M. Csörgő and Révész (1981) and an elementary bound for the tail of the standard normal distribution [cf. Feller (1968)] that for positive constants a, b and λ,

$$P(B_n(j,p) > z/2) < a \exp\{b(p-j) - \lambda z\}.$$

Together with (27) this proves the lemma. □

We are now in a position to prove Inequality 2. Choose any $n \ge 1$, $1 \le z \le n$, p so that $n2^{-(p+1)} < z/32 \le n2^{-p}$ and $0 \le j \le p$. Since for every $0 \le s \le 2^{-j}$, there exists an integer $k \geq 0$ such that $(2k+1)2^{-(p+1)} \leq 2^{-j}$ and

$$|s - (2k+1)2^{-(p+1)}| \leq 2^{-p},$$

we see that

$$P\Big(\sup_{0 \leq s \leq 2^{-j}} n^{1/2}|\tilde{\alpha}_n(s) - B(s)| > z\Big) \leq P(\alpha_n(j,p) + B_n(j,p) > z/2)$$
$$+ P\Big(\max_{k \in \mathscr{I}_{j,p}} n^{1/2}|\tilde{\alpha}_n((2k+1)2^{-(p+1)})$$
$$- B((2k+1)2^{-(p+1)})| > z/2\Big),$$

with $\mathscr{I}_{j,p}$ as in Lemma 6. As $z/2$ satisfies the assumptions of Lemma 6 as well as those of Lemma 7, application of these lemmas completes the proof of Inequality 2 and also of Theorem 1. □

Acknowledgements. The authors are indebted to Professor F. Götze for a very helpful discussion and to Professor S. Csörgő for pointing out an oversight in our original statement of (3). They thank Professor Gaenssler for his interest and the Associate Editor and the referee for some useful comments.

REFERENCES

BENNETT, G. (1962). Probability inequalities for the sum of independent random variables. *J. Amer. Statist. Assoc.* **57** 33–45.

CSÖRGŐ, M. (1983). *Quantile Processes with Statistical Applications.* SIAM, Philadelphia.

CSÖRGŐ, M., CSÖRGŐ, S., HORVÁTH, L. and MASON, D. M. (1986). Weighted empirical and quantile processes. *Ann. Probab.* **14** 31–85.

CSÖRGŐ, M. and RÉVÉSZ, P. (1978). Strong approximations of the quantile process. *Ann. Statist.* **6** 882–894.

CSÖRGŐ, M. and RÉVÉSZ, P. (1981). *Strong Approximations in Probability and Statistics.* Academic, New York–Akadémai Kiadó, Budapest.

DVORETZKY, A., KIEFER, J. and WOLFOWITZ, J. (1956). Asymptotic minimax character of the sample distribution function and of the multinomial estimator. *Ann. Math. Statist.* **27** 642–669.

FELLER, W. (1968). *An Introduction to Probability Theory and Its Applications* **1**, 3rd ed. Wiley, New York.

KIEFER, J. (1970). Deviations between the sample quantile process and the sample df. In *Nonparametric Techniques in Statistical Inference* (M. L. Puri, ed.) 299–319. Cambridge Univ. Press, Cambridge.

KOMLÓS, J., MAJOR, P. and TUSNÁDY, G. (1975). An approximation of partial sums of independent rv's and the sample df. I. *Z. Wahrsch. verw. Gebiete* **32** 111–131.

KOMLÓS, J., MAJOR, P. and TUSNÁDY, G. (1976). An approximation of partial sums of independent rv's and the sample df. II. *Z. Wahrsch. verw. Gebiete* **34** 33–58.

YURINSKIĬ, V. (1976). Exponential inequalities for sums of random vectors. *J. Multivariate Anal.* **6** 473–499.

DEPARTMENT OF MATHEMATICAL SCIENCES
501 EWING HALL
UNIVERSITY OF DELAWARE
NEWARK, DELAWARE 19716

UNIVERSITY OF LEIDEN
DEPARTMENT OF MATHEMATICS AND COMPUTER SCIENCE
P.O. BOX 9512
2300 RA LEIDEN
THE NETHERLANDS

The Asymptotic Distribution of Point Charges on a Conducting Sphere

Willem R. van Zwet

University of Leiden and University of North Carolina

Abstract. Consider n point charges, each with charge $\frac{1}{n}$, in electrostatic equilibrium on the surface S of a conducting sphere. It is shown that as n tends to infinity, the distribution of the total charge 1 on S tends to the uniform distribution on S. Though this is an entirely deterministic result, the proof is probabilistic in nature.

1 Introduction

Consider n point charges (electrons), each with charge $\frac{1}{n}$, in equilibrium position on the surface S of a conducting unit sphere in $\mathbf{R}^3$. If d denotes Euclidean distance in $\mathbf{R}^3$, these charges will be located at points $\xi_{n1}, \xi_{n2}, \ldots, \xi_{nn}$ on S for which the potential energy

$$\sum_{i \neq j} \frac{1}{d(\xi_{ni}, \xi_{nj})}$$

is an absolute minimum. Let P_n denote the measure that assigns measure $\frac{1}{n}$ to each of the points $\xi_{n1}, \ldots, \xi_{nn}$ so that for any $E \subset S$, $P_n(E)$ denotes the charge situated in E. Let λ be Lebesgue measure on S and $\Pi = \lambda/(4\pi)$ the uniform probability measure on S. Is it true that P_n converges weakly to Π as $n \to \infty$? In other words, is the macroscopic model where the electrical charge is viewed as a continuous phenomenon compatible with the microscopic description in terms of point charges?

The problem of providing a rigorous proof of this intuitively obvious conjecture was raised by Korevaar (1972) and Robbins (1975). Two different proofs were given independently by Korevaar (1976) and van Zwet (1976), the former preceding the latter by some months. At the time, however, neither proof was published. Sixteen years later, matters of priority don't seem terribly relevant any more and since the present author's probabilistic proof is simple and perhaps somewhat amusing, it is presented here. We shall prove

Theorem 1 *P_n converges weakly to Π, so $\lim_{n \to \infty} P_n(B) = \Pi(B)$ for every Borel set $B \subset S$ whose boundary relative to S has Π-measure zero.*

In fact we prove more. It will be shown that the result remains valid if S is replaced by an arbitrary compact set $K \subset \mathbf{R}^3$ with positive capacity and Π by the so-called minimizing measure P_0 on K. For a review of relevant literature see Korevaar (1976).

S. van de Geer and M. Wegkamp (eds.), *Selected Works of Willem van Zwet*, Selected Works in Probability and Statistics, DOI 10.1007/978-1-4614-1314-1_27, © Springer Science+Business Media, LLC 2012

2 Charges on a Compact Set

We begin by dealing with the general case. Let K be an infinite compact set in $\mathbf{R}^3$ and let $\mathcal{P}_n$ denote the class of probability measures that assign measure $\frac{1}{n}$ to each of n distinct points of K. Consider n identical point charges situated at points $x_1, x_2, \ldots, x_n$ in K. For convenience we take these charges to be $\frac{1}{n}$ so that the total charge is equal to 1. If the potential energy of this configuration is finite, the points $x_1, \ldots, x_n$ must be distinct and the charge distribution may be described by a measure $P \in \mathcal{P}_n$ assigning measure $\frac{1}{n}$ to $x_1, \ldots, x_n$. In this case the energy may be written as

$$\tilde{\psi}(P) = \frac{1}{n^2} \sum_{i \neq j} \frac{1}{d(x_i, x_j)} = \int_K \int_K \frac{1}{d(x,y)} 1_{\{x \neq y\}} \, dP(x) \, dP(y) . \tag{2.1}$$

Now let $\xi_{n1}, \ldots, \xi_{nn}$ be a configuration of the n point charges for which the energy is an absolute minimum. Because K is infinite and compact such configurations exist and have finite energy, and the points $\xi_{n1}, \ldots, \xi_{nn}$ are distinct. Let $P_n \in \mathcal{P}_n$ be the corresponding probability measure which puts mass $\frac{1}{n}$ at $\xi_{n1}, \ldots, \xi_{nn}$. Then clearly

$$\tilde{\psi}(P_n) = \int_K \int_K \frac{1}{d(x,y)} 1_{\{x \neq y\}} \, dP_n(x) \, dP_n(y) = \min_{P \in \mathcal{P}_n} \tilde{\psi}(P) < \infty . \tag{2.2}$$

Instead of the above discrete model with indivisible point charges which are not subject to internal forces that would make them explode, one can also consider a model where charge is viewed as a "continuous" phenomenon. In this model the distribution of a total charge 1 on K is given by a measure P in the class $\mathcal{P}$ of all probability measures on the Borel sets in K, where $P(B)$ denotes the charge in the Borel set B. For $P \in \mathcal{P}$ and $x \in \mathbf{R}^3$ one defines the potential of P by

$$U(P, x) = \int_K \frac{1}{d(x,y)} dP(y) , \tag{2.3}$$

the energy of P by

$$\psi(P) = \int_K \int_K \frac{1}{d(x,y)} dP(x) \, dP(y) = \int_K U(P, x) \, dP(x) \tag{2.4}$$

and the capacity of the set K by

$$C(K) = \left[\inf_{P \in \mathcal{P}} \psi(P) \right]^{-1} . \tag{2.5}$$

Note that under this model the presence of a point charge implies infinite energy.

If $C(K) = 0$, then clearly $\psi(P) = \infty$ for every $P \in \mathcal{P}$. On the other hand, if $C(K) > 0$, then there exists a unique measure $P_0 \in \mathcal{P}$ for which ψ assumes its absolute minimum on $\mathcal{P}$ (c.f. Landkof (1972) p.131–133). P_0 is called the minimizing measure on K and represents the equilibrium (i.e. minimum energy) distribution of a charge 1 on K under the continuous model. Note that since

$$\psi(P_0) = \min_{P\in\mathcal{P}} \psi(P) = \frac{1}{C(K)}, \tag{2.6}$$

the assumption $C(K) > 0$ ensures that P_0 assigns measure zero to single points and that the compact set K must therefore be non–denumerable.

Theorem 2 *Let* K *be a compact set in* $\mathbf{R}^3$ *with positive capacity. Then* P_n *converges weakly to* P_0, *so* $\lim_{n\to\infty} P_n(B) = P_0(B)$ *for every Borel set* $B \subset K$ *whose boundary relative to* K *has* P_0*–measure zero.*

Proof. Let $X_1,\ldots,X_n$ be independent random points in K that are identically distributed according to the probability measure P_0. Since P_0 assigns probability zero to single points, the points $X_1,\ldots,X_n$ are distinct with probability 1. Hence (2.1) and (2.2) imply that with probability 1

$$\tilde{\psi}(P_n) \le \frac{1}{n^2}\sum_{i\neq j}\frac{1}{d(X_i,X_j)}.$$

Taking the expectation on the right we find

$$\tilde{\psi}(P_n) \le \frac{n-1}{n}\int_K\int_K \frac{1}{d(x,y)}dP_0(x)\,dP_0(y) = \frac{n-1}{n}\psi(P_0)\,. \tag{2.7}$$

Now let $Q_n = P_n \times P_n$ denote the product measure on $K\times K$. Because $K\times K$ is compact, the set $\{Q_n : n = 1,2,\ldots\}$ is relatively compact in the topology of weak convergence ($\int f dQ_n \to \int f dQ$ for bounded continuous f). To show that Q_n converges weakly to a probability measure Q on $K\times K$ it is therefore sufficient to show that every weakly convergent subsequence has limit Q. Let Q_{n_k}, $k = 1,2,\ldots$, denote such a weakly converging subsequence with limit Q_0 and define a bounded and continuous function f_c on $K\times K$ by $f_c(x,y) = \min(c, 1/d(x,y))$ for $c > 0$. Noting that Q_n assigns probability $\frac{1}{n}$ to the set $\{(x,y) : x = y\}$ we see that for every $c > 0$,

$$\begin{aligned}
\liminf_{k\to\infty}\tilde{\psi}(P_{n_k}) &= \liminf_{k\to\infty}\int_{K\times K}\frac{1}{d(x,y)}1_{\{x\neq y\}}dQ_{n_k}(x,y)\\
&\ge \liminf_{k\to\infty}\int_{K\times K} f_c(x,y)1_{\{x\neq y\}}dQ_{n_k}(x,y) = \liminf_{k\to\infty}\Big[\int_{K\times K} f_c dQ_{n_k} - \frac{c}{n_k}\Big]\\
&= \liminf_{k\to\infty}\int_{K\times K} f_c dQ_{n_k} = \int_{K\times K} f_c dQ_0,
\end{aligned}$$

so that the monotone convergence theorem implies that

$$\liminf_{k\to\infty}\tilde{\psi}(P_{n_k}) \ge \int_{K\times K}\frac{1}{d(x,y)}dQ_0(x,y). \tag{2.8}$$

For every n, Q_n is the product of two identical probability measures on K and Q_0 must therefore have the same structure, say $Q_0 = P_0' \times P_0'$ with $P_0' \in \mathcal{P}$. Hence (2.7) and (2.8) yield

$$\psi(P_0') \le \liminf_{k\to\infty} \tilde{\psi}(P_{n_k}) \le \liminf_{k\to\infty} \frac{n_k - 1}{n_k} \psi(P_0) = \psi(P_0)$$

and since P_0 is the unique measure in $\mathcal{P}$ minimizing ψ, we find that $P_0' = P_0$ so that $Q_0 = P_0 \times P_0$. Since the limit Q_0 is independent of the weakly convergent subsequence Q_{n_k} we have chosen, it follows that Q_n converges weakly to $P_0 \times P_0$ and hence that P_n converges weakly to P_0.

3 Charges on the Surface of a Sphere

It remains to consider the special case where $K = \mathcal{S}$. Clearly $\mathcal{S}$ is compact and one easily verifies that the uniform probability measure $\Pi = \lambda/(4\pi)$ on $\mathcal{S}$ has finite energy $\psi(\Pi)$ and a constant potential $U(\Pi, x)$ for $x \in \mathcal{S}$. But this implies that $\mathcal{S}$ has positive capacity and that Π is the unique minimizing measure on $\mathcal{S}$ (see Landkof (1972, p. 137). Theorem 1 is therefore an immediate consequence of Theorem 2.

References

Korevaar, J. (1972). Prijsvraag wiskundig genootschap 1972–2. *Nieuw Arch. Wiskunde*, (3), **20**, p. 73.

Korevaar, J. (1976). Problems of equilibrium points on the sphere and electrostatic fields. Technical Report 76-03, Department of Mathematics, University of Amsterdam.

Landkoff, N.S. (1972). *Foundations of Modern Potential Theory*. Springer-Verlag, Berlin.

Robbins, H.E. (1975). Lecture presented at the Fourth Lunteren Meeting on Probability and Statistics.

van Zwet, W.R. (1976). Solution Prijsvraag Wiskundig Genootschap, Amsterdam.

This article was processed using the LaTeX macro package with LLNCS style

The Annals of Probability
2004, Vol. 32, No. 1A, 380–423

WEAK CONVERGENCE RESULTS FOR THE KAKUTANI INTERVAL SPLITTING PROCEDURE

BY RONALD PYKE AND WILLEM R. VAN ZWET

University of Washington and University of Leiden

This paper obtains the weak convergence of the empirical processes of both the division points and the spacings that result from the Kakutani interval splitting model. In both cases, the limit processes are Gaussian. For the division points themselves, the empirical processes converge to a Brownian bridge as they do for the usual uniform splitting model, but with the striking difference that its standard deviations are about one-half as large. This result gives a clear measure of the degree of greater uniformity produced by the Kakutani model. The limit of the empirical process of the normalized spacings is more complex, but its covariance function is explicitly determined. The method of attack for both problems is to obtain first the analogous results for more tractable continuous parameter processes that are related through random time changes. A key tool in their analysis is an approximate Poissonian characterization that obtains for cumulants of a family of random variables that satisfy a specific functional equation central to the K-model.

1. Introduction. We are interested in comparisons between two probability models for the random subdivision of the unit interval. The first is the usual model in which the division points are independent Unif(0, 1) random variables (r.v.'s). We refer to this as the U-model. The second model will be referred to as the K-model (for Kakutani) in which the first division point, X_1, is a Unif(0, 1) r.v., and then thereafter the nth division point, X_n, conditionally given the preceding $n-1$ points $\{X_1, \dots, X_{n-1}\}$, is uniformly distributed over the largest subinterval formed by $0, 1, X_1, X_2, \dots, X_{n-1}$. The K-model was suggested by Kakutani (1975) who conjectured that the empirical distribution function (d.f.) of the first n subdivision points converges to the uniform d.f. on [0, 1], just as is well known to be the case for the U-model. This Glivenko–Cantelli result for the K-model was shown to be true by van Zwet (1978).

The K-method of interval splitting, however, should by its very nature result in "more uniform" spacings than those of the U-method. This is intuitively clear since in the K-model the largest spacing is always the one that is being split, whereas in the U-model, the largest spacing may remain untouched for several iterations while at the same time the smaller intervals are consequently being divided into even smaller ones. This difference between the two models was clarified in Pyke (1980)

Received December 2001; revised December 2002.
AMS 2000 subject classifications. Primary 60F17; secondary 60G99, 62G30.
Key words and phrases. Empirical processes, Kakutani interval splitting, spacings, weak convergence, cumulants, self-similarity.

S. van de Geer and M. Wegkamp (eds.), *Selected Works of Willem van Zwet*, Selected Works in Probability and Statistics, DOI 10.1007/978-1-4614-1314-1_28, © Springer Science+Business Media, LLC 2012

where it was shown that for the K-model the empirical d.f. of the normalized spacings converges uniformly with probability one to the uniform d.f. on $[0, 2]$. This is in sharp contrast to the U-model where the limit is an exponential d.f. over $(0, \infty)$; a result of Blum [cf. the footnote in Weiss (1955)].

The purpose of this paper is to study the weak convergence under the Kakutani model of the empirical processes for both the division points and their spacings. The results and their proofs clarify further the differences between the U- and the K-model. The differences are rather striking. In particular, the difference between the two interval-splitting models is summarized by the fact that although the empirical processes for the division points converge in law to Brownian bridges under both the U- and K-models, the standard deviations in the latter case are approximately *half* what they are for the former; see Theorem 4.1.

To be more precise we introduce the following notation. Let $\{X_n : n \geq 1\}$ be the sequence of r.v.'s with values in $(0, 1)$ that represent the successive division points of the unit interval. Let $X_{n1} \leq X_{n2} \leq \cdots \leq X_{nn}$ be the ordered values of $\{X_1, \ldots, X_n\}$. Define the *spacings*

$$(1.1) \qquad D_{ni} = X_{ni} - X_{n,i-1}, \qquad 1 \leq i \leq n+1 \text{ with } X_{n0} = 0, X_{n,n+1} = 1,$$

and let $D^*_{ni} := (n+1)D_{ni}$, $1 \leq i \leq n+1$, denote the *normalized spacings*. Since under the K-model, the maximum normalized spacing converges a.s. to 2 [see (1.12)], it is expedient to introduce the *relative* spacings, $\{D_{ni}/M_n; 1 \leq i \leq n+1\}$ in which $M_n := \max\{D_{n1}, \ldots, D_{nn+1}\}$.

Let F_n, G_n and G^*_n denote, respectively, the empirical d.f.'s of the division points $\{X_1, \ldots, X_n\}$, the spacings $\{D_{n1}, \ldots, D_{n,n+1}\}$ and the normalized spacings $\{D^*_{n1}, \ldots, D^*_{n,n+1}\}$. Let F be the Unif(0, 1) d.f., G be the Unif(0, 2) d.f., and H be the exponential d.f. with mean 1. Then the Glivenko–Cantelli results reviewed above can be stated as follows, where $\|\cdot\|$ is the supremum norm in $\mathbb{R}^1$: with probability 1 under the U-model,

$$(1.2) \qquad \|F_n - F\| \to 0 \quad \text{and} \quad \|G^*_n - H\| \to 0$$

whereas under the K-model

$$(1.3) \qquad \|F_n - F\| \to 0 \quad \text{and} \quad \|G^*_n - G\| \to 0.$$

Thus no differentiation between the two models shows up at this level for the division points, though it does for the spacings. However, Theorem 4.1 shows dramatically that differences are in fact present for the division points in the orders of $n^{1/2}\|F_n - F\|$.

Before introducing the notation for the processes to be studied, we recall that the key method of proofs for results under the K-model involves a random time change from the discrete index $n \in \mathbb{Z}^+$ to the continuous parameter $s > 0$ defined by

$$(1.4) \qquad N_s = \min\{n \in \mathbb{Z}^+ : M_n \leq s\}, \qquad s > 0,$$

where $M_0 = 1$. Interpret $\min \varnothing = +\infty$. Note that $N_s = 0$ when $s \geq 1$. Thus N_s denotes the smallest sample size n for which no spacing exceeds s. The method relies essentially upon a stochastic recursion relationship [(1.9) or (1.10)] that holds in the continuously indexed case. Results are first proved for this case and then an argument is provided to show that the results desired for the original quantities (indexed by n) follow as corollaries.

In terms of the parameter s, the analogous functions to those introduced above are

$$\begin{aligned} F(x, s) &= F_{N_s}(x), \\ N_s(x) &= N_s F(x, s) = {}^{\#}\{j : X_j \leq x, 1 \leq j \leq N_s\}, \\ G(x, s) &= G_{N_s}(x), \\ K(x, s) &= (N_s + 1)G(x, s) = {}^{\#}\{j : D_{N_s j} \leq x, 1 \leq j \leq N_s + 1\}. \end{aligned} \tag{1.5}$$

The following results from van Zwet (1978) and Pyke (1980) are used extensively throughout the paper:

$$\mu(t) := EN_t = \begin{cases} 2/t - 1, & \text{for } 0 < t < 1, \\ 0, & \text{for } t \geq 1, \end{cases} \tag{1.6}$$

$$v(t) := \operatorname{var} N_t = c/t \qquad \text{for } 0 < t \leq 1/2, \tag{1.7}$$

$$\mu(x, s) := EK(x, s) = \begin{cases} 2x/s^2, & \text{if } 0 < x \leq s < 1, \\ 2/s, & \text{if } 0 < s < x \leq 1, \\ \varepsilon(x - 1), & \text{if } s \geq 1, \end{cases} \tag{1.8}$$

where $\varepsilon(u) = 0$ or 1 according as $u < 0$ or $u \geq 0$. The constant $c = v(1/2)/2$ in (1.7) is evaluated in Lemma 3.2. A key result in this paper is Theorem 2.2 that shows in particular that all of the remaining cumulants of N_t are also proportional to t^{-1} in intervals of the form $(0, 1/k)$. Central to the study of these and all other results about the continuous parameter version of the Kakutani method are the recursive representations that come directly from the iterative nature of the Kakutani procedure. In particular, one may check that N_t satisfies the relationship

$$N_t \stackrel{L}{=} N_{t/U} + N^*_{t/(1-U)} + 1, \qquad 0 < t < 1, \tag{1.9}$$

where N and N^* are independent identically distributed processes and U is a Unif(0, 1) r.v. independent of N and N^*. More generally, one can show that

$$K(x, t) \stackrel{L}{=} K\left(\frac{x}{U}, \frac{t}{U}\right) + K^*\left(\frac{x}{1-U}, \frac{t}{1-U}\right), \qquad 0 < t < 1, \tag{1.10}$$

where again $K \stackrel{L}{=} K^*$ and K, K^* and U are independent. Of course, U represents the first (uniform) partition point of the unit interval. Since $K(1, t) = K(t, t) = N_t + 1$, (1.9) is seen to be a special case of (1.10).

Throughout the paper we also need the following limit results from van Zwet (1978) and Pyke (1980) which are contained in their proofs of the Glivenko–Cantelli results of (1.3):

(1.11) $sN_s \to 2$ a.s. as $s \to 0$;

(1.12) $nM_n \to 2$ a.s. as $n \to \infty$ where $M_n = \max\{D_{ni} : 1 \le i \le n+1\}$;

(1.13) $s^{-1}M_{N_s} \to 1$ a.s. as $s \to 0$ [from (1.11) and (1.12)];

(1.14) $sK(ys, s) \to 2y$ uniformly for $0 \le y \le 1$, a.s. as $s \to 0$.

The purpose of this paper is to study under the K-model the weak convergence of the empirical processes associated with the division points and the spacings. We denote these processes of interest as follows:

(i) empirical processes of the division points: for $0 \le x \le 1$,

(1.15)
$$\text{parameter } n \ge 1, \quad U_n(x) = n^{1/2}\{F_n(x) - x\},$$
$$\text{parameter } s > 0, \quad U(x, s) = (s/2)^{1/2}\{N_s(x) - xN_s\};$$

(ii) empirical processes of the normalized spacings: for $0 \le y \le 1$,

(1.16)
$$\text{parameter } n \ge 1, \quad V_n^*(y) = n^{1/2}\{G_n^*(2y) - y\},$$
$$\text{parameter } s > 0, \quad V^*(y, s) = (s/2)^{1/2}\left\{K\left(\frac{2y}{N_s+1}, s\right) - y(N_s+1)\right\};$$

(iii) empirical processes of the relative spacings: for $0 \le y \le 1$,

(1.17)
$$\text{parameter } n \ge 1, \quad V_n(y) = \sqrt{n+1}\,\{G_n(M_n y) - y\},$$
$$\text{parameter } s > 0, \quad V(y, s) = (2/s)^{1/2}\left\{\frac{K(ys, s)}{N_s+1} - y\right\}.$$

For convenience, we will refer to processes indexed by continuous parameters as *stopped* processes, referring thereby to the random stopping times N_s involved in their definitions.

Central to the study of these processes is the related stopped process defined by

$$W(y, s) = (2/s)^{1/2}\left\{\frac{s}{2}K(ys, s) - y\right\}, \qquad 0 \le y < \infty, s > 0, \tag{1.18}$$

since as we now show, V and V^* are expressible in terms of W and W is simpler to study. Observe that in view of (1.8), $W(\cdot, s)$ is a centered process only for $0 \le y \le 1$ and $s < 1$. Since $N_s + 1 = K(s, s)$,

$$W(1, s) = (2/s)^{1/2}\left\{\frac{s}{2}(N_s + 1) - 1\right\}$$

and one can check that with $\delta_s := 2/(s(N_s+1))$, the two stopped spacings processes satisfy

$$V^*(y,s) = W(y\delta_s, s) - W(1,s)(\delta_s+1)y \tag{1.19}$$

and

$$\begin{aligned} V(y,s) &= (2/s)^{1/2}\left\{\frac{K(ys,s)}{N_s+1} - y\right\} \\ &= \frac{(2/s)^{1/2}}{N_s+1}\{K(ys,s) - y(N_s+1)\} \\ &= \delta_s(2/s)^{1/2}\left\{\frac{s}{2}K(ys,s) - y - \frac{s}{2}y\left(K(s,s) - \frac{2}{s}\right)\right\} \\ &= \delta_s\{W(y,s) - yW(1,s)\} \end{aligned} \tag{1.20}$$

for $0 < s \leq 1$. Since $\delta_s \to 1$ a.s., by (1.11), the limiting behaviors of $V^*(\cdot, s)$ over $0 < y < 1$ and $V(\cdot, s)$ over $0 < y \leq 1$ will follow from that of $W(\cdot, s)$ in $D[0,1]$. Notice that although V^* may appear to be a type of "tied-down" version of W, it is not actually zero at $y = 1$, as is V. Moreover, the support interval of significance for $V^*(\cdot, s)$ is random, namely, $[0, \delta_s]$. This is a result of the fact that the normalized maximum spacing has a finite limit; see (1.12). Since the limiting distribution of the maximum spacing may be obtained separately [see (6.6) and the discussion following] it suffices to place our emphasis here upon the processes of the relative spacings, namely, V_n and $V(\cdot, s)$, which we do in Section 6.

The limiting behaviors of the empirical processes U_n and V_n^* under the U-model are well known. Essentially due to Donsker [(1952); cf. Billingsley (1968)] is the fact that $U_n \to_L U$, where U is the standard Brownian bridge with representation $U(t) = B_0(t) := B(t) - tB(1), 0 \leq t \leq 1$, in which B is the standard Brownian motion with $B(0) = 0$ and $\operatorname{var} B(1) = 1$. For the spacings' empirical process, weak convergence was obtained in Pyke [(1965), Theorem 6.4]. Here the definition must be modified to $V_n^*(y) = n^{1/2}\{G_n^*(H^{-1}(y)) - y\}$ to keep the process on $[0,1]$ since by (1.2) and (1.3) the a.s. limit for G_n^* is the exponential H rather than the uniform G over $(0,2)$. Hence $H^{-1}(y) = -\ln(1-y)$. [In (1.16) observe that $G^{-1}(y) = 2y$.] With this notational change, the U-model's weak convergence result for the spacings' empirical process is that $V_n^* \to_L V^*$ where V^* is a mean zero, Gaussian process with

$$\operatorname{Cov}\{V^*(x), V^*(y)\} = x(1-y) - m(x)m(y), \qquad 0 \leq x \leq y \leq 1,$$

where $m(y) = -(1-y)\ln(1-y)$.

NOTE. Although the functions introduced above are point indexed, we will use the same symbol to represent their corresponding set-indexed functions whenever they are well defined. For example, since $K(\cdot, s)$ is nondecreasing, it determines a Lebesgue–Stieltjes measure which we will write as $K(B,s)$. In particular, (1.8) implies

$$EK(sJ,s) = \frac{2}{s}|J| \tag{1.21}$$

for any Borel subset J of [0, 1] and $s < 1$. Here, $|J|$ denotes the Lebesgue measure of J.

The outline of the paper is as follows. The key result about the eventual simple form of the cumulants is proved in Section 2. The weak convergence of the empirical processes for the division points and for the normalized spacings are obtained, respectively, in Sections 4 and 6. The corresponding preliminary results for the convergence of the stopped processes are given, respectively, in Sections 3 and 5. Finally, in Section 7, the covariance function for the limiting Gaussian processes in the spacings case is derived, thereby characterizing those processes completely.

2. Cumulants of functions of the stopped process. As mentioned above, there is a fundamental recursive structure present in the Kakutani interval-splitting procedure that is central to its study. Recall that N_s is the number of partition points that are necessary to get all spacings $\le s$. The first splitting point, X_1, is a Unif(0, 1) r.v. For simplicity, write $U = X_1$. After the first split, there are two intervals, $(0, U)$ and $(U, 1)$ of lengths U and $1 - U$, respectively. Once U is observed, the procedure is equivalent to watching two *independent* Kakutani procedures taking place on these two intervals until both of them result in spacings smaller than s. Moreover, the number of division points needed to partition an interval of length U according to the K-model until no subinterval exceeds s has the same distribution as the number of points needed to divide (0, 1) so that no subinterval exceeds s/U. From this, the representations (1.9) and (1.10) follow. These relations are really of the same type. For if one sets $x = yt$ in (1.10), then for fixed y, the resulting recursion for K is of the same form as that which (1.9) gives for $N_t + 1$. To emphasize this general nature, let $\{D(t) : t > 0\}$ be a real-valued process satisfying

$$D(t) \stackrel{L}{=} D(t/U) + D^*\big(t/(1-U)\big) \qquad \text{for } 0 < t < 1 \tag{2.1}$$

where $D =_L D^*$, U is Unif(0, 1) and D, D^* and U are independent.

LEMMA 2.1. *If D satisfies* (2.1) *and, for a positive integer m, $E|D(t)|^m$ is bounded for $t \ge 1$, then for every $t_0 > 0$, $E|D(t)|^m$ is bounded for $t \ge t_0$.*

PROOF. Fix a positive integer r and choose $t \in [2^{-r}, 2^{-r+1})$. Define independent $D_0, D_1, \dots$ and $U_1, U_2, \dots$ with $D_i =_L D$ and $U_i =_L U$ for all i. Let

$V_i = U_i \vee (1 - U_i)$, $W_i = U_i \wedge (1 - U_i)$ and

$$\nu_t = \min\left\{n : \prod_{i=1}^{n} V_i \le t2^{r-1}\right\}.$$

By iterating (2.1) until the arguments of the D_i are all $\geq 2^{-r+1}$ we find

$$\begin{aligned} D(t) &\stackrel{L}{=} D_0(t/W_1) + D_1(t/V_1) \\ &\stackrel{L}{=} D_0(t/W_1) + \mathbb{1}_{\{\nu_t=1\}} D_1(t/V_1) \\ &\quad + \mathbb{1}_{\{\nu_t>1\}}\left[D_1(t/(V_1 W_2)) + D_2(t/(V_1 V_2))\right] \\ &\stackrel{L}{=} \cdots \stackrel{L}{=} \sum_{k=0}^{\nu_t-1} D_k\left(t\Big/\left(\left(\prod_{i=1}^{k} V_i\right) W_{k+1}\right)\right) + D_{\nu_t}\left(t\Big/\prod_{i=1}^{\nu_t} V_i\right). \end{aligned}$$

Conditioning first on $\{U_1, U_2, \ldots\}$ (and hence on ν_t), Minkowski's inequality implies

$$E|D(t)|^m \le E(\nu_t + 1)^m \sup\{E|D(s)|^m : s \ge 2^{-r+1}\}.$$

Now Markov's inequality yields

$$\begin{aligned} P(\nu_t > n) &= P\left(\prod_{i=1}^{n} V_i > t2^{r-1}\right) \le P\left(\prod_{i=1}^{n} V_i > \tfrac{1}{2}\right) \\ &\le 2E\prod_{i=1}^{n} V_i = 2\left(\tfrac{3}{4}\right)^n, \end{aligned}$$

so that

$$E(\nu_t + 1)^m = \sum_{n=2}^{\infty} n^m P(\nu_t = n - 1) \le 2\sum_{n=2}^{\infty} n^m (3/4)^{n-2} = A_m < \infty.$$

Because $A_m > 1$, this yields

$$\begin{aligned} \sup\{E|D(t)|^m : t \ge 2^{-r}\} &\le A_m \sup\{E|D(t)|^m : t \ge 2^{-r+1}\} \\ &\le A_m^r \sup\{E|D(t)|^m : t \ge 1\} \end{aligned}$$

by recursion over r, which gives the desired result. □

We now establish that the structure of D that is implicit in the representation (2.1) forces the process D to have a pseudo-Poissonian nature (in terms of cumulants) as is made precise in the following theorem. Here and throughout, we denote the mth cumulant and the mth central moment of a r.v. Z by $\kappa_m(Z)$ and $\mu_m(Z)$, respectively.

THEOREM 2.2. *Suppose that* D *satisfies* (2.1) *and that, for* $m = 1, 2, \ldots$, $E|D(t)|^m$ *is bounded for* $t \geq 1$. *There then exist constants* $c_1, c_2, \ldots$ *such that*

$$ED(t) = \frac{c_1}{t} \qquad \text{for } 0 < t < 1, \tag{2.2}$$

and for $m \geq 2$,

$$\kappa_m(D(t)) = \frac{c_m}{t} \qquad \text{for } 0 < t \leq 1/m. \tag{2.3}$$

It follows that $c_1 = \lim_{t\uparrow 1} ED(t)$ *and* $c_m = m^{-1}\kappa_m(D(1/m))$.

PROOF. We write $\kappa_m(t) = \kappa_m(D(t))$ and $\mu(t) = \kappa_1(t) = ED(t)$. For $0 < t < 1$, (2.1) implies that

$$\mu(t) = \int_0^1 \{\mu(t/u) + \mu(t/(1-u))\}\,du = 2\int_0^1 \mu(t/u)\,du = 2t\int_t^\infty \frac{\mu(y)}{y^2}\,dy.$$

By Lemma 2.1, $\sup\{\mu(y) : y \geq t\}$ is bounded for $t > 0$, so that μ is first of all continuous on (0, 1), and therefore also differentiable on (0, 1) with

$$\left(\frac{\mu(t)}{t}\right)' = \frac{\mu'(t)}{t} - \frac{\mu(t)}{t^2} = -2\frac{\mu(t)}{t^2}.$$

Hence $\mu(t) + t\mu'(t) = 0$ on (0, 1) and (2.2) follows.

Define

$$\psi(t, w) = \log\bigl(E\exp\{iwD(t)\}\bigr) = \sum_{j=1}^{\infty} \kappa_j(t)\frac{(iw)^j}{j!}.$$

The right-hand side is an asymptotic expansion in the sense that if we truncate the sum after r terms, the remainder is $O(|w|^{r+1})$ as $w \to 0$, uniformly for $t \geq t_0 > 0$. Of course (2.1) implies

$$\begin{aligned}\exp\{\psi(t, w)\} &= \int_0^1 \exp\{\psi(t/u, w) + \psi(t/(1-u), w)\}\,du \\ &= 2\int_0^{1/2} \exp\{\psi(t/u, w) + \psi(t/(1-u), w)\}\,du.\end{aligned} \tag{2.4}$$

Fix $m \geq 2$, $t \in (0, 1/m]$, and assume that $\kappa_j(t) = c_j/t$ for $t < 1/j$ and $j = 1, 2, \ldots, m-1$. To prove (2.3) we shall show that this implies that $\kappa_m(t) = c_m/t$ for $t \leq 1/m$. Define $n = [m/2]$ and note that:

(i) if $u \in (nt, 1/2)$, then $t/(1-u) < t/u < 1/n$, so that $\kappa_j(t/u) = c_j u/t$ and $\kappa_j(t/(1-u)) = c_j(1-u)/t$ for $j = 1, 2, \ldots, n$;

(ii) if $u \in ((k-1)t, kt)$ for some $k = 1, \ldots, n$, then $t/u < 1/(k-1)$ and $t/(1-u) < 1/(m-k)$, so that $\kappa_j(t/u) = c_j u/t$ for $j = 1, \ldots, k-1$ and $\kappa_j(t/(1-u)) = c_j(1-u)/t$ for $j = 1, \ldots, m-k$.

Multiplying (2.4) by

$$\exp\left\{-\sum_{j=1}^{m-1}\kappa_j(t)\frac{(iw)^j}{j!}\right\}=\exp\left\{-t^{-1}\sum_{j=1}^{m-1}c_j\frac{(iw)^j}{j!}\right\}$$

we find

$$\exp\left\{\sum_{j=m}^{\infty}\kappa_j(t)\frac{(iw)^j}{j!}\right\}$$

$$=2\int_{nt}^{1/2}\exp\left\{\sum_{j=n+1}^{\infty}\left(\kappa_j\left(\frac{t}{u}\right)+\kappa_j\left(\frac{t}{1-u}\right)\right)\frac{(iw)^j}{j!}-\frac{1}{t}\sum_{j=n+1}^{m-1}c_j\frac{(iw)^j}{j!}\right\}du$$

$$+2\sum_{k=1}^{n}\int_{(k-1)t}^{kt}\exp\left\{\sum_{j=k}^{m-k}\left(\kappa_j\left(\frac{t}{u}\right)-c_j\frac{u}{t}\right)\frac{(iw)^j}{j!}\right.$$

$$+\sum_{j=m-k+1}^{m-1}\left(\kappa_j\left(\frac{t}{u}\right)+\kappa_j\left(\frac{t}{1-u}\right)-\frac{c_j}{t}\right)\frac{(iw)^j}{j!}$$

$$\left.+\sum_{j=m}^{\infty}\left(\kappa_j\left(\frac{t}{u}\right)+\kappa_j\left(\frac{t}{1-u}\right)\right)\frac{(iw)^j}{j!}\right\}du.$$

Now we expand both sides in powers of (iw) and equate the coefficients of $(iw)^m/m!$. Note that in the first integral only terms containing κ_m contribute to this coefficient and that

$$\int_{(k-1)t}^{kt}f(t/u)\,du=t\int_{k-1}^{k}f(1/y)\,dy=Ct$$

where $C=C(k,f)$ is constant in t. Hence we find after some reflection that

$$\kappa_m(t)=\int_0^1\big(\kappa_m(t/u)+\kappa_m(t/(1-u))\big)\,du+Ct=2\int_0^1\kappa_m(t/u)\,du+Ct$$

$$=2t\int_t^{\infty}\kappa_m(y)y^{-2}\,dy+Ct\qquad\text{for } 0<t\le 1/m.$$

By Lemma 2.1, $\kappa_m(y)$ is bounded on (t,∞), so that κ_m is continuous on $(0,1/m)$ and differentiable on $(0,1/m)$ with

$$\left(\frac{\kappa_m(t)}{t}\right)'=\frac{\kappa_m'(t)}{t}-\frac{\kappa_m(t)}{t^2}=-2\frac{\kappa_m(t)}{t^2}.$$

It follows that $\kappa_m(t)+t\kappa_m'(t)=0$ on $(0,1/m)$ and so $\kappa_m(t)=c_m/t$ for $t\in(0,1/m]$. □

In view of (1.9) and (1.10), two special examples of D-processes to which this theorem applies are N_t+1 and $K(\alpha t,t)$. Since these examples are central in what follows, we summarize their structure as follows.

COROLLARY 2.3. *For $m = 2, 3, \ldots$ and $0 < t \le 1/m$, the cumulants of N_t and $K(\alpha t, t)$ for $0 < \alpha \le 1$ are given by*

$$\kappa_m(N_t) = \kappa_m(N_t + 1) = c_m/t \qquad \text{with } c_m = \frac{1}{m}\kappa_m(N_{1/m}),$$

(2.5)

$$\kappa_m(K(\alpha t, t)) = c_{m,\alpha}/t \qquad \text{with } c_{m,\alpha} = \frac{1}{m}\kappa_m(K(\alpha/m, 1/m))$$

and $E(N_t) = \mu(t) = 2/t - 1$, $E(K(\alpha t, t)) = 2\alpha/t$ for $0 < t < 1$.

In particular, this corollary shows that the variance of N_t is c/t if $0 < t \le 1/2$ (with $c = c_2$), as given previously in (1.7), and the fourth central moment is

$$\mu_4(t) := E[N_t - \mu(t)]^4 = c_4/t + 3c^2/t^2 \qquad \text{if } 0 < t \le 1/4. \tag{2.6}$$

The latter is needed several times in what follows.

The main result above generalizes straightforwardly to the case of vector-valued $\mathbf{D}(t) = (D_1(t), D_2(t), \ldots, D_r(t))$. In this paper, only the bivariate case $r = 2$, is needed (in Sections 5 and 7) so we will restrict our discussion to this case for notational convenience. In analogy with the univariate case, multivariate cumulants are the coefficients in the multivariate Taylor expansion of the logarithm of the joint characteristic function. Thus in particular, if $Z = (X, Y)$ is a r.v. with $E|X|^m|Y|^n < \infty$ for all $m, n \ge 1$, the (m, n)th cumulants, $\kappa_{m,n} \equiv \kappa_{m,n}(X, Y)$ are defined by

$$\log E \exp(ivX + iwY) = \sum_{\substack{m=0\, n=0 \\ m+n\ge 1}}^{\infty}\sum^{\infty} \kappa_{mn}\frac{(iv)^m}{m!}\frac{(iw)^n}{n!}.$$

Clearly, the joint cumulants $\{\kappa_{mn}\}$ are determined by the univariate cumulants of $vX + wY$; for $l \ge 1$,

$$\kappa_l(vX + wY) = \sum_{m=0}^{l}\binom{l}{m} v^m w^{l-m}\kappa_{m,l-m}. \tag{2.7}$$

Now, if we take $X = D_1(t)$, $Y = D_2(t)$ and assume that for every v, w, $vD_1(t) + wD_2(t)$ satisfies the conditions of Theorem 2.2 so that

$$\kappa_l(t) \equiv \kappa_l(vD_1(t) + wD_1(t)) = \frac{c_l(v, w)}{t} \qquad \text{for } 0 < t \le 1/l,$$

it follows from the identity in (2.7) that the coefficients $\kappa_{m,l-m}(t)$, now depending upon t, must satisfy $\kappa_{m,l-m}(t) = c_{m,l-m}/t$ for $0 < t \le 1/l$ for some constants $c_{m,l-m}$. This verifies:

THEOREM 2.4. *Let $\mathbf{D}(t) = (D_1(t), D_2(t)), t > 0$, satisfy (2.1), with $E|D_1(t)|^m$ and $E|D_2(t)|^m$ bounded in $t \ge 1$ for each $m \ge 1$. Then there exist constants $\{c_{mn}\}$ such that*

$$ED_1(t) = \frac{c_{10}}{t}, \qquad ED_2(t) = \frac{c_{01}}{t} \qquad \text{for } 0 < t < 1$$

and for $m \geq 0, n \geq 0, m+n \geq 2,$

$$\kappa_{mn}(\mathbf{D}(t)) = \frac{c_{m,n}}{t} \quad \textit{for } 0 < t \leq \frac{1}{m+n}.$$

Note that in the above, $\kappa_{10}(\mathbf{D}(t)) = \kappa_1(D_1(t)) = ED_1(t)$, with a similar identity for D_2.

3. Weak convergence of the $U(\cdot, s)$ processes. In this section, we prove that the stopped empirical process of the division points, $U(\cdot, s)$, as defined in (1.15), converges weakly to a nonstandard Brownian bridge, σB_0, as $s \to 0$ in which the constant $\sigma = (4 \ln 2 - 5/2)^{1/2} \approx 0.5221003$. [It turns out that $\sigma^2 = c/2$ with c defined in (1.7).] In the following section, we show that U_n, the ordinary empirical process for the partition points, inherits this same limit. Consequently, even though the K-model is indistinguishable from the U-model with regard to the Glivenko–Cantelli result for division points, when one considers weak convergence the two cases are quite different. The K-model results in a limiting process that has only *about half* of the variation as does the limit under the U-model.

Consider the definitions of the empirical processes of partition points given in (1.15). With $U_n = n^{1/2}(F_n - F)$, the stopped version of the process would be U_{N_s}. But

$$U_{N_s}(x) = (N_s)^{1/2}\left\{\frac{N_s(x)}{N_s} - x\right\}, \qquad 0 \leq x \leq 1, 0 < s < 1,$$

$$= (N_s)^{-1/2}\{N_s(x) - xN_s\}.$$

Since $sN_s \to 2$ a.s. by (1.11), this process is asymptotically equivalent to $U(x, s)$. But one can expand

$$U(x, s) = (s/2)^{1/2}\{N_s(x) - xN_s\}$$

$$= (s/2)^{1/2}\left\{N_s(x) - \frac{2x}{s}\right\} - x(s/2)^{1/2}\left\{N_s - \frac{2}{s}\right\}.$$

Just as for the usual U-model, this representation suggests the study of the non-tied-down process

(3.1) $$Z(x, s) := (s/2)^{1/2}\{N_s(x) - 2x/s\}, \qquad 0 \leq x \leq 1, 0 < s < 1,$$

in terms of which $U(x, s) = Z(x, s) - xZ(1, s)$. The proof of the following theorem is therefore a proof of the convergence of $Z(\cdot, s)$, from which that of $U(\cdot, s)$ follows directly.

THEOREM 3.1. *As* $t \to 0$, $U(\cdot, t) \to_L \sigma B_0(\cdot)$, *where* B_0 *is standard Brownian bridge and* $\sigma^2 = \frac{1}{4}\,\mathrm{var}(N_{1/2}) = 4 \ln 2 - 5/2$ *so that* $\sigma = 0.5221003$.

PROOF. For $0 < s < 1$, introduce the notation $0 = X_{s0} \le X_{s1} \le \cdots \le X_{s,N_s} \le X_{s,N_s+1} = 1$ to represent the N_s division points and write $D_{si} = D_{N_s i} = X_{si} - X_{s,i-1}$, $1 \le i \le N_s + 1$, for the associated spacings. For $x \in [0, 1]$, define $X_s(x) = X_{s,N_s(x)}$ and $X_s^+(x) = X_{s,N_s(x)+1}$, so that $X_s(x) \le x < X_s^+(x)$ are the division points that straddle x. Write $D_s(x) = X_s^+(x) - X_s(x)$.

The following representation is key. For any $0 < t < s < 1$,

$$N_t(x) \overset{L}{=} \sum_{i=1}^{N_s(x)} (N_{t/D_{si}}^{(i)} + 1) + N_{t/D_s(x)}^*\left(\frac{x - X_s(x)}{D_s(x)}\right) \tag{3.2}$$

where $\{N_\cdot^{(i)}\}$ are independent processes with the same laws as $N_\cdot$, $N_\cdot^*(\cdot) =_L N_\cdot(\cdot)$ and all of these processes are independent of each other and of $N_s(\cdot)$. Thus, conditionally given $\mathcal{F}_s = \sigma(D_{si} : 1 \le i \le N_s + 1) = \sigma(X_1, X_2, \ldots, X_{N_s})$, $N_t(x)$ is a sum of independent r.v.'s. More to the point is the observation that $N_t(\cdot)$ is essentially a partial-sum process, the difference being the N^*-term in (3.2).

Our approach, suggested by (3.2), is to apply standard weak convergence results to this partial-sum process, and then show that the difference term is negligible. Actually, there are two partial-sum processes involved. The one suggested by (3.2) has jumps of $N_{t/D_{si}}^{(i)} + 1$ at the times X_{si}. [Remember that we will be studying these processes *conditional on* $\mathcal{F}_s$ and with $t = t(s) < s$ going to zero appropriately with s.] The more standard time scale for plotting partial-sum processes is to plot the ith sum at its variance. We will therefore first use this standard time scale to get weak convergence, then show that the difference between the two time scales converges uniformly to zero, and finally prove that the contribution due to the extra N^* term in (3.2) is negligible.

Write

$$S_t(x; s) := (t/2)^{1/2} \sum_{i : X_{si} \le x} \{N_{t/D_{si}}^{(i)} - \mu(t/D_{si})\}, \tag{3.3}$$

where $\mu(s) = EN_s$ is given in (1.6). Thus $S_t(\cdot; s)$ is a partial sum process with increments $(t/2)^{1/2}[N_{t/D_{si}}^{(i)} - \mu(t/D_{si})]$ plotted at X_{si}. Let $S_t^*(\cdot; s)$ be the related partial-sum process whose increments are the same but which are plotted at the cumulative proportional variances $\tau_i = (\sigma_1^2 + \cdots + \sigma_i^2)/(\sigma_1^2 + \cdots + \sigma_{N_s+1}^2)$ with $\sigma_i^2 = \operatorname{var} N_{t/D_{si}}^{(i)}$. Before obtaining the limit of this $S_t^*(\cdot; s)$ process it is necessary to determine the limiting behavior of the time scale given by $\{\tau_i\}$. For this, we first need to complete the evaluation of $v(u) = \operatorname{var}(N_u)$.

From (1.7) and the definition of N_u, it is known that $v(u) = 0$ if $u \ge 1$ and $= c/u$ for $0 < u \le 1/2$ where $c = v(1/2)/2$. It remains to compute $v(u)$ for $1/2 \le u < 1$ and thereby evaluate c.

LEMMA 3.2. *For $1/2 \le u < 1$, the distribution of N_u is given by*

$$P[N_u > k] = 2^k u \sum_{j=k}^{\infty} \left(\ln \frac{1}{u}\right)^j \Big/ j! = 2^k P\left[\mathcal{P}\left(\ln \frac{1}{u}\right) \ge k\right] \tag{3.4}$$

for $k = 0, 1, 2, \ldots$ where $\mathcal{P}(\lambda)$ denotes a Poisson r.v. of mean λ. Moreover, the variance of N_u is

$$(3.5)\qquad v(u) = \frac{1}{u}\left(8\ln\frac{1}{u} + 2\right) - \frac{4}{u^2} + 2, \qquad \frac{1}{2} \le u < 1.$$

In particular, $c = v(1/2)/2 = 8\ln 2 - 5$.

PROOF. For $1/2 \le u < 1$, the splitting points $X_1, X_2, \ldots, X_{N_u}$ never return to an interval they have left, so that as in the proof of Lemma 2.1,

$$P[N_u > k] = P\left[\prod_{i=1}^{k} V_i > u\right]$$

where $V_1, V_2, \ldots$ are independent Unif(1/2, 1) r.v.'s. Hence the $-\ln V_i$ are distributed as independent standard exponential random variables Z_i, each conditioned on being smaller than $\ln 2$. For $1/2 \le u < 1$, $\sum_{i=1}^{k} Z_i < \ln(1/u)$ implies $Z_i < \ln(1/u) \le \ln 2$ for $i = 1, \ldots, k$ and

$$\begin{aligned} P[N_u > k] &= P\left(\sum_{i=1}^{k} Z_i < \ln\frac{1}{u} \,\Big|\, \max Z_i < \ln 2\right) \\ &= 2^k P\left(\sum_{i=1}^{k} Z_i < \ln\frac{1}{u}\right) \\ &= 2^k P\left(\mathcal{P}\left(\ln\frac{1}{u}\right) \ge k\right), \end{aligned}$$

which proves (3.4). This in turn implies

$$\begin{aligned} EN_u^2 &= \sum_{k=0}^{\infty}(2k+1)P[N_u > k] = u\sum_{j=0}^{\infty}\frac{(\ln 1/u)^j}{j!}\sum_{k=0}^{j}(2k+1)2^k \\ &= u\sum_{j=0}^{\infty}\frac{(\ln 1/u)^j}{j!}\{(2j-1)2^{j+1} + 3\} \\ &= \frac{1}{u}\left(8\ln\frac{1}{u} - 2\right) + 3. \end{aligned}$$

Since $EN_u = 2/u - 1$ by (1.6), the expression (3.5) follows by direct calculation. □

To establish the weak convergence of the $S_i^*(\cdot, s)$ partial-sum process, it suffices [cf. Gihman and Skorokhod (1974), page 411] to show [with E_s and P_s denoting the conditional quantities, $E(\text{—}|\mathcal{F}_s)$ and $P[\text{—}|\mathcal{F}_s]$, respectively, that

$$(3.6)\qquad \lim_{s\to 0}\sum_{i=1}^{N_s+1} E_s t\{N_{t/D_{si}}^{(i)} - \mu(t/D_{si})\}^2 \mathbb{1}_{[t^{1/2}|N_{t/D_{si}}^{(i)} - \mu(t/D_{si})| > \varepsilon]} = 0,$$

for a suitable choice of $t = t(s) < s$ going to zero with s. By the Cauchy–Schwarz and Chebyshev inequalities, the sum in (3.6) is bounded by

$$\sum_{i=1}^{N_s+1} t\{\mu_4(t/D_{si})tv(t/D_{si})\varepsilon^{-2}\}^{1/2} \tag{3.7}$$

in which $\mu_4(u) = E[N_u - \mu(u)]^4$ and $v(u) = \mathrm{var}(N_u)$. By definition and by Theorem 2.2, $u^2\mu_4(u)$ and $uv(u)$ are bounded for all $u > 0$. Hence the bound in (3.7) is

$$\varepsilon^{-1}\sum_{i=1}^{N_s+1} D_{si}^{3/2}\{(t/D_{si})^2\mu_4(t/D_{si})\}^{1/2}\{(t/D_{si})v(t/D_{si})\}^{1/2} \le C_0(M_{N_s})^{1/2} \tag{3.8}$$

for some constant C_0 where M_n is the maximum spacing at the nth stage. But by (1.13) this bound goes to zero, which establishes (3.6) and hence the desired weak convergence result. The limit process must be a mean zero Brownian motion and it remains only to determine its variance at $x = 1$. By the discussion following (3.3),

$$(t/2)^{1/2}\sum_{i=1}^{N_s+1}\{N_{t/D_{si}}^{(i)} - \mu(t/D_{si})\} = S_t^*(1, s).$$

By (1.7) and (3.5) the (conditional) variance of this sum, with $\sigma^2 = c/2$, is equal to

$$\begin{aligned}\frac{t}{2}\sum_{i=1}^{N_s+1} v(t/D_{si}) &= t\sum_{i\,:\,D_{si}>2t}\sigma^2 D_{si}/t + \frac{t}{2}\sum_{i\,:\,t<D_{si}\le 2t} v(t/D_{si})\\ &= \sigma^2 + \frac{t}{2}\sum_{i\,:\,D_{si}\le 2t}\{v(t/D_{si}) - 2\sigma^2 D_{si}/t\}\\ &= \sigma^2 + \frac{1}{2}\sum_{i\,:\,D_{si}\le 2t} D_{si}\{(t/D_{si})v(t/D_{si}) - 2\sigma^2\}.\end{aligned} \tag{3.9}$$

But since $uv(u)$ is bounded for all $u > 0$ (see Theorem 2.2) and since $\Sigma\{D_{si} : D_{si} \le 2t\} \le 2tK(2t, s)$, the second term in (3.9) is bounded for some constant C by $CtK(2t, s) = C(\frac{t}{s})sK(\frac{2t}{s}s, s)$. By (1.14) this is $O(1)$ with probability 1 and, moreover, is $o(1)$ if $t = o(s)$. This proves that when $t/s \to 0$, $\mathrm{var}\, S_t^*(1, s) \to \sigma^2 = c/2 = v(1/2)/4 = 4\ln 2 - 5/2$ by Lemma 3.2. Thus

$$S_t^*(\cdot, s) \xrightarrow{L} \sigma B(\cdot) \tag{3.10}$$

as $s \to 0$ with $t = o(s)$.

REMARK. Let us clarify how the unconditional weak convergence follows from conditional applications of limit theorems. Our approach is to use two parameter values, $t < s$ with $t = t(s)$, and express a process, $X_t(\cdot)$ say, in such a way that conditionally given $\mathcal{F}_s$ a limit result holds. For example, if g is any bounded continuous real-valued function defined on the range of $X_t(\cdot)$, suppose $E\{g(X_t(\cdot))|\mathcal{F}_s\} \to Eg(X(\cdot))$ a.s. Then by Lebesgue's dominated convergence theorem, $Eg(X_t(\cdot)) \to Eg(X(\cdot))$. This example suffices for our purposes, since it shows how conditional weak convergence a.s. proves unconditional weak convergence; all of our examples are for $D[0,1]$ processes with limits in $C(0,1)$.

We now compare the time scales of $S_t(\cdot;s)$ and $S_t^*(\cdot;s)$. By (3.3), the increments of $S_t(\cdot;s)$ are the same as for $S_t^*(\cdot;s)$ but they occur at X_{si} rather than at τ_i. The differences between the two time scales are

$$X_{sj} - \tau_j = \sum_{i=1}^{j} \{D_{si} - v(t/D_{si})/\bar{\sigma}_s^2\}$$

where $\bar{\sigma}_s^2 = \sigma_1^2 + \cdots + \sigma_{N_s+1}^2$. Note that $t\bar{\sigma}_s^2/2$ is equal to (3.9) and therefore $t\bar{\sigma}_s^2/2 = \sigma^2 + o(1)$ a.s. if $t = o(s)$, as was shown following (3.9).

By means of the same partition used in (3.9),

$$\begin{aligned} \max_{1\le j\le N_s+1} |X_{sj} - \tau_j| &\le (t\bar{\sigma}_s^2)^{-1} t \sum_{i=1}^{N_s+1} \left|\bar{\sigma}_s^2 D_{si} - v\left(\frac{t}{D_{si}}\right)\right| \\ &\le (t\bar{\sigma}_s^2)^{-1} \Bigg\{ |t\bar{\sigma}_s^2 - 2\sigma^2| + 2\sigma^2 \sum_{D_{si}\le t} D_{si} \\ &\qquad + \sum_{t<D_{si}\le 2t} D_{si} \left| 2\sigma^2 - \left(\frac{t}{D_{si}}\right) v\left(\frac{t}{D_{si}}\right)\right| \Bigg\} \\ &\le o(1) + C_0 2t K(2t, s) \end{aligned} \tag{3.11}$$

for some constant C_0. Thus as before, (1.14) implies that this converges a.s. to zero as $s \to 0$ provided $t = o(s)$. Since this proves that the difference between the time scales converges uniformly to zero with probability 1, it follows from (3.10) and the above remark that

$$S_t(\cdot;s) \xrightarrow{L} \sigma B(\cdot) \tag{3.12}$$

as $s \to 0$ with $t = o(s)$.

It remains to show that the extra N^* term in (3.2) and the centering differences of the $Z(\cdot,t)$ and $S_t(\cdot;s)$ processes are asymptotically negligible. Observe first that the N^* term is bounded by

$$\sup_{0\le x\le 1} \left| N_t(x) - \sum_{i=1}^{N_s(x)} (N_{t/D_{si}}^{(i)} + 1) \right| = \max_{1\le i\le N_s+1} \{N_{t/D_{si}}^{(i)} + 1\} := M_{s,t}. \tag{3.13}$$

[The difference on the left-hand side is actually nonnegative by (3.2) and approaches $N^{(i)}_{t/D_{s,i+1}}+1$ as $x \nearrow X_{s,i+1}$.] To show that $t^{1/2}M_{s,t} \to 0$ in probability, given $\mathcal{F}_s$, compute

$$P_s[M_{s,t} > \varepsilon t^{-1/2}] \le \sum_{i=1}^{N_s+1} P_s[N^{(i)}_{t/D_{si}}+1 > \varepsilon t^{-1/2}]$$
$$\le (N_s+1)P[N_{t/s}+1 > \varepsilon t^{-1/2}]$$

since each $D_{si} \le s$. Thus in particular, each $N^{(i)}_{t/D_{si}}$ is stochastically (given $\mathcal{F}_s$) less than $N_{t/s}$. Suppose $t/s \le 1/4$. Then by (2.6), Markov's inequality with fourth moments gives

$$P[N_{t/s}+1 > \varepsilon t^{-1/2}] \le \mu_4(t/s)/(\varepsilon t^{-1/2} - 2s/t)^4$$
$$= \{c_4 s/t + 12\sigma^4 s^2/t^2\}/(\varepsilon t^{-1/2} - 2s/t)^4$$

so that

$$P_s[M_{s,t} > \varepsilon t^{-1/2}] \le s(N_s+1)\frac{c_4 t + 12\sigma^4 s}{(\varepsilon - 2s/t^{1/2})^4}.$$

Since $sN_s \to 2$ a.s. by (1.11), this bound converges to zero a.s. provided $s = o(\sqrt{t}\,)$. In view of (3.12) we also need $t = o(s)$, so choose $t = s^{3/2}$. In this case then, this proves that almost surely, $t^{1/2}M_{s,t} \to 0$ in probability conditionally given $\mathcal{F}_s$.

Now by (3.1)–(3.3) and (3.13),

$$\sup_{0\le x\le 1} |Z(x,t) - S_t(x;s)|$$
(3.14)
$$\le t^{1/2}M_{s,t} + t^{1/2}\sum_{i=1}^{N_s} |\mu(t/D_{si}) + 1 - 2D_{si}/t| + 2t^{-1/2}M_{N_s}$$

where we have written $x = \sum_{i=1}^{N_s(x)} D_{si} + (x - X_{si})$ and used $|x - X_{si}| \le D_s(x)$ and so $\max_i |x - X_{si}| \le \max_i D_{si} = M_{N_s}$. We have just shown that the first term is $o_p(1)$ a.s. if $s = o(\sqrt{t}\,)$. By (1.13) the third term is $o(1)$ under the same proviso. By (1.6) the middle term is bounded by

$$t^{1/2} \sum_{i\,:\,D_{si}\le t} |2D_{si}/t - 1| \le t^{1/2}K(t,s).$$

Since $t < s < 1$, $E\{t^{1/2}K(t,s)\} = 2t^{3/2}s^{-2}$ by (1.8). Thus if $t = o(s^{4/3})$ this converges to zero and so the middle term on the right-hand side of (3.14) converges to zero in probability. We have thus established that if $s = s(t) = o(\sqrt{t}\,)$ and $t = o(s^{4/3})$, as is the case if $s = t^{2/3}$ for example, then the left-hand side of (3.14) is bounded by the sum of three terms, $T_1(t) + T_2(t) + T_3(t)$ say, in which

T_2 and T_3 are measurable $\mathcal{F}_s$ and each converge in probability to zero, while with probability 1, T_1 converges to zero in probability conditional on $\mathcal{F}_s$. It follows that T_1 also converges to zero in probability, thus showing that $Z(\cdot, t)$ and $S_t(\cdot; s)$ converge weakly to the same limiting process when $s = t^{2/3}$. In view of (3.12) the proof of Theorem 3.1 is complete. □

An important consequence of the above proof is that it gives the limiting distribution for N_t. Since $Z(1, t) = (t/2)^{1/2}\{N_t - 2/t\}$, we have the following corollary.

COROLLARY 3.3. *As* $t \to 0$, $(t/2)^{1/2}\{N_t - 2/t\}$ *converges in law to a* $N(0, \sigma^2)$ *random variable, with* $\sigma = 4\ln 2 - 5/2$ *as in Theorem* 3.1.

4. Weak convergence of the U_n process. As discussed at the start of Section 3, the stopped process $U(\cdot, s)$ is asymptotically equivalent to $U_{N_s}(\cdot)$, the regular empirical process of the division points computed at the random sample size N_s. We show in this section that the process U_n inherits from $U(\cdot, s)$ and U_{N_s} the same weak convergence. Thus the limiting process for U_n, which is B_0 under the U-model, becomes σB_0 for the K-model.

Random sample size central limit theorems were first considered in a general setting by Anscombe (1952) who studied the case of sums of independent r.v.'s. The weak convergence of uniform empirical processes under random sample size was studied in Pyke (1968); see also Csörgő (1974) and Klaassen and Wellner (1992). The situation here is quite different in that we will deduce the convergence of the fixed sample size process from that of the random sample size case. Of course, one can reverse this formally by defining random times s_n so that $U_n = U(\cdot, s_n)$; simply use $s_n = M_n$.

The result to be proved is the following:

THEOREM 4.1. *As* $n \to \infty$, $U_n \to_L \sigma B_0$, *where* B_0 *is standard Brownian bridge and* $\sigma = (4\ln 2 - 5/2)^{1/2} = 0.52210$.

PROOF. The proof essentially is by moments but entails a coupling argument in a critical spot. Some technical results are needed that are presented first in a series of lemmas. The first three involve the cumulants κ_m and central moments μ_m of differences $N_s - N_t$.

LEMMA 4.2. *For* $m = 2, 3, \ldots$ *and* $0 < s < t \leq 1/m$,

$$\kappa_m(N_s - N_t) = \kappa_m(N_{s/tm} - N_{1/m})\frac{1}{mt}. \tag{4.1}$$

PROOF. It is easy to check that $D(t) = N_{\alpha t} - N_t$ for $0 < \alpha < 1$ satisfies the relationship (2.1). Thus (4.1) follows from Theorem 2.2 with $\alpha = s/t$. □

LEMMA 4.3. *For every* $0 < \varepsilon < 1$ *and* $m = 2, 3, \ldots$, *there exists a positive number* $c_m(\varepsilon)$ *such that*

$$|\kappa_m(N_s - N_t)| \le c_m(\varepsilon)\left(\frac{1}{s} - \frac{1}{t}\right) \qquad \text{for } \varepsilon t \le s < t < 1, \tag{4.2}$$

$$|\mu_m(N_s - N_t)| \le c_m(\varepsilon)\left\{\left(\frac{1}{s} - \frac{1}{t}\right)^{m/2} + \left(\frac{1}{s} - \frac{1}{t}\right)\right\} \qquad \text{for } \varepsilon t \le s < t < 1. \tag{4.3}$$

PROOF. First take $1/m \le t < 1$. Let $M = K((s,t],t) := K(t,t) - K(s,t)$, be the number of intervals with length $l \in (s,t]$ at the first time when all intervals are less than or equal to t. Since $[M = 0] = [N_s - N_t = 0]$ and $N_s - N_t \ge 0$ a.s.,

$$P(M > 0) = P(N_s - N_t \ge 1) \le E(N_s - N_t) = 2(1/s - 1/t).$$

Now, for $l \in (s,t]$, $l/t \ge s/t \ge \varepsilon$ and since $M \le 1/s$ a.s.,

$$N_s - N_t \le \sum_{j=1}^{M} N_{s/t}^{(j)} \le \mathbb{1}_{\{M>0\}} \sum_{j=1}^{[1/s]} N_\varepsilon^{(j)}$$

where the $N^{(j)}$ are independent copies of N which are also independent of M. Hence, for $k = 1, 2, \ldots, m$ and $1/m \le t < 1$, Minkowski's inequality implies

$$\begin{aligned} E(N_s - N_t)^k &\le 2\left(\frac{1}{s} - \frac{1}{t}\right)s^{-k} E N_\varepsilon^k \\ &\le 2\left(\frac{m}{\varepsilon}\right)^m E N_\varepsilon^m \left(\frac{1}{s} - \frac{1}{t}\right) = \tilde{c}_m(\varepsilon)\left(\frac{1}{s} - \frac{1}{t}\right), \end{aligned}$$

where $\tilde{c}_m(\varepsilon)$ is finite since $E N_\varepsilon^m < \infty$ [see van Zwet (1978)]. For $t \ge 1/m$, $1/s - 1/t \le 1/\varepsilon t \le m/\varepsilon$, and this yields (4.2). Insertion of this into (4.1) suffices to cover the case of $t < 1/m$ so that

$$|\kappa_m(N_s - N_t)| \le c'_m(\varepsilon)(1/s - 1/t) \qquad \text{for } \varepsilon t \le s < t < 1.$$

But this yields (4.3) for some $c_m(\varepsilon) > c'_m(\varepsilon)$ and the proof is complete. □

Recall that $N_t(x)$ denotes the number of points among $X_1, \ldots, X_{N_t}$ which fall in $(0, x]$, $0 \le x \le 1$. Suppose that among the $N_t(x)$ points in $(0, x]$, ξ is the *first* point in $[x - t, x]$ and that we delete all points in $(\xi, x]$ that follow it. Let $N'_{t/x}$ denote the number of points remaining. Clearly $N_t(x) \ge N'_{t/x}$ and $N'_{t/x}$ is distributed like $N_{t/x}$. Note that for fixed x the processes $\{N'_{t/x} : 0 < t < x\}$ and $\{N_{t/x} : 0 < t < x\}$ also have the same distribution.

Furthermore, $(N_t(x) - N'_{t/x})$ is stochastically smaller than $N_{1/2} - 2$. To see this, note that among the points $X_1, \ldots, X_{N_t}$, the first points ξ and ξ' in $[x - t, x]$ and $(x, x + t]$, respectively, plus all points in (ξ, ξ') form a Kakutani splitting of the interval $(x-t, x+t)$ into intervals of length $\le t$, which is half the width of the interval.

Summarizing, we have a process $N'_{t/x}$ distributed like $N_{t/x}$ and such that

$$N_t(x) = N'_{t/x} + R(x,t) \tag{4.4}$$

where, for every x and t,

$$0 \le R(x,t) \overset{\text{st}}{\le} N_{1/2} - 2. \tag{4.5}$$

LEMMA 4.4. *For every $\varepsilon > 0$ and $m = 2, 3, \ldots$, there exists a positive number $\tilde{c}_m(\varepsilon)$ such that*

$$|\mu_m(N_s(x) - N_t(x))| \le \tilde{c}_m(\varepsilon)\left\{\left(\frac{x}{s} - \frac{x}{t}\right)^{m/2} + 1\right\} \tag{4.6}$$

for $\varepsilon t \le s < t < x \le 1$.

PROOF. In view of (4.4) and (4.3),

$$\begin{aligned}
&|\mu_m(N_s(x) - N_t(x))| \\
&\quad = |\mu_m(N'_{s/x} - N'_{t/x} + R(x,s) - R(x,t))| \\
&\quad = \left|\sum_{k=0}^{m}\binom{m}{k}\mu_k(R(x,s) - R(x,t))\mu_{m-k}(N'_{s/x} - N'_{t/x})\right| \\
&\quad \le c_m(\varepsilon)\left\{\left(\frac{x}{s} - \frac{x}{t}\right)^{m/2} + \left(\frac{x}{s} - \frac{x}{t}\right)\right\} + |\mu_m(R(x,s) - R(x,t))| \\
&\qquad + \sum_{k=2}^{m-2}\binom{m}{k}|\mu_k(R(x,s) - R(x,t))|c_{m-k}(\varepsilon) \\
&\qquad \times \left\{\left(\frac{x}{s} - \frac{x}{t}\right)^{(m-k)/2} + \left(\frac{x}{s} - \frac{x}{t}\right)\right\}.
\end{aligned}$$

Now (4.5) implies that

$$|\mu_k(R(x,s) - R(x,t))| \le 2^{2k-1}\{E|R(x,s)|^k + E|R(x,t)|^k\} \le 2^k E(N_{1/2} - 2)^k$$

and since $N_{1/2}$ has finite moments of every order, the proof is complete. □

LEMMA 4.5. *For every $0 < a < A$,*

$$\lim_{n\to\infty} \sup_{\substack{0 \le x \le 1 \\ an^{1/2} \le |\tau^{-1} - n/2| \le An^{1/2}}} \left|\frac{N_{2/n}(x) - N_\tau(x)}{n - 2/\tau} - x\right| = 0 \qquad a.s.$$

PROOF. Fix $x \in (0,1]$. Consider two sequences s_n and t_n such that $\varepsilon \max(s_n, t_n) \le \min(s_n, t_n) < \max(s_n, t_n) < x \le 1$ and $n^{-\delta}|\frac{1}{s_n} - \frac{1}{t_n}| \to \infty$ for some positive ε and δ. Then (4.6) implies that for every $m = 1, 2, \dots$,

$$P\left(|N_{s_n}(x) - N_{t_n}(x) - EN_{s_n}(x) + EN_{t_n}(x)| \ge n^{-\delta/4}\left|\frac{x}{s_n} - \frac{x}{t_n}\right|\right) = o(n^{-\delta m/4})$$

as $n \to \infty$. Hence, by choosing $m > 4/\delta$,

$$\lim_n \frac{|N_{s_n}(x) - N_{t_n}(x) - EN_{s_n}(x) + EN_{t_n}(x)|}{|x/s_n - x/t_n|} = 0 \qquad \text{a.s.}$$

For $s, t < x$, (4.4), (4.5) and (1.6) insure that $|E(N_s(x) - N_t(x)) - 2x/s + 2x/t| \le E(N_{1/2} - 2) = 1$ and hence

$$\text{(4.7)} \qquad \lim_{n\to\infty} \frac{N_{s_n}(x) - N_{t_n}(x)}{2x/s_n - 2x/t_n} = 1 \qquad \text{a.s.}$$

for every pair of sequences satisfying the above requirements.

Take $s_n = 2/n$ and define $t_{n,k}^{-1} = n/2 + k\eta n^{1/2}$ where

$$k \in \mathcal{J} = \left\{\pm\left[\frac{a}{\eta}\right], \pm\left(\left[\frac{a}{\eta}\right]+1\right), \pm\left(\left[\frac{a}{\eta}\right]+2\right), \dots, \pm\left(\left[\frac{A}{\eta}\right]+1\right)\right\}$$

and η is a fixed (small) positive number. For each $k \in \mathcal{J}$, $s_n = 2/n$ and $t_{n,k}$ satisfy the requirements for (4.7), so

$$\text{(4.8)} \qquad \lim_{n\to\infty} \max_{k\in\mathcal{J}} \left|\frac{N_{2/n}(x) - N_{t_{n,k}}(x)}{n - 2/t_{n,k}} - x\right| = 0 \qquad \text{a.s.}$$

Let τ_n be a sequence with $an^{1/2} \le |\tau_n^{-1} - n/2| \le An^{1/2}$. Then there exist $k_n, k_n + 1 \in \mathcal{J}$ such that $t_{n,k_n}^{-1} \le \tau_n^{-1} \le t_{n,k_n+1}^{-1}$. Now $t_{n,k_n+1}^{-1} - t_{n,k_n}^{-1} = \eta n^{1/2}$ and $|t_{n,k_n}^{-1} - n/2| \ge [a/\eta]\eta n^{1/2}$, so for $\eta \le a/2$,

$$\text{(4.9)} \qquad \begin{aligned} &\limsup_n \left|\frac{N_{\tau_n}(x) - N_{t_{n,k_n}}(x)}{n - 2/t_{n,k_n}}\right| \\ &\quad\le \limsup_n \left|\frac{N_{t_{n,k_n+1}}(x) - N_{t_{n,k_n}}(x)}{2/t_{n,k_n+1} - 2/t_{n,k_n}}\right| \frac{1}{[a/\eta]} \\ &\quad\le \frac{2\eta}{a} \limsup_n \max_{k\in\mathcal{J}} \left|\frac{N_{t_{n,k+1}}(x) - N_{t_{n,k}}(x)}{2/t_{n,k+1} - 2/t_{n,k}}\right| \\ &\quad\le \frac{2\eta x}{a} \le \frac{2\eta}{a} \qquad \text{a.s.,} \end{aligned}$$

since the pair of sequences $t_{n,k}$ and $t_{n,k+1}$ satisfy the requirements for (4.7). Also

$$\text{(4.10)} \qquad \limsup_n \left|\frac{n - 2/\tau_n}{n - 2/t_{n,k_n}} - 1\right| \le \frac{1}{[a/\eta]} \le \frac{2\eta}{a}.$$

Combining (4.8)–(4.10) and noting that $\eta > 0$ may be taken arbitrarily small, we find that for fixed $x \in (0, 1]$,

$$(4.11)\qquad \lim_{n\to\infty} \sup_{an^{1/2}\le|\tau_n^{-1}-n/2|\le An^{1/2}} \left|\frac{N_{2/n}(x) - N_{\tau_n}(x)}{n - 2/\tau_n} - x\right| = 0 \qquad \text{a.s.}$$

Since $(N_{2/n}(x) - N_{\tau_n}(x))/(n - 2/\tau_n)$ is nondecreasing in x and equals 0 for $x = 0$, a standard argument completes the proof. □

Let $n(x) = \sum_{i=1}^n \mathbb{1}_{(0,x]}(X_i)$ be the number of points among $X_1, \ldots, X_n$ that fall in $(0, x]$, so that $F_n(x) = \frac{n(x)}{n}$ is the empirical d.f. of $X_1, \ldots, X_n$.

LEMMA 4.6. *As $n \to \infty$,*

$$\sup_{0\le x\le 1} n^{1/2}|F_{N_{2/n}}(x) - F_n(x)| \to 0$$

in probability.

PROOF. We have

$$n^{1/2}|F_{N_{2/n}}(x) - F_n(x)| = \frac{n^{1/2}|N_{2/n} - n|}{N_{2/n}}\left|\frac{N_{2/n}(x) - n(x)}{N_{2/n} - n} - F_n(x)\right|.$$

By (1.3), (1.6), (1.7) and (1.11), it suffices to show that

$$(4.12)\qquad n^{-1/2}|N_{2/n} - n| \sup_{0\le x\le 1}\left|\frac{N_{2/n}(x) - n(x)}{N_{2/n} - n} - x\right| \xrightarrow{P} 0.$$

The definition of M_n following (1.1) implies that$N_{M_n} = n$, so $N_{M_n}(x) = n(x)$. Applying Lemma 4.5 twice, once for general x and once for $x = 1$ and substituting $\tau = M_n$, we find that, for every $0 < a < A$,

$$\lim_{n\to\infty} \sup_{0\le x\le 1}\left|\frac{N_{2/n}(x) - n(x)}{N_{2/n} - n} - x\right| \mathbb{1}_{[a,A]}(n^{-1/2}|M_n^{-1} - n/2|) = 0 \qquad \text{a.s.}$$

and since $n^{-1/2}|N_{2/n} - n|$ is bounded in probability by (1.6) and (1.7), we have, for every $0 < a < A$,

$$(4.13)\qquad \begin{aligned} &n^{-1/2}|N_{2/n} - n| \sup_{0\le x\le 1}\left|\frac{N_{2/n}(x) - n(x)}{N_{2/n} - n} - x\right| \\ &\qquad \times \mathbb{1}_{[a,A]}(n^{-1/2}|M_n^{-1} - n/2|) \to 0 \qquad \text{a.s.} \end{aligned}$$

We have $\{M_n > t\} = \{N_t > n\}$ and hence, if $n - 2/t + 1 > 0$ and $0 < t < 1$,

$$(4.14)\qquad \begin{aligned} P(M_n > t) &= P\left(\frac{N_t - EN_t}{\sigma(N_t)} > \frac{n - 2/t + 1}{c^{1/2}} t^{1/2}\right) \\ &\le \frac{c}{t(n - 2/t + 1)^2}, \end{aligned}$$

and therefore, for sufficiently large n,

$$P(n^{-1/2}(M_n^{-1} - n/2) < -A)$$
$$= P(M_n > (n/2 - An^{1/2})^{-1}) \le \frac{c(n/2 - An^{1/2})}{(2An^{1/2}+1)^2} \le \frac{c}{8A^2}.$$

This probability can be made arbitrarily small by taking A large and the same is true for $P(n^{-1/2}(M_n^{-1} - n/2) > A)$, so (4.13) can be extended to

$$n^{-1/2}|N_{2/n} - n| \sup_{0\le x\le 1} \left| \frac{N_{2/n}(x) - n(x)}{N_{2/n} - n} - x \right| \tag{4.15}$$
$$\times \mathbb{1}_{[a,\infty)}(n^{-1/2}|M_n^{-1} - n/2|) \xrightarrow{P} 0.$$

Finally we consider the set $B = \{n^{-1/2}|M_n^{-1} - n/2| \le a\}$. Writing $s_n^{-1} = (n/2 + an^{1/2})$ and $t_n^{-1} = (n/2 - an^{1/2})$ we see that on the set B, $N_{t_n} \le n \le N_{s_n}$ and since $s_n < 2/n < t_n$, we have $|N_{2/n} - n| \le |N_{s_n} - N_{t_n}|$ on B. Hence, by Lemma 4.3, we have for sufficiently large n and any $\delta, \varepsilon > 0$,

$$P\left(n^{-1/2}|N_{2/n} - n| \sup_{0\le x\le 1} \left| \frac{N_{2/n}(x) - n(x)}{N_{2/n} - n} - x \right| \mathbb{1}_B \ge \delta\right)$$
$$\le P(n^{-1/2}|N_{s_n} - N_{t_n}| \ge \delta) \le \frac{E(N_{s_n} - N_{t_n})^2}{\delta^2 n} \tag{4.16}$$
$$\le \frac{2c_2(1/2)an^{1/2} + 16a^2 n}{\delta^2 n} \le 17\frac{a^2}{\delta^2} \le \varepsilon,$$

if we take a sufficiently small. Together (4.15) and (4.16) imply (4.12) and the lemma. □

To complete the proof of Theorem 4.1, it suffices now to take $s = 2/n$ in $U(\cdot, s)$ to see that

$$\{(n/2)^{1/2}(F_{N_{2/n}}(x) - x); 0 < x < 1\} \xrightarrow{L} \sigma B_0$$

by Theorem 3.1. It then follows from Lemma 4.6 that the proof is complete. □

5. Weak convergence of the $W(\cdot, t)$ and $V(\cdot, t)$ processes. We now prove that the stopped empirical processes of the relative spacings, $V(\cdot, t)$, converge weakly on [0, 1] to a Gaussian process $V(\cdot)$ as $t \to 0$. The proof concentrates in fact upon establishing the weak convergence on [0, 1] of the related processes $W(\cdot, s)$ of which the V-processes are tied-down versions; see (1.17), (1.18) and (1.20). The proof is based on a representation of $W(\cdot, t)$ as a sum of independent processes. As in Section 3, let $\mathcal{F}_s = \sigma(D_{si} : 1 \le i \le N_s + 1)$ be the σ-field of the partitions at level s. For any $0 < t < s < 1$, we may write

$$K(x,t) = \sum_{i=1}^{N_s+1} K^{(i)}(x/D_{si}, t/D_{si}) \tag{5.1}$$

where $K^{(i)}$, $1 \le i \le N_s+1$, are independent copies of K that are independent also of $\mathcal{F}_s$. From this and the definition of $W(\cdot,t)$ in (1.18), we get the following key representation of $W(\cdot,t)$ as a sum of conditionally independent processes, namely, for any $0<t<s<1$,

$$W(\cdot,t)=\sum_{i=1}^{N_s+1} D_{si}^{1/2} W^{(i)}(\cdot,t/D_{si}) \tag{5.2}$$

in which $W^{(i)}$, $1 \le i \le N_s+1$, are independent copies of W that are also independent of $\mathcal{F}_s$. The visual simplicity of this representation is due to the definition (1.18) in which the centering for $(r/2)K(yr,r)$ is chosen to be y rather than its mean when $y>1$ or $r\ge 1$. Since we are only concerned with the processes $W(y,t)$ for $0\le y\le 1$, the case of $y>1$ plays no role in (5.2), but since t/D_{si} may exceed 1, the case of $r\ge 1$ does. When only terms centered at expectations are used, the expression (5.2) becomes

$$\begin{aligned} W(y,t)=&\sum_{i\,:\,D_{si}>t} D_{si}^{1/2} W^{(i)}(y,t/D_{si}) \\ &+(t/s)^{1/2}W(yt/s,s)+2y(s/t)^{1/2}\int_{t/s}^{1} zW(dz,s), \end{aligned} \tag{5.3}$$

for $0\le y\le 1$ and $0<t<s<1$. It is in this form that the recursion is used in Section 6. Note that the full integral over [0, 1] in the above is zero.

We first use the representation (5.2) to prove the limiting normality of the finite-dimensional distributions. We do this by applying the Lindeberg central limit theorem to the sum in (5.2) *conditionally given* $\mathcal{F}_s$. [The remark following (3.10) should be noted for this section as well.] Since we need to compute moments using Theorem 2.2, we split the summation of (5.2) into two parts according as $t/D_{si}>1/4$ or $\le 1/4$. Write $W(\cdot,t)=W_s^-(\cdot,t)+W_s^+(\cdot,t)$, where W_s^- represents the summation over those i for which $D_{si}<4t$. Recall from (1.8) that for $0\le y\le 1$ and $0<u<1$, $EW(y,u)=0$. Then, conditionally given $\mathcal{F}_s$, W_s^- and W_s^+ are independent and W_s^+ has mean zero, so that for $0\le y\le 1$ and $0<t<s<1$,

$$\mathrm{var}_s\{W(y,t)\}=\mathrm{var}_s\{W_s^-(y,t)\}+E_s[W_s^+(y,t)]^2,$$

where var_s, E_s, P_s indicate the conditional quantities given $\mathcal{F}_s$. Also, for $0<t\le 1/2$ and $m=2$, Corollary 2.3 implies

$$E_s[W_s^+(y,t)]^2=\sum_{i\,:\,D_{si}\ge 4t} D_{si}E_s[W(y,t/D_{si})]^2=\tfrac{1}{2}c_{2,y}{\sum}^+ D_{si}\le \tfrac{1}{2}c_{2,y} \tag{5.4}$$

where $c_{2,y}$ is the constant of Corollary 2.3 for $m=2$ and $\alpha=y$, and $\sum^+$ denotes summation over $\{i : D_{si}\ge 4t\}$. [The actual covariance function for the $W(\cdot,t)$ processes, and hence for the limiting $W(\cdot)$ process, is derived in (7.10) where in particular, $c_{2,y}=\alpha(y,y)$ with α defined in (7.9).]

Next we deal with $W^-(y,t)$. Since $E_s[W(y,t/D_{si})] = 0$ whenever $D_{si} > t$, we find in view of (1.8) and (1.14) that with probability 1,

$$|E_s[W_s^-(y,t)]| = \left|\sum_{i:D_{si}\le t} D_{si}^{1/2} E_s W(y,t/D_{si})\right|$$

$$= (t/2)^{1/2}\left|\sum_{i:D_{si}\le t} [E_s K(yt/D_{si}, t/D_{si}) - (2y/t)D_{si}]\right| \tag{5.5}$$

$$\le (t/2)^{1/2} \sum_{i:D_{si}\le yt} 1 + y(2/t)^{1/2} \sum_{i:D_{si}\le t} D_{si}$$

$$\le (t/2)^{1/2} K(yt,s) + y(2/t)^{1/2} t K(t,s) = O(t^{3/2}/s^2).$$

To handle the variance of $W^-(y,t)$, observe first that $W(y,u)$ is nonrandom when $u \ge 1$. Thus

$$\mathrm{var}_s\{W^-(y,t)\} = \sum_{i:D_{si}<4t} D_{si}\, \mathrm{var}_s\{W(y,t/D_{si})\}$$

$$\le \sum_{t<D_{si}<4t} D_{si} E_s[W(y,t/D_{si})]^2 \tag{5.6}$$

$$\le 4tK(4t,s) \sup_{1/4<u<1} E[W(y,u)]^2.$$

By (1.14), $tK(4t,s) \to 0$ a.s. as $s \to 0$ if $t = t(s)$ is chosen so that $t/s \to 0$. Moreover, the quantity $C(y) := \sup_{1/4<u<1} E[W(y,u)]^2$ is finite, since, for $1/4 \le u < 1$,

$$E[W(y,u)]^2 \le E\left[\frac{u}{2}K^2(uy,u)\right] \le \frac{u}{2}E(N_u+1)^2 \le \frac{1}{2}E(N_{1/4}+1)^2$$

which is finite; see van Zwet (1978). Thus

$$W_s^-(y,t) \xrightarrow{P} 0 \qquad \text{as } s \to 0 \text{ and } t^3/s^4 \to 0.$$

Consider now the limiting finite-dimensional distributions of $W_s^+(\cdot,t)$. The Lindeberg criterion (conditional given $\mathcal{F}_s$) for the one-dimensional case involves

$$L_t(\varepsilon) := \sum\nolimits^+ D_{si} E_s\left\{[W^{(i)}(y,t/D_{si})]^2 \mathbb{1}_{[D_{si}^{1/2}|W^{(i)}(y,t/D_{si})| \ge \varepsilon B_t^{1/2}]}\right\}$$

for $\varepsilon > 0$ and $B_t = \mathrm{var}_s(W_s^+(y,t))$. From Hölder's and Chebyshev's inequalities we obtain

$$L_t(\varepsilon) \le \sum\nolimits^+ D_{si}\left\{E_s[W^{(i)}(y,t/D_{si})]^4 P_s[D_{si}^{1/2}|W^{(i)}(y,t/D_{si})| \ge \varepsilon B_t^{1/2}]\right\}^{1/2}$$

$$\le \sum\nolimits^+ D_{si}^{3/2}\left\{E_s[W^{(i)}(y,t/D_{si})]^4 E_s[W^{(i)}(y,t/D_{si})]^2\right\}^{1/2} / \varepsilon B_t^{1/2}.$$

Since $t/D_{si} \leq 1/4$, the moments in the summation are bounded by constants (for fixed y) by Corollary 2.3. Also, B_t converges to the nonzero constant $c_{2,y}/2$ by (5.4) and the sentence following (5.6). Since each $D_{si} \leq s$ by definition, it follows from the above that $L_t(\varepsilon)/B_t = O(s^{1/2}) = o(1)$ as $s \to 0$ provided only that $t \to 0$ as well. Thus Lindeberg's condition is satisfied, and therefore the one-dimensional distributions of $W(\cdot, t)$ converge to those of $W(\cdot)$. For higher dimensions, the proof is similar requiring only that finite linear combinations $\sum a_j W(y_j, t)$ be considered. We have therefore established the following lemma.

LEMMA 5.1. *As $t \to 0$, the finite-dimensional distributions of $\{W(y, t) : 0 \leq y \leq 1\}$ converge to those of $W(\cdot)$, a mean zero Gaussian process on* [0, 1] *with covariance given in* (7.9) *and* (7.10).

To complete the proof of weak convergence, we will apply a standard sufficient condition for tightness in $D[0, 1]$ that is based on a moment bound for adjacent increments of the process; see Theorem 15.6 of Billingsley (1968). The appropriate bound is given in the following lemma. (We gratefully acknowledge our appreciation to Christian Genest for pointing out an error in an earlier attempt to prove this result based only on a moment bound for a single interval.)

Write $J_1 = (x, y]$, $J_2 = (y, z]$ for $0 \leq x < y < z \leq 1$, and recall that we write, for example, $W(J_1, t) = W(y, t) - W(x, t)$. The adjacency and interval structure of J_1 and J_2 is not required in the moment bound that we now derive, and so we state it for general disjoint Borel sets. Observe that although $K(\cdot, t)$ and $W(\cdot, t)$ are defined as point functions, since they are clearly equivalent to (signed) measures, this enables us to write $K(B, t)$ and $W(B, t)$ unambiguously for any Borel set B as well as for intervals.

LEMMA 5.2. *There exists a constant C such that for all $t \in (0, 1]$ and any disjoint Borel subsets J_1 and J_2 of* [0, 1],

$$E[W(J_1, t)W(J_2, t)]^2 \leq C|J_1||J_2|. \tag{5.7}$$

PROOF. Assume first of all that $0 < t \leq 1/4$. [The reader should note that for its application to tightness, the bound of (5.7) is only needed for t in some interval of the form $(0, t_0)$.] By applying Theorem 2.4 to the pair $D_i(t) = K(tJ_i, t)$, $i = 1, 2$, we get for $j, k = 0, 1, 2$, that

$$\kappa_{j,k}(t) \equiv \kappa_{j,k}(D_1(t), D_2(t)) = \frac{c_{j,k}}{t} \qquad \text{for } 0 < t \leq \frac{1}{4}. \tag{5.8}$$

This joint cumulant may be expressed in terms of central moments; specifically, a straightforward computation shows that if $EX = \mu$, $EY = \nu$,

$$\begin{aligned}
&\kappa_{2,0}(X, Y) = \sigma_X^2, \\
&\kappa_{0,2}(X, Y) = \sigma_Y^2, \\
&\kappa_{1,1}(X, Y) = \mathrm{Cov}(X, Y) \quad \text{and} \\
&\kappa_{2,2}(X, Y) = E(X-\mu)^2(Y-\nu)^2 - \sigma_X^2\sigma_Y^2 - 2[\mathrm{Cov}(X, Y)]^2.
\end{aligned} \tag{5.9}$$

Thus, using $W(J_1, t) = X$ and $W(J_2, t) = Y$, this means by (1.8) and the scalar homogeneity evident in (5.9), that for $0 < t \le 1/4$,

$$\begin{aligned}
E[W(J_1, t)W(J_2, t)]^2 &= \left(\frac{t}{2}\right)^2 \kappa_{2,2}(t) + E[W(J_1, t)]^2 E[W(J_2, t)]^2 \\
&\quad + 2\{EW(J_1, t)W(J_2, t)\}^2 \\
&= \left(\frac{t}{2}\right)^2 \{\kappa_{2,2}(t) + \kappa_{2,0}(t)\kappa_{0,2}(t) + 2[\kappa_{1,1}(t)]^2\} \\
&= \frac{t}{4}c_{2,2} + \frac{1}{4}c_{2,0}c_{0,2} + \frac{1}{2}c_{1,1}^2,
\end{aligned} \tag{5.10}$$

with the last equation following from (5.8). It remains, then, to obtain bounds for the constants c_{22}, c_{20} and c_{11} in terms of $|J_1|$ and $|J_2|$ so as to verify (5.7) when $0 < t \le 1/4$.

By definition, and in view of (5.9) and (1.21), it follows that for any $0 < t \le 1/4$,

$$\begin{aligned}
c_{2,2} &= t\kappa_{2,2}(t) = t\kappa_{2,2}(D_1(t), D_2(t)) \\
&\le tE(\{D_1(t) - ED_1(t)\}^2\{D_2(t) - ED_2(t)\}^2) \\
&= tE(\{D_1(t) - 2t^{-1}|J_1|\}^2\{D_2(t) - 2t^{-1}|J_2|\}^2) \\
&\le t\Big(E\{D_1(t)D_2(t)\}^2 + (2t^{-1}|J_1|)^2E[D_2(t)]^2 + (2t^{-1}|J_2|)^2E[D_1(t)]^2 \\
&\quad + 16t^{-2}|J_1|\,|J_2|E(D_1(t)D_2(t)) + 16t^{-4}|J_1|^2|J_2|^2\Big) \\
&\le tE\{D_1(t)D_2(t)\}^2 + C_t\{|J_1|E[D_2(t)]^2 + |J_2|E[D_1(t)]^2 + |J_1|\,|J_2|\}
\end{aligned} \tag{5.11}$$

for some constant C_t depending on the chosen t but not on the J_i's. For this, recall that $D_i(t) \le (N_t + 1)$ and N_t has finite moments by Lemma 2.1. Thus to obtain the desired bound for $c_{2,2}$, it suffices to bound $E\{D_1(t)D_2(t)\}^2$ and $E[D_i(t)]^2$ appropriately for some value of $t \le 1/4$. To bound the other two terms of (5.10) observe first that by Cauchy–Schwarz, $c_{1,1}^2 \le c_{2,0}c_{0,2}$. Hence it suffices to establish an appropriate bound for the right-hand side of

$$c_{2,0} = tE\left\{K(tJ_1, t) - \frac{2}{t}|J_1|\right\}^2 \le tE[K(tJ_1, t)]^2$$

for some $0 < t \le 1/4$. Such bounds are contained in the following:

LEMMA 5.3. *For any Borel set $J \subset [0,1]$, $0 < t < 1$ and $k = 1, 2, \ldots,$*

$$E[K(tJ,t)]^k \le 2^k |J| E N_t^k, \tag{5.12}$$

and for any disjoint Borel sets J_1 and J_2 in $[0,1]$ and $0 < t < 1$,

$$E[K(tJ_1,t)K(tJ_2,t)]^2 \le C|J_1||J_2| E N_t^4 \tag{5.13}$$

for some constant C.

PROOF. The first inequalities are the simplest to prove since they involve only one set J. (Although only the case $k = 2$ is needed here, we give it for general k since this requires no new ideas.) For this single J, write $D(t) = K(tJ, t)$. We use the representation

$$0 \le D(t) = \sum_{j=1}^{N_t} \eta_j$$

where the r.v. η_j equals the number of new spacings with length in tJ that originate with the jth splitting, which occurs at X_j. Since $\eta_j = 0, 1$ or 2, we have for any $k \ge 1$ that

$$\begin{aligned} E[D(t)]^k &= E\left[\sum_{j=1}^{N_t} \eta_j\right]^k = \sum_{j_1=1}^{\infty} \cdots \sum_{j_k=1}^{\infty} E\left(\prod_{i=1}^{k} \eta_{j_i} \mathbb{1}_{[N_t \ge \max_i j_i]}\right) \\ &\le 2^{k-1} \sum_{j_1=1}^{\infty} \cdots \sum_{j_k=1}^{\infty} E(\eta_{\max j_i} \mathbb{1}_{[N_t \ge \max j_i]}) \\ &= 2^{k-1} \sum_{j=1}^{\infty} (j^k - (j-1)^k) E(\eta_j \mathbb{1}_{[N_t \ge j]}) \\ &= 2^{k-1} \sum_{j=1}^{\infty} (j^k - (j-1)^k) P[N_t \ge j] E[\eta_j | N_t \ge j]. \end{aligned}$$

On the event $[N_t \ge j]$, X_j splits an interval whose length exceeds t. Hence, since the splitting is done uniformly, $E[\eta_j | N_t \ge j] \le 2|J|$ and so, for any $0 < t \le 1$,

$$E[D(t)]^k \le 2^k |J| \sum_{j=1}^{\infty} j^k P[N_t = j] = 2^k |J| E N_t^k.$$

To prove the second inequality (5.13), let η_{ij} equal the number of the two new spacings that originate with the ith splitting, whose lengths are in tJ_j. Thus, each $\eta_{ij} \in \{0, 1, 2\}$, and

$$D_j(t) = K(tJ_j, t) = \sum_{i=1}^{N_t} \eta_{ij} = \sum_{i=1}^{\infty} \eta_{ij} \mathbb{1}_{[N_t \ge i]}. \tag{5.14}$$

Recall that M_n is the maximum spacing after n splittings. Write

$$
\begin{aligned}
&EK^2(tJ_1,t)K^2(tJ_2,t)\\
&\quad = E\sum_{i=1}^{\infty}\sum_{k=1}^{\infty}\sum_{j=1}^{\infty}\sum_{l=1}^{\infty}\eta_{i1}\eta_{k1}\eta_{j2}\eta_{l2}\mathbb{1}_{[N_t\ge\max\{i,j,k,l\}]}
\end{aligned}
\tag{5.15}
$$

$$
\le 4\sum_{n=1}^{\infty}\sum_{i\le n}\sum_{\substack{j\le n\\ i\vee j=n}}(i^2-(i-1)^2)(j^2-(j-1)^2)E\big(\eta_{i1}\eta_{j2}\mathbb{1}_{[N_t\ge n]}\big).
$$

Set

$$
\begin{aligned}
A_i &= \frac{tJ_1}{M_{i-1}}\cup\left(1-\frac{tJ_1}{M_{i-1}}\right) \quad\text{and}\\
B_j &= \frac{tJ_2}{M_{j-1}}\cup\left(1-\frac{tJ_2}{M_{j-1}}\right),
\end{aligned}
\tag{5.16}
$$

where for $J\subset\mathbb{R}$ and $a,b\in\mathbb{R}$, $a-bJ=\{a-bx:x\in J\}$. Note that the measures of A_i and B_j satisfy

$$
|A_i|\le\frac{2t}{M_{i-1}}|J_1|,\qquad |B_j|\le\frac{2t}{M_{j-1}}|J_2|
\tag{5.17}
$$

for all i and j. By definition,

$$
\begin{aligned}
\eta_{m1} &= \mathbb{1}_{[(1-U_m)M_{m-1}\in tJ_1]}+\mathbb{1}_{[U_mM_{m-1}\in tJ_1]}\\
&\le 2\mathbb{1}_{[U_m\in(tJ_1/M_{m-1})\cup(1-tJ_1/M_{m-1})]}=2\mathbb{1}_{[U_m\in A_m]},
\end{aligned}
\tag{5.18}
$$

with a similar bound holding for η_{m2}. Substitution of this into (5.15) yields

$$
\begin{aligned}
&EK^2(tJ_1,t)K^2(tJ_2,t)\\
&\quad\le 64\sum_{n=1}^{\infty}\sum_{i\le n}\sum_{\substack{j\le n\\ i\vee j=n}} ijP[U_i\in A_i,U_j\in B_j,N_t\ge n].
\end{aligned}
\tag{5.19}
$$

There are two cases to consider in evaluating the inner summations of (5.19), namely, $i=j=n$ and $i<j=n$ (with $i=n>j$ being similar).

Consider $i=j=n$, for which the summands involve $A_n\cap B_n$. Because of the disjointness of J_1 and J_2, it follows from their definitions in (5.16) that

$$
\begin{aligned}
&A_n\cap B_n\\
&\quad=\left\{\left(\frac{t}{M_{n-1}}J_1\right)\cap\left(1-\frac{t}{M_{n-1}}J_2\right)\right\}\cup\left\{\left(\frac{t}{M_{n-1}}J_2\right)\cap\left(1-\frac{t}{M_{n-1}}J_1\right)\right\}.
\end{aligned}
$$

Notice that the two sets in parentheses are disjoint and have the same Lebesgue measure. Thus

$$
\begin{aligned}
&P[U_n \in A_n \cap B_n, N_t \geq n] \\
&\quad = 2P\left[U_n \in \frac{t}{M_{n-1}}\left\{J_1 \cap \left(\frac{M_{n-1}}{t} - J_2\right)\right\}, M_{n-1} > t\right] \\
&\quad = 2E\left(\frac{t}{M_{n-1}}\left|J_1 \cap \left(\frac{M_{n-1}}{t} - J_2\right)\right| \mathbb{1}_{[M_{n-1}>t]}\right) \\
&\quad \leq 2E\left(\left|J_1 \cap \left(\frac{M_{n-1}}{t} - J_2\right)\right| \mathbb{1}_{[M_{n-1}>t]}\right).
\end{aligned} \tag{5.20}
$$

However, for any $v > 1$,

$$|J_1 \cap (v - J_2)| = \int_0^1 \mathbb{1}_{J_1}(x)\mathbb{1}_{J_2}(v-x)\,dx = (\mathbb{1}_{J_1} * \mathbb{1}_{J_2})(v),$$

the convolution of two indicator functions. By Fubini,

$$
\begin{aligned}
&P[U_n \in A_n \cap B_n, N_t \geq n] \\
&\quad \leq 2\int_0^1 \mathbb{1}_{J_1}(x) E\left\{\mathbb{1}_{J_2}\left(\frac{M_{n-1}}{t} - x\right)\mathbb{1}_{[M_{n-1}>t]}\right\} dx.
\end{aligned} \tag{5.21}
$$

Since $J_2 \subset [0,1]$ and $0 < x \leq 1$, the expectation in the integrand is equal to $P[M_{n-1} \in tL_x]$ where $L_x = (x + J_2) \cap (1, 2]$. But

$$P[M_{n-1} \in tL_x] = \int_{tL_x} d_s P[M_{n-1} \leq s] = \int_{tL_x} d_s(-P[N_s \geq n]).$$

Thus (5.21) yields

$$P[U_n \in A_n \cap B_n, N_t \geq n] \leq 2\int_0^1 \mathbb{1}_{J_1}(x)\int_{tL_x} d_s(-P[N_s \geq n])\,dx. \tag{5.22}$$

Substitution of this into (5.19) shows that the part of the summation for which $i = j = n$ satisfies

$$
\begin{aligned}
&64\sum_{n=1}^{\infty} n^2 P[U_n \in A_n \cap B_n, N_t \geq n] \\
&\quad \leq 128\int_0^1 \mathbb{1}_{J_1}(x)\int_{tL_x} d_s\left(-\sum_{n=1}^{\infty} n^2 P[N_s \geq n]\right) dx.
\end{aligned} \tag{5.23}
$$

However, the mass function

$$\begin{aligned}\Phi(s) &:= \sum_{n=1}^{\infty} n^2 P[N_s \ge n] \\ &= \sum_{n=1}^{\infty} n^2 \sum_{k=n}^{\infty} P[N_s = k] = \sum_{k=1}^{\infty} \frac{k(k+1)(2k+1)}{6} P[N_s = k] \qquad (5.24) \\ &= \frac{1}{3} EN_s^3 + \frac{1}{2} EN_s^2 + \frac{1}{6} EN_s.\end{aligned}$$

Since in the integration of (5.23), $s \in (t, 2t]$, it follows that $s \le 1/3$ whenever $t \le 1/6$. In this case, then, the key Corollary 2.3 implies that

$$\begin{aligned}\Phi(s) &= \frac{1}{3}\{\kappa_3(N_s) + 3\kappa_2(N_s)EN_s + (EN_s)^3\} + \frac{1}{2}\{\kappa_2(N_s) + (EN_s)^2\} + \frac{1}{6}EN_s \\ &= \frac{c_3}{3s} + \frac{c_2}{s}\left(\frac{2}{s} - 1\right) + \frac{1}{3}\left(\frac{2}{s} - 1\right)^3 + \frac{c_2}{2s} + \frac{1}{2}\left(\frac{2}{s} - 1\right)^2 + \frac{1}{6}\left(\frac{2}{s} - 1\right) \\ &\equiv a_0 + a_1 s^{-1} + a_2 s^{-2} + a_3 s^{-3},\end{aligned}$$

say, and so Φ is differentiable with $-\Phi'(s) \le bs^{-4}$ for some constant b for $s \in (t, 2t]$. Substitution into (5.23) shows that the part of the summation in (5.19) with $i = j = n$ satisfies

$$64 \sum_{n=1}^{\infty} n^2 P[U_n \in A_n \cap B_n, N_t \ge n] \le 128bt^{-3}|J_1||J_2| \le Ct^{-3}|J_1||J_2| \qquad (5.25)$$

as desired whenever $0 < t \le 1/6$. Note that since $t(N_t + 1) \ge 1$ always, $t^{-3} \le E(N_t + 1)^3 \le 8EN_t^4$.

Consider now $t > 1/6$. To show that the measure determined by Φ remains dominated by Lebesgue measure over $(t, 1)$, it suffices to show that for any $k \ge 1$ and any $1/6 < s < r < 1$, $E(N_s^k - N_r^k) \le b(r - s)$ for some constant b. To this end, for $1/6 < s < r < 1$ consider

$$\begin{aligned}N_s^k - N_r^k &= [N_r + (N_s - N_r)]^k - N_r^k \\ &= \sum_{l=0}^{k-1} \binom{k}{l} N_r^l (N_s - N_r)^{k-l} \le C_k N_r^{k-1} (N_s - N_r)^k \qquad (5.26)\end{aligned}$$

for some constant C_k. Moreover, by (5.1),

$$\begin{aligned}N_s - N_r &= \sum_{i : D_{ri} > s} N_{s/D_{ri}}^{(i)} \\ &= \sum_{i : D_{ri} > s} (N_{s/D_{ri}}^{(i)} - 1) + K((s, r], r) \qquad (5.27)\end{aligned}$$

in which the superscript i indexes independent processes as in (5.1); see also the proof of Lemma 4.3. Thus, by Minkowski's inequality,

$$\begin{aligned} E\big((N_s - N_r)^k | \mathcal{F}_r\big) &\le \left\{ \sum_{i:\, D_{ri}>s} \{E(N^{(i)}_{s/D_{ri}} - 1)^k\}^{1/k} + K\big((s,r],r\big) \right\}^k \\ &\le \left\{ K\big((s,r],r\big)\{E(N_{s/r}-1)^k\}^{1/k} + K\big((s,r],r\big) \right\}^k \\ &\le \{K\big((s,r],r\big)\}^k \{(E(N_{s/r}-1)^k)^{1/k} + 1\}^k, \end{aligned}$$

in which $\mathcal{F}_r = \sigma(D_{ri} : 1 \le i \le N_r + 1)$ as introduced in Section 3. Since to show the a.e. differentiability of Φ it suffices to consider $r - s$ small, assume without loss of generality that $r - s < 1/6$, implying that $s/r > 1/2$; recall that $1/6 < s < r < 1$. Hence, the second factor in the last expression is (by Lemma 2.1) bounded, and so

$$\begin{aligned} E\big((N_s - N_r)^k | \mathcal{F}_r\big) &\le C_k \{K\big((s,r],r\big)\}^k \\ &\le C_k N_r^k \mathbb{1}_{[N_s > N_r]}, \end{aligned}$$

where, here and in the following, C_k is used generically to denote constants depending only upon k. Therefore, (5.26) yields

$$EN_s^k - EN_r^k \le C_k E\big(N_r^{2k-1} \mathbb{1}_{[N_s > N_r]}\big). \tag{5.28}$$

The event $[N_s > N_r] = [K((s,r],r) > 0]$ is the event that at least one of the first N_r splits resulted in a spacing in $(s,r]$. As for the proof of (5.12) above but with r in place of t, let η_j be the number of spacings formed by the jth splitting that have lengths in $rJ \equiv (s,r]$ with $J = (s/r, 1]$. Then

$$[N_s > N_r] \subset \bigcup_{j=1}^{\infty} [\eta_j > 0, N_r \ge j]$$

so that

$$E\big(N_r^{2k-1} \mathbb{1}_{[N_s > N_r]}\big) \le \sum_{j=1}^{\infty} E\big(N_r^{2k-1} \mathbb{1}_{[\eta_j > 0, N_r \ge j]}\big). \tag{5.29}$$

However, conditionally given $\{D_{j-1,i} : 1 \le i \le j\}$ and $\mathbb{1}_{[N_r \ge j]}$, N_r is stochastically dominated as follows:

$$N_r \overset{L}{=} j + \sum_{i:\, D_{j,i} > r} N^{(i)}_{r/D_{j,i}} \overset{\text{stoch}}{\preceq} j + \sum_{i=1}^{[1/r]} N_r^{(i)}, \tag{5.30}$$

since after j splittings, the splitting process continues independently in the intervals whose lengths, $D_{j,i}$, still exceed r, and since the number of such intervals does not exceed $1/r$. Note in particular that this stochastic bound does not depend upon U_j, the jth splitting uniform r.v., nor, therefore, upon η_j. Thus

$$\begin{aligned}
&E(N_r^{2k-1}\mathbb{1}_{[\eta_j>0,N_r\geq j]})\\
&\leq E\left(j+\sum_{i=1}^{[1/r]}N_r^{(i)}\right)^{2k-1}P[\eta_j>0,N_r\geq j]\\
&\leq\left(j+\frac{1}{r}(EN_r^{2k-1})^{1/(2k-1)}\right)^{2k-1}P[\eta_j>0,N_r\geq j]\\
&\leq C_k j^{2k-1}P[\eta_j>0,N_r\geq j],
\end{aligned}\tag{5.31}$$

in which the second inequality utilizes Minkowski's inequality and Lemma 2.1; recall $r>1/6$. Moreover, as in the proof of (5.12),

$$\begin{aligned}
P[\eta_j>0,N_r\geq j]&\leq 2P[U_jM_{j-1}\in(s,r],N_r\geq j]\\
&=2E((r/M_{j-1})(1-s/r)\mathbb{1}_{[N_r\geq j]})\\
&\leq\frac{2}{r}(r-s)P[N_r\geq j].
\end{aligned}\tag{5.32}$$

Thus (5.29), (5.31) and (5.32) applied to (5.28) yields

$$\begin{aligned}
EN_s^k-EN_r^k&\leq C_k(r-s)\sum_{j=1}^{\infty}j^{2k-1}P[N_r\geq j]\\
&\leq C_k(r-s)EN_r^{2k}\leq C_k(r-s)
\end{aligned}\tag{5.33}$$

by Lemma 2.1 for all $k\geq 1$ and $1/6<s<r<1$ with $r-s<1/6$. By (5.24) this shows that a bounded Φ' exists a.e. over $(1/6,1)$. Hence (5.23) yields

$$\begin{aligned}
64\sum_{n=1}^{\infty}n^2P[U_n\in A_n\cap B_n,N_t\geq n]&\leq C\int_0^1\mathbb{1}_{J_1}(x)|tL_x|\,dx\\
&\leq C|J_1|\,|J_2|
\end{aligned}\tag{5.34}$$

as desired when $t>1/6$. This and (5.25) complete the bounding of the terms in (5.19) with $i=j=n$.

To compute a bound for the other terms of (5.19) in which $i<j=n$, observe that

$$\begin{aligned}
&P[U_i\in A_i,U_n\in B_n,N_t\geq n]\\
&\quad=E\{E(\mathbb{1}_{[U_n\in B_n]}|M_{n-1},U_i\in A_i,N_t\geq n)\mathbb{1}_{[U_i\in A_i]}\mathbb{1}_{[N_t\geq n]}\}\\
&\quad\leq E\{(2t/M_{n-1})|J_2|\mathbb{1}_{[U_i\in A_i,N_t\geq n]}\}\\
&\quad\leq 2|J_2|P[U_i\in A_i,N_t\geq i,N_t\geq n]
\end{aligned}\tag{5.35}$$

where the insertion of $[N_t \geq i]$ changes nothing since $n > i$. To compute the remaining probability, notice that when $U_i \in A_i$, (5.16) implies that at least one of $U_i M_{i-1}$ or $(1 - U_i)M_{i-1}$ is in tJ_1, and thus is less than or equal to t. Hence, at least one of the two spacings formed by the ith splitting is never split again during the next $N_t - i$ splittings.

Given $\{D_{i-1,k} : 1 \leq k \leq i\}$, $[N_t \geq i]$ and $[U_i \in tJ_1/M_{i-1}]$, the conditional distribution of N_t is, as in (5.30), that of

$$i + \sum_{k:\, D_{i-1,k} < M_{i-1}} N^{(k)}_{t/D_{i-1,k}} + N^{(0)}_{t/(1-U_i)M_{i-1}} \tag{5.36}$$

where the $N^{(k)}$ are independent copies of N. Clearly the last term of (5.36) does not exceed $N^{(0)}_{t/M_{i-1}}$ so that (5.36) is stochastically smaller than

$$i + \sum_{k=1}^{i} N^{(k)}_{t/D_{i-1,k}}.$$

This in turn has the same distribution as the conditional distribution of $N_t + 1$ given $\{D_{i-1,k} : 1 \leq k \leq i\}$ and $[N_t \geq i]$. The same result holds if we had assumed that $1 - U_i \in tJ_1/M_{i-1}$, and hence if $U_i \in A_i$. It follows that

$$\begin{aligned}
&P[U_i \in A_i, N_t \geq i, N_t \geq n] \\
&\quad = E\{P[N_t \geq n | \{D_{i-1,k} : 1 \leq k \leq i\}, U_i \in A_i, N_t \geq i] \mathbb{1}_{[U_i \in A_i, N_t \geq i]}\} \\
&\quad \leq E\{P[N_t \geq n-1 | \{D_{i-1,k} : 1 \leq k \leq i\}, N_t \geq i] \mathbb{1}_{[U_i \in A_i, N_t \geq i]}\} \\
&\quad = EP[N_t \geq n-1 | N_t \geq i, M_{i-1}] \mathbb{1}_{[N_t \geq i]} P[U_i \in A_i | N_t \geq i, M_{i-1}].
\end{aligned}$$

However,

$$P[U_i \in A_i | N_t \geq i, M_{i-1}] \leq 2(t/M_{i-1})|J_1| \leq 2|J_1|.$$

Thus, for $1 \leq i \leq n-1$,

$$\begin{aligned}
P[U_i \in A_i, N_t \geq i, N_t \geq n] &\leq 2|J_1| P[N_t \geq n-1 | N_t \geq i] P[N_t \geq i] \\
&\leq 2|J_1| P[N_t \geq n-1].
\end{aligned}$$

In view of (5.35) this implies that the sum of the terms of (5.19) with $i < j = n$ is bounded by

$$\begin{aligned}
&256|J_1||J_2| \sum_{n=1}^{\infty} \sum_{i=1}^{n-1} ni P[N_t \geq n-1] \\
&\qquad = 128|J_1||J_2| \sum_{n=1}^{\infty} n^2(n-1) P[N_t \geq n-1] \\
&\qquad = 128|J_1||J_2| E\left\{ \sum_{n=1}^{N_t+1} n^2(n-1) \right\} \leq 64|J_1||J_2| E(N_t+1)^4.
\end{aligned}$$

Together with (5.25) and (5.34) this completes the proof of (5.13) and hence of Lemma 5.3. □

Return now to the proof of Lemma 5.2 in the remaining case of $t > 1/4$, or more generally, when t is bounded away from 0. Observe that since for any t, $W(J_i, t) = (t/2)^{1/2}[K(tJ_i, t) - 2|J_i|/t]$, direct expansion in (5.7) yields

$$
\begin{aligned}
&E[W(J_1,t)W(J_2,t)]^2 \\
&\quad = \frac{t^2}{4} E\left\{K^2(tJ_1,t) - \frac{4}{t}|J_1|K(tJ_1,t) + \frac{4}{t^2}|J_1|^2\right\} \\
&\qquad \times \left\{K^2(tJ_2,t) - \frac{4}{t}|J_2|K(tJ_2,t) + \frac{4}{t^2}|J_2|^2\right\} \qquad (5.37)\\
&\quad \le \frac{t^2}{4} E\left\{[K(tJ_1,t)K(tJ_2,t)]^2 + \frac{16}{t^2}|J_1|\,|J_2|[K(tJ_1,t)K(tJ_2,t)]\right. \\
&\qquad \left. + \frac{16}{t^4}|J_1|^2|J_2|^2 + \frac{4}{t^2}|J_1|^2K^2(tJ_2,t) + \frac{4}{t^2}|J_2|^2K^2(tJ_1,t)\right\}.
\end{aligned}
$$

Thus (5.7) follows directly from Lemma 5.3 whenever t is bounded away from 0. This completes the proof of Lemma 5.2. □

Tightness clearly follows from Lemma 5.2; Chebyshev's inequality implies that for any $\lambda > 0$ and disjoint adjacent intervals J_1 and J_2 in $[0, 1]$,

$$
\begin{aligned}
&P[|W(J_1,t)| > \lambda, |W(J_2,t)| > \lambda] \\
&\qquad \le P[|W(J_1,t)W(J_2,t)| > \lambda^2] \le \lambda^{-4}C|J_1|\,|J_2|.
\end{aligned}
$$

Theorem 15.6 of Billingsley (1968), then suffices, together with the finite-dimensional limits established earlier in Lemma 5.1, to prove the following main result.

THEOREM 5.4. *The stopped processes $W(\cdot, t)$ converge weakly in $D[0, 1]$ as $t \to 0$ to a mean zero Gaussian process $W(\cdot)$ with covariance given by* (7.10). *Moreover, by* (1.20), *the stopped empirical processes of the relative spacings satisfy $V(\cdot, t) \to_L V$ where $V(y) = W(y) - yW(1)$, a mean zero Gaussian process with covariance given by* (7.12).

6. Weak convergence of the V_n-processes. Consider the empirical process of the relative spacings, $\{D_{ni}/M_n; 1 \le i \le n+1\}$, defined in (1.17) by

$$V_n(y) = \sqrt{n+1}\{G_n^*((n+1)M_ny) - y\}, \qquad 0 \le y \le 1. \qquad (6.1)$$

The weak convergence of the related stopped process

(6.2) $$V(y,s)=\frac{2}{s(N_s+1)}\{W(y,s)-yW(1,s)\}$$

for $0\le y\le 1$ and $0<s\le 1$ is given in Theorem 5.4, namely,

(6.3) $$V(\cdot,s)\stackrel{L}{\to}W(\cdot)-(\cdot)W(1)=V(\cdot)$$

in $D[0,1]$ as $s\to 0$, with the limit process V being a mean zero Gaussian process with covariance given in (7.12).

By the above definitions, one may write

(6.4) $$V_n(y)=(M_n(n+1)/2)^{1/2}V(y,M_n),$$

which means that the desired weak convergence of V_n will be established once it is established for $W(\cdot,M_n)$.

To show the latter, we will establish that as $n\to\infty$,

(6.5) $$W(\cdot,2/n)-W(\cdot,M_n)\stackrel{L}{\to}0$$

in $D[0,1]$. From this, the limit process for $W(\cdot,M_n)$ is seen to be that of $W(\cdot,2/n)$, namely, W.

The first step is to show that for any $\varepsilon>0$, there exists n_ε and $L=L_\varepsilon>0$ such that

(6.6) $$P\left(\left|M_n-\frac{2}{n}\right|\ge Ln^{-3/2}\right)\le\varepsilon$$

for all $n\ge n_\varepsilon$. To see this, observe that the expression following (4.14) states that for any $A>0$,

$$P\left(M_n-\frac{2}{n}>\left(\frac{n}{2}-An^{1/2}\right)^{-1}-\frac{2}{n}\right)=P\left(M_n>\left(\frac{n}{2}-An^{1/2}\right)^{-1}\right)\le\frac{c}{8A^2}$$

for all n sufficiently large. But, for $A=L/8$,

$$\left(\frac{n}{2}-An^{1/2}\right)^{-1}-\frac{2}{n}=\frac{2}{n}[(1-2An^{-1/2})^{-1}-1]<8An^{-3/2}=Ln^{-3/2}$$

for all n sufficiently large, implying that

$$P\left(M_n-\frac{2}{n}\ge Ln^{-3/2}\right)\le\frac{8c}{L^2}$$

for all n sufficiently large. Together with (4.14)'s analogous bound for small values of M_n-2/n, this proves (6.6). Alternatively, one may use a standard renewal theory argument to obtain the limit law for $n^{3/2}(M_n-2/n)$ from that of N_s since $(M_n>s)=(N_s>n)$; specifically, one obtains from Corollary 3.3 that $n^{1/2}\{nM_n/2-1\}$ has the same asymptotic normal distribution $N(0,\sigma^2)$ as does $(2/t)^{1/2}(tN_t/2-1)$.

In view of (6.6), the proof of (6.5) will be complete if we can show that for every integer $L > 0$

$$\sup_{s:|t-s|\le Lt^{3/2}} \sup_{0\le y\le 1} |W(y,s) - W(y,t)| \xrightarrow{P} 0 \tag{6.7}$$

as $t \to 0$. To handle the supremum over s in the above, equip the interval $I_t = [t - Lt^{3/2}, t + Lt^{3/2}]$ with the grid of $2L^2 + 1$ equally spaced points $s_i \equiv s_i(t) = t + (-L + i/L)t^{3/2}$, $i = 0, 1, \ldots, 2L^2$. Note that $s_{i+1} - s_i = t^{3/2}/L$ for each i. Assume without loss of generality that $t < (2L)^{-2}$ to insure that each $s_i \in (0, 1)$ and that $t/s_i < 2$ for each i. To prove (6.7) it obviously suffices to show that for every positive integer $L \ge 2$,

$$\sup_{0\le y\le 1} |W(y,t) - W(y,s_i)| \xrightarrow{P} 0 \qquad \text{for } i = 0, 1, \ldots, 2L^2 \tag{6.8}$$

as $t \to 0$, and that for every $\varepsilon > 0$, $\delta > 0$, there exist $L > 0$ and $t^* > 0$ such that for $t < t^*$,

$$P\left(\max_{0\le i<2L^2} \sup_{s_i\le s\le s_{i+1}} \sup_{0\le y\le 1} |W(y,s) - W(y,s_i)| > \delta\right) < \varepsilon. \tag{6.9}$$

To prove (6.8), observe first that known characterizations of tightness for $D[0,1]$-valued processes imply that the processes formed by summing two tight families of processes are also tight; compare Theorem 15.2 of Billingsley (1968) and check that the $D[0,1]$ modulus $w'_x(\delta)$ satisfies the following: For any $\varepsilon > 0$ and $\delta > 0$, there exists $\delta^* \equiv \delta^*(\varepsilon, \delta)$ for which

$$w'_{f+g}(\delta^*) \le w'_f(\delta) + w'_g(\delta) + 2\varepsilon.$$

This follows by first choosing δ -partitions that approximate the moduli $w'_f(\delta)$ and $w'_g(\delta)$ to within ε, and then use the refinement of these two partitions to obtain an upper bound for $w'_{f+g}(\delta^*)$ in which δ^* is the span of this refinement. In view of Theorem 5.4 this means that the family of processes $\{W(\cdot,t) - W(\cdot,s_i) : t \in (0,1]\}$ is tight. Thus, to show $W(\cdot,t) - W(\cdot,s_i) \to_L 0$ as $t \to 0$, and hence the uniform convergence in probability to zero that is expressed in (6.8), it suffices to show $W(y,t) - W(y,s_i) \to_P 0$ for each *fixed* $y \in [0,1]$. This we do by establishing the following lemma.

LEMMA 6.1. *For* $0 < s \le 1/2$ *and* $0 < \lambda < 1$,

$$\sup_{0\le y\le 1} E[W(y,s) - W(y,\lambda s)]^2 = O(1-\lambda).$$

PROOF. From (7.9) and (7.10), we have, for any $0 \le y, z \le 1$ and $0 < s \le 1/2$,

$$\operatorname{Cov}(W(y,s), W(z,s)) = \tfrac{1}{2}\alpha(y \wedge z, y \vee z) \equiv A(y,z) + y \wedge z \tag{6.10}$$

in which the non-Brownian portion of the covariance,

$$A(y,z) = -6yz + 2yz\{(1+y)^{-1} + (1+z)^{-1}\} + 2y\ln(1+z) + 2z\ln(1+y) + (y+z-1)^{+}(1-(y+z)^{-1}), \tag{6.11}$$

is a symmetric function with uniformly bounded partial derivatives over $[0,1]^2$. By (5.3),

$$\begin{aligned} &E\{W(y,\lambda s)W(y,s)\} \\ &\quad = \lambda^{1/2}E\{W(\lambda y,s)W(y,s)\} \\ &\qquad + 2y\lambda^{-1/2}E\left\{\left[W(1,s) - \lambda W(\lambda,s) - \int_\lambda^1 W(x,s)\,dx\right]W(y,s)\right\} \end{aligned}$$

where integration by parts has been applied to the last term of (5.3). Consequently, (6.10) implies that

$$\begin{aligned} &E\{W(y,\lambda s)W(y,s)\} \\ &\quad = \sqrt{\lambda}\{A(\lambda y,y) + \lambda y\} + \frac{2y}{\sqrt{\lambda}}A(y,1) \\ &\qquad - 2y\sqrt{\lambda}A(y,\lambda) - \frac{2y}{\sqrt{\lambda}}\int_\lambda^1 A(x,y)\,dx \\ &\qquad + \frac{2y^2}{\sqrt{\lambda}} - 2y\sqrt{\lambda}(y\wedge\lambda) - \frac{2y}{\sqrt{\lambda}}\int_\lambda^1 (x\wedge y)\,dx. \end{aligned} \tag{6.12}$$

Hence,

$$\begin{aligned} &E[W(y,s) - W(y,\lambda s)]^2 \\ &\quad = 2A(y,y) + 2y - 2E\{W(y,\lambda s)W(y,s)\} \\ &\quad = 2\{A(y,y) - \sqrt{\lambda}A(\lambda y,y)\} + \frac{4y}{\sqrt{\lambda}}\{\lambda A(y,\lambda) - A(y,1)\} \\ &\qquad + \frac{4y}{\sqrt{\lambda}}\int_\lambda^1 A(x,y)\,dx + 2y(1-\lambda^{3/2}) \\ &\qquad + \frac{4y}{\sqrt{\lambda}}\{\lambda(y\wedge\lambda) - y\} + \frac{4y}{\sqrt{\lambda}}\int_\lambda^1 (x\wedge y)\,dx. \end{aligned} \tag{6.13}$$

It follows from (6.11) that $A(\cdot,\cdot)$ and its first-order partials are bounded, thereby insuring that the first three terms above are of order $O(1-\lambda)$. The last three terms, not involving $A(\cdot,\cdot)$, are easily checked to be of the desired order as well, thereby completing the proof. □

By taking $s = s_i$ and $\lambda = t/s_i$ for $i = 0, 1, \ldots, 2L^2$, it follows from Lemma 6.1 that $W(y,t) - W(y,s_i) \to_P 0$ for each y as $t \to 0$; note that for each fixed $L > 0$,

$$1 - \lambda \equiv \frac{s_i - t}{s_i} \leq \frac{L\sqrt{t}}{1 - L\sqrt{t}} = O(\sqrt{t}) \to 0$$

as $t \to 0$. This completes the proof of (6.8).

The proof of (6.9) uses the following inequalities. For $0 < u \leq s \leq v \leq 1$, we have [cf. (1.5)]

$$K(yu, v) \leq K(ys, v) \leq K(ys, s) \leq K(ys, u) \leq K\big((yv) \wedge u, u\big) \tag{6.14}$$

and so by (1.18), with $\rho \equiv u/v$ and $0 \leq y \leq 1$,

$$\begin{aligned} &\sqrt{\rho}W(\rho y, v) - \sqrt{\frac{2}{v}}\frac{1}{\sqrt{\rho}}(1-\rho^2) \\ &\quad \leq W(y,s) \\ &\quad \leq \frac{1}{\sqrt{\rho}}W\Big(\frac{y}{\rho} \wedge 1, \rho v\Big) + \sqrt{\frac{2}{v}}\Big\{\frac{1}{\rho}\Big(\frac{y}{\rho} \wedge 1\Big) - y\Big\} \\ &\quad \leq \frac{1}{\sqrt{\rho}}W\Big(\frac{y}{\rho} \wedge 1, \rho v\Big) + \sqrt{\frac{2}{v}}(\rho^{-2} - 1). \end{aligned} \tag{6.15}$$

For application of these bounds to (6.9), take $v = s_{i+1}$ and $u = s_i$ so that $\rho = s_i/s_{i+1}$. Observe that by definition,

$$0 \leq 1 - \rho = \frac{s_{i+1} - s_i}{s_{i+1}} = \frac{\sqrt{t}/L}{1 + (-L + (i+1)/L)\sqrt{t}} \leq \frac{\sqrt{t}/L}{1 - L\sqrt{t}}.$$

Hence, since we have assumed $t < (2L)^{-2}$ and $L \geq 2$, then $1 - \rho < 2\sqrt{t}/L$ and $\rho > 3/4$. It follows that the nonrandom terms in the bounds of (6.15) are bounded in absolute value by $16/L$. Consequently, for $s_i \leq s \leq s_{i+1}$ and $0 \leq y \leq 1$, we obtain the following uniform bounds for (6.9), in which we write $\|\cdot\|$ for the supremum over $[0, 1]$ and $w_f(\delta) = \sup\{|f(u) - f(v)| : 0 \leq u \leq v \leq u + \delta \leq 1\}$ for the usual modulus of continuity:

$$\begin{aligned} W(y,s) - W(y,s_i) &\leq \frac{1}{\sqrt{\rho}}W\Big(\frac{y}{\rho} \wedge 1, s_i\Big) - W(y, s_i) + \frac{16}{L} \\ &= (\rho^{-1/2} - 1)W\Big(\frac{y}{\rho} \wedge 1, s_i\Big) + W\Big(\Big(y, \frac{y}{\rho} \wedge 1\Big], s_i\Big) + \frac{16}{L} \\ &\leq (1-\rho)\|W(\cdot, s_i)\| + w_{W(\cdot,s_i)}(\rho^{-1} - 1) + \frac{16}{L} \\ &\leq \frac{2\sqrt{t}}{L}\|W(\cdot, s_i)\| + w_{W(\cdot,s_i)}\Big(\frac{2\sqrt{t}}{L}\Big) + \frac{16}{L} \end{aligned}$$

and

$$
\begin{aligned}
W(y,s) - W(y,s_i) &\geq \sqrt{\rho}\, W(\rho y, s_{i+1}) - W(y,s_i) - \frac{16}{L} \\
&= (\rho^{1/2} - 1) W(\rho y, s_{i+1}) - W(y(\rho,1], s_{i+1}) \\
&\quad + W(y, s_{i+1}) - W(y, s_i) - \frac{16}{L} \\
&\geq -(1-\rho)\|W(\cdot, s_{i+1})\| - w_{W(\cdot,s_{i+1})}(1-\rho) \\
&\quad - \|W(\cdot, s_{i+1}) - W(\cdot, s_i)\| - \frac{16}{L} \\
&\geq -\frac{2\sqrt{t}}{L}\|W(\cdot, s_{i+1})\| - w_{W(\cdot,s_{i+1})}\left(\frac{2\sqrt{t}}{L}\right) \\
&\quad - \|W(\cdot, s_{i+1}) - W(\cdot, s_i)\| - \frac{16}{L}.
\end{aligned}
$$

Since the maximum discontinuity of $W(\cdot, s)$ is $(s/2)^{1/2}$, the limiting process, W, in Theorem 5.4 is continuous so that for every i and L, $w_{W(\cdot,s)}(2\sqrt{t}/L) \to_P 0$ and $\|W(\cdot, s_i)\| = O_p(1)$. Moreover, a similar argument to that used earlier to prove (6.9) suffices to show that $\|W(\cdot, s_{i+1}) - W(\cdot, s_i)\| \to_P 0$ for each i. Therefore, for each fixed L,

$$
\max_{0 \leq i < 2L^2} \sup_{s_i \leq s \leq s_{i+1}} \|W(\cdot, s) - W(\cdot, s_i)\| = o_p(1) + \frac{16}{L},
$$

which establishes (6.9). This completes the proof of (6.8) and (6.9), and hence of (6.7) and (6.5), thereby proving:

THEOREM 6.2. *The empirical processes of the relative spacings, $V_n : n \geq 1$, converge weakly in $D(0,1)$ to the mean zero Gaussian process V with covariance function given in* (7.12).

7. The covariance of the spacings processes. The covariance functions of the limiting empirical processes for the normalized spacings are only given implicitly in the above in the sense that they are expressible in terms of constants from Theorem 2.2 for $D(t) = K(tJ, t)$. An explicit expression for these covariances is now derived, thereby completing the characterization of the limiting process.

The basic function is the covariance of $K(\cdot, s)$. For $0 \leq x \leq y \leq s$ and $0 \leq s \leq 1$, set

$$
\begin{aligned}
c(x,y,s) &= \mathrm{Cov}(K(x,s), K(y,s)), \\
C(u,v,s) &= c(us, vs, s) \qquad \text{for } 0 \leq u \leq v.
\end{aligned} \tag{7.1}
$$

Since K satisfies the representation (1.10), we already know much of the structure of the covariance because of Theorem 2.4; note that $\mathbf{D}(t) = (K(ut,t), K(vt,t))$ satisfies the hypotheses of Theorem 2.4. Thus for $0 \le t \le 1/2$ the second mixed cumulant, $\kappa_{1,1}(t) \equiv \kappa_{1,1}(K(ut,t), K(vt,t)) = C(u,v,t)$, is proportional to t^{-1} for each u and v. Specifically,

$$C(u,v,t) = (2t)^{-1} C(u,v,\tfrac{1}{2}) \qquad \text{for } 0 < t \le \tfrac{1}{2}. \tag{7.2}$$

What remains to be done is to compute the actual proportionality constants and this is what is done below.

In view of (1.10), with K^* denoting an independent copy of K,

$$\begin{aligned} c(x,y,s) &= \operatorname{Cov}\left(K\left(\frac{x}{U}, \frac{s}{U}\right) + K^*\left(\frac{x}{1-U}, \frac{s}{1-U}\right), \right. \\ &\qquad\qquad \left. K\left(\frac{y}{U}, \frac{s}{U}\right) + K^*\left(\frac{y}{1-U}, \frac{s}{1-U}\right)\right) \\ &= 2\int_0^1 c(x/u, y/u, s/u)\,du + 2\int_0^1 \mu(x/u, s/u)\mu(y/u, s/u)\,du \\ &\qquad + 2\int_0^1 \mu(x/u, s/u)\mu(y/(1-u), s/(1-u))\,du - \mu(x,s)\mu(y,s) \end{aligned} \tag{7.3}$$

where $\mu(x,s) = EK(x,s)$. This uses the following easy computation: if U is a Unif(0, 1) r.v. independent of processes $X(\cdot)$ and $Y(\cdot)$, then whenever the integrals make sense,

$$\begin{aligned} &\operatorname{Cov}(X(U), Y(U)) \\ &\quad = \int_0^1 EX(u)Y(u)\,du - EX(U)EY(U) \\ &\quad = \int_0^1 \{\operatorname{Cov}(X(u), Y(u)) + EX(u)EY(u)\}\,du - EX(U)EY(U). \end{aligned}$$

[In applying this to (7.3) we use the conditional independence of the summands making up $X(u)$ and $Y(u)$.]

In view of the evaluation of μ in (1.8), the second integral in (7.3) is

$$\begin{aligned} \int_0^1 \mu(x/u, s/u)\mu(y/u, s/u)\,du &= \int_0^s \varepsilon(x-u)\varepsilon(y-u)\,du + \int_s^1 \frac{2xu}{s^2}\frac{2yu}{s^2}\,du \\ &= x + 4xy/3s^4 - 4xy/3s. \end{aligned}$$

To evaluate the third integral of (7.3), note first that by (1.8), the integrand is zero when either $x < u \le s$ or $1 - s \le u < 1 - y$. Upon applying (1.8) appropriately to the integrand over the remaining intervals of integration $0 < u \le x \wedge (1-s)$, $s \vee (1-y) < u \le 1$, $1 - y < u \le x$ (when $x + y > 1$) and $s < u \le (1-s)$ (when

$s < 1/2$), one obtains

$$
\begin{aligned}
&\int_0^1 \mu(x/u, s/u)\mu\left(\frac{y}{1-u}, \frac{s}{1-u}\right) du \\
&\quad = (x/s^2)(1 - \{s \vee (1-y)\}^2) \\
&\qquad + (y/s^2)(1 - \{s \vee (1-x)\}^2) + (x+y-1)^+ \\
&\qquad + \mathbb{1}_{[s<1/2]} \frac{2xy}{3s^4}(4s^3 - 6s^2 + 1).
\end{aligned} \tag{7.4}
$$

Together with (1.8) this allows us to evaluate $c(x, y, s)$ for $0 \le x \le y \le s$ and $0 < s \le 1$ by means of (7.3). Substituting $x = us$ and $y = vs$ we find that, for $0 \le u \le v \le 1$ and $0 < s \le 1$,

$$
\begin{aligned}
C(u, v, s) &= 2s \int_s^1 C(u, v, w) w^{-2} \, dw + 2us - 4uv/3s^2 - 8uvs/3 \\
&\quad + (2u/s)[1 - \{s \vee (1 - vs)\}^2] + (2v/s)[1 - \{s \vee (1 - us)\}^2] \\
&\quad + 2[(u+v)s - 1]^+ + \mathbb{1}_{[s<1/2]} \frac{4uv}{3s^2}(4s^3 - 6s^2 + 1).
\end{aligned} \tag{7.5}
$$

This expression shows that $C(u, v, \cdot)$ is continuous on $(0, 1)$ (recall that there is a discontinuity at $s = 1$) and is differentiable for all s except possibly at the five values: $1/2 \le (1+v)^{-1} \le (1+u)^{-1} \le [(u+v) \wedge 1]^{-1} \le 1$. Dividing by s, then differentiating with respect to s and finally multiplying by s^2 shows that

$$
\begin{aligned}
&\frac{d}{ds}\{sC(u, v, s)\} \\
&\quad = \frac{4uv}{s^2} - \frac{4u}{s}\mathbb{1}_{[s(1+v) \ge 1]} - \frac{4v}{s}\mathbb{1}_{[s(1+u) \ge 1]} - 4uv\mathbb{1}_{[s(1+v)<1]} \\
&\qquad - 4uv\mathbb{1}_{[s(1+u)<1]} + 2\mathbb{1}_{[s(u+v) \ge 1]} + 4uv(2 - s^{-2})\mathbb{1}_{[s<1/2]}
\end{aligned} \tag{7.6}
$$

for all but those five exceptional points. [The reader may note that the right-hand side of this expression is zero for $0 < s < 1/2$, thereby leading to an alternate proof that $sC(u, v, s)$ is constant over that range as stated in (7.2).] To obtain the proportionality constant of (7.2), we integrate the above from $s = 1/2$ to $s = 1-$ to obtain

$$
\begin{aligned}
&C(u, v, 1-) - \tfrac{1}{2}C(u, v, 1/2) \\
&\quad = 4uv - 4u \ln(1+v) - 4v \ln(1+u) \\
&\qquad - 4uv[(1+u)^{-1} + (1+v)^{-1} - 1] + 2(u+v-1)^+/(u+v).
\end{aligned} \tag{7.7}
$$

Thus, to complete the computation it remains only to determine $C(u, v, 1-)$. For this, we need the distribution of $K(x, 1-)K(y, 1-)$ which is deducible from the following in which we represent the two ordered spacings by $(1-U)/2$ and $(1+U)/2$ with U being a Unif(0, 1) r.v. For $0 \le x \le y \le 1$,

$$P[K(x,1-)=2,\ K(y,1-)=2] = P[(1+U)/2 \le x],$$
$$P[K(x,1-)=1,\ K(y,1-)=2] = P[(1-U)/2 < x < (1+U)/2 < y]$$
$$= P[(1-2x) \vee (2x-1) < U < 2y-1]$$

and

$$P[K(x,1-)=1,\ K(y,1-)=1] = P[(1-U)/2 < x \le y < (1+U)/2]$$
$$= P[U > (1-2x) \vee (2y-1)].$$

Since $EK(x,1-) = \mu(x,1-) = 2x$, straightforward computations lead to

$$C(u,v,1-) = 2u - 4uv + 2(u+v-1)^+. \tag{7.8}$$

A combination of (7.2), (7.7) and (7.8) completes the proof of the following theorem.

THEOREM 7.1. *For* $0 < u \le v \le 1$,

$$\mathrm{Cov}\big(K(us,s), K(vs,s)\big) = C(u,v,s) = \frac{\alpha(u,v)}{s}, \qquad 0 < s \le 1/2,$$

where

$$\alpha(u,v) = \tfrac{1}{2}C(u,v,1/2)$$
$$= -12uv + 4uv\{(1+v)^{-1} + (1+u)^{-1}\} + 4u\ln(1+v) \tag{7.9}$$
$$+ 4v\ln(1+u) + 2u + 2((u+v-1)^+)^2(u+v)^{-1},$$

with $(x)^+ = \max(0,x)$.

As a corollary of this result, the covariance of the process $W(\cdot,s)$, which is defined in (1.18) and has mean zero on [0, 1] by (1.8), is

$$\mathrm{Cov}\big(W(u,s), W(v,s)\big) = (s/2)C(u,v,s) = \tfrac{1}{2}\alpha(u,v) \tag{7.10}$$

for $0 \le u \le v \le 1$ and $0 < s \le 1/2$. Note also that $\alpha(1,1)/2 = 4\ln 2 - 5/2 = \sigma^2$; see Theorem 4.1. Clearly, the limiting process $W(\cdot)$ has the same covariance. By (1.19) this would mean that the covariance function for a limiting V^*-process for the normalized spacings would become

$$\mathrm{Cov}\big(V^*(u), V^*(v)\big)$$
$$= u(1-v) + (8\ln 2 - 3)uv - 2u\ln(1+v) - 2v\ln(1+u) \tag{7.11}$$
$$+ \{(u+v-1)^+\}^2/(u+v)$$

for $0 \le u \le v \le 1$. More importantly, however, the covariances for the limiting V-process of Theorem 6.1 for the relative spacings becomes, by (7.10) and (1.20),

$$\begin{aligned}\operatorname{Cov}(V(u), V(v)) &= u(1-v) - uv\left(\frac{3}{2} - \frac{1}{1+u} - \frac{1}{1+v}\right)\\ &\quad + \{(u+v-1)^+\}^2/(u+v)\end{aligned} \tag{7.12}$$

for $0 \le u \le v \le 1$. Observe that this latter covariance is zero at $V = 1$ since V is a tied-down process, whereas the untied process V^* has variance at $v = 1$ of $\sigma^2 = 4\ln 2 - 5/2$; compare Theorem 3.1 and the difference between V and V^*, which is seen through (1.19) and (1.20) to be $V(1) - V^*(1) = W(1)$.

REMARK. The focus of this paper has been solely upon the interval-splitting procedure of Kakutani (1975), and the methodologies required to obtain the weak convergence limits for the two main empirical processes under the particular dependence structure determined by this procedure. The paper therefore extends in a natural way the strong law or Glivenko–Cantelli results previously obtained for the Kakutani model; compare Lootgieter (1977), van Zwet (1978) and Pyke (1980).

Generalizations of the Kakutani procedure have been proposed. For example, the splitting random variables, $\{U_i\}$ in this paper, could have distributions other than uniform. Alternatively, procedures could allow for random selection of the interval to be split, rather than restricting it to be always the longest interval. Glivenko–Cantelli results for generalized procedures of these types have been studied in Brennan and Durrett (1987) and papers referenced therein. It is an open question whether weak convergence results for the analogous empirical processes can also be derived by the methodologies of this paper.

Other related references are Sibuya and Itoh (1987) and Komaki and Itoh (1992).

During the preparation of this paper, the authors had discussions with P. Diaconis and M. Shahshahani about their interests in this and related work. In particular, correspondence from P. Diaconis described calculations involving moments of the trace of a random $n \times n$ permutation matrix on the one hand, and of a random $n \times n$ orthogonal matrix on the other hand, for which the first n (resp., $2n + 1$) moments are exactly the moments of a Poisson (resp. normal) random variable. The connection with our work lies in the loose similarity with the type of result contained in our Theorem 2.2 in which an increasing number of moments become constant as a parameter, $1/t$ increases.

REFERENCES

ANSCOMBE, P. (1952). Large-sample theory of sequential estimation. *Proc. Cambridge Philos. Soc.* **45** 600–607.

BILLINGSLEY, P. (1968). *Convergence of Probability Measures*. Wiley, New York.

BRENNAN, M. D. and DURRETT, R. (1987). Splitting intervals. II. Limit laws for lengths. *Probab. Theory Related Fields* **75** 109–127.

CSÖRGŐ, S. (1974). On weak convergence of the empirical process with random sample size. *Acta Sci. Math. Szeged* **36** 17–25.

DONSKER, M. (1952). Justification and extension of Doob's heuristic approach to the Kolmogorov–Smirnov theorems. *Ann. Math. Statist.* **23** 277–281.

GIHMAN, I. I. and SKOROHOD, A. V. (1974). *The Theory of Stochastic Processes* I. Springer, New York.

KAKUTANI, S. (1975). A problem of equidistribution on the unit interval [0, 1]. *Proceedings of Oberwolfach Conference on Measure Theory. Lecture Notes in Math.* **541** 369–376. Springer, Berlin.

KLAASSEN, C. A. J. and WELLNER, J. A. (1992). Kac empirical processes and the bootstrap. In *Proceedings of the Eighth International Conference on Probability in Banach Spaces* (M. Hahn and J. Kuebs, eds.) 411–429. Birkhäuser, Boston.

KOMAKI, F. and ITOH, Y. (1992). A unified model for Kakutani's interval splitting and Renyi's random packing. *Adv. in Appl. Probab.* **24** 502–505.

LOOTGIETER, J. C. (1977). Sur la répartition des suites de Kakutani (I). *Ann. Inst. H. Poincaré Ser. B* **13** 385–410.

PYKE, R. (1965). Spacings. *J. Roy. Statist. Soc. Ser. B* **27** 395–449.

PYKE, R. (1968). The weak convergence of the empirical process with random sample size. *Proc. Cambridge Philos. Soc.* **64** 155–160.

PYKE, R. (1980). The asymptotic behavior of spacings under Kakutani's model for interval subdivision. *Ann. Probab.* **8** 157–163.

SIBUYA, M. and ITOH, Y. (1987). Random sequential bisection and its associated binary tree. *Ann. Inst. Statist. Math.* **39** 69–84.

VAN ZWET, W. R. (1978). A proof of Kakutani's conjecture on random subdivision of longest intervals. *Ann. Probab.* **6** 133–137.

WEISS, L. (1955). The stochastic convergence of a function of sample successive differences. *Ann. Math. Statist.* **26** 532–536.

DEPARTMENT OF MATHEMATICS
UNIVERSITY OF WASHINGTON
SEATTLE, WASHINGTON 98195
USA
E-MAIL: pyke@math.washington.edu

DEPARTMENT OF MATHEMATICS
UNIVERSITY OF LEIDEN
P.O. BOX 9512
2300 RA LEIDEN
THE NETHERLANDS
E-MAIL: vanzwet@math.leidenuniv.nl

Chapter 29
A discussion of Willem van Zwet's probability papers

David M. Mason

Abstract I discuss five papers of Willem van Zwet: [19], [20], [13], [21] and [15].

I shall begin my discussion of Willem's probability papers with his 1978 paper on the Kakutani conjecture. Willem tells me that this is his favorite paper. I can see why. It is not only a fine piece of mathematics; it also displays very well a common feature of many of Willem's best papers, namely, it begins with a key insight, which lights the way to the solution of a knotty problem.

The full story of Willem's involvement with the Kakutani conjecture appears in the *Statistical Science* interview of him by Rudy Beran and Nick Fisher (Beran and Fisher (2009)). Therefore I shall not repeat it here. I met Kakutani in Lunteren in 1984, when he was a speaker and I was a participant. By that time his conjecture was solved. He did not lecture about it and we did not discuss it. On the other hand, I have heard Willem talk about it a number of times at Oberwolfach Meetings and other occasions.

Here is a description of the conjecture and a sketch of Willem's beautiful solution. The Kakutani interval splitting is as follows: Let V_1 be a Uniform $(0,1)$ random variable. Now proceed sequentially, choosing for each integer $n \geq 1$, V_{n+1} uniformly from the largest of the $n+1$ subintervals into which the previously constructed variables $V_1, \ldots, V_n$ partition $(0,1)$. Consider the empirical distribution function F_n based on the sample $V_1, \ldots, V_n$ at stage n. Kakutani conjectured, that with probability 1, as $n \to \infty$,

$$\sup_{0 \leq t \leq 1} |F_n(t) - t| \to 0. \tag{29.1}$$

He was, of course, motivated by the Glivenko-Cantelli theorem, which implies that if G_n is the empirical distribution function of $U_1, \ldots, U_n$, i.i.d. Uniform $(0,1)$, then, with probability 1, as $n \to \infty$,

David M. Mason
Statistics Program, University of Delaware, 213 Townsend Hall Newark, DE 19716
e-mail: davidm@udel.edu This research was partially supported by an NSF Grant.

S. van de Geer and M. Wegkamp (eds.), *Selected Works of Willem van Zwet*, Selected Works in Probability and Statistics, DOI 10.1007/978-1-4614-1314-1_29,

$$\sup_{0\le t\le 1} |G_n(t) - t| \to 0. \tag{29.2}$$

It turns out that the Kakutani conjecture is correct.

In van Zwet (1978), Willem devised an ingenious proof of (29.1). Essential to his approach was a stopping time formed as follows: For each integer $n \ge 1$ let $0 = V_{0,n} \le V_{1,n} \le \dots \le V_{n,n} \le V_{n+1,n} = 1$ denote the order statistics of $V_1, \dots, V_n$, and introduce the $n+1$ spacings $D_{i,n} = V_{i,n} - V_{i-1,n}$, $i = 1, \dots, n+1$. For $t > 0$, define the stopping time $N_t = \min\{n : \max_{1\le i\le n+1} D_{i,n} \le t\}$. We shall write N for the process N_t, $t > 0$. Notice that N_t is the smallest integer such that all of the subintervals have length at most t. Crucial to his proof was the distributional identity

$$N_t =_d N^{(1)}_{t/V} + N^{(2)}_{t/(1-V)} + 1, \tag{29.3}$$

where $N^{(1)}, N^{(2)}$ and V are independent, with $N^{(1)} =_d N^{(2)} =_d N$ and V being Uniform $(0,1)$. The distributional identity (29.3), in turn, leads to differential equations whose solutions give that for some $c > 0$,

$$\mu(t) := EN_t = 2/t - 1 \text{ for } 0 < t < 1 \text{ and } \sigma^2(t) := Var(N_t) = c/t \text{ for } 0 < t < 1/2. \tag{29.4}$$

Here is a synopsis of Willem's proof of (29.1). For any $x \in (0,1]$ let $N_t(x) = N_t F_{N_t}(x)$. It is easy to argue that for any $x \in (0,1]$ there exist distributionally equivalent versions of $N'_{t/x}$ of $N_{t/x}$ and $N'_{t/(1-x)}$ of $N_{t/(1-x)}$ such that

$$N'_{t/x} \le N_t(x) \le N_t - N'_{t/(1-x)}. \tag{29.5}$$

Set $t_m = m^{-2}$ for $m \ge 1$. Using Chebyshev's inequality and the identities in (29.4) he shows that for any sequences $\{N'_{t_m}\}$, $\{N'_{t_m/x}\}$ and $\{N'_{t_m/(1-x)}\}$, where for $m \ge 1$, $N'_{t_m} =_d N_{t_m}$, $N'_{t_m/x} =_d N_{t_m/x}$ and $N'_{t_m/(1-x)} =_d N_{t_m/(1-x)}$, with probability 1,

$$t_m N'_{t_m}/2 \to 1,\ t_m N'_{t_m/x}/2 \to x \text{ and } t_m N'_{t_m/(1-x)}/2 \to 1 - x.$$

This implies by the inequalities in (29.5) that, with probability 1, $F_{N_{t_m}}(x) \to x$, from which (29.1) follows by a routine argument, noting that, with probability 1, $N_{t_m}/N_{t_{m+1}} \to 1$.

In Pyke and van Zwet (2004), Ron Pyke and Willem considered the question of the weak convergence of the *Kakutani empirical process*,

$$U_n(x) = \sqrt{n}\{F_n(x) - x\},\ x \in [0,1].$$

They prove that U_n converges weakly to σB, written $U_n \to_d \sigma B$, where B is a standard Brownian bridge on $[0,1]$ with $\sigma = \sqrt{4\ln 2 - 5/2} = .5221003$, which is approximately one half as large as in the i.i.d. case. Recall that the *uniform empirical process*

$$\alpha_n(x) = \sqrt{n}\{G_n(x) - x\},\ x \in [0,1], \tag{29.6}$$

converges weakly to B. This shows that though the Kakutani sampling procedure does not differ from the usual i.i.d. situation at the Glivenko-Cantelli level in the sense that both (29.2) and (29.1) hold, it does at the asymptotic distributional level. Roughly speaking, this says that the Kakutani scheme leads to more regularly spaced points than in the i.i.d. uniform model.

Once again the stopping time N_t is a basic ingredient in the proof. Now higher moments are required and these are obtained through the cumulants of $N_t + 1$. They extend the method in van Zwet (1978) used to derive (29.4) to show that for $m \geq 1$, the m-th cumulant of $N_t + 1$ is

$$k_m(t) = c_m/t \text{ for } 0 < t \leq 1/m, \text{ with } mc_m = k_m(1/m).$$

It was also necessary to ferret out the underlying process that drives the weak convergence of U_n, which they found to be the following: For $0 < t < s < 1$ let

$$S_t(x;s) = (t/2)^{1/2} \sum_{V_{i,N_s} \leq x} \left\{ N^{(i)}_{t/D_{i,N_s}} - \mu(t/D_{i,N_s}) \right\}, \; x \in [0,1],$$

where $N^{(1)}, N^{(2)}, \ldots,$ are i.i.d. N.

Conditioned on D_{i,N_s}, $i = 1, \ldots, N_s + 1$, $S_t(\cdot;s)$ is clearly a partial sum process on $[0,1]$ based on independent summands, with jumps at the V_{i,N_s}. By a careful analysis, applying a classical weak convergence result, they prove that as long as $t(s) = o(s)$ as $s \searrow 0$ at a proper rate, the process $S_t(\cdot;s)$ converges weakly, both conditionally and unconditionally, to σW, where W is a standard Brownian motion. After that is established, they complete the proof by verifying that random jump times can be replaced by deterministic ones and the appropriate pieces can be fit together properly to conclude that $U_n \to_d \sigma B$.

Pyke (1980) had shown that the empirical distribution of the normalized spacings $(n+1)D_{i,n}$, $i = 1, \ldots, n+1$, of the Kakutani sample converges to the Uniform $(0,2)$ distribution, whereas in the i.i.d. case it goes to the exponential distribution with mean 1. Pyke and van Zwet (2004) establish that the corresponding empirical process converges weakly to a mean zero Gaussian process with a complicated covariance function. They also treat the empirical process of the relative spacings $D_{i,n}/M_n$, where $M_n = \max\{D_{i,n} : i = 1, \ldots, n+1\}$. They do this by adapting the methods that they used to prove $U_n \to_d \sigma B$. Sadly this was Ron Pyke's last paper.

Willem and I first met at an *N.S.F. Regional Conference* in 1982 in Eugene, Oregon, where he was the featured speaker. However, my first professional contact with Willem was in 1979, when I wrote him for a preprint of his paper van Zwet (1980), which is on a strong law of large numbers for linear functions of orders statistics. These are statistics of the form

$$\mathbb{L}_n = \sum_{i=1}^{n} g(U_{i,n}) \int_{(i-1)/n}^{i/n} J_n(u)\, du,$$

where g is a measurable function on $(0,1)$ and $\{J_n\}$ is a sequence of measurable functions on $(0,1)$ that converge in an appropriate sense to a function J. At that time, proving central limit theorems and strong laws for $\mathbb{L}_n$ was still in vogue. The game was to balance smoothness conditions on g with those on J.

Willem came up with an elegant solution to the strong law problem for $\mathbb{L}_n$ in van Zwet (1980). To state his main results, introduce the centering constants

$$\mu_n = \int_0^1 J_n g d\lambda \text{ and } \mu = \int_0^1 J g d\lambda.$$

(We denote Lebesgue measure by λ and assume here whatever conditions needed to insure that the integrals are finite.) The following result is a combination of Theorem 2.1 and Corollary 2.1 of van Zwet (1980).

Theorem 1 *Let* $1 \le p \le \infty$, $p^{-1} + q^{-1} = 1$, *and suppose that* $J_n \in L_p(d\lambda)$ *for* $n \ge 1$, *and* $g \in L_q(d\lambda)$ *. If either*

(i) $1 < p \le \infty$ *and* $\sup_n \|J_n\| < \infty$, or

(ii) $p = 1$ *and* $\{J_n\}$ *is uniformly integrable,*

then with probability 1

$$\mathbb{L}_n - \mu_n \to 0. \tag{29.7}$$

Moreover, if there is a $J \in L_p(d\lambda)$ *such that* $\int_0^t J_n g d\lambda \to \int_0^t J g d\lambda$ *for every* $t \in (0,1)$. *Then we can replace* μ_n *by* μ *in* (29.7).

The key technical result used in his proof was the following lemma based on Lusin's theorem: With $[x]$ denoting the integer part of x, define for each integer $n \ge 1$,

$$g_n(t) = g\left(U_{[tn]+1,n}\right) \text{ for } t \in (0,1).$$

Lemma 1 *With probability* 1, g_n *converges to* g *in Lebesgue measure* λ *on* $(0,1)$, *i.e. for all* $\delta > 0$,

$$\lim_{n\to\infty} \lambda\{t : |g_n(t) - g(t)| > \delta\} = 0.$$

Armed with this result, the proof of (29.7) is short and sweet. Some classical criteria for weak convergence in L_p, $1 \le p < \infty$, and weak* convergence in L_∞ play a role. Willem then shows how this result and its extensions imply the strong laws for $\mathbb{L}_n$ of Helmers (1977), Wellner (1977) and Sen (1978).

I published my own strong law for $\mathbb{L}_n$ in Mason (1982), in which for certain subclasses of linear functions of order statistics I provided necessary and sufficient conditions for (29.7) to hold. I also considered $\mathbb{L}_n$ whose g functions are not in $L_q(d\lambda)$ for any $q \ge 1$.

I next discuss the most significant of my two joint papers with Willem. This is Mason and van Zwet (1987), where we obtained the following refinement of the Komlós, Major and Tusnády [KMT] (1975) Brownian bridge approximation to the uniform empirical process α_n. Recall the definition of α_n given in (29.6).

Theorem 2 *There exists a probability space* (Ω, A, P) *with independent Uniform* $(0,1)$ *random variables* $U_1, U_2, \ldots$, *and a sequence of Brownian bridges* $B_1, B_2, \ldots$, *such that for all* $n \ge 1$, $1 \le d \le n$, *and* $-\infty < x < \infty$,

$$P\left\{\sup_{0\le t\le d/n}|\alpha_n(t)-B_n(t)|\ge n^{-1/2}(a\log d+x)\right\}\le b\exp(-cx) \qquad (29.8)$$

and with the same inequality holding with the supremum is taken over $1-d/n\le t\le 1$, *where* a,b *and* c *are suitable positive constants independent of* n, d *and* x.

Setting $d=n$ into these inequalities yields the original KMT inequality. Our result drew a fair amount of interest and has been applied in a number of papers. I note that as of this writing google scholar lists 54 citations to our paper, which is pretty good for a theoretical paper. Rio (1994) has computed values for the constants in (29.8). Later, Castelle, and Laurent-Bonvalot (1998) obtained a similar refinement to the KMT Kiefer process approximation to α_n.

M. Csörgő, S. Csörgő, Horváth and Mason [Cs-Cs-H-M](1986) had earlier constructed a probability space on which an analog to these inequalities holds with α_n replaced by β_n (the uniform quantile process). We shall not define β_n in this discussion.

Willem and I did much of the work on our paper during the month of March 1985, while I was visiting him in Leiden. At the time, I was on extended leave from the University of Delaware and was in the middle of the first year of a two year stay at the University of Munich supported by an Alexander von Humboldt stipend.

Our progress was severely hampered by the fact that KMT (1975) had only provided a bare sketch of the proof of their original inequality. We spent an enormous amount of time filling in the missing details. This became an obsession to me. After two weeks in Leiden watching me consumed with this project night and day, my wife at the time got fed up with me and returned to Munich.

Complete proofs are now available. Mason and van Zwet (1987) combined with Mason (2001a), which contains additional details based on Leiden notes, provides a proof. Also consult Bretagnolle and Massart (1989), Major (1999) and Dudley (2000). Bretagnolle and Massart (1989) determined values of the constants in the original KMT inequality. For more on the history of proofs refer to Mason (2007).

Mason and van Zwet (1987) pointed out that their inequality leads to the following useful weighted approximation: For any $0\le\nu<1/2$, $n\ge 1$, and $1\le d\le n-1$ let

$$\Delta_{n,\nu}(d):=\sup_{d/n\le t\le 1-d/n}\frac{n^{\nu}|\alpha_n(t)-B_n(t)|}{(t(1-t))^{1/2-\nu}}. \qquad (29.9)$$

On the probability space of Theorem 2, one has

$$\Delta_{n,\nu}(1)=O_p(1). \qquad (29.10)$$

Cs-Cs-H-M (1986) had obtained a version of this result on their probability space under the restriction that $0\le\nu<1/4$. Later, motivated by a question of Evarist Giné, I derived the following improved version of the Mason and van Zwet weighted approximation (29.10). I published this in the van Zwet Festschrift (see Mason (2001b)).

Theorem 3 *On the probability space of Theorem 2, for every* $0 \le \nu < 1/2$ *there exist positive constants* A_ν *and* C_ν *such that for all* $n \ge 2$, $1 \le d \le n-1$ *and* $0 \le x < \infty$,

$$P\{\Delta_{n,\nu}(d) \ge x\} \le 2A_\nu \exp(d^{1/2-\nu} C_\nu) \exp\Big(-d^{1/2-\nu} cx/4\Big). \qquad (29.11)$$

One of the key inequalities in Mason and van Zwet (1987) that was essential to derive the refinement of KMT was the following: Let $(X_1, \ldots, X_{k+1})$ with $k \ge 1$ be a multinomial random vector with parameters n, $p_1, \ldots, p_{k+1}$. Assume that $p_i > 0$ for $i = 1, \ldots, k$ and let $s = \sum_{i=1}^k p_i \in (0,1]$.

Lemma 2 *For every* $C > 0$ *and* $\delta > 0$, *there exist positive numbers* a, b *and* λ *such that for all* $n \ge 1$, $k \ge 1$ *and positive* s, $p_1, \ldots, p_k$ *satisfying* $z \le Cn \min\{p_1, \ldots, p_k\}$ *and* $s \le 1 - \delta$

$$P\left(\sum_{i=1}^k \frac{(X_i - np_i)^2}{np_i} > z\right) < a\exp(bk - \lambda z). \qquad (29.12)$$

This is Lemma 3 of Mason and van Zwet (1987). It was just what we needed to complete the proof of inequality (29.8) and it required us some time to formulate. (The final result was Willem's doing.) Therefore I was surprised when I found it quoted in the 1996 monograph on empirical processes by Aad van der Vaart and Jon Wellner (van der Vaart and Wellner (1996)), with a remark that it follows from Talagrand's general empirical process inequality.

Willem and I also published a nice paper on the strong approximation to the renewal process. It was also begun during my Leiden visit. Unfortunately it had a controversial history, which I will not discuss here.

Finally I must say something about the remaining paper in the collection of Willem's papers that are classified as 'probability'. This is van Zwet (1994). In this little gem can be found an elegant solution to the following problem:

Consider n point charges, each with charge $1/n$, in electrostatic equilibrium on the surface S of a conducting sphere. Show that, as n tends to infinity, the distribution of the total charge 1 on S tends to the uniform distribution on S.

Though this is an entirely deterministic result, the proof is probabilistic in nature. I will conclude this discussion with some informative and personal remarks by Willem about the history of this paper:

Let me also say something about the paper classified as 'probability', to wit the paper on point charges on a conducting sphere. In 1975 Herbert Robbins spoke at Lunteren. At the end of his talk he proposed a few problems, one of these being: prove that n electrons become uniformly distributed on the sphere as n tends to infinity. He said this problem had been bothering him for years.

Some time later I saw the same problem again, this time as a prize problem of the Dutch Mathematical Society. I found a simple proof that is a funny mixture of potential theory and probability, and sent it to the mathematics society. They awarded me the prize, but the jury report mentioned that the problem had been solved before by a well-known Dutch analyst called Jaap Korevaar.

It turned out that Korevaar had suggested the problem to the mathematics society. (He also heard it from Robbins.) They had used it for a prize problem for a number of years without receiving a solution. Then Korevaar found a solution himself, but forgot to tell the mathematics society about this, so they continued running the problem as a prize question.

They were pretty embarrassed, and though Korevaar's proof was totally different from mine (lots of calculations), we both buried our proofs in a desk drawer. However, 17 years later I still liked my proof and thought 'what the hell' and spoke about it at the Purdue Symposium in 1992, mentioning Korevaar's proof too, of course. David Siegmund was present and after the meeting he told Robbins about my talk. Robbins' – as usual cynical –comment was 'My god, did it take him 20 years to solve this?

This is Willem, the inimitable storyteller. It has been a pleasure writing these pages.

References

1. BERAN, R. J. and FISHER, N. I. (2009). An evening spent with Bill van Zwet. *Statistical Science* **24** 87–115.
2. BRETAGNOLLE, J. and MASSART, P. (1989). Hungarian constructions from the nonasymptotic viewpoint. *Ann. Probability* **17** 239–256.
3. CASTELLE, N. and LAURENT-BONVALOT, F (1998). Strong approximations of bivariate uniform empirical processes. *Ann. Inst. Henri Poincaré* **34** 425–480.
4. CSÖRGŐ, M., CSÖRGŐ, S., HORVÁTH, L. and MASON, D. M. (1986). Weighted empirical and quantile processes. *Ann. Probability* **14** 31–85.
5. DUDLEY, R. M. (2000). An exposition of Bretagnolle and Massart's proof of the KMT (1975) theorem for the uniform empirical process. In: Notes on empirical processes, lectures notes for a course given at Aarhus University, Denmark, August 2000.
6. HELMERS, R. (1977). A strong law of large numbers for linear combinations of order statistics. Report SW 50/77, Mathematisch Centrum, Amsterdam.
7. KOMLÓS, J., MAJOR, P. and TUSNÁDY, G. (1975). An approximation of partial sums of independent RV's and the sample DF. I. *Z. Wahrscheinlichkeitstheorie und Verw. Gebiete* **32** 111–131.
8. MAJOR, P. (1999). The approximation of the normalized empirical distribution function by a Brownian bridge, http://www.renyi.hu/~major/probability/empir.html
9. MASON.D. M. (1982) Some characterizations of strong laws for linear functions of order statistics. *Ann. Probability* **10** 1051–1057.
10. MASON, D. M. (2001a) Notes on the KMT Brownian bridge approximation to the uniform empirical process. *Asymptotic methods in probability and statistics with applications* (St. Petersburg, 1998), 351–369, Stat. Ind. Technol., Birkhäuser, Boston, Boston, MA.
11. MASON, D. M. (2001b) An exponential inequality for a weighted approximation to the uniform empirical process with applications. *State of the art in probability and statistics* (Leiden, 1999), 477–498, IMS Lecture Notes Monogr. Ser., 36, Inst. Math. Statist., Beachwood, OH.
12. MASON, D. M. (2007). Some observations on the KMT dyadic scheme. (Special issue on Nonparametric Statistics and Related topics in honor of M. L. Puri.) *J. Statist. Plann. Inf.* **137** 895–906.
13. MASON, D. M. and VAN ZWET, W. R. (1987). A refinement of the KMT inequality for the uniform empirical process. *Ann. Probability* **15** 871–884.
14. PYKE, R. (1980). The asymptotic behavior of spacings under Kakutani's model for interval subdivision. *Ann. Probability* **8** 157–163.

15. Pyke, R. and van Zwet, W. R. (2004). Weak convergence results for the Kakutani interval splitting procedure *Ann. Probability* **32** 380–423
16. Rio, E. (1994). Local invariance principles and their application to density estimation. *Probab. Theory Related Fields* **98** 21–45.
17. Sen, P. K. (1978). An invariance principle for linear combinations of order statistics. *Z. Wahrscheinlichkeitstheorie und Verw. Gebiete* **42** 327–340.
18. Vaart, van der, A.W. and Wellner, J. A. (1996). *Weak Convergence and Empirical Processes*. Springer, New York.
19. van Zwet, W. R. (1978). A proof of Kakutani's conjecture on random subdivision of longest intervals. *Ann Probability* **6** 133–137.
20. van Zwet, W. R. (1980). A strong law for linear functions of uniform order statistics. *Ann Probability* **8** 986–990.
21. van Zwet, W. R. (1994). The asymptotic distribution of point charges on a conducting sphere. *Statistical Decision Theory and Related Topics,* V, S. S. Gupta and J. O. Berger editors, Springer New York 427–430.
22. Wellner, J. A. (1977). A Glivenko-Cantelli theorem and strong laws of large numbers for functions of order statistics. *Ann. Statist.* **5** 473–480.

Printed in the United States
By Bookmasters